QUANTUM CHEMISTRY

An Introduction

QUANTUM CHEMISTRY

An Introduction

WALTER KAUZMANN

Department of Chemistry

Princeton University

1957

ACADEMIC PRESS INC · PUBLISHERS · NEW YORK

By

ACADEMIC PRESS INC.

111 Fifth Avenue, New York 3, N.Y.

Library of Congress Catalog Card Number: 56-6606

Printed in the United States of America

Preface

The primary aim of this book is to help chemists understand the more important concepts that quantum mechanics has introduced into chemistry. A secondary aim is to examine critically the limitations of many of these concepts.

Unfortunately the theory of quantum mechanics can be satisfactorily understood only if it is presented in a mathematical language that is likely to be unfamiliar to chemists. Therefore chemists are tempted to be satisfied with the word pictures and loose analogies that are often used when quantum mechanical concepts are first introduced to students. If an adequate and critical comprehension of these concepts is to be achieved, however, the mathematical formulation of quantum mechanics must be mastered in some detail. Without such mastery the chemist who tries to make quantum mechanical interpretations of experimental observations is apt to find himself swimming in treacherous waters.

Several excellent texts on quantum chemistry are available, but the author has reason to believe that many students and teachers will welcome a book in which the necessary mathematics is presented in a more extended and more understandable form. In this volume more space than usual is also devoted to discussions of the nature and seriousness of the mathematical approximations one is forced to use in solving many quantum chemical problems.

Although relatively few chemists have had formal training in the fields of mathematics that are extensively employed in quantum chemistry, a great many chemists do have the mathematical aptitude that is needed in order to understand the subject. This book was written with such people in mind. The typical reader is assumed to have been exposed at some time to introductory courses in calculus and physics. It is assumed that the memory of the basic elements of these subjects can be revived and that the reader is willing to think in mathematical terms, once these terms have been fully explained to him. A special effort has been made to give an understandable presentation of the more advanced topics in mathematics. In particular, the theory of partial differential equations has been discussed separately and at some length, using the classical theory of vibrations as a means of developing physical insight into this important branch of mathematics. Thus the student does not have to contend with the strange new physical concepts of quantum mechanics at the same time that he is grappling with the solution of some rather complicated differential equations. Special effort has also been made to bring the theory into frequent contact with experimental observations, especially in Parts III, IV, and V, so that the reader can gain a feeling for the quantitative importance and adequacy of the concepts that are discussed.

Chemists have always made extensive use of light and optical instruments in investigating the properties of matter. It is felt that this justifies a somewhat more detailed discussion of optical phenomena than is usually given in texts on quantum chemistry. Since the quantum theory of optics is closely related to the classical electron theory of optics both in its terminology and in the formal appearance of its results, an extensive discussion of the classical theory has been included. The recent widespread interest in light scattering as a tool for studying the sizes and shapes of macromolecules has made the scattering phenomenon more familiar than ever to chemists. It has long been known that many optical effects can be interpreted as consequences of light scattering by atoms. The concept of light scattering has therefore been used to account for optical phenomena that are usually explained in textbooks in a more formal, less physical manner by the solution of Maxwell's equations.

It is hoped that a path of learning has been provided that can be traveled by the average chemist and chemist-to-be. If the path seems to be long and at times steep and rough, this is probably unavoidable. The value of following the path to its end ought to be self-evident in view of the many basic contributions that quantum mechanics has made to chemistry. Indeed, it is safe to say that no one can fully understand chemistry without a good grounding in quantum mechanics.

The book has been organized according to a logical arrangement of topics, which can be read in the sequence in which they appear. This entails a rather lengthy preliminary excursion into the classical theory of vibration, however, and many readers will not want to wait so long before getting their teeth into quantum mechanics. Such readers may follow the first eleven chapters according to the plan given in the accompanying table (page viii). (Other methods of attack are, of course, possible, especially in an organized course where suitable lecture material may be interpolated.) The last four chapters can be read in any order, depending on the interests of the student, except that Chapters 14 and 15 should precede Chapter 16. Chapter 15 may be read independently of the rest of the book, although Section D of Chapter 2 should prove useful to those who know little about vector analysis.

Because of limitations of space it has not been possible to discuss certain topics in quantum chemistry that are of interest to many chemists. These include the crystalline state, the nucleus and nuclear reactions, nuclear magnetic relaxation times and line widths, the Raman effect, and quantum statistics. With the background provided by this book, however, it should not be very difficult to learn about these subjects from the general literature.

I am grateful to students in the Princeton Chemistry Department who have gone over various versions of the earlier chapters and who have offered many useful suggestions. Dr. John Weil has carefully read most of the later chapters and his criticisms have been extremely helpful. It has been a pleasure to work with the staff of Academic Press in the troublesome task of putting the book into print. It is also a pleasure to have this opportunity for mentioning my indebtedness to Dean Henry Eyring, from whom I first learned about quantum mechanics and all that it means to chemistry. Finally, I must express my deep gratitude to my wife, Elizabeth Flagler Kauzmann, who has worked hard, long, and effectively at typing and editing the manuscript, and at reading the proof.

Princeton, New Jersey

October 1, 1956 WALTER KAUZMANN

Suggested Plan of Study

Topic	Chapters and Sections
Introduction	1
Essential mathematical background	2 A–C, E, G, H; 3 A a–g
Concepts and laws of quantum mechanics	5 A, B, C a–e, E
Precise solutions of Schroedinger's equation	6 A a; 2 D (pp. 22–25); 6 A b i; 3 B f–h, C; 6 A b ii, B–E, G
Uncertainty relations	5 C f, g; 7 A–C
Atomic structure	9 A, B, C a; 2 F; 3 A h–j; 4 A, C a; 9 C b (to p. 288); 4 B; 9 C b (to p. 290); 2 I; 3 B a–e; 4 C b, D, E; 5 C h, D; 2 D; 8; 9 C c–e, D; 10 A, B, C
Molecules and the chemical bond	6 J; 7 D; 6 H; 11

Contents

PART II
GENERAL PRINCIPLES OF QUANTUM MECHANICS

PART III
ATOMIC SYSTEMS

PART IV
MOLECULAR SYSTEMS

PART V

SYSTEMS IN NON-STATIONARY STATES

APPENDIXES

Chapter 1

Introduction

There is every reason to believe that all of chemistry should be deducible from the laws of quantum mechanics. The purpose of this book is to show how quantum mechanical deductions are made and to describe the achievements and limitations of the application of quantum mechanics to chemical problems.

Unfortunately, in spite of its great interest to chemists, quantum mechanics is not very easy to understand. For one thing, it is inherently mathematical, even though definite physical ideas can always be detected behind its equations. The student might wish that he could avoid this difficulty by learning quantum mechanics without its mathematics. It is, however, impossible to acquire a satisfactory grasp of the subject without some understanding of its mathematical structure. Therefore this book does not avoid the essential mathematical details of the theory. On the other hand, an effort has been made to reduce the mathematical difficulties by explaining from a rather elementary point of view the features that are likely to be new to the student and by emphasizing the physical ideas expressed by the mathematics.

Quite aside from the difficulties that the student might have with the mathematics, the physical content of quantum mechanics and its methods of "explaining" chemistry are likely to seem strange. Part of this feeling of strangeness can be overcome if, at the outset, the student understands something of the point of view of modern physical science – why the atomic theory necessarily reduced chemistry to a problem in mechanics, how scientific theories are put together, and what can and cannot be expected of them. This chapter gives a brief discussion of these points.

A. Chemistry as a Branch of Mechanics

Twenty five hundred years ago certain Greek philosophers suggested that matter is made up of indivisible particles, which they called atoms. According to this shrewdly conceived idea, the properties of matter result from the interactions and motions of elementary particles. The theory was highly developed in classical times and the detailed exposition given by the Roman poet-philosopher Lucretius in *De rerum natura* ("On the Nature of Things") contains much that is remarkably close to the modern version of the atomic theory–as well as a great deal that is amusingly different. The following account of the properties of sea water is typical.

> Liquids must owe their fluid consistency to component atoms that are smooth and round. Substances hurtful to the senses but not solid are sharp-pointed but without projections. Some things are both bitter and fluid as, for instance, sea water. This, being fluid, consists of smooth round atoms. It causes pain because of the admixture of many rough ones. There is a way of separating the two ingredients and viewing them in isolation by filtering the sweet fluid through many layers of earth so that it flows into a pit and loses its tang. It leaves behind the atoms of unpalatable brine because owing to their roughness they are more apt to stick fast in the earth (1, p. 73).

In the past 150 years a vast number of experimental observations have been interpreted successfully in terms of the notion of the atomic constitution of matter. We have, of course, become much more sophisticated than Lucretius in our conceptions of atomic behavior and structure, but most of us believe in the existence of atoms almost as firmly as we believe in the existence of the chair on which we sit.

Chemistry deals with the properties of matter and with their changes. According to the modern version of the atomic theory, these properties result from the motions of electrons and atomic nuclei. The motions are determined by the interaction of the elementary particles with each other's electric and magnetic fields. The principles of chemistry must therefore be a consequence of two more fundamental sciences:

(1) The science of *electrodynamics*, which deals with the behavior of electric and magnetic fields and with the forces that result when these fields interact with the elementary particles of matter.

(2) The science of *mechanics*, which deals with the actions of forces on material objects and especially with the motions that they cause.

If we seek a really penetrating understanding of chemistry, it is clear that we must begin by studying mechanics and electrodynamics. Fortunately it is possible to go a long way with a rather elementary understanding of electrodynamics. We shall therefore not have to dwell on this complicated field, which is, in fact, still not completely understood. Mechanics is quite another matter, however, and we must be prepared to make an extensive pursuit of this subject.

We have at our disposal three different systems of mechanics, each suited to limited sets of circumstances:

(1) *Classical mechanics* (that is, the mechanics of Newton) usually gives an accurate description of the motions of particles that are considerably heavier than individual atoms and that move with velocities much smaller than the velocity of light.

(2) *Quantum mechanics* is an improvement on classical mechanics that may be used in studying the motions of atomic and subatomic particles. When applied to the motions of heavier particles it gives the same results as classical mechanics, so that classical mechanics is really an approximate form of the more general system of quantum mechanics. The limitation to particles moving slowly compared to the velocity of light continues to apply in most of quantum mechanics.

(3) The *relativistic mechanics* of Einstein is an improvement on classical mechanics adapted to heavy particles moving with velocities approaching that of light. Physicists have not yet been completely successful in formulating *relativistic quantum mechanics*, universally valid for all sizes of particles moving at any speed.

The ordinary chemical properties of matter are caused by electrons and atomic nuclei moving at velocities much less than that of light. Therefore the mechanical basis of chemistry is to be found in quantum mechanics. This means that if we are to achieve what the Greek atomists and Lucretius set out to do, such concepts as the roughness and smoothness of atoms must be reinterpreted in quantum mechanical terms. Aside from this, the spirit of the present-day approach is essentially the same as that of the ancient atomists: we try to analyze the properties of matter in terms of the behavior of its elementary parts.

An engineer who designs aeroplane engines must know something about the classical mechanical theory that determines the motions occurring inside the engine. For similar reasons the chemist of today and tomorrow will find it profitable to learn about the principles of quantum mechanics, which ultimately determine the motions of atoms in chemical systems.

B. The Structure and Aims of Scientific Theories

a. The two components of a scientific theory

The aim of any scientific theory is the reasonable and systematic description of some aspect of nature. In accomplishing this, the scientist invariably proceeds in two steps. First he notices that his primary measurements of the world tend to be rather cumbersome to use directly in systematizing observations. He therefore invents a set of concepts that are related in some definite way to the original observations, but are better suited for logical analysis.

For instance, if the thermal effects of a chemical reaction are being studied in a calorimeter, the raw data consist of (among other things) the weights of the reactants, a series of thermometer and clock readings before, during, and after the reaction, and similar thermometer readings when an electric current is passed through a coil in the calorimeter at a definite voltage for a definite time. The very process of writing down these data is already something of an abstraction from direct observation. The experimenter then proceeds to "rephrase" his data in terms of concepts that thermodynamicists have found to be appropriate. After performing a series of manipulations on his raw data he produces a number called the "heat of reaction," expressed perhaps as calories per mole of one of the reactants. This "heat of reaction" is a long way from the original observations, but insofar as logically sound procedures were followed in going from one to the other, the heat of reaction has a right to be called an experimentally measured quantity.

The second step that the scientist makes in developing a theory is to find universal relationships between his concepts. These relationships are called laws.

Scientific theories are thus made up of two equally important parts: first, a set of defined *concepts* in terms of which we express our experience; and second, a set of *laws* relating these concepts to one another. The importance of concepts in science often tends to be forgotten because of a confusion of the meanings of the word "law." A rule that society sets up to regulate our lives is also called a "law." Man-made laws have such a great influence on our everyday affairs that it is natural, though unjustifiable, to impose a similar meaning on the "laws of nature." We tend to think of scientific laws as rules by which nature "regulates" our existence. Just as our relationship to society is dominated by man-made laws, we assume that our relationship to nature is dominated by natural laws and we come to believe that the discovery of these laws is the chief function of science. We are likely to forget that our relationship to nature depends just as much on the concepts we use in observing nature.

In a practical sense concepts have often played a more critical role than laws in the history of scientific theories, because it is usually more difficult to recognize useful concepts than it is to find the laws relating these concepts. There are many examples of this. The laws of thermodynamics are expressed in terms of the concepts of heat, energy, work, and entropy. The laws of classical mechanics use the concepts of force, velocity, and acceleration. The laws of chemistry lean heavily on a particular criterion for a chemical element. Before most of these concepts were correctly formulated, there had to be painstaking groping by some of the greatest minds of human history for hundreds or even thousands of years. Once the concepts were formulated, however, the laws relating them usually followed without much difficulty. Modifications of laws have also sometimes become necessary, because subtle weaknesses and inconsistencies have been recognized in the definitions of certain fundamental concepts. A striking example was Heisenberg's discovery of the uncertainty relations (Chapter 7), which necessitated the modification of the treatment of position and velocity. This was one of the factors leading to the revision of the laws of classical mechanics with which we shall be concerned in this book.

All students are familiar with the difficulties in understanding the concepts employed in most scientific theories, because in learning a theory they have to rediscover for themselves both the concepts and the laws that scientists have developed, and it is usually the concepts that cause them the most trouble. For instance, Newton's law of motion, $f = ma$, is easy to comprehend once the concepts of force, friction, and the relationship of time, position, velocity, and acceleration are understood. Similarly, in learning thermodynamics, the real battle is with the concepts of heat, work, energy, entropy, and free energy; the laws of thermodynamics are not at all difficult to understand once these concepts have been mastered.

It is interesting and important that quantum mechanics differs from most theories in this respect. Many of the concepts that are used in quantum mechanics are based on the familiar concepts used in classical mechanics, and since the student has usually been exposed to them earlier in his career, they do not give him any difficulty. The manner in which the laws of quantum mechanics relate these concepts to one another, however, is quite remarkable and puzzles most students. This fact seems to be one of the major difficulties that beset the student of quantum mechanics. He is subconsciously accustomed to having trouble with concepts, but to being able to perceive the laws more or less intuitively. When quantum mechanics presents him with difficulties in the reverse order, he is inclined to view it with disfavor, if not with skepticism. Yet there is no reason why the laws of nature should be intuitively obvious, and if the student is aware of this situation before he begins his study of quantum mechanics, he will perhaps not be so much disturbed by it.

b. The possibility of alternative formulations of scientific theories

A particular set of natural events may often be dealt with in a number of different but equivalent ways, depending on the set of concepts with which one chooses to begin. Thus there are several equivalent theories of classical mechanics, each starting with a different set of fundamental variables and utilizing different laws relating these variables. In classical mechanics, Newton began with forces, masses and accelerations; Lagrange, with the so-called Lagrangian function; Hamilton, with the Hamiltonian function. In thermodynamics there are the methods of Planck, Gibbs, Lewis, Caratheodory, and Brønsted; each is based on different laws relating different sets of concepts, but in the end they all agree in their description of nature.

In quantum mechanics, too, we find that several different formulations have been put forward, based on different sets of laws but resulting in equivalent descriptions of nature. The most important of these formulations are Heisenberg's "matrix mechanics," Schroedinger's "wave mechanics," and Dirac's "symbolic method." Of these three formulations, those of Dirac and Heisenberg have certain advantages, but the Schroedinger formulation is easier to understand. Fortunately nearly all of the phenomena of chemistry can be satisfactorily interpreted in terms of the Schroedinger method, which is therefore very widely used in quantum chemistry. It will be the method followed in this book.

The Schroedinger method has a pedagogical advantage in that one is often able to turn to analogies with vibrating systems in order to clarify the mathematical arguments. (It is these analogies that have given the name "wave mechanics" to the Schroedinger method.) In this way one can often (though not always!) set up physical pictures that assist the student considerably in understanding this otherwise strange theory. There are pitfalls in the use of analogies, however, and the student will do well to bear them in mind.

c. Analogies: their use and abuse

It is often found that two theories, describing two different kinds of phenomena, have close similarities in form. For instance, the differential equations describing the mechanical oscillations of a weight on a spring can be made to look the same as the equations describing the electrical oscillations of a condenser connected across an inductance. Systems of this kind are said to be "analogues" of one another.

Analogies are very helpful to most people, both in understanding old theories and in developing new ones. It is always easier to arrange new ideas in the framework of what one already knows than it is to build up an entirely new system of ideas from bare logical essentials. Furthermore, analogies provide "insight"–a very valuable thing in itself.

Unfortunately it can be dangerous to rely too heavily on analogies. Analogies are never complete. Two analogous systems always show differences at one point or another. (If they did not, they would be identical!) In making use of analogies, it is therefore important to know where to stop.

The full importance of this simple fact was not realized until the early years of this century. From Maxwell's laws of electromagnetic fields, one can derive an equation that has the same form as the equations for the elastic vibrations of a solid medium. The analogy being thus established, one soon finds oneself speaking of "electromagnetic vibrations." But if there are vibrations, something must be vibrating, and this leads one to postulate a medium for the electromagnetic vibrations–the so-called "ether." The "ether" turns out to have some remarkable properties. For instance, the famous Michelson-Morley experiment shows that the ether does not seem to realize that the earth is moving around the sun. Thus we have the famous "ether problem," which worried scientists a half century ago. Of course the ether problem was really not a problem at all, but arose because too much reliance was placed on the analogy that could be drawn between two equations.

The failure of a simple physical analogy to go beyond a certain point in explaining a new set of facts can be the source of acute discomfort if we are not prepared for it. There is something tangible and "real" about a familiar physical model, and when it fails we tend to blame nature for it and to accuse her of being unreal, complicated, paradoxical, and lacking in common sense. This, in fact, seems to be the contemporary popular attitude toward the developments that have taken place in physics in the last fifty years.

d. "Explaining" natural phenomena

In scientific literature one continually finds statements such as this: "The results of this experiment can be explained by such and such a theory." What does such a statement really mean?

The word "explain" can have a number of different meanings, depending on the way in which it is used. It is important that the student realize this.

In a limited sense we always feel that we can "explain" a situation if we are able to construct an analogy between it and something that is more familiar to us. The more extended the analogy is, the better we feel about it. But as we have just seen, analogies often exact a bitter tribute: the better an analogy is, the more we are troubled by those inevitable features that fail to correspond.

Another sense of the word "explain" can be understood by the following example. It is observed that if there is gasoline in the tank of an automobile, it is possible for the automobile to travel from one place to another. When we are asked to explain this observation, we answer by giving a description of the movements of the valves, pistons, gasoline vapors, air, ignition system, exhaust gases, and gears. If the description agrees with what we find on taking the automobile apart, the observation is said to be explained. In this sense, "explanation" is synonomous with *description*.

On the other hand, we may be asked for an explanation of the motions of the planets through the sky. The best answer would be a demonstration that the motions are consistent with the laws of mechanics and Newton's gravitational law. Here "explanation" does not imply description, but *interpretation* in terms of general laws.

For some reason, we tend to be more easily satisfied by descriptive explanations than by interpretive ones. Most of us have a keener intuitive sense of the "reality" of the workings of a gasoline engine than we do of the workings of the Newtonian system of mechanics, and we might be rather pleased if someone were to replace Newton's laws by some more tangible mechanism governing the motions of the heavenly bodies. In this respect, Aristotle's and Ptolemy's systems of astronomy are more satisfying than Newton's.

The satisfaction one obtains from descriptive explanations is, however, bound to be ephemeral. One inevitably reaches a point beyond which it is impossible to continue to give descriptive explanations, so that one is forced to turn to interpretive ones in the end. Even the operation of an automobile rests ultimately on Newton's laws.

In speaking of "the quantum mechanical explanation of the phenomena of chemistry" we therefore mean interpretive explanations, not descriptive ones. This is sometimes forgotten and some confusion and misunderstanding results. For instance, it is often said that the concept of quantum mechanical resonance is merely a mathematical trick about which nature knows nothing. In a sense this is very true. But in the same sense, Newton's laws, the laws of thermodynamics, and all of the other generalizations of science are also notions that exist only in the minds of scientists. They are, however, no less useful or significant because of this deficiency in "reality."

C. Quantum Mechanics as a Tool and as a Language

A theory may be proven in a particularly spectacular way by making predictions about future events that eventually turn out to be true. The ability of a theory to make successful predictions can be very useful, since it is often desirable to know properties of systems that are not easily accessible for experimental study. This quality in a theory is much admired, especially by experimenters, for whom theories in this way provide a *tool* for research, useful to them in much the same way as is the equipment in their laboratories. A theory can be useful in another important way, however, which is not so generally appreciated. We have seen that one of the components of a theory is a system of concepts abstracted from direct sense observations. These concepts are extremely useful to the working scientist because they provide a *language* in terms of which he can describe and arrange his observations.

Some theories are used most often as tools in supplying information that is otherwise difficult to obtain. Other theories are useful primarily because they supply a framework for the logical arrangement and discussion of past observations. Still other theories are useful in both ways. The Newtonian theory of mechanics is an example of a theory that has been primarily useful as a tool in research (e.g., the location of new planets and the calculation of stresses in a rapidly accelerating component of an engine). The Darwinian theory of evolution has been more useful as a framework or language (e.g., as a basis for the classification of living things and for the interpretation of human history). Thermodynamics may be said to be useful both as a tool (e.g., the use of heats of reaction to find the temperature dependence of equilibrium constants) and as a language (e.g., the use of activity coefficients to describe the properties of solutions). In which of these respects is quantum mechanics likely to be most useful to the average chemist?

In the ten years during which quantum mechanics was first applied to chemistry, many fundamental problems of long standing were solved. The covalent bond, the periodic system of the elements, the aromatic character of benzene, the stereochemistry of carbon and other elements, the mechanism of bimolecular reactions, the existence of free radicals, van der Waals forces, the magnetic properties of matter, the conduction of electricity by metals–these are only a few of the phenomena that were finally explained in atomic terms using quantum mechanics. These striking successes led many people to think that a new age of chemistry was at hand, in which quantum mechanics would become an important tool, making unnecessary much difficult and uncertain work in the laboratory. There was the pleasant prospect that a chemist who had run into experimental difficulties in measuring some molecular property in his laboratory could resolve his difficulties by stepping into his office and spending a few hours or days with a calculating machine. This hope has not been fulfilled. Accurate quantum mechanical computations on any but the most simple systems have turned out to be far beyond the capacities of the best computing systems. Furthermore, many problems that one would have expected would be vulnerable to quantum mechanical methods have had to await other methods for their resolution. For instance, the structure of the boron hydrides and the existence of restricted rotation about single bonds have been demonstrated by other means; we must still go into the laboratory to find the dissociation energy of the carbon–carbon bond in ethane; the absolute configurations of optical isomers have not been reliably calculated.

On the other hand, the concepts introduced by quantum mechanics have given the chemist a new and powerful language with which to interpret chemical observations. This language has continued to grow in volume and usefulness. The chemist in his everyday work makes constant use of such concepts as resonance, electronegativity, hybridization, zero point energies, tunnelling through energy barriers, atomic and molecular orbitals, nonadiabatic reactions, and the classification of the states of molecules and atoms according to symmetry and angular momentum. None of these concepts can be adequately understood without a good grasp of the principles of quantum mechanics. Indeed, the chemical bond itself (surely the fundamental entity of chemical science) is essentially a quantum mechanical phenomenon.

Unfortunately there are inherent practical and theoretical limitations on the precision with which certain quantum mechanical concepts can be defined. (This is true, for instance, of the resonance energy.) It is important that the chemist be aware of these limitations and that he understand how they arise, so that the success that these very useful concepts have had does not delude him into thinking that they are better than they really are.

Therefore, in studying quantum mechanics, the average chemist will have as his chief purpose the development of an understanding of the quantum mechanical language. This should include an appreciation of both the possibilities and the limitations of this language. In order to achieve his purpose, the chemist will find it necessary to follow the detailed mathematical discussions of various basic chemical phenomena, but in learning to make these mathematical manipulations, he should not think that he is necessarily acquiring a tool for solving experimental problems.

REFERENCE CITED

1. Lucretius, "The Nature of the Universe" (Translation by R. Latham), Penguin Books, New York, 1951.

GENERAL REFERENCE

H. Margenau, "The Nature of Physical Reality," McGraw-Hill, New York, 1950. This book contains an extensive analysis of some of the points mentioned in this chapter, such as the nature of concepts ("constructs") and the meaning of scientific explanations. There are also many references for further reading.

PART I

MATHEMATICAL BACKGROUND

Chapter 2

Some Basic Mathematical Concepts

Quantum mechanics makes use of a number of mathematical concepts which, though they are not familiar to the average chemist, are not difficult for him to grasp. This chapter gives a brief and simplified summary of some of these concepts. It is intended to serve as an introduction for those readers who do not already know them, but it may also refresh the memory and clarify the understanding of the reader who has at one time been confronted with them. The discussion in this chapter is not particularly detailed, comprehensive, or rigorous. The primary aim is to strike up or renew an acquaintance; familiarity will come later as these concepts are actually used.

A. Operators

a. Definition and examples

An *operator* is a symbol for a mathematical procedure that changes one function into another. For instance, the operation of taking a square root is represented by the operator, $\sqrt{\ }$. When this operator is applied to the function $x^2 + a$, we obtain a new function, $(x^2 + a)^{1/2}$. A list of some typical operators, along with the results of their operation on the function $x^2 + a$, is given in Table 2-1.

TABLE 2-1
SOME TYPICAL OPERATORS

Operation	Operator	Result of operation on $x^2 + a$
Multiplication by the constant, c	$c\cdot$	$cx^2 + ca$
Taking the square	$(\quad)^2$	$x^4 + 2ax^2 + a^2$
Differentiation with respect to x	d/dx	$2x$
Integration with respect to x	$\int (\quad) dx$	$x^3/3 + ax + C$
Addition of x	$x +$	$x^2 + x + a$

Operations such as these can always be written in the form of an equation,

$$\text{(operator) (function)} = \text{(new function)}$$

For instance, $(d/dx)\ (u(x)) \quad = \quad du/dx$

Symmetry operators form an important class which is best illustrated by means of examples. Suppose that we have a function, $f(x,y,z)$, where x, y and z are the usual Cartesian coordinates of a point in space. It is helpful to think of $f(x,y,z)$ as representing some physical property that varies from point to point in space, such as the temperature in a solid body into which heat is introduced nonuniformly, or the concentration of salt in a solution that is not uniform. If the body or the solution is rotated by 180° about the z-axis, the x, y, and z axes themselves being held fixed in space, then $f(x,y,z)$ will be changed into a new function, $f'(x,y,z)$. The rotation is therefore an operation in the sense described above. (It changes one function into another.) It is generally denoted by the symbol, C_{2z}, the subscript 2 referring to the number of successive rotations needed to give a complete rotation ($2 \times 180° = 360°$), and the subscript z referring to the axis of rotation. The symbol C_{2z} is thus an operator, and we can write

$$C_{2z}f(x,y,z) = f'(x,y,z)$$

But this operation is equivalent to replacing x in $f(x,y,z)$ by $-x$, replacing y by $-y$, and leaving z unchanged. That is,

$$C_{2z}f(x,y,z) = f(-x, -y, z)$$

Thus if

$$f(x,y,z) = (x-a)^2 + (y-b)^2 + (z-c)^2$$

then

$$C_{2z}f = (-x-a)^2 + (-y-b)^2 + (z-c)^2$$

Another symmetry operator is σ_{xy}, which represents the reflection of a body in the $x-y$ plane (i.e., the body is replaced by its image in a mirror located in the $x-y$ plane). It is easy to see that

$$\sigma_{xy}f(x,y,z) = f(x, y, -z)$$

The operation of inversion through the origin is represented by the symbol i. It means that a line is to be drawn from each point in the body through the origin of the coordinates and the point is to be moved along this line to a point equidistant from the origin on the other side of the origin.

This means that

$$if(x,y,z) = f(-x,-y,-z)$$

Exercises. (1) The operator S_{nz} means rotation by $360/n$ degrees about the z-axis, followed by reflection in the x–y plane. Show that

$$S_{2z}f(x,y,z) = if(x,y,z)$$
$$S_{4z}f(x,y,z) = f(-y, x, -z)$$

(2) Show that

$$C_{6z}f(x,y,z) = f\left(\frac{1}{2}x + \frac{\sqrt{3}}{2}y, -\frac{\sqrt{3}}{2}x + \frac{1}{2}y, z\right)$$

(3) Show that if

$$f = 2x^2 + y^2 + z^2$$

then

$$C_{2y}f = C_{2z}f = C_{2x}f = if = \sigma_{xy}f = f$$
$$C_{4x}f = f$$
$$C_{4y}f = x^2 + y^2 + 2z^2$$

(4) Suppose that your right hand is placed on top of a table so that it covers the origin of the coordinate system in which x and y are in the plane of the table. Describe the results of the following operations: C_{2z}, S_{4z}, σ_{xy}, i. Which of these operations transforms a right hand into a left hand?

b. The algebra of operators

In a sense operators are meaningless when they stand by themselves. Yet it is possible to formulate an algebra of operators in which the operator symbols are manipulated in a manner analogous to the manipulation of number symbols in ordinary algebra. In doing this, certain simple conventions are generally followed.

Multiplication of operators: Let α, β, and γ represent three different operations. Then the expression

$$\alpha\beta\gamma f(x)$$

means that we first operate on $f(x)$ with γ, to obtain a new function, $f'(x)$

$$\gamma f(x) = f'(x)$$

Then $f'(x)$ is operated on by β

$$\beta f'(x) = f''(x)$$

and finally $f''(x)$ is operated on by α

$$\alpha f''(x) = f'''(x)$$

If the same operator is applied several times in succession, it is written as a power. Thus

$$\alpha\alpha\alpha f(x) = \alpha^3 f(x)$$

The order of application of the operators is *always* from right to left as they are written. This is very important because the result of a series of operations is often different, depending on the sequence in which the operations are performed. For instance, let α represent "take the square root" and let β represent "multiply by four". Then

$$\alpha\beta f = \sqrt{4f} = 2\sqrt{f}$$

whereas

$$\beta\alpha f = 4\sqrt{f}$$

The two results are not at all the same.

If the result of two operations is the same regardless of the order in which the operations are performed, the two operators are said to *commute*. Thus the two operators just mentioned do not commute, whereas the operations, "multiply by two" and "multiply by four" do commute. The fact that many operators do not commute with one another gives operator algebra a different form from that of the ordinary algebra we learned in school.

Parentheses are often used in operator expressions in the following way. Consider the expression $(\alpha\,\beta\,f)\,(\gamma\,g)$, where α, β, and γ are operators, and f and g are functions. This expression means that the entire function $(\alpha\,\beta\,f)$ is to be multiplied (using ordinary multiplication) by the function $(\gamma\,g)$. On the other hand, in the expression $(\alpha\,\beta\,f\,\gamma\,g)$ it is to be understood that f acts as an operator that multiplies the function $\gamma\,g$, the result being operated on first by β and then by α. Thus the parentheses may be said to "insulate" operators from each other. Obviously it is always true that $(\alpha\,\beta\,f)\,(\gamma\,g) = (\gamma\,g)\,(\alpha\,\beta\,f)$, whereas it may not be true that $(\alpha\,\beta\,f\,\gamma\,g) = (\gamma\,g\,\alpha\,\beta\,f)$.

Addition and subtraction of operators: New operators can be formed by adding and subtracting operators. Thus if α and β are operators, then the new operators $\alpha + \beta$ and $\alpha - \beta$ are defined in the following way

$$(\alpha + \beta)f = \alpha f + \beta f$$

$$(\alpha - \beta)f = \alpha f - \beta f$$

Consider the function $(\alpha\,\beta\,f) - (\beta\,\alpha\,f)$, which can be represented by the expression $(\alpha\,\beta - \beta\,\alpha)f$. Then $(\alpha\,\beta - \beta\,\alpha)$ is an operator called the *commutator* of the pair α and β. If two operators commute, their commutator is "multiply by zero". The commutator of "take the square root" and "multiply by four" is $\sqrt{4} - 4\sqrt{\ } = -2\sqrt{\ }$.

If in operating on the sum of any two functions an operator gives the same result as the sum of the operations on the two functions separately, the operator is said to be *linear*. That is, the operator α is linear if for any two functions f and g,

$$\alpha(f+g) = (\alpha f) + (\alpha g)$$

An example of a linear operator is d/dx, since

$$d/dx(f+g) = df/dx + dg/dx$$

A nonlinear operator is "take the square root," since $\sqrt{f+g} \neq \sqrt{f} + \sqrt{g}$. Linear operators have the property that they always commute with multiplication by any number

$$\alpha n f = n \alpha f \quad (n = \text{a constant; } \alpha \text{ is linear})$$

Exercises. (1) Is it always true that $\alpha + \beta = \beta + \alpha$ and that $\alpha - \beta = -\beta + \alpha$?

(2) Let $\alpha = d/dx$, $\beta =$ "multiply by four", $\gamma =$ "add six", $f = x$ and $g = x^2$. Show that

$$(\alpha\beta f)(\gamma g) = 24 + 4x^2 = (\gamma g)(\alpha\beta f)$$

$$(\alpha\beta f\gamma g) = 24 + 12x^2$$

and

$$(\gamma g\alpha\beta f) = 6 + 4x^2$$

(3) Which of the following pairs of operators commute?

(a) $\sqrt{\ }$ and $(\quad)^2$

(b) d/dx and $\int_0^x (\quad)dx$

(c) d/dx and $3\,+$

(d) d/dx and $x\cdot$

(e) C_{4x} and σ_{xy}

(f) C_{2x} and i

(g) d/dx and C_{2x}

(h) d/dx and C_{2z}

(4) What is the commutator of $x\cdot$ and d/dx?

(5) Show that, if for two linear operators, α and β, it is true that $\alpha\beta - \beta\alpha = 1$, then $\alpha\beta^2 - \beta^2\alpha = 2\beta$. Extend this proof to show that $\alpha\beta^n - \beta^n\alpha = n\beta^{n-1}$.

(6) Which of the following are linear operators?

$$x\cdot,\ C_{2z},\ 6\,+,\ \int_0^x (\quad)dx,\ (\quad)^n,\ i.$$

(7) Show that $\alpha^2 + \beta^2 = (\alpha + i\beta)(\alpha - i\beta) - i\gamma$, where γ is the commutator of α and β and $i = \sqrt{-1}$.

c. Eigenfunctions and eigenvalues

If for a function, f, and an operator, α, we have

$$\alpha f = af$$

where a is a number, then f is said to be an *eigenfunction* (or *characteristic function*, or *proper function*) of the operator α, and a is said to be its *eigenvalue* (or *characteristic value* or *proper value*). For instance, e^{2x} is an eigenfunction of d/dx with an eigenvalue 2 because

$$(d/dx)e^{2x} = 2e^{2x}$$

Similarly, $\sin 3x$ is an eigenfunction of d^2/dx^2 with an eigenvalue of -9 and x^n, y^n, and z^n are eigenfunctions of σ_{xy} with eigenvalues 1, 1 and $(-1)^n$ respectively.

Exercises. (1) Find the value of the constant a that makes e^{-ax^2} an eigenfunction of the operator $d^2/dx^2 - Bx^2$, where B is a constant. What is the corresponding eigenvalue?

(2) Show that the following are eigenfunctions of the operator in (1) if the constant a is properly chosen, and find the eigenvalues (expressed in terms of the constant B):

(a) xe^{-ax^2}

(b) $(4ax^2 - 1)e^{-ax^2}$

(c) $(4ax^3 - 3x)e^{-ax^2}$

(3) Find a function of the form $(bx^4 + cx^2 + d)e^{-kx^2}$ that is an eigenfunction of the operator in (1). Express the eigenvalue and the four constants, b, c, d, and k, in terms of B.

(4) Make rough graphs of the five eigenfunctions obtained in exercises (1), (2) and (3), above, all to the same scale (let $B = 1$). Make a table of the eigenvalues and note the simple relation between them. These eigenfunctions and eigenvalues are basic to the quantum mechanical theory of vibrating systems, as we shall see in Chapter 6.

d. Differential operators. The Laplacian

Operators that involve differentiation are called differential operators. Several examples of differential operators have already been mentioned; another one that we shall have frequent occasion to use is the Laplacian operator, usually denoted by either of the symbols ∇^2 (called "del squared" or "nabla squared") or Δ. It is defined as

$$\nabla^2 \equiv \Delta \equiv \frac{\partial^2}{\partial x^2} + \frac{\partial^2}{\partial y^2} + \frac{\partial^2}{\partial z^2} \tag{A-1}$$

where x, y, and z are Cartesian coordinates. It is called the Laplacian operator from its occurrence in Laplace's differential equation

$$\frac{\partial^2 f}{\partial x^2} + \frac{\partial^2 f}{\partial y^2} + \frac{\partial^2 f}{\partial z^2} = 0$$

or

$$\nabla^2 f = 0$$

If one transforms the Cartesian coordinates into polar coordinates, (r, θ, φ), setting (see Fig. 2-1)

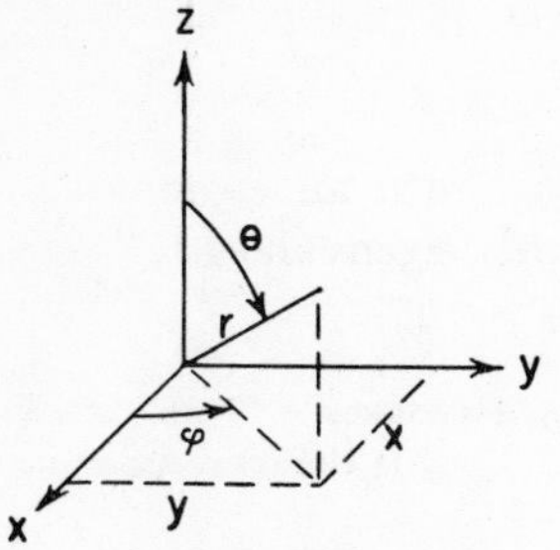

FIG. 2-1. Relation of Cartesian coordinates to polar coordinates.

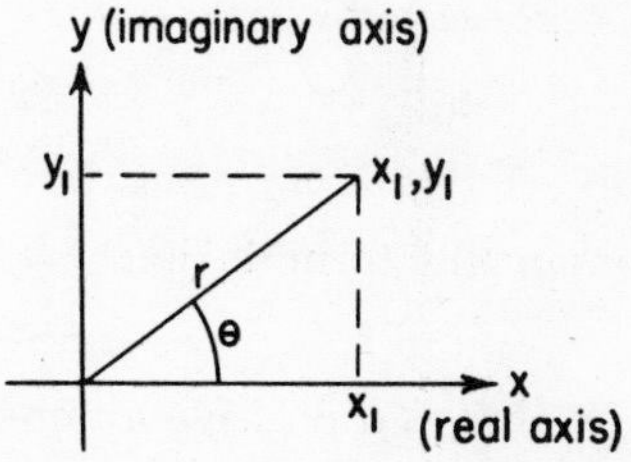

FIG. 2-2. The Argand diagram.

$$\begin{aligned} x &= r \sin\theta \cos\varphi & \qquad r &= \sqrt{x^2 + y^2 + z^2} \\ y &= r \sin\theta \sin\varphi \quad \text{or} & \qquad \theta &= \cos^{-1}(z/r) \\ z &= r \cos\theta & \qquad \varphi &= \tan^{-1}(y/x) \end{aligned}$$

the Laplacian operator takes the form

$$\begin{aligned} \nabla^2 &= \frac{1}{r^2}\frac{\partial}{\partial r}\left(r^2 \frac{\partial}{\partial r}\right) + \frac{1}{r^2 \sin\theta}\frac{\partial}{\partial\theta}\left(\sin\theta \frac{\partial}{\partial\theta}\right) + \frac{1}{r^2 \sin^2\theta}\frac{\partial^2}{\partial\varphi^2} \\ &= \frac{\partial^2}{\partial r^2} + \frac{2}{r}\frac{\partial}{\partial r} + \frac{1}{r^2}\frac{\partial^2}{\partial\theta^2} + \frac{\cot\theta}{r^2}\frac{\partial}{\partial\theta} + \frac{1}{r^2 \sin^2\theta}\frac{\partial^2}{\partial\varphi^2} \qquad \text{(A-2)} \end{aligned}$$

(This transformation will be discussed further when we consider the vibrations of gases in Chapter 3.)

Exercises. Show that the following are eigenfunctions of the Laplacian and find the eigenvalues:

(a) $\sin kx \sin my \sin nz$ (b) $1/r$ (c) $\frac{1}{r} \sin kr$

(d) $\left[\frac{\sin kr}{k^2 r^2} - \frac{\cos kr}{kr}\right] \cos\theta$ (e) $\frac{1}{r^2} \cos\theta$ (f) $\frac{1}{r^2} \sin\theta \sin\varphi$

(g) $\frac{1}{r^3} \sin^2\theta \sin 2\varphi$ (h) $\frac{1}{r^3} \sin\theta \cos\theta \cos\varphi$

B. Complex Numbers

The student will recall that the quantity $\sqrt{-1}$ is known as the *imaginary quantity* and is ordinarily denoted by the symbol i or j. It gives rise to the so-called complex numbers, that is, numbers that can be written as

$$z = x + iy$$

x and y being ordinary real numbers. The *conjugate complex* of the complex number z is simply the number with a minus sign written in front of the i; it is represented by the symbol z^* or $\bar{z}$. Thus, if

$$z = x + iy$$

then

$$z^* \equiv \bar{z} \equiv x - iy$$

It is important to note that

$$zz^* = x^2 + y^2$$

which is always a positive number. The quantity

$$|z| = \sqrt{zz^*}$$

is known as the *absolute value of* z; it is always positive and real.

Our comprehension of the behavior of real numbers is immensely aided by looking at them from a geometrical point of view. If we draw a straight line of infinite length, mark off equal increments of length and then proceed to number each mark, starting at $-\infty, \cdots, -2, -1, 0, 1, 2, \cdots$, and ending at $+\infty$, then it is possible to locate every real number on this line. If we try to do the same thing for complex numbers we are faced with the difficulty that every complex number has two parts–a real part, x in the expression $z = x + iy$, and an imaginary part, y. This means that we need two degrees of freedom to represent a complex number, and it is natural to choose a plane for this purpose. One draws two perpendicular axes on a piece of paper, labelling the horizontal axis the x-axis or real axis, and the vertical one the y-axis, or imaginary axis (Fig. 2-2). Then every conceivable complex number is represented by a point in the x-y plane. The number $z_1 = x_1 + iy_1$, for instance, corresponds to the point (x_1, y_1) in Fig. 2-2. Such a means of representing complex numbers is called an *Argand diagram.*

It is not necessary, however, to use Cartesian coordinates, x, y, in such a representation; it is also possible to represent a complex number by its polar coordinates, r, θ. Thus,

$$z = x + iy = r\cos\theta + ir\sin\theta = r(\cos\theta + i\sin\theta)$$

Now it happens that

$$\cos\theta + i\sin\theta = e^{i\theta} \tag{B-1}$$

a very important relation which is proven below. Thus in the polar coordinate representation we have

$$z = re^{i\theta} \tag{B-2}$$

It is also easy to see that

$$z^* = re^{-i\theta}$$

and

$$zz^* = r^2 e^{i\theta - i\theta} = r^2$$

Thus

$$r = |z| \tag{B-3}$$

The quantity r is therefore simply the absolute value of the complex number; it is also sometimes called the *modulus* of z. The angle θ is sometimes called the *amplitude* or *phase* of z.

Derivation of equation (B-1). It is well known that the following series are valid for all values of x

$$\begin{aligned} e^x &= 1 + x + (x^2/2!) + (x^3/3!) + (x^4/4!) + \cdots \\ \sin x &= x - (x^3/3!) + (x^5/5!) - \cdots \\ \cos x &= 1 - (x^2/2!) + (x^4/4!) - \cdots \end{aligned}$$

If we write $x = i\theta$ and remember that $i^2 = -1$, $i^3 = -i$, $i^4 = 1$, etc., then we find

$$\begin{aligned} e^{i\theta} &= 1 + i\theta - \theta^2/2! - i\theta^3/3! + \theta^4/4! + i\theta^5/5! - \theta^6/6! - \cdots \\ &= (1 - \theta^2/2! + \theta^4/4! - \cdots) + i(\theta - \theta^3/3! + \theta^5/5! - \cdots) \\ &= \cos\theta + i\sin\theta \end{aligned}$$

Exercises. (1) Show that

$$\sin x = \frac{e^{ix} - e^{-ix}}{2i} \quad \text{and} \quad \cos x = \frac{e^{ix} + e^{-ix}}{2}$$

(2) Using the result of (1), show that if m and n are integers,

$$\int_{-\pi}^{\pi} \cos mx \cos nx \, dx = \int_{-\pi}^{\pi} \sin mx \sin nx \, dx = \begin{cases} 0 \text{ if } m \neq n \\ \pi \text{ if } m = n \neq 0 \end{cases}$$

$$\int_{-\pi}^{\pi} \cos mx \sin nx \, dx = 0$$

for all integral values of m and n.

(3) Show that (a) $(2 + 3i)/(1 + i) = \frac{5}{2} + \frac{1}{2}i$

(b) $\sqrt{1 + i} = 2^{1/4} (\cos \pi/8 + i \sin \pi/8)$

(c) $\log (x + iy) = \log |x + iy| + i \tan^{-1}(y/x)$

(d) $\sin (x + iy) = \sin x \cosh y + i \cos x \sinh y$

Note: $\cosh y =$ "hyperbolic cosine of y" $= \frac{1}{2}(e^{y} + e^{-y})$
$\sinh y =$ "hyperbolic sine of y" $= \frac{1}{2}(e^{y} - e^{-y})$

(4) Give a simple rule for finding the location of the complex conjugate of a complex number on the Argand diagram.

C. Well Behaved Functions

A function $\psi(x, y, \cdots)$ is said to be well behaved if three conditions are satisfied:

(1) It is *single valued* (i.e., for each value of the variables $x, y, \cdots$; there is only one value of the function ψ). This means that the function must not "bend over" on itself, as $f(x)$ does in the vicinity of x' in Fig. 2-3.

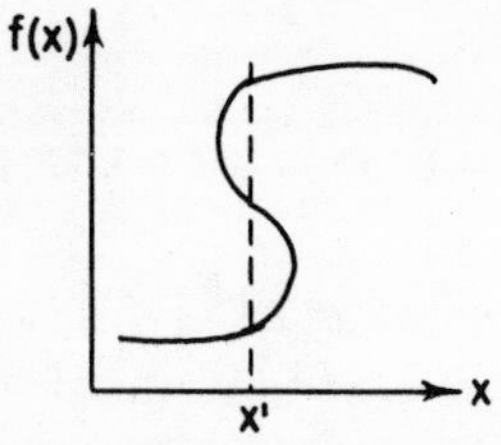

FIG. 2-3. A function of x that is not single valued at $x = x'$.

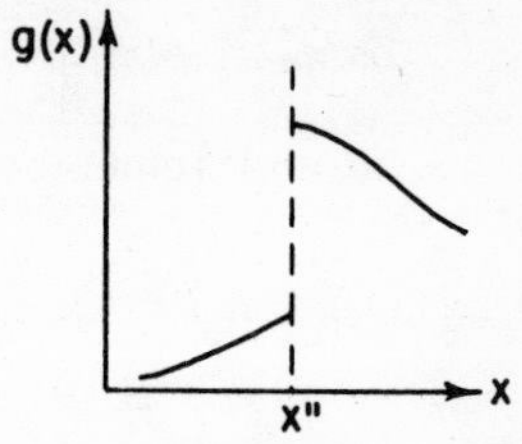

FIG. 2-4. A function of x that is not continuous at $x = x''$.

(2) It is *continuous* (i.e., there are no sudden changes in ψ as its variables are changed. Thus Fig. 2-4 shows a function $g(x)$ that is discontinuous at x'').

(3) *The integral of the square of the absolute value of the function is finite*, the limits of integration being the entire range of all of the variables. That is, if x can vary from a to a', y can vary from b to b', etc., then $\int_a^{a'} \int_b^{b'} \ldots . \psi\psi^* \, dx \, dy \cdots =$ a finite number. An example of a well behaved function is $e^{-x^2/2}$, where x can have values between $-\infty$ and $+\infty$, because $\int_{-\infty}^{\infty} e^{-x^2} \, dx = \sqrt{\pi}$. On the other hand, the function x^2 is not well behaved in this range of values of x because $\int_{-\infty}^{\infty} x^4 \, dx$ is infinite. Sometimes we are concerned with variables which can have only a limited range of values. For instance θ in $e^{i\theta}$ can vary only from 0 to 2π if it represents a polar coordinate. Thus $e^{i\theta}$ is well behaved

since $\int_0^{2\pi} e^{i\theta}\, e^{-i\theta}\, d\,\theta = 2\,\pi$. On the other hand, e^x and e^{ix} are not well behaved if x can vary from $-\infty$ to $+\infty$.

D. Vectors

Quantities which have only magnitude are known as *scalars* (for instance, mass, volume, time, wave length, temperature, entropy). Quantities which have a direction in space as well as magnitude are called *vectors* (e.g., force, velocity, concentration gradient, flow of heat through a body). The symbols for scalars and vectors are usually distinguished by underlining the vectors (or in printed books, by writing them in bold faced type). Vectors can be represented by arrows whose length is equal to their magnitude and whose direction is parallel to the direction of the vector.

a. Addition of vectors

Two vectors can be added together to give a new vector. This is done by the so-called parallelogram law: the sum of two vectors, **A** and **B**, is the diagonal of the parallelogram of which **A** and **B** are the adjacent sides (Fig. 2-5).

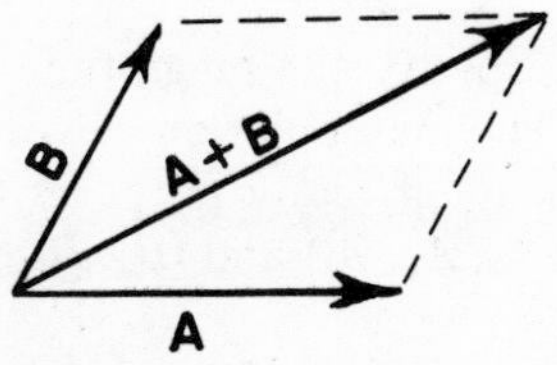

FIG. 2-5. The addition of vectors.

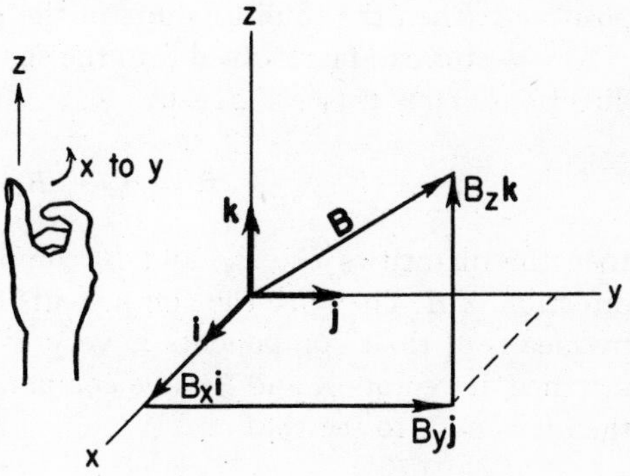

FIG. 2-6. The resolution of a vector into components along the x, y, and z axes.

The vector $-$ **B** is simply the vector **B** turned about in space by 180°. Obviously the sum of **B** and $-$**B** is a vector of zero length, or

$$\mathbf{B} + (-\mathbf{B}) = 0$$

A *unit vector* is one whose magnitude is unity. Any vector can be expressed as the product of its magnitude (a scalar generally represented by placing vertical lines on either side of the vector) and a unit vector in the direction of the vector. Thus if $\boldsymbol{b}$ is a unit vector in the direction of $\mathbf{B}$, we can write

$$\mathbf{B} = |\mathbf{B}|\, \boldsymbol{b}$$

b. Resolution of vectors into components

Just as any group of vectors can be added together to give a single vector, so any one vector can be considered to be the sum of a group of other vectors, each having a specified direction. The latter are called *components* of the given vector in these directions. The most common system of resolution of a vector into components is the following, and when one speaks of "the components" of a vector, it is usually this system which is implied.

A Cartesian coordinate system is set up as shown in Fig. 2-6, with unit vectors, $\boldsymbol{i}$, $\boldsymbol{j}$, and $\mathbf{k}$ directed respectively along the positive x, y, and z axes. Unless otherwise stated, it is always understood that x, y, and z form a *right-handed system*; that is, if y and z in Fig. 2-6 are in the plane of the paper, x sticks *out of* the paper toward the reader; or, if one grasps the z-axis with one's right hand, with fingers pointing from positive x toward positive y, then the thumb points in the positive z direction.

Any vector can be resolved into the sum of three vectors in the $\boldsymbol{i}$, $\boldsymbol{j}$, and $\mathbf{k}$ directions. Thus if (see Fig. 2-6)

$$\mathbf{B} = B_x\boldsymbol{i} + B_y\boldsymbol{j} + B_z\mathbf{k}$$

then the quantities B_x, B_y, and B_z completely specify the magnitude and direction of $\mathbf{B}$. They are the components of $\mathbf{B}$. The manipulation of vectors by means of their components is very convenient for many purposes. For instance, if vectors $\mathbf{A}$ and $\mathbf{B}$ have components (A_x, A_y, A_z) and (B_x, B_y, B_z), then it is easy to see that

$$\mathbf{A} + \mathbf{B} = (A_x + B_x)\boldsymbol{i} + (A_y + B_y)\boldsymbol{j} + (A_z + B_z)\mathbf{k}$$

Thus the components of the sum of two vectors are simply the sums of the corresponding pairs of components. It is much easier and more accurate to add vectors in this way than to add them by drawing parallelograms.

Exercise. A cube has sides of length 2. A right-handed Cartesian coordinate system is set up with its origin at the center of the cube and with axes parallel to the sides of the cube (see Fig. 2-7). Write down the following vectors in terms of the components parallel to $\boldsymbol{i}$, $\boldsymbol{j}$ and $\boldsymbol{k}$: $\boldsymbol{OA}$, $\boldsymbol{AB}$, $\boldsymbol{OB}$, $\boldsymbol{OC}$. Show that $\boldsymbol{OB} = \boldsymbol{OA} + \boldsymbol{AB}$, both according to the parallellogram law and by the addition of components.

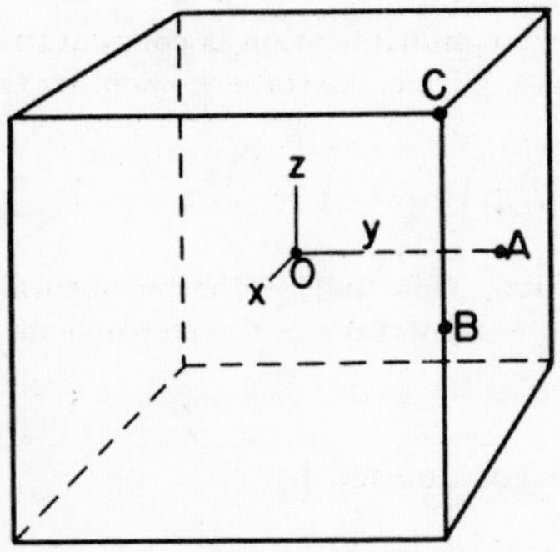

FIG. 2-7.

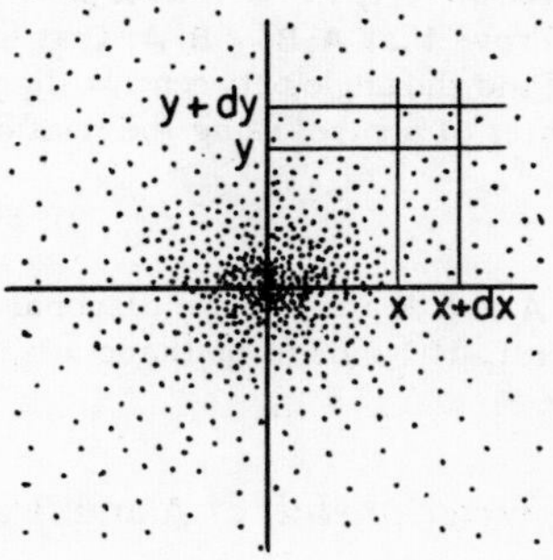

FIG. 2-8. A two-dimensional probability distribution resulting from shooting at a target.

c. Multiplication of vectors

There are two different methods of combining vectors by multiplication. They are known as the scalar product and the vector product.

The *scalar product* of two vectors, $\boldsymbol{A}$ and $\boldsymbol{B}$, is a scalar quantity defined by

$$\boldsymbol{A}\cdot\boldsymbol{B} = |\boldsymbol{A}|\,|\boldsymbol{B}| \cos \theta \tag{D-1}$$

where θ is the angle between the two vectors. The scalar product, $\boldsymbol{A}\cdot\boldsymbol{B}$, is also sometimes written as $(\boldsymbol{A},\boldsymbol{B})$. It is evidently equal to the length of one of the two vectors ($|\boldsymbol{A}|$) multiplied by the projection of the other vector on it ($|\boldsymbol{B}| \cos \theta$). An important example is the work done in moving against a force. Displacement and force are both vectors, and work is a scalar defined by the product of the displacement times the component of the force in the direction of the displacement. If $\boldsymbol{F}$ is the force, $\boldsymbol{s}$ is the displacement, and θ is the angle between $\boldsymbol{F}$ and $\boldsymbol{s}$, then it is clear that the work done in moving $\boldsymbol{F}$ a distance $\boldsymbol{s}$ is

$$\text{Work} = (|\boldsymbol{F}| \cos \theta)\,|\boldsymbol{s}| = \boldsymbol{F}\cdot\boldsymbol{s}$$

It is easy to see that

$$\boldsymbol{i}\cdot\boldsymbol{i} = \boldsymbol{j}\cdot\boldsymbol{j} = \boldsymbol{k}\cdot\boldsymbol{k} = 1$$

$$\boldsymbol{i}\cdot\boldsymbol{j} = \boldsymbol{j}\cdot\boldsymbol{i} = \boldsymbol{j}\cdot\boldsymbol{k} = \boldsymbol{k}\cdot\boldsymbol{j} = \boldsymbol{k}\cdot\boldsymbol{i} = \boldsymbol{i}\cdot\boldsymbol{k} = 0$$

Furthermore $\quad \boldsymbol{A}\cdot\boldsymbol{B} = (A_x\boldsymbol{i} + A_y\boldsymbol{j} + A_z\boldsymbol{k})\cdot(B_x\boldsymbol{i} + B_y\boldsymbol{j} + B_z\boldsymbol{k})$

$$= A_xB_x + A_yB_y + A_zB_z \tag{D-2}$$

Exercises. (1) Prove that $\boldsymbol{A}\cdot\boldsymbol{A} = |\boldsymbol{A}|^2$ for any vector.

(2) Prove that $\boldsymbol{A}\cdot\boldsymbol{B} = \boldsymbol{B}\cdot\boldsymbol{A}$; that is, that scalar multiplication is commutative.

(3) Find the angle between two diagonals passing from one corner to another through the center of a cube, using the relation

$$\cos\theta = \boldsymbol{A}\cdot\boldsymbol{B}/(|\boldsymbol{A}|\ |\boldsymbol{B}|)$$

where $\boldsymbol{A}$ and $\boldsymbol{B}$ are the two diagonals in question. This angle is the tetrahedral angle between C–H bonds in methane, etc., which is so important in the stereochemistry of carbon.

The *vector product* of $\boldsymbol{A}$ and $\boldsymbol{B}$ is the vector defined by

$$\boldsymbol{A}\times\boldsymbol{B} = |\boldsymbol{A}|\ |\boldsymbol{B}|\ \boldsymbol{c}\sin\theta \tag{D-3}$$

where θ is the angle between $\boldsymbol{A}$ and $\boldsymbol{B}$ and $\boldsymbol{c}$ is a unit vector perpendicular to $\boldsymbol{A}$ and $\boldsymbol{B}$ such that $\boldsymbol{A}$, $\boldsymbol{B}$, $\boldsymbol{c}$ forms a right-handed system. (If the fingers of the right hand point from $\boldsymbol{A}$ to $\boldsymbol{B}$, the thumb points along $\boldsymbol{c}$.) The vector product, $\boldsymbol{A}\times\boldsymbol{B}$, is also sometimes written $[\boldsymbol{A},\boldsymbol{B}]$. The length of $\boldsymbol{A}\times\boldsymbol{B}$ is numerically equal to the area of the parallelogram whose sides are $\boldsymbol{A}$ and $\boldsymbol{B}$, and its direction is perpendicular to the plane of this parallelogram.

It is easy to see that

$$\begin{aligned}
&\boldsymbol{i}\times\boldsymbol{i} = \boldsymbol{j}\times\boldsymbol{j} = \boldsymbol{k}\times\boldsymbol{k} = 0\\
&\boldsymbol{i}\times\boldsymbol{j} = \boldsymbol{k} \qquad \boldsymbol{j}\times\boldsymbol{i} = -\boldsymbol{k}\\
&\boldsymbol{j}\times\boldsymbol{k} = \boldsymbol{i} \qquad \boldsymbol{k}\times\boldsymbol{j} = -\boldsymbol{i}\\
&\boldsymbol{k}\times\boldsymbol{i} = \boldsymbol{j} \qquad \boldsymbol{i}\times\boldsymbol{k} = -\boldsymbol{j}
\end{aligned} \tag{D-4}$$

Furthermore, it is always true that

$$\boldsymbol{A}\times\boldsymbol{B} = -\boldsymbol{B}\times\boldsymbol{A}$$

That is, vector multiplication is not commutative.

Also

$$\begin{aligned}
\boldsymbol{A}\times\boldsymbol{B} &= (A_x\boldsymbol{i} + A_y\boldsymbol{j} + A_z\boldsymbol{k}) \times (B_x\boldsymbol{i} + B_y\boldsymbol{j} + B_z\boldsymbol{k})\\
&= A_xB_x\boldsymbol{i}\times\boldsymbol{i} + A_xB_y\boldsymbol{i}\times\boldsymbol{j} + A_xB_z\boldsymbol{i}\times\boldsymbol{k}\\
&\quad + A_yB_x\boldsymbol{j}\times\boldsymbol{i} + A_yB_y\boldsymbol{j}\times\boldsymbol{j} + A_yB_z\boldsymbol{j}\times\boldsymbol{k}\\
&\quad + A_zB_x\boldsymbol{k}\times\boldsymbol{i} + A_zB_y\boldsymbol{k}\times\boldsymbol{j} + A_zB_z\boldsymbol{k}\times\boldsymbol{k}\\
&= (A_yB_z - A_zB_y)\boldsymbol{i} + (A_zB_x - A_xB_z)\boldsymbol{j} + (A_xB_y - A_yB_x)\boldsymbol{k}
\end{aligned} \tag{D-5a}$$

This can also be written in the form of the determinant (see page 46)

$$\mathbf{A} \times \mathbf{B} = \begin{vmatrix} \mathbf{i} & \mathbf{j} & \mathbf{k} \\ A_x & A_y & A_z \\ B_x & B_y & B_z \end{vmatrix} \tag{D-5b}$$

An example of the occurrence of the vector product that is of considerable importance in quantum mechanics is the angular momentum of a particle moving about a fixed point. Let the mass of the particle be m and suppose that it moves with a velocity $\mathbf{v}$ (a vector) at a distance $\mathbf{r}$ (also a vector) from the point. Then the angular momentum $\mathbf{M}$ is defined as the vector whose direction is perpendicular to both $\mathbf{v}$ and $\mathbf{r}$, and whose sense is such that $\mathbf{r}$, $\mathbf{v}$, $\mathbf{M}$ form a right-handed system. Its magnitude is the product of the component of linear momentum perpendicular to $\mathbf{r}$ with the distance from the point. Thus

$$|\mathbf{M}| = m|\mathbf{v}|\ |\mathbf{r}| \sin \theta$$

It is evident that the angular momentum is simply the vector product of the linear momentum and the radius vector, or

$$\mathbf{M} = m\,\mathbf{r} \times \mathbf{v} \tag{D-6}$$

E. Probability Functions and Average Values

If one shoots a gun at a target many times a pattern such as that shown in Fig. 2-8 can be expected. Let N be the number of shots hitting the target. The location of a point on the target can be specified by giving its coordinates (x, y) in a suitable two dimensional Cartesian coordinate system. Then if N is large and the marksmanship is random it is possible to set up a function, $f(x, y)$, such that

$$Nf(x, y)\, dx\, dy$$

is the number of hits in the rectangular region of the target between x, $x + dx$, y, and $y + dy$. The function $f(x, y)$ is called a *probability distribution* (or *probability function*).

It is obvious that since the total number of shots hitting someplace on the target is N, then

$$\iint_{\text{target}} Nf\, dx\, dy = N$$

or

$$\iint_{\text{target}} f\, dx\, dy = 1 \tag{E-1}$$

The function f defined in this way is said to be *normalized.*

Suppose that a prize of $g(x,y)$ dollars is given for each shot hitting the point (x,y). Then the total prize won with N shots will be

$$G = \iint Ngf\,dx\,dy$$

the integration being performed over the entire target. The average prize per shot will be

$$\bar{g} = \frac{G}{N} = \frac{\iint gf\,dx\,dy}{\iint f\,dx\,dy} \tag{E-2a}$$

or if f is normalized

$$\bar{g} = \iint gf\,dx\,dy \tag{E-2b}$$

It is easy to generalize this expression so as to be able to calculate the average value of any property associated with a system for which a probability function can be written.

Exercises. (1) Suppose that the probability function for hitting the point (x,y) on a target is $f = A\exp(-(x^2 + y^2)/\sigma^2)$ and that a prize awarded is

$$g = \$\,2.00 \text{ if } \sqrt{x^2 + y^2} < a$$

$$g = 0 \qquad \text{if } \sqrt{x^2 + y^2} > a$$

The constant σ is a measure of the accuracy of the marksmanship of the shooter, a small value indicating good marksmanship. Find the value of the constant A which normalizes the function f. If a charge of one dollar is made for each shot fired, what is the largest value of σ that the marksman can have if he is not to lose money in the shooting match? (Answer: $\sigma = 1.20\,a$).

(2) The probability function for a molecular velocity c in an ideal gas at temperature T is

$$f(c) = Ac^2 \exp(-mc^2/2kT)$$

where m is the molecular mass, k is the Boltzmann gas constant, and A is the normalization constant. If the kinetic energy of a molecule moving with velocity c is $\frac{1}{2}mc^2$, find the mean kinetic energy per molecule in the gas. Note that

$$\int_0^\infty x^{2n} e^{-ax^2}\,dx = [(2n)!/(2^{2n+1}\,n!a^n)]\sqrt{\pi/a}$$

F. Series Expansions of Functions

Certain sets of functions have the remarkable property that they may be used as the basis for the expansion of other arbitrary functions. Such sets of functions play a very important part in quantum mechanics, and in order to familiarize the student with the notion of series expansions we shall consider two special types of series.

a. Power series

The student is undoubtedly already familiar with the Taylor-Maclaurin series, which is based on the integral powers of $(x - x_0)$, where x_0 is a constant. In this series we have

$$f(x) = \sum_{n=0}^{\infty} a_n (x - x_0)^n$$

where the constant coefficients a_n are given by

$$a_n = (1/n!)\,(d^n f/dx^n)_{x=x_0}$$

Such a series is often valid over only a limited range of values of x. For instance, the expansion of the function $1/(1 - x)$ about $x_0 = 0$ gives

$$1/(1 - x) = 1 + x + x^2 + x^3 + \cdots$$

which is valid only in the range $-1 < x < 1$, since outside of this range the series fails to converge. The expansions

$$e^x = 1 + x + (x^2/2!) + (x^3/3!) + \cdots$$

$$\sin x = x - (x^3/3!) + (x^5/5!) - \cdots$$

$$\cos x = 1 - (x^2/2!) + (x^4/4!) - \cdots$$

are, on the other hand, valid for all values of x. Many functions cannot be expanded at all about certain values of x_0, because some or all of their derivatives are infinite. An example is $f(x) = x^{1/2}$, which cannot be expanded about $x_0 = 0$.

b. Fourier series

A more powerful basic set of expansion functions was unearthed in 1807 by Fourier, who showed that in the range $-\pi < x < \pi$ almost any function, $f(x)$, can be expressed in the form

$$f(x) = \sum_{n=0}^{\infty} (a_n \cos nx + b_n \sin nx) \tag{F-1}$$

where a_n and b_n are constants whose values will be given below. A function can always be expanded in terms of sines and cosines in this way if it satisfies the following conditions: (1) If it has only a finite number of discontinuities. (2) If it is single valued except at the discontinuities. (3) If it oscillates only a finite number of times. (Can the function sin $(1/x)$ be expanded in a Fourier series?) For a proof of the validity of the Fourier expansion, see *1*, Chapter 1; or *2*, p. 175.

The Fourier series differs from the power series in several important respects. The range of convergence of the power series is different for different functions; for the Fourier series (F-1) on the other hand, the range of convergence is always $-\pi$ to $+\pi$. Furthermore, the power series expansion is not possible at all for many unremarkable functions, whereas the Fourier series will represent all but the most extraordinary functions.

The validity of the Fourier expansion is by no means self-evident. It is interesting that in the half-century before Fourier made his discovery, several mathematicians (notably Lagrange and Euler) had encountered these expansions in specific problems, but had failed to generalize them because they seemed intuitively unreasonable. Fourier encountered some difficulty in having his theory accepted for the same reason.

The constants a_n and b_n in equation (F-1) are readily evaluated because of the fact that, as is shown in Exercise (2) on page 21

$$\int_{-\pi}^{\pi} \cos mx \sin nx \, dx = 0 \tag{F-2}$$

m and n being integers, and

$$\left.\begin{aligned} &\int_{-\pi}^{\pi} \cos mx \cos nx \, dx = 0 \qquad \text{(F-3)} \\ &\int_{-\pi}^{\pi} \sin mx \sin nx \, dx = 0 \qquad \text{(F-4)} \end{aligned}\right\} (m \neq n)$$

If the integral of the product of two functions vanishes in this way the functions are said to be *orthogonal*. (Of course, the orthogonality of a pair of functions depends on the limits of the integration, and these must be stated if they are not obvious from context.) If m and n in equations (F-3) and (F-4) are equal, we have for $m \neq 0$

$$\int_{-\pi}^{\pi} \sin^2 mx\, dx = \int_{-\pi}^{\pi} \cos^2 mx\, dx = \pi \tag{F-5}$$

Therefore, if we multiply equation (F-1) on both sides by cos kx and integrate from $-\pi$ to π, we find that all terms on the right other than the kth term vanish, giving

$$a_k = (1/\pi) \int_{-\pi}^{\pi} f(x) \cos kx\, dx \tag{F-6a}$$

The coefficient a_0 is, however, a special case, and we find

$$a_0 = (1/2\pi) \int_{-\pi}^{\pi} f(x)\, dx \tag{F-6b}$$

Similarly, multiplying by sin kx, we find

$$b_k = (1/\pi) \int_{-\pi}^{\pi} f(x) \sin kx\, dx \tag{F-7}$$

Let us expand the function $f(x) = x$ as an illustration of the Fourier series. Then

$$a_k = (1/\pi) \int_{-\pi}^{\pi} x \cos kx\, dx = 0$$

as is easily shown by making a graph of the integrand, from which it will be seen that the area under the curve to the left of $x = 0$ is equal and opposite in sign to that on the right, so that the two areas cancel.*

$$b_k = (1/\pi) \int_{-\pi}^{\pi} x \sin kx\, dx = (1/\pi k^2) \int_{-\pi k}^{\pi k} y \sin y\, dy$$

where $y = kx$. Integrating by parts

$$b_k = (1/\pi k^2) \left[-y \cos y + \int \cos y\, dy \right]_{-k\pi}^{k\pi} = (-1)^{k-1}\,(2/k)$$

Thus

$$x = 2 \sin x - \sin 2x + (2/3) \sin 3x - (1/2) \sin 4x + \cdots$$

* If a function $f(x)$ is unchanged in value when x is replaced by $-x$, the function is said to have *even* symmetry about the origin. If the function merely changes its sign on such a replacement, it is said to have *odd* symmetry about the origin. The functions x, $\sin x$, x^3, and $\tan^{-1} x$ are odd functions, while x^2 and $\cos x$ are even. It is easy to see that the integral of the product of any even function with any odd function must vanish if the limits of integration lie on either side of the origin and at equal distances from the origin. This is a simple example of the use of symmetry operations in evaluating integrals. We shall see that in quantum mechanics extensive use is made of many kinds of operators for this purpose.

If a series of drawings is made (Fig. 2-9) showing, successively, the first term in this series, the sum of the first two terms, the sum of the first three terms,

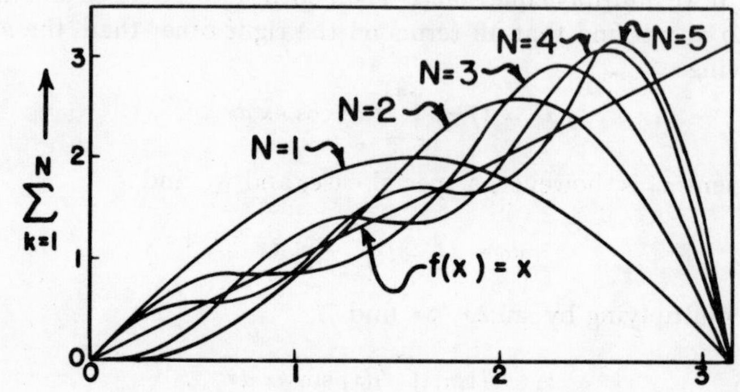

FIG. 2-9. Comparison of the incomplete Fourier series, $-\sum_{k=1}^{N}(-1)^k(2/k)\sin kx$, with the function $f(x) = x$. Note the improvement in the approximation as N increases from 1 to 5. (Curves are plotted only for x between 0 and π; similar curves are obtained if x lies between $-\pi$ and 0, but the ordinates are negative.)

etc., we can gain some insight into the fascinating versatility of sines and cosines, which are able to cooperate in reproducing almost any conceivable function. It is especially interesting that discontinuous functions can be represented in this way (cf. Exercises (3) and (4), below).

Other kinds of Fourier series are also possible. In the interval, $0 < x < \pi$, arbitrary functions can be expanded in terms of either sines or cosines alone

$$f(x) = \sum_{n=0}^{\infty} a_n \cos nx \tag{F-8}$$

$$f(x) = \sum_{n=0}^{\infty} b_n \sin nx \tag{F-9}$$

The student should show that for these two kinds of series,

$$a_n = (2/\pi)\int_0^{\pi} f(x)\cos nx\,dx; \quad a_0 = (1/\pi)\int_0^{\pi} f(x)\,dx \tag{F-10}$$

$$b_n = (2/\pi)\int_0^{\pi} f(x)\sin nx\,dx \tag{F-11}$$

The interval over which the series is valid can also easily be changed merely by introducing a scaling factor. Thus the expansion

$$f(x) = \sum_{n=0}^{\infty} a_n \cos(n\pi x/l) + b_n \sin(n\pi x/l) \tag{F-12}$$

is valid in the range $-l < x < l$.

c. Other series

A great many other sets of mutually orthogonal functions can serve as a basis for the expansion of arbitrary functions over various intervals. Such groups of functions are called *complete sets,* and will be discussed in more detail in Chapter 4.

Exercises. Find the Fourier expansions of some of the following functions and show by means of a graph how in each instance the sum of the first few terms begins to approach the true function.

(1) $f(x) = \sin^2 x$,

(*i*) in the range $-\pi$ to π

(*ii*) as a sine series (cf. Equation (F-9)) in the range 0 to π

(2) $f(x) = x$ as a cosine series (cf. Equation (F-8)) in the range 0 to π

(3) $f(x) = \begin{cases} x + 1 \text{ from } -\pi \text{ to } 0 \\ x - 1 \text{ from } 0 \text{ to } \pi \end{cases}$

(4) $f(x) = \begin{cases} 1 \text{ from } -\pi \text{ to } 0 \\ x - 1 \text{ from } 0 \text{ to } \pi \end{cases}$

(5) $f(x) = x^2$

(*i*) from $-\pi$ to π

(*ii*) as a sine series in the range 0 to π

d. The Fourier integral

Because of the relationship between $e^{\pm i\theta}$ and $\cos\theta$ and $\sin\theta$ (Section *B* of this chapter), the Fourier series can be written in the equivalent form

$$f(x) = \sum_{n=-\infty}^{\infty} A_n e^{inx} \tag{F-13}$$

where

$$A_n = \frac{1}{2\pi} \int_{-\pi}^{\pi} f(x) e^{-inx}\, dx \tag{F-14}$$

Similarly Equation (F-12) can be written

$$f(x) = \sum_{n=-\infty}^{\infty} A_n e^{in\pi x/l} \tag{F-15}$$

where

$$A_n = \frac{1}{2l} \int_{-l}^{l} f(x) e^{-in\pi x/l}\, dx \tag{F-16}$$

It is interesting to see what happens to this representation when we let l approach infinity, so that the function $f(x)$ is represented over the entire range of the variable x, from $-\infty$ to ∞. Let us make the substitution $k = n\pi/l$. Note that if l is large, the new quantity k changes in small increments, $\Delta k = \pi/l$, when n is changed by the increment $\Delta n = 1$. Thus as $l \to \infty$, k becomes a continuous variable and we can write the coefficients A_n as ordinary functions of k, $A(k)$. It is also convenient to write

$$lA(k)/\pi = g(k)/\sqrt{2\pi}$$

If, therefore, we set $1 = \Delta n$ in Equation (F-15), we may write

$$f(x) = \lim_{l \to \infty} \sum_{k=-\infty}^{\infty} A_n e^{in\pi x/l} \Delta n$$

$$= \lim_{\Delta k \to 0} \sum_{k=-\infty}^{\infty} \frac{lA(k)}{\pi} e^{ikx} \Delta k$$

or

$$f(x) = \frac{1}{\sqrt{2\pi}} \int_{-\infty}^{\infty} g(k) e^{ikx}\, dk \tag{F-17}$$

where

$$g(k) = \frac{1}{\sqrt{2\pi}} \int_{-\infty}^{\infty} f(x) e^{-ikx}\, dx \tag{F-18}$$

A rigorous proof of the validity of equations (F-17) and (F-18) may be found in standard treatises (e.g. *3, 4*). Equation (F-17) is known as a *Fourier integral* and $f(x)$ and $g(k)$ are said to be *Fourier transforms* of one another.

It is of interest to mention that there are devices (called harmonic analyzers) which automatically compute the coefficients of the Fourier expansion of a function. A spectroscope is, in fact, one example of just such a device. The electric field in a beam of polychromatic light is usually a complicated function of the distance along the beam, and the spectroscope resolves this function into its harmonic components. In order to see this, consider a light wave made up of a superposition of monochromatic waves having wave lengths $\lambda_1, \lambda_2, \lambda_3, \cdots$. The resulting wave can be represented by the function

$$f(x) = A_1 \sin 2\pi x/\lambda_1 + A_2 \sin 2\pi x/\lambda_2 + A_3 \sin 2\pi x/\lambda_3 + \cdots \qquad \text{(F-19)}$$

where $A_1, A_2, A_3, \cdots$ are the amplitudes of the respective monochromatic components. The student can readily show that $g(k)$, the Fourier transform of the function $f(x)$ in (F-19), is very large when $k = 2\pi/\lambda_1, 2\pi/\lambda_2, 2\pi/\lambda_3, \cdots$, and that it is negligible at other values of k. Furthermore, the values of $g(k)$ at these respective values of k are in the ratio $A_1 : A_2 : A_3 : \cdots$. That is, the Fourier transform is proportional to the amplitude of the corresponding monochromatic component in the incident light beam. But a spectroscope would merely split the light beam represented by Equation (F-19) into the same monochromatic components, and the observed intensities would be in the ratio

$$A_1^2 : A_2^2 : A_3^2 : \cdots = g^2(2\pi/\lambda_1) : g^2(2\pi/\lambda_2) : g^2(2\pi/\lambda_3) : \cdots$$

Thus the Fourier transform, Equation (F-18), and the spectroscope do essentially the same thing to $f(x)$. The only difference is that, as ordinarily used, the spectroscope gives the square of the Fourier transform rather than its first power.

Exercises. (1) Prove by direct substitution of (F-19) into (F-18) that $g(k)$, the Fourier transform of $f(x)$ in Equation (F-19), is very large when $k = 2\pi/\lambda_1, 2\pi/\lambda_2, 2\pi/\lambda_3, \cdots$, and that the corresponding values of $g(k)$ are in the ratio $A_1 : A_2 : A_3 : \cdots$.

(2) Find the Fourier transform of the function

$$f(x) = \begin{cases} 0; & x < -\pi \\ x; & -\pi < x < \pi \\ 0; & x > \pi \end{cases}$$

[Answer: $g(k) = 2(k\pi \cos k\pi + \sin k\pi)/(i\sqrt{2\pi}\, k^2)$]

(3) Find the Fourier transform of the function

$$f(x) = \begin{cases} 0; & x < -n\lambda \\ \sin 2\pi x/\lambda; & -n\lambda < x < n\lambda \\ 0; & x > n\lambda \end{cases}$$

[Answer: $g(k) = -2\sqrt{2\pi}\,\lambda(k^2\lambda^2 - 4\pi^2)^{-1} \sin nk\lambda$]

In Exercise (3), $f(x)$ represents a train of monochromatic waves of finite duration. The Fourier transform in this instance shows that such a wave does not appear to be monochromatic when viewed through a spectroscope. It is seen, however, that the spectral intensity (as given by $|g(k)|^2$) is large only when $k\lambda \approx 2\pi$. If we write $k = 2\pi(1 + e)/\lambda$, where e is a new variable, we find that

$$|g(k)|^2 \propto \frac{\sin^2 n\pi e/2}{(2e + e^2)^2}$$

This function has a series of peaks at $e = 0, \pm 3/n, \pm 5/n, \cdots$, whose heights are respectively in the ratio $\pi^2/16:1/36:1/100:\cdots$ (see Fig. 2-10). Therefore most of the

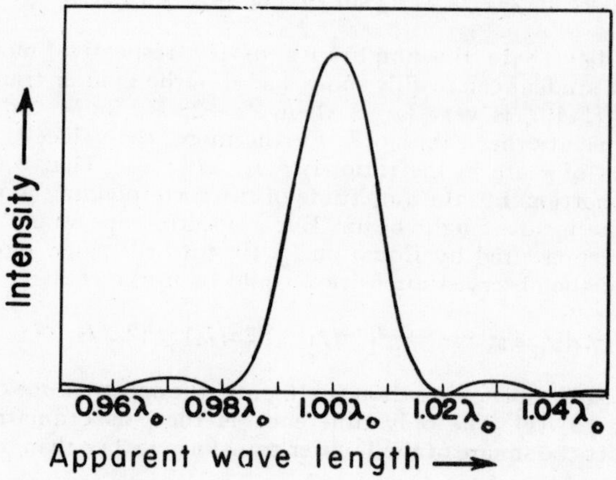

FIG. 2-10. Square of the Fourier transform of a wave train containing 200 oscillations whose wave length is λ_0 .If this finite wave train is passed through a spectroscope, the intensities observed at different settings of the wave length scale of the spectroscope will be proportional to the ordinate.

spectral intensity lies in the first peak, corresponding to k values lying within about $\pm (1/n)(2\pi/\lambda)$ of the mean value of k at $2\pi/\lambda$. The wave lengths in the finite "monochromatic" wave train, as measured in a spectroscope, are thus mostly distributed over a range between

$$\frac{\lambda}{1 + \dfrac{1}{n}} \quad \text{and} \quad \frac{\lambda}{1 - \dfrac{1}{n}}$$

If $n = 100$, the light (as analyzed by the spectroscope) will appear to be made up of a distribution of wave lengths differing by as much as $\pm 1\%$ from the mean value. It is clear that no light beam of finite duration can, on spectroscopic analysis, act as if it were truly monochromatic. This result will be useful to us in Chapters 7, 15, and 16.

G. Ordinary Differential Equations

Since some of the laws of quantum mechanics are expressed in the form of differential equations, it is important that we have a clear idea of just what a differential equation represents. A differential equation is simply an equation in which the derivatives of a function appear. If the function has only one independent variable, the differential equation is called an *ordinary differential equation*. For instance

$$dy/dx = x - y \tag{G-1}$$

is an ordinary differential equation in which $y(x)$ is a function that depends on the independent variable, x. A differential equation that contains at most first derivatives is called a *first order differential equation*, while if the highest derivative it contains is a second derivative, it is called a *second order differential equation*. Most of the differential equations that we shall encounter in quantum chemistry are second order differential equations.

A differential equation is said to have been solved when all of the functions that satisfy it have been found. Every differential equation may be satisfied by a vast family of functions and it is often not easy to find out what these functions are. Furthermore, in many practical applications we seek a particular member of this family that satisfies certain specified conditions. Therefore the solution of differential equations can require a high order of mathematical skill and perseverance. Nevertheless, it is important to realize that there is nothing inherently obscure about what a differential equation stands for. One can, in fact, often gain some idea of the general appearance of the solutions of a differential equation by relatively simple methods.

a. Ordinary first order differential equations

We shall now describe a simple method by which approximate graphical solutions of first order ordinary differential equations can usually be drawn. These equations can almost always be thrown into the form

$$dy/dx = f(x, y) \tag{G-2}$$

where $f(x, y)$ is a known function. According to this equation, for every value of x and y there is a definite value of the slope, dy/dx. If, therefore, we set up x and y axes in a plane, a slope can be assigned to each of the points in this plane: that is, a short, straight line segment having the slope $f(x, y)$ and passing through the point (x, y) is a piece of one solution of the differential equation. If a sufficient number of such short line segments are drawn on the (x, y) plane, it should be possible to join them and so obtain all of the smooth curves that are the solutions of the differential equation.

Let us apply this method to the solution of Equation (G-1). We shall set

$$x - y = a \tag{G-3}$$

where $a = 0, \pm\frac{1}{2}, \pm 1, \pm\frac{3}{2}, \cdots$. Equation (G-3) represents a series of parallel straight lines in the (x, y) plane, and along each of these lines we may draw a set of short line segments all having the same slope, a. The result is shown in Fig. 2-11A. It is easy to see that these segments can be joined to give a family of smooth curves as shown in Fig. 2-11B.

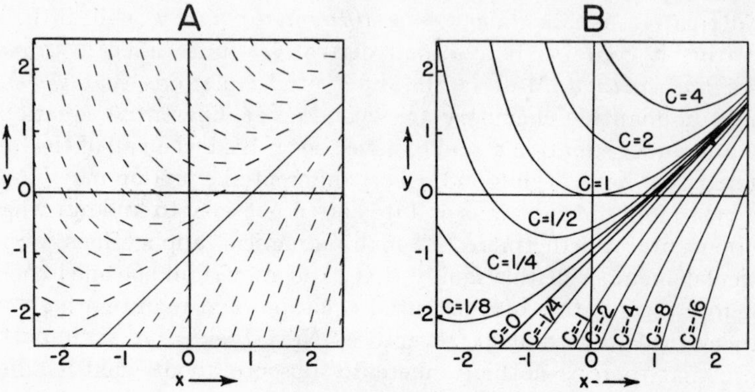

FIG. 2-11. Solutions of the differential equation $dy/dx = x - y$.
A. Segments of the solutions passing through various points in the x-y plane as derived directly from the differential equation. Note that the slopes are identical along the lines $x = y + a$.
B. Exact solutions, $y = x - 1 + Ce^{-x}$, corresponding to various values of the constant C. The general shapes of these solutions could be deduced directly from the segments drawn in A.

Exercises. (1) The exact solution of Equation (G-1) is found to be $y = x - 1 + Ce^{-x}$, where C is an arbitrary constant. Verify that these solutions have the shapes shown in Fig. 2–11B, different values of the constant C giving different members of the family.

(2) Using the graphical method just described, find the general appearance of the family of solutions of the following differential equations:

(a) $dy/dx = x - \sin y$

(b) $dy/dx = xy + e^{xy}$

(c) $(dy/dx)^2 - (x + y \sin y)dy/dx + xy \sin y = 0$

b. Ordinary second order differential equations

A similar, though somewhat more qualitative, graphical method can be used with ordinary second order differential equations. Equations of this type can usually be thrown into the form

$$d^2/ydx^2 = f(x, y, dy/dx) \tag{G-4}$$

The second derivative of a function is merely the rate at which the slope of the function changes when the independent variable is varied. If the slope of a function is not constant, the graph of the function must be curved; the value of d^2y/dx^2 is thus a measure of the "curvature" of the solution. Therefore a second order differential equation can be regarded as an expression for the *curvature* of its solutions, just as a first order differential equation may be regarded as an expression for the *slope* of its solutions.

Second order differential equations give us no information concerning the slope of the solution passing through a given point (x,y). Therefore this slope can be chosen arbitrarily. Accordingly, an infinite number of solutions, all differing in slope, can pass through each point in the x-y plane. Evidently a second order differential equation has an even greater multiplicity of solutions than we found for first order differential equations, for which only one solution passes through each point in the x-y plane. This greater multiplicity expresses itself through the occurrence of *two* arbitrary constants in the solutions of all second order differential equations.

Fortunately we shall find that quantum mechanics is concerned only with second order differential equations having the special form,

$$P(x)d^2y/dx^2 + Q(x)dy/dx + R(x)y = 0 \qquad \text{(G-5)}$$

where $P(x)$, $Q(x)$ and $R(x)$ are specified functions of x. Such an equation is called a *linear* second order differential equation, because the operator $P(x)d^2/dx^2 + Q(x)d/dx + R(x)$ is a linear operator (see page 17). Linear differential equations have the useful property that, if $y = f(x)$ is a solution, then $y = Af(x)$ must also be a solution, where A is any constant. (The student can verify this by direct substitution in (G-5).) Therefore, one of the two arbitrary constants appearing in the general solution of a linear second order differential equation merely multiplies the function as a whole.

Now let us apply the graphical method to the linear second order differential equation

$$d^2y/dx^2 = -y \qquad \text{(G-6)}$$

According to this equation, when y is positive the solution has a slope that decreases as x increases. On the other hand, if y is negative, the slope increases with increasing x. In short, all of the solutions are curved downward in the upper half of the x-y plane, and they are curved upward in the lower half of the x-y plane. Furthermore, the curvature of the solution passing through a given point is greater, the greater the value of the ordinate of the point. And when the solutions pass through the x-axis ($y = 0$) they are not curved at all; that is, points lying on the x-axis are always points of inflection.

It is easy to see that the solutions of (G-6) must oscillate from one side of the x-axis to the other as x is varied, because as long as we are on one side of the x-axis, the curvature is such that the solution tries to get to the other side, and once it gets to the other side, the solution will not be happy until it returns to the original side. Thus, by simply looking at the differential equation, it is possible to deduce the general behavior of the solutions.

In Fig. 2-12 a series of solutions of equation (G-6) is drawn, all of which pass through the point (0,1), but with different slopes. It is readily verified

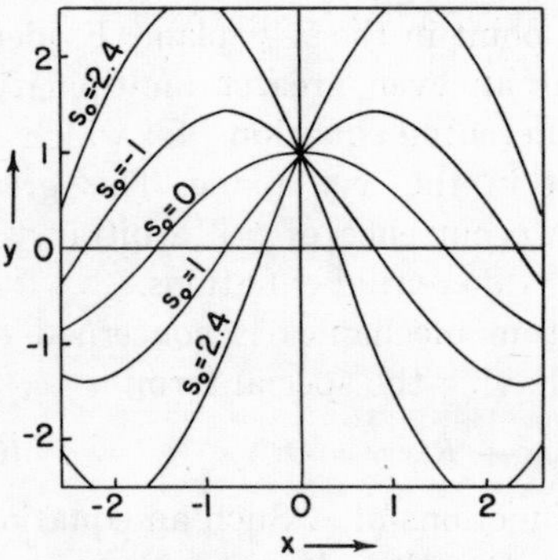

FIG. 2-12. Solutions of the differential equation $d^2y/dx^2 = -y$ passing through the point (0,1) and having the indicated slopes s_0 at this point. Note tendency of the solutions to curve toward the x-axis.

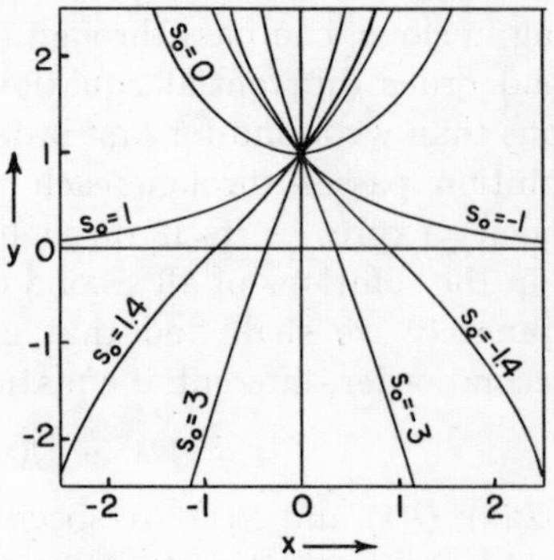

FIG. 2-13. Solutions of the differential equation $d^2y/dx^2 = y$ passing through the point (0,1) and having the indicated slopes s_0 at this point. Note tendency of the solutions to curve away from the x-axis.

by substitution in (G-6) that the family of solutions passing through this point has the general form $y = A \sin (x + b)$ where A and b are constants which are related by the condition $A \sin b = 1$. These solutions are, of course, exactly of the form inferred by means of the arguments in the previous paragraph.

Next let us consider the equation

$$d^2y/dx^2 = y \tag{G-7}$$

Using arguments similar to those just given, we can see that the solutions of this equation do not oscillate at all. In fact, they tend to "explode" since a large positive value of y tends to make the slope increase as x increases, so that if dy/dx is positive, y will tend to assume even larger values with increasing x, whereas if dy/dx is negative, y will increase without limit as x decreases. Similar explosive behavior is found for negative values of y, regardless of the sign of dy/dx. If we restrict ourselves to solutions which pass through the point (0,1) we can easily see that nearly every solution explodes with both increasing and decreasing values of x (see Fig. 2-13); in some cases the explosion is toward positive values of y; in others, it is toward negative values. In only two instances do the solutions fail to explode in *both* the positive and negative x-directions. In these solutions y (and hence the curvature) approaches zero in such a way that the solution becomes asymptotic to either the positive or the negative x-axis. It is readily verified that the exact solutions of (G-7) which pass through the point (0,1) are $y = Ae^{-x} + (1 - A)e^{x}$. The two asymptotic solutions result when $A = 0$ and 1. All other values of A result in "explosions" in both positive and negative x-directions.

The oscillatory and explosive behaviors described above are prominent features of the solutions of the differential equations of quantum mechanics. We shall find in Chapter 6 that energy quantization results from the fact that explosive solutions are not acceptable and that the asymptotic behavior is required at great distances from most systems of chemical interest.

Exercises. (1) Describe the general appearance of the solutions of the differential equations

$$d^2y/dx^2 + (A - x^2)y = 0$$

which pass through the point (0,0) and have slope unity at that point. (Do this by assuming initially that $A = 0$, and then show what happens to the solutions as A is increased. It will be found that for certain values of A the solutions will approach the x-axis asymptotically as $x \to \infty$ and as $x \to -\infty$. For all other values of A the solutions are of the explosive variety. This equation occurs in the quantum mechanical treatment of the harmonic oscillator and the asymptotic behavior occurs only when the energy is quantized according to the well-known rule, $E = h\nu(n + 1/2)$. (See page 205).)

(2) Legendre's equation of order one is

$$d^2y/dx^2 = [2x(dy/dx) - 2y]/(1 - x^2)$$

(a) Show that the only solutions of this equation that pass through the origin are those having the form $y = Ax$, where A is an arbitrary constant.

(b) Show that the solution that passes through the point (0,1) and has $dy/dx = 0$ becomes negatively infinite at $x = \pm 1$.

(c) Show that all other solutions passing through the point (0,1) also become negatively infinite at $x = \pm 1$.

(d) Show that all solutions passing through $(0,a)$ become negatively infinite at $x = \pm 1$ if $a > 0$ and positively infinite at $x = \pm 1$ if $a < 0$.

H. Partial Differential Equations

When there is more than one independent variable in a system, differential equations containing partial derivatives will generally be obtained. In quantum mechanics we must frequently deal with linear second order partial differential equations; if there are two independent variables, these equations have the general form

$$P(x,y)\frac{\partial^2 f}{\partial x^2} + Q(x,y)\frac{\partial^2 f}{\partial x \partial y} + R(x,y)\frac{\partial^2 f}{\partial y^2}$$

$$+ S(x,y)\frac{\partial f}{\partial x} + T(x,y)\frac{\partial f}{\partial y} + U(x,y)f = 0 \qquad \text{(H-1)}$$

where P, Q, R, S, T, and U are given functions of the independent variables. If there are more than two independent variables the differential equations are of similar form except that there are additional terms involving first and second partial derivatives of the other variables.

Exercise. Why is such an equation called linear? If $f(x,y)$ is a solution of (H-1), show that $Af(x,y)$ is also a solution, where A is an arbitrary constant.

a. Geometrical interpretation of partial differential equations

It is possible to discuss the geometrical significance of partial differential equations in a manner similar to that outlined above for ordinary differential equations. Evidently the curvature of the solutions of (H-1) with respect to any one of the variables is determined by the other curvatures and slopes appearing in the equation. An enormous variety of solutions is possible, because even if there are only two variables, x and y, acceptable solutions will be obtained on giving arbitrary values to the function, to two of its slopes, $\partial f/\partial x$ and $\partial f/\partial y$, and to two of its curvatures, say $\partial^2 f/\partial y^2$ and $\partial^2 f/\partial x \partial y$.

For instance, consider the relatively simple second order equation

$$\frac{\partial^2 u}{\partial x^2} + \frac{\partial^2 u}{\partial y^2} = 0 \qquad \text{(H-2)}$$

Let us make a three dimensional plot of the solutions, $u(x,y)$, against the values of x and y. This plot will give a surface. Consider the intersection of this surface with the plane $x = 0$. The intersection of the plane with the surface results in a curve, $u(0,y)$, which we shall denote by $f(y)$. According to Equation (H-2), the solutions have merely to satisfy the condition

$$\partial^2 u/\partial x^2 = -\partial^2 u/\partial y^2 \tag{H-3}$$

in the immediate vicinity of the y-u plane. The function $f(y)$ can have *any shape whatsoever*, as long as it has a second derivative, and furthermore the values of $\partial u/\partial x$, which give the slope of the surface in a direction normal to the y-u plane, are *completely arbitrary*. Thus an enormous number of different surfaces are consistent with Equation (H-2). In fact, as we shall see in studying the shapes of vibrating membranes (page 72), all of the shapes that can be assumed by a stationary stretched elastic membrane are consistent with Equation (H-2).

The student might gain the impression from the above discussion that, because of the enormous diversity in the shapes of their solutions, partial differential equations do not tell us very much about a system. This, however, is not the case. In practical applications of partial differential equations, we almost always know something about the solutions before we begin. For instance, in vibrating bodies we know that certain portions are held rigidly in place, or that certain portions have no stresses acting on them. In quantum mechanics we know that the solutions are well behaved. These conditions—which are called *boundary conditions*—are usually sufficient to limit the solutions very drastically, so that the solutions provide us with detailed descriptions of the possible behavior of the systems.

Exercises. (1) Show that the following functions are solutions of Equation (H-2)

(a) $u = \log(x^2 + y^2)$

(b) $u = x/(x^2 + y^2)$

(c) $u = Af(x + iy) + Bg(x - iy)$

where A and B are arbitrary constants, $i = \sqrt{-1}$, and f and g are any functions that have second derivatives. Show that the solutions (a) and (b) can be written in the form (c).

(2) Show that the differential equation

$$\frac{\partial u}{\partial x} + \frac{\partial u}{\partial y} = 0$$

has the general solution $u(x,y) = Af(x - y)$, where A is an arbitrary constant and f is any function in which x and y always appear in the combination $(x - y)$ (e.g., $u = (x - y)^3$; $u = \sin(x - y)$; $u = \exp(x - y)$). Draw a sketch of the solution that, for $y = 0$, has the shape $u(x,0) = \exp(-x^2)$.

b. The solution of partial differential equations by the method of separation of variables

Consider the differential equation

$$\frac{\partial u}{\partial x} - \frac{\partial u}{\partial y} = 0 \qquad \text{(H-4)}$$

Let us see what happens if we assume that the solution has the form

$$u(x, y) = X(x)\ Y(y) \qquad \text{(H-5)}$$

where $X(x)$ is a function involving only x and $Y(y)$ involves only y. On differentiating we find

$$\partial u/\partial x = Y\, dX/dx;\ \partial u/\partial y = X\, dY/dy \qquad \text{(H-6)}$$

Total derivatives are written on the right hand sides of these two equations because each of the differentiated functions contains only one variable. On substituting (H-6) into (H-4) we find

$$Y\, dX/dx - X\, dY/dy = 0 \qquad \text{(H-7)}$$

Dividing through by $u = XY$, we obtain an equation in which the variables have been "separated" into different terms

$$\frac{1}{X(x)}\ \frac{dX(x)}{dx} - \frac{1}{Y(y)}\ \frac{dY(y)}{dy} = 0 \qquad \text{(H-8)}$$

It is easy to see that each of the two terms appearing on the left hand side of (H-8) is constant independent of both x and y. The first term, $(1/X)dX/dx$, clearly depends only on x and is independent of y, whereas the second term, $(1/Y)dY/dy$, does not depend on x. Therefore, if we vary x in Equation (H-8), the second term on the left must remain constant. Since the equation as a whole must be true for all values of x, this means that the first term on the left is also independent of x. Similarly, as y is varied, the first term on the left cannot vary, so the second term must also be independent of y. In other words, both $(1/X)dX/dx$ and $(1/Y)dY/dy$ are constant. Furthermore, the two terms must be equal to one another in order that their difference vanish. Thus we arrive at the two ordinary differential equations

$$(1/X)dX/dx = c \tag{H-9a}$$

$$(1/Y)dY/dy = c \tag{H-9b}$$

where c is an arbitrary constant. These equations have the solutions

$$X = Ae^{cx} \tag{H-10a}$$

$$Y = Be^{cy} \tag{H-10b}$$

where A and B are arbitrary constants. Combining A and B into a single arbitrary constant, $D = AB$, we obtain for the solution of (H-4)

$$u(x, y) = De^{c(x+y)} \tag{H-11}$$

The method outlined above does not always work. In some instances the solution obtained does not fit the boundary conditions that have been set for the problem. More often the coefficients P, Q, R, S, T, or U of Equation (H-1) contain the variables in such combinations that they cannot be separated from each other. In either case it is sometimes possible to transform the differential equation by using a new coordinate system and so obtain an equation which can be separated. Unfortunately, as Robertson and Eisenhart have shown (see *5*, p. 171), separation of variables is possible for only a few general types of equations occurring in quantum mechanics. Almost all of the more complex chemically interesting quantum mechanical problems give equations that do not belong to one of these types, so that the method of separation of variables is not applicable. It is this fact more than any other that stands in the way of progress in the rigorous application of quantum mechanics to chemistry and in the precise *a priori* calculation of the properties of chemical systems. Nevertheless we shall find frequent occasion to use this method in the solution of simpler problems. Examples will be found on pages 60, 73, 86, 161, and 214.

Exercise. Can the method of separation of variables be used with the equation

$$\frac{\partial^2 u}{\partial x^2} + \frac{\partial^2 u}{\partial y^2} + \frac{u}{\sqrt{x^2 + y^2}} + u = 0$$

Show that the transformation $x = r \cos \theta$, $y = r \sin \theta$ will give a differential equation that can be separated, and find the resulting total differential equations.

I. Determinants

A *determinant* is an arrangement of quantities or *elements* A_{ij} in rows and columns in which the number of rows is equal to the number of columns

$$\begin{vmatrix} A_{11} & A_{12} & A_{13} & \cdots & A_{1n} \\ A_{21} & A_{22} & A_{23} & \cdots & A_{2n} \\ A_{31} & A_{32} & A_{33} & \cdots & A_{3n} \\ \vdots & \vdots & \vdots & & \vdots \\ A_{n1} & A_{n2} & A_{n3} & \cdots & A_{nn} \end{vmatrix} \tag{I-1}$$

The number of rows or columns, n, is called the *order* of the determinant.

Each element in a determinant has a *minor*, which is defined as the determinant remaining when the row and column containing the given element are removed from the original determinant. The minors of the elements in an nth order determinant are therefore determinants of order $(n-1)$.

The *value* of a determinant is defined as the quantity obtained in the following way. There are $n!$ different ways of choosing one and only one element from each row and each column in any n^{th} order determinant. Form the $n!$ different products each containing n elements chosen in this way. Arrange the elements A_{ij} in each of these products so that their first subscripts are in their natural order, $1, 2, 3, \cdots, n$. In all but one of the products the second subscripts will not occur in their natural order; find out how many permutations of the elements in each product are required in order to bring the second subscripts into their natural order, $1, 2, 3, \cdots, n$. If this number of permutations is even, give the product a positive sign, and if it is odd, give it a negative sign. Then the value of the determinant is simply the algebraic sum of all of the products.

In order to illustrate this procedure, consider the third order determinant

$$\begin{vmatrix} A_{11} & A_{12} & A_{13} \\ A_{21} & A_{22} & A_{23} \\ A_{31} & A_{32} & A_{33} \end{vmatrix} \tag{I-2}$$

The $3! = 6$ different products obtained from this determinant are, when written so that their first subscripts fall in the order 1, 2, 3

$$A_{11}A_{22}A_{33} \ (p = 0) \qquad A_{11}A_{23}A_{32} \ (p = 1)$$
$$A_{12}A_{21}A_{33} \ (p = 1) \qquad A_{12}A_{23}A_{31} \ (p = 2)$$
$$A_{13}A_{22}A_{31} \ (p = 1) \qquad A_{13}A_{21}A_{32} \ (p = 2)$$

The number of permutations, p, required to bring the second subscripts into the order 1, 2, 3 in each case is indicated beside each product. The value of the determinant is thus

$$A_{11}A_{22}A_{33} - A_{12}A_{21}A_{33} - A_{13}A_{22}A_{31} - A_{11}A_{23}A_{32} + A_{12}A_{23}A_{31} + A_{13}A_{21}A_{32} \tag{I-3}$$

In general we may write for the value of a determinant

$$\sum (-1)^p A_{1k_1}A_{2k_2}A_{3k_3}\cdots A_{nk_n} \tag{I-4}$$

where p is the number of permutations required to bring $k_1, k_2, k_3, \cdots, k_n$ into the order $1, 2, 3, \cdots, n$, and the sum is over all permutations of the k_i's.

It is easy to show that the value of a determinant can also be found in the following way: Multiply each of the elements in the first row or column by its minor. Give the product a plus sign if the element is first, third, fifth, $\cdots$ in the row or column, and a minus sign if it is second, fourth, sixth, $\cdots$. Add all of the products. The result is the value of the determinant. Thus we have for the determinant (I-2)

$$A_{11}\begin{vmatrix} A_{22} & A_{23} \\ A_{32} & A_{33} \end{vmatrix} - A_{21}\begin{vmatrix} A_{12} & A_{13} \\ A_{32} & A_{33} \end{vmatrix} + A_{31}\begin{vmatrix} A_{12} & A_{13} \\ A_{22} & A_{23} \end{vmatrix} \tag{I-5}$$

which gives the same result as (I-3).

The evaluation and manipulation of determinants is greatly simplified by taking note of the following properties of determinants (all of which are easily proven by means of the definitions given above).

(1) The value of a determinant merely changes sign when two rows or two columns are interchanged.

(2) If every element in a row or column is zero, the value of the determinant is zero.

(3) If two rows are equal, or if two columns are equal, the determinant is zero.

(4) If all of the elements of the ith row are written as the sums of two numbers, $A_{ij} = B_{ij} + C_{ij}$, then the determinant can be written in the form

$$\begin{vmatrix} A_{11} & A_{12} & \cdots & A_{1n} \\ \vdots & \vdots & & \vdots \\ A_{i1} & A_{i2} & \cdots & A_{in} \\ \vdots & \vdots & & \vdots \\ A_{n1} & A_{n2} & \cdots & A_{nn} \end{vmatrix} = \begin{vmatrix} A_{11} & A_{12} & \cdots & A_{1n} \\ \vdots & \vdots & & \vdots \\ B_{i1} & B_{i2} & \cdots & B_{in} \\ \vdots & \vdots & & \vdots \\ A_{n1} & A_{n2} & \cdots & A_{nn} \end{vmatrix} + \begin{vmatrix} A_{11} & A_{12} & \cdots & A_{1n} \\ \vdots & \vdots & & \vdots \\ C_{i1} & C_{i2} & \cdots & C_{in} \\ \vdots & \vdots & & \vdots \\ A_{n1} & A_{n2} & \cdots & A_{nn} \end{vmatrix}$$

A similar relationship holds for the columns of a determinant.

(5) If one multiplies all terms in one row or one column by a constant, the determinant in multiplied by the same constant.

(6) If one multiplies all members of one row or one column by a constant and adds the result to another row or column, the value of the determinant is unchanged.

A frequently used practical procedure for evaluating determinants of high order is to make all but one of the elements in one row equal to zero by means of property (6). The order of the determinant is in this way effectively reduced by one. The same reduction procedure can then be employed on the minor of the remaining nonzero element, and so on.

In the quantum chemical treatment of aromatic and conjugated hydrocarbons we shall encounter so-called *cyclic determinants*. An nth order cyclic determinant, C, is formed from n quantities $c_1, c_2, c_3, \cdots, c_n$ by rearranging their order in different rows in the following way

$$C = \begin{vmatrix} c_1 & c_2 & c_3 & \cdots & c_n \\ c_2 & c_3 & c_4 & \cdots & c_1 \\ c_3 & c_4 & c_5 & \cdots & c_2 \\ \cdot & \cdot & \cdot & \cdot & \cdot \\ c_n & c_1 & c_2 & \cdots & c_{n-1} \end{vmatrix}$$

It can be shown (*6*, p. 106) that the value of this determinant is given by

$$C = (-1)^{(n-1)(n-2)/2} f_1 f_2 f_3 \cdots f_n \tag{I-6}$$

where

$$f_k = \sum_{j=1}^{n} c_j x_k^{j-1}$$

$$x_k = \exp(2k\pi i/n)$$

and

$$i = \sqrt{-1}$$

Other properties of determinants will be found in Kowalewski (*6*).

Exercises. (1) Evaluate the determinant

$$\begin{vmatrix} 4 & 1 & 2 & 3 \\ 1 & 2 & 3 & 4 \\ 2 & 3 & 4 & 1 \\ 3 & 4 & 1 & 2 \end{vmatrix}$$

using the method of minors or Equation (I-6). [Answer: –160.]

(2) Show that
$$\begin{vmatrix} 1 & 2 & 3 & \cdots & n \\ 2 & 3 & 4 & \cdots & 1 \\ 3 & 4 & 5 & \cdots & 2 \\ \cdot & \cdot & \cdot & \cdot & \cdot \\ n & 1 & 2 & \cdots & (n-1) \end{vmatrix} = (-1)^{n(n-1)/2} \frac{n+1}{2} n^{n-1}$$

(3) Consider the equation

$$C_n \equiv \begin{vmatrix} x & 1 & 0 & 0 & \cdots & 0 & 0 & 1 \\ 1 & x & 1 & 0 & \cdots & 0 & 0 & 0 \\ 0 & 1 & x & 1 & \cdots & 0 & 0 & 0 \\ 0 & 0 & 1 & x & \cdots & 0 & 0 & 0 \\ \cdot & \cdot & \cdot & \cdot & \cdot & \cdot & \cdot & \cdot \\ 0 & 0 & 0 & 0 & \cdots & 1 & x & 1 \\ 1 & 0 & 0 & 0 & \cdots & 0 & 1 & x \end{vmatrix} = 0 \tag{I-7a}$$

where C_n is an nth order cyclic determinant in which all but three elements in each row vanish. Show that the n roots of this equation are given by

$$x = -2 \cos (2\pi k/n) \tag{I-7b}$$

where $k = 1, 2, 3, \cdots, n$.

(4) Let D_n be the nth order non-cyclic determinant defined by

$$D_n \equiv \begin{vmatrix} x & 1 & 0 & 0 & \cdots & 0 & 0 & 0 \\ 1 & x & 1 & 0 & \cdots & 0 & 0 & 0 \\ 0 & 1 & x & 1 & \cdots & 0 & 0 & 0 \\ 0 & 0 & 1 & x & \cdots & 0 & 0 & 0 \\ \cdot & \cdot & \cdot & \cdot & \cdot & \cdot & \cdot & \cdot \\ 0 & 0 & 0 & 0 & \cdots & 1 & x & 1 \\ 0 & 0 & 0 & 0 & \cdots & 0 & 1 & x \end{vmatrix} \tag{I-8}$$

(Note that this determinant is identical with C_n in Exercise (3) except in the first and last rows, where all but *two* of the elements vanish.)

(a) Show that

$$D_n = xD_{n-1} - D_{n-2} \tag{I-9a}$$

where

$$D_1 = x \tag{I-9b}$$

and

$$D_2 = \begin{vmatrix} x & 1 \\ 1 & x \end{vmatrix} = x^2 - 1 \tag{I-9c}$$

(b) Show that if one sets $x = 2\cos\theta = e^{i\theta} + e^{-i\theta}$, then

$$D_n = Ae^{in\theta} + Be^{-in\theta} \tag{I-10}$$

where A and B may depend on θ but not on n. [Hint: Use the method of induction, showing that if the equation holds for D_n it also holds for D_{n+1}, and that it must therefore hold for all values of n.]

(c) Show that $A = e^{i\theta}/(e^{i\theta} - e^{-i\theta})$ and $B = -e^{-i\theta}/(e^{i\theta} - e^{-i\theta})$. [Hint: Compare (I-10) with (I-9b) and (I-9c).]

(d) Using these values of A and B in (I-10) show that

$$D_n = \frac{\sin(n+1)\theta}{\sin\theta} \tag{I-11}$$

(e) Show that the roots of the equation $D_n = 0$ occur at

$$x = 2\cos[k\pi/(n+1)] \tag{I-12}$$

where $k = 1, 2, 3, \cdots, n$.

REFERENCES CITED

1. T. M. MacRobert, "Spherical Harmonics," 2nd ed., Dover, New York, 1948.
2. E. T. Whittaker and G. N. Watson, "A Course of Modern Analysis," 4th ed., Cambridge U.P., New York, 1927.
3. E. C. Titmarsh, "Introduction to the Theory of Fourier Integrals," Oxford, New York, 1937.
4. H. S. Carslaw, "Introduction to the Theory of Fourier Series and Integrals," 3rd ed., Macmillan, New York, 1930. (Reprinted by Dover, New York).
5. H. Margenau and G. M. Murphy, "The Mathematics of Physics and Chemistry," Van Nostrand, New York, 1943.
6. G. Kowalewski, "Determinententheorie," 3rd ed., Chelsea, New York, 1943.

GENERAL REFERENCES

I. S. and E. S. Sokolnikoff, "Higher Mathematics for Engineers and Physicists," McGraw-Hill, New York, 1934. An excellent introduction to vectors, differential equations, determinants, Fourier series, and probability. Generally less difficult than the references given below.

H. Margenau and G. M. Murphy, reference 5. Excellent general reference.

P. M. Morse and H. Feshbach, "Methods of Theoretical Physics," McGraw-Hill, New York, 1953. More advanced and detailed, but relatively readable. Especially good on differential equations. Contains numerous further references.

Chapter 3

The Classical Theory of Vibrations I
Some Typical Vibrating Systems

Quantum mechanics can be adequately grasped only by underst
mathematical structure. This is the cause of much difficulty for th
chemist, since the mathematics required ranges somewhat beyond
course in calculus, and the physical interpretation of the mathe
often unusual, if not weird. To reduce the pain of learning thi
subject, it is therefore helpful to learn some of the mathemati
framework of a more familiar physical problem. The mathemati
been mastered in this way, it will then be less of a shock to enco
new notions of the behavior of matter that quantum mechanics a
accept. Fortunately, the classical theory of the vibrations of elas
furnishes an excellent training field on which to play these practi
Not only is the mathematical treatment of vibrations closely simil
of a large segment of quantum mechanics, but the theory of vibrat
interest and value in its own right. Before going into the quantum m
theory we shall therefore find it profitable to devote some time to
of ordinary vibrations.

We are all familiar with the great variety of oscillations that occ
hanging chain or a dish of water is disturbed. All material objects a
able to vibrate, though the frequencies are often too high and the a
too low for us to notice the motion directly. Our aim will be to d
systematic method of describing these oscillations for variou
systems. In every instance the procedure is to derive the different
tion of motion and then to look for appropriate solutions of this

A. Vibrations in One Dimension

a. The wave equation for a stretched string

One of the simplest vibrating systems is a flexible string held in
as in a violin or in the device shown in Fig. 3-1A. When the string
it is assumed to be straight (dashed line in Fig. 3-1A). Let x be the

measured along the string when it is at rest. Let ρ be the mass of the string per unit length (which may or may not be constant along the string), and let T be the tension in the string (which will be assumed constant along the string). We shall consider the motion of the string in only one plane, say the vertical plane. The vertical displacement of the string from its rest position

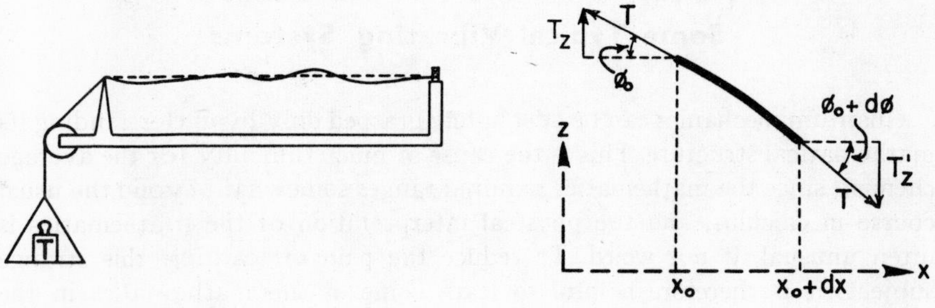

FIG. 3-1. The vibrating string.

will be denoted by z. This displacement is different at different times and at different points along the string; that is, z is a function of x and of the time, t ($z = z(x,t)$). The problem is to find this function, which describes all of the possible vibrations of a stretched string.

The function is found by solving the differential equation of motion (called the "wave equation") obtained in the following way. Mentally cut an infinitesimal segment of length dx out of the string. To the two cut ends apply forces that act on the segment in the same way as the rest of the string did–that is, as tensions of magnitude T (see Fig. 3-1B). Since the string is flexible, the tension T is directed everywhere *along* the string. The two substitute forces therefore act in the direction of tangents to the segment at its two ends. If the segment happens to be straight, the substitute forces will exactly balance and there will be no net force on the segment. If the segment is curved, however, the substitute forces will not balance and a net force will result. According to Newton's laws of motion, the element will accelerate in the direction of this force, and the greater the curvature of the segment, the greater its acceleration will be.

Let us calculate the z-component of this unbalanced force. Suppose that the string is nowhere sharply kinked and that z is everywhere small (this means that only oscillations of small amplitude are considered in what follows). Then the tension is always nearly parallel to the x-axis and $\sin\varphi \simeq \tan\varphi = \partial z/\partial x$, where φ is the angle the string makes with the

x-axis. At one end of the segment ($x = x_0$), the z-component of the tension is

$$T_z = -T \sin \varphi_0 \cong -T(\partial z/\partial x)_{x=x_0}$$

(the sign is negative because the force is downward if $\partial z/\partial x$ is positive). At the other end of the segment ($x = x_0 + dx$) it is

$$T_z' = T \sin(\varphi_0 + d\varphi) \cong T(\partial z/\partial x)_{x=x_0+dx}$$

$$\cong T\,[(\partial z/\partial x)_{x=x_0} + (\partial^2 z/\partial x^2)_{x=x_0} dx]$$

(the partial derivatives here refer to the values of the slopes, etc., at a definite value of the time, t). The net force in the z-direction is thus

$$T_z + T_z' = T(\partial^2 z/\partial x^2)dx$$

as long as the oscillations are small. According to Newton's law, this force must be equated to the mass of the element, ρdx, times its acceleration in the z-direction, $(\partial^2 z/\partial t^2)$. Thus we find that

$$T \frac{\partial^2 z(x,t)}{\partial x^2} = \rho \frac{\partial^2 z(x,t)}{\partial t^2} \tag{A-1}$$

which is called the *wave equation in one dimension*. This equation is valid if ρ depends on x, but is not valid if T depends on x. Both ρ and T may, however, depend on the time.

Exercises. (1) Show that if the oscillations are not small, so that the tension, though constant, is not always very nearly horizontal, the wave equation has the form

$$\{T/[1 + (\partial z/\partial x)^2]^{3/2}\} \frac{\partial^2 z}{\partial x^2} = \rho \frac{\partial^2 z}{\partial t^2} \tag{A-2}$$

Can you think of any other factors that ought to be taken into account in a real string undergoing large oscillations?

(2) Show that if the tension, T, is not constant along the string, the wave equation becomes

$$\frac{\partial}{\partial x}\left(T(x) \frac{\partial z}{\partial x}\right) = \rho \frac{\partial^2 z}{\partial t^2} \tag{A-3a}$$

or

$$T \frac{\partial^2 z}{\partial x^2} + \frac{\partial T}{\partial x} \frac{\partial z}{\partial x} = \rho \frac{\partial^2 z}{\partial t^2} \tag{A-3b}$$

b. The wave equation for a liquid in a shallow trough*

Consider a long, narrow trough containing a shallow layer of an incompressible liquid (Fig. 3-2A). Let

x = distance along the length of the trough,
A = cross-sectional area of the trough, which may vary with x,
z = vertical displacement of the liquid surface from the rest position. It is assumed that z is constant across the narrow dimension of the trough; that is, we consider only waves running along the long dimension.
ρ = density of the liquid.

If the tank is sufficiently long and shallow, oscillations of the liquid require the transport of liquid horizontally from one part of the tank to another, so the motion of the liquid will be mostly in the x-direction.† If friction at the walls is neglected, the velocity will be constant across any given cross-section. Call the component of the velocity in the x-direction u.

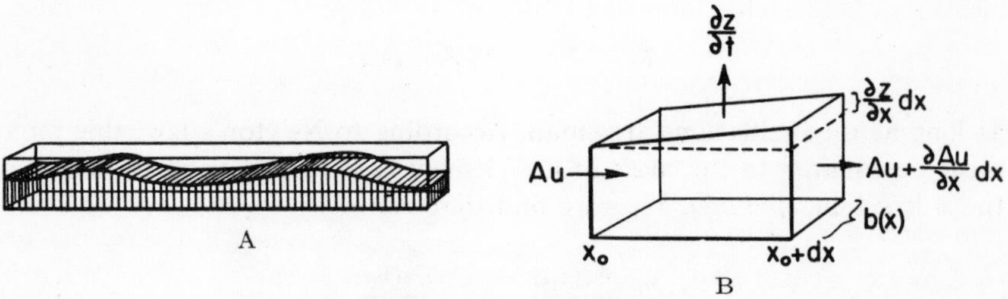

FIG. 3-2. Oscillations of a liquid in a trough.

First we apply the law of conservation of matter. Consider a slice of the trough lying between $x = x_0$ and $x_0 + dx$ (Fig. 3-2B). In unit time the volume of fluid flowing into this slice through the cross-section at $x = x_0$ is Au, and the amount flowing out through the cross-section at $x = x_0 + dx$ is

$$Au + (\partial Au/\partial x)dx$$

* See Coulson (*1*, pp. 63-70) and Lamb (*2*, Chapter 8)

† Waves of this kind are known as *tidal waves*, since they are characteristic of the movements of the oceans that are responsible for the tides and for the catastrophic waves caused by submarine earthquakes, which are popularly known as "tidal waves". Waves for which vertical motions cannot be neglected are called *surface waves*; these are the wind-engendered waves responsible for the rocking of ocean liners, etc. In contrast with tidal waves, their wave lengths are short compared with the depth of the liquid. If the wave length is less than one or two centimeters, the effect of surface tension becomes predominant in controlling the wave motion in nearly all liquids; such waves are called *capillary waves*, or *ripples*. (See Lamb or Coulson, *loc. cit.*, for details.)

The difference between these two quantities is equal to the rate at which liquid accumulates in the slice between the two cross-sections. This accumulation causes the surface of the liquid in the slice to rise at a rate $(\partial z/\partial t)$. If $b(x)$ is the width of the trough at x, we see that, assuming the liquid to be incompressible

$$-\frac{\partial A u}{\partial x} = b(x)\,\frac{\partial z}{\partial t} \tag{A-4}$$

This is known as the *equation of continuity*.

Next we consider the forces acting on the slice of the liquid. If the free surface of the liquid is not level, the pressure on one face of the slice will be greater than that on the other by an amount $g\rho(\partial z/\partial x)dx$, where g is the acceleration due to gravity. This difference in pressure, acting over the area A, produces a net force on the slice in the negative x-direction amounting to $Ag\rho(\partial z/\partial x)dx$, which in turn results in an acceleration $\partial u/\partial t$ of the liquid in the slice. Applying Newton's law and setting the mass of the liquid in the slice equal to $\rho A dx$ we find

$$g\,\frac{\partial z}{\partial x} = -\frac{\partial u}{\partial t} \tag{A-5}$$

This is known as the *equation of motion*.

Equations (A-4) and (A-5) are now combined in the following manner: (a) Equation (A-4) is differentiated partially with respect to the time; (b) Equation (A-5) is multiplied by A and differentiated partially with respect to x. (c) $\partial^2 Au/\partial t\partial x$ is eliminated from the resulting pair of equations. This leads to the equation

$$g\,\frac{\partial}{\partial x}\left(A\,\frac{\partial z}{\partial x}\right) = b(x)\,\frac{\partial^2 z}{\partial t^2} \tag{A-6a}$$

or

$$A\,\frac{\partial^2 z}{\partial x^2} + \frac{\partial A}{\partial x}\,\frac{\partial z}{\partial x} = \frac{b}{g}\,\frac{\partial^2 z}{\partial t^2} \tag{A-6b}$$

which have the same form as the wave equations for a string of variable tension and density, Equations (A-3a) and (A-3b).

If the cross-section of the tank, A, is independent of x, then since $A = bh$, we have

$$\frac{\partial^2 z}{\partial x^2} = \frac{1}{gh}\,\frac{\partial^2 z}{\partial t^2} \tag{A-7}$$

where h is the depth of the tank (the depth h may depend on x as long as the width, b, also varies in such a way that the area of the cross-section is constant). Equation (A-7) has the same form as the wave equation for a stretched string of uniform tension, Equation (A-1).

c. General solutions of the wave equations

Equations (A-1) and (A-7) have the general form

$$\partial^2 z/\partial x^2 = (1/c^2)\partial^2 z/\partial t^2 \tag{A-8}$$

where c^2 may or may not depend on x and t; we shall, however, for the moment assume that c^2 is a constant. Then by direct substitution we can verify that the following are solutions of Equation (A-8),

$$z = F(x + ct) \tag{A-9a}$$

$$z = G(x - ct) \tag{A-9b}$$

where F and G are any functions capable of being differentiated.* Furthermore, any linear combination of F and G is also a solution

$$z = AF(x + ct) + BG(x - ct) \tag{A-9c}$$

where A and B are any constants. We have thus solved the problem that was set before us, but the solution is so general that it is almost disconcerting. When we recall the complexity of the oscillations of, say, a stretched clothesline, however, we should not be surprised at the failure of the solution to commit itself more definitely.

On the other hand, these solutions do have a very simple physical interpretation. Equation (A-9a) states that, if at time $t = 0$ the string has a shape $z = F(x)$, then at any subsequent time it will have the same shape except that the function is shifted in the negative x-direction by an amount ct. That is to say, Equation (A-9a) corresponds to a wave of constant shape moving with velocity c in the negative x-direction. Similarly, Equation

* To prove that (A-9a) is a solution of (A-8) set $u = x + ct$. Then

$$\frac{\partial F}{\partial x} = \frac{dF(u)}{du}\,\frac{\partial u}{\partial x} = \frac{dF(u)}{du}$$

$$\frac{\partial^2 F}{\partial x^2} = \frac{\partial}{\partial x}\left(\frac{\partial F}{\partial x}\right) = \frac{\partial u}{\partial x}\,\frac{d}{du}\left(\frac{\partial F}{\partial x}\right) = \frac{d^2F(u)}{du^2}$$

$$\frac{\partial F}{\partial t} = \frac{dF}{du}\,\frac{\partial u}{\partial t} = c\,\frac{dF}{du}$$

$$\frac{\partial^2 F}{\partial t^2} = \frac{\partial}{\partial t}\left(\frac{\partial F}{\partial t}\right) = \frac{\partial u}{\partial t}\,\frac{d}{du}\left(\frac{\partial F}{\partial t}\right) = c^2\,\frac{d^2F(u)}{{}^2du}$$

It is evident that $\partial^2F/\partial x^2 = (1/c^2)\,\partial^2F/\partial t^2$ so that $F(x + ct)$ satisfies Equation (A-8) regardless of the shape of the function F. The student can readily show by the same equations that $G(x - ct)$ is also a solution for any function G.

(A-9b) represents a wave of constant shape moving in the positive x-direction with the same velocity, c, and (A-9c) represents a "superposition" of these two waves in the proportions $A:B$. The constant c is thus equal to the "wave velocity." For the string

$$c = (T/\rho)^{1/2} \tag{A-10}$$

whereas for the liquid in the shallow trough

$$c = (gh)^{1/2} \tag{A-11}$$

Interestingly enough, the latter velocity is independent of the density of the liquid in the trough.

Exercise. On November 11, 1922, there was an earthquake off the coast of Chile at Carrizal, which caused seismic waves in the ocean. Fifteen hours later the wave was noticed at Honolulu, 5850 miles away, as a series of oscillations superimposed on the normal tidal motions. What is the average depth of the Pacific Ocean between Carrizal and Honolulu?

Solutions of the form (A-9a, b, c) will occur only if c is independent of x and t. On the other hand, even if T and ρ, or A and h, vary with x and t, we may still use Equations (A-10) and (A-11) to give a "local wave velocity" which will retain some physical significance if waves of small wave length are used. In such a wave the slope $\partial z/\partial x$ changes rapidly with x, so that $\partial^2 z/\partial x^2$ is large, and the terms in $\partial T/\partial x$ and $\partial A/\partial x$ in Equations (A-3b) and (A-6b) can be neglected. Under these circumstances waves will be propagated with nearly constant form over fairly long distances, and (A-10) and (A-11) give a local velocity of wave propagation that can be given some physical significance. We can in this approximation think of such strings and troughs as having wave velocities that vary from point to point. We may also define an "index of refraction" at each point of the system as the ratio

$$n = c_0/c \tag{A-12}$$

where c_0 is the wave velocity under some standard set of conditions and c is the wave velocity at the point in question. Strings of variable tension and density, and troughs of variable depth and cross-section may in this sense be regarded as having a variable index of refraction.

d. The effect of boundary conditions

Equation (A-9c) is the most general solution for systems not subject to any external restraints. In real systems, however, the solution must always meet certain additional requirements. For instance, the ends of the string may be rigidly fixed, so that z is always equal to zero at these two points. Or the two ends of the trough may be closed by rigid walls, so that $u = 0$ here at all times, which by Equation (A-5) gives $\partial z/\partial x = 0$ at either end. These ever-present requirements are known as *boundary conditions*. Of course, an infinite variety of solutions is possible even when the boundary conditions are satisfied, but it is not so "big" an infinity as that allowed by the most general solution.

e. Normal modes of vibration

When an object is set in vibration by some momentary jarring, the resulting oscillations usually seem very complex. For instance, if a length of chain suspended from a fixed point is given a sudden push it will squirm in an amazingly complicated way which it might seem hopelessly difficult to describe mathematically. On the other hand, it is possible, with care, to excite certain kinds of motion that are relatively simple. For instance, if the lower end of the chain is pulled off to one side while exerting the proper amount of tension on the chain, and if it is then released, a particularly simple motion will occur (see Fig. 3-3). In this motion, all points of the chain move so that their displacements vary with time in the same way, the frequency and phase being everywhere the same (i.e., all points are at their maximum displacement at the same time, and have their maximum velocity at the same time). Mathematically we can say that these motions are described by functions of the form

$$z(x, t) = f(x)\, \phi(t) \tag{A-13}$$

where ϕ is independent of x, and f is independent of t. Such motions are known as *normal modes of vibration*.

Elastic bodies are generally capable of many different normal modes of different frequencies (there would be an infinite number of normal modes for a string or a trough were it not for the atomic composition of matter). For instance, the instantaneous shapes of the first four normal modes of the hanging chain are illustrated in Fig. 3-4. (They are related to Bessel functions of zero order. See Morse (*3*, p. 149).) Notice that successive modes differ in the number of points which remain at rest ("nodes"). The number and arrangement of nodes is often a means of classifying the normal modes of elastic bodies.

A convenient method of exciting normal modes of the chain in fairly pure form is to move the point of suspension back and forth over a small distance with a frequency equal to that of the normal mode desired. The student can use this method to verify the appearance of some of the normal modes of a hanging chain illustrated in Fig. 3-4.

Molecules also have normal modes of vibration. The number of these is, however, finite, being equal to $3N-5$ for linear molecules and $3N-6$ otherwise, where N is the number of atoms in the molecule. It is the frequencies of these normal modes that are observed in infra-red and Raman spectra. Although the methods of calculating these normal modes seem at first sight quite different from those used with macroscopic bodies (see Wilson, Decius and Cross, *4*), they must ultimately amount to the same thing, since macrosopic bodies are really only molecules containing very many atoms. The relationship between the two methods of calculation is discussed in treatments of the statistical mechanical theory of the specific heats of crystals (e.g., *5*, pp. 246 ff.; *6*, pp. 241 ff.; (*7*)).

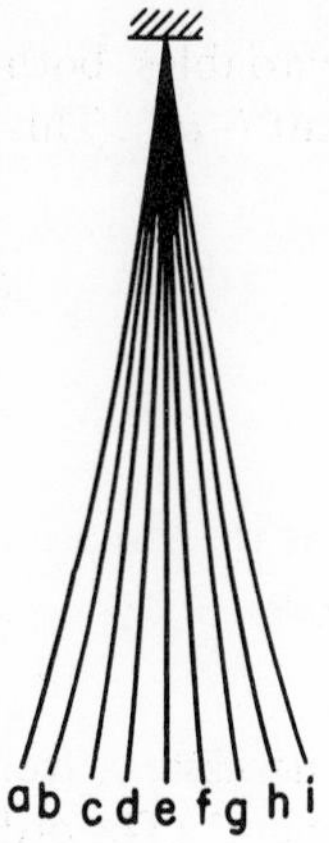

Fig. 3-3

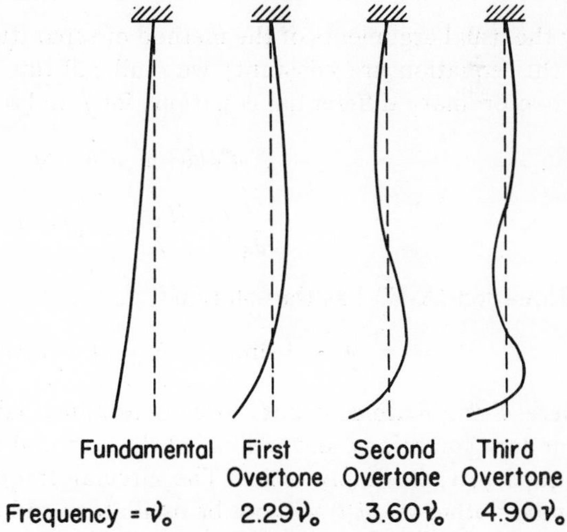

Fig. 3-4

FIG. 3-3. Successive positions of a hanging chain during one cycle of its fundamental normal mode of oscillation. The chain is initially stationary in position *a*. It subsequently moves to the right through positions *b*, *c*, *d*, *e*, *f*, *g*, and *h*, reversing its motion at *i* and returning to *a* in reverse order of positions *h* through *b*. Since the motion is a normal mode, all points on the chain have their maximum displacements at the same instant (positions *a* and *i*) and all points have their maximum velocities at the same instant (position *e*).

FIG. 3-4. Shapes of the first four normal modes of the hanging chain. The normal mode frequencies are given in terms of ν_0, the frequency of the fundamental mode.

f. The differential equations of normal modes

We have seen that a normal mode may be defined as an oscillation in which the displacements are of the form

$$z(x,t) = f(x)\,\phi(t) \tag{A-14}$$

and which obey the equation

$$\partial/\partial x(T\partial z/\partial x) = \rho\,\partial^2 z/\partial t^2 \tag{A-15}$$

where T and ρ may depend on x. Let us substitute (A-14) in (A-15). This is equivalent to the application of the method of separation of variables as described on page 44. We find

$$\frac{1}{f(x)\rho(x)}\frac{d}{dx}\left(T(x)\frac{df(x)}{dx}\right) = \frac{1}{\phi(t)}\frac{d^2\phi(t)}{dt^2} \tag{A-16}$$

By the usual arguments of the method of separation of variables, both sides of this equation are constant; we shall call this constant $(-\omega^2)$. This gives us two ordinary differential equations for f and ϕ

$$d^2\phi/dt^2 + \omega^2\phi = 0 \tag{A-17}$$

$$\frac{d}{dx}\left(T\frac{df}{dx}\right) + \omega^2\rho f = 0 \tag{A-18}$$

Equation (A-17) has the solution

$$\phi = A\sin(\omega t + \delta) = A\sin(2\pi\nu t + \delta) \tag{A-19}$$

where A and δ are constants to be determined later, if necessary, from the boundary conditions, and ω is called the "circular frequency." (The ordinary frequency, ν, is simply $\omega/2\pi$. The circular frequency, ω, must be a real number, otherwise $\phi(t)$ will not be oscillatory and (A-19) will not correspond to a vibration (cf. page 40). Physically this will be seen to mean that T must be positive – that is, the string must be under tension, not compression. Equation (A-19) tells us that the motions in a normal mode are "harmonic" – that is, all parts of the body move sinusoidally with respect to time. They also move with the same phase, δ, and the same frequency, ν, as required by the definition in Section e, above.

The shape of the normal mode is given by the solution of (A-18) and will depend on T, ρ and the boundary conditions. Equation (A-18) will be called the *normal mode equation*. It involves only the coordinates, as compared with the *wave equation*, (A-15), which contains the time as well as the coordinates.

It is interesting to note that the normal mode solutions, $f(x)$, are eigenfunctions (see page 18) of the operator

$$W = -\frac{1}{\rho}\frac{d}{dx}\left(T\frac{d}{dx}\right) \tag{A-20}$$

and that the squares of the circular frequencies are the eigenvalues

$$Wf = \omega^2 f \tag{A-21}$$

g. The normal modes of a uniform stretched string

For a stretched string in which T and ρ are constant and the ends are held fixed we have the normal mode equation

$$\frac{d^2f}{dx^2} + \frac{4\pi^2\nu^2}{c^2} f = 0 \tag{A-22}$$

and the boundary conditions

$$(i)\ f = 0 \text{ at } x = 0$$

$$(ii)\ f = 0 \text{ at } x = a \tag{A-23}$$

where a is the length of the string, c is the wave velocity in the string, and ν is the frequency of the normal mode. Setting

$$\lambda = c/\nu \tag{A-24}$$

we find the general solution of (A-22) to be

$$f = C\sin(2\pi x/\lambda) - D\cos(2\pi x/\lambda) \tag{A-25}$$

where C and D are constants. From the boundary condition (*i*) we see that $D = O$ and from condition (*ii*), $C \sin (2\pi a/\lambda) = 0$, which, if we are to have any solution at all, must mean that

$$\sin(2\pi a/\lambda) = 0, \text{ or } 2a/\lambda = n \tag{A-26}$$

where n is a positive integer (negative integers merely correspond to normal modes whose phases are 180° behind those of the modes with positive integers). Thus $n\lambda/2 = a$ and

$$\nu = nc/2a = n\nu_0 \tag{A-27}$$

The normal modes are thus stationary sine waves whose wave lengths (λ) are such that the length of the string is an integral number of half waves (see Fig. 3–5). The frequencies are integral multiples of the so-called "fundamental" frequency, $\nu_0 = c/2a$. The complete solution for a normal mode is

$$z(x,t) = C \sin(n\pi x/a) \sin(2\pi n\nu_0 t + \delta) \tag{A-28}$$

where C and δ are called the *amplitude* and *phase*, respectively.*

Insofar as a violin corresponds to a stretched string with fixed ends, these are the only frequencies that a violinist can extract from his instrument with a given fingering.† The simple numerical relationship between the

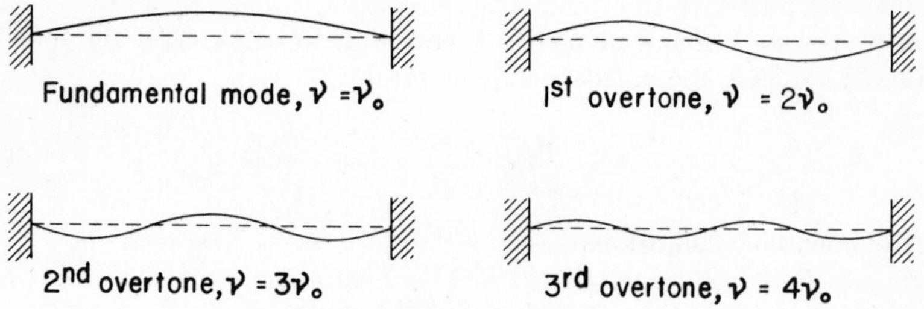

FIG. 3-5. Normal modes of a stretched string.

frequencies accounts for the excellent musical properties of stringed instruments. Using musical notation to express frequencies, the lowest overtones of the *G*-string of a violin are *g*, *d'*, *g'*, b'', *d''*, *f''*, *g''*, *a'''*, *b'''*, etc. Only at the sixth overtone (*f''*) do we begin to find frequencies that do not harmonize with the fundamental. This "harmonic" property of vibrating strings and other one-dimensional systems is not at all common in other vibrating systems, as we shall see.

* Equation (A-28) may be written

$$z = C\left[\cos\left(\frac{n\pi}{a}(x - 2a\nu_0 t) - \delta\right) - \cos\left(\frac{n\pi}{a}(x + 2a\nu_0 t) + \delta\right)\right]$$
$$= C\left[\cos\left(\frac{n\pi}{a}(x - ct) - \delta\right) - \cos\left(\frac{n\pi}{a}(x + ct) + \delta\right)\right]$$

which has the same form as the general solution of the wave equation, Equation (A-9c). The normal modes of the stretched string may thus be regarded as a superposition of two waves having the same phase and equal and opposite amplitude, which move in opposite directions along the string.

† He can, however, alter the relative proportions of the different frequencies–that is, change the "quality" of the note–by a change in his bowing. Thus, by moving the bow nearer to the bridge, a "shrill" tone, containing more of the higher overtones, is obtained.

Exercises. (1) Try to apply the normal mode treatment to the uniform string with fixed ends in which there is a uniform compression instead of a tension (that is, T is negative). Will such a string oscillate?

(2) Find the normal modes and frequencies of the motion of a liquid in a long, shallow, rectangular trough of constant depth and cross-section whose length is a cm. and whose ends are closed. Will all liquids give the same normal mode frequencies in a given trough filled to a given depth?

h. Fourier series and the analysis of vibrations in terms of normal modes

If a string is plucked or struck with a hammer, it will not vibrate in a normal mode, but will take on some other shape which will vary in a more complicated way with the time. At any moment, t, the displacement will be given by $z(x,t)$. Let us expand this function as a Fourier series (see page 30) of sines in the range $x = 0$ to $x = a$.

$$z(x,t) = \sum_{n=1}^{\infty} C_n(t) \sin (n\pi x/a) \tag{A-29}$$

The cœfficients, C_n, will of course, depend on the time, and their time dependence can be found by substituting (A-29) directly into the original wave equation, (A-8). This gives

$$\sum \left[(n\pi/a)^2 C_n + (1/c^2)\,(d^2C_n/dt^2)\right] \sin n\pi x/a = 0 \tag{A-30}$$

Equation (A-30) must be valid for all values of x. This is only possible if each of the coefficients in brackets is separately zero

$$d^2C_n/dt^2 + (n\pi c/a)^2 C_n = 0 \tag{A-31}$$

The solution of (A-31) is

$$C_n = A_n \sin \left[(n\pi ct/a) + \delta_n\right] \tag{A-32}$$

where A_n and δ_n are constants. But $nc/2a$ is the frequency of the nth normal mode, ν_n. Thus we find

$$z(x,t) = \sum A_n \sin (n\pi x/a) \sin (2\pi\nu_n t + \delta_n) \tag{A-33}$$

On comparing (A-33) with (A-28) we see that each term in this sum describes a normal mode of vibration. *Because of Fourier's expansion theorem, any oscillation of the string can be considered to be the resultant of the simultaneous excitation of all of the normal modes. Any arbitrary motion can be completely described by specifying the amplitudes, A_n, and phases, δ_n, of the normal modes into which it can be resolved.*

Exercise. A stretched string (e.g., a string of a harp) is pulled off to one side at its center and then released. At the moment of release its velocity is everywhere zero and each half is straight. Give the amplitudes and phases of all of the normal modes thus excited.

(At $t = 0$,

$$z(x, 0) = \begin{cases} sx/a \text{ for } 0 \leqq x \leqq (1/2)a \\ s[1 - (x/a)] \text{ for } (1/2)a \leqq x \leqq a \end{cases}$$

$$dz/dt = 0 \text{ for all } x$$

Using these relations with (A-33), A_n and δ_n may be evaluated.)

i. Perturbations and the concept of hybridization

We have seen that a string of length a, uniform tension T, and uniform density ρ, which is fixed at both ends, has a set of normal modes

$$f_{n0} = \sin(n\pi x/a) \tag{A-34}$$

$$z_{n0} = f_{n0} \sin(2\pi\nu_{n0}t + \delta) \tag{A-35}$$

where n is an integer, the frequency of the fundamental mode is $\nu_0 = (1/2a)(T/\rho)^{1/2}$, and the frequency of the nth mode is $\nu_{n0} = n\nu_0$. Suppose that we distribute a *small* extra load nonuniformly along the string, say by wrapping a fine wire around the string. Let this extra load have a mass $\rho e(x)$ per unit length and assume that $e(x)$ is much less than unity. As a result of this "perturbation" of the system, the shapes and frequencies of the normal modes will be slightly changed. It is not difficult to obtain useful approximate expressions for these changes.

The differential equation for the perturbed mode that corresponds to the nth unpertbured one is (*cf.* (A-18))

$$\frac{d^2 f_n}{dx^2} + \frac{\rho(1 + e(x))}{T} 4\pi^2 \nu_n^2 f_n = 0 \tag{A-36}$$

Let us express the perturbed normal modes in terms of the unperturbed ones, using Fourier's theorem

$$f_n = A_n f_{n0} + \sum_{j \neq n} A_j f_{j0} \tag{A-37}$$

and define a quantity α_n through the relation

$$\nu_n^2 = \nu_{n0}^2(1 + \alpha_n) \tag{A-38}$$

If α_n and all of the A_j's are known, the shape and frequency of the perturbed mode will be known; the problem is to find these quantities. As long as the perturbation is small, we may expect that each perturbed normal mode will resemble one of the unperturbed ones. This means that A_n in (A-37) is large, while the rest of the A_j's as well as α_n are small. Substituting (A-38) and (A-37) into (A-36), and recalling that

$$\frac{d^2 f_{j0}}{dx^2} = -(4\pi^2\rho/T)\nu_{j0}^2 f_{j0} \tag{A-39}$$

and neglecting terms that involve the products of two or more small quantities ($\alpha_n e$; $\alpha_n A_j$, where $j \neq n$; etc.) we find

$$A_n\left[\frac{d^2 f_{n0}}{dx^2} + \frac{4\pi^2\rho}{T}\nu_{n0}^2 f_{n0}\right] + \frac{4\pi^2\nu_{10}^2\rho}{T}\left[A_n n^2(e+\alpha_n)f_{n0} + \sum_j A_j(n^2-j^2)f_{j0}\right] = 0 \tag{A-40}$$

The first term in brackets is zero because of equation (A-39). Therefore, the second term in brackets must also vanish. This leads to

$$n^2 A_n(e+\alpha_n)f_{n0} + \sum_{j\neq n} A_j(n^2-j^2)f_{j0} = 0 \tag{A-41}$$

Multiplying through by f_{n0} and integrating from $x = 0$ to $x = a$, we obtain

$$n^2 A_n\left[\int_0^a e(x)f_{n0}^2(x)dx + \alpha_n\int_0^a f_{n0}^2(x)dx\right] + \sum_{j\neq n} A_j(n^2-j^2)\int_0^a f_{j0}f_{n0}dx = 0 \tag{A 42}$$

But all of the terms in the sum in this equation vanish, because if $j \neq n$

$$\int_0^a f_{j0}f_{n0}dx = \int_0^a \sin(j\pi x/a)\sin(n\pi x/a)dx = 0 \tag{A-43}$$

(this follows from Exercise 2 on page 21). Therefore we are left with

$$\alpha_n = -\int_0^a e(x)f_{n0}^2(x)dx \bigg/ \int_0^a f_{n0}^2(x)dx \tag{A-44}$$

which gives, to a first approximation, the relative change in the frequency due to the perturbation. It is seen that α_n is the average relative change in density, weighted according to the square of the amplitude of the unperturbed normal mode. If the perturbation is large where the normal mode has a large amplitude, it will cause a big change in the frequency of that mode. If the perturbation is localized near nodes, there will be a correspondingly small effect on the frequency.

The values of the A_j's in Equation (A-37) are found by multiplying (A-41) by f_{k0} and integrating. As before, nearly everything drops out because of (A-43), and we are left with

$$n^2 A_n \int_0^a e(x) f_{k0} f_{n0} dx + A_k(n^2 - k^2) \int_0^a f_{k0}^2 dx = 0 \tag{A-45}$$

$$A_k/A_n = \frac{n^2}{k^2 - n^2} \frac{\int_0^a e(x) f_{k0} f_{n0} dx}{\int_0^a f_{k0}^2 dx} \tag{A-46}$$

The ratio A_k/A_n has a simple physical interpretation. Suppose that the perturbed string oscillates in the nth mode and that the perturbation is then suddenly removed. The string will no longer oscillate in a pure normal mode, but the motion will contain other normal modes in proportion to the A_k's in the expansion

$$z(x, t) = \sum A_k f_{k0} \sin(2\pi\nu_{k0} t + \delta) \tag{A-47}$$

This is true because (A-47) satisfies the wave equation for the unperturbed string and with the appropriate value of the constant δ it also gives the shape of the string at the instant the perturbation was removed. The A_k's thus represent the "contamination" of the original normal mode caused by the perturbation. This contamination will be large for those modes for which $e(x)$, f_{k0}, and f_{n0} are all large in the same regions. If $e(x)$ is added near the nodes of either the kth or the nth modes, the contamination of the nth mode by the kth mode will be small.

This "mixing" of normal modes under the influence of a perturbation (as illustrated by Equation (A-37)) will be called *hybridization* because it is the exact analogue of an important quantum chemical concept known by the same name. The concept of hybridization thus makes use of the normal modes of simple systems in order to describe the normal modes of more complex systems. It is important to realize that the concept has meaning only in terms of this use of simple systems in describing more complex ones; it depends entirely on the initial use of Equation (A-37) to describe the perturbed system.

As an example of the perturbation theory, let us find the effect of distributing a small mass uniformly along one half of a stretched string. Then

$$e(x) = \begin{cases} e = \text{const. if } 0 < x < a/2 \\ 0 \qquad \text{if } a/2 < x < a \end{cases}$$

which gives

$$a_n = -e \frac{\int_0^{a/2} \sin^2 (n\pi x/a) dx}{\int_0^a \sin^2 (n\pi x/a) dx} = -e/2$$

To a first approximation all frequencies are reduced by the factor $(1 - \frac{1}{2}e)^{1/2}$. For the coefficients in the Fourier expansion of the perturbed normal mode function,

$$A_k/A_n = \frac{n^2 e}{k^2 - n^2} \int_0^{a/2} \sin (k\pi x/a) \sin (n\pi x/a) dx \Big/ \int_0^a \sin^2 (k\pi x/a) dx$$

$$= \frac{n^2 e}{(k^2 - n^2)^2 \pi} \{(k + n) \sin [(k - n)\pi/2] - (k - n) \sin [(k+n)\pi/2]\}$$

The shape of the lowest mode is given by taking $n = 1$, with A_1 very large. Then it is found on substitution that

$$A_2/A_1 = 4e/9\pi; \qquad A_3/A_1 = A_5/A_1 = 0;$$

$$A_4/A_1 = -8e/225\pi; \; A_6/A_1 = 12e/1225\pi; \text{ etc.}$$

These values give for the shape of the first perturbed mode

$$f_1 = A_1[\sin(\pi x/a) + 0.14e\sin(2\pi x/a) - 0.011e\sin(4\pi x/a) + 0.0031e\sin(6\pi x/a) - \cdots]$$

The effect of the added weight on the fundamental is thus to make the heavy side move with a greater amplitude (see Fig. 3-6A). Similarly, for the first overtone we find

$$f_2 = A_2[\sin(2\pi x/a) - 0.57e\sin(\pi x/a) + 0.20e\sin(3\pi x/a) - 0.012e\sin(5\pi x/a) + \cdots]$$

showing that the added load makes the heavy half move with a smaller amplitude (see Fig. 3-6B).

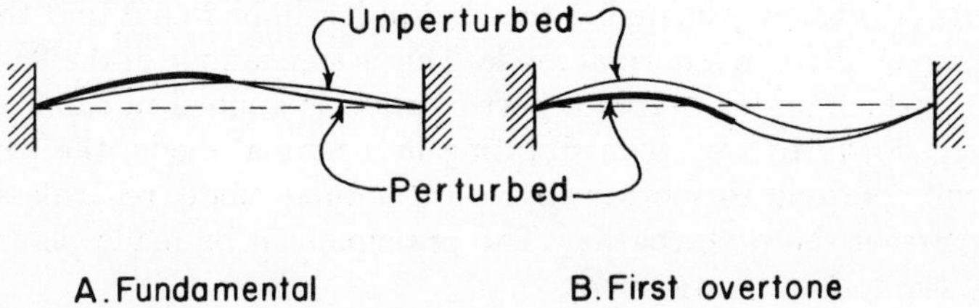

FIG. 3-6. Fundamental and first overtone of a stretched string perturbed by uniformly loading the left half of the string.

As a result of the addition of the extra load e to the left half of the string we can say that the fundamental mode has been changed in shape by being hybridized with the first overtone to the extent of $0.14e$, with the second overtone to the extent of $-0.011e$, etc. Similarly, the shape of the first overtone has been changed by hybridization to the extent of $-0.57e$ with the fundamental, $0.20e$ with the second overtone, etc.

Exercises. (1) Show that if the perturbation consists of a change in the tension by an amount $Tg(x)$, as well as of a change in the density by $\rho e(x)$, the relative change in the frequency of the nth mode is, to a first approximation, given by Equation (A-38), with

$$a_n = \int_0^a (g - e) f_{n0}^2 dx \Big/ \int_0^a f_{n0}^2 dx \qquad \text{(A-48)}$$

(2) Suppose that $e(x)$ is everywhere a constant, independent of x. How does the approximate change in frequency, as given by equations (A-44) and (A-38), compare with the exact value of the change when $e = 0.01$, 0.1 and 0.5? How do the shapes of the normal modes, as given by equations (A-37) and (A-46) compare with the exact shapes?

(3) Find the first approximation to the percentage change in the frequency of the nth normal mode of a uniform stretched string of length a if the density of the portion between $x = 0.45a$ and $0.55a$ is increased by 10%.

(4) Find the approximate change in the fundamental frequency of a stretched string of length a and density ρ to which a perturbing load with density $0.1\ \rho[x - (\tfrac{1}{2})a]$ has been added.

j. The continuity of the normal modes in perturbed systems: the adiabatic principle

If a hanging chain which has been oscillating in a normal mode is suddenly perturbed by adding a weight to its lower end or by changing its length, it will vibrate in a complicated fashion that can be regarded as a superposition of many normal modes of the perturbed system. On the other hand, if the same perturbation is applied gradually, (by adding the weight in small increments or by slowly changing the length) it will be found that the chain continues to oscillate in a normal mode. This is an example of the "adiabatic principle," which states that if a perturbation is applied in small steps or sufficiently slowly to a system vibrating in a normal mode, the perturbed system will continue to vibrate in a pure normal mode, regardless of the ultimate size of the perturbation. The principle can be made plausible by means of the following argument.

Suppose that a system vibrating in its nth normal mode is slowly subjected to a perturbation (e.g., an increase in the tension). How much will the

n^{th} mode be contaminated by the mth mode after the perturbation has been applied? If the frequencies of the two modes are ν_n and ν_m, respectively, it is clear that if any of the mth mode did happen to be excited at some stage during the perturbation, it will go into and out of phase with the nth mode at a rate of $\nu_m - \nu_n$ times a second. Now if the perturbation happens to increase the contamination when the two modes are in one phase relative to one another, it would be expected to decrease the contamination when the two are half a cycle out of phase–i.e., $[2(\nu_m - \nu_n)]^{-1}$ sec. later. The contamination should therefore increase and decrease rather than continually build up, as long as the time required for the application of the perturbation is large compared with $1/(\nu_m - \nu_n)$. The more slowly the perturbation is applied, the smaller will this ebb and flow of contamination be.

It is often possible to change one vibrating system into another by gradually applying some perturbation. As a result of the adiabatic principle, one can relate the normal modes of the two systems to each other. This is a useful result, since it often offers a means of characterizing the normal modes of a complicated object in terms of the normal modes of similar but simpler objects. It is also sometimes possible to make rough estimates of shapes and frequencies of normal modes in this way. Later in this chapter several illustrations will be given of this important method of discussing vibrations; the method is even more useful in quantum chemistry than it is in the theory of vibrations. A simple illustration of the method will now be considered.

Suppose that a particle of mass m is attached to the center of a string which is vibrating in its nth normal mode. If n is even, there will be no effect on the motion of the string, since the center is a node and does not move anyway. But if n is odd, the motion of the center of the string, and therefore of the string as a whole, will be modified by the particle. In the odd normal modes of the perturbed system, each half of the string will still have the form of a sine wave, but the wave length will be increased, giving a kink in the string where the particle is located. It is not hard to show (see *1*, p. 37) that the frequencies of the resulting odd-numbered normal modes are given by the roots of the transcendental equation,

$$\frac{\pi}{2}\frac{m}{M}\frac{\nu_n}{\nu_0} = \cot\left(\frac{\pi}{2}\frac{\nu_n}{\nu_0}\right) \tag{A-47}$$

where ν_n is the frequency of the nth mode in the presence of the particle, ν_0 is the fundamental frequency of the unperturbed string and M is the total mass of the string. If $m/M \ll 1$, we find roots of Equation (A-47) at approximately

$$\nu_n/\nu_0 = 1, 3, 5, 7, \cdots$$

If $m/M \gg 1$ we find roots at approximately

$$\nu_n/\nu_0 = 0, 2, 4, 6, \cdots$$

The modes of even n, which are unaffected by the perturbation, have, of course, the frequencies

$$\nu_n/\nu_0 = 2, 4, 6, \cdots$$

The approximate dependence on m/M of the frequencies and the shapes of the normal modes are indicated in Fig. 3-7. We see that for $m/M = \infty$ all modes except the fundamental come in pairs having the same frequency. Such modes are said to be *doubly degenerate* (in general, if N normal modes happen to have the same frequency, they are said to be *N-fold degenerate*). The physical reason for the degeneracy in the present instance is easy to

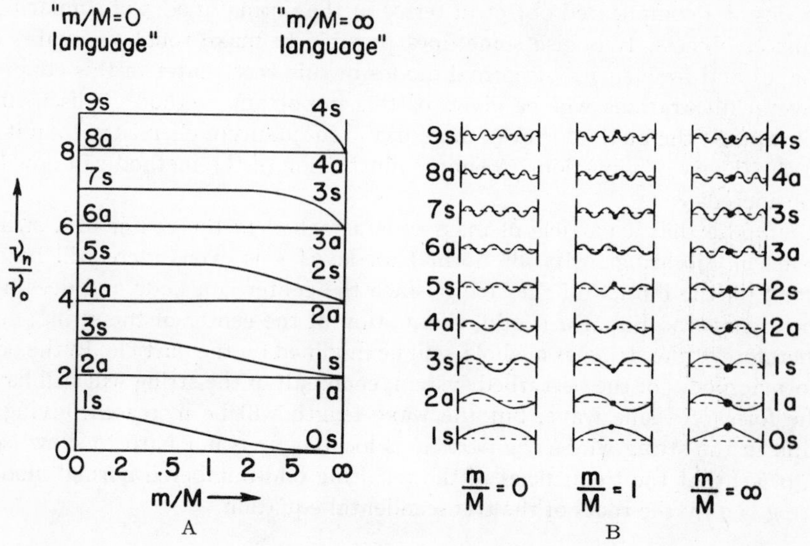

FIG. 3-7.

A. Variation of the normal mode frequencies of a stretched string of mass M caused by the addition of a point mass m at its center. The fundamental frequency of the unloaded string is ν_0.

B. Appearance of some of the normal modes of the loaded string for three values of the added mass.

see: when the central particle is very heavy it remains at rest in all modes except the fundamental mode, and the two halves of the string can vibrate independently of one another. The frequencies are those of a string half the original length–a fact that could have been guessed initially.

The diagram giving ν_n/ν_0 as a function of m/M is called a *correlation diagram,* since it correlates the vibrational states of all systems of this type. It is important because it provides a basis for rational systems of nomenclature with which to label the normal modes. If we start at the $m/M = 0$ side of the diagram, it is natural to label the modes according to the value of ν_n/ν_0 at $m/M = 0$. Thus the fundamental would be called mode No. 1, the first overtone No. 2, etc. Starting from the $m/M = \infty$ side of the diagram, however, the numbering would naturally run 0, 1, 2, 3, $\cdots$, but with two kinds of vibration for each overtone. In order to distinguish these, it is convenient to consider the symmetry of the vibrations. It will be seen that, regardless of the value of m/M, one of each pair of normal modes is "symmetric" with respect to reflection in a plane through the center of the string. (That is, all displacements from equilibrium remain unchanged on such a reflection.) The other member of the pair is "antisymmetric." (That is, the displacements all change their signs on reflection.) Thus, in the "$m/M = \infty$ language," we can denote the symmetric 1-mode by $1s$ and the antisymmetric 1-mode by $1a$. Furthermore, since the s and a character is preserved at all points across the diagram, we can, if we wish, add the letters a and s to the "$m/M = 0$ language;" this, of course, removes no ambiguities, but it does help to make the "$m/M = 0$ language" more descriptive.

We thus have two sets of symbols with which to denote the normal modes of any string pertubed by a weight at its center, and there is a one-to-one correspondence between them as indicated by the following "dictionary":

$$\text{“}m/M = 0 \text{ language:”}\quad 1s,\ 2a,\ 3s,\ 4a,\ 5s,\ \cdots$$

$$\updownarrow\ \ \updownarrow\ \ \updownarrow\ \ \updownarrow\ \ \updownarrow$$

$$\text{“}m/M = \infty \text{ language:”}\quad 0s,\ 1a,\ 1s,\ 2a,\ 2s,\ \cdots$$

Which of these languages we choose to use in describing any particular system of this type is, strictly speaking, immaterial. If we wish, we may even use the "$m/M = \infty$ language" to describe the $m/M = 0$ states. It is, however, more sensible to restrict the use of each language more or less to its half of the correlation diagram.

B. Vibrations of Two-Dimensional Systems

a. The wave equation of a membrane under uniform tension

Consider an elastic membrane of constant density, ρ gm./cm.^{2}, in which there is a uniform tension of T dynes per cm. (that is, if we make a slit one centimeter long anywhere in the membrane, we shall have to exert a pull of T dynes on either side of the slit in order to keep it closed). Let x and y be position coordinates measured in the plane of the membrane, and let z be the vertical displacement of the membrane from its equilibrium position. Of course z will be a function of x and y, as well as of the time, t. Let us find the forces acting in the z-direction on an infinitesimal rectangular element of the membrane located between $x, x + dx, y, y + dy$. Forces of magnitude Tdx and Tdy will pull tangentially outward on the two pairs of parallel edges of the element (see Fig. 3-8). If the element is curved, these

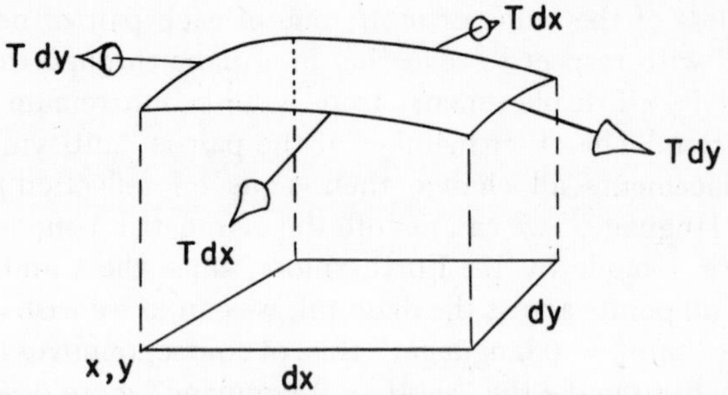

FIG. 3-8. Forces acting on an element of a membrane under uniform tension (Cartesian coordinates).

forces will not balance one another. By arguments similar to those used in the derivation of the wave equation of the string, the curvature in the y-direction will, if the displacements and the slopes are everywhere small, result in a net force in the z-direction amounting to

$$Tdx(\partial^2 z/\partial y^2)dy$$

Similarly, the curvature in the x-direction will result in a net force

$$Tdy(\partial^2 z/\partial x^2)dx$$

These two forces act independently, and thus must be added and equated to the inertial force

$$\rho dxdy(\partial^2 z/\partial t^2)$$

giving finally

$$\partial^2 z/\partial x^2 + \partial^2 z/\partial y^2 = (\rho/T)\partial^2 z/\partial t^2 \qquad \text{(B-1)}$$

which is the *wave equation for a membrane.*

b. The normal modes of a square membrane; degenerate normal modes

Consider a square membrane of uniform density and under constant tension, whose edges are held rigidly fixed. If the origin of the coordinate system is placed at one corner, the boundary conditions are

$$z = 0 \text{ when } \begin{cases} x = 0 \text{ or } x = a, \text{ for all values of } y \text{ and } t \\ y = 0 \text{ or } y = a, \text{ for all values of } x \text{ and } t \end{cases} \tag{B-2}$$

where a is the length of a side of the square.

In order to find the normal modes of such a membrane we must find solutions of Equation (B-1) of the form

$$z = f(x, y)\phi(t) \tag{B-3}$$

By the usual arguments of the method of separation of variables, the substitution of (B-3) into (B-1) gives

$$\phi(t) = C \sin(2\pi\nu t + \delta) \tag{B-4}$$

and

$$\partial^2 f/\partial x^2 + \partial^2 f/\partial y^2 + (4\pi^2\nu^2/c^2)f = 0 \tag{B-5}$$

where ν is the frequency of the vibration and $c = \sqrt{T/\rho}$. We must now find a solution of (B-5) that will satisfy the boundary conditions, (B-2). Let us try the method of separation of variables once again, setting

$$f(x, y) = X(x)\,Y(y) \tag{B-6}$$

Then we find

$$\frac{1}{X}\frac{d^2X}{dx^2} + \frac{1}{Y}\frac{d^2Y}{dy^2} + \frac{4\pi^2\nu^2}{c^2} = 0 \tag{B-7}$$

Each term on the left side of this equation must be constant, independent of x and y. Let us set

$$\frac{1}{X}\frac{d^2X}{dx^2} = -4\pi^2 k_1^2 \tag{B-8a}$$

$$\frac{1}{Y}\frac{d^2Y}{dy^2} = -4\pi^2 k_2^2 \tag{B-8b}$$

where k_1 and k_2 are constants satisfying the relation

$$k_1^2 + k_2^2 = (\nu/c)^2 \tag{B-8c}$$

which follows from Equation (B-7). The solutions of Equations (B-8a) and (B-8b) are

$$X = A \sin(2\pi k_1 x) + B \cos(2\pi k_1 x) \tag{B-9a}$$

$$Y = C \sin(2\pi k_2 y) + D \cos(2\pi k_2 y) \tag{B-9b}$$

where A, B, C, and D are arbitrary constants. The boundary conditions at $x = 0$ and $y = 0$ may be satisfied by setting $B = D = 0$, while those at $x = a$ and $y = a$ make it necessary that $k_1 = n/2a$ and $k_2 = m/2a$, where m and n are positive integers. Thus the method of separation of variables gives an acceptable solution, and the normal modes have the form

$$f(x,y) = A'\sin(n\pi x/a)\sin(m\pi y/a) \tag{B-10}$$

while the normal mode frequencies are

$$\nu = c\sqrt{k_1^2 + k_2^2} = (c/2a)\sqrt{m^2 + n^2} \tag{B-11}$$

The fundamental mode occurs when $m = n = 1$, and has the frequency

$$\nu_0 = (\sqrt{2}/2)c/a \tag{B-11a}$$

It contains no nodes. The frequencies and appearances of this and some of the other modes are shown in Fig. 3-9. Most of these modes are evidently

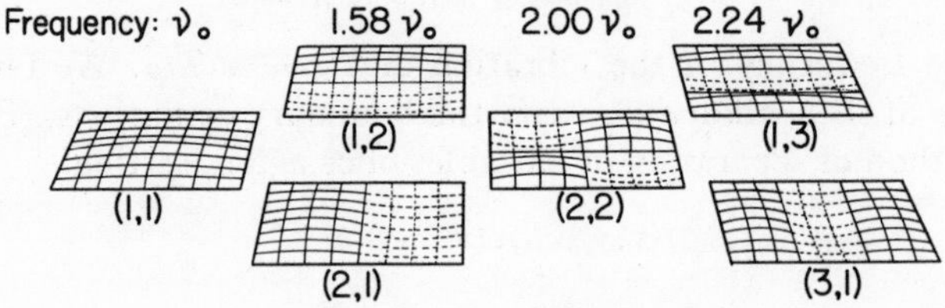

FIG. 3-9. The normal modes of a square membrane. Numbers in parenthesis give the values of the integers m and n in Equations (B-10) and (B-11). The frequencies are those given by Equation (B-11), with ν_0 defined in (B-11a). The normal modes (2,1) and (1,2) have the same frequency, as do also (3,1) and (1,3).

doubly degenerate. It will also be noticed that the frequencies of most of the overtones are not whole multiples of the fundamental.

Degenerate modes have the important property that any superposition is also a normal mode. Thus, substitution into the wave equation (B-1) will verify that

$$z(x,y,t) = [A\sin(m\pi x/a)\sin(n\pi y/a) + B\sin(n\pi x/a)\sin(m\pi y/a)]\cos(2\pi\sqrt{m^2+n^2}\,\nu_0 t) \tag{B-12}$$

is a solution regardless of the values of A and B. It also obviously has the normal mode form.

The results of combining some degenerate modes of a square membrane in various ways are shown in Fig. 3-10. The changing of the shapes of normal modes by combining them in this way is an example of hybridization (see page 66). When the modes are degenerate, however, hybridization is possible even in the absence of a perturbation. Evidently degenerate modes can hybridize with each other much more readily than can modes having different frequencies.

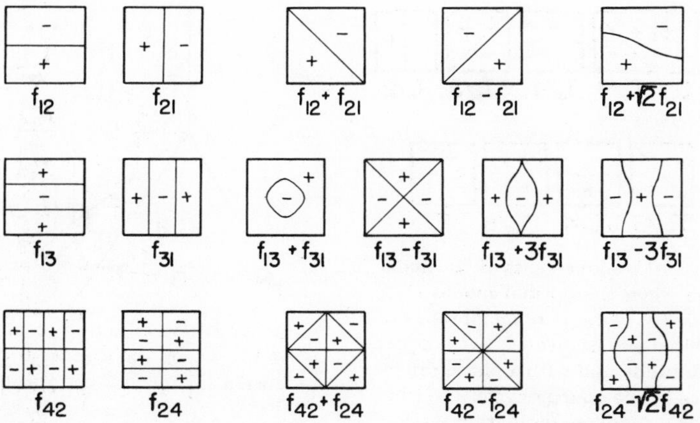

FIG. 3-10. Hybridization of some degenerate normal modes of a square membrane. The plus and minus signs indicate that at a given instant the portions of the membrane near the signs are displaced above (+) or below (−) the equilibrium position. The lines separating the + and − regions are nodal lines (where the membrane is at rest at all times).

Degenerate normal modes can also give interesting shapes if they are combined with different phases. The nodal lines are then no longer stationary, but move about on the vibrating object. For instance, let us superimpose the modes f_{12} and f_{21} of the square membrane, letting them be one quarter cycle out of phase with each other, so that

$$z(x, y, t) = f_{21}(x, y)\sin(2\pi t/T_0) + f_{12}(x, y)\cos(2\pi t/T_0)$$

where T_0 is the period of the oscillation. The nodal line will then move about on the membrane in the manner shown in Fig. 3-11. These motions are, of course, repeated every T_0 seconds.

It is evident, then, that when there is degeneracy one can construct an infinite number of normal modes having different shapes, but the same frequency. Of these only a small number are really independent modes, however. The degeneracy is in fact defined as the number of *linearly in-*

dependent modes of a given frequency; that is, the largest number of modes for which it is not possible to find a set of constants $c_1, c_2, \cdots, c_n$, all differing from zero, such that

$$c_1 f_1 + c_2 f_2 + c_3 f_3 + \cdots + c_n f_n = 0 \tag{B-13}$$

where the f's are normal modes having the same frequency.

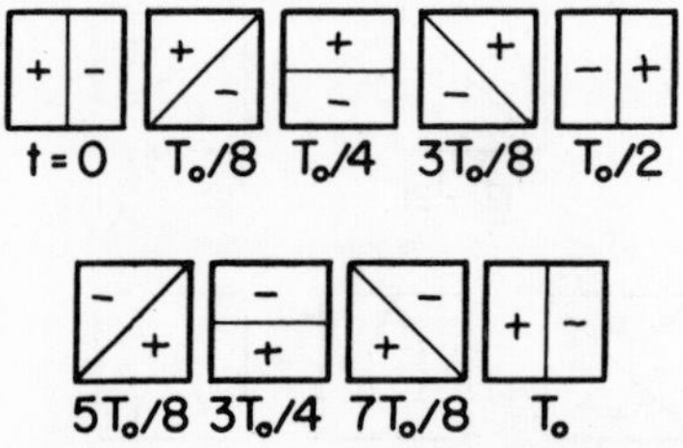

FIG. 3-11. Movements of the nodal line when the normal modes (1,2) and (2,1) of a square membrane are excited simultaneously with equal amplitude and with a phase difference of one quarter cycle. The time for a complete oscillation is T_0.

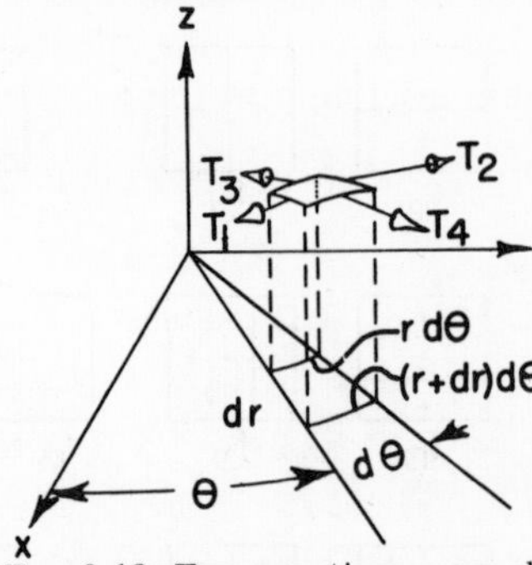

FIG. 3-12. Forces acting on an element of a membrane under uniform tension (cylindrical coordinates).

Exercises. (1) Find the normal modes and frequencies of a liquid in a shallow square dish of constant depth. If possible, put some mercury or water into a square dish, excite some of the normal modes, and compare their frequencies with those you calculate. Particular normal modes can usually be excited by placing a finger where a maximum amplitude is desired and moving it up and down with the frequency of the mode.

(2) What are the normal modes and frequencies of a rectangular membrane whose length is a and whose width is b?

(3) Draw a series of sketches showing the movements of the nodal lines when the following pairs of degenerate modes of a square membrane are combined with a phase difference of one quarter cycle:

(a) f_{13} and f_{31}

(b) f_{42} and f_{24}

c. The wave equation for a membrane in cylindrical coordinates

Let us derive the wave equation for a membrane using cylindrical coordinates, (r, θ), instead of Cartesian coordinates, (x, y). These are defined by the equations

$$x = r \cos\theta \tag{B-14a}$$

$$y = r \sin\theta \tag{B-14b}$$

It would be possible to transform the wave equation, (B-1), using the rules of differential calculus, but it is more instructive to rederive the equation from first principles.

Consider the infinitesimal sector of the membrane contained between $r, r + dr$, $\theta, \theta + d\theta$ (see Fig. 3-12). Evidently the tension, T_3, pulling on the side $r = r$, is not equal to the tension, T_4, pulling on the side $r = r + dr$, since these two sides are not of equal length. This must be taken into account in computing the z-components of the tension on the element. These components are found in the usual way to be

$$\begin{aligned}
z\text{-component of } T_1 &= -T\,dr(1/r)\,(\partial z/\partial\theta)\\
z\text{-component of } T_2 &= T\,dr(1/r)\,[(\partial z/\partial\theta) + (\partial^2 z/\partial\theta^2)d\theta]\\
z\text{-component of } T_3 &= -T\,r d\theta(\partial z/\partial r)\\
z\text{-component of } T_4 &= T(r + dr)d\theta\,[(\partial z/\partial r) + (\partial^2 z/\partial r^2)dr]
\end{aligned}$$

The net force in the z-direction is thus

$$T\,dr\,d\theta\,[r(\partial^2 z/\partial r^2) + (\partial z/\partial r) + (1/r)\,(\partial^2 z/\partial\theta^2)]$$

In contrast to what we found with Cartesian coordinates, the net force does not depend only on the curvature in the r-direction, but also on the slope in this direction. This is a result of the inequality of the opposite sides of the sector noted above.

Equating the net force to the inertial force, given by the mass of the element, $\rho r d\theta dr$, multiplied by the acceleration, $\partial^2 z/\partial t^2$, we find, writing $c^2 = \sqrt{T/\rho}$,

$$\frac{\partial^2 z}{\partial r^2} + \frac{1}{r}\frac{\partial z}{\partial r} + \frac{1}{r^2}\frac{\partial^2 z}{\partial\theta^2} = \frac{1}{c^2}\frac{\partial^2 z}{\partial t^2} \qquad \text{(B-15)}$$

which is the *wave equation in cylindrical coordinates*.

d. The normal modes of a circular membrane

Consider a circular membrane (e.g., the head of a drum) whose radius is a and whose edges are held fixed. The boundary condition is

$$z = 0 \text{ at } r = a \text{ at all times} \qquad \text{(B-16)}$$

It is obvious that Equations (B-6), (B-9a), and (B-9b) cannot be made to fit this condition, so that the method of separation of variables using the wave equation in Cartesian coordinates is not suited to this problem. It seems natural, however, to apply the method to the wave equation in cylindrical coordinates. We therefore first assume a solution of the form

$$z(r, \theta, t) = f(r, \theta)\phi(t) \tag{B-17}$$

Substituting this into Equation (B-15), we find in the usual way that

$$\phi(t) = C \sin(2\pi\nu t + \delta) \tag{B-18}$$

and

$$\frac{\partial^2 f}{\partial r^2} + \frac{1}{r}\frac{\partial f}{\partial r} + \frac{1}{r^2}\frac{\partial^2 f}{\partial \theta^2} + \frac{4\pi^2\nu^2}{c^2} f = 0 \tag{B-19}$$

Applying the method of separation of variables to Equation (B-19) we set

$$f(r, \theta) = R(r)\, S(\theta) \tag{B-20}$$

giving

$$\frac{d^2R}{dr^2} + \frac{1}{r}\frac{dR}{dr} + [k^2 - (n/r)^2]\, R = 0 \tag{B-21}$$

$$\frac{d^2S}{d\theta^2} + n^2 S = 0 \tag{B-22}$$

where n^2 is a constant and $k^2 = 4\pi^2\nu^2/c^2$. Equation (B-22) gives

$$S(\theta) = A \sin n\theta + B \cos n\theta \tag{B-23}$$

where A and B are constants. Now $S(\theta)$ must return to its original value every time θ is increased by 2π. That is,

$$S(\theta) = S(\theta + 2\pi) \tag{B-24}$$

which is possible only if n is an integer. Thus $n = 0, 1, 2, \cdots$.

Equation (B-21) is known as *Bessel's equation of order n*. It has as one of its solutions the Bessel function of order n, for which the symbol $J_n(kr)$ is usually used. These functions are simply a new type of transcendental function belonging in the same category as the trigonometric functions, sin nx and cos nx. They arise in many problems and rival the trigonometric functions in practical importance. Their numerical values have been tabulated (*8, 9*), and their mathematical properties have been thoroughly investigated (*10*). The general appearance of several of these functions is shown in Fig. 3-13. It is worth noting that if r is large, the second and fourth terms in Equation (B-21) can be neglected and the remaining differential equation is satisfied by $R(r) \simeq A \sin k(r + r_0)$, where A and r_0

are constants. A more careful analysis shows that at large values of r the Bessel functions are given approximately by

$$J_n(kr) \simeq \sqrt{\frac{2}{\pi kr}} \sin\left[kr - (n - \tfrac{1}{2})\frac{\pi}{2}\right] \tag{B-25}$$

Thus at large values of r, Bessel functions approximate to sine functions whose amplitude is "damped" by a factor proportional to $r^{-1/2}$.

Since each of these functions repeatedly passes through zero as its argument increases, the boundary conditions (B-16) can be fulfilled. The value of the constant k must be such that

$$J_n(ka) = 0 \tag{B-26}$$

This condition determines the values of the normal mode frequencies. If x_{nl} is the value of x for which $J_n(x)$ crosses the abscissa in Fig. 3-13 for the lth time, the normal mode frequencies are given by

$$\nu_{nl} = ck_{nl}/2\pi = cx_{nl}/2\pi a \tag{B-27}$$

Values of x_{nl} are tabulated in Jahnke and Emde (*8*, page 168).

The normal modes themselves are given by

$$f(r, \theta) = J_n(2\pi\nu_{nl}r/c)\,(A\sin n\theta + B\cos n\theta) \tag{B-28}$$

A section of the vibrating membrane along any radius (θ = a constant) therefore has the appearance of one of the curves in Fig. 3-13, the center

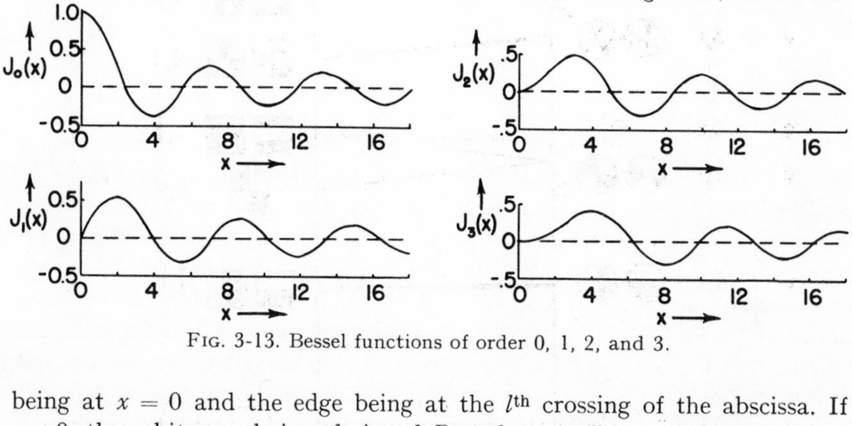

FIG. 3-13. Bessel functions of order 0, 1, 2, and 3.

being at $x = 0$ and the edge being at the lth crossing of the abscissa. If $n \neq 0$, the arbitrary choice of A and B makes possible an infinite number of normal modes for each frequency ν_{nl}. Of these, however, only two are linearly independent. Therefore the degeneracy is two-fold when $n \neq 0$, and if $n = 0$ there is no degeneracy. It is convenient to choose as the two linearly independent degenerate modes for each frequency the functions

$$f(r, \theta) = A\, J_n(2\pi\nu_{nl}r/c)\cos n\theta \quad \text{(B-29a)}$$

$$f(r, \theta) = A\, J_n(2\pi\nu_{nl}r/c)\sin n\theta \quad \text{(B-29b)}$$

The nodal patterns and relative frequencies of the first few of these normal modes are shown in Fig. 3-14. None of the overtones is an integral multiple of the fundamental. This is why the circular drum produces little more than noise.

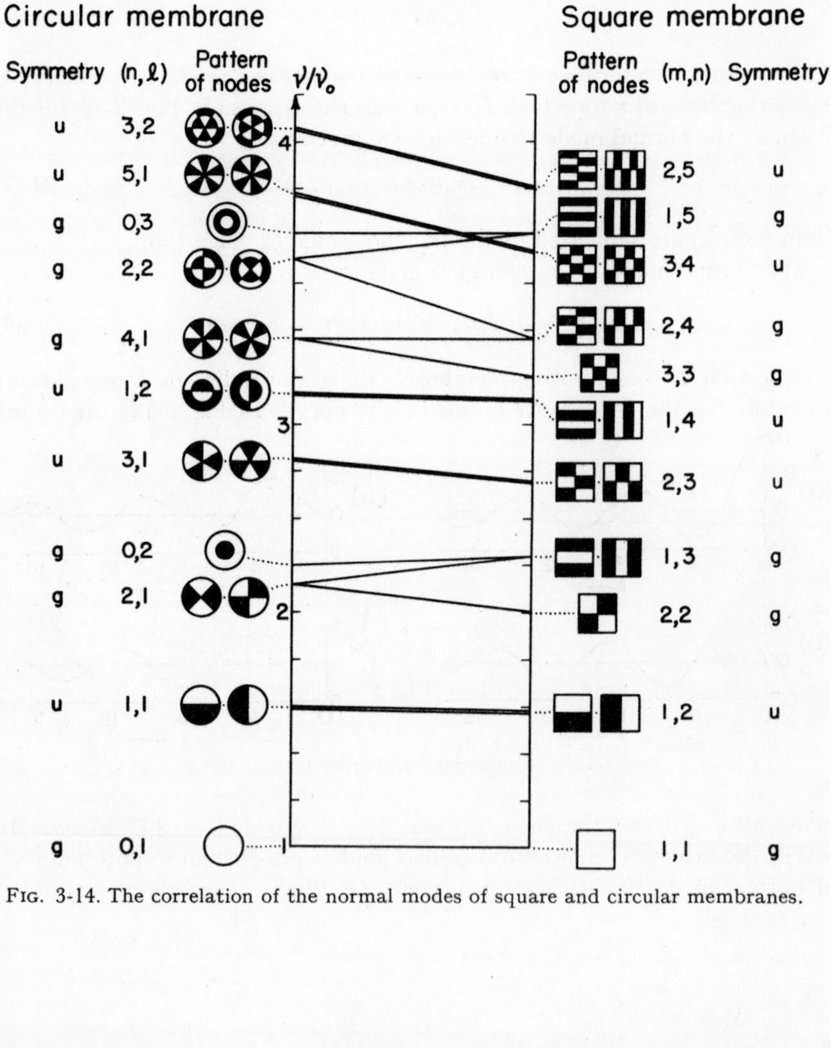

FIG. 3-14. The correlation of the normal modes of square and circular membranes.

Exercises. (1) Setting $x = kr$, show that (B-25) satisfies (B-21) if terms involving powers of r more negative than -2 are neglected in (B-21).

(2) Draw a series of sketches showing the movements of the nodal lines if the two members of the degenerate mode with $n = 1$, $l = 1$ are superimposed with a phase difference of a quarter cycle.

(3) Show that a solution of the differential equation

$$\frac{d^2f}{dx^2} + xf = 0 \tag{B-30a}$$

is

$$f = \sqrt{x}\, J_{1/3}\left(\tfrac{2}{3}x^{3/2}\right) \tag{B-30b}$$

[Hint: Transform the variables in equation (B-30a) by setting

$$x = \left(\tfrac{3}{2}y\right)^{2/3} \text{ and } f = y^{1/3}g(y).]$$

e. The correlation of the normal modes of the square and circular membranes

It should be possible to convert the square membrane into a circular one by the gradual application of a suitable perturbation to the corners of the vibrating membrane—for instance, by increasing the value of the density, ρ, in the shaded regions shown in Fig. 3-15. As the density of these regions

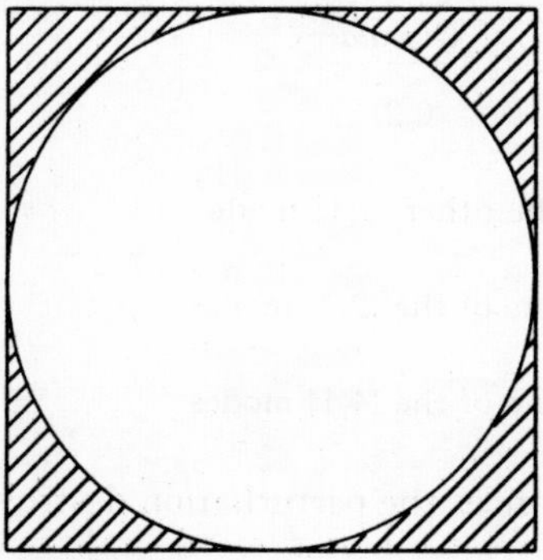

FIG. 3-15. Method of perturbing the square membrane to convert its normal modes into those of the circular membrane. A uniform load is gradually applied to the shadedregions.

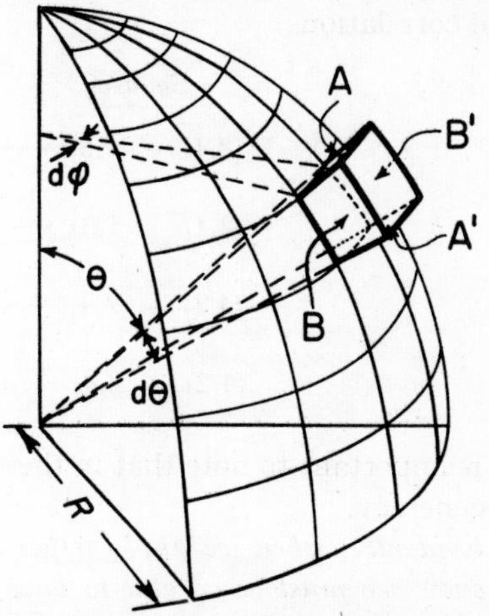

FIG. 3-16. Element of volume in the ocean on a flooded planet (polar coordinates).

is raised, the amplitude of their motion will decrease until, in the limit $\rho \to \infty$, they will not move at all and the membrane will behave like a circular one. According to the adiabatic principle, if the perturbation is applied sufficiently slowly to a square membrane vibrating in a pure normal mode, the membrane will continue to vibrate in a pure normal mode. It should therefore be possible to correlate the normal modes of square and circular membranes in an unequivocal way. In comparing the normal modes of Fig. 3-9 with those shown on the left hand side of Fig. 3-14 there is little doubt that the following correlations will occur under the application or removal of the slowly applied perturbation illustrated in Fig. 3-15:

Square		*Circular*
(1,1)	⟷	(0,1)
(1,2)	⟷	(1,1)
(2,1)	⟷	(1,1)
(2,2)	⟷	one of the (2,1) modes

The relationships between the remaining modes are unfortunately anything but obvious. If, however, we look at some of the hybrid modes of Fig. 3-10 on page 75, we can readily perceive some similarities. Thus we may expect the correlations:

Square		*Circular*
(3,1) + (1,3)	⟷	(0,2)
(3,1) − (1,3)	⟷	the other (2,1) mode
(4,2) + (2,4)	⟷	one of the (2,2) modes
(4,2) − (2,4)	⟷	one of the (4,1) modes

It is important to note that in these instances the perturbation destroys the degeneracy.

Evidently, when we apply a perturbation to the modes of a degenerate frequency, we must be careful to have the system excited in a particular linear combination of modes, otherwise the perturbed system may not remain in a normal mode. Whenever perturbations destroy degeneracy in this way, care must be taken to begin with the proper hybrid. A general method of finding these proper hybrids will be dealt with in the next chapter.

The correlation for square and circular membranes is shown in Fig. 3-14.

f. The wave equation for tidal waves on a flooded planet

Suppose that all of the continents could be leveled off into the seas so that the earth would be covered everywhere by an ocean of constant depth. Let us find the wave equation that determines the oscillations of this idealized ocean. This wave equation is especially important in quantum chemistry, since its normal mode solutions appear in one way or another in most chemical problems. In particular they determine the form of the periodic table and they are basic to the understanding of stereochemistry and molecular structure. Although the process of finding these solutions is rather lengthy, there are no essential difficulties. The student should make a special effort to follow the manipulations described on the next few pages.

Let the location of a point on the earth's surface be given in terms of the longitude, φ, and the co-latitude, θ (which is merely the latitude measured from the north pole instead of from the equator). Let the displacement of the surface of the ocean above its equilibrium position be given by z, let R be the radius of the planet, and let the depth of the ocean be h. The velocity component of the water in the direction of the parallels of latitude (direction of increasing φ) is denoted by v_φ, and that in the direction of a meridian (increasing θ) is v_θ. Consider a small elementary column of the ocean defined by the arcs θ, $\theta + d\theta$, φ, $\varphi + d\varphi$ (see Fig. 3-16). Let us apply Newton's law of motion and the principle of the conservation of matter to the water flowing through this column. The arguments used here are similar to those employed in the discussion of tidal waves in one dimension (page 54). It is assumed that the component of velocity normal to the surface of the planet is negligible in comparison with the component parallel to the surface.

(i) The Equation of Continuity. The volume of water flowing into each side of the column in unit time is equal to the area of the side multiplied by the component of velocity of the water normal to the side. For side A (Fig. 3-16) the area is equal to $h\,R \sin\theta\, d\varphi$ and the velocity normal to the side is v_θ. Thus we have

$$\text{Rate of flow into side } A = v_\theta h\, R \sin\theta\, d\varphi$$

Similarly we find

$$\text{Rate of flow into side } A' = - [v_\theta + (\partial v_\theta/\partial\theta)d\theta]hR[\sin\theta + (d\sin\theta/d\theta)d\theta]d\varphi$$

$$\text{Rate of flow into side } B = v_\varphi\, hR\, d\theta$$

$$\text{Rate of flow into side } B' = - [v_\varphi + (\partial v_\varphi/\partial\varphi)d\varphi]hR\, d\theta$$

The sum of these four rates is the rate of accumulation of water in the elementary column. Since water is essentially incompressible, this accumulation must result in a change in the depth of the water in the column such that the rate of accumulation is equal to

$$(\text{area of base of column}) \times (\text{rate of change of depth of water}) =$$
$$(R \sin\theta \, d\varphi \, R d\theta) \times (\partial z/\partial t)$$

These relationships lead to the equation of continuity (*cf.* page 55)

$$\frac{h}{R}\frac{\partial v_\theta}{\partial \theta} + \frac{h}{R}\frac{\cos\theta}{\sin\theta} v_\theta + \frac{h}{R}\frac{1}{\sin\theta}\frac{\partial v_\varphi}{\partial \varphi} = -\frac{\partial z}{\partial t} \tag{B-31}$$

(ii) The Equation of Motion. The mean difference in hydrostatic pressure on faces A and A' is $g\rho(\partial z/\partial\theta)d\theta$; therefore a net force $[hR \sin\theta \, d\varphi] \, [g\rho(\partial z/\partial\theta)d\theta]$ acts on the liquid between these two faces. This force results in a deceleration of the fluid in the element in the meridional direction. According to Newton's law we have

$$(\text{volume of element}) \times (\text{density of liquid}) \times (\partial v_\theta/\partial t) = \text{net force}$$

or

$$-hR^2 \sin\theta \, d\theta \, d\varphi \, \rho(\partial v_\theta/\partial t) = hR \sin\theta \, d\theta \, d\varphi \, g\rho(\partial z/\partial\theta)$$

or

$$R \, \partial v_\theta/\partial t = -g \, \partial z/\partial\theta \tag{B-32}$$

Similarly, the difference in the forces on B and B' is $hR \, d\theta(\partial z/\partial\varphi)d\varphi \, g\rho$ and the net force results in a deceleration of the fluid in the direction of the parallels of latitude; thus (*cf.* page 55)

$$-hR^2 \sin\theta \, d\theta \, d\varphi \, \rho(\partial v_\varphi/\partial t) = hR \, d\theta \, d\varphi \, g\rho(\partial z/\partial\varphi)$$

or

$$R \sin\theta \, \partial v_\varphi/\partial t = -g \, \partial z/\partial\varphi \tag{B-33}$$

Equations (B-32) and B-33) are the equations of motion.

Differentiating (B-31) with respect to the time, (B-32) with respect to θ, and (B-33) with respect to φ, we find on eliminating all terms containing v_θ and v_φ that

$$\frac{1}{R^2}\frac{\partial^2 z}{\partial\theta^2} + \frac{1}{R^2}\frac{\cos\theta}{\sin\theta}\frac{\partial z}{\partial\theta} + \frac{1}{R^2 \sin^2\theta}\frac{\partial^2 z}{\partial\varphi^2} = \frac{1}{gh}\frac{\partial^2 z}{\partial t^2} \tag{B-34}$$

which may be written in the more usual form (setting $c = (gh)^{1/2}$ = velocity of propagation of the wave as shown on page 56)

$$\frac{1}{R^2 \sin\theta} \frac{\partial}{\partial\theta}\left(\sin\theta \frac{\partial z}{\partial\theta}\right) + \frac{1}{R^2 \sin^2\theta} \frac{\partial^2 z}{\partial\varphi^2} = \frac{1}{c^2} \frac{\partial^2 z}{\partial t^2} \tag{B-35}$$

The terms on the left are seen to be the angular parts of the Laplacian operator in polar coordinates (page 19). This derivation shows why a term in $(\partial z/\partial\theta)$ occurs in the Laplacian: if v_θ = constant, less fluid will run into face A than runs out of face A' because of the larger area of the latter caused by the convergence of the meridians toward the poles. Thus in the equation of continuity $(\partial z/\partial t)$ contains a term proportional to v_θ, so that $(\partial^2 z/\partial t^2)$ depends on $(\partial v_\theta/\partial t)$, which according to (B-32), is in turn proportional to $(\partial z/\partial\theta)$.

Exercise. Show that if the depth of the ocean is not everywhere the same (i.e., $h = h(\theta,\varphi)$), then

$$\frac{1}{R^2 \sin\theta} \frac{\partial}{\partial\theta}\left(h \sin\theta \frac{\partial z}{\partial\theta}\right) + \frac{1}{R^2 \sin^2\theta} \frac{\partial}{\partial\varphi}\left(h \frac{\partial z}{\partial\varphi}\right) = \frac{1}{g} \frac{\partial^2 z}{\partial t^2} \tag{B-36}$$

This equation could be used to investigate the tides in the ocean on the earth as it actually is, without having to perform any bulldozing operations on the continents.

g. The normal modes of an ocean of uniform depth

In this example there are no boundary conditions. It is only necessary that the solution be everywhere finite and single valued. Since, in previous examples, it was the boundary conditions that led to a discrete set of frequencies for the normal modes, we might anticipate a continuous set of frequencies here. This is not the case, however, for as we shall see, the requirement that the solution be everywhere finite and single-valued is enough to permit only certain frequencies.

To obtain the normal mode solution of (B-35) we set

$$z(\theta,\varphi,t) = f(\theta,\varphi)T(t) \tag{B-37}$$

and find in the usual way

$$T(t) = A \sin(2\pi\nu t + \delta) \tag{B-38}$$

$$\frac{1}{R^2 \sin\theta} \frac{\partial}{\partial\theta}\left(\sin\theta \frac{\partial f}{\partial\theta}\right) + \frac{1}{R^2 \sin^2\theta} \frac{\partial^2 f}{\partial\varphi^2} + \frac{4\pi^2\nu^2}{c^2} f = 0 \tag{B-39}$$

In order to solve Equation (B-39) we try once again the method of separation of variables, setting

$$f(\theta,\varphi) = F(\theta)G(\varphi) \tag{B-40}$$

and obtaining

$$d^2G/d\varphi^2 + m^2G = 0 \tag{B-41}$$

and

$$\frac{1}{\sin\theta}\frac{d}{d\theta}\left(\sin\theta\frac{dF}{d\theta}\right) + \left[\frac{4\pi^2\nu^2R^2}{c^2} - \frac{m^2}{\sin^2\theta}\right]F = 0 \tag{B-42}$$

where m^2 is a constant.

The solution of (B-41) is

$$G = A\sin m\varphi + B\cos m\varphi \tag{B-43}$$

and the condition that $G(\varphi) = G(\varphi + 2\pi)$ leads (*cf.* page 78) to the requirement that m be an integer ($m = 0, 1, 2, \cdots$). A and B are arbitrary constants but only two of the infinite number of possible solutions are linearly independent. It is convenient to choose as these two solutions

$$G_{m1} = \sin m\varphi \tag{B-44a}$$

$$G_{m2} = \cos m\varphi \tag{B-44b}$$

In solving Equation (B-42) it is convenient to make the following substitutions

$$\beta = (2\pi\nu R/c)^2 = [(2\pi R)/(c/\nu)]^2$$

$$y = \cos\theta$$

$$d/d\theta = -\sqrt{1-y^2}\,d/dy \tag{B-45a}$$

$$\sin\theta = \sqrt{1-y^2}$$

$$F(\theta) = P(y)$$

Note that $\sqrt{\beta}$ is the ratio of the circumference of the planet $(2\pi R)$ to the wave length associated with the frequency $\nu\,(\lambda = c/\nu)$. The substitutions give

$$(1-y^2)^2\frac{d^2P}{dy^2} - 2y(1-y^2)\frac{dP}{dy} + [\beta(1-y^2) - m^2]\,P = 0 \tag{B-45b}$$

This is known as the *associated Legendre equation.* When $y = \pm 1$, every term but the last in this equation vanishes, giving

$$m^2 P(\pm 1) = 0 \tag{B-46}$$

showing that $P(\pm 1)$ must vanish when $m \neq 0$. The condition is fulfilled if P is written in the form

$$P(y) = (1 - y^2)^{am} K(y) \tag{B-47}$$

where a is a constant which, as will be shown below, it is convenient to set equal to ½. $K(y)$ is a new function which need not vanish at $y = \pm 1$. Substituting (B-47) into (B-45) we find

$$(1 - y^2)^2 \frac{d^2K}{dy^2} - 2y(1 - y^2)\,(2am + 1) \frac{dK}{dy} +$$

$$[(4a^2m^2 + 2am)y^2 - (m^2 + 2am) + \beta(1 - y^2)]\,K = 0 \tag{B-48}$$

If we set $a = ½$, the coefficient of K in (B-48) is simplified to $(1 - y^2)$ $[\beta - m(m + 1)]$, and (B-48) reduces to

$$(1 - y^2) \frac{d^2K}{dy^2} - 2y(m + 1) \frac{dK}{dy} + [\beta - m(m + 1)]\,K = 0 \tag{B-49}$$

Let us now express $K(y)$ as a power series in y

$$K(y) = y^k(a_0 + a_1 y + a_2 y^2 + \cdots) \tag{B-50}$$

This series may be differentiated, giving

$$dK/dy = k a_0 y^{k-1} + (k + 1)a_1 y^k + (k + 2)a_2 y^{k+1} + \cdots +$$

$$(k + n)a_n y^{k+n-1} + \cdots \tag{B-50a}$$

$$d^2K/dy^2 = k(k - 1)a_0 y^{k-2} + k(k + 1)a_1 y^{k-1} + (k + 1)\,(k + 2)a_2 y^k +$$

$$\cdots + (k + n)\,(k + n - 1)a_n y^{k+n-2} + \cdots \tag{B-50b}$$

The derivatives may be substituted into (B-49), giving (after grouping together terms having the same power of y)

$$\begin{aligned}
&k(k-1)a_0y^{k-2} + k(k+1)a_1y^{k-1}\\
&+ (k+1)(k+2)a_2y^k + \cdots + (k+n+1)(k+n+2)a_{n+2}y^{k+n} + \cdots\\
&\quad -(k-1)ka_0y^k - \cdots \qquad -(k+n-1)(k+n)a_ny^{k+n} - \cdots\\
&\quad -2k(m+1)a_0y^k - \cdots \qquad -2(k+n)(m+1)a_ny^{k+n} - \cdots\\
&+ [\beta - m(m+1)]a_0y^k - \cdots \qquad + [\beta - m(m+1)]a_ny^{k+n} + \cdots = 0
\end{aligned} \tag{B-51}$$

If this expression is to be valid for all values of y, the coefficient of each power of y must vanish. From the coefficients of y^{k-2} and y^{k-1} we find

$$k(k-1)a_0 = 0 \tag{B-52a}$$

$$k(k+1)a_1 = 0 \tag{B-52b}$$

which require that $k = 0$. Thus the series in $K(y)$, Equation (B-50), evidently begins with at least a constant term, no negative powers of y being permitted. The remaining coefficients in (B-50) give, for y^0

$$1\cdot 2\, a_2 + [\beta - m(m+1)]a_0 = 0 \tag{B-53a}$$

and for the general term, y^n

$$(n+1)(n+2)a_{n+2} - [(n-1)n + 2(m+1)n - \beta + m(m+1)]a_n = 0 \tag{B-53b}$$

The expression in the brackets may be written

$$\begin{aligned}
n^2 - n + 2mn + 2n + m^2 + m - \beta &= n^2 + (2m+1)n + m(m+1) - \beta\\
&= (n+m)(n+m+1) - \beta
\end{aligned} \tag{B-54}$$

We thus obtain the *recursion formula* relating the coefficients of alternate powers of y in (B-50)

$$a_{n+2} = \frac{(n+m)(n+m+1) - \beta}{(n+1)(n+2)}a_n \tag{B-55}$$

By means of this relation we may obtain the coefficients of all even powers of y in (B-50) in terms of a_0, and the coefficients of all odd powers of y in terms of a_1. Therefore we may write

$$K(y) = a_0 S_{\text{even}} + a_1 S_{\text{odd}} \tag{B-56}$$

where S_{even} is an infinite sum involving only even powers of y, and S_{odd} is one involving only odd powers of y. We have at our disposal only two arbitrary constants, a_0 and a_1, as we should expect for a second order differential equation such as (B-49).

Let us examine the behavior of S_{even} and S_{odd} when $y = \pm 1$. We can see from the recursion formula (B-55) that for large values of n

$$a_{n+2} \simeq a_n \tag{B-57}$$

Thus when $y = \pm 1$, both S_{even} and S_{odd}, and hence $K(y)$, become infinite. Of course

$$P(y) = (1 - y^2)^{m/2} K(y) = (\text{for } y = \pm 1)\ 0 \cdot \infty$$

so the behavior of $P(\pm 1)$ is not clear. Closer analysis reveals, however, that the "infiniteness" of $K(y)$ is stronger than the "zeroness" of $(1 - y^2)^{m/2}$ (see *11*, vol. I, p. 281), and $P(\pm 1)$ is therefore infinite. Since $y = \pm 1$ at the north and south poles, this leads to the ridiculous result that the amplitude of the wave motion at the poles is infinite in a normal mode. The only way out of this dilemma is to require that β have only those values that cause the numerator in one of the recursion formulae, (B-55), to vanish. That is, for some value of n, say $n = s$, we must have

$$\beta = (s + m)(s + m + l) \tag{B-58}$$

Then the corresponding series breaks off after the sth term, becoming a polynomial of degree s which remains finite at $y = \pm 1$. The other series will still be infinite, but it can be disposed of by setting its initial cofficient (a_0 or a_1 as the case may be) equal to zero.

Both s and m are positive integers, so β must have the form

$$\beta = l(l + 1) \tag{B-59}$$

where $l = 0, 1, 2, 3, \cdots$, and $l \geqq m$. But through the definition (B-45a), β determines the frequencies of the normal modes. The permitted frequencies are therefore

$$\nu_l = (c/2\pi R)\sqrt{l(l + 1)} \quad (l = 0, 1, 2, 3, \cdots) \tag{B-60}$$

A discrete set of normal mode frequencies is thus obtained in spite of the absence of boundary conditions, the requirements of single valuedness and a finite amplitude at all points being all that is needed. These frequencies are again not integral multiples of the fundamental frequency.

Since $\sqrt{\beta}$ is the ratio of the circumference of the planet to the wave length associated with the frequency ν (that is, the length that a wave of the same frequency would have in a long trough whose depth is the same as that of the ocean), it is interesting that $\sqrt{\beta}$ is not an integer (although it approximates to one when l is large). The frequencies of the normal modes of an ocean of uniform depth are somewhat larger than would be obtained in a canal of the same depth encircling the earth at the equator.

The well behaved solutions $P(y)$ which we have obtained each contain a single arbitrary constant multiplying the entire function. If this constant is chosen in such a way that

$$K(1) = \frac{(l+m)!}{2^m\, m!\, (l-m)!} \tag{B-61}$$

then $P(y)$ is called the *associated Legendre function of degree l and order m* and it is given the symbol $P_l^m(y)$. The solutions of Equation (B-39) for given l and m will be written in terms of these functions, using the notation

$$f_{lm}(\theta,\varphi) = P_l^m(\cos\theta)G_m(\varphi) \tag{B-62}$$

where $G_m(\varphi)$ is defined by Equations (B-44). The functions $f_{lm}(\theta,\varphi)$ are known as *surface spherical harmonics*. They occur in many important quantum mechanical problems. The corresponding normal mode solutions of the wave equation, (B-35), will be denoted by $z_{lm}(\theta,\varphi,t)$. These solutions will contain a multiplicative constant whose value is related to the amplitude of the normal mode, but which has no effect on the shape.

Let us now find the solutions corresponding to some of the values of l and m that we are free to choose.

(*i*) If $l = 0$, then $\beta = 0$ because of (B-59), and the only possible value of m is zero. Therefore, from (B-55),

$$a_{n+2}/a_n = [n(n+1)]/[(n+1)(n+2)]$$

It is easy to see that the odd series, beginning with $n = 1$, does not break off. Therefore we must set $a_1 = 0$. For the even series,

$$a_2 = 0 \cdot a_0 = 0$$

Therefore $K(y)$ is a constant, and by the convention (B-61), this constant is unity. Thus

$$P_0^0(\cos\theta) = 1$$

Furthermore, from (B-44b)

$$G_0(\varphi) = 1$$

so that

$$f_{00}(\theta, \varphi) = 1$$

Since $\beta = 0$, the frequency of the oscillation is zero because of (B-45a). Thus

$$z_{00}(\theta, \varphi, t) = \text{constant}$$

and in this "normal mode" the ocean is quiet everywhere.

(ii) Next we will let $l = 1$, so that $\beta = 2$. There are then two possible values of m: either $m = 0$ or $m = 1$.

(a) First let $m = 0$. Then

$$a_{n+2}/a_n = [n(n+1) - 2]/[(n+1)(n+2)]$$

In this case the even series is infinite, so we set $a_0 = 0$. Then

$$a_3 = \frac{1 \cdot 2 - 2}{2 \cdot 3} a_1 = 0$$

Thus $K(y) = a_1 y$, and by the convention (B-61), $a_1 = 1$ so that

$$P_1^0(y) = y, \; f_{10}(\theta, \varphi) = \cos\theta$$

$$z_{10}(\theta, \varphi, t) = A \cos\theta \sin(\sqrt{2}\, ct/R + \delta)$$

(b) Now select $m = 1$

$$a_{n+2} = \frac{(n+1)(n+2) - 2}{(n+1)(n+2)} a_n$$

The odd series is infinite, and $a_2 = 0 \cdot a_0$ so that

$$K(y) = a_0 = 1$$

$$P_1^1(y) = (1 - y^2)^{1/2} = \sin\theta$$

$$G_1(\varphi) = \begin{cases} \sin\varphi \\ \cos\varphi \end{cases}$$

$$f_{11} = \begin{cases} \sin\theta \sin\varphi \\ \sin\theta \cos\varphi \end{cases}$$

$$z_{11} = \begin{cases} A \sin\theta \sin\varphi \sin(\sqrt{2}\, ct/R + \delta) \\ A \sin\theta \cos\varphi \sin(\sqrt{2}\, ct/R + \delta) \end{cases}$$

The frequency, $\nu_1 = \sqrt{2}\, c/2\pi R$, is thus three-fold degenerate. The associated Legendre functions and spherical harmonics for $l = 2$ are shown in Table 3-1. The student should verify that these functions follow from Equations (B-40), (B-44), (B-45a), (B-47, with $a = \frac{1}{2}$), (B-50), (B-55), (B-59), and (B-61).

In general, the frequency associated with a given l is $(2l + 1)$-fold degenerate, because for each value of l, m can have values from 0 to l, $m = 0$ giving one mode and each value of m greater than zero giving two modes.

TABLE 3-1

SURFACE SPHERICAL HARMONICS

l	m	$P_l^m(y)$	Spherical harmonics: In polar coordinates	Spherical harmonics: In Cartesian coordinates
0	0	1	$f_{00} = 1$	$s = 1$
1	0	y	$f_{10} = \cos\theta$	$p_z = z/R$
1	1	$(1-y^2)^{1/2}$	$f_{11} = \begin{cases} \sin\theta \sin\varphi \\ \sin\theta \cos\varphi \end{cases}$	$p_y = y/R$ $p_x = x/R$
2	0	$\frac{1}{2}(3y^2 - 1)$	$f_{20} = 3\cos^2\theta - 1$	$d_{z^2} = (3z^2 - R^2)/R^2$
2	1	$3y(1-y^2)^{1/2}$	$f_{21} = \begin{cases} \sin\theta \cos\theta \sin\varphi \\ \sin\theta \cos\theta \cos\varphi \end{cases}$	$d_{yz} = yz/R^2$ $d_{xz} = xz/R^2$
2	2	$3(1-y^2)$	$f_{22} = \begin{cases} \sin^2\theta \sin 2\varphi \\ \sin^2\theta \cos 2\varphi \end{cases}$	$d_{xy} = xy/R^2$ $d_{x^2-y^2} = (x^2-y^2)/R^2$
3	0	$\frac{1}{2}(5y^2 - 3y)$	$f_{30} = 5\cos^3\theta - 3\cos\theta$	$f_{z^3} = (5z^3 - 3R^2 z)/R^3$
3	1	$\frac{3}{2}(1-y^2)^{1/2}(5y^2-1)$	$f_{31} = \begin{cases} \sin\theta\,(5\cos^2\theta - 1)\sin\varphi \\ \sin\theta\,(5\cos^2\theta - 1)\cos\varphi \end{cases}$	$f_{yz^2} = y(5z^2-R^2)/R^3$ $f_{xz^2} = x(5z^2-R^2)/R^3$
3	2	$15(1-y^2)y$	$f_{32} = \begin{cases} \sin^2\theta \cos\theta \sin 2\varphi \\ \sin^2\theta \cos\theta \cos 2\varphi \end{cases}$	$f_{xyz} = xyz/R^3$ $f_{z(x^2-y^2)} = z(x^2-y^2)/R^3$
3	3	$15(1-y^2)^{3/2}$	$f_{33} = \begin{cases} \sin^3\theta \sin 3\varphi \\ \sin^3\theta \cos 3\varphi \end{cases}$	$f_{y^3} = y(y^2-3x^2)/R^3$ $f_{x^3} = x(x^2-3y^2)/R^3$

The appearance of the surface of the ocean when it is vibrating in the first few normal modes is shown in Fig. 3-17, the nodal lines and the sign of the displacements from equilibrium being indicated. For brevity we have set $\nu_0 = c/2\pi R$ which is the reciprocal of the time required for a disturbance to travel in a canal around the equator of the planet.

Evidently the number of nodal lines is equal to l.

For reasons which will become clear later in the book, it is convenient to denote the normal mode having $l = 0$ as an "s-mode" (actually it is not a

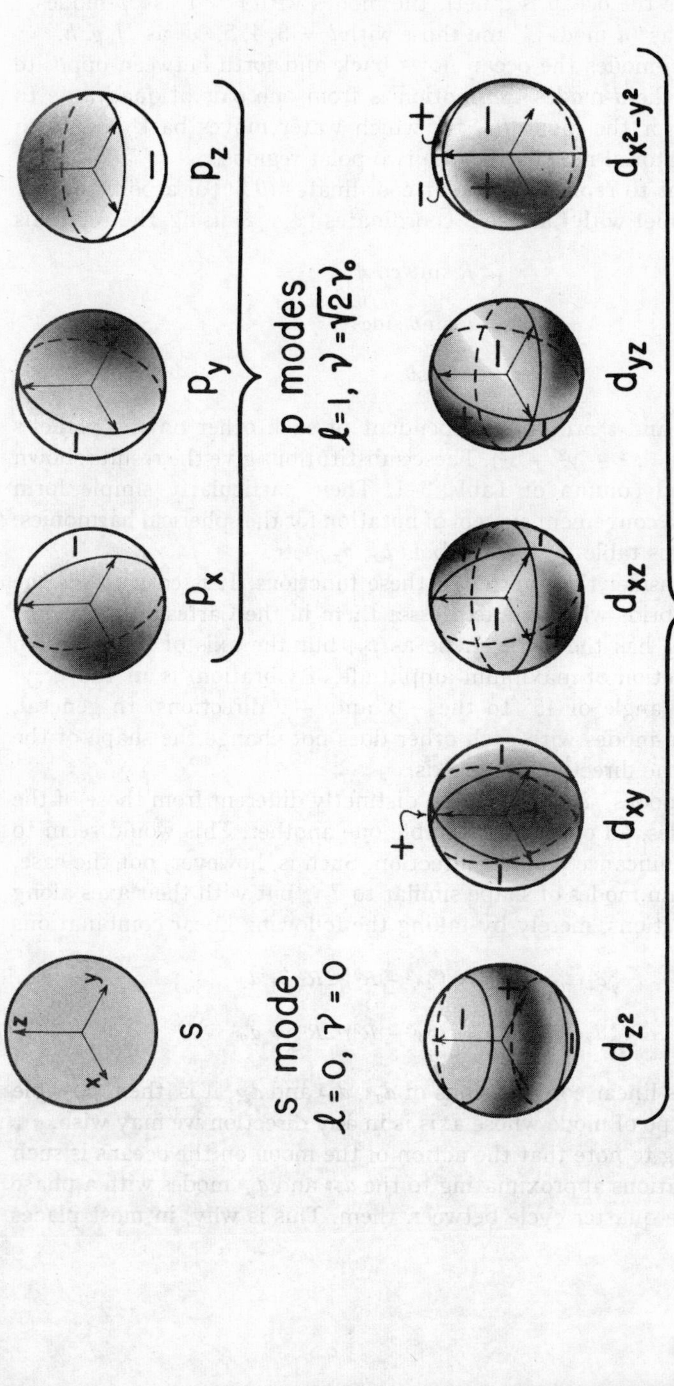

FIG. 3-17. Arrangement of nodal lines in various normal modes of a uniformly flooded planet. The plus and minus signs indicate that the surface of the ocean is respectively above or below mean sea level in the regions bounded by the nodal lines.

mode at all, since the ocean is quiet), the modes with $l = 1$ as "p-modes," those with $l = 2$ as "d-modes," and those with $l = 3, 4, 5, \cdots$ as "$f, g, h, \cdots$ modes." In the p-modes the ocean flows back and forth between opposite hemispheres. In the d-modes the motion is from one pair of quadrants to another, except for the mode f_{20}, for which water moves back and forth between the equatorial regions and the two polar regions.

It is convenient to replace the polar coordinates (θ, φ) of a point on the surface of the planet with Cartesian coordinates (x, y, z) using the relations

$$x = R \sin\theta \cos\varphi$$

$$y = R \sin\theta \sin\varphi$$

$$z = R \cos\theta$$

(of course, x, y, and z are not independent of each other on the planet's surface, since $R^2 = x^2 + y^2 + z^2$). These substitutions give the results shown in the right-hand column of Table 3–1. Their particularly simple form provides us with a convenient system of notation for the spherical harmonics, as indicated in this table by the symbols p_x, d_{xy}, etc.

Let us now consider the hybrids of these functions. It is easy to see the shapes of the hybrids when one expresses them in the Cartesian form. For instance, $p_x + p_y$ has the same shape as p_x, but the axis of the function (that is, the direction of maximum amplitude of vibration) is in the (x, y) plane and at an angle of 45° to the $+x$ and $+y$ directions. In general, hybridizing the p-modes with each other does not change the shape of the mode, but only the direction of its axis.

Among the d-modes, d_{z^2} has a shape distinctly different from those of the remaining d-modes, all of which resemble one another. This would seem to give a special significance to the z-direction. Such is, however, not the case, for one may obtain modes of shape similar to d_{z^2}, but with their axes along the z- and y-directions, merely by taking the following linear combinations

$$\tfrac{3}{2}d_{x^2-y^2} - \tfrac{1}{2}d_{z^2} = (3x^2 - R^2)/2R^2 = d_{x^2}$$

$$-\tfrac{3}{2}d_{x^2-y^2} - \tfrac{1}{2}d_{z^2} = (3y^2 - R^2)/2R^2 = d_{y^2}$$

By taking various linear combinations of d_{x^2}, d_{y^2} and d_{z^2}, it is, then, possible to obtain a d_{z^2} type of mode whose axis is in any direction we may wish.

It is interesting to note that the action of the moon on the oceans is such as to excite vibrations approximating to the d_{x^2} and d_{y^2} modes with a phase difference of one quarter cycle between them. This is why, in most places on the earth, the period of the tides is twelve hours in spite of the fact that the earth rotates under the moon with a period of about 24 hours.

Because their frequencies are not the same, it is not possible to hybridize, say, the *p*-modes of the ocean very extensively with the *d*-modes without applying a fairly strong perturbation. Certain types of vibrating systems do, however, yield *s*-, *p*-, and *d*-modes of nearly the same frequency. In these systems interesting shapes can occur by hybridization from relatively weak perturbations. Hybrids of this kind are of great importance in chemistry, as we shall see.

Exercises. (1) If the mean depth of the oceans is 13,000 feet, what are the natural frequencies of the *p, d, f* and *g* tidal modes, and which most nearly approximates the period of the earth's rotation?

(2) Suppose that a low submarine ridge is added to the planet we have been considering, extending as a great circle around the planet and making an angle of 45° with the equator. The perturbation thus introduced will remove some of the degeneracy. Make a guess as to which hybrids of the *d*-modes and of the *p*-modes of a uniform planet would correlate with the normal modes of the perturbed planet.

(3) Find the seven *f*-modes, express them in Cartesian coordinates and draw sketches.

(4) Draw sketches showing the movements of the nodal lines when the following pairs of degenerate modes are superimposed with a phase difference of one quarter cycle:

(a) p_x and p_y

(b) d_{xy} and d_{xz}

(c) d_{xy} and d_{z^2}

h. Note on the role of surface spherical harmonics in electrostatics

Another very helpful way of looking at the functions just described arises in connection with the electric field associated with a cluster of positive and negative charges. Poisson's equation states that, outside of such a cluster,

$$\nabla^2 V = 0$$

where V is the electrostatic potential and ∇^2 is the Laplacian operator. Applying the method of separation of variables, with

$$V(r, \theta, \varphi) = R(r) f(\theta, \varphi)$$

we find

$$\frac{d}{dr}\left(r^2 \frac{dR}{dr}\right) - l(l+1)R = 0$$

$$\frac{1}{\sin\theta}\frac{\partial}{\partial\theta}\left(\sin\theta\frac{\partial f}{\partial\theta}\right) + \frac{1}{\sin^2\theta}\frac{\partial^2 f}{\partial\varphi^2} + l(l+1)f = 0$$

where the separation constant has been written in the form $l(l + 1)$. The solution of the first equation that remains finite at infinity is

$$R(r) = A/r^{l+1}$$

as can be verified by direct substitution, A being an arbitrary constant. The second equation is the associated Legendre equation, which, as we already know, has well behaved solutions only if l is a positive integer. These solutions are, of course, the surface spherical harmonics described in the previous section. The general solution of Poisson's equation can therefore be written in the form

$$\begin{aligned} V = A_0/r &+ (1/r^2)(A_{1x}p_x + A_{1y}p_y + A_{1z}p_z) \\ &+ (1/r^3)(A_{2z^2}d_{z^2} + A_{2xz}d_{xz} + A_{2yz}d_{yz} + A_{2xy}d_{xy} + A_{2x^2-y^2}d_{x^2-y^2}) \\ &+ (1/r^4)(\text{sum of } f\text{-modes}) + \cdots \end{aligned}$$

It can be shown that the constant A_0 is equal to the net charge of the cluster. The first term therefore represents the contribution to the field due to this net charge. This term dies away with distance the least rapidly, so that at great distances the cluster appears to be a single charged point.

The second term represents the contribution to the field due to the "dipole moment" of the cluster. This dipole moment can be represented by a single pair of oppositely charged points joined by a line whose direction cosines are proportional to A_{1x}, A_{1y}, and A_{1z}. If there is no net charge on the cluster, this term will be the most important at great distances, so that the cluster appears from far off to consist only of a single dipole oriented as mentioned.

The third term represents the contribution to the field due to the "quadrupole moment" of the cluster. The quadrupole moment evidently has five independent components, and therefore does not behave like a vector. As far as its field at large distances is concerned, any cluster of four charges that has neither a dipole moment nor a net charge can be represented by a set of five quadrupoles, whose fields are proportional to the d-modes.

Similarly, any cluster of eight charges that has neither a net charge, a dipole moment nor a quadrupole moment can be represented by a set of seven elementary "octupoles" whose fields are proportional to the f-modes.

The surface spherical harmonics therefore represent the angular dependence of the electrostatic fields of the various multipoles encountered in electrostatic theory. The multipoles associated with each of the surface harmonics are indicated in Fig. 3-18 for the first few harmonics.

i. Some mathematical properties of the associated Legendre functions

We shall give here a brief survey of some important properties of the associated Legendre functions, many of which will be utilized in later chapters. This survey will take the form of a series of exercises which will give the student practice in manipulating these functions.

Exercises. (1) Show that the functions $P_l^0(y)$ are polynomials in which the highest power of y is y^l. Also show that

$$P_l^0(1) = 1 \tag{B-63}$$

for all values of l.

(2) Show that $P_l^0(y)$ satisfies the differential equation

$$(1 - y^2)\frac{d^2P_l^0}{dy^2} - 2y\frac{dP_l^0}{dy} + l(l+1)P_l^0 = 0 \tag{B-64}$$

This is known as *Legendre's differential equation.*

(3) The *Legendre polynomials* may be defined by

$$P_l(y) = \frac{1}{2^l l!}\frac{d^l(y^2 - 1)^l}{dy^l} \tag{B-65}$$

Show that these polynomials satisfy Legendre's differential equation and that $P_l(1) = 1$, thereby proving that $P_l(y)$ is identical with $P_l^0(y)$. Equation (B-65) is known as *Rodrigue's formula.* [Hint: Note that the function $f = (y^2 - 1)^l$ satisfies the differential equation $(1 - y^2)df/dy + 2lyf = 0$. Differentiate this equation $l+1$ times and compare the result with Legendre's equation.]

(4) Show from Rodrigue's formula that

$$\begin{aligned} P_0(y) &= 1 & P_3(y) &= (5y^3 - 3y)/2 \\ P_1(y) &= y & P_4(y) &= (35y^4 - 30y^2 + 3)/8 \\ P_2(y) &= (3y^2 - 1)/2 & P_5(y) &= (63y^5 - 70y^3 + 15y)/8 \end{aligned} \tag{B-66}$$

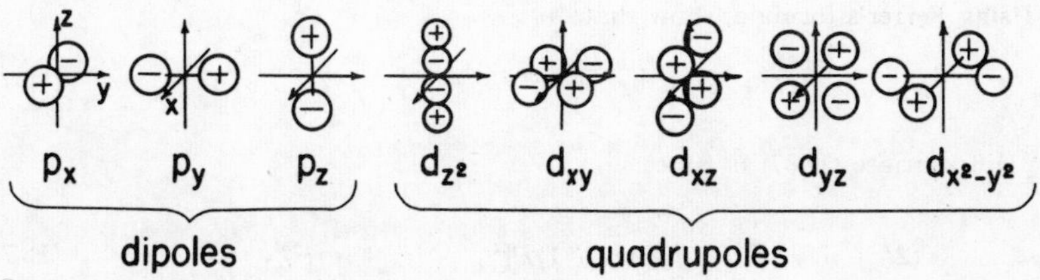

FIG. 3-18. Arrangements of electric charges giving electrostatic fields whose angular dependence is similar to the normal modes in Fig. 3-17.

(5) Using Rodrigue's formula, show that

$$(2l + 1)P_l = dP_{l+1}/dy - dP_{l-1}/dy \tag{B-67}$$

[Hint: Note that

$$\frac{d^l}{dy^l}\left[y(y^2-1)^l\right] = \frac{d^{l-1}}{dy^{l-1}}\left[\frac{d}{dy}\, y(y^2-1)^l\right]$$
$$= \frac{d^{l-1}}{dy^{l-1}}\left\{\left[(2l+1)\,(y^2-1) + 2l\right](y^2-1)^{l-1}\right\}$$

(6) Using Rodrigue's formula, show that

$$lP_l = y\, dP_l/dy - dP_{l-1}/dy \tag{B-68}$$

and

$$(l + 1)P_l = dP_{l+1}/dy - y\, dP_l/dy \tag{B-69}$$

[Hint: Note that

$$\frac{d^l}{dy^l}\, y(y^2-1)^n = l\,\frac{d^{l-1}}{dy^{l-1}}\,(y^2-1)^n + y\,\frac{d^l}{dy^l}\,(y^2-1)^n\ ]$$

(7) Show that

$$(y^2 - 1)\, dP_l/dy = lyP_l - lP_{l-1}$$
$$= -\,(l+1)\,(yP_l - P_{l+1}) \tag{B-70}$$

(8) Show that the function $P_l^m(y)$ defined by the relation

$$P_l^m(y) = (1-y^2)^{m/2}\frac{d^m P_l}{dy^m} = \frac{(1-y^2)^{m/2}}{2^l l!}\frac{d^{l+m}(y^2-1)^l}{dy^{l+m}} \tag{B-71}$$

satisfies the associated Legendre equation (B-45b) and that this relation gives the same functions as those derived on pages 90–91. Equation (B-71) is *Ferrer's formula* for the associated Legendre functions.

(9) Using Ferrer's formula, show that

(a)
$$(2l + 1)\sqrt{1-y^2}\, P_l^m = P_{l+1}^{m+1} - P_{l-1}^{m+1} \tag{B-72}$$

[Hint: Differentiate (B-67) m times.]

(b)
$$(2l + 1)yP_l^m = (l - m + 1)P_{l+1}^m + (l + m)P_{l-1}^m \tag{B-73}$$

[Hint: Differentiate (B-69) $(m - 1)$ times and make use of (B-72).]

(c) $$2myP_l^m = \sqrt{1-y^2}\,[P_l^{m+1} + (l+m)(l-m+1)P_l^{m-1}] \quad \text{(B-74)}$$

(d) $$(2l+1)\sqrt{1-y^2}\,P_l^m = (l+m)(l+m-1)P_{l-1}^{m-1} - (l-m+1)(l-m+2)P_{l+1}^{m-1} \quad \text{(B-75)}$$

(e) $$(y^2-1)dP_l^m/dy = (l-m+1)P_{l+1}^m - (l+1)yP_l^m \quad \text{(B-76)}$$

(f) $$(y^2-1)dP_l^m/dy = myP_l^m - \sqrt{1-y^2}\,P_l^{m+1} \quad \text{(B-77)}$$

(g) $$d/dy[(1-y^2)^{m/2}P_l^m] = -(l+m)(l-m+1)(1-y^2)^{\frac{1}{2}(m-1)}\,P_l^{m-1} \quad \text{(B-78)}$$

(10) Show that
$$\int_{-1}^{1}[P_l^m(y)]^2 dy = \frac{(l+m)!}{(l-m)!}\frac{2}{2l+1} \quad \text{(B-79)}$$

[Hint: Integrate repeatedly by parts, using Ferrer's formula and (B-78), obtaining finally
$$\int_{-1}^{1}[P_l^m(y)]^2 dy = (l+m)(l-m+1)\int_{-1}^{1}[P_l^{m-1}(y)]^2 dy = \frac{(l+m)!}{(l-m)!}\int_{-1}^{1}[P_l(y)]^2 dy$$
Then integrate $[P_l(y)]^2$ by parts l times using Rodrigue's formula for $P_l(y)$.]

C. Vibrations of Three-Dimensional Systems

a. The wave equation for sound waves in a gas

Since it is impossible to have three-dimensional membranes and three-dimensional tidal waves, we must go on to a new type of system in order to illustrate waves in three dimensions. Sound waves in a gas are most convenient for this, although crystals and electromagnetic radiation also offer many good examples.

Suppose that sound waves pass through a sample of gas that is confined to some container. Let us define the following variables:

x, y, z = Cartesian coordinates of a point in the container
t = time
$\rho(x,y,z,t)$ = density of the gas at point x,y,z and at time t
$p(x,y,z,t)$ = pressure of the gas
$u(x,y,z,t)$ = component of velocity of the gas in the x-direction
$v(x,y,z,t)$ = component of velocity of the gas in the y-direction
$w(x,y,z,t)$ = component of velocity of the gas in the z-direction

The quantities ρ, p, u, v and w all fluctuate in the sound wave.

It is convenient to write

$$\rho(x,y,z,t) = \rho_0[1 + s(x,y,z,t)] \tag{C-1}$$

where ρ_0 is the mean density of the gas over a long time. It is possible for ρ_0 to depend on x, y, and z, but it does not depend on t.

Consider a small region of the gas bounded by the planes x, $x + dx$, y, $y + dy$, z, $z + dz$. The motion of the gas through this element must be consistent with Newton's laws and with the law of conservation of matter.

Applying the law of conservation of matter, it must be true that the rate of increase of the mass of gas in the element is equal to the net rate of influx of gas.

$$\text{Rate of increase of mass of the element} = (\partial\rho/\partial t)dxdydz$$

$$\text{Rate of influx through plane } x = u\rho dydz$$

$$\text{Rate of influx through plane } x + dx = -[u\rho + (\partial u\rho/\partial x)dx]dydz$$

$$\text{Net rate of influx through the two planes} = -(\partial u\rho/\partial x)dxdydz$$

Similarly,

$$\text{Net rate of influx through } y \text{ and } y + dy = -(\partial v\rho/\partial y)dxdydz$$

$$\text{Net rate of influx through } z \text{ and } z + dz = -(\partial w\rho/\partial z)dxdydz$$

Thus we have

$$\partial\rho/\partial t + \partial u\rho/\partial x + \partial v\rho/\partial y + \partial w\rho/\partial z = 0 \tag{C-2}$$

This is the *equation of continuity*.

In applying Newton's laws, we see that there is a net force on the element in the x-direction due to the variation of pressure in this direction. The force on the face at x is $pdydz$, while the force on the face at $x + dx$ is $-[p + (\partial p/\partial x)dx]dydz$ (negative sign because it acts in the opposite direction). The net force in the x-direction is thus $-(\partial p/\partial x)dxdydz$. The mass of material in the element is $\rho dxdydz$, and this is accelerated in the x-direction by an amount $(\partial u/\partial t)$. Applying Newton's second law we see that

$$\partial p/\partial x = -\rho\partial u/\partial t \tag{C-3}$$

Similarly

$$\partial p/\partial y = -\rho\partial v/\partial t \tag{C-4}$$

$$\partial p/\partial z = -\rho\partial w/\partial t \tag{C-5}$$

These are the *equations of motion*.

Inserting Equation (C-1) in the equation of continuity, we find

$$\rho_0 \partial s/\partial t + \partial u\rho_0/\partial x + \partial v\rho_0/\partial y + \partial w\rho_0/\partial z + \partial su\rho_0/\partial x + \partial sv\rho_0/\partial y + \partial sw\rho_0/\partial z = 0 \quad \text{(C-6)}$$

If we restrict ourselves to waves of small amplitude (small velocities and small changes in pressure and density), the last three terms in (C-6) can be neglected in comparison with the first four since they involve the products of small quantities. Thus

$$\rho_0(\partial s/\partial t) + \partial u\rho_0/\partial x + \partial v\rho_0/\partial y + \partial w\rho_0/\partial z = 0 \quad \text{(C-7)}$$

The pressure and the density of the gas are related by the equation of state. If the vibrations are rapid the changes in local density will occur adiabatically, there being insufficient time for heat conduction to take place during the cycles of rarefaction and compression. (This turns out to be the case in air down to much shorter wavelengths than are commonly encountered; see Rayleigh (*12*, p. 23).) Therefore we must use the adiabatic equation of state

$$p/p_0 = (\rho/\rho_0)^\gamma = (1+s)^\gamma \simeq (1+\gamma s) \quad \text{(C-8)}$$

where γ is the ratio of the specific heat at constant pressure to the specific heat at constant volume. Thus

$$\partial s/\partial t = (1/\gamma p_0)\,(\partial p/\partial t) \quad \text{(C-9)}$$

Substituting this in (C-7) and differentiating partially with respect to the time, we find

$$(\rho_0/\gamma p_0)\,(\partial^2 p/\partial t^2) + (\partial^2 \rho_0 u/\partial x \partial t) + (\partial^2 \rho_0 v/\partial y \partial t) + (\partial^2 \rho_0 w/\partial z \partial t) = 0 \quad \text{(C-10)}$$

But setting $\rho = \rho_0$ in (C-3, C-4, C-5), and differentiating (C-3) with respect to x, (C-4) with respect to y, and (C-5) with respect to z, we find on substituting into (C-10)

$$\partial^2 p/\partial x^2 + \partial^2 p/\partial y^2 + \partial^2 p/\partial z^2 = (\rho_0/\gamma p_0)\,(\partial^2 p/\partial t^2) \quad \text{(C-11)}$$

or

$$\nabla^2 p = (1/C^2)\,(\partial^2 p/\partial t^2) \quad \text{(C-11a)}$$

which is the wave equation for sound waves,* with C the velocity of sound

$$C = \sqrt{\gamma p_0/\rho_0} \quad \text{(C-12)}$$

* From Maxwell's equations it is easy to derive the wave equation

$$\nabla^2 E = (n^2/c^2)\,(\partial^2 E/\partial t^2)$$

for the electric field in electromagnetic radiation (see *13*, p. 315); or *14*, Chap. 8). Here n is the refractive index, related to the dielectric constant and magnetic permeability, and c is the velocity of light in a vacuum. It is possible for n to be a function of x, y, and z.

We note that in (C-12), C may be a function of x, y, z. For ideal gases, the ratio $(p_0/\rho_0) = RT/M$, where M is the molecular weight, so that

$$C = \sqrt{\gamma RT/M}$$

and C may be made to vary by changing the temperature or composition from point to point in the container.

b. The normal modes of sound waves in a room

Consider a room filled with air at constant temperature and having the shape of a rectangular parallelepiped the lengths of whose sides are a, b, and c. If the walls are perfectly rigid, no motion will be possible in a direction perpendicular to them in the gas immediately adjacent to them. We must therefore solve Equation (C-11) subject to the conditions, C = constant and

$$\partial u/\partial t = 0 \text{ at } x = 0 \text{ and } a, \text{ which, from (C-3) gives } \partial p/\partial x = 0 \qquad \text{(C-13a)}$$

$$\partial v/\partial t = 0 \text{ at } y = 0 \text{ and } b, \text{ which, from (C-4) gives } \partial p/\partial y = 0 \qquad \text{(C-13b)}$$

$$\partial w/\partial t = 0 \text{ at } z = 0 \text{ and } c, \text{ which, from (C-5) gives } \partial p/\partial z = 0 \qquad \text{(C-13c)}$$

Since we are looking for normal modes we begin by assuming

$$p(x,y,z,t) - p_0 = P(x,y,z)\phi(t) \qquad \text{(C-14)}$$

which, on substitution in (C-11) gives in the usual way

$$\phi = A \sin(\omega t + \delta) \qquad \text{(C-15)}$$

$$\nabla^2 P + (\omega/C)^2 P = 0 \qquad \text{(C-16)}$$

where $\omega = 2\pi\nu$ = a constant. Now let us try to apply the method of separation of variables, setting

$$P(x,y,z) = F(x)G(y)H(z) \qquad \text{(C-17)}$$

This gives

$$d^2F/dx^2 + k_1^2 F = 0 \qquad \text{(C-18a)}$$

$$d^2G/dy^2 + k_2^2 G = 0 \qquad \text{(C-18b)}$$

$$d^2H/dz^2 + k_3^2 H = 0 \qquad \text{(C-18c)}$$

$$k_1^2 + k_2^2 + k_3^2 = (\omega/C)^2 \qquad \text{(C-19)}$$

which give in turn

$$F = A_1 \sin k_1x + B_1 \cos k_1x \tag{C-20a}$$

$$G = A_2 \sin k_2y + B_2 \cos k_2y \tag{C-20b}$$

$$H = A_3 \sin k_3z + B_3 \cos k_3z \tag{C-20c}$$

Applying the boundary conditions (C-13a) we see that when $x = 0$ or a

$$dF/dx = 0 = k_1A_1 \cos k_1x - k_1B_1 \sin k_1x$$

This means that $A_1 = 0$ and $\sin k_1a = 0$, or $k_1a = n_1\pi$ $(n_1 = 0,1,2,\cdots)$ so that

$$k_1 = n_1\pi/a \qquad F(x) = B_1 \cos n_1\pi x/a \tag{C-21a}$$

Similarly for the other functions we find

$$k_2 = n_2\pi/b \qquad G(y) = B_2 \cos n_2\pi y/b \tag{C-21b}$$

and

$$k_3 = n_3\pi/c \qquad H(z) = B_3 \cos n_3\pi z/c \tag{C-21c}$$

and

$$p - p_0 = C \cos (n_1\pi x/a) \cos (n_2\pi y/b) \cos (n_3\pi z/c) \sin (2\pi\nu t + \delta) \tag{C-22}$$

where

$$\nu = (\tfrac{1}{2})C\ [(n_1/a)^2 + (n_2/b)^2 + (n_3/c)^2]^{\frac{1}{2}} \tag{C-23}$$

and n_1, n_2 and n_3 are integers *including zero*. There are three fundamental modes, (1,0,0), (0,1,0) and (0,0,1), corresponding to waves in which the motion is in the x-, y-, and z-directions, respectively. If the room is a cube, with $a = b = c$, the fundamental mode is triply degenerate.

Exercise. Calculate the first ten or so normal mode frequencies at 0°C. in a room 3 meters high by 10 meters wide by 15 meters long. The normal modes of ordinary rooms evidently have frequencies in the audio range. As may be expected, this causes complications in the acoustical design of auditoriums. ($C = 3.32 \times 10^4$ cm./sec. in air at 0°C.)

If a room could be designed with light, movable walls, so that the pressure exerted by them on the gas is always the same, the boundary conditions would be $p - p_0 = 0$ at $x = 0$ and a, $y = 0$ and b, $z = 0$ and c. This gives

$$\nu = (\tfrac{1}{2})C\ [(n_1/a)^2 + (n_2/b)^2 + (n_3/c)^2]^{\frac{1}{2}} \tag{C-24}$$

$$p - p_0 = A \sin (n_1\pi x/a) \sin (n_2\pi y/b) \sin (n_3\pi z/c) \sin (2\pi\nu t + \delta) \tag{C-25}$$

Obviously, if *any* of the n's is zero there will be no variation in pressure at all, so that the fundamental mode here is (1,1,1). The correlation of these modes with those given in Equation (C-22) is best done in the light of the results of pages 138ff.

c. The normal modes of sound waves in a spherical cavity

For a spherical cavity of radius a whose walls are rigid the radial component of the velocity must be zero at the wall. This means that $(\partial p/\partial r) = 0$ at $r = a$, where r is the distance from the center of the sphere. Such a condition can certainly not be fulfilled with solutions of the type (C-17) and (C-20), so it is natural to transform to polar coordinates in this case. The normal mode equation in polar coordinates is* (*cf.* Equation (*A*-2) of Chapter 2)

$$\frac{1}{r^2}\frac{\partial}{\partial r}\left(r^2\frac{\partial P}{\partial r}\right) + \frac{1}{r^2 \sin\theta}\frac{\partial}{\partial \theta}\left(\sin\theta\frac{\partial P}{\partial \theta}\right) + \frac{1}{r^2\sin^2\theta}\frac{\partial^2 P}{\partial \varphi^2} + \left(\frac{\omega}{c}\right)^2 P = 0 \tag{C-26}$$

Let us try the solution

$$P(r, \theta, \varphi) = R(r)\, S(\theta)\, T(\varphi) \tag{C-27}$$

This gives, by the usual arguments†

$$\frac{1}{r^2}\frac{d}{dr}\left(r^2\frac{dR}{dr}\right) + \left(\frac{\omega^2}{c^2} - \frac{n(n+1)}{r^2}\right) R = 0 \tag{C-28}$$

$$\sin\theta\frac{d}{d\theta}\left(\sin\theta\frac{dS}{d\theta}\right) + [n(n+1)\sin^2\theta - m^2]\, S = 0 \tag{C-29}$$

$$\frac{d^2T}{d\varphi^2} + m^2 T = 0 \tag{C-30}$$

where $n(n+1)$ and m^2 are the separation constants. The solution of (C-30) is

$$T(\phi) = A\sin m\varphi + B\cos m\varphi \tag{C-31}$$

* The first derivatives of P with respect to r and θ appear in the polar coordinate form of the Laplacian operator, ∇^2, because of the noncubical shape of the element of volume enclosed by the surfaces, r, $r + dr$, θ, $\theta + d\theta$, φ, $\varphi + d\varphi$. This causes gas to disappear from the element when the velocity components in the θ and r directions are constant inside the element.

† It is important to note that Equations (C-27, -28, and -29) are obtained even if the sound velocity, C, varies with r. If, however, C also depends on θ and φ, separation of variables is not in general possible.

and in order to have $T(0) = T(2\pi)$, so that there will be no discontinuities in pressure, m must be an integer ($m = 0, 1, 2, \cdots$).

Equation (C-29) has already arisen in an earlier problem. It is the associated Legendre equation (*cf.* Equation (B-42)), and we have found that acceptable solutions occur only when n is an integer ($n = 0, 1, 2, \cdots$). Furthermore, n must be equal to or greater than m. The acceptable solutions of (C-29) are associated Legendre functions of degree n and order m,

$$S(\theta) = P_n^m(\cos\theta) \tag{C-32}$$

Equation (C-28) can be thrown into recognizable form by making the substitution

$$R(r) = r^{-\frac{1}{2}}\, V(r) \tag{C-33}$$

Then

$$\frac{d^2V}{dr^2} + \frac{1}{r}\frac{dV}{dr} + \left[\frac{\omega^2}{c^2} - \frac{(n + \frac{1}{2})^2}{r^2}\right] V = 0 \tag{C-34}$$

Substituting $z = (\omega/c)r$, we obtain Bessel's equation of half integral order (*cf.* Equation (B-21)),

$$\frac{d^2V}{dz^2} + \frac{1}{z}\frac{dV}{dz} + \left[1 - \frac{(n + \frac{1}{2})^2}{z^2}\right] V = 0 \tag{C-35}$$

The first two acceptable solutions have the form

$$n = 0 \quad V(z) = J_{\frac{1}{2}}(z) = (2/\pi z)^{\frac{1}{2}} \sin z$$

$$n = 1 \quad V(z) = J_{3/2}(z) = \left[\frac{1}{z}\sin z - \cos z\right](2/\pi z)^{\frac{1}{2}}$$

(See Margenau and Murphy (*15*, p. 114) for other solutions.)

The radial part of the solution of this problem is thus

$$R(r) = J_{n+\frac{1}{2}}\left(\frac{\omega}{c}r\right)\Big/ r^{\frac{1}{2}} \tag{C-36}$$

The boundary condition, $\partial P/\partial r = 0$ at $r = a$ means that

$$\frac{dR}{dr} = \frac{1}{a^{\frac{1}{2}}}\left[\frac{\omega}{c}\frac{dJ}{dz} - \frac{1}{2a}J\right] = 0 \tag{C-37}$$

The zeros of this expression may be found with the help of the tables and other information in Jahnke and Emde (*8*, Chapter 8). It is clear that only certain values of ω will satisfy (C-37). These give the possible normal mode frequencies, but the equation for these frequencies is obviously rather complicated, and we shall leave the problem at this point.

If the walls of the cavity were made of very light material exerting constant pressure on the gas, the boundary condition is $p - p_0 = 0$ at $r = a$ at all times, which leads to the requirement

$$R(a) = a^{-1/2} J_{n+1/2}\left(\frac{\omega}{c} a\right) = 0 \tag{C-38}$$

The roots of this equation are much easier to find. It is interesting that this condition would also apply to a gaseous star whose temperature and composition are everywhere the same (giving a constant sound velocity) and which is held together by gravitational forces.

For $n = 0$ the roots of (C-38) are at $\omega a/c = l\pi$, where l is an integer. This gives frequencies $\nu = \omega/2\pi = lc/2a$. These frequencies are harmonic.

For $n = 1$ the roots of (C-38) are at $\omega a/c = 4.4934, 7.72, 10.89, 14.08, \cdots$, which give frequencies that are not harmonic.

For $n = 2$ the roots are at $\omega a/c = 5.77, 9.11, 12.34, 15.52, \cdots$, which also fail to give harmonic frequencies*.

The appearance of these normal modes is difficult to represent in two dimensions, but the sketches in Fig. 3–19 should be of some help. The frequencies given refer to the cavity with light walls, but the general arrangement of the nodes is more or less the same for light and heavy walls. The modes are labeled according to their angular dependence (denoted by $s, p, d, \cdots$, as with the modes of an ocean) and according to the number of nodal surfaces (denoted by a number in front of s, p, d, etc., which is one greater than the total number of nodes).

It will be noticed that here, especially among the higher modes, the s-modes with l radial nodes tend to have the same frequencies as the d-modes with $l - 1$ radial nodes. Thus ns tends to have the same frequency as $(n + 1)d$. The same is found to be true of np and $(n + 1)f$. In fact, it is found that:

$s, d, g, i, \cdots$ modes tend to have frequencies which are whole integral multiples of the fundamental,

$p, f, h, j, \cdots$ modes tend to have frequencies which are half integral multiples of the fundamental.

* It should be clear by this time that harmonic vibrations generally arise only in one-dimensional systems. This is why all musical instruments of any pretension are based on linear vibrators – such as strings, columns of air and bars of metal.

The higher *s*-modes can therefore be "approximately" hybridized with the *d*-modes even without a perturbation. For instance the frequencies of 3*s* and 4*d* differ by about one part in thirty, so that if they are excited together the resulting hybrid will act like a normal mode for about five cycles–that is, they will maintain the same phase relative to each other. During this time the 3s–4*d* "hybrid" will have an appearance entirely different from that of any other normal mode. It is evident that certain

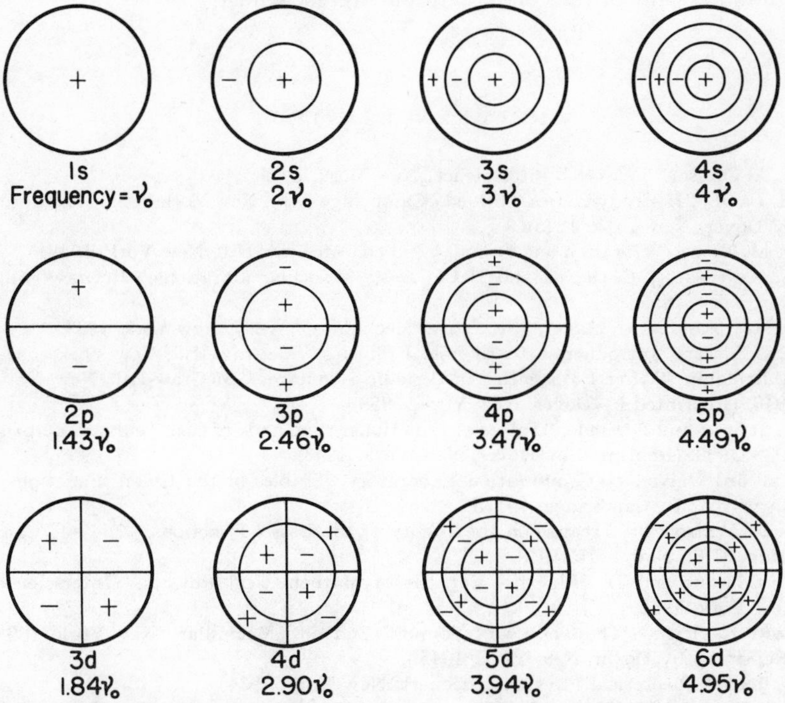

FIG. 3-19. Cross-sections of the nodal surfaces of some normal modes of a gas in a spherical cavity with very light walls that exert a constant pressure on the gas. The plus and minus signs indicate that the pressure at some instant is respectively greater than or less than the mean pressure in the cavity.

types of weak perturbation can induce rather drastic changes in the shapes of some of the normal modes of this system.

If one were to allow the sound velocity to be a function of $r, C(r)$, inside a spherical cavity, the shapes of the radial parts of the normal modes would be changed, but their angular dependence would be unaffected. The normal mode frequencies would also be affected, and by the choice of a suitable function for $C(r)$, we might expect to be able to establish degeneracies, or at least approximate degeneracies, between, say, some of the *s*- and *p*-modes, or even between *s*-, *p*-, and *d*-modes. In fact, this happens in the hydrogen atom, where all "modes" having the same number of nodes turn out to have the same energy (which, as we shall see, is the analogue of the frequency). The possibilities for hybridization therefore become rather great, and this accounts for some of the complexities of stereochemistry.

REFERENCES CITED

1. C. A. Coulson, "Waves," Interscience, New York, 1949.
2. H. Lamb, "Hydrodynamics," 6th ed., Cambridge U.P., New York, 1932. (Reprinted by Dover, New York, 1945).
3. P. M. Morse, "Vibration and Sound," 2nd ed., McGraw-Hill, New York, 1948.
4. E. B. Wilson, J. C. Decius, and P. C. Cross, "Molecular Vibrations," McGraw-Hill, New York, 1955.
5. J. Mayer and M. G. Mayer, "Statistical Mechanics," Wiley, New York, 1939.
6. J. C. Slater, "Introduction to Chemical Physics," McGraw-Hill, New York, 1939.
7. L. Brillouin, "Wave Propagation in Periodic Structures," McGraw-Hill, New York, 1946. (Reprinted by Dover, New York, 1953).
8. E. Jahnke and F. Emde, "Tables of Functions," 2nd and 3rd eds., Teubner, Leipzig, 1933, 1938. (Reprinted by Dover, New York, 1943).
9. Harvard University Computation Laboratory, "Tables of the Bessel Functions," Harvard U.P., Cambridge, 1947ff.
10. G. N. Watson, "A Treatise on the Theory of the Bessel Functions," 2nd ed., Cambridge U.P., New York, 1948.
11. R. Courant and D. Hilbert, "Methods of Mathematical Physics," Interscience, New York, 1953.
12. Lord Rayleigh, "The Theory of Sound," 2nd ed., Macmillan, New York, 1894. (Reprinted by Dover, New York, 1945.)
13. G. Joos, "Theoretical Physics," Stechert, New York, 1934.
14. J. C. Slater and N. H. Frank, "Electromagnetism," McGraw-Hill, New York, 1947.
15. H. Margenau and G. M. Murphy, "The Mathematics of Physics and Chemistry," Van Nostrand, New York, 1943.

GENERAL REFERENCES

C. A. Coulson, reference 1. A short, readable introduction.

P. M. Morse, reference 3. Somewhat more detailed and advanced than Coulson's book.

Lord Rayleigh, reference 12. Extensive, detailed, and readable account of the classical theory of vibrations.

Chapter 4

The Classical Theory of Vibrations II
Approximate Methods for Complex Systems

In finding the normal modes of vibrating strings, tides, etc., we have had to solve some second order differential equations subject to certain boundary conditions. Because of the mathematical difficulties that so often intrude, this method of direct attack on the differential equation is usually not feasible in dealing with more complicated systems. Fortunately several methods exist for finding approximate solutions, and these methods will concern us in this chapter. They depend on the so-called Sturm-Liouville theory, which deals with the general mathematical properties of equations of the type arising in vibrating systems. We shall first give a brief discussion of this theory. For a somewhat more detailed discussion see Margenau and Murphy (*1*, p. 253 ff.). A still more detailed discussion will be found in Courant and Hilbert (*2*, Vol. I, p. 280 ff.).

In this chapter we shall use the abbreviation SL for "Sturm-Liouville."

A. The Sturm-Liouville Theory

a. The SL operator and the SL eigenvalue problem

The *SL operator* is a general type of differential operator defined as

$$L = \frac{d}{dx}\left(p(x)\frac{d}{dx}\right) - q(x) \tag{A-1}$$

where $p(x)$ and $q(x)$ are functions of x. The *SL differential equation* is

$$Lu(x) + \lambda w(x)u(x) = 0 \tag{A-2a}$$

or

$$(1/w)Lu = -\lambda u \tag{A-2b}$$

where $w(x)$ is a function (called the weighting function) which is always positive or zero, and λ is a constant. The function $u(x)$ may be said to be the eigenfunction of the operator $(1/w)L$, with the eigenvalue $-\lambda$.

It is easy to see that many of the equations we have so far encountered are of the form of the SL differential equation. For example, the normal mode equation for the vibrating string with constant tension and density is

$$\frac{d}{dx}\left(1\,\frac{du}{dx}\right) + \left(\frac{\omega}{c}\right)^2 u = 0 \tag{A-3}$$

giving $p = 1$, $q = 0$, $\lambda = (\omega/c)^2$, $w = 1$. For the string with variable tension and density,

$$\frac{d}{dx}\left(T(x)\,\frac{du}{dx}\right) + \omega^2 \rho(x) u = 0 \tag{A-4}$$

giving $p = T(x)$, $q = 0$, $\lambda = \omega^2$ and $w(x) = \rho(x)$. For the associated Legendre equation,

$$\frac{d}{dx}\left((1 - x^2)\,\frac{du}{dx}\right) + \left[n(n+1) - \frac{m^2}{1 - x^2}\right] u = 0 \tag{A-5}$$

we have $p = 1 - x^2$, $q = m^2/(1 - x^2)$, $\lambda = n(n+1)$, $w = 1$. Bessel's equation may be written in the SL form,

$$\frac{d}{dx}\left(x\,\frac{du}{dx}\right) + \left[a^2 x - \frac{n^2}{x}\right] u = 0 \tag{A-6}$$

with $p = x$, $q = n^2/x$, $\lambda = a^2$, $w = x$.

In fact any linear second order differential equation,

$$f(x)u'' + g(x)u' + h(x)u = 0 \tag{A-7a}$$

(where u'' stands for d^2u/dx^2, u' for du/dx) can be transformed into the SL form as follows

$$\frac{d}{dx}\left[f \exp\left(\int \frac{g - f'}{f}\,dx\right)\frac{du}{dx}\right] + h \exp\left(\int \frac{g - f'}{f}\,dx\right) u = 0 \tag{A-7b}$$

For instance, the differential equation $u'' + u' + u = 0$ (where $f = g = h = 1$, so that $\int (g - f')/f\,dx = \int dx = x$) can be written

$$\frac{d}{dx}\left(e^x\,\frac{du}{dx}\right) + e^x u = 0$$

Exercises. (1) Show that (A-7b) reduces to (A-7a).

(2) Transform the following linear second order differential equations into the SL form:

$$\text{(a) } u'' + xu' + x^2u = 0$$

$$\text{(b) } xu'' + u' + xu = 0$$

$$\text{(c) } x^n u'' + ax^{n-1}u' + x^m u = 0$$

$$\text{(d) } x^r u'' + ax^s u' + x^t u = 0$$

The boundary conditions of vibrating systems are usually of only a few types. Either the solution is required to vanish at the boundaries ($u = 0$, as in the vibrating string), or the coefficient $p(x)$ vanishes at one or both limits of the variable (as in the associated Legendre equation, where $p = 1 - x^2$ vanishes at the limits of x, *viz.*, $x = \pm 1$), or the first derivative of the solution vanishes ($du/dx = 0$, as with tidal waves in a trough or sound waves in a rigid container). These boundary conditions, as well as others not normally encountered in vibration problems, can easily be satisfied by the following requirement:

If $u(x)$ and $v(x)$ are any two solutions of the SL differential equation, then at the two limits $x = a$ and $x = b$,

$$vp \frac{du}{dx}\bigg|_{x=a} = vp \frac{du}{dx}\bigg|_{x=b} \tag{A-8a}$$

and

$$up \frac{dv}{dx}\bigg|_{x=a} = up \frac{du}{dx}\bigg|_{x=b} \tag{A-8b}$$

These are called the *SL boundary conditions.*

From what we have seen so far in the treatment of vibrations, it is clear that finding the shapes of normal modes is equivalent to the determination of the eigenfunctions of a SL operator that satisfy certain SL boundary conditions, whereas the problem of the normal mode frequencies consists of finding the corresponding eigenvalues, λ. Normal mode problems are for this reason often spoken of as "eigenvalue problems." The SL eigenvalue problem is of fundamental importance in the theory of vibrations and in quantum chemistry. We shall now investigate certain properties of the SL operator, its eigenfunctions, and its eigenvalues. Sections b, c, and d are concerned with the four concepts summarized in the following equations:

Hermitian character of the SL operator:

$$\int_a^b uLv\,dx = \int_a^b vLu\,dx$$

Orthogonality of the SL eigenfunctions:

$$\int_a^b u_m u_n w\,dx = 0 \quad \text{if } n \neq m$$

Normalized SL eigenfunctions:

$$\int_a^b u_n^2 w\,dx = 1$$

Expansion of arbitrary functions $z(x)$ in terms of a complete set of SL eigenfunctions:

$$z(x) = \sum_{\substack{\text{all}\\ \text{eigenfunctions}}} A_n u_n(x)$$

$$\text{with} \quad A_n = \int_a^b z(x) u_n(x) w(x)\,dx$$

These four relatively simple concepts play an important role in the two methods of approximation on which a large part of quantum chemistry is based. The two methods will be discussed in Sections B and C.

b. The Hermitian character of the SL operator

A real operator, α, is said to be Hermitian in the range (a, b) with respect to some class of functions if

$$\int_a^b \psi\alpha\phi dx = \int_a^b \phi\alpha\psi dx \tag{A-9}$$

where ψ and ϕ are any real functions of the class.*

* If the operator is complex, it is said to be Hermitian if

$$\int_a^b \psi\alpha\phi dx = \int_a^b \phi\alpha^*\psi dx$$

where α^* is the conjugate complex (page 20) of α.

It is not hard to show that *the SL operator, L, is Hermitian in the range (a, b) with respect to any pair of functions that are in the class obeying the SL boundary conditions.* That is

$$\int_a^b uLv\,dx = \int_a^b vLu\,dx \tag{A-10}$$

for any pair of functions u and v that satisfy the conditions (A-8). (It is *not* necessary that u and v be solutions of the SL equation.)

Proof: Using the notation $du/dx = u'$, and the definition (A-1)

$$\int_a^b uLv\,dx = \int_a^b u(pv')'dx - \int_a^b uqv\,dx = I - \int_a^b uqv\,dx \tag{A-11}$$

Integrating by parts

$$I = \int_a^b u(pv')'dx = upv'\Big|_a^b - \int_a^b pv'u'dx \tag{A-12}$$

The first term on the right is zero because of the condition (A-8). Integrating the second term by parts again

$$I = -vpu'\Big|_a^b + \int_a^b v(pu')'dx \tag{A-13}$$

Again the first term on the right is zero because of (A-8). Substituting (A-13) in (A-11) we find

$$\int_a^b uLv\,dx = \int_a^b v(pu')'dx - \int_a^b uqv\,dx = \int_a^b vLu\,dx \tag{A-14}$$

c. The orthogonality and normalization of the SL eigenfunctions

If u and v in equation (A-10) are eigenfunctions, u_m and u_n respectively, of L with eigenvalues λ_m and λ_n in the sense of Equation (A-2), then it is easy to show that

$$\int_a^b u_m u_n w\,dx = 0 \quad \text{if} \quad \lambda_m \neq \lambda_n \tag{A-15}$$

Any such pair of functions is said to be *orthogonal* relative to the weighting function, $w(x)$. Nondegenerate eigenfunctions of L are therefore always orthogonal.

Proof: Since

$$Lu_m = -\lambda_m w u_m \tag{A-16a}$$

and

$$Lu_n = -\lambda_n w u_n \tag{A-16b}$$

we have

$$\int_a^b u_m \, Lu_n dx = -\lambda_n \int_a^b u_m u_n w \, dx \tag{A-17a}$$

and

$$\int_a^b u_n L u_m dx = -\lambda_m \int_a^b u_n u_m w \, dx \tag{A-17b}$$

But because L is Hermitian with respect to functions of the type u_m and u_n, we find on subtracting (A-17a) from (A-17b)

$$-(\lambda_m - \lambda_n) \int_a^b u_m u_n w \, dx = \int_a^b u_m L u_n \, dx - \int_a^b u_n L u_m \, dx = 0 \tag{A-18}$$

Therefore if $\lambda_m \neq \lambda_n$, Equation (A-15) must be true.

Even if there is degeneracy, with $\lambda_m = \lambda_n$, so that u_m and u_n are not orthogonal, *it is always possible to form combinations of u_m and u_n which are orthogonal*. Thus, if for a pair of degenerate functions we have

$$\int_a^b u_m u_n w \, dx = A \tag{A-19a}$$

then we may define a new function,

$$u'_n = u_n - A \frac{u_m}{\int_a^b u_m^2 w \, dx} \tag{A-19b}$$

which is orthogonal to u_m, since

$$\int_a^b u_m u'_n w \, dx = \int_a^b u_m u_n w \, dx - A = 0$$

If for all functions, $u_1, u_2, \cdots, u_m, \cdots$ of a set of functions we have

$$\int_a^b u_m^2 w \, dx = 1 \tag{A-20}$$

then the set is said to be *normalized*. It is always possible to normalize any function U simply by multiplying it by a constant c defined by

$$c = \left[\int_a^b U^2 w \, dx\right]^{-1/2} \tag{A-21}$$

Then the function cU is normalized since

$$\int_a^b (cU)^2 w \, dx = c^2 \int_a^b U^2 w \, dx = 1$$

If a set of functions is normalized and orthogonal, we can write

$$\int_a^b u_m u_n w \, dx = \delta_{mn} \tag{A-22}$$

where δ_{mn} is the so-called "Kronecker delta" or "Weierstrass delta." It is by definition equal to zero if $m \neq n$ and equal to unity if $m = n$.

Evidently we can make an orthogonal normalized set of functions out of the complete collection of normal mode solutions of any vibrating system whose wave equation is reducible to the SL form. This is evidently true not only for one-dimensional systems, but also for many-dimensional systems whose solutions are expressible as products of functions of individual variables–for instance, the ocean on a flooded planet, the circular membrane, and gases in spherical and rectangular cavities. Thus suppose that for a three-dimensional vibrating body we have normal mode functions f_{lmn}, where l, m, and n are indices used in identifying the modes. Then if q_1, q_2, and q_3 are the coordinates used in separating the normal mode equation, we may write

$$\int f_{l'm'n'} f_{lmn} w_1 w_2 w_3 \, dq_1 dq_2 dq_3 = \delta_{ll'} \delta_{mm'} \delta_{nn'} \tag{A-23}$$

where the integral is over the entire volume occupied by the body, and w_1, w_2 and w_3 are the weighting functions which appear in the separated normal mode equations. It usually turns out that

$$d\tau = w_1 w_2 w_3 \, dq_1 dq_2 dq_3 \tag{A-24}$$

where $d\tau$ is the "volume element" in the coordinate system q_1, q_2, q_3. That is, $d\tau$ is the volume of the element defined by the surfaces q_1, $q_1 + dq_1$, q_2, $q_2 + dq_2$, q_3, $q_3 + dq_3$. Thus we may write

$$\int f_{l'm'n'} f_{lmn} \, d\tau = \delta_{ll'} \delta_{mm'} \delta_{nn'} \tag{A-25}$$

Exercises. (1) Show that
for Cartesian coordinates $d\tau = dxdydz$
for polar coordinates $d\tau = r^2 \sin\theta \, drd\theta d\varphi$
for cylindrical coordinates $d\tau = rdrd\theta dz$

(2) Prove by direct integration that $P_0^0(x)$, $P_1^0(x)$, and $P_2^0(x)$ are mutually orthogonal, using the weighting function $w = 1$ and the limits of integration $x = \pm 1$.

(3) Find the normalizing factor for $\sin n\pi x/a$, using the limits of integration $x = 0$ and $x = a$ and the weighting function $w = 1$.

Find the normalizing factors for $P_0^0(x)$, $P_1^0(x)$, $P_2^0(x)$, and $P_1^1(x)$, using the limits of integration $x = \pm 1$ (cf. Equation (B-79) of Chapter 3).

(4) Show that the surface spherical harmonics f_{00}, f_{10}, f_{11}, f_{20}, f_{21}, and f_{22} given in Table 3-1 are mutually orthogonal when the integration covers the complete range of the variables θ and φ.

(5) Consider the following polynomials:

$$y_0 = 1$$

$$y_1 = x + a_1$$

$$y_2 = a_2x^2 + b_2x + c_2$$

$$y_3 = a_3x^3 + b_3x^2 + c_3x + d_3$$

$$y_4 = a_4x^4 + b_4x^3 + c_4x^2 + d_4x + e_4$$

Find the values of the constants that make all of these functions orthogonal to one another in the range of integration $x = \pm 1$; that is,

$$\int_{-1}^{1} y_i y_j \, dx = 0 \quad \text{for} \quad i \neq j$$

The resulting functions turn out to be proportional to the Legendre polynomials, $y_n = AP_n^0(x)$ (cf. page 97).

d. The completeness of the SL eigenfunctions

In Chapter 3 (page 63) we have seen that as a result of Fourier's theorem, all shapes of a uniform, stretched string can be described in terms of its normal mode functions:

$$z(x) = \sum A_n \sin n\pi x/a \quad (0 \leq x \leq a)$$

where the coefficients A_n are independent of x. Similarly, the set of all eigenfunctions of any particular SL eigenvalue problem can serve as a basis for the expansion of arbitrary functions. If $u_1, u_2, \cdots, u_n, \cdots$ are the eigenfunctions of a SL operator satisfying the SL boundary conditions at $x = a$ and $x = b$, and if $z(x)$ is an arbitrary function that satisfies the same boundary conditions,* then it is possible to express $z(x)$ in terms of the $u(x)$'s for any value of x in the range $a \leqq x \leqq b$:

$$z(x) = \sum_{\substack{\text{all eigen-}\\ \text{functions}}} A_n u_n(x) \tag{A-26}$$

where the A_n's are constants independent of x. A proof of the possibility and adequacy of such an expansion will be found in Margenau and Murphy (*1*, p. 262 ff.). A set of functions that is able to act as a basis for the expansion of arbitrary functions is said to be *complete*. Any family of SL eigenfunctions is therefore complete. (One must, however, be careful to include in the family *all* of the eigenfunctions obtainable from the problem; if even one of them is omitted, the expansion (A-26) will be imperfect.) Much of quantum chemistry depends on the possibility of expansions of this kind.

The concept of expansions in terms of complete sets has a simple physical interpretation. It means that if a body can be bent elastically into some shape, then the same shape can be obtained momentarily by exciting the appropriate mixture of the normal modes of vibration of the body.

If the u_n's have been orthogonalized and normalized, it is easy to see that the coefficients in the expansion (A-26) are given by

$$A_n = \int_a^b z(x) u_n(x) w(x)\, dx \tag{A-27}$$

Exercise. Prove this. [Hint: Multiply both sides of (A-26) by wu_n and integrate.]

Similar expansions are possible in terms of the normal modes of two- and three-dimensional systems

$$v(x, y, z) = \sum_{\substack{\text{all}\\ \text{modes}}} A_m u_m(x, y, z) \tag{A-28}$$

where the u_m's are solutions of a normal mode equation and v is an arbitrary function having the same boundary conditions as the u_m's. If the u_m's have been orthogonalized and normalized, we may find the constants A_n by means of the relation (*cf.* (A-25))

$$A_m = \int v u_m d\tau \tag{A-29}$$

* The function $z(x)$ must also be single valued and can have only a finite number of discontinuities in the range (a,b).

Exercises. (1) Find the numerical values of the coefficients A_0, A_1, A_2, and A_3 in the expansion

$$f(x) = \sum_{n=0}^{\infty} A_n P_n^0(x) \quad (-1 < x < 1)$$

where $P_n^0(x)$ are the Legendre polynomials (see page 97) and where $f(x)$ is one of the following functions:

(a) $\sin x$ (b) $\cos x$

(c) $f(x) = 0$ for $-1 \leqq x < 0$ and $f(x) = 1$ for $0 < x \leqq 1$

(d) $f(x) = -x$ for $-1 \leqq x \leqq 0$ and $f(x) = x$ for $0 \leqq x \leqq 1$

Plot the sum of the first four terms in the expansion and compare it with $f(x)$.

(2) Show that if $z(x,t)$ satisfies the wave equation, $Lz = \rho\partial^2 z/\partial t^2$, then the time dependence of $z(x,t)$ may be included in the expansion (A-26) if

$$A_n = b_n \sin(\sqrt{\lambda_n}t + \delta_n)$$

where b_n is independent of the time and of x, δ_n is a constant, and λ_n is the SL eigenvalue. This means that the general motion can be expressed as a superposition of normal modes.

B. The Variation Method

In the problems that have been considered in Chapter 3, it has been possible to find exact solutions of the wave equation by a more or less direct attack on the equation. Only a limited number of vibration problems have succumbed to this procedure, however, since mathematical difficulties have a way of becoming insurmountable when even a slight degree of complexity is introduced into the wave equation or the boundary conditions. Therefore it is very fortunate that we have at our disposal other less direct means by which we are able to obtain at least approximate solutions to such problems. In this section and in the next section we shall discuss two of these methods. A third method has already been mentioned–that of the correlation diagram, in which we try to relate the system to other, simpler systems whose precise solutions we know, and attempt to interpolate the properties of the complex system from the known properties of the simpler systems. All of these methods are of great importance in quantum chemistry, where few problems have been solved exactly.

a. The variation principle

The variation method (also often called the Rayleigh-Ritz method) consists essentially in making a shrewd guess at the solution of the normal mode equation. The method depends for much of its usefulness on the validity of the so-called variation principle.

Suppose that a SL differential equation,

$$Lu(x) = -\lambda w(x)u(x) \tag{B-1}$$

has a set of solutions, $u_0, u_1, \cdots, u_n, \cdots$ that satisfy the SL boundary conditions at $x = a$ and at $x = b$. Let the respective eigenvalues be $\lambda_0, \lambda_1, \cdots, \lambda_n, \cdots$. Suppose further that $\phi(x)$ is any function (to be called the "trial function") that also satisfies the boundary conditions at a and b. Let us define the quantity

$$W_\phi = -\int_a^b \phi L\phi \, dx \Big/ \int_a^b \phi^2 w \, dx \tag{B-2}$$

Then the variation principle states that

$$W_\phi \geqq \lambda_0 \tag{B-3}$$

where λ_0 is algebraically the lowest of the eigenvalues.

Proof of the variation principle: Assume that we have a set of eigenfunctions, u_0, u_1, etc., satisfyingEquation (B-1) and the SL boundary conditions at $x = a$ and b. According to Section Ad these u_n's form a complete set, and we can expand the "trial function," ϕ, in terms of this set

$$\phi = \sum a_n u_n$$

Also,

$$L\phi = \sum a_n L u_n = -\sum a_n \lambda_n u_n w$$

Substituting these two expressions into (B-2) we find

$$W_\phi = \frac{\sum\sum a_n a_m \lambda_n \int_a^b u_n u_m w dx}{\sum\sum a_n a_m \int_a^b u_n u_m w dx} = \frac{\sum a_n^2 \lambda_n}{\sum a_n^2} \tag{B-4}$$

Let λ_0 be the least of all the λ_n's in the set, and subtract it from both sides of equation (B-4)

$$W_\phi - \lambda_0 = \frac{\sum a_n^2 (\lambda_n - \lambda_0)}{\sum a_n^2} \tag{B-5}$$

Everything on the right hand side of this equation is greater than zero: all of the a_n^2 are inherently positive, and we have deliberately chosen λ_0 so that $(\lambda_n - \lambda_0)$ must be positive. Therefore,

$$W_\phi - \lambda_0 \geqq 0$$

or

$$W_\phi \geqq \lambda_0$$

Furthermore, the only way in which the right side of (B-5) can vanish is to have all $a_n = 0$ except for a_0. Therefore, if

$$W_\phi = \lambda_0$$

then

$$\phi = a_0 u_0$$

and ϕ is precisely the normal mode having the lowest λ_0. We also note that the closer W_ϕ is to λ_0, the closer ϕ is to u_0 (i.e., the smaller all of the other a_n's are relative to a_0).

b. Application of the variation principle to vibrating systems

When the normal mode equations of vibration theory are thrown into the form of the SL equation, the eigenvalue λ_n is usually equal to $4\pi^2\nu_n^2$, where ν_n is the frequency of the nth normal mode. Therefore,

$$\lambda_0 = 4\pi^2\nu_0^2 \tag{B-6}$$

where ν_0 is the lowest frequency of the system–that is, the "fundamental" frequency. The variation principle therefore states that

$$\nu_0 \leqq W_\phi^{1/2}/2\pi \tag{B-7}$$

Furthermore, the closer ϕ approaches to the correct solution for the fundamental mode, the more nearly (B-7) will approach to an equality.

If, therefore, we are willing to make a guess as to the shape of the fundamental mode, this remarkable principle gives us a means of estimating the frequency of this mode. The better the guess is, the more accurate the estimated frequency will be, and the estimated frequency will always be greater than the correct one.

Let us now illustrate this by means of a simple example. Consider the uniform stretched string. Here the normal mode equation and boundary conditions are

$$\frac{d}{dx}\left(T\frac{dz}{dx}\right) = -4\pi^2\nu^2\rho z; \quad z = 0 \text{ at } x = 0 \text{ and } a$$

T is a constant. The boundary conditions are consistent with the SL boundary conditions, and we have

$$L = T\, d^2/dx^2 \quad (w = \rho = \text{constant})$$

We know, of course, that the solution corresponding to the lowest frequency is $z_0 = A \sin \pi x/a$, but let us see what happens if we use as a trial function a parabola that passes through the fixed end points of the string:

$$\phi = x(x - a)$$

On substituting this into (B-2) we find

$$W_\phi = 10T/\rho a^2$$

and from this, using (B-7), we can estimate the lowest frequency of the string:

$$\nu_{0\ \text{est}} = (10^{1/2}/2\pi a)\,(T/\rho)^{1/2} = \frac{1.0066}{2a}(T/\rho)^{1/2}$$

The correct fundamental frequency is, of course,

$$\nu_0 = \frac{1}{2a}(T/\rho)^{1/2}$$

The parabola was therefore a very good guess, the estimated frequency being only 0.66 % too high.

The variation method allows us to go one step further than this. Suppose that the trial function is chosen with one or more arbitrary parameters, $\alpha, \beta, \cdots$. Then on evaluating the integrals in (B-2) we will find that W_ϕ is a function of these parameters

$$W_\phi = W_\phi(\alpha, \beta, \cdots) \tag{B-8}$$

If we now select the values of the parameters that give W_ϕ the lowest possible value, we will have the best estimate that can be made with a trial function of the form originally chosen. This makes the variational method something better than a mere matter of guessing. The more parameters one introduces into the trial function, the closer one is likely to get to the correct solution and frequency. Furthermore, an approximation is usually improved by adding more trial functions, with parameters, to a given set of trial functions.

The minimizing of W_ϕ $(\alpha, \beta, \cdots)$ is very easily accomplished in principle by taking the partial derivatives of W_ϕ with respect to each of the parameters and setting them equal to zero

$$\partial W_\phi/\partial\alpha = 0, \quad \partial W_\phi/\partial\beta = 0, \cdots \tag{B-9}$$

This results in just as many equations as there are unknown parameters, so that all of the parameters can be determined. In practice, however, these equations are often not easily solved, and one of the tricks of the variational method is to select trial functions containing parameters that can easily be determined when the minimizing condition is applied. (Particularly well suited in this respect is the linear trial function

$$\phi = \alpha f_1 + \beta f_2 + \gamma f_3 + \cdots$$

where $f_1, f_2, f_3, \cdots$ are chosen on the basis of common sense. The treatment of trial functions of this kind will be discussed in more detail in Section c.)

As an illustration of this minimizing procedure, let us calculate W_ϕ for the uniform stretched string using

$$\phi = x(x - a) + (\alpha/a^2)x^2(x - a)^2$$

where a is the length of the string and α is the parameter whose value is to be found. Substitution into (B-2) gives

$$W_\phi = 6 \cdot \frac{35 - 14\alpha + 2\alpha^2}{21 - 9\alpha + \alpha^2} (T/\rho a^2)$$

$\partial W_\phi/\partial\alpha$ is evaluated and equated to zero, giving

$$4\alpha^2 - 14\alpha - 21 = 0$$

which has the roots

$$\alpha = 4.63 \text{ and } -1.133143$$

Using $\alpha = 4.63$ we find $W_\phi = 97.5\ T/\rho a^2$, and using $\alpha = -1.133143$ we find $W_\phi = 9.869748\ T/\rho a^2$. Since the latter is the smaller value of W_ϕ, it is obviously the one we are looking for. Using (B-7), the estimated frequency of the fundamental mode is found to be

$$\nu_{0\text{ est}} = \frac{(9.869748)^{1/2}}{2\pi a}(T/\rho)^{1/2} = \frac{1.000006}{2a}(T/\rho)^{1/2}$$

which is only 0.0006 % above the correct value. The corresponding estimated shape of the fundamental mode is

$$u_{0\ \text{est}} = x(x-a)\left[1 - 1.13\frac{x(x-a)}{a^2}\right]$$

The variational method is not limited to the calculation of the fundamental frequency, but may be extended to the calculation of successive overtone frequencies, once the fundamental has been accurately determined. One merely uses as trial functions ϕ's that are orthogonal to the final fundamental solutions, $\phi_0 = u_0$:

$$\int \phi_1 \phi_0 w\, dx = 0$$

Then in the expansion of ϕ_1 in terms of the complete set of u_n's,

$$\phi_1 = \sum a_n u_n$$

the coefficient a_0 will be zero, since from Equation (A-27)

$$a_0 = \int \phi_1 u_0 w\, dx = \int \phi_1 \phi_0 w\, dx = 0$$

Thus

$$W_{\phi_1} = \frac{a_1{}^2 \lambda_1 + a_2{}^2 \lambda_2 + \cdots}{a_1{}^2 + a_2{}^2 + \cdots}$$

and if λ_1 is the eigenvalue next larger than λ_0

$$W_{\phi_1} - \lambda_1 = \frac{a^2{}_2\,(\lambda_2 - \lambda_1) + a^2{}_3\,(\lambda_3 - \lambda_1) + \cdots}{a^2{}_1 + a^2{}_2 + \cdots}$$

The quantity on the right is positive, so that

$$\lambda_1 \leqq W_{\phi_1}$$

or

$$\nu_1 \leqq W_{\phi_1}{}^{1/2}/2\pi \qquad \text{(B-10)}$$

Thus one can estimate the frequency of the first overtone, ν_1. This estimate can then be improved by making better and better guesses using variable parameters. Having finally attained the exact value of ν_1, one may then make use of trial functions orthogonal to both ϕ_0 and ϕ_1 in order to find the frequency of the next overtone, and so on.

For instance the two functions, $\phi_0 = x(x - a)$ and $\phi_1 = x(x - \frac{1}{2}a)(x - a)$, are orthogonal; and ϕ_0, as we have just seen, gives an estimated value for the fundamental frequency of the stretched string that is close to the correct value. Therefore we should expect that ϕ_1 will yield a fair approximation to the frequency of the first harmonic of the stretched string. It is found that $W_{\phi_1} = 42\ T/\rho a^2$, or $\nu_{\text{est}} = (2.06/2a)\ (T/\rho)^{1/2}$, which is only 3 % above the correct value.

In using the variation method to estimate frequencies other than that of the fundamental mode there is no assurance that the estimated frequency will be greater than the correct value unless one knows with certainty that the trial function is orthogonal to the correct normal mode functions of the lower states. In many applications, however, it is possible to choose such trial functions even if the exact shape of the fundamental function is not known. For instance, in the stretched string we know that the fundamental mode is symmetric with respect to reflection in a plane passing through the midpoint of the string. Any trial function that changes sign in such a reflection must be orthogonal to the fundamental mode (*cf.* footnote on page 31). Therefore any antisymmetric trial function must yield an estimated frequency greater than the true frequency of the lowest antisymmetric mode. This mode happens to be the first harmonic. The function ϕ_1 discussed in the previous paragraph is antisymmetric. This is why the frequency estimated from ϕ_1 turned out to be greater than the true frequency of this mode.

So far, all that we have said refers only to one-dimensional systems, but it is easy to generalize to systems of more than one dimension in the light of what has been said on page 115 about the orthogonality of normal mode functions in general and on page 117 about the completeness of these functions. Wave equations, regardless of the dimensionality of the problem, always take the form

$$KU = \partial^2 U/\partial t^2 \tag{B-11}$$

where K is a differential operator involving only the coordinates and not the time, and U is some property that oscillates in the wave. If U is written in the normal mode form

$$U = u(x, y, z)T(t) \tag{B-12}$$

then it is clear that we will have

$$T = A \sin(\sqrt{\lambda}\, t + \delta) \tag{B-13}$$

and

$$Ku = -\lambda u \tag{B-14}$$

Such being the case, if for some specific boundary conditions we have a complete orthogonal set of solutions, u_0, $u_1, \cdots$, of (B-14), having eigenvalues λ_0, $\lambda_1, \cdots$, then the student can readily show by arguments analogous to those used above that

$$W_\phi = -\frac{\int \phi K \phi d\tau}{\int \phi^2 d\tau} \geqq \lambda_0 = 4\pi^2 \nu_0^2 \tag{B-15}$$

where λ_0 is the smallest eigenvalue of the set, ν_0 is the fundamental frequency of the system, and ϕ is any function that satisfies the boundary conditions of the problem. The integrations here are carried out over the entire vibrating system, and $d\tau$ is the volume element as defined on page 115 (and also subject to the same qualifications as those inferred there).

c. Variation method with a linear trial function

Suppose that one uses as a trial function a linear combination of l functions,

$$\phi = a_1 f_1 + a_2 f_2 + \cdots + a_l f_l \tag{B-16}$$

where a_1, a_2, $a_3, \cdots$, a_l are parameters whose values are to be chosen so as to minimize W_ϕ, and where the f's are any functions satisfying the SL boundary conditions. Then

$$W_\phi = \frac{\sum_{i=1}^{l} \sum_{j=1}^{l} a_i a_j L_{ij}}{\sum_{i=1}^{l} \sum_{j=1}^{l} a_i a_j w_{ij}} \tag{B-17}$$

where (recalling the Hermitian character of L)

$$L_{ij} = -\int f_i L f_j \, dx = L_{ji} \tag{B-17a}$$

$$w_{ij} = \int f_i w f_j \, dx = w_{ji} \tag{B-17b}$$

In order to find the values of the parameters a_i that minimize W_ϕ, we differentiate successively with respect to the a_i's and set each derivative equal to zero. This gives

$$\frac{1}{2}\frac{\partial W_\phi}{\partial a_1} = \frac{a_1L_{11} + a_2L_{12} + \cdots + a_lL_{1l}}{\sum\sum a_ia_jw_{ij}}$$

$$- \frac{\sum\sum a_ia_jL_{ij}}{\sum\sum a_ia_jw_{ij}}\left[\frac{a_1w_{11} + a_2w_{12} + \cdots + a_lw_{1l}}{\sum\sum a_ia_jw_{ij}}\right]$$

$$= \frac{a_1(L_{11} - W_mw_{11}) + a_2(L_{12} - W_mw_{12}) + \cdots + a_l(L_{1l} - W_mw_{1l})}{\sum\sum a_ia_jw_{ij}}$$

$$= 0$$

or, more generally,

$$\frac{1}{2}\frac{\partial W_\phi}{\partial a_i} = \frac{\sum_{j=1}^{l} a_j(L_{ij} - W_mw_{ij})}{\sum_{j=1}^{l}\sum_{k=1}^{l} a_ja_kw_{jk}} = 0 \tag{B-18}$$

where W_m is the minimum value of W_ϕ appropriate to the optimum choice of the a_i's. This leads to a set of l linear equations in the a_i's,

$$(L_{11} - W_mw_{11})a_1 + (L_{12} - W_mw_{12})a_2 + \cdots + (L_{1l} - W_mw_{1l})a_l = 0$$
$$(L_{21} - W_mw_{21})a_1 + (L_{22} - W_mw_{22})a_2 + \cdots + (L_{2l} - W_mw_{2l})a_l = 0$$
$$\cdots\cdots\cdots\cdots\cdots\cdots\cdots\cdots\cdots\cdots\cdots\cdots$$
$$(L_{l1} - W_mw_{l1})a_1 + (L_{l2} - W_mw_{l2})a_2 + \cdots + (L_{ll} - W_mw_{ll})a_l = 0 \tag{B-19}$$

which can be solved for the a_i's by the method of determinants (often called Cramer's rule; see Margenau and Murphy (*1*, page 299))

$$a_1 = \frac{\begin{vmatrix} 0 & L_{12} - W_mw_{12} & \cdots & L_{1l} - W_mw_{1l} \\ 0 & L_{22} - W_mw_{22} & \cdots & L_{2l} - W_mw_{2l} \\ \vdots & \vdots & \cdots & \vdots \\ 0 & L_{l2} - W_mw_{l2} & \cdots & L_{ll} - W_mw_{ll} \end{vmatrix}}{\begin{vmatrix} L_{11} - W_mw_{11} & L_{12} - W_mw_{12} & \cdots & L_{1l} - W_mw_{1l} \\ L_{21} - W_mw_{21} & L_{22} - W_mw_{22} & \cdots & L_{2l} - W_mw_{2l} \\ \vdots & \vdots & & \vdots \\ L_{l1} - W_mw_{l1} & L_{l2} - W_mw_{l2} & \cdots & L_{ll} - W_mw_{ll} \end{vmatrix}} \tag{B-20}$$

Similar expressions are obtained for the other a_i's, except that the ith column in the numerator is in each case made up of zeros. Now any determinant having a column of zeros is itself zero (cf. page 47). Therefore, unless the denominator in (B-20) is also zero, all of the a_i's must be zero. This gives

$$\begin{vmatrix} L_{11} - W_m w_{11} & L_{12} - W_m w_{12} & \cdots & L_{1l} - W_m w_{1l} \\ L_{21} - W_m w_{21} & L_{22} - W_m w_{22} & \cdots & L_{2l} - W_m w_{2l} \\ \vdots & \vdots & & \vdots \\ L_{l1} - W_m w_{l1} & L_{l2} - W_m w_{l2} & \cdots & L_{ll} - W_m w_{ll} \end{vmatrix} = 0 \qquad \text{(B-21)}$$

Equation (B-21) is a polynomial of the lth degree in W_m, and since all of the L_{ij}'s and w_{ij}'s are known, it can be solved for W_m. There are, of course, l roots, of which the smallest is the one we seek. Using this value of W_m we can then go back to (B-19), divide through by a_l and solve for the $l-1$ ratios a_i/a_l by Cramer's rule. Substituting in (B-16) we find for our "best" normal mode approximation of this form

$$\phi = a_l \sum_{i=1}^{l} \left(a_i/a_l\right)_{W_m} f_i \qquad \text{(B-22)}$$

If ϕ is very close to the true fundamental normal mode, then it can be shown that the set of (a_i/a_l) obtained from each of the other roots, W_{m2}, $W_{m3}, \cdots, W_{ml}$, of (B-21) will also be close to actual normal modes, and the roots themselves will be close to the true eigenvalues, λ_n, of the corresponding SL equation.

Exercises. (1) A chain of length a and density ρ grams per unit length is suspended from one end, the other end being free so that the chain, when at rest, hangs vertically.

(a) Show that the tension on the chain at a distance x from the fixed end is $T = g\rho(a-x)$, where g is the acceleration due to gravity.

(b) Show that the SL operator derived from the normal mode equation describing the lateral oscillations of such a chain is

$$L = d/dx\,[g\rho(a-x)d/dx]$$

and that the weighting function is $w = \rho$.

(c) Using the trial wave function $\phi_0 = x$ and the variation principle, estimate the frequency of the fundamental mode. [Answer: $\nu_{0\,\text{est}} = (1/2\pi)\,(3g/2a)^{1/2} = 0.195(g/a)^{1/2}$. Note that the correct value for ν_0 is $0.1914\,(g/a)^{1/2}$.]

(d) Determine the value of the constant α in $\phi_1 = x(x-\alpha)$ that makes ϕ orthogonal to the trial function $\phi_0 = x$ used in the previous calculation. Use the resulting trial function to estimate the frequency of the first overtone of the hanging chain. [Answer: $\alpha = 3a/4$. $\nu_{1\,\text{est}} = (1/2\pi)\,(55g/6a)^{1/2} = 0.4819\,(g/a)^{1/2}$. The correct value for ν_1 is $0.4385\,(g/a)^{1/2}$.]

(e) Use the method of linear trial functions with the function

$$\phi = a_1 x + a_2 x^2$$

to estimate the frequencies and shapes of the first two normal modes of the hanging chain. [Answer: $\nu_{0\ \mathrm{est}} = 0.1916\,(g/a)^{1/2}$; $\nu_{1\ \mathrm{est}} = 0.4833\,(g/a)^{1/2}$; in the fundamental mode $a_2/a_1 = 0.5873/a$ and in the first overtone $a_2/a_1 = -1.3051/a$.]

(f) Show that the functions obtained with the two sets of values of a_1 and a_2 in the previous example are orthogonal.

(2) On page 77 we have derived the wave equation for a membrane in cylindrical coordinates.

(a) Show by comparison of Equation (B-15) of Chapter 3 with Equation (B-11) of the present chapter that the operator K for such a membrane may be written

$$K = c^2 \left[\frac{\partial}{\partial r} r \frac{\partial}{\partial r} + \frac{1}{r} \frac{\partial^2}{\partial \varphi^2}\right]$$

(b) Use the trial wave function $\phi_0 = (r - a)$ with the operator K given above to estimate the frequency of the fundamental mode of a circular drumhead of radius a, in which the wave velocity is c. [Answer: $\nu_{0\ \mathrm{est}} = \sqrt{6}\,c/2\pi a = 2.4495c/2\pi a$. The correct value is $2.4048c/2\pi a$.]

(3) Using the trial wave function $\phi = r - a$, estimate the fundamental frequency of a gas-filled spherical cavity of radius a, whose walls are of negligible mass and exert a constant pressure on the gas. [Hint: Find the operator K by referring to Equation (C-11a) or (C-26) of Chapter 3. Note that the volume element is

$$d\tau = r^2 \sin\theta\, dr d\theta d\varphi$$

Answer: $\nu_{0\ \mathrm{est}} = \sqrt{10}\,c/2\pi a = 0.504\,c/a$ where c is the sound velocity in the gas. This is less than 1% above the correct value, $\nu_0 = 0.500c/a$.]

C. The Perturbation Method

A second method of handling problems not soluble by the direct method of attack has been mentioned on page 64, where we have shown that it is possible to calculate the approximate shapes and frequencies of the normal modes of a stretched string that has been subjected to a slight perturbation in the form of a change in density or tension of the string. The theory presented there is easily generalized and extended on the basis of the results of the SL theory. The method of attack depends, however, on whether or not the perturbed normal mode is degenerate. We shall consider first the perturbation theory for nondegenerate modes.

a. Perturbation theory for nondegenerate modes

Suppose that we know the solutions, u_{k0}, for the normal mode equation of a vibrating object

$$L_0 u_{k0} = \lambda_{k0} w_0 u_{k0} \tag{C-1}$$

where L_0 is a differential operator of the SL type, λ_{k0} is a constant related to the frequency of the normal mode, and w_0 is the weighting function. The solution u_{k0} is subject to the SL boundary conditions, and we shall assume that all of the u_{k0}'s have been orthogonalized and normalized; that is

$$\int_a^b u_{k0} u_{m0} w_0 \, dx = \delta_{km} \tag{C-2}$$

Now suppose that the vibrating object is perturbed in such a way that L_0 is changed to $(L_0 + \gamma L)$ (which need no longer be of the SL type), and w_0 is changed to $(w_0 + \gamma w)$, where γ is a parameter which is independent of x, but which will be considered as variable. The new wave equation becomes

$$(L_0 + \gamma L) u_n = \lambda_n (w_0 + \gamma w) u_n \tag{C-3}$$

The parameter γ is introduced because in principle any perturbation can be applied in small steps, and the wave functions and eigenvalues for one value of γ must pass continuously into those belonging to another value of γ. In this way a "family tree" can be traced out for each level of the perturbed system. The original ancestor can be formed by allowing γ to approach zero; then the solutions and eigenvalues of (C-3) approach the solutions and eigenvalues of (C-1). If γ is small, we should expect a given perturbed solution u_n to resemble a particular one of the unperturbed solutions of (C-1). Let us suppose that this unperturbed solution is u_{n0} and that the eigenvalue corresponding to λ_n is λ_{n0}. Then it is natural to expand u_n and λ_n in a power series in γ, as follows

$$u_n = u_{n0} + \gamma u'_n + \gamma^2 u''_n + \cdots \tag{C-4}$$

$$\lambda_n = \lambda_{n0} + \gamma \lambda'_n + \gamma^2 \lambda''_n + \cdots \tag{C-5}$$

where u'_n and u''_n are functions of x, and λ'_n and λ''_n are constants. We shall assume that the mode u_{n0} is nondegenerate (the theory for degenerate modes will be given below).

It is convenient to make use of the fact that the u_{n0}'s form a complete set, so that we can expand u'_n, u''_n, etc., in terms of the u_{k0}'s

$$u'_n = \sum A_k u_{k0} \tag{C-6}$$

$$u''_n = \sum B_k u_{k0} \tag{C-7}$$

The problem then becomes simply one of finding the values of the constants A_k, B_k, etc., as well as those of λ'_n, λ''_n, etc.* This is done by substituting Equations (C-4) through (C-7) into (C-3), making use of (C-1), and collecting coefficients of the various powers of γ. This leads to the following result:

$$\begin{aligned}&(L_0 u_{n0} - \lambda_{n0} w_0 u_{n0})\gamma^0\\ &+ [\sum A_k(\lambda_{k0} - \lambda_{n0}) w_0 u_{k0} + (L - \lambda_{n0} w) u_{n0} - \lambda'_n w_0 u_{n0}]\gamma^1\\ &+ [\sum B_k(\lambda_{k0} - \lambda_{n0}) w_0 u_{k0} + \sum A_k(L - \lambda_{n0} w - \lambda'_n w_0) u_{k0} - (\lambda''_n w_0 + \lambda'_n w) u_{n0}]\gamma^2\\ &+ \cdots = 0\end{aligned} \tag{C-8}$$

Since this result must be true for all values of γ, the coefficients of each power of γ must vanish. The coefficient of γ^0 vanishes because of (C-1). From the coefficient of γ^1,

$$\sum A_k(\lambda_{k0} - \lambda_{n0}) w_0 u_{k0} + (L - \lambda_{n0} w) u_{n0} - \lambda'_n w_0 u_{n0} = 0 \tag{C-9}$$

Multiplying (C-9) by u_{n0} and integrating, we find, because of the orthogonality of the u_{k0}'s, Equation (C-2), that the *first order perturbation of the eigenvalue* is given by

$$\lambda'_n = \int_a^b u_{n0}(L - \lambda_{n0} w) u_{n0}\, dx = \mathscr{L}_{nn} \tag{C-10}$$

where the symbol $\mathscr{L}_{km}$ is defined by

$$\mathscr{L}_{km} = \int_a^b u_{k0}(L - \lambda_{n0} w) u_{m0}\, dx \tag{C-11}$$

Multiplying (C-9) by u_{j0} and integrating, we find similarly that *the first order perturbation of the nth normal mode* is determined by

$$A_j = \frac{1}{\lambda_{n0} - \lambda_{j0}} \int_a^b u_{j0}(L - \lambda_{n0} w) u_{n0} dx = \mathscr{L}_{jn}/(\lambda_{n0} - \lambda_{j0}) \text{ if } j \neq n \qquad \text{(C-12(}$$

* Equations (C-6) and (C-7) are readily seen to be generalized expressions of the hybridization concept discussed on page 66, the coefficients A_k, B_k, $\cdots$ giving the degree of mixing of the kth level with the nth level as a result of the perturbation.

It will be shown below that if u_n is to be normalized, then

$$A_n = -\tfrac{1}{2}\int_a^b u_{n0}^2 w dx \tag{C-12a}$$

From the coefficient of γ^2 we find

$$\sum B_k(\lambda_{k0} - \lambda_{n0})w_0 u_{k0} + \sum A_k(L - \lambda_{n0}w - \lambda'_n w_0)u_{k0} - (\lambda''_n w_0 + \lambda'_n w)u_{n0} = 0 \tag{C-13}$$

Multiplying this by u_{n0} and integrating as above, we find for the *second order perturbation of the eigenvalue*

$$\lambda''_n = \sum_{k \neq n} \frac{\mathscr{L}_{kn}\mathscr{L}_{nk}}{\lambda_{n0} - \lambda_{k0}} - \mathscr{L}_{nn}w_{nn} \tag{C-14}$$

where w_{km} is defined as

$$w_{km} = \int_a^b u_{k0} w u_{m0}\, dx \tag{C-15}$$

Similarly, on multiplying (C-13) by u_{j0} and integrating, we find that the *second order perturbation of the normal mode* is determined by

$$B_j = \sum_{k \neq n} \frac{\mathscr{L}_{kn}\mathscr{L}_{jk}}{(\lambda_{n0} - \lambda_{k0})(\lambda_{n0} - \lambda_{j0})} - \mathscr{L}_{nn}w_{jn} - \tfrac{1}{2}\mathscr{L}_{jn}w_{nn} - \frac{\mathscr{L}_{jn}\mathscr{L}_{nn}}{\lambda_{n0} - \lambda_{j0}} \quad \text{if } j \neq n \tag{C-16}$$

If u_n is to be normalized, then we shall find below that

$$B_n = -(\tfrac{1}{2})\sum_k A_k\,[A_k + 2w_{kn}] \tag{C-16a}$$

The higher order perturbations can be found, if necessary, by continuing the procedure indicated above.

To prove that A_n and B_n are given by (C-12a) and (C-16a) we evaluate the integral

$$\int_a^b u_n^2(w_0 + \gamma w)\, dx = 1 + \gamma[w_{nn} + 2A_n] + \gamma^2[\sum A_k(A_k + 2w_{kn}) + 2B_n] + \cdots \tag{C-17}$$

If this integral is to be normalized for all values of γ, the coefficients of γ and γ^2 must vanish. This gives (C-12a) and (C-16a).

In order to illustrate the application of these results, let us consider a stretched string originally uniform in density, ρ_0, and tension, T_0. A perturbation in the form of a uniform increase in density, ρ', and a uniform increase in tension, T', is applied. Then the normal mode equation for the unperturbed string is (*cf.* Equation (A-22) of Chapter 3)

$$T_0 \frac{d^2 f_{n0}}{dx^2} = -4\pi^2 \nu_{n0}^2 \rho_0 f_{n0}$$

whereas in the presence of the perturbation, it is

$$(T_0 + T') \frac{d^2 f_n}{dx^2} = -4\pi^2 \nu_n^2 (\rho_0 + \rho') f_n$$

We see that

$$L_0 = T_0 \, d^2/dx^2$$

$$\gamma L = T' \, d^2/dx^2$$

$$w_0 = \rho_0$$

$$\gamma w = \rho'$$

$$\lambda_{n0} = -4\pi^2 \nu_{n0}^2 = -\pi^2 n^2 T_0 / a^2 \rho_0$$

$$\lambda_n = -4\pi^2 \nu_n^2$$

$$f_{n0} = N_n \sin n\pi x/a$$

where N_n is a normalization constant such that

$$\int_0^a f_{n0}^2 w_0 \, dx = 1 = N_n^2 \int_0^a \sin^2 n\pi x/a \; \rho_0 \, dx = N_n^2 \, \rho_0 \, a/2$$

This gives

$$f_{n0} = (2/a\rho_0)^{1/2} \sin n\pi x/a$$

Substituting these functions into the appropriate integrals, we find

$$\gamma \lambda'_n = \frac{n^2 \pi^2}{a^2} \frac{T_0}{\rho_0} \left(\frac{\rho'}{\rho_0} - \frac{T'}{T_0} \right)$$

$$\gamma A_k = 0 (k \neq n); \; \gamma A_n = -\tfrac{1}{2} \rho'/\rho_0$$

$$\gamma^2 \lambda''_n = \frac{n^2 \pi^2}{a^2} \frac{T_0}{\rho_0} \left[\frac{T' \rho'}{T_0 \rho_0} - \left(\frac{\rho'}{\rho_0} \right)^2 \right]$$

$$\gamma^2 B_k = 0 (k \neq n); \; \gamma^2 B_n = \tfrac{3}{8} \left(\frac{\rho'}{\rho_0} \right)^2$$

Up to terms in the second order of the perturbation, we thus have

$$\nu_n^2 = \nu_{n0}^2 \, [1 + T'/T_0 - \rho'/\rho_0 + (\rho'/\rho_0)^2 - \rho' T'/\rho_0 T_0]$$

$$f_n = \sqrt{\frac{2}{a\rho_0}} \, [1 - \tfrac{1}{2}\rho'/\rho_0 + {}^3/_8 (\rho'/\rho_0)^2] \sin n\pi x/a$$

The exact solution is (*cf.* pages 61-62)

$$\nu_n^2 = \frac{n^2}{4a^2}\frac{T_0 + T'}{\rho_0 + \rho'}$$

$$f_n = \sqrt{\frac{2}{a(\rho_0 + \rho')}}\sin n\pi x/a$$

which on Taylor's series expansion in powers of ρ' and T' is found to agree with the perturbation theory up to terms of the second order.

Exercises. (1) Find the third order corrections to λ_n and u_n.

(2) Find the first order change in the frequency of a stretched string of length a if the density of the string is changed from ρ_0 to ρ_0 $(1 - ex)$, where $e \ll 1/a$ and x is the distance along the string measured from one end.

b. Perturbation theory for degenerate modes

We have seen on page 82 that when a perturbation is applied to a degenerate group of modes, the correlation can only be carried through if we choose a particular combination of the degenerate modes out of an infinite number of possible combinations. The problem for degenerate modes is thus more complicated than that treated above since the proper linear combination of the degenerate unperturbed functions must be determined, in addition to the functions u'_n, u''_n, $\cdots$, and the constants λ'_n, λ''_n, $\cdots$, in Equations (C-4) and (C-5).

Suppose that the unperturbed normal mode equation is

$$L_0 u_{k0} = \lambda_{k0} w_0 u_{k0} \tag{C-18}$$

and that a group of l functions among the u_{k0}'s have the same eigenvalue, λ_{n0}. We shall refer to these l-fold degenerate modes as $u_{10}, u_{20}, \cdots, u_{l0}$, and we shall assume that they have been orthogonalized and normalized. The perturbed case then gives the normal mode equation

$$(L_0 + \gamma L)v_n = \lambda_n(w_0 + \gamma w)v_n \tag{C-19}$$

As γ approaches zero, v_n approaches a particular linear combination or hybrid of the u_{j0}'s, say

$$v_{n0} = \sum_{j=1}^{l} c_j u_{j0} \tag{C-20}$$

(There are, in fact, l different combinations of this sort, each corresponding to one of the modes into which the degenerate group is split by the perturbation.) Thus we may write

$$v_n = \sum_{j=1}^{l} c_j u_{j0} + \gamma V'_n + \cdots \tag{C-21}$$

$$\lambda_n = \lambda_{n0} + \gamma \lambda'_n + \cdots \tag{C-22}$$

$$V'_n = \sum_{\substack{\text{all}\\\text{modes}}} A_k u_{k0} \tag{C-23}$$

Substituting these in (C-19) and arranging terms according to powers of γ as before, we find for the coefficient of γ^0,

$$L_0 \sum_{j=1}^{l} c_j u_{j0} - \lambda_{n0} w_0 \sum_{j=1}^{l} c_j u_{j0} = \sum_{j=1}^{l} c_j (L_0 - \lambda_{n0} w_0) u_{j0} = 0 \tag{C-24}$$

which we already know to be true from (C-18).

From the coefficient of γ^1 we find

$$\sum_{\substack{\text{all}\\\text{modes}}} (\lambda_{k0} - \lambda_{n0}) w_0 A_k u_{k0} + L \sum_{j=1}^{l} c_j u_{j0} - (\lambda_{n0} w + \lambda'_n w_0) \sum_{j=1}^{l} c_j u_{j0} = 0 \tag{C-25}$$

If this is multiplied by one of the degenerate modes, say u_{i0}, and integrated we find

$$\sum_{j=1}^{l} c_j (\mathscr{L}_{ij} - \delta_{ij} \lambda'_n) = 0 \tag{C-26}$$

where δ_{ij} is the Kronecker delta function and $\mathscr{L}_{ij}$ is defined by Equation (C-11). Repeating this procedure with each of the l degenerate modes, we obtain a set of l linear equations in $l + 1$ unknowns (*viz.*, the l coefficients, c_j, plus λ'_n). Written out, these have the form

$$\begin{aligned}
(\mathscr{L}_{11} - \lambda'_n) c_1 + \mathscr{L}_{12} c_2 + \mathscr{L}_{13} c_3 + \cdots + \mathscr{L}_{1l} c_l &= 0 \\
\mathscr{L}_{21} c_1 + (\mathscr{L}_{22} - \lambda'_n) c_2 + \mathscr{L}_{23} c_3 + \cdots + \mathscr{L}_{2l} c_l &= 0 \\
\vdots \qquad\qquad \vdots \qquad\qquad \vdots \qquad\qquad \vdots \quad & \\
\mathscr{L}_{l1} c_1 + \mathscr{L}_{l2} c_2 + \mathscr{L}_{l3} c_3 + \cdots + (\mathscr{L}_{ll} - \lambda'_n) c_l &= 0
\end{aligned} \tag{C-27}$$

By the argument used on page 127, we see that nonvanishing values of the

c_j's can be obtained only if the determinant of the coefficients of the c_j's in (C-27) vanishes,

$$\begin{vmatrix} (\mathscr{L}_{11} - \lambda'_n) & \mathscr{L}_{12} & \mathscr{L}_{13} \cdots & \mathscr{L}_{1l} \\ \mathscr{L}_{21} & (\mathscr{L}_{22} - \lambda'_n) & \mathscr{L}_{23} \cdots & \mathscr{L}_{2l} \\ \vdots & \vdots & \vdots & \vdots \\ \mathscr{L}_{l1} & \mathscr{L}_{l2} & \mathscr{L}_{l3} & (\mathscr{L}_{ll} - \lambda'_n) \end{vmatrix} = 0 \tag{C-28}$$

In this determinant everything is known except λ'_n. Equation (C-28) *is therefore a means of determining λ'_n, which in turn determines the first order correction to the energy*. It is known as the *secular equation*. The determinant, on expansion, will give a polynomial of the lth degree in λ'_n, and therefore has l roots, which we shall denote by $\lambda'_{n1}, \lambda'_{n2}, \lambda'_{n3}, \cdots, \lambda'_{nl}$. This corresponds to the removal of the degeneracy by the perturbation. (If some of the roots are equal, some degeneracy of course remains, to a first approximation.)

We can now return to (C-27) and solve for the c_j's. The value of one of the roots, λ'_{ni}, is substituted in (C-27) and all equations are divided by c_l. This gives l equations in the $l - 1$ ratios (c_j/c_l) which can be solved with no essential difficulties, giving, from (C-20), for one of the correct "zero order" linear combinations of degenerate modes

$$v_{ni0} = c_l \sum_{j=1}^{l} (c_j/c_l)_{\lambda'_{ni}} u_{j0} \tag{C-29}$$

The value of c_l itself may be found from the normalization condition

$$\int_a^b v_{ni0}^2 w_0 dx = \sum_{j=1}^{l} c_j^2 = 1 = c_l^2 \sum_{j=1}^{l} (c_j/c_l)^2 \tag{C-30}$$

This procedure is repeated for each of the l roots, λ'_{ni}, of the secular equation, giving l sets of the coefficients c_j. In this way all of the correct zero order combinations may be found.

It is important to note that if the secular equation should happen to be "diagonalized," that is, if it happens to have the form

$$\begin{vmatrix} (\mathscr{L}_{11} - \lambda'_n) & 0 & 0 \cdots & 0 \\ 0 & (\mathscr{L}_{22} - \lambda'_n) & 0 \cdots & 0 \\ \vdots & \vdots & \vdots & \vdots \\ 0 & 0 & 0 \cdots & (\mathscr{L}_{ll} - \lambda'_n) \end{vmatrix} = 0 \tag{C-31}$$

then the roots are given directly by $\lambda'_n = \mathscr{L}_{11}, \mathscr{L}_{22}, \mathscr{L}_{33}, \cdots, \mathscr{L}_{ll}$. Furthermore, substitution of, say, $\lambda'_n = \mathscr{L}_{11}$ into the Equations (C-27) gives in this case

$$\begin{array}{lllll}
(\mathscr{L}_{11} - \mathscr{L}_{11})c_1 + & 0\, c_2 & + \cdots + & 0\, c_l & = 0 \\
0\, c_1 & + (\mathscr{L}_{22} - \mathscr{L}_{11})c_2 & + \cdots + & 0\, c_l & = 0 \\
\vdots & \vdots & & \vdots & \\
0\, c_1 \quad + & 0\, c_2 & + \cdots + & (\mathscr{L}_{ll} - \mathscr{L}_{11})c_l & = 0
\end{array}$$

which is satisfied by $c_1 \neq 0$ and all other c_i's $= 0$ as long as $\mathscr{L}_{11}$ is different from all other $\mathscr{L}_{ii}$. Similarly, each of the other nondegenerate roots $\lambda'_n = \mathscr{L}_{jj}$ gives $c_j \neq 0$ and the other c_i's $= 0$. Thus we find that we have in this case been lucky enough to have stumbled onto the correct zero order functions at the very beginning. Only if some of the roots happen to be equal do we have any possibility that for a given value of λ'_n more than one c_i is not zero. But in that case some degeneracy remains after perturbation and there is no unique combination of the persistently degenerate group, anyway.

Evidently everything is made much simpler if we choose the original degenerate functions so that the secular equation is as nearly diagonal as possible, if not completely diagonal. Some systematic methods of accomplishing this will be described in the next section of this chapter.

The values of the coefficients A_k in (C-23) may be found from (C-25) by multiplying by u_{k0} and integrating, and higher order perturbations may also be found by methods similar to those used in the nondegenerate case. (The student can do this as an exercise.)

The perturbation method that has been outlined above for both nondegenerate and degenerate modes is really a generalized procedure for ascertaining the extent of hybridization of normal modes resulting from a perturbation. It is interesting to note that, because of the occurence of quantities such as $(\lambda_{no}-\lambda_{jo})$ in the denominators of Equations (C-12) and (C-16), hybridization tends to be most extensive (A_k and B_k large) between modes whose eigenfrequencies are close together (λ_{no} close to λ_{jo} and λ_{ko}). In the limit $\lambda_{no} = \lambda_{jo} = \lambda_{ko}$ we have the case of degeneracy, and hybridization among the degenerate levels is already complete (Equation (C-20)) when the perturbation is still infinitesimally small.

Exercises. (1) Suppose that the depth, h, of the ocean on a flooded planet varies with the co-latitude, θ, according to the relation

$$h = h_0(1 + e \sin \theta)$$

where e is a constant small compared to unity.

(a) From Equation (B-36) of Chapter 3 show that the normal mode equation for such an ocean can be written in the form

$$(L_0 + L)f(\theta, \varphi) = -\lambda w_0 f$$

where

$$L_0 = (gh_0/R^2)[\partial/\partial\theta(\sin\theta\,\partial/\partial\theta) + (1/\sin\theta)\partial^2/\partial\varphi^2]$$

$$L = (egh_0/R^2)[\partial/\partial\theta(\sin^2\theta\,\partial/\partial\theta) + \partial^2/\partial\varphi^2]$$

$$w_0 = \sin\theta$$

$$\lambda = 4\pi^2\nu^2$$

$$\nu = \text{normal mode oscillation frequency}$$

Note that L_0 may be regarded as an SL operator with respect to each of the two coordinates, θ and φ.

(b) For the unperturbed case ($e = 0$), the fundamental mode is known to be triply degenerate with a frequency given by

$$4\pi^2\nu^2 = \lambda_0 = 2gh_0/R^2$$

(see page 89). We may take for the three normal mode functions corresponding to this frequency the normalized orthogonal set, $u_{10} = (3/4\pi)^{1/2}\cos\theta$, $u_{20} = (3/4\pi)^{1/2}\sin\theta\cos\varphi$, $u_{30} = (3/4\pi)^{1/2}\sin\theta\sin\varphi$. Show that

$$\mathscr{L}_{11} = \int_0^{\pi}\int_0^{2\pi} u_{10}Lu_{10}d\varphi d\theta = -\frac{9\pi gh_0}{16R^2}e$$

and that, similarly

$$\mathscr{L}_{12} = \mathscr{L}_{21} = \mathscr{L}_{13} = \mathscr{L}_{31} = \mathscr{L}_{23} = \mathscr{L}_{32} = 0$$

and

$$\mathscr{L}_{22} = \mathscr{L}_{33} = -\frac{15\pi gh_0}{32R^2}e$$

(c) Use these integrals to set up the secular equation. Show that the perturbation $h_0 e\sin\theta$ removes part of the degeneracy of the fundamental mode and that to a first approximation we obtain one nondegenerate mode that resembles u_{10}, for which

$$4\pi^2\nu^2 = \left(2 - \frac{9\pi}{16}e\right)gh_0/R^2$$

and a degenerate pair resembling u_{20} and u_{30}, for which

$$4\pi^2\nu^2 = \left(2 - \frac{15\pi}{32}e\right)gh_0/R^2$$

(2) Find the first approximation to the frequencies of the three lowest normal modes of a flooded planet if the depth of the ocean is given by

$$h = h_0(1 + e \cos 2\varphi)$$

Begin with the degenerate functions u_{10}, u_{20}, and u_{30} given in the previous exercise.

D. The Use of Symmetry and Commutation Properties in the Variation and Perturbation Methods

In Sections B and C of this chapter we have obtained many expressions involving integrals of the type

$$\int_a^b \psi \alpha \phi d\tau$$

where ψ and ϕ are functions, and α is an operator. The secular equation, (C-28), contains a great number of these integrals, and they also occur in (B-2), (B-15), (B-21), (C-10), (C-12), (C-14), and (C-16). In some systems, especially those containing a certain amount of geometrical symmetry, many of these integrals either vanish or can be made to vanish by the proper selection of ψ and ϕ. Since this may result in a significant simplification of the problem, it is worth looking into some of the factors that can make these integrals vanish. We shall here consider two of these factors, symmetry and commutation.

a. Effect of symmetry

Consider the integral

$$\int_a^b \psi \alpha \phi d\tau \tag{D-1}$$

Any definite integral such as this is a number whose value is independent of the coordinate system used in its evaluation. Therefore, *any transformation of coordinates in (D-1) must leave the value of the integral unchanged.* (Theorem I.)

The boundary conditions of many vibrating systems show a certain amount of symmetry. For instance, the edges of a square membrane are unchanged by the following operations:

$$C_2 = \text{rotation by } 180° \text{ about the axis } a \text{ in Fig. 4-1}$$

$$C_4 = \text{rotation by } 90° \text{ about } a$$

$$C_4{}^3 = \text{rotation by } 270° \text{ about } a$$

$$C'_{21}, C'_{22} = \text{rotation by } 180° \text{ about } b \text{ and } c\text{, respectively}$$

$$C''_{21}, C''_{22} = \text{rotation by } 180° \text{ about } d \text{ and } e\text{, respectively}$$

$$\sigma_v, \sigma'_v, \sigma_d, \sigma'_d = \text{reflection in the vertical planes including } a \text{ and } b,\ a \text{ and } c,\ a \text{ and } d\text{, and } a \text{ and } e\text{, respectively}$$

$$i = \text{inversion through the center}$$

Each of these symmetry operations, when applied to a function, corresponds to a transformation of the coordinates in the function (*cf.* discussion of

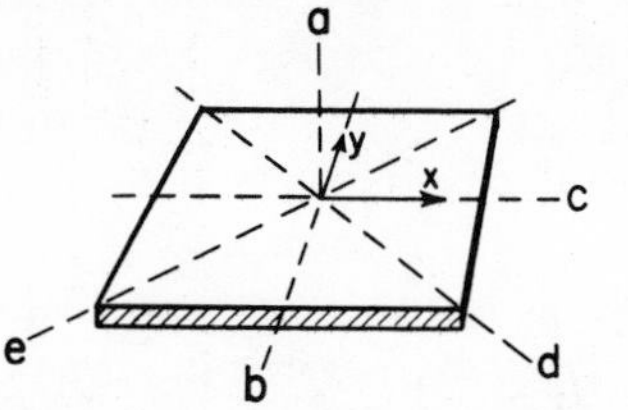

FIG. 4-1. Symmetry elements for a square.

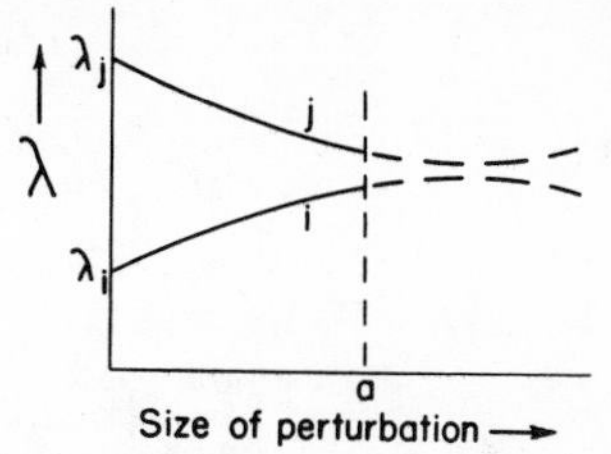

FIG. 4-2. The noncrossing rule.

symmetry operations on pp. 14-15). For instance, if we begin with Cartesian coordinates having their origin at the center of the membrane and the x and y axes in the plane of the membrane and parallel to the edges, we see that

C_4 means "$x \to y,\ y \to -x$"

σ_v means "$x \to -x,\ y \to y$"

C'_{21} means "$x \to -x,\ y \to y$"

i means "$x \to -x,\ y \to -y$," etc.

Exercise. What are the transformations of cylindrical coordinates corresponding to C_4, C_2, C''_{21}, σ_d, i?

For the square membrane we have, because of Theorem I,

$$\int_{-a/2}^{a/2}\int_{-a/2}^{a/2} (C_4\psi)\,(C_4\alpha)\,(C_4\phi)dxdy = \int_{-a/2}^{a/2}\int_{-a/2}^{a/2} \psi\alpha\phi d\tau$$

Now suppose that we are interested in the integral

$$I = \int_{-a/2}^{a/2}\int_{-a/2}^{a/2} f_{22}\mathscr{L} f_{33} dxdy$$

where f_{22} and f_{33} are the two normal modes illustrated in Fig. 3-10, having $(m,n) = (2,2)$ and $(3,3)$, respectively, and where $\mathscr{L}$ represents the perturbation mentioned on page 81 – namely, the addition of weight symmetrically to the four corners of the membrane. ($\mathscr{L}$ is the function of x and y illustrated in Fig. 3-15. It is constant in the shaded regions in the corners and zero elsewhere.) Inspection of Figs. 3-10 and 3-15 will show that

$$C_4 f_{22} = - f_{22}$$
$$C_4 \mathscr{L} = \mathscr{L}$$
$$C_4 f_{33} = f_{33}$$

Therefore,

$$I = \int_{-a/2}^{a/2}\int_{-a/2}^{a/2} (C_4 f_{22})\,(C_4\mathscr{L})\,(C_4 f_{33})\,dxdy$$

$$= - \int_{-a/2}^{a/2}\int_{a-/2}^{a/2} f_{22}\mathscr{L} f_{33}\,dxdy = - I$$

Clearly the integral I must vanish.

This result may be generalized as follows. *If for any symmetry operator R that leaves the boundary conditions unchanged, we have*

$$R\psi = A_1\psi$$
$$R\alpha = A_2\alpha$$
$$R\phi = A_3\phi$$

then

$$\int_a^b \psi\alpha\phi\,d\tau = 0$$

unless

$$A_1 A_2 A_3 = +1$$

The integral is zero unless the integrand remains unchanged by all operations that have no effect on the limits of the integral. (Theorem II.)

The usefulness of this result may be illustrated by an example. In the perturbation of the square membrane leading to a rounding of its corners and ending ultimately in a round membrane, the selection of the proper zero order combination of degenerate modes is not always obvious. The degenerate modes (1,3) and (3,1) are an example. We could apply the perturbation theory for degenerate modes to these two functions, but it is possible to proceed more directly. We note on looking at these modes (Fig. 3-9) that

$$C_4 f_{31} = f_{13}$$

$$C_4 f_{13} = f_{31}$$

Also, as stated above

$$C_4 \mathscr{L} = \mathscr{L}$$

Thus neither f_{31} nor f_{13} are eigenfunctions of C_4, but we see that

$$g_1 = f_{31} + f_{13}$$

and

$$g_2 = f_{31} - f_{13}$$

are eigenfunctions

$$C_4 g_1 = g_1$$

$$C_4 g_2 = -g_2$$

Therefore if we use g_1 and g_2 instead of f_{31} and f_{13} as our initial wave functions we will be able to apply theorem II. In fact

$$\int g_1 \mathscr{L} g_2 \, d\tau = \int g_2 \mathscr{L} g_1 \, d\tau = 0$$

since the product of the eigenvalues is not $+1$. Therefore our secular equation is

$$\begin{vmatrix} \mathscr{L}_{11} - \lambda'_n & 0 \\ 0 & \mathscr{L}_{22} - \lambda'_n \end{vmatrix} = 0$$

the roots are

$$\lambda'_{n1} = \int g_1 \mathscr{L} g_1 \, d\tau$$

$$\lambda'_{n2} = \int g_2 \mathscr{L} g_2 \, d\tau$$

and g_1 and g_2 are the correct zero order wave functions.

Symmetry properties are not only useful in solving secular equations, but they also provide a simple clue to the unravelling of the correlation diagram in many instances. This is most easily seen by referring to the correlation diagram for round and square membranes (Fig. 3-14). Evidently all of the normal modes of both the circular and the square membranes are eigenfunctions of C_2, the operation of rotating the membrane by 180° about an axis through the center of the membrane and perpendicular to the plane of the membrane. The eigenvalues are either $+1$ or -1 in all cases; the value is indicated in the columns labelled "Symmetry" on either side of the diagram, the letter g being used if the eigenvalue is $+1$, and u if it is -1.

The perturbation is also symmetrical with respect to C_2 (in fact $C_2\mathscr{L} = \mathscr{L}$), so we may expect that all perturbed modes will also be eigenfunctions of C_2. Since symmetry properties cannot change abruptly on the gradual application of a perturbation, the eigenvalue of C_2 must remain the same all of the way across the diagram. Thus g-states always go to g-states and u-states to u-states. In general, *each line in the correlation diagram preserves its eigenvalue with respect to any symmetry operation R, if R has no effect on the perturbation or on the boundary conditions.* (Theorem III.)

Another principle that is of great importance in drawing correlation diagrams is the following: *if in the correlation diagram the modes on either side are arranged in order according to their eigenvalues, then, with rare exceptions, no two correlation lines will ever cross if they belong to modes having the same symmetry properties.* (Theorem IV–called the "noncrossing rule.") This is not difficult to prove. Consider two nondegenerate correlation lines, i and j, (see Fig. 4-2, page 139) which are about to cross (i.e., two modes whose eigenvalues approach one another as the perturbation is changed). Let the perturbation at this point, a, be given by L_a, w_a, the complete normal mode equation being

$$(L_0 + L_a)u = \lambda(w_0 + w_a)u$$

Suppose that this equation is somehow solved precisely, giving a set of solutions and eigenvalues u_n and λ_n of which u_i and u_j and λ_i and λ_j belong to the two modes in question. Now let the perturbation be increased by L', w', and apply the perturbation theory for nondegenerate levels. The new eigenvalues will, according to Equations (C-5), (C-10), and (C-14), be given by

$$\lambda'_i = \lambda_i + (\mathscr{L}_i)_{ii} + (\mathscr{L}_i)_{ij}(\mathscr{L}_i)_{ji}/(\lambda_i - \lambda_j) + \text{(terms involving other states)} \quad \text{(D-2a)}$$

$$\lambda'_j = \lambda_j + (\mathscr{L}_j)_{jj} + (\mathscr{L}_j)_{ij}(\mathscr{L}_j)_{ji}/(\lambda_j - \lambda_i) + (\text{terms involving other states}) \quad \text{(D-2b)}$$

where

$$(\mathscr{L}_i)_{ij} = \int u_i(L' - \lambda_i w')u_j \, d\tau$$

and

$$(\mathscr{L}_j)_{ij} = \int u_i(L' - \lambda_j w')u_j \, d\tau$$

If the operator L' is Hermitian (and perturbations can always be found that are Hermitian and that will transform the system from one type to another), then

$$(\mathscr{L}_i)_{ij} = (\mathscr{L}_i)_{ji}$$

and

$$(\mathscr{L}_j)_{ij} = (\mathscr{L}_j)_{ji}$$

This means that the numerators of the third terms on the right of (D-2a) and (D-2b) are positive. But the denominator of this term in (D-2a) is negative, whereas that in (D-2b) is positive (since $\lambda_i < \lambda_j$ by definition). Therefore the perturbation will tend to decrease λ'_i and increase λ'_j, as far as these terms are concerned. It is clear that the closer λ_i and λ_j approach one another, the more this term will dominate the proceedings, so that unless the $(\mathscr{L})_{ij}$'s are zero the lines cannot cross. In fact, the lines can be said to "repel" one another in inverse proportion to their "distance" apart. One condition that the $(\mathscr{L})_{ij}$'s vanish is, however, that u_j and u_i have different symmetry properties. It is possible for the $(\mathscr{L})_{ij}$'s to vanish even though the symmetry properties of u_i and u_j are the same, but this is unusual. If this should happen, there is generally some property other than a symmetry property that can be used in establishing the correlation (see below).

Theorem IV makes it possible to correlate the states of two systems unequivocally, since there is only one way in which states of the same symmetry can be joined so that correlation lines of the same symmetry fail to cross. This is how the correlation lines were drawn in Fig. 3-14. In this case the behavior with respect to the operation C_2 alone is sufficient to fill in the figure, but the student can verify that each correlation joins modes identical in their behavior toward all of the symmetry operations that were mentioned on page 139. Although some of the correlation lines do cross, these never involve two g- or two u-states.

It should be evident that the effects of symmetry operations can be useful in certain aspects of vibration theory. Any further discussion of these important matters leads one into group theory, however, and will not be considered in this book. (See *3*, Chapter 10).

b. Effect of commuting operators

If in connection with the evaluation of the integral $\int \psi \alpha \phi d\tau$, we can find any Hermitian operator β, such that

$$\beta\psi = b_1\psi$$

$$\beta\phi = b_2\phi$$

$$\beta\alpha = \alpha\beta$$

then unless $b_1 = b_2$,

$$\int \psi\alpha\phi \, d\tau = 0$$

That is, if α and β commute, if β is Hermitian and if ψ and ϕ are eigenfunctions of β, the integral $\int \psi\alpha\phi d\tau$ will vanish unless the eigenvalues are the same. (Theorem V.)

The *proof* of this theorem is as follows. Because of the Hermitian character of β

$$\int \psi\beta\alpha\phi \, d\tau = \int (\alpha\phi)\beta\psi \, d\tau$$

But

$$\int (\alpha\phi)\beta\psi \, d\tau = b_1 \int (\alpha\phi)\psi \, d\tau = b_1 \int \psi\alpha\phi \, d\tau$$

and because α and β commute

$$\int \psi\beta\alpha\phi \, d\tau = \int \psi\alpha\beta\phi \, d\tau = b_2 \int \psi\alpha\phi \, d\tau$$

Thus

$$b_1 \int \psi\alpha\phi \, d\tau = b_2 \int \psi\alpha\phi \, d\tau$$

and unless $b_1 = b_2$ the integral must vanish. This theorem is of much more importance in quantum mechanics than in vibration theory, so we shall delay its further discussion.

Exercises. (1) Using the symmetry properties of a cube, correlate the first dozen normal modes of a gas in a cubical cavity with those of a gas in a spherical cavity, assuming that the walls in all instances are weightless.

(2) Correlate the first dozen normal modes of a gas in a cubical cavity which has infinitely heavy walls with those of a gas in a cubical cavity having weightless walls exerting a constant pressure on the gas adjacent to them.

E. The Interaction of Vibrating Systems

a. The effect of coupling on normal modes and frequencies

The frequencies and shapes of the normal modes of two completely independent violin strings are, of course, simply those of the separate systems (see Fig. 4-3). If the two strings are identical, each normal fre-

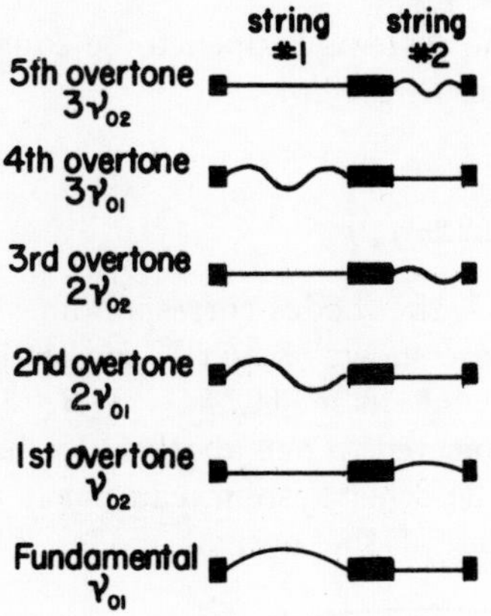

FIG. 4-3. Normal modes of a pair of stretched strings of unequal length.

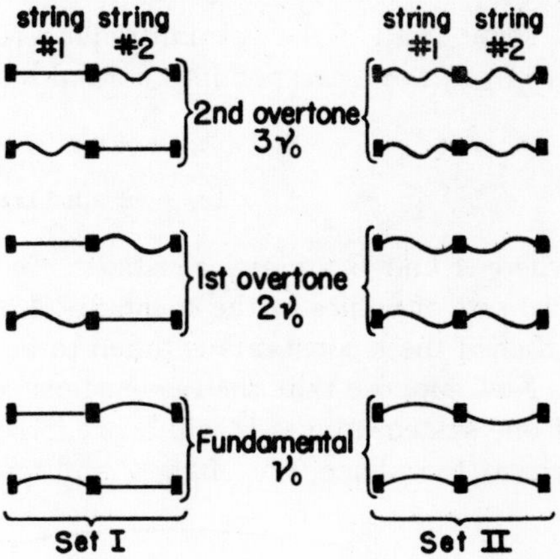

FIG. 4-4. Normal modes of an identical pair of stretched strings. Because of the pairwise degeneracy of the modes, various linear combinations of Set I can be taken, of which Set II is one.

quency of the combined system will be doubly degenerate and any combination of their normal modes will still be a normal mode (see Fig. 4-4). These results can be stated in mathematical terms as follows: let the set of functions

$$z_1 = \sin(n\pi x_1/a_1) \sin 2\pi\nu_{1n}t$$

represent the normal modes of system # 1, and

$$z_2 = \sin(n\pi x_2/a_2) \sin 2\pi\nu_{2n}t$$

represent those of system # 2 (here z_1 is the displacement of string # 1 and

z_2 is that of string # 2). Then the normal modes of the combined system are

$$\begin{Bmatrix} z_1 = 0 \\ z_2 = \sin(n\pi x_2/a_2) \sin 2\pi \nu_2{}^{nt} \end{Bmatrix}$$

and

$$\begin{Bmatrix} z_1 = \sin(n\pi x_1/a_1) \sin 2\pi \nu_{1n}t \\ z_2 = 0 \end{Bmatrix}$$

If some or all of the normal frequencies of the systems happen to be equal, however, the corresponding normal modes will be given by

$$\begin{Bmatrix} z_1 = A \sin(n\pi x_1/a_1) \sin 2\pi \nu_{1n}t \\ z_2 = B \sin(n\pi x_2/a_2) \sin 2\pi \nu_{1n}t \end{Bmatrix}$$

where A and B are any constants. In Fig. 4-4 the modes corresponding to two sets of values of the constants A and B are shown. In set I, one or the other of these constants is taken to be zero, while in set II, $A = \pm B$.

Now suppose that the two systems are connected to one another, so that if one system moves it will exert forces on the other system, causing it to move, too. Then the shapes and frequencies of the normal modes will

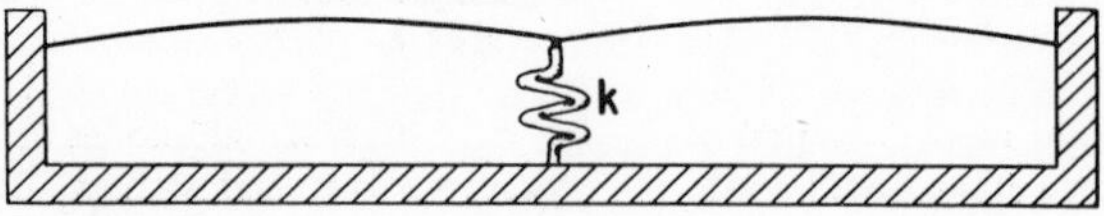

FIG. 4-5. Coupling of a pair of stretched strings by means of a spring whose force constant is k.

usually be changed. Such systems are said to be *coupled*. An example of two coupled strings is shown in Fig. 4-5, where the ends of the two strings are tied to a spring whose force constant is k. The strength of the coupling can be varied by varying k; a very stiff spring (large k) gives "weak coupling" (that is, the movement of one string has only a slight effect on the other), whereas a soft spring (small k) gives "strong coupling." What will the normal modes and frequencies of such a coupled system be?

This question is best considered from the point of view of the correlation diagram, since we already know the exact solutions for very soft and very stiff springs ($k = 0$ and ∞). A zero force constant in the coupling spring is equivalent to simply tying the two ends of the string together and results in normal modes that are those of a string whose length is the sum of the lengths of the original strings. The correlation diagram for a coupled pair of strings of unequal length is shown in Fig. 4-6, whereas that for a coupled pair of identical strings is shown in Fig. 4-7. In both instances the frequency of the fundamental mode is decreased by coupling. This is a general rule for any system of coupled vibrators. Suppose that one has a group of separate oscillators of which the lowest normal mode frequency is ν_0. If one then couples these oscillators in any way, the lowest frequency among the normal modes of the coupled system will be smaller than ν_0.

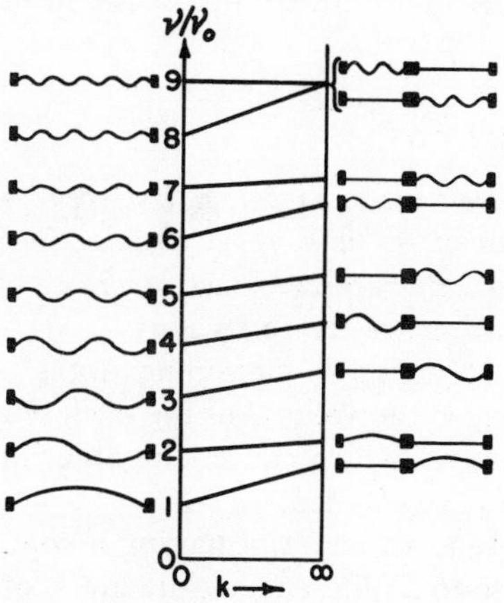

FIG. 4-6. Correlation diagram for the normal modes of a pair of non-identical stretched strings coupled by means of a spring as shown in Fig. 4-5 (force constant of spring = k). The lengths of the two strings are in the ratio 5:4.

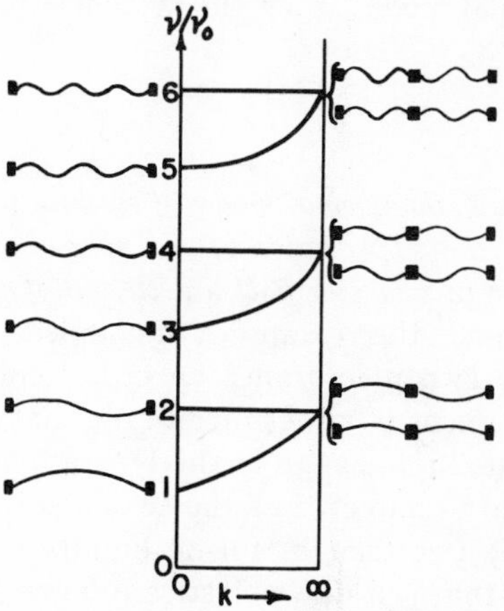

FIG. 4-7. Correlation diagram for the normal modes of a pair of identical stretched strings coupled by a spring as in Fig. 4-5 (force constant of spring = k).

The phenomenon of "resonance" in quantum mechanics is very similar to this behavior of coupled oscillators. Indeed the well-known stabilization of the ground state of a system by "resonance" between several structures arises mathematically in exactly the same way as the lowering of the fundamental frequency by the coupling of several oscillators.

In Fig. 4-7 it is obvious from the appearance of the modes on the left hand ($k = 0$) side that when two identical strings are coupled, the correct combinations of the degenerate uncoupled modes are

$$\text{Mode } a \begin{cases} z_1 = \sin(n\pi x_1/a)\sin 2\pi n\nu_0 t \\ z_2 = \sin(n\pi x_2/a)\sin 2\pi n\nu_0 t \end{cases}$$

$$\text{Mode } b \begin{cases} z_1 = \sin(n\pi x_1/a)\sin 2\pi n\nu_0 t \\ z_2 = -\sin(n\pi x_2/a)\sin 2\pi n\nu_0 t \end{cases}$$

These modes are illustrated on the right in Fig. 4-7. The frequencies of those modes that are antisymmetric with respect to reflection through the center of the system are unaffected by the coupling because such a mode exerts no force on the spring. The frequencies of the other, symmetrical modes are decreased by the coupling.*

The combinations of uncoupled degenerate modes that approximate to normal modes when the systems are coupled are invariably of the form

$$z_1 = \quad f_{1n} \sin 2\pi\nu_n t$$

$$z_2 = \pm f_{2n} \sin 2\pi\nu_n t$$

That is, *the normal modes of coupled identical objects have, to a first approximation, the appearance of the normal modes of the uncoupled objects vibrating either in phase or half a cycle out of phase with respect to each other.*

Some other examples of coupled vibrators are the following:

(*i*) Two membranes on either end of a cylinder containing a gas. Any movement of either membrane that changes the volume of the enclosed gas results in a change in the pressure of the gas, which causes the other membrane to move; thus the two membranes are coupled.

(*ii*) Two troughs full of liquid, joined end to end and having a common wall that can move elastically in response to a difference in the levels of the liquids on either side.

(*iii*) Two spheres full of gas, connected by a tube so that the changes in pressure of the gas at the mouth of the tube in one sphere are communicated to the other sphere.

(*iv*) Two overlapping spheres full of gas, whose walls are cut away in the region of overlap. (This provides an analogue to a diatomic molecule.)

Exercises. (1) Work out the correlation diagram for two identical troughs coupled by an elastic wall as in (*ii*), above. Allow the flexibility of the wall to vary from complete rigidity to complete freedom of motion (that is, no wall at all).

(2) Work out the correlation diagram for some of the lower modes of the gas in two

* The exact expression for the symmetric normal frequencies of the identical coupled strings can be shown to be given by the roots of the equation,

$$\tan \pi(\nu/\nu_0) = \sigma\pi(\nu/\nu_0)$$

where $\sigma = T/ka$, T is the tension on the string, a is the length of the string, and ν_0 is the fundamental frequency of the uncoupled strings.

It will be noticed that the correlations obtained here are not the same as those found in the similar problem of the addition of a weight to the center of the string (page 69). Thus the nature of the correlation can depend on the kind of perturbation by which one goes from one system to the other.

identical overlapping spheres as in (*iv*), above, considering the distance between the centers of the spheres to vary between 0 and $2a$, where a is the radius of one of the spheres. Assume that the walls of the spheres are very light, so that the pressure they exert on the gas is constant (that is, the case considered on page 106). Make use of the symmetry properties of the coupled system, especially with respect to inversion through the center of the line joining the centers of the spheres. The frequencies in an isolated sphere are as follows:

s-, *p*-, and *d*-modes: see Fig. 3-19

f-modes: $4f$, $2.22\ \nu_0$; $5f$, $3.33\ \nu_0$; $6f$, $4.35\ \nu_0$

g-modes: $5g$, $2.62\ \nu_0$; $6g$, $3.73\ \nu_0$; $7g$, $4.79\ \nu_0$

h-modes: $6h$, $2.99\ \nu_0$; $7h$, $4.13\ \nu_0$

s-, *d*-, *g*-, *i*-, etc., modes are symmetric, and *p*-, *f*-, *h*-, *j*-, etc., modes are antisymmetric with respect to inversion through the center of the sphere.

(3) Work out the correlation diagram for three identical strings coupled by means of a pair of springs as shown in Fig. 4-8.

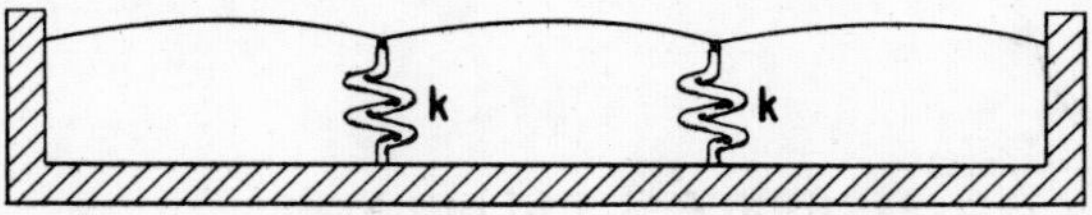

FIG. 4-8. Three identical stretched strings coupled by springs (see Exercise 3).

(4) Suppose that two identical oscillators are coupled and that the two lowest normal modes are simultaneously excited with equal amplitude, giving for the displacements of the oscillators,

$$z_1 = A(\sin 2\pi\nu_0 t + \sin 2\pi\nu'_0 t)$$
$$z_2 = A(\sin 2\pi\nu_0 t - \sin 2\pi\nu'_0 t)$$

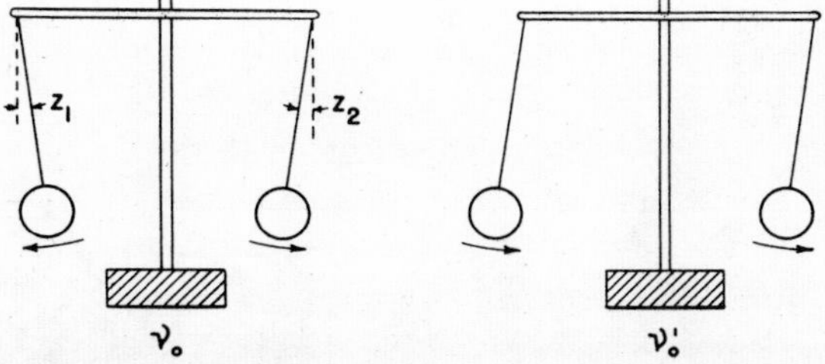

FIG. 4-9. Normal modes of an identical pair of pendulums coupled by means of a slightly flexible support.

where ν_0 and ν'_0 are the frequencies of the two normal modes, and A is a constant. Show that energy flows back and forth from one oscillator to the other with a frequency

$1/(\nu'_0 - \nu_0)$. This phenomenon is called "resonance" and is the source of the term as used in quantum chemistry, though quantum mechanical resonance is not strictly comparable. It is nicely demonstrated by hanging a pair of weights from a slightly flexible support, weak coupling being provided in this way (Fig. 4-9). The two normal modes are as shown in Fig. 4-9; this can be verified by the fact that, once excited, these motions will persist indefinitely. If, however, only one of the weights is set in motion at first, there is a gradual transfer of energy from this weight to the other. Eventually the other weight gets all of the energy, and the first weight becomes motionless. Then the energy gradually returns to the first weight. This shuttling of energy back and forth continues indefinitely.

REFERENCES CITED

1. H. Margenau and G. M. Murphy, "The Mathematics of Physics and Chemistry," Van Nostrand, New York, 1943.
2. R. Courant and D. Hilbert, "Methods of Mathematical Physics," Interscience, New York, 1953.
3. H. Eyring, J. Walter, and G. E. Kimball, "Quantum Chemistry," Wiley, New York, 1944.

GENERAL REFERENCES

See general references listed for Chapters 2 and 3.

R. Courant and D. Hilbert, reference *2*, also gives an extensive discussion of the methods described in this chapter.

PART II

GENERAL PRINCIPLES OF QUANTUM MECHANICS

Chapter 5

The Schroedinger Formulation of Quantum Mechanics

A. Some Fundamental Concepts used in Quantum Mechanics

In Chapter 1 it was mentioned that one proceeds in two stages when looking at nature from a scientific point of view. In the first stage one selects the concepts that are most convenient in describing the natural world. In the second stage one seeks the laws relating these concepts to one another. The discussion of quantum mechanics is therefore logically begun with a description of the concepts that it uses. These quantum mechanical concepts are of three types, involving the notions of dynamical variables, state functions, and operators.

a. The dynamical variables

The old classical mechanical concepts of time, position, velocity, mass, linear momentum, angular momentum, energy, etc., retain a central position in quantum mechanics. As in classical mechanics, they are regarded as measurable properties of any system. The behavior of these properties is, however, quite different from that predicted by classical mechanics. In addition, certain new properties appear that do not occur at all in classical mechanics – for instance, spin and symmetry with respect to interchanges of identical particles. These properties will be discussed later in the book.

It is interesting and significant, but hardly surprising, that quantum mechanics is so largely constructed out of concepts borrowed from the *ancien régime* of classical mechanics. Quantum mechanics therefore does not represent a completely new way of looking at nature; it is a historical development of classical mechanics. As with all revolutions, the quantum mechanical revolution is really an evolution, strongly conditioned by its antecedents.

b. The state function

Extremely important in quantum mechanics is the concept of a state function of a system. Suppose that we have a group of *n* particles, identified by numbers as particle 1, particle 2 , $\cdots$, particle n, and whose Cartesian coordinates are respectively $x_1, y_1, z_1; x_2, y_2, z_2; \cdots ; x_n, y_n, z_n$. Then we can conceive of a function $\Psi (x_1, y_1, z_1, x_2 \cdots , z_n, t)$ (which may be a complex function, that is, involving $\sqrt{-1}$), such that its absolute value squared is a measure of the probability of finding the particles in a particular configuration at a given time t. That is,

$$| \Psi (x_1, y_1, z_1, x_2, \cdots, z_n, t) |^2 \, dx_1 dy_1 dz_1 dx_2 \cdots dz_n \tag{a}$$

is the relative probability that at time t, particle 1 will be found between x_1 and $x_1 + dx_1$, y_1 and $y_1 + dy_1$, z_1 and $z_1 + dz_1$; particle 2 will be found between x_2 and $x_2 + dx_2$, etc. This means operationally that if we have many systems each containing n particles and represented by the same state function Ψ, and if we observe the locations of the particles in each system, then the relative frequencies of the different configurations are given by (a). Such a group of many systems, each containing n particles, is known as an *assembly of systems.*

The function $| \Psi |^2$ is a probability function of the type described in Chapter 2, Section E. It expresses all that we know about the configuration of any system in the assembly. The function Ψ can even be constructed so as to include uncertainties in our knowledge of the systems.

We begin to see that quantum mechanics is essentially statistical in outlook, and we shall not be surprised to find ourselves speaking of probabilities and average values, where classical mechanics dealt with much more precise descriptions of mechanical systems. This does not mean, however, that certainties cannot occur in quantum mechanics.

The student may be puzzled by our introduction of the state function, Ψ; its square, the probability function, is physically much more "real," so why must we insist on using Ψ? The reason is simply that the laws of quantum mechanics are more readily expressed in terms of the state function than in terms of the probability function. That is, nature seems to be constructed in such a way that the Ψ concept is particularly useful to us.

It is important to note that there is nothing exclusively quantum mechanical about the notion of a state function. For instance, we could devise a classical state function for the solar system that would contain predictions of the classical motions of the planets and that could include the effects of uncertainties in the experimental observations on which the predictions are based. The results of classical mechanical calculations are not, however, conveniently expressed in this way. Furthermore, an assembly of solar systems would be rather difficult to prepare.

In quantum mechanics, on the other hand, we usually deal with atomic and molecular systems. Observations are generally made on large numbers of systems, and the detailed motions of the particles in each atom or molecule are rarely subject to measurement. (We shall see in Chapter 7 that even when these detailed motions are followed, the very process of measurement so disturbs the system that the observer has to be included as a part of the system.) Thus it is most convenient that quantum mechanics makes use of a statistical approach.

Some characteristics of the state function. The state function Ψ is never observed directly; only its square can be measured in any experiment.

In defining Ψ we have left room for the occurrence of that peculiar quantity, $\sqrt{-1}$. This quantity does not occur in classical mechanics except where introduced for convenience in performing certain mathematical operations; whenever it is so introduced it can always be avoided if desired. This is not the case in quantum mechanics, where we shall find that $\sqrt{-1}$ is firmly embedded in both the concepts and the laws.

In systems containing a finite number of particles, it is obvious that the integral

$$I = \int_{\substack{\text{all}\\ \text{configurations}}} |\Psi|^2 \, dx_1 dy_1 dz_1 dx_2 \cdots dz_n \qquad \text{(A-1)}$$

must be a finite constant independent of the time, because the particles making up the system must always be somewhere in space. This means that *if there is a finite number of particles in a system, Ψ must vanish at infinity.* (For nuclear and cosmic processes in which particles are created or destroyed, the integral I need not be independent of the time, but such processes will not concern us.)

If normalized probabilities are used, then $I = 1$, and Ψ is said to be *normalized.*

In order to make any physical sense, Ψ must always be single-valued. That is, for any choice of the coordinates $x_1, \cdots, z_n$ there must be only one value of Ψ. This means that we cannot use such functions as $\Psi = \sin^{-1}x$ without including extra conditions that tell us which value of the function to choose for a given x.

In other words, *Ψ must be well behaved,* in the sense defined in Chapter 2, Section C.

c. Operators

The third type of concept used in quantum mechanics is that of operators associated with the classical dynamical variables. These associations are made in a very definite way as indicated in the following table:

Dynamical variable	Operator
(i) Coordinates: x,	"Multiplication by x," or $x\cdot$
y,	"Multiplication by y," or $y\cdot$
z,	"Multiplication by z," or $z\cdot$
(ii) Linear momentum:	
in x-direction, p_x	$(\hbar/i)\ \partial/\partial x$
in y-direction, p_y	$(\hbar/i)\ \partial/\partial y$
in z-direction, p_z	$(\hbar/i)\ \partial/\partial z$

where $i = \sqrt{-1}$ (again revealing its strange presence) and $\hbar$ is written for $h/2\pi$, h being Planck's constant, 6.625×10^{-27} erg. sec.

According to classical mechanics, all other dynamical variables may be expressed in terms of x, y, z, p_x, p_y, and p_z. The rule for setting up their quantum mechanical operators is simply to take the classical expression and replace x, y, z, p_x, p_y, p_z by the operators mentioned above. If the variable occurs raised to the nth power, this is taken to mean that the corresponding operation is to be repeated n times. For example,

$$x^2 \rightarrow \text{x}\cdot\text{x}\cdot = x^2\cdot$$

$$p_x{}^2 \rightarrow \left(\frac{\hbar}{i}\frac{\partial}{\partial x}\right)\left(\frac{\hbar}{i}\frac{\partial}{\partial x}\right) = -\hbar^2\frac{\partial^2}{\partial x^2}$$

(We can evidently use only expressions in which integral powers of the momenta occur.) Furthermore, only linear, Hermitian operators are permissible (see page 112 for the definition of a Hermitian operator). That is, for all quantum mechanical operators, α

$$\alpha(\phi+\psi) = \alpha\phi + \alpha\psi \quad (\alpha \text{ is linear})$$

and

$$\int_{\text{all space}} \phi\alpha\psi d\tau = \int_{\text{all space}} \psi\alpha^*\phi d\tau \quad (\alpha \text{ is Hermitian})$$

where ϕ and ψ are any functions that, with their derivatives, go to zero at infinity, and the star means that we are to take the conjugate complex of the quantity immediately to the left of the star.

The operator $x\cdot$ is obviously Hermitian.

That $p_x = (\hbar/i)\, \partial/\partial x$ is Hermitian is shown by the following

$$\int_{-\infty}^{\infty} \phi(x)\, \frac{\hbar}{i}\, \frac{\partial \psi(x)}{\partial x}\, dx = \frac{\hbar}{i}\, \phi\psi \Big|_{-\infty}^{\infty} - \int_{-\infty}^{\infty} \psi\, \frac{\hbar}{i}\, \frac{\partial \phi}{\partial x}\, dx$$

by the rules of partial integration. The first term on the right goes to zero since both ϕ and ψ vanish at infinity. The second term on the right is simply

$$\int \psi \left(\frac{\hbar}{i}\, \frac{\partial}{\partial x}\right)^{*} \phi\, dx$$

Two particularly important dynamical variables are the angular momentum about some axis (say the z-axis) and the energy. The classical expression for the angular momentum about the z-axis is

$$M_z = xp_y - yp_x$$

This leads to the *operator for the component of angular momentum* about the z-axis

$$M_z = \frac{\hbar}{i}\left(x\frac{\partial}{\partial y} - y\frac{\partial}{\partial x}\right) \tag{A-2}$$

which is easily seen to be both linear and Hermitian. Similar operators for angular momentum about the x- and y-axes are easily constructed.

The classical (so-called Hamiltonian) expression for the energy in terms of coordinates and linear momenta is, for a single particle of mass m

$$H = \frac{1}{2m}\,(p_x^2 + p_y^2 + p_z^2) + V(x, y, z, t) \tag{A-3}$$

where V is the potential energy of the particle.

The corresponding *energy operator* (or *Hamiltonian operator*)

$$H = -\frac{\hbar^2}{2m}\left(\frac{\partial^2}{\partial x^2} + \frac{\partial^2}{\partial y^2} + \frac{\partial^2}{\partial z^2}\right) + V(x\cdot, y\cdot, z\cdot, t\cdot) = -\frac{\hbar^2}{2m}\,\nabla^2 + V \tag{A-3a}$$

which is linear and Hermitian. If there are several particles, then

$$H = -\frac{\hbar^2}{2} \sum_{\substack{\text{all}\\ \text{particles}}} \nabla_j^2/m_j + V \tag{A-3b}$$

where m_j is the mass of the jth particle and ∇_j^2 is the Laplacian operator containing the coordinates of the jth particle.

Exercise. Show that H is linear and Hermitian. (Hint: Use partial integration twice repeated. *Cf.* page 113.)

Evidently the operators for x and p_x do not commute; that is

$$(x\cdot)\left(\frac{\hbar}{i}\frac{\partial}{\partial x}\right) \neq \left(\frac{\hbar}{i}\frac{\partial}{\partial x}\right)(x\cdot) \tag{A-4}$$

This, as we shall see, has many important consequences.

It should be mentioned that there is another way of setting up operators, called the "momentum method," in which the operator for momentum is p_x and the operator for position is $i\hbar\, \partial/\partial p_x$. This method will not be considered here (see *1*, Chapter 7).

For a much more general discussion of quantum mechanical operators, see Kemble (*2*, Chapter 7) or Dirac (*3*, Chapter 2).

B. The Laws of Quantum Mechanics

Scientific laws are relationships between concepts. In contrast to concepts, they are given to us by nature and we can at most merely rephrase them by expressing them in terms of different sets of concepts. The following are the laws of quantum mechanics in a form suitable for our purposes.

Law I. The possible state functions, Ψ, of a system are given by the solutions of the differential equation,

$$\boxed{H\Psi = -\frac{\hbar}{i}\frac{\partial \Psi}{\partial t}} \tag{I}$$

where H is the energy operator ("Hamiltonian operator") for the system. Written out for a system containing a single particle, we have

$$\boxed{-\frac{\hbar^2}{2m}\nabla^2\Psi + V(x, y, z, t)\Psi = -\frac{\hbar}{i}\frac{\partial \Psi}{\partial t}} \tag{Ia}$$

Equation I is called the *time-dependent Schroedinger equation.*

Law II. ***The only possible values that can be observed for a dynamical variable in an assembly of systems are the eigenvalues, λ, in the equation***

$$\boxed{\alpha\phi = \lambda\phi} \tag{II}$$

where α is the operator for the variable, and ϕ is any well-behaved eigenfunction of α.

Law III. ***When a great many measurements of any dynamical variable are made on an assembly of systems whose state function is Ψ, the average result obtained will be***

$$\boxed{\bar{a} = \int \Psi^* \alpha \Psi d\tau \Big/ \int \Psi^* \Psi d\tau} \tag{III}$$

where α is the operator corresponding to the dynamical variable, and the integration is over all configurations accessible to the system.

The symbol $d\tau$ stands for the volume element of the system. Using Cartesian coordinates, for a system containing n particles

$$d\tau = dx_1 dy_1 dz_1 dx_2 \cdots dx_n dy_n dz_n \tag{B-1}$$

Unless otherwise stated, all integrations in the rest of this book will be over all configurations of the system.

Law IV is the so-called Pauli Principle; it will be presented later (Chapter 9).

All of chemistry and much of physics are merely mathematical consequences of these four laws. The rest of this book will indicate how this comes about.

C. Some Important Corollaries of the Laws of Quantum Mechanics

a. Corollary I

Suppose that an assembly of systems has a state function Ψ which happens to be the eigenfunction for some dynamical variable, with eigenvalue λ. Then every measurement of that variable on each system in the assembly will yield precisely the value λ for the variable.

Proof. If α is the operator for the variable in question, then

$$\alpha\Psi = \lambda\Psi$$

From Law III the average value of the property for all systems is

$$\bar{a} = \int \Psi^* \alpha \Psi d\tau \bigg/ \int \Psi^* \Psi d\tau = \lambda \tag{C-1}$$

Furthermore

$$\alpha^2\Psi = \alpha(\alpha\Psi) = \alpha(\lambda\Psi) = \lambda(\alpha\Psi) = \lambda^2\Psi$$

because λ is a constant, and for any power n

$$\alpha^n\Psi = \lambda^n\Psi \tag{C-2}$$

Since the operator corresponding to the nth power of the variable is the operation for the variable repeated n times, the average value of the nth power of the variable for all systems in the assembly is

$$\overline{a^n} = \int \Psi^* \alpha^n \Psi d\tau \bigg/ \int \Psi^* \Psi d\tau = \lambda^n = (\bar{a})^n \tag{C-3}$$

for all integral values of n. This result is only possible if every measurement gave the same result, as can be seen from the following examples.

Consider the series of three numbers, 1, 1, 4. The average value of these numbers is $(1 + 1 + 4)/3 = 2$. The average of their squares is $(1 + 1 + 16)/3 = 6$. The average of their cubes is $(1 + 1 + 64)/3 = 22$. Clearly $6 \neq (2)^2$ and $22 \neq (2)^3$. It is easy to see that for any group of unequal numbers the average value of any power of the numbers is not equal to the average of the numbers raised to that power.

On the other hand, with any series of n equal numbers $a, a, a, a, \cdots, a$, the average is $(a + a + a + \cdots + a)/n = a = \bar{a}$. The average of the squares is $(a^2 + a^2 + a^2 + \cdots + a^2)/n = a^2$, which is the same as $(\bar{a})^2$. The average of the kth powers is $(a^k + a^k + a^k + \cdots + a^k)/n = a^k$, which is the same as $(\bar{a})^k$.

A system whose state function is an eigenfunction of the operator of some dynamical variable is said to be in an *eigenstate* of that variable. We shall be particularly interested in *energy eigenstates*, defined by

$$H\Psi = E\Psi \tag{C-4}$$

where E is the energy.

b. Steady states and the steady state Schroedinger equation

Suppose that for some system the potential energy, V, does not depend explicitly on the time.

Examples of potential energies which do not depend explicitly on the time:
1. Two charges e_1 and e_2 at a distance r: $V = e_1e_2/r$.
2. A spring (force constant k) stretched by an amount x from its equilibrium length: $V = (1/2)kx^2$.
3. A mass m a distance h above the earth's surface: $V = mgh$.

Example of a potential energy that does depend explicitly on the time: A charge e in a light wave polarized in the x-direction: $V = eE_0x \sin 2\pi\nu t$, where $E_0 \sin 2\pi\nu t$ is the electric field strength in the wave and ν is the frequency of the wave. The position of the charge is given by x.

Then we may apply the method of separation of variables (page 44) to Equation (I), writing

$$\Psi(x, y, z, t) = \psi(x, y, z)\phi(t) \tag{C-5}$$

On substitution in Equation (I), dividing by $\psi\phi$, and making use of the fact that since V is independent of the time, H is also independent of the time, we find

$$H\psi = \lambda\psi \tag{C-6}$$

$$\phi = A \exp(-i\lambda t/\hbar) \tag{C-7}$$

where λ is the separation constant and A is an arbitrary constant which may be made equal to unity with no loss in generality. Therefore,

$$H\Psi = H(\psi\phi) = \lambda\psi\phi = \lambda\Psi \tag{C-8}$$

and from Corollary I (Equation C-4) we see that the separation constant is equal to E, the energy of the system. Furthermore,

$$\Psi = \psi(x, y, z) \exp(-iEt/\hbar) \tag{C-9}$$

so that the probability distribution is

$$\Psi\Psi^* = \psi\psi^* \tag{C-10}$$

which is independent of the time. Also, if α is the operator for any variable that does not depend explicitly on the time, the average value of the variable is

$$\bar{a} = \int \Psi^*\alpha\Psi d\tau \Big/ \int \Psi^*\Psi d\tau = \int \psi^*\alpha\psi d\tau \Big/ \int \psi^*\psi d\tau \tag{C-11}$$

which is also independent of the time. Such a system is said to be in a *steady state* or *stationary state*, and all of its properties may be found by solving the differential equation,

$$\boxed{H\psi(x, y, z) = E\psi(x, y, z)} \tag{IV}$$

We can thus make the following important observation:

Corollary II. *If the potential energy of a system depends only on the coordinates and not on the time explicitly, it is possible to construct a time-independent probability distribution and to find all of the properties of the system from the solution of Equation (IV).*

Equation (IV) is a restricted form of Law (I) and is known as the *steady state Schroedinger equation.* It is the basis of the greater part of quantum chemistry and will be the starting point for most of the problems considered in this book.

c. The steady state Schroedinger equation as a normal mode equation. The "wave-like" properties of matter. The de Broglie relation. The idea of complementarity

When written out explicitly for one-, two-, and three-dimensional systems, the steady state Schroedinger equation is seen to have the form of some of the normal mode equations we have encountered in the classical theory of vibrations.

For one-dimensional systems containing single particles Equation (IV) becomes

$$\frac{d^2\psi(x)}{dx^2} + \frac{2m}{\hbar^2}(E - V(x))\psi(x) = 0 \tag{IVa}$$

This resembles the normal mode equation for a string of constant tension and nonuniform density

$$\frac{d^2z(x)}{dx^2} + 4\pi^2\nu^2\frac{\rho(x)}{T}z(x) = 0$$

The analogue of the density in (IVa) evidently varies with x in proportion to $(1 - V(x)/E)$, and the energy E is analogous to the square of the frequency.

For a single particle moving in two dimensions

$$\frac{\partial^2\psi}{\partial x^2} + \frac{\partial^2\psi}{dy^2} + \frac{2m}{\hbar^2}(E - V(x, y))\psi = 0 \tag{IVb}$$

which has the same form as the normal mode equation of a membrane of variable density.

For a single particle moving in three dimensions

$$\nabla^2\psi + \frac{2m}{\hbar^2}(E - V(x, y, z))\psi = 0 \qquad \text{(IVc)}$$

which has the same form as the normal mode equation for sound waves in a system in which the sound velocity is inversely proportional to $\sqrt{1 - V/E}$.

In regions in which $V > E$, $(1 - V/E)$ becomes negative, and the curvature of the solutions changes sign. In such regions ψ cannot oscillate. We do not encounter such a behavior in classical vibration theory because it implies the existence of negative mass, which has never been observed. This does not destroy the analogy, however, since negative mass is quite conceivable in the limited sense that its effects on vibrations can be surmised from the classical theory of vibrations.

Quantum mechanics is often called "wave mechanics" because of this similarity between the steady state Schroedinger equation and the normal mode equation of vibration theory. Matter can in this sense be said to have "wave-like" properties. It should be noted, however, that the validity of the wave analogy is restricted, because the time-dependence of $\Psi(q,t)$ is not given by a true wave equation. (The time-dependent Schroedinger equation has only a first derivative with respect to the time, whereas a true wave equation involves a second derivative.)

The normal mode equation of vibration theory can often (but not always) be written in the form

$$\nabla^2 u + \frac{4\pi^2\nu^2}{c^2} u = 0 \qquad \text{(C-12)}$$

where c is the wave velocity and u is the amplitude. Now the ratio c/ν is of the nature of a wave length

$$\lambda = c/\nu$$

If we compare (C-12) with the steady state Schroedinger equation, we see that

$$4\pi^2/\lambda^2 \text{ (or } 4\pi^2\nu^2/c^2\text{)}$$

takes the place of

$$\frac{2m}{\hbar^2}(E - V) = \frac{8\pi^2 m}{h^2}(E - V)$$

Thus we are led to the identification

$$\frac{1}{\lambda} = \frac{\sqrt{2m(E-V)}}{h} \tag{C-13}$$

and the ψ function corresponds to an amplitude of vibration, u. According to classical mechanics, for a single particle

$$E = \text{kinetic energy} + \text{potential energy}$$

$$= (\tfrac{1}{2})mv^2 + V$$

$$= p^2/2m + V$$

or

$$\sqrt{2m\,(E-V)} = p \tag{C-14}$$

where v is the linear velocity of the particle and p is its linear momentum ($p = mv$).

Substituting (C-14) in (C-13) we see that

$$\boxed{\lambda = h/p} \tag{V}$$

That is, Schroedinger's steady state equation can be interpreted as a normal mode equation in which the motion of a particle is "regulated" by a wave of length $\lambda = h/p$. The probability of finding the particle at any given point depends on the square of the amplitude of this wave at that point. We are thus led to the following statement: **Corollary III.** *Particles in stationary states act as though "guided" by a wave whose amplitude is the wave function, ψ, and whose wave length is $\lambda = h/p$.* The exact meaning of the word "guided" is to be set by the student from the context of the proof of this corollary.

Relation (V) was hypothesized by de Broglie shortly before the discovery of quantum mechanics, in order to account for the stability of the Bohr orbits of the hydrogen atom. Bohr, in developing his theory of the hydrogen atom, had postulated that only those orbits are possible for which the angular momentum, M, of the electron about the nucleus is an integral multiple of $\hbar$,

$$M = n\hbar \quad (n = 1, 2, 3, \cdots)$$

Now M = (moment of inertia) × (angular velocity). For a particle of mass m moving with linear velocity v in a circular orbit of radius r

$$M = (mr^2)\cdot(v/r) = mvr = pr = n\hbar = nh/2\pi$$

so that

$$r = nh/2\pi p$$

De Broglie wrote this as

$$\text{Circumference of orbit} = 2\pi r = n(h/p) = n\lambda$$

and pointed out that the only orbits observed in the hydrogen atom are those for which the circumference is an integral multiple of the wave length, $\lambda = h/p$. Every time the electron moves around the nucleus it traverses an integral number of wave lengths of some hypothetical oscillating medium. A standing wave therefore results. In any other orbit the waves would somehow interfere destructively as the electron circles about, and a stable orbit cannot exist.

De Broglie also pointed out that a similar relation holds for photons. The Compton effect shows that photons "act as though" they have a mass $m = h\nu/c^2$, where c is the velocity of light and ν is the frequency of the photon (this is also a consequence of the Einstein mass-energy relation, since $h\nu$ is the energy of a photon, so it must have a mass such that $mc^2 = h\nu$). Since the velocity of a photon is c, its momentum is $p = mc = h\nu/c = h/\lambda$, or $\lambda = h/p$, which is of the same form as Equation (V). Thus we can say that photons as well as electrons are particles that have a wave length, $\lambda = h/p$, somehow "associated" with them, and we may suspect that this is a general property of all particles.

The existence of a wave length and of interference phenomena of the type postulated by de Broglie for electrons was soon verified experimentally by Davisson and Germer and by G. P. Thomson, who showed that beams of electrons are diffracted by crystals according to exactly the same geometrical laws as are X-rays of the same wave length. Similar diffraction effects have also been observed with protons, neutrons and α particles; in each instance the wave length observed was that given by the de Broglie relation, (V). The experimental difficulties of proving its validity for particles heavier than atoms have been so great that this has not been done, but there is every reason to believe that the de Broglie relation is valid for all particles. If we could make a fine enough grating, we could diffract baseballs, people and planets. Diffraction by electrons has become an important laboratory tool in studying the structures of molecules and of surfaces, and structure determination by neutron diffraction is now in an advanced stage of development.

Exercise. What is the de Broglie wave length of a person weighing 150 pounds and moving with a velocity of 3 miles per hour? Compare this wave length with the diameter of a proton (of the order of 10^{-13} cm.). What frequency would a gamma ray have if its wave length were this small? Express the energy of such a gamma ray in electron volts. Can such energies be attained in machines now available in nuclear research?

The de Broglie relation was instrumental in leading Schroedinger to the discovery of his formulation of quantum mechanics; in fact, Schroedinger went through more or less the same arguments as those just given, but in reverse order. He did this guided by certain analogies between optics and ordinary mechanical motions, that had been noticed earlier by mathematicians and classical theoretical physicists such as Hamilton. When, in tracing these analogies, it became necessary to assign a "wave length" to the moving particles, the classical theoretical physicists automatically took this wave length to be zero in order to avoid diffraction effects such as are observed with light, but would have seemed outrageous at that time for material particles. The analogy is, in fact, not quite complete, and Schroedinger was forced to make some assumptions in carrying it through. Thus it cannot be said that quantum mechanics can be "derived" from these considerations, and for our purposes it is better simply to accept the quantum mechanical laws as a universally valid structure, without trying to reconstruct the scaffolding that was used in erecting it. The student who wishes to pursue this interesting question should read Schroedinger (*4*, pp. 13-30, or *4a*) and Bohm (*5*, Part I).

On the basis of de Broglie's considerations one is tempted to postulate a "super ether" in which vibrations occur governing the motions of material particles, the state functions being the amplitudes of these vibrations. This idea leads to many difficulties, however, and is a good example of the danger awaiting anyone who tries to ride an analogy too far. It is better simply to take the equations as written and not ask too many questions about what is behind them.

There is no denying the fact, however, that matter seems to exhibit characteristics of both waves and particles, and the same is true of light. This aspect of nature has been developed by Bohr into a most interesting point of view known as the *Principle of Complementarity*. According to this, nature has provided us with two complete, equivalent, and distinct languages, one characteristic of particles, the other characteristic of waves, quantum mechanics being the means of translating from one language into the other. It is possible to set up a consistent description of the behavior and interaction of matter and radiation in either of these languages. The results are equally valid; neither language can be said to be better than the other. The supposed paradoxes of "particles that act like waves" and "beams of light waves that act like particles" appear only when the two languages are mixed – which is not admissable. The quantum mechanical translation of the languages involves the use of the momentum method mentioned on page 158. We cannot go into any detail about this interesting and important principle. It is discussed extensively in the works of Bohr (*6*), Landé (*7*) and Bohm (*5*). Some of its philosophical implications are considered by Oppenheimer (*8*) and Pauli (*9*).

d. Possible values of position, linear momentum, and angular momentum. Derivation of Bohr's angular momentum postulate

According to Law (II), it is only possible to observe values of a property that are the eigenvalues of the operator corresponding to the property. Let us see what restrictions this places on the properties of position, linear momentum, and angular momentum. This merely involves finding all of the well behaved eigenfunctions of these operators.

(*i*) *Position.* Law II states that it is possible to observe a particular value, a, for the x-coordinate of a system only if there exists a function ϕ such that

$$x\phi = a\phi \tag{C-15}$$

where ϕ is single valued. This relation is satisfied for any value of a by the functions defined in the following way. Consider the function

$$f(x-a) = \lim_{\varepsilon \to 0} \frac{1}{\sqrt{\varepsilon}} \exp\left[-(x-a)^2/\varepsilon^2\right] \tag{C-16}$$

This function is zero everywhere except at $x = a$, where it is infinite. Furthermore, the integral of its square,

$$\int_{-\infty}^{\infty} f^2\, dx = \lim_{\varepsilon \to 0} \frac{1}{\varepsilon} \int_{-\infty}^{\infty} \exp\left[-2(x-a)^2/\varepsilon^2\right] dx = \sqrt{\pi/2}$$

is finite, so the function is well behaved. It is evident that the function f satisfies the equation

$$x \cdot f(x-a) = a\, f(x-a) \tag{C-17}$$

for all values of x, because if $x \neq a$ both sides of the equation are zero, while if $x = a$ then $a \cdot f = a\,f$. Obviously a function of the type $f(x-a)$ can be constructed for any value of a, so that Law II brings about no restrictions on the possible values of the positions of the particles of a system.

The function

$$f^2(x-a) = \lim_{\varepsilon \to 0} \frac{1}{\varepsilon} \exp\,[-2(x-a)^2/\varepsilon^2]$$

is an example of a *Dirac delta function*. For further discussion of these functions, see Bohm (*5*, pages 212 ff.).

(*ii*) *Momentum*. In order to find the possible values of the momentum in the x-direction we must find solutions of

$$\frac{\hbar}{i}\frac{\partial \phi}{\partial x} = \lambda \phi \tag{C-18}$$

These are seen to be

$$\phi = \exp\,(i\lambda x/\hbar) \tag{C-19}$$

and in the absence of other restrictions, eigenfunctions may be constructed for all values of the constant, λ. Therefore there is no limitation on the possible values of x-momentum.

(*iii*) *Angular Momentum*. The equation here is (taking the z-axis for convenience as the axis of rotation and referring to Equation (A-2))

$$M_z\phi = \frac{\hbar}{i}\left(x\frac{\partial \phi}{\partial y} - y\frac{\partial \phi}{\partial x}\right) = \lambda\phi \tag{C-20}$$

where λ is a possible value of the component of the angular momentum about the z-axis. If we transform to cylindrical coordinates, for which $x = r\cos\theta$, $y = r\sin\theta$, the rules of calculus tell us that

$$\left(\frac{\partial}{\partial x}\right)_y = \left(\frac{\partial r}{\partial x}\right)_y\left(\frac{\partial}{\partial r}\right)_\theta + \left(\frac{\partial \theta}{\partial x}\right)_y\left(\frac{\partial}{\partial \theta}\right)_r = \frac{x}{r}\left(\frac{\partial}{\partial r}\right)_\theta - \frac{y}{r^2}\left(\frac{\partial}{\partial \theta}\right)_r \tag{C-21a}$$

$$\left(\frac{\partial}{\partial y}\right)_x = \left(\frac{\partial r}{\partial y}\right)_x\left(\frac{\partial}{\partial r}\right)_\theta + \left(\frac{\partial \theta}{\partial y}\right)_x\left(\frac{\partial}{\partial \theta}\right)_r = \frac{y}{r}\left(\frac{\partial}{\partial r}\right)_\theta + \frac{x}{r^2}\left(\frac{\partial}{\partial \theta}\right)_r \tag{C-21b}$$

and substitution into (C-20) gives

$$M_z = \frac{\hbar}{i}\frac{\partial}{\partial\theta} \tag{C-22}$$

$$M_z\phi = \frac{\hbar}{i}\frac{\partial\phi}{\partial\theta} = \lambda\phi \tag{C-23}$$

$$\phi(\theta) = \exp\,(i\lambda\theta/\hbar) = \cos\,(\lambda\theta/\hbar) + i\sin\,(\lambda\theta/\hbar) \tag{C-24}$$

The function $\phi(\theta)$ must be single valued; that is

$$\phi(\theta) = \phi(\theta + 2\pi) \tag{C-25}$$

which is possible only if λ is zero or an integer times $\hbar$

$$\lambda = n\hbar \tag{C-26}$$

This result is Bohr's angular momentum postulate: **Corollary IV.** *The component of angular momentum about any axis must be an integral multiple of* $\hbar$.

e. Corollary V

Measurable quantities are always real numbers.

Although the disconcerting quantity, $\sqrt{-1}$, is woven into the quantum mechanical formalism, the structure of quantum mechanics is such that, whenever a number is extracted that can be measured in an experiment, it never contains $\sqrt{-1}$. This is assured in the following way. Observables appear in the laws as λ in Law II and as $\bar{a}$ in Law III. Consider Equation III and its conjugate complex

$$\bar{a} = \int \Psi^*\alpha\Psi d\tau \Big/ \int \Psi^*\Psi d\tau \tag{C-27a}$$

$$\bar{a}^* = \int \Psi\alpha^*\Psi^* d\tau \Big/ \int \Psi\Psi^* d\tau \tag{C-27b}$$

Since α is always a Hermitian operator

$$\int \Psi^*\alpha\Psi d\tau = \int \Psi\alpha^*\Psi^* d\tau \tag{C-28}$$

so that $\bar{a} = \bar{a}^*$, and $\bar{a}$ must be real.

Next consider (II) and its conjugate complex

$$\alpha\phi = \lambda\phi \tag{C-29a}$$

$$\alpha^*\phi^* = \lambda^*\phi^* \tag{C-29b}$$

Multiply the first by ϕ^* and the second by ϕ and integrate over the entire system

$$\int \phi^* \alpha \phi d\tau = \lambda \int \phi^* \phi d\tau \tag{C-30a}$$

$$\int \phi \alpha^* \phi^* d\tau = \lambda^* \int \phi \phi^* d\tau \tag{C-30b}$$

The two quantities on the left are equal because α is Hermitian. Thus $\lambda = \lambda^*$, and λ is always real.

Evidently the requirement that all quantum mechanical operators be Hermitian is sufficient to prevent $\sqrt{-1}$ from occurring in any observable quantity.

f. Corollary VI

It is possible to set up a system having precisely specified values of two dynamical variables if, and only if, the operators for these variables commute. For instance, according to this principle it is not possible to locate a particle precisely and at the same time to know its energy precisely, since the operators for location, $x\cdot$, $y\cdot$, and $z\cdot$, do not commute with the operator for energy, H. That is, for example

$$Hx\cdot = -\frac{\hbar^2}{2m}\nabla^2 x\cdot + Vx\cdot = -\frac{\hbar^2}{2m}x\nabla^2 + xV - \frac{\hbar^2}{m}\frac{\partial}{\partial x}$$

$$= xH\cdot - \frac{\hbar^2}{m}\frac{\partial}{\partial x} \neq xH\cdot \tag{C-31}$$

Similarly, the location and linear momentum of a particle in any one direction cannot be precisely known at the same time, since

$$p_x x\cdot = \frac{\hbar}{i}\frac{\partial}{\partial x}x\cdot = \frac{\hbar}{i}\left(x\frac{\partial}{\partial x} + 1\cdot\right) = xp_x + \frac{\hbar}{i} \neq xp_x \tag{C-32}$$

The momentum of a particle in some direction and its energy can only be known precisely if the potential energy does not vary in that direction, since

$$p_x H = \frac{\hbar}{i}\frac{\partial}{\partial x}\left(-\frac{\hbar^2}{2m}\nabla^2 + V\right) \tag{C-33}$$

and $(\partial/\partial x)V = V(\partial/\partial x)$ only if V is independent of x, whereas it is always true that $(\partial/\partial x)\nabla^2 = \nabla^2(\partial/\partial x)$. These conclusions are in direct contradiction to classical mechanics, where any dynamical variable may be specified to any degree of accuracy.

The *proof* of Corollary (VI) is as follows. Let α and β be the operators for the two properties in question. Then if both properties are to be known precisely, there must be some state function, Ψ, for which

$$\alpha\Psi = a\Psi \tag{C-34a}$$

and

$$\beta\Psi = b\Psi \tag{C-34b}$$

where a and b are constants. Operate on the first of these with β and the second with α. Since all quantum mechanical operators are linear, they commute with constants, so that

$$\beta\alpha\Psi = \beta a\Psi = a\beta\Psi = ab\Psi \tag{C-35a}$$

and

$$\alpha\beta\Psi = \alpha b\Psi = b\alpha\Psi = ab\Psi \tag{C-35b}$$

That is

$$\beta\alpha\Psi = \alpha\beta\Psi \tag{C-36a}$$

or

$$\beta\alpha = \alpha\beta \tag{C-36b}$$

The converse of this corollary – that if two operators commute, systems can always be prepared in which the corresponding variables are precisely known – can also be proven. (See Eyring, Walter, and Kimball (*10*, page 35) and Kemble (*2*, page 284).)

g. Corollary VII

Let a and b be two properties whose operators α and β fail to commute by an amount γ (the "commutator" of α and β, see page 16), *i.e.*

$$\gamma \equiv \alpha\beta - \beta\alpha$$

Then in a series of measurements of the values of a and b in an assembly of systems having a state function Ψ, the standard deviations, δa and δb, of a and b from their average values will be limited by the condition*

$$\delta a \delta b \geqq \frac{1}{2}\left| \int \Psi^* \gamma \Psi d\tau \Big/ \int \Psi^* \Psi d\tau \right| \tag{C-37}$$

For a proof of this corollary, see Margenau and Murphy (*11*, page 332).

* The "standard deviation" of a series of n numbers, $a_1, a_2, a_3, \cdots, a_n$, from their mean value is defined by

$$(\delta a)^2 = \sum_{j=1}^{n} (a_j - \bar{a})^2/n$$

where $\bar{a} = \sum a_j/n$ = average value of the numbers.

If the two properties a and b are x, the x-coordinate of position, and p_x, the linear momentum in the x-direction, then we have seen in the discussion of Corollary (VI) that

$$p_x x - x p_x = \hbar/i \tag{C-38}$$

so that $\gamma = \hbar/i$. Substituting this in (C-37), we find

$$\delta p_x \cdot \delta x \geqq \hbar/2 \tag{C-39}$$

This is one of the so-called *Heisenberg uncertainty relations,* which will be discussed in more detail in Chapter 7.

h. Corollary VIII

If ψ_1 and ψ_2 are two eigenfunctions of a Hermitian operator α, with eigenvalues a_1 and a_2, then

$$\int \psi_1{}^* \psi_2 d\tau = 0$$

if $a_1 \neq a_2$. That is, ψ_1 and ψ_2 are orthogonal.

Proof. By hypothesis

$$\alpha\psi_1 = a_1\psi_1 \tag{i}$$

$$\alpha\psi_2 = a_2\psi_2 \tag{ii}$$

and

$$\int \psi_1{}^* \alpha \psi_2 d\tau = \int \psi_2 \alpha^* \psi_1{}^* d\tau$$

Take the conjugate complex of (i), multiply it by ψ_2 and integrate (remembering that since α is Hermitian, a_1 is real, so $a_1{}^* = a_1$)

$$\int \psi_2 \alpha^* \psi_1{}^* d\tau = a_1 \int \psi_2 \psi_1{}^* d\tau$$

Multiply (ii) by $\psi_1{}^*$ and integrate

$$\int \psi_1{}^* \alpha \psi_2 d\tau = a_2 \int \psi_1{}^* \psi_2 d\tau$$

Thus

$$a_1 \int \psi_1{}^* \psi_2 d\tau = a_2 \int \psi_1{}^* \psi_2 d\tau$$

and unless $a_1 = a_2$, the integral must vanish.

A theorem very similar to this has already been encountered in the classical theory of vibrations (page 113). Here, as there, it is always possible to construct orthogonal functions out of two nonorthogonal ones even if $a_1 = a_2$ (see page 114 for the method).

D. The Quantum Mechanical Treatment of Chemical Systems

As far as chemical applications are concerned, the energy is by far the most important dynamical variable for atoms and molecules. Furthermore, unless light or other radiation happens to be shining on the atom or molecule, its potential energy, and hence its Hamiltonian operator, will not depend explicitly on the time. Because of Corollary II we shall therefore be particularly interested in the solutions and eigenvalues of the steady state Schroedinger equation

$$H\psi = E\psi \tag{D-1}$$

since the eigenvalues, E, of this equation are the sole values of the energy that the atom or molecule will ever be found to have.

In this equation the operator H is the Hamiltonian for the particular system under investigation. It will vary from one system to another because the number of electrons and the number and kind of atomic nuclei vary in different systems. Therefore Equation (D-1) must be solved separately for each system in which one may happen to be interested.

As has been mentioned on pages 162 to 167, Equation (D-1) bears a strong resemblance to the normal mode equations of classical vibration theory, and a useful, though limited, physical analogy between the classical mechanics of waves and the quantum mechanics of particles can be developed on the basis of this resemblance. The mathematical similarity between the classical theory of normal modes and the quantum mechanical theory of steady states is, however, much more complete than this limited physical analogy, and it will be alluded to constantly in what follows. We shall find that in Chapter 3 we have, indeed, already solved several important quantum mechanical problems, and we have at our disposal the means for solving many others. It will be useful, therefore, to summarize the results of vibration theory in the form in which they are used in solving the steady state Schroedinger equation.

In one-dimensional problems, (D-1) can be written

$$H\psi = \left[-\frac{\hbar^2}{2m}\frac{d^2}{dx^2} + V(x)\right]\psi(x) = E\psi(x) \tag{D-2}$$

This has the same form as the Sturm-Liouville differential equation

$$Lu = \left[\frac{d}{dx}T(x)\frac{d}{dx} - q(x)\right]u(x) = -\lambda w(x)u(x) \tag{D-3}$$

Evidently we may always take

$$T(x) = \text{constant} = \hbar^2/2m \tag{D-4a}$$

$$q(x) = \text{potential energy} = V(x) \tag{D-4b}$$

$$\lambda = \text{energy} = E \tag{D-4c}$$

$$w(x) = 1 \tag{D-4d}$$

$$u(x) = \psi(x) \tag{D-4e}$$

and everything we have said about the Sturm-Liouville theory in Chapter 4 can be applied immediately to the quantum mechanical treatment of one-dimensional systems.

We have seen (page 155) that ψ must vanish at infinity. Therefore, the Sturm-Liouville boundary conditions (page 111) are satisfied by taking $a = -\infty$ and $b = \infty$.

The concept of Hermitian functions (page 112) remains useful but needs to be modified to include the possibility of complex operators (footnote on page 112).

We may expect that the solutions ψ will obey the boundary conditions only if the energy E has certain values. Energy quantization therefore arises for the same reason that the normal mode frequencies in vibrating systems are discrete.

We have already shown (Corollary VIII) that two wave functions corresponding to different energies will be orthogonal regardless of how many independent variables they contain. If two eigenfunctions belong to the same eigenvalue, they can be made orthogonal in the usual way. It will also be useful to work with normalized wave functions.

It will be assumed that the normalized and orthogonalized eigenfunctions of a given Hamiltonian, H, form a complete set – that is, they may serve as a basis for the expansion of other functions (see page 116)

$$f = \sum A_n \psi_n \tag{D-5}$$

where f is any function that vanishes at infinity, and the coefficients are given by

$$A_n = \int_{\text{all space}} \psi_n{}^* f d\tau \tag{D-6}$$

The variation method may be used in order to find the lowest possible energy of a system with a given Hamiltonian operator (see page 118). The basic relation is

$$E_0 \leqq \int \phi^* H \phi d\tau \bigg/ \int \phi^* \phi d\tau \tag{D-7}$$

where ϕ is some trial wave function that is usually a sensible guess at the true solution for the lowest energy state. The closer ϕ is to the true solution for this state, the more nearly (D-7) approaches to an equality. The energies of higher states may be found by using trial wave functions that are orthogonal to the exact solutions for lower states.

The perturbation method (pages 128ff) is simpler in quantum mechanics than in vibration theory, since the weight function $w(x)$ is always unity. Thus if

$$H_0\psi_{n0} = E_{n0}\psi_{n0} \tag{D-8}$$

corresponds to the unperturbed case and

$$(H_0 + \gamma H)\psi_n = E_n\psi_n \tag{D-9}$$

corresponds to the perturbed case, then writing

$$\psi_n = \psi_{n0} + \gamma\psi'_n + \gamma^2\psi''_n + \cdots \tag{D-10}$$

$$E_n = E_{n0} + \gamma E'_n + \gamma^2 E''_n + \cdots \tag{D-11}$$

$$\psi'_n = \sum A_k\psi_{k0} \tag{D-12}$$

$$\psi''_n = \sum B_k\psi_{k0} \tag{D-13}$$

we find for nondegenerate levels

$$E'_n = H_{nn} = \int \psi_{n0}{}^*H\psi_{n0}d\tau \tag{D-14}$$

$$E''_n = \sum_{k\neq n} H_{kn}H_{nk}/(E_{n0} - E_{k0}) \tag{D-15}$$

$$A_k = H_{kn}/(E_{n0} - E_{k0}) \quad (n \neq k) \tag{D-16}$$

$$B_k = \frac{\sum_{j\neq n} H_{jn}H_{nj}/(E_{n0} - E_{j0})}{E_{n0} - E_{k0}} - \frac{H_{kn}H_{nn}}{E_{n0} - E_{k0}} \quad (n \neq k) \tag{D-17}$$

$$A_n = 0 \tag{D-18}$$

$$B_n = -\tfrac{1}{2}\sum A_k{}^2 \tag{D-19}$$

where

$$H_{jk} = \int \psi_{j0}{}^*H\psi_{k0}d\tau \tag{D-20}$$

For degenerate levels we have instead of (D-10)

$$\psi_n = \sum_{\substack{\text{degenerate}\\ \text{states}}} c_j \psi_{j0} + \gamma \psi'_n + \cdots \tag{D-21}$$

and the secular equation is

$$\begin{vmatrix} (H_{11} - E'_n) & H_{12} & \cdots & H_{1l} \\ H_{21} & (H_{22} - E'_n) & \cdots & H_{2l} \\ \vdots & \vdots & & \vdots \\ H_{l1} & H_{l2} & \cdots & (H_{ll} - E'_n) \end{vmatrix} = 0 \tag{D-22}$$

where l is the degeneracy of the level.

Symmetry and commutation properties will prove extremely useful in helping to diagonalize the secular equation (see pages 138ff), as well as in evaluating integrals of the general type $\int \phi \alpha \psi d\tau$.

The adiabatic law is valid in quantum mechanics. That is, if a perturbation is applied sufficiently slowly to a system in an eigenstate, it will remain in an eigenstate. Since the time dependence of quantum mechanical state functions is different from that governing vibrating systems, the proof of this adiabatic law is somewhat altered, however (see page 529).

Correlation diagrams continue to be useful as the basis for a descriptive terminology of states.

All that was said about coupled systems (pages 145ff) can be directly applied to interacting electrons, atoms, molecules, etc. In particular, we can expect that the energy of the lowest state of a system ought to be lowered by coupling two systems together. In actual fact this is often found to be true only when the coupling is weak because of features in quantum mechanics that are not encountered in ordinary vibrating systems (*viz.*, the Pauli principle and the occurence of a term corresponding to negative mass in the quantum mechanical equations).

E. Atomic Units

For atoms and molecules, with which we shall be chiefly concerned in this book, the potential energy in the Hamiltonian is made up of terms representing the electrostatic interactions of the electrons and the atomic nuclei

$$V = -\sum_{\substack{\text{all electrons, } i\\ \text{all nuclei, } k}} Z_k e^2 / r_{ik} + \sum_{\substack{\text{all pairs}\\ \text{of electrons}}} e^2 / r_{ij} + \sum_{\substack{\text{all pairs}\\ \text{of nuclei}}} Z_k Z_l e^2 / r_{kl} \tag{E-1}$$

where r_{ik} is the distance between the ith electron and the kth nucleus, r_{kl} is the distance between the kth and l^{th} nuclei, r_{ij} is the distance between the ith and j^{th} electrons, Z_k is the atomic number of the kth nucleus, and e is the charge on an electron. This expression is to be substituted into the Schroedinger equation. In many problems it is justifiable to disregard the kinetic energy of the nuclei, and hence their Laplacian operators. We then obtain

$$\sum_{\text{electrons}} \nabla_i^2 \psi + \frac{2m_e}{\hbar^2} (E - V) \psi = 0 \tag{E-2}$$

where ∇_i^2 is the Laplacian operator involving the coordinates of the ith electron and m_e is the electronic mass.

Equation (E-2) assumes a much simpler form if we change the units of length and of energy. Let us measure length in units of *Bohr radii*, $a_0 = \hbar^2/m_e e^2$. (This is the radius of the innermost orbit of the electron in the old Bohr theory of the hydrogen atom.) Let us measure the energy in units of e^2/a_0, which is the potential energy of the electron in the first Bohr orbit.* Then Equation (E-2) becomes

$$\sum_{\text{electrons}} \nabla_i^2 \psi + 2 \left(E - \sum \frac{1}{r_{ij}} - \sum \frac{Z_k Z_l}{r_{kl}} + \sum \frac{Z_k}{r_{ik}} \right) \psi = 0 \tag{E-3}$$

Similar simplifications of other equations of quantum chemistry are obtained if a new system of units is used instead of the more familiar c.g.s. system. These new units are called *atomic units* (also sometimes known as *quantum units* and *Hartree units*). The unit of length is the Bohr radius defined above. The unit of mass is that of the electron. The unit of time is that required by an electron in the lowest Bohr orbit to travel a distance equal to the radius of the orbit. Appendix I gives the factors for converting atomic units into c.g.s. units. Appendix II gives the conversion factors for energy in the various systems of units used by physicists and chemists.

The student will find it useful to memorize the following approximate conversion factors:

1 e.v. $\simeq$ 23 kcal./mole

1 atomic unit $\simeq$ 27 e.v.

300° K $\simeq$ 1/40 e.v.

10,000 cm^{-1} $\simeq$ 10,000 A. $\simeq$ 1.24 e.v.

$\simeq$ 28.6 kcal./mole

1 e.v. $\simeq$ 8,000 cm^{-1}.

* Some workers use $e^2/2a_0$ as the atomic unit of energy.

REFERENCES CITED

1. R. C. Tolman, "The Principles of Statistical Mechanics," Oxford U.P., New York, 1938.
2. E. C. Kemble, "The Fundamental Principles of Quantum Mechanics," McGraw-Hill, New York, 1937.
3. P. A. M. Dirac, "The Principles of Quantum Mechanics," 2nd ed., Oxford U.P., New York, 1935.
4. E. Schroedinger, "Collected Papers on Wave Mechanics," Blackie and Son, London, 1928.

4a. E. Schroedinger, *Ann. Physik* **79,** 489 (1926).

5. D. Bohm, "Quantum Theory," Prentice-Hall, New York, 1951.
6. N. Bohr, "Atomic Theory and the Description of Nature," Cambridge U.P., New York, 1934.
7. A. Landé, "Quantum Mechanics," Cambridge U.P., New York, 1937.
8. J. R. Oppenheimer, "Science and the Common Understanding," Simon and Schuster, New York, 1954.
9. W. Pauli, *Experientia* **6,** 72 (1950).
10. H. Eyring, J. Walter, and G. E. Kimball, "Quantum Chemistry," Wiley, New York 1944.
11. H. Margenau and G. M. Murphy, "The Mathematics of Physics and Chemistry," Van Nostrand, New York, 1943.

GENERAL REFERENCES

For a more extensive discussion of the physical basis of quantum mechanics, see D. Bohm, reference 5.

The following references discuss the Heisenberg formulation of quantum mechanics, which is an alternative to the Schroedinger method used in this book:

V. Rojansky, "Introductory Quantum Mechanics," Chapter 10, Prentice-Hall, New York, 1938.

L. Schiff, "Quantum Mechanics," Chapter 6, McGraw-Hill, New York, 1949.

D. Bohm, reference 5, Chapter 16.

Another alternative formulation of quantum mechanics, that of Dirac, is described in the following references:

V. Rojansky, *ibid.*, Chapter 11.

E. U. Condon and G. Shortley, "The Theory of Atomic Spectra," rev. ed., Chapter 2, Cambridge U.P., New York, 1950.

P. A. M. Dirac, reference 3.

Chapter 6

Some Solutions of the Steady State Schroedinger Equation

All the properties and all the possible motions of a particle of mass m moving in a potential energy field $V(x, y, z)$ can be found from the well behaved solutions and eigenvalues, E, of the second order partial differential equation

$$-\frac{\hbar^2}{2m}\nabla^2\psi + V(x, y, z)\psi = E\psi \tag{I}$$

Evidently it is only the mass and especially the potential energy function that distinguish one system from another. We shall now discuss some exact solutions of Equation (I) and other equations of closely similar form for a number of simple systems having different potential energy functions.

A chemist might think that these simple systems are so different from the complex systems with which he usually has to deal that they can hardly be of any value to him. This, however, is not the case. Since exact solutions have not been obtained for the more complex chemical problems, we must rely exclusively on approximate methods of solution, and these methods are generally based on the solutions of the simpler, soluble problems. Furthermore, even if precise solutions were available for the complex systems, they would be so elaborate that we would undoubtedly have to use the approximate methods and the simple, inexact solutions in order to understand and apply them. Thus the solutions of simple problems provide us with insights that are essential for the interpretation of the behavior of complex systems.

I. SYSTEMS WITH CONSTANT POTENTIAL ENERGY

A. Free Particles

a. One-dimensional case

A free particle is one whose potential energy is everywhere the same; it is convenient to assume that $V = 0$. If the particle is somehow constrained to move in only one direction (a specific example would be a bead sliding on a wire) the Schroedinger equation becomes

$$-(\hbar^2/2m)d^2\psi/dx^2 = E\psi \tag{A-1}$$

This equation is analogous to the normal mode equation for a vibrating string (page 61), and its solutions can be written

$$\psi = A \sin kx + B \cos kx \tag{A-2}$$

where A and B are arbitrary constants and $k^2 = 2mE/\hbar^2$. Since there are no boundary conditions, all of the constants, including k, can have any value one chooses to give them. Therefore the energy is not quantized in this case.

Since the integral of $|\psi|^2$ over all values of x from $-\infty$ to ∞ is infinite, the wave function ψ given in (A-2) cannot be normalized. To avoid this difficulty it is convenient to regard the system as a beam of noninteracting particles in which there are n_0 particles per unit length along the beam. Then the wave function can be normalized by merely applying the condition

$$\int_x^{x+L} |\psi|^2 \, dx = n_0 L \tag{A-3}$$

where L is a length such that $kL \gg 2\pi$. This leads to the requirement that

$$|A|^2 + |B|^2 = 2n_0 \tag{A-4}$$

Because of the fact that sines and cosines can be expressed in the form of complex exponentials (see Exercise 1 on page 21) we may write the solution (A-2) in the equivalent form

$$\psi = Ce^{ikx} + De^{-ikx} \tag{A-5}$$

where C and D are arbitrary constants. The normalization condition (A-4) gives

$$|C|^2 + |D|^2 = n_0 \tag{A-6}$$

The functions e^{ikx} and e^{-ikx} are both eigenfunctions of the linear momentum operator, $p_x = (\hbar/i)d/dx$; the eigenvalues are $k\hbar$ and $-k\hbar$, respectively, which are equal to $\pm\sqrt{2mE}$. These quantities are the classical values of the momentum of a free particle having an energy E; $2\pi/k$ is the de Broglie wave length associated with the particles. Thus the physical interpretation of Equation (A-5) is as follows: suppose that we set up a state in which $D = 0$ in Equation (A-5). We would then have a beam of particles, all of which move in the positive x-direction with momentum $+\sqrt{2mE}$. On the other hand, if we set $C = 0$, we would have a beam of particles moving to the left with the momentum $-\sqrt{2mE}$. The probability density, $|\psi|^2$, would be a constant independent of x in both instances, which means that we would have equal probabilities of finding the particles at every value of x.

On the other hand, either of the functions $A\sin kx$ or $B\cos kx$ corresponds to a superposition of two beams of equal intensity moving in opposite directions (*cf.* Exercise 1 on page 21). The probability density, $|\psi|^2$, is then not independent of x, but varies sinusoidally along the beam. This variation occurs because of the formation of "standing waves" through the interference of the de Broglie waves associated with the beams moving in opposite directions. It is, of course, a completely nonclassical effect. If the student is made unhappy by it, we can only assure him that all evidence points to the conclusion that the classical theory is wrong in not having told us about it.

We note that in this system the value of p_x is known precisely, whereas the position of a given particle is completely unknown. Thus the uncertainties in p_x and x are given by $\Delta p_x = 0$ and $\Delta x = \infty$, and the Heisenberg product, $\Delta p_x \Delta x$, is indeterminate, so that the Uncertainty Principle (pages 172 and 238) is not violated.

b. Three-dimensional case

If the free particle moves in three dimensions, the Schroedinger equation is

$$-\frac{\hbar^2}{2m}\nabla^2\psi = E\psi \tag{A-7}$$

and the form of the solution depends on the coordinate system we choose in writing the Laplacian operator, ∇^2.

(*i*) *Plane Wave Solution.* If we choose Cartesian coordinates, the Laplacian takes the form

$$\nabla^2 = \partial^2/\partial x^2 + \partial^2/\partial y^2 + \partial^2/\partial z^2$$

and the eigenfunctions can be written

$$\psi = Ae^{i\mathbf{K}\cdot\mathbf{r}} = Ae^{i(x\cos\alpha + y\cos\beta + z\cos\gamma)K_0} \tag{A-8}$$

where

$$\mathbf{r} = \mathbf{i}x + \mathbf{j}y + \mathbf{k}z \tag{A-8a}$$

$$\mathbf{K} = (\mathbf{i}\cos\alpha + \mathbf{j}\cos\beta + \mathbf{k}\cos\gamma)K_0 \tag{A-8b}$$

$$K_0 = \sqrt{2mE}/\hbar \tag{A-8c}$$

$\mathbf{i}$, $\mathbf{j}$, and $\mathbf{k}$ being unit vectors (page 24), and A, α, β, and γ being arbitrary constants. It is evident that this wave function is an eigenfunction of the operator for the total linear momentum, $\mathbf{p} = p_x\mathbf{i} + p_y\mathbf{j} + p_z\mathbf{k}$, having the eigenvalues $\mathbf{K}\hbar$. Thus $K_0\hbar$ is the magnitude of the linear momentum, and $\cos\alpha$, $\cos\beta$, and $\cos\gamma$ give the direction cosines of the particles' motion. The de Broglie waves in this case are simply plane waves whose wave front is normal to the direction of **K** and whose wave length is $2\pi/K_0$. For the free particle there are no boundary conditions, so K_0, and hence E, can assume any positive value. That is, the energy is not quantized.

(*ii*) *Spherical Wave Solutions.* If we express the Laplacian in polar coordinates, (r, θ, φ), we obtain from Equation (A-7)

$$\frac{1}{r^2}\frac{\partial}{\partial r}r^2\frac{\partial\psi}{\partial r}+\frac{1}{r^2\sin\theta}\frac{\partial}{\partial\theta}\sin\theta\frac{\partial\psi}{\partial\theta}+\frac{1}{r^2\sin^2\theta}\frac{\partial^2\psi}{\partial\varphi^2}=-K_0^2\psi \tag{A-9}$$

where K_0 has the value given in Equation (A-8c). If we write

$$\psi(r, \theta, \varphi) = R(r)Y(\theta, \varphi) \tag{A-10}$$

and apply the method of separation of variables (page 44) we find

$$\frac{1}{\sin\theta}\frac{\partial}{\partial\theta}\sin\theta\frac{\partial Y}{\partial\theta}+\frac{1}{\sin^2\theta}\frac{\partial^2 Y}{\partial\varphi^2}+l(l+1)\,Y=0 \tag{A-11a}$$

$$\frac{1}{r^2}\frac{d}{dr}r^2\frac{dR}{dr}+\left[K_0^2-\frac{l(l+1)}{r^2}\right]R=0 \tag{A-11b}$$

where we have written the separation constant in the form $l(l + 1)$, l being a constant to be determined presently. Equation (A-11a) has the same form as the differential equation obtained in the discussion of the normal modes of the ocean on a flooded planet (Equation (B-39) of Chapter 3). Its solutions are the surface spherical harmonics, which have been described in detail in Chapter 3,

$$Y = P_l^m(\cos\theta)\begin{cases}\sin m\varphi\\ \cos m\varphi\end{cases} \tag{A-12}$$

Making use of the complex exponential forms of the sine and cosine terms, we can write these solutions in the alternative form

$$Y = P_l^{|m|}(\cos\theta)\,e^{im\varphi} \tag{A-13}$$

where m can have either positive or negative values. These solutions are well behaved only if l is a positive integer, and if m is an integer lying in the range $-l \leq m \leq l$.

Equation (A-11b) has the same form as Equation (C-28) of Chapter 3, which was encountered in discussing the normal modes of a gas in a spherical cavity. Its solutions are

$$R(r) = (1/\sqrt{r})\,[AJ_{l+\frac{1}{2}}(K_0r) + BJ_{-l-\frac{1}{2}}(K_0r)] \tag{A-14}$$

where $J_{l+\frac{1}{2}}$ is Bessel's function of order $(l+\frac{1}{2})$. These functions have been discussed on page 105. In the case of a free particle there are no boundary conditions, so once again K_0 is arbitrary and the energy is not quantized.

If we take $l = 0$ we obtain

$$Y(\theta, \varphi) = 1 \tag{A-15}$$

$$R(r) \quad = (1/\sqrt{r}) (A \sin K_0 r + B \cos K_0 r)$$

Setting $A = -iB$ we obtain

$$\psi = (B/\sqrt{r})e^{-iK_0 r} \tag{A-15a}$$

which corresponds to a system of particles converging toward the origin from all directions with momentum $K_0 \hbar$. Similarly, by taking $A = iB$ we obtain a solution corresponding to particles flying away from the origin in all directions with momentum $K_0 \hbar$. Furthermore, at large distances from the origin (A-14) becomes approximately

$$R(r) \cong (1/\sqrt{r}) (A'e^{iK_0 r} + B'e^{-iK_0 r}) \tag{A-15b}$$

which also shows that these solutions correspond to particles moving approximately toward and away from the origin.

We shall find in Chapter 8 that the solutions we have just obtained for Equation (A-9) are eigenfunctions of the operator for the square of the total angular momentum about the origin of the coordinates, the eigenvalue being $l(l + 1)\hbar^2$. Therefore the solutions having $l \neq 0$ correspond to systems in which particles do not pass through the origin. It can be shown that the solutions with $l = 1$ correspond to beams of particles coming from different directions and passing a distance $\lambda/2\pi$ to one side or the other of the origin, where λ is the de Broglie wave length, $2\pi/K_0$. Solutions with $l = 2$ miss the origin by twice this distance, and so on.

These solutions are important in discussing the scattering of a beam of particles by centers of force that interact with the particles. The most important example is the scattering of high energy particles by atomic nuclei. Much of our knowledge of the nucleus comes from the analysis of scattering experiments using the wave functions described above.

B. Particles in Boxes

A box is a system in which the potential energy is zero when the particle is within a closed region and infinite everywhere else.

a. One-dimensional boxes

Suppose that a particle moving in only the x-direction is restricted to the region between $x = 0$ and $x = a$. (We might think of a bead moving on a wire with obstacles attached to the wire at $x = 0$ and a.) Outside this region the potential energy will be taken to be infinite, and within the region it will be taken to be zero.

Outside of the box the Schroedinger equation is

$$-(\hbar^2/2m)d^2\psi/dx^2 + \infty\,\psi = E\psi \tag{B-1}$$

This equation can be satisfied only if $\psi = 0$ at all points outside the box. That is, the particle cannot exist outside the region $0 < x < a$–which, of course, is just what we should expect.

Inside the box the Schroedinger equation is

$$-(\hbar^2/2m)d^2\psi/dx^2 = E\psi \tag{B-2}$$

which is identical in form to the normal mode equation for a vibrating string (page 61) and has solutions that may be written

$$\psi = A \sin kx + B \cos kx \tag{B-3}$$

where $k = \sqrt{2mE}/\hbar$. There is a further very important restriction on this solution. According to the laws of quantum mechanics, ψ must be a continuous function. Therefore it must join the solution found for the outside of the box at the edges of the box–that is, the solution inside the box must go to zero at $x = 0$ and $x = a$. The problem is identical with that of finding the normal modes of a string held at both ends. Using the same arguments as those applied in that problem (page 61) we see that $B = 0$ and $k = n\pi/a$, where n is an integer. Thus the permitted solutions are

$$\psi = A \sin n\pi x/a \tag{B-4}$$

and the energy can only have the values

$$E = n^2h^2/8ma^2 \tag{B-5}$$

Equation (B-4) could also be written

$$\psi = (A/2)\,(e^{ikx} - e^{-ikx}) \tag{B-4a}$$

showing that the solution corresponds to a particle moving back and forth within the box, the two directions occurring with equal probability and the momentum having the values $k\hbar$ and $-k\hbar$. This, of course, agrees with the classical picture, except for the fact that the energy is quantized.

There is one further difference between this result and the result expected on the basis of the classical theory. If a particle were bouncing back and forth inside a box, we should expect classically that the probability of finding it at a given location would be the same everywhere inside the box. Thus the classical picture leads us to expect that $|\psi|^2$ = a constant. According to Equation (B-4), however, the probability distribution is

$$|\psi|^2 = A^2 \sin^2 n\pi x/a \tag{B-6}$$

which varies between 0 and A^2 at different points in the box, the distance between successive null points being half a de Broglie wave length. For macroscopic particles moving at reasonable speeds the de Broglie wave length is so small, however (6×10^{-27} cm. for a one gram mass moving with a velocity of one cm./sec.), that the oscillation in probability density is far beyond the possibility of observation. Therefore Equation (B-6) cannot be said to disagree with our everyday experience, even though it seems to be quite at variance with it. On the other hand, for boxes of atomic dimensions and for particles whose mass is close to that of the electron, these non-classical effects become very noticeable.

Let us now calculate the average values of some of the properties of a particle in a box, using Law III (page 159). For the momentum we obtain

$$\bar{p} = \frac{\int_0^a \psi^* \left(\frac{\hbar}{i}\frac{d}{dx}\right) \psi\, dx}{\int_0^a \psi^* \psi\, dx} = \frac{\hbar}{i} \frac{\int_0^a \sin(n\pi x/a)\cos(n\pi x/a)\, dx}{\int_0^a \sin^2(n\pi x/a) dx} = 0 \tag{B-7}$$

and for the square of the momentum

$$\overline{p^2} = \frac{\int_0^a \psi^* (-\hbar^2\, d^2/dx^2)\psi\, dx}{\int_0^a \psi^* \psi dx} = n^2\pi^2\hbar^2/a^2 = 2mE \tag{B-8}$$

Both of these values correspond exactly to the classical picture of a particle moving back and forth in the box with constant velocity.

The mean position of the particle is

$$\bar{x} = \int_0^a x \sin^2(n\pi x/a) dx \Big/ \int_0^a \sin^2(n\pi x/a) dx = a/2 \tag{B-9}$$

which also agrees with the classical result, since if $|\psi|^2 = c$, where c is a constant, we obtain (*cf.* page 28)

$$\bar{x} = \int_0^a xcdx \Big/ \int_0^a cdx = a/2 \tag{B-9a}$$

On the other hand we find for the mean value of x^2

$$\overline{x^2} = \frac{\int_0^a x^2 \sin^2 (n\pi x/a)\, dx}{\int_0^a \sin^2 (n\pi x/a)\, dx} = \frac{a^2}{3}\left(1 - \frac{3}{2n^2\pi^2}\right) \tag{B-10}$$

whereas the classical probability function gives the somewhat different result

$$\overline{x^2} = \int_0^a x^2 cdx \Big/ \int_0^a cdx = a^2/3 \tag{B-10a}$$

We note, however, that for large values of n the two results agree with one another. This is an example of Bohr's *correspondence principle,* according to which classical mechanics and quantum theory tend to give the same results when the systems are in highly excited quantum states.

b. Three-dimensional boxes

(*i*) *Rectangular Cavities.* Suppose that (x, y, z) are the Cartesian coordinates of the particle and that the potential energy is infinite everywhere except inside the rectangular parallelepiped whose walls are the planes $x = 0$, $x = a$, $y = 0$, $y = b$, $z = 0$, $z = c$. We shall set the potential energy equal to zero inside this box. The Schroedinger equation within the box is then

$$-(\hbar^2/2m)(\partial^2\psi/\partial x^2 + \partial^2\psi/\partial y^2 + \partial^2\psi/\partial z^2) = E\psi \tag{B-11}$$

For reasons analogous to those given in the previous example, the wave function is zero everywhere outside the box. The wave function must, therefore, be zero at all six walls, and the solution inside the box must also be zero at the walls for reasons of continuity. This problem is mathematically identical with that of finding the normal modes of a gas in a room whose walls are weightless and exert a constant pressure on the gas. We have seen on page 103 that the solutions are

$$\psi = A \sin k_1 x \sin k_2 y \sin k_3 z \tag{B-12a}$$

where

$$k_1 = n_1\pi/a, \; k_2 = n_2\pi/b, \text{ and } k_3 = n_3\pi/c \tag{B-12b}$$

and

$$E = (h^2/8m)\,[(n_1/a)^2 + (n_2/b)^2 + (n_3/c)^2] \tag{B-12c}$$
$$= (\hbar^2/2m)\,(k_1^2 + k_2^2 + k_3^2)$$

The integers n_1, n_2, and n_3 each have the values unity or greater. The energy level diagram in this case can be quite complex, as can be seen from Fig. 6-1 where the lower levels for the example $c = 2b = 3a$ are shown. The

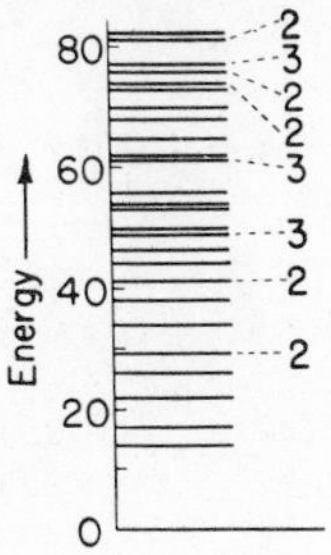

FIG. 6-1. Energy levels of a particle of mass m in a box whose dimensions are c by $\frac{1}{2}c$ by $\frac{1}{3}c$. Units of energy are $h^2/8mc^2$. Integers to the right of the levels indicate their degeneracies.

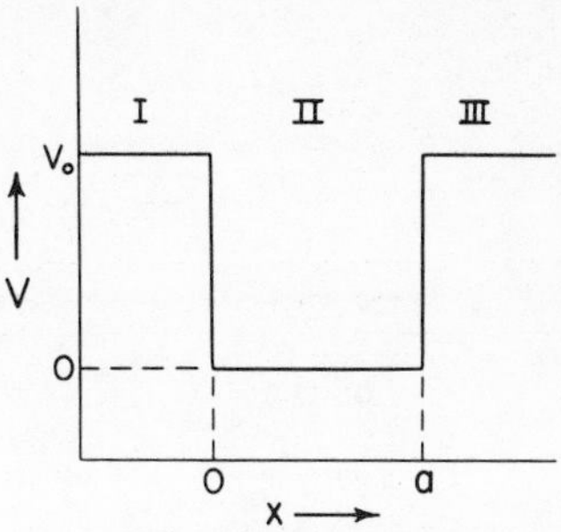

FIG. 6-2. Potential energy function for a particle in a well.

levels generally tend to become closer together as the energy increases, and there may be groups of several states having the same energy (for instance, the states with $(n_1, n_2, n_3) = (1, 1, 6)$, $(1, 3, 2)$ and $(2, 1, 3)$ all have the energy $49\, h^2/8mc^2$).

(*ii*) *Spherical Cavities.* If the particle moves inside a spherical box whose radius is R, and outside of which the potential energy is infinite, it is easy to find acceptable solutions by transforming the Schroedinger equation to polar coordinates, (r, θ, φ). The problem is identical with that discussed on page 106, and the solutions are

$$\psi = AP_l^{|m|}(\cos\theta)\, e^{im\varphi}\, \frac{1}{r}\, J_{l+\frac{1}{2}}(Kr) \tag{B-13}$$

where $K = \sqrt{2mE}/\hbar$, $P_l^{|m|}$ is an associated Legendre function and $J_{l+1/2}$ is a Bessel function. Since we must satisfy the boundary condition $\psi = 0$ when $r = R$, we have the condition on K (and hence on the energy, E) that

$$J_{l+1/2}\,(KR) = 0 \tag{B-14}$$

The numerical values of the roots of Equation (B-14) can be found readily from the discussion on page 106 and from Jahnke and Emde's "Tables" (*1*, pp. 154, 167–9). Each value of l leads to an infinite number of roots and hence an unlimited, but discrete, set of energy levels. The degeneracy of each level is $(2l + 1)$-fold, corresponding to the choice of the integer m, whose values lie between $-l$ and l.

The energies of the lower levels are given in Table 6-1. In this table, $n = 1$ denotes the state derived from the smallest root of Equation (B-14), $n = 2$ denotes the next highest state, etc. The number of radial nodes is thus $(n - 1)$, and as we have seen from our discussion of the spherical harmonics (page 92), the number of angular nodes is l. Evidently states of high angular momentum (that is, high l) are relatively stable, radial nodes in the wave function requiring considerably more energy than angular nodes. We shall find that this is quite different from the situation in atoms, where radial nodes in the wave function tend to confer stability on the system.

TABLE 6-1

ENERGY LEVELS FOR A PARTICLE IN A SPHERICAL CAVITY

(energy given in units of $h^2/8mR^2$)

l	$n = 1$	$n = 2$	$n = 3$	$n = 4$
0	1.00 (1*s*)	4.00 (2*s*)	9.00 (3*s*)	16.00 (4*s*)
1	2.04 (2*p*)	6.04 (3*p*)	12.00 (4*p*)	20.00 (5*p*)
2	3.37 (3*d*)	8.40 (4*d*)	15.45 (5*d*)	24.45 (6*d*)
3	4.95 (4*f*)	11.0 (5*f*)	19.0 (6*f*)	29.0 (7*f*)
4	6.8 (5*g*)	13.9 (6*g*)	22.8 (7*g*)	34.0 (8*g*)
5	8.9 (6*h*)	17.1 (7*h*)	27.1 (8*h*)	
6	11.2 (7*i*)	20.5 (8*i*)		

C. Systems involving Potential Walls of Finite Height

a. Particle in a potential well

Let a particle of mass m move in one dimension subject to a potential energy that varies with the position, x, as shown in Fig. 6-2. Between $x = 0$ and $x = a$ the potential energy is zero, and elsewhere it has the constant value V_0. In the present instance we shall assume that $V_0 > 0$; later in this chapter we shall consider the case in which $V_0 < 0$. (The two systems defined in this way may be called *potential energy wells* and *potential energy barriers*, respectively.) Referring to Fig. 6-2, we see that three regions are defined by this potential function. In regions I and III, where $V = V_0$, the Schroedinger equation has the form

$$d^2\psi/dx^2 - k_1^2\psi = 0 \tag{C-1}$$

where

$$k_1 = \sqrt{2m\,(V_0 - E)}/\hbar \tag{C-1a}$$

In region II, where $V = 0$, the Schroedinger equation is

$$d^2\psi/dx^2 + k_2^2\psi = 0 \tag{C-2}$$

where

$$k_2 = \sqrt{2mE}/\hbar \tag{C-2a}$$

We must distinguish between two states of affairs in solving these equations. In one case, the energy, E, is less than the height of the potential wall, V_0; classically, the particle then does not have enough energy to escape from the well, so it should be restricted to the region between $x = 0$ and $x = a$. We shall call this the *bound particle case*. The other possibility is that the particle has an energy larger than V_0, so that classically it is able to escape from the well; this will be called the *free particle case*. These two cases will now be considered separately.

(*i*) *Bound Particle Case*. If $E < V_0$, then k_1^2 is a positive number so k_1 itself is real. The solution of Equation (C-1) is then

$$\psi = Ae^{k_1x} + Be^{-k_1x} \tag{C-3}$$

where A and B are arbitrary constants. In region I, however, where x is negative, the second term on the right in (C-3) increases without limit as x goes to $-\infty$. In order that ψ be well-behaved, this part of the solution must be thrown away in region I—which is accomplished by setting $B = 0$. Similarly, in region III it is necessary to set $A = 0$ in order that ψ remain finite as x goes to $+\infty$. Thus we arrive at the following solutions:

$$\text{In region I:} \quad \psi_{\text{I}} = Ae^{k_1x} \tag{C-3a}$$

$$\text{In region III:} \quad \psi_{\text{III}} = Be^{-k_1x} \tag{C-3b}$$

It will be convenient in later steps to replace B in (C-3b) by $B'e^{-k_1a}$, giving

$$\text{In region III: } \psi_{III} = B'e^{-k_1(x-a)} \tag{C-3c}$$

Turning now to the solution in region II, we see that the constant k_2 in Equation (C-2) is real, so the solution in region II may be written

$$\psi_{II} = C \sin (k_2x + \delta) \tag{C-3d}$$

where C and δ are arbitrary constants whose values are to be determined presently.

In order that the solutions (C-3a), (C-3c), and (C-3d) be acceptable, they must join each other continuously at $x = 0$ and $x = a$. Furthermore, since the potential energy is everywhere finite, the Schroedinger equation tells us that $d^2\psi/dx^2$ must be finite everywhere. Therefore, there can be no sudden changes in slope at $x = 0$ and $x = a$. This means that the constants A, B, C, δ, k_1, and k_2 must be chosen so that at $x = 0$,

$$\psi_I = \psi_{II} \tag{C-4a}$$

$$d\psi_I/dx = d\psi_{II}/dx \tag{C-4b}$$

and at $x = a$,

$$\psi_{II} = \psi_{III} \tag{C-4c}$$

$$d\psi_{II}/dx = d\psi_{III}/dx \tag{C-4d}$$

It is easy to see that these relations lead to the conditions

$$A = C \sin \delta \tag{C-5a}$$

$$k_1A = k_2 C \cos \delta \tag{C-5b}$$

$$B' = C \sin (k_2a + \delta) \tag{C-5c}$$

$$-k_1B' = k_2C \cos (k_2a + \delta) \tag{C-5d}$$

On rearranging these equations we find

$$\tan \delta = k_2/k_1 = \sqrt{E/(V_0 - E)} \tag{C-6a}$$

$$A/C = \sqrt{E/V_0} \tag{C-6b}$$

$$\tan k_2a = 2 \tan \delta/(\tan^2 \delta - 1) \tag{C-6c}$$

$$\tan \sqrt{2mEa^2/\hbar^2} = 2\sqrt{E(V_0 - E)}/(2E - V_0) \tag{C-6d}$$

In (C-6d) the quantities m, a and V_0 are fixed by the nature of the problem. This equation therefore imposes a condition on the energy; in general only a discrete and limited set of values of the energy will fulfill this condition, as we shall now show.

Let us measure the energy and V_0 in units of $\hbar^2/2ma^2$. Then (C-6d) takes the form

$$\tan\sqrt{E} = 2\,\frac{\sqrt{\dfrac{E}{V_0}\left(1 - \dfrac{E}{V_0}\right)}}{2\dfrac{E}{V_0} - 1} \tag{C-6e}$$

This is most easily solved graphically. In Fig. 6-3 we have superimposed two graphs; one of the graphs is a plot of E *vs.* the left side of (C-6e); the other is a set of curves showing E *vs.* the right side of (C-6e) for various values of V_0. The permissible values of E are those for which the two graphs cross. (The intersection at E = 0 is of no interest, since it requires that

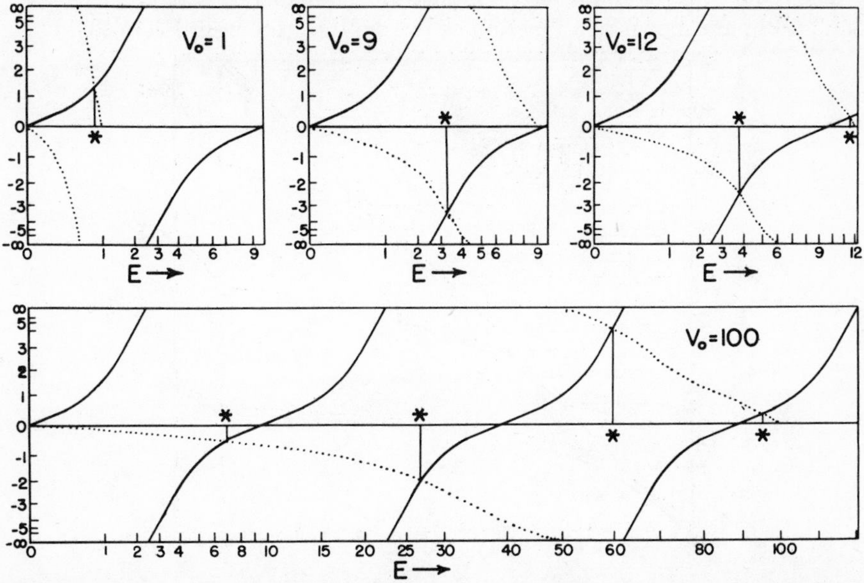

Fig. 6-3. Determination of the roots of Equation (C-6e) for $V_0 = 1$, 9, 12, and 100. The solid curve is a plot of the left side of the equation *vs.* the energy, and the dotted curve is a plot of the right side of the equation. Intersections of the two curves occur at the roots of the equation and are indicated by an asterisk. Vertical and horizontal scales have been distorted so as to simplify the drawing. Units of energy are $\hbar^2/2ma^2$.

$\psi = 0$ everywhere.) It is seen that if $V_0 < \pi^2$, there is only one intersection aside from the one at $E = 0$. This means that if the box is too narrow or too shallow, only one bound state is possible. If $\pi^2 < V_0 < 4\pi^2$, only two bound states are possible, and if $4\pi^2 < V_0 < 9\pi^2$ there are three bound states, etc. If V_0 is very large compared with E, we have

$$2\sqrt{(E/V_0)(1 - E/V_0)} \Big/ \left(2\frac{E}{V_0} - 1\right) \cong 0 \tag{C-6f}$$

or

$$\tan\sqrt{E} \cong 0 \tag{C-6g}$$

which means that $\sqrt{E} \cong n\pi$, or $E \cong n^2\pi^2$, where n is an integer. If we recall that E in this expression is measured in units of $\hbar^2/2ma^2$, we see that in this case ($V_0 \gg E$)

$$E \cong n^2h^2/8ma^2 \tag{C-6h}$$

This is identical with the expression (B-5) for the energy of a particle in a box with infinitely high walls, as of course we should expect when $V_0 \gg E$.

The roots of Equation (C-6e), obtained in the manner outlined in the last paragraph, are shown as a function of the depth of the well, V_0, in Fig. 6-4.

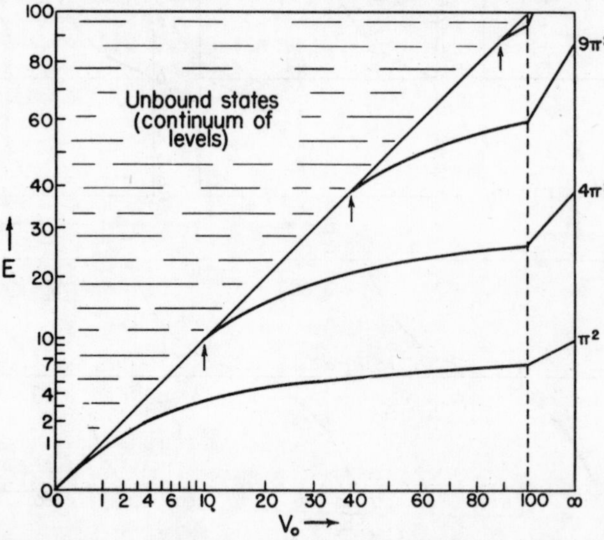

FIG. 6-4. Energy levels of a particle in a potential energy well of depth V_0. Arrows indicate points of appearance of new bound levels as V_0 is increased. Levels for a well of infinite depth are shown on the right. Vertical and horizontal scales are distorted so that a greater range of values could be shown. Units of energy are the same as in Fig. 6-3.

When E is less than V_0 (that is, in the portion of Fig. 6-4 lying below the line $V_0 = E$), there is a discrete set of energy levels. We shall show below that a continuum of levels exists above the line $V_0 = E$.

Let us now investigate the wave functions themselves. Here we find a most remarkable phenomenon. According to Equation (C-6b), if there is a particle in region II (that is, if $C \neq 0$) then there is a finite probability that the particle will also be found in region I (since the constant A cannot vanish; and if $A \neq 0$, then according to Equation (C-3a) ψ_1 cannot vanish). In other words, *quantum mechanics tells us that there is a definite possibility of finding particles in regions in which the total energy is less than the potential energy–that is, in regions in which the kinetic energy of the particles must be negative.* It is impossible to give a reasonable classical interpretation to this astonishing result. Yet there is much experimental evidence that particles can exist in such regions. Some of this evidence will be mentioned in Section D, below.

It is interesting to ask what this result means in terms of the oscillations of material bodies, as discussed in Chapters 3 and 4. The crux of the problem is the occurrence of the minus sign in Equation (C-1). If we look at the normal mode equation for a vibrating string (Equation (A-18) of Chapter 3), we see that a minus sign can be introduced if the tension is negative–that is, if the string is under compression. (It would also result if the string had a negative density, but this is physically difficult to conceive.) In accordance with the arguments given on pages 39ff, the string would in this case always curve away from the line of zero displacement. Only if at least one part of the string is under tension (so that the string can curve toward the line of zero displacement over part of its length) are normal modes conceivable in this case. If the string were under constant compression for $x < 0$ and $x > a$, and if it were under constant tension between $x = 0$ and $x = a$, it could theoretically execute perfectly respectable normal mode vibrations whose shapes are identical with the wave functions found for the bound particle in a potential energy well. These normal modes would be unstable.

Because of the close analogy between the steady state Schroedinger equation and the normal mode equations of vibrating bodies, it has probably occurred to the student that eigenvalues might be found for complex molecular systems by constructing suitable mechanical models and finding their normal mode frequencies (that is, by the use of so-called "analogue methods"). We shall find, however, that molecular systems invariably give rise to terms requiring negative tensions or negative densities in their analogous mechanical models. The difficulty of constructing models having such terms is one of the reasons that analogue methods for solving Schroedinger's equation are not feasible. Another reason is the difficulty in introducing the analogue of the electrostatic interactions between electrons and nuclei into a mechanical model.

It should be mentioned that there are also certain optical analogues to the behavior described above. For instance, it is found that if light passes from a medium of high refractive index onto an interface with a medium of low index of refraction, total reflection from the interface will occur if the angle of incidence is greater than a critical value. On solving the electromagnetic wave equation for this phenomenon, however, it is found that before it is reflected, the light does penetrate a short distance (of the order of one wave length) into the medium of low index of refraction.

(*ii*) *Free Particle Case.* If the total energy of the particle is greater than the height of the potential wall, V_0, then k_1 in Equation (C-1) becomes imaginary and the solutions in regions I and III are oscillatory. If we define the real constant k'_1 as

$$k'_1 = \sqrt{2m(E - V_0)}/\hbar \qquad \text{(C-7a)}$$

the Schroedinger equation (C-1) becomes

$$d^2\psi/dx^2 + k'^2_1\psi = 0 \qquad \text{(C-7)}$$

The solution of this equation could, of course, be written as the sum of a sine term and a cosine term, but instead we shall write it in the equivalent complex form

$$\psi_{\text{I}} = Ae^{ik'_1x} + Be^{-ik'_1x} \qquad \text{(C-7b)}$$

$$\psi_{\text{III}} = Fe^{ik'_1x} + Ge^{-ik'_1x} \qquad \text{(C-7c)}$$

where A, B, F, and G are arbitrary constants. These solutions correspond to a particle moving freely outside the box, and therefore do not change in amplitude as $x \to \pm\infty$. There is therefore no reason to discard any of the four constants. This gives us two more arbitrary constants with which to satisfy the continuity conditions, and the result is that it is no longer necessary to quantize the energy in order to obtain acceptable solutions. The particle can therefore have any energy greater than V_0. This is indicated in Fig. 6-4 by the continuum of levels lying above the line $E = V_0$.

Let us simplify this problem by assuming that in region III particles move only to the right, away from the well. This means that the constant G in (C-7c) must be set equal to zero (cf. page 180). Let us also write the solution in region II in the complex form

$$\psi_{\text{II}} = Ce^{ik_2x} + De^{-ik_2x} \qquad \text{(C-7d)}$$

where C and D are constants whose values will be determined.

The continuity requirements (C-4c) and (C-4d) at $x = a$ give the relations

$$C = \tfrac{1}{2} F e^{i(k'_1 - k_2)a} \left(1 + \frac{k'_1}{k_2}\right) \tag{C-7e}$$

$$D = \tfrac{1}{2} F e^{i(k'_1 - k_2)a} \left(1 - \frac{k'_1}{k_2}\right) \tag{C-7f}$$

Since $k'_1 \neq k_2$, neither C nor D can vanish. Therefore ψ_{II} contains a term involving e^{-ik_2x}. This means that some of the particles in region II move toward the left, although the number moving toward the right is larger than the number moving toward the left because $|D|^2$ is smaller than $|C|^2$. Physically this means that some of the particles moving toward the right in region II do not continue into region III when they strike the potential wall at $x = a$; that is, some of the particles are reflected back into region II by the wall. Such a partial reflection is not consistent with the classical treatment of a free particle passing a potential well, since a classical particle, moving through a region in which the potential energy is low, merely changes its velocity in this region without reversing its direction. Reflections of this type are, however, observed in optics when a light wave passes an interface between two media having different indices of refraction. The phenomenon is, in short, another result of the wave-like character of matter.

Exercises. (1) Consider the free particle case of the potential well problem. Assume that there are equal numbers of particles moving from right to left and from left to right in region I. (That is, assume that $|A|^2 = |B|^2$.) Show that in this case $|C|^2 = |D|^2$ and $|A|^2 = |F|^2 = |G|^2$; that is, the numbers of particles moving in the two directions is equal in regions II and III. Also show that $|C|^2 + |D|^2$ is less than $|A|^2 + |B|^2$; that is, the number of particles per unit length of path is smaller in region II than in regions I and III. Give a classical interpretation of this result.

(2) Approximately how many bound states are there for a 75-kg. man walking back and forth in a narrow passageway 10 meters long with walls 2 meters high at either end?

(3) Nuclei have dimensions of the order of 10^{-13} cm., and protons and neutrons are bound to the nucleus by energies of the order of 10^6 electron volts. Approximately how many bound states would you expect a typical nucleon in a nucleus to have?

b. "Tunnelling" through potential energy barriers

Suppose that a particle moves in a potential energy field for which $V = V_0$ between $x = 0$ and $x = a$, and $V = 0$ everywhere else, V_0 being a positive quantity (see Fig. 6-5). In this way a potential energy barrier may be made to intercept a beam of particles. According to classical mechanical ideas, if the energy of the particles in the beam is less than the height of the barrier, none of the particles will be able to get past the barrier. We shall now show, however, that according to quantum mechanics some of the particles are able to penetrate the barrier.

Let us suppose that the energy of the particles, E, is less than V_0. Then the Schroedinger equation for regions I and III of Fig. 6-5 may be written

$$d^2\psi/dx^2 + k_2{}^2\psi = 0 \tag{C-8a}$$

and for region II

$$d^2\psi/dx^2 - k_1{}^2\psi = 0 \tag{C-8b}$$

where $k_1 = \sqrt{2m(V_0 - E)}/\hbar$ and $k_2 = \sqrt{2mE}/\hbar$. We shall look for the solution in the case that a beam of particles moves only to the right in region III, whereas particles will be permitted to move in both directions

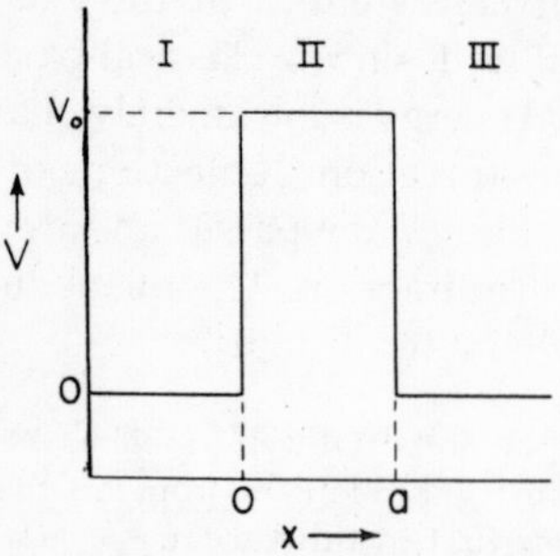

FIG. 6-5. Potential energy of a particle at a barrier.

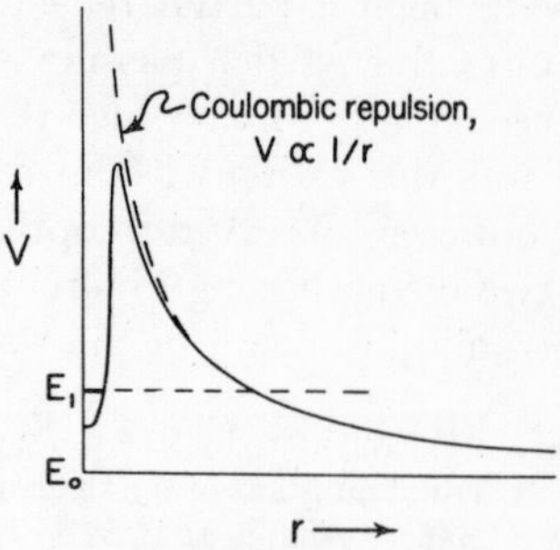

FIG. 6-6. Potential energy of an alpha particle near a nucleus.

in regions I and II. This is equivalent to the situation in which a beam of particles in region I impinges on the barrier from the left, a certain fraction of them being able to get through while the remainder are merely reflected and move off toward the left in region I. The solution we seek has the form,

$$\psi_{\rm I} = Ae^{ik_2x} + Be^{-ik_2x} \quad \text{in region I} \tag{C-8c}$$

$$\psi_{\rm II} = Ce^{k_1x} + De^{-k_1x} \quad \text{in region II} \tag{C-8d}$$

$$\psi_{\rm III} = Fe^{ik_2x} \quad \text{in region III} \tag{C-8e}$$

The number of particles impinging on the barrier from the left is proportional to $|A|^2$, and the number that get through the barrier is proportional to

$|F|^2$. The probability that a particle will get past the barrier, the *transmission coefficient* of the barrier, is therefore given by

$$T = |F|^2/|A|^2 \tag{C-9a}$$

The value of the transmission coefficient may be found by applying the continuity relations (C-4a, b, c, d) at $x = 0$ and $x = a$ to Equations (C-8c, d, e). It is found that these continuity relations may be satisfied regardless of the value of the energy; therefore the energy is not quantized. They result in the following expression

$$T = \left[1 + \frac{V_0^2 \sinh^2 (\sqrt{2m(V_0 - E)a^2/\hbar})}{4E(V_0 - E)}\right]^{-1} \tag{C-9b}$$

If $2m(V_0 - E)a^2/\hbar^2 >> 1$, this reduces to

$$T \simeq \frac{16E(V_0 - E)}{V_0^2} e^{-2\sqrt{2m(V_0-E)a^2}/\hbar} \tag{C-9c}$$

There is much evidence that tunnelling phenomena of this type actually occur. The most direct evidence is found in the radioactive disintegration of atomic nuclei by alpha decay. It is well known, for instance, that radium decays by the process

$$\text{Ra} \longrightarrow \text{Rn} + \alpha$$

If an alpha particle is brought up to a radon nucleus, it is strongly repelled by the strong Coulombic forces arising from the large nuclear charge of the radon. At very short distances, however, nuclear binding forces come into play and result in attraction and binding of the alpha particle by the nucleus. These interactions are described by the potential energy curve shown in Fig. 6-6. Suppose now that an alpha particle inside a radium nucleus has an energy E_1, which is greater than the energy E_0 of the particle far from the nucleus, but not sufficient for the alpha particle to climb over the top of the Coulombic repulsion barrier. In spite of this, the particle is still able to leave the radium nucleus by quantum mechanical tunnelling through the barrier. This picture of alpha decay has been worked out in considerable detail and is able to account for experimental laws such as the Geiger-Nuttal relationship between the half life and the kinetic energy of the alpha particle (see *2*, Chapter 6; *3*, Chapter 11).

Tunnelling is also important whenever electrons move from one atom to another because of the small mass of the electron (note that the mass occurs in the exponent of the transmission coefficient in Equation (C-9c)). It is therefore unquestionably important in many chemical processes in which electron transfer takes place–such as redox reactions and reactions at electrodes. The emission of electrons from metals at low temperatures and high electric field strengths ("cold emission") is also controlled by tunnelling.

Except in nuclear disintegrations, tunnelling is much less important for particles other than electrons, because of the strong influence of the mass. The protons in ammonia are able to tunnel through the barrier that maintains the pyramidal form of the ammonia molecule, and this process is responsible for the very strong microwave absorption by ammonia at a wave length of 1.26 cm. Tunnelling has also been suggested as a factor in reactions involving motions of hydrogen atoms (e.g., the electrolytic conductance by hydrogen and hydroxyl ions in acidic and basic solutions), but the importance of this factor is still not clear.

Exercises. (1) Prove Equation (C-9b).

(2) Calculate the probability that you would "tunnel" through a wall quantum mechanically if the wall were 20 ft. high and one foot thick and if you were moving at 20 miles per hour. (Answer: $T \simeq 10^{-10^{36}}$.) Would you say that the failure to observe tunnelling on a macroscopic scale is evidence against the Schroedinger equation?

(3) Calculate the probability that an electron moving with thermal energy, kT, at 300°K will tunnel through a barrier 10 A. thick and 4 e.v. high.

D. Rotating Bodies

a. The rotator with fixed axis of rotation (one-dimensional rotator)

Suppose that a particle of mass m is constrained to move in a circular path of fixed radius about a fixed axis. To a crude approximation this may be said to describe the motion of the so-called "pi electrons" around the benzene ring (page 428). It also represents the motion of any body whose axis of rotation is rigidly fixed in space by stationary bearings. The position of such a body is completely determined by a single angle, φ, Classically the energy of such a system is entirely kinetic, and its Hamiltonian takes the form

$$H = M_{\varphi}^2/2I \tag{D-1}$$

where I is the moment of inertia about the axis and M_{φ} is the angular momentum. The Schroedinger equation thus has the form

$$-(\hbar^2/2I)\, d\psi^2/d\varphi^2 = E\psi \tag{D-2}$$

which is similar to Equation (A-1) for the free particle. The well behaved solutions of this equation must, however, obey the condition

$$\psi(\varphi) = \psi(\varphi + 2\pi) \tag{D-3}$$

(*cf.* page 169). As a result we have for the acceptable solutions

$$\psi = e^{im\varphi} \tag{D-4}$$

where $m = 0, \pm 1, \pm 2, \cdots$. The permissible energy values are

$$E = m^2\hbar^2/2I \tag{D-5}$$

Each energy level except that for $m = 0$ is doubly degenerate.

These solutions are eigenfunctions of the operator for angular momentum about the chosen axis, $M_\varphi = (\hbar/i)\,\partial/\partial\varphi$, the eigenvalues being $m\hbar$. The solutions given in (D-4) therefore correspond to states in which the body moves in a definite direction about the axis, and with a definite rate of rotation. It is also possible to write the solutions of (D-2) in the form

$$\psi = A \sin m\varphi + B \cos m\varphi \tag{D-6}$$

but in general these are not eigenfunctions of M_φ, since they result from superposition of two solutions of the form (D-4) having positive and negative values of m.

b. The rotating dumb-bell or linear molecule

More important and somewhat more complex is the system typified by the diatomic molecule, consisting of two or more point masses arranged along a straight line free to rotate in any direction. The moment of inertia about the line through the masses is zero. Assume that the center of gravity of the system is fixed. Then its position in space may be completely specified by means of two polar coordinates, θ and φ, which give the orientation of the line through the masses relative to three fixed, mutually perpendicular directions in space, as shown in Fig. 2-1. The classical expression for the energy is

$$H = M^2/2I \tag{D-7}$$

where M is the total angular momentum and I is the moment of inertia about an axis passing through the center of gravity and normal to the line through the masses. We shall show in Chapter 8 that the eigenfunctions of this operator are the surface spherical harmonics

$$S_{l,m}(\theta,\varphi) = P_l^{|m|}(\cos\theta)\, e^{im\varphi} \tag{D-8}$$

where l and m are integers and $P_l^{|m|}(\cos\theta)$ is the associated Legendre function of order $|m|$ and degree l, which has been discussed on pages 90*ff* in connection with the oscillations of the ocean on a flooded planet. The energy can have the values

$$E = (\hbar^2/2I)\, l(l+1) \tag{D-9}$$

For each value of E there are $2l + 1$ values of m, running from $-l$ to l. As will be shown in Chapter 8, the quantum number l determines the rate of rotation of the molecule, whereas m is concerned only with the spatial orientation of the axis of rotation.

c. The symmetrical top

A *symmetrical top* is a rigid body for which two of the principle moments of inertia are equal, but differ from the third. Examples are the CH_3Cl and benzene molecules, the conventional children's top, and any other object with cylindrical symmetry. The quantum mechanical treatment is somewhat involved, so we shall merely give the eigenvalues. A more complete treatment will be found in Margenau and Murphy (*4*, p. 352) and Pitzer (*5*, p. 488).

Let the two identical moments of inertia be A, and the third moment of inertia be C. Then the possible energy values are

$$E = \tfrac{1}{2}\hbar^2 \left[\frac{J(J+1)}{A} + K^2\left(\frac{1}{C} - \frac{1}{A}\right)\right] \tag{D-10}$$

where

$$J = 0, 1, 2, \cdots$$
$$K = 0, \pm 1, \pm 2, \cdots, \pm J$$

There is also a third quantum number, M, whose value has no influence on the energy. It can take all integral values from $-J$ to $+J$ for each value of J and K.

The total angular momentum of the body is $\hbar\sqrt{J(J+1)}$, so J is called the total angular momentum quantum number. The component of angular momentum about the C-axis is $K\hbar$, and the spatial orientation of the axis of total angular momentum is determined in the usual way by M. Thus for a given value of J, if $K = 0$ the molecule rotates about an axis normal to the C-axis, whereas if $K = \pm J$ it rotates essentially (though not exactly, see page 250) about the C-axis.

If the two moments of inertia, A and C, are equal, we have what is known as the *spherical top*, exemplified by such molecules as CH_4 and SF_6, and by such objects as the cube and the sphere. In this case the second term on the right in Equation (D-10) vanishes and we have energy levels that are identical with those of the linear molecule, (D-9). The degeneracy is, however, $(2J + 1)^2$ instead of $(2J + 1)$, because both the quantum numbers K and M can assume integral values from $-J$ to $+J$ independently.

d. The unsymmetrical top

If all of the principle moments of inertia of a molecule are unequal, we have what is called the *unsymmetrical top*. Most molecules are of this type, e.g., the water and toluene molecules. The wave functions and energy levels are rather complex, and will not be described here (see Pitzer (*5*, p. 437) and articles by King, Hainer, and Cross (*6*)).

II. SYSTEMS FOR WHICH THE POTENTIAL ENERGY IS NOT CONSTANT

E. The One-Dimensional Harmonic Oscillator

Consider a particle of mass m attached to a weightless spring and restricted in some way so that it can move only in the x-direction. The force acting on this particle is

$$f = -kx \tag{E-1}$$

where k is a characteristic constant of the spring, known as the *force constant*, and x is the displacement of the particle from its equilibrium position. The potential energy of the particle is

$$V = -\int f\,dx = kx^2/2 \tag{E-2}$$

the zero of energy being taken as the particle at rest at $x = 0$.

Let us first investigate the classical description of the motion of the particle. This follows immediately from Newton's law of motion

$$md^2x/dt^2 = -kx \tag{E-3}$$

which has the general solution

$$x = x_0 \sin (\sqrt{k/m}\, t + \delta) \tag{E-4}$$

where x_0 and δ are arbitrary constants, x_0 being the amplitude of the oscillation. According to the classical theory, therefore, the particle oscillates from $x = x_0$ to $x = -x_0$ sinusoidally with the time at a frequency $\nu = (\frac{1}{2\pi}) \sqrt{k/m}$. The energy thus changes back and forth from the kinetic energy form to the potential energy form, being exclusively kinetic when $x = 0$ and exclusively potential when $x = x_0$. The total energy is a constant given by $kx_0^2/2$. It can have any positive value, there being no limit on the value of x_0.

Now let us consider the properties of such a system according to the laws of quantum mechanics. The Schroedinger equation is

$$d^2\psi/dx^2 + (2m/\hbar^2)(E - \tfrac{1}{2}kx^2)\psi = 0 \tag{E-5}$$

Inspection of this equation in the manner described in Section Gb of Chapter 2 (see especially Exercise 1 on page 41) will show that well behaved solutions are possible only for certain values of the energy, E. It is therefore immediately evident that the energy of the harmonic oscillator is quantized. Unfortunately none of the vibrating systems discussed in Chapter 3 give normal mode equations resembling (E-5), so we must expect solutions different from any yet encountered. Furthermore, it is difficult to conceive a real system that would yield a normal mode equation of this type. If we write the second term on the left of (E-5) in the form $(2mE/\hbar^2)\,[1 - (kx^2/2E)]$ and compare this with the normal mode equation for a string, (A-18) of Chapter 3, we see that a string obeying (E-5) would have to have a negative density when $x^2 > 2E/k$. This, of course, is not realizable in a real system, although it is possible to imagine such a system.

Equation (E-5) may be simplified by writing

$$\alpha = 2mE/\hbar^2; \quad \beta = \sqrt{mk}/\hbar; \quad \xi = \sqrt{\beta}\, x \tag{E-6}$$

the new variable ξ being dimensionless. We then obtain

$$d^2\psi/d\xi^2 - [(\alpha/\beta) - \xi^2]\psi = 0 \tag{E-7}$$

In order to find the possible values of the energy and the shapes of the corresponding wave functions, let us first investigate this equation when ξ has large values, so that the factor (α/β) can be neglected in (E-7). We then obtain

$$d^2\psi/d\xi^2 - \xi^2\psi \simeq 0 \tag{E-8a}$$

which has the approximate solution

$$\psi \simeq Ae^{-\xi^2/2} - Be^{\xi^2/2} \tag{E-8b}$$

(Equation (E-8b) is actually the exact solution of the differential equation $d^2\psi/d\xi^2 = (\xi^2 - 1)\psi$, but this reduces to (E-8b) if $\xi >> 1$.) In this solution ψ will increase without limit at large positive and negative values of x unless we set $B = 0$. Thus we may conclude that for large values of ξ the solution of (E-7) must vary approximately as $e^{-\xi^2/2}$. This suggests that for the precise solution we assume that ψ has the form

$$\psi(\xi) = H(\xi)e^{-\xi^2/2} \tag{E-9}$$

where $H(\xi)$ is a function that varies in such a way that its product with $e^{-\xi^2/2}$ vanishes as ξ becomes large. Substitution of (E-9) into (E-7) gives us the following differential equation for $H(\xi)$

$$d^2H/d\xi^2 - 2\xi dH/d\xi + [(\alpha/\beta) - 1]H = 0 \tag{E-10}$$

In order to solve this equation we shall invoke the method of the power series used in treating the normal modes of the flooded planet (page 87). We shall assume that

$$H(\xi) = \sum_{m=0}^{\infty} a_m\xi^m \tag{E-11}$$

where the coefficients a_m are constants to be determined.

Substituting (E-11) into (E-10) we find

$$dH/d\xi = \sum ma_m\xi^{m-1} \tag{E-11a}$$

$$d^2H/d\xi^2 = \sum m(m-1)a_m\xi^{m-2} \tag{E-11b}$$

giving, on substitution into (E-10)

$$\sum m(m-1)a_m\xi^{m-2} - \sum 2ma_m\xi^m + [(\alpha/\beta) - 1]\sum a_m\xi^m = 0 \tag{E-12}$$

or

$$\sum \{(m+2)(m+1)a_{m+2} - [2m + 1 - (\alpha/\beta)]a_m\}\xi^m = 0 \tag{E-12a}$$

In order for this equation to be true for all values of ξ, the coefficients of each of the powers of ξ must vanish. Thus we obtain the recursion formula

$$\frac{a_{m+2}}{a_m} = \frac{2m + 1 - (\alpha/\beta)}{(m + 2)\,(m + 1)} \tag{E-13}$$

Since this formula relates alternate coefficients, and since it gives merely the ratios of coefficients, it is clear that the first two coefficients of the series, a_0 and a_1, may be chosen arbitrarily. Therefore the general solution can be expressed as any linear combination of two series: one series involving only odd powers of ξ and the other involving only even powers of ξ

$$H_{\text{even}} = a_0 \sum_{m=0}^{\infty} (a_{2m}/a_0)\,\xi^{2m} \tag{E-14a}$$

$$H_{\text{odd}} = a_1 \sum_{m=0}^{\infty} (a_{2m+1}/a_1)\,\xi^{2m+1} \tag{E-14b}$$

The constants a_0 and a_1 are the two arbitrary constants that always appear in the general solution of a second order differential equation.

We must now investigate the behavior of these solutions at large values of ξ. Since they are power series, it is clear that this behavior must be determined by the terms involving the higher powers of ξ. From the recursion formula (E-13) we see that for large values of the exponent m

$$a_{m+2}/a_m \cong 2/m \tag{E-15}$$

Now the Taylor expansion of the function e^{ξ^2} shows that

$$e^{\xi^2} = 1 + \xi^2 + \xi^4/2! + \xi^6/3! + \cdots + \xi^m/(\tfrac{1}{2}m)! + \cdots \tag{E-15a}$$

showing that the successive coefficients in the expansion of this function are in the ratio $(2/m)$, just as was found for the higher power terms of the function $H(\xi)$. Thus for large values of ξ the functions $H(\xi)$ and e^{ξ^2} vary in approximately the same way, so that we may write

$$H(\xi) \sim e^{\xi^2} \quad (\xi >> 1) \tag{E-15b}$$

If we substitute this result into the expression (E-9) for the wave function itself, we see that

$$\psi \sim e^{\xi^2} e^{-\xi^2/2} = e^{\xi^2\,2} \quad (\xi >> 1) \tag{E-15c}$$

This function is, however, not well behaved, since it increases without limit as ξ increases. Therefore it is clear that we do not obtain well behaved solutions if $H(\xi)$ is an infinite series. There is only one way out of this difficulty. One of the two series, either H_{even} or H_{odd}, must break off after a finite number of terms. (The other series can then be eliminated by setting its first coefficient, a_0 or a_1 as the case may be, equal to zero.) The series may be made to break off by choosing the ratio (α/β) so that for some one value of m (say $m = n$) we have

$$\alpha/\beta = 2n + 1 \tag{E-16}$$

Then according to the recursion formula (E-13), all coefficients after the nth coefficient must vanish. At worst our wave function will then vary as $\xi^n e^{-\xi^2/2}$ for large values of ξ, but this is still a well behaved function.

Thus if we are to have well behaved solutions, Equation (E-16) must be true. On substituting for α and β from (E-6), we find that this means that the energy can only have the values

$$E = \hbar\sqrt{k/m}\,(n + \tfrac{1}{2}) \tag{E-17a}$$

or, since the frequency ν of the oscillator is equal to $(\frac{1}{2\pi})\sqrt{k/m}$, we have

$$E = h\nu\,(n + \tfrac{1}{2}) \tag{E-17b}$$

This is the well known rule for the quantized energy levels of an oscillator.

For each value of n in (E-17) we can derive a polynomial in which the highest power of ξ is ξ^n. The polynomial thus obtained is given the symbol $H_n(\xi)$, and is called the *Hermite polynomial of degree n*. It is readily shown from the recursion formula (E-13) that

$$H_0 = 1$$

$$H_1 = 2\xi$$

$$H_2 = 4\xi^2 - 2$$

$$H_3 = 8\xi^3 - 12\xi$$

$$H_4 = 16\xi^4 - 48\xi^2 + 12$$

In these expressions, the choice of the initial coefficient has been made in accordance with a convention to be described below.

On introducing (E-16) into (E-10) we obtain the following differential equation for the Hermite functions

$$d^2H_n/d\xi^2 - 2\xi dH_n/d\xi + 2nH_n = 0 \tag{E-18}$$

It may be shown by direct substitution that this equation, which is known as *Hermite's equation*, has as a solution

$$H_n = Ce^{\xi^2}d^ne^{-\xi^2}/d\xi^n \tag{E-19}$$

where C is an arbitrary constant. It is conventional to choose $C = (-1)^n$; this results in the functions given above.

The wave functions for the harmonic oscillator thus have the form

$$\psi(x) = H_n(\sqrt{\beta}x)e^{-\beta x^2/2} \tag{E-20}$$

These functions* are plotted in Fig. 6-7 for various values of the quantum

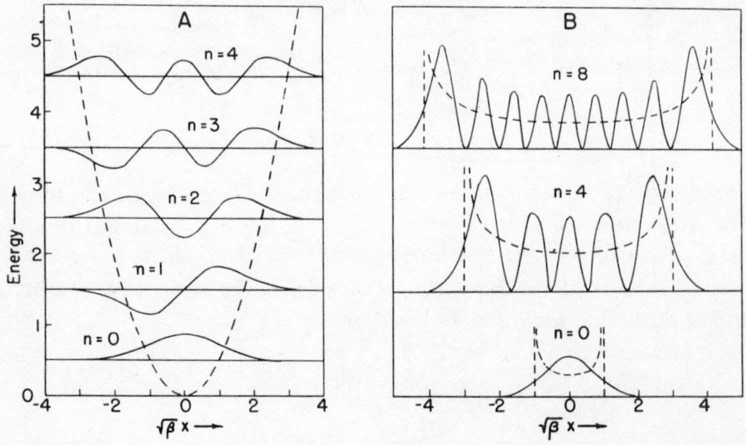

FIG. 6-7.

A. Energy levels and wave functions of the harmonic oscillator. The horizontal lines represent the energy levels (energy being measured in units of $h\nu$). The dashed line is the potential energy function. The wave function is superimposed on each level.

B. Probability distribution for the harmonic oscillator in levels $n = 0$, 4, and 8. The solid lines are the quantum mechanical probability functions and the dashed lines are the classical probability functions corresponding to the same energy.

* Some of these functions have already been encountered in the Exercises on page 18.

number n. Evidently the particle is not necessarily to be found at the point $x = 0$ when it is in its lowest energy state, and furthermore its energy is not zero in the lowest energy state. These results are important from the standpoint of the uncertainty principle and will be considered further in Chapter 7.

It is interesting to compare the probability distribution $|\psi|^2$ with the classical probability of finding the oscillating particle at any given position on its path. Classically the probability of finding a particle at a point x is inversely proportional to its velocity at the point. Since

$$E = kx^2/2 + mv^2/2 \tag{E-20}$$

where v is the velocity, we obtain

$$v = \sqrt{(2E/m) - (kx^2/m)} \tag{E-21}$$

Thus the classical probability distribution has the form

$$p_c \propto 1/\sqrt{(2E/m) - (kx^2/m)} \tag{E-22a}$$

for values of x in the range $-x_0 < x < x_0$, where $x_0 = \sqrt{2E/k}$; and of course we also have

$$p_c = 0 \tag{E-22b}$$

outside of this range.

Quantum mechanically we find, on the other hand, that the probability distribution is given by

$$p_q \propto H_n^2(\sqrt{\beta}x)e^{-\beta x^2} \tag{E-23}$$

In Fig. 6-7, p_c and p_q are plotted for various energies of the harmonic oscillator. It is evident that the envelope of p_q tends to approach p_c at higher energies; this is another example of the correspondence principle (see page 186). It is evident also that quantum mechanics gives the particle a finite probability of appearing in regions which, according to the classical theory, are completely forbidden. This, of course, is another example of the barrier penetration effects mentioned earlier (page 193).

Exercises. (1) Show that aside from an arbitrary multiplier, Equations (E-19) and (E-11) (with (E-13 and 16) give the same function for $H_4(\xi)$.

(2) Find the wave functions and energy levels for an isotropic three-dimensional harmonic oscillator, that is, for a particle that can move about the origin with a potential energy $V = kr^2/2$, where r is the distance from the particle to the origin. [Hint: Set $r = \sqrt{x^2 + y^2 + z^2}$ and separate the variables in the resulting Schroedinger equation.]

(3) Find the wave functions and energy levels for the anisotropic oscillator, for which $V = (k_x x^2 + k_y y^2 + k_z z^2)/2$, where k_x, k_y, and k_z are three unequal constants.

(4) Prove the following mathematical relations between the Hermite polynomials, starting with the relation (E-19):

$$\text{(a) } dH_n(\xi)/d\xi = 2n\, H_{n-1}(\xi) \tag{E-24}$$

$$\text{(b) } H_n - 2\xi\, H_{n-1} + 2(n-1)H_{n-2} = 0 \tag{E-25}$$

(5) Show that the functions

$$\psi_n = \frac{(\beta/\pi)^{1/4}}{\sqrt{2^n\, n!}} H_n(\sqrt{\beta}\, x)e^{-\beta x^2/2} \tag{E-26}$$

are normalized. (That is, $\int_{-\infty}^{\infty} \psi_n^2 dx = 1$) [Hint: Make use of (E-19) and (E-24) and integrate repeatedly by parts.]

(6) ***The Particle in a Uniform Field.*** Suppose that a constant force, $-g$, acts on a mass m moving in the x-direction. This will be the case for a charged particle in a uniform electric field or for an object subject to the earth's gravitational field and not too far from the surface of the earth.

(a) Starting with Newton's law of motion, show that the classical probability distribution function is

$$p_c \propto 1/\sqrt{x_0 - x}$$

where x_0 is the value of x at which the kinetic energy of the particle vanishes ($x_0 = E/g$ where E is the total energy of the particle and the potential energy is $V = gx$).

(b) The Schroedinger equation in this case is

$$-(\hbar^2/2m)d^2\psi/dx^2 + gx\psi = E\psi$$

Making the substitutions

$$z = \frac{2}{3}\sqrt{2mg/\hbar^2}(x_0 - x)^{3/2} \text{ and } \psi = z^{1/3}f(z)$$

show that one obtains the equation

$$d^2f/dz^2 + (1/z)df/dz + [1 - (1/9z^2)]f = 0$$

having as solutions Bessel's functions of order $1/3$, $f = J_{1/3}(z)$. [Hint: *Cf.* Equation (B-21) of Chapter 3.]

(c) For large, real, positive values of z, it can be shown that $J_{1/3}(z) \cong \sqrt{2/\pi z} \cos (z - 5\pi/12)$. Show that for large values of x the envelope of the quantum mechanical distribution function, $|\psi|^2$, has the same shape as the classical distribution function derived in (a), above.

(d) For a ball bouncing on an infinitely hard surface under the influence of gravity $V = \infty$ if $x < 0$ and $V = gx$ if $x > 0$. Show that the energy of such a ball is quantized with the values

$$E = (3z_n/2a)^{2/3}g$$

where $a = (2mg/h^2)^{1/3}$ and z_n is any one of the roots of the equation $J_{1/3}(z) = 0$. Also, show from the behavior of $J_{1/3}$ for large values of z that the higher energy levels are given approximately by

$$E \simeq \left[\frac{3h^2g^{1/2}}{16\pi m}\left(n - \frac{1}{12}\right)\right]^{2/3}$$

where n is an integer.

F. The Sinusoidal Potential

The potential

$$V = -V_0 \cos nx \qquad \text{)F-1)}$$

is believed to be a good representation for the energy changes that take place in the ethane molecule when the two methyl groups are rotated with respect to one another about the carbon–carbon bond. In this case the constant n is three and the variable x is the angle between the two planes that include the C–C bond and one of the C–H bonds on each methyl group. V_0 is a constant typical of the molecule (approximately 1.5 kcal. per mole for ethane). A potential of this type also arises in the problem of the pendulum (see *7*, p. 77), and it is of some interest in connection with crystals (see *8*).

In the case of the ethane molecule x is an angular variable, and the Schroedinger equation becomes

$$-(\hbar^2/2I)d^2\psi/dx^2 - V_0 \cos nx\psi = E\psi \qquad \text{(F-2)}$$

where I is the moment of inertia of a methyl group about the carbon–carbon bond. If we make the substitutions

$$\phi = nx/2 \qquad a = 2IE/n^2\hbar^2$$

$$f(\phi) = \psi(x) \qquad q = -IV_0/2n^2\hbar^2$$

this equation is transformed into Mathieu's equation

$$\frac{1}{4}\, d^2f/d\phi^2 + (a - 4q \cos 2\phi)f = 0 \qquad \text{(F-3)}$$

The solutions of this equation are described in Jahnke and Emde (*1*, pp. 283ff.). Since x varies from 0 to 2π we must have

$$\psi(0) = \psi(2\pi) \quad \text{or} \quad f(0) = f(n\pi)$$

in order that the solutions be well behaved. For a given value of the constant q (that is of n, I and V_0), this is possible only for a discrete set of values of a (that is, of the energy E). Thus the energy is quantized. The relationship of a to q is, however, rather complex; it is more readily understood if we consider the two extremes, q very small and q very large.

(i) *q very small.* If q can be neglected, (F-3) takes the form

$$d^2f/d\phi^2 + 4af = 0 \tag{F-4}$$

which has well behaved solutions only if $a = m^2$, where $m = 0, \pm 1, \pm 2, \cdots$. If we write $M = mn$ we obtain the energy levels

$$E = M^2\hbar^2/2I \tag{F-5}$$

and the eigenfunction

$$\psi = e^{iMx/2} \tag{F-6}$$

These are similar in form to those of a rotator with fixed axis (Section Da, above), as one would expect. The difference here is in the fact that M takes on only the values 0, $\pm n$, $\pm 2n$, $\cdots$. Thus only one out of every n levels of the rotator appear here.

(ii) *q very large.* If q is very large we should expect that the system will be found mostly in configurations in which x is close to 0, $2\pi/n$, $4\pi/n$, $\cdots$, 2π. If we make the approximation $\cos 2\phi \simeq 1 - 2\phi^2$, Equation (F-3) becomes

$$\frac{1}{4}\, d^2f/d\phi^2 + (a - 4q + 8q\phi^2)f = 0 \tag{F-7}$$

If we write

$$\chi = \sqrt[4]{-32q}\,\phi;\, A = (a - 4q)/\sqrt{-2q}$$

Equation (F-7) becomes

$$d^2f/d\chi^2 - (A - \chi^2)f = 0 \tag{F-8}$$

which has the same form as the Schroedinger equation for a harmonic oscillator (Equation (E-7)). The solutions are well behaved only if $A = 2K + 1$, where K is an integer, so that

$$a = 4q + (2K + 1)\sqrt{-2q}$$

and the energy levels are

$$E = -V_0 + n\hbar\sqrt{V_0/I}\,(K + \tfrac{1}{2}) \tag{F-9}$$

The dependence of E on V_0 for intermediate values of V_0 is shown in Fig. 6-8.

G. The Hydrogen Atom and Hydrogen-Like Ions

a. The classical treatment of the Rutherford planetary atom

Rutherford showed in 1911 that atoms consist of a concentrated, heavy positive charge, the nucleus, about which the much less massive electrons move. The nature of this motion is most easily seen by considering the simplest atom–a nucleus plus a single electron. This system is found in the hydrogen atom, and also in the He^+, Li^{++}, Be^{+++}, etc., ions. The potential energy function in this case is readily derivable from Coulomb's law of force between a pair of charge particles. The force is in the direction of the line joining the two charged particles and has the magnitude $F = -e \cdot Ze/r^2$ where $-e$ is the charge on the electron, Ze is the charge on the nucleus (that is, Z is the atomic number), and r is the distance between the two particles. The potential energy resulting from this force is

$$V = -\int_{\infty}^{r} F\,dr = -Ze^2/r \tag{G-1}$$

where the zero of energy is taken at $r = \infty$.

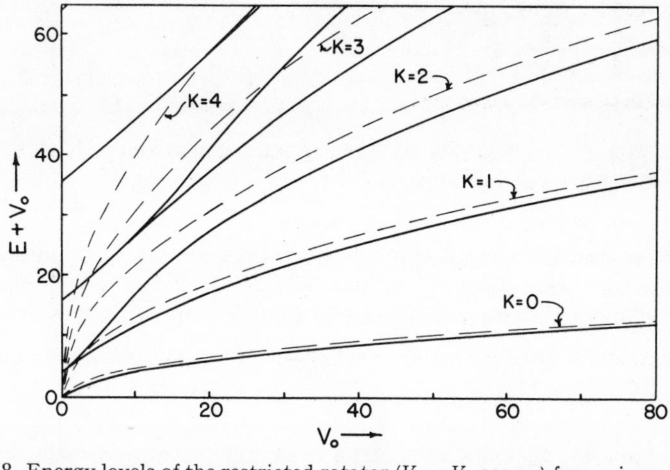

FIG. 6-8. Energy levels of the restricted rotator ($V = V_0 \cos nx$) for various values of V_0. Dashed lines give the energy levels according to the approximate expression, Equation (F-9). Energy and V_0 are expressed in units of $n^2\hbar^2/8I$. Note that the ordinate, $E + V_0$, is the energy measured from the minimum of the potential energy function.

Classically it can be shown (see, for example, *9*, pp. 36*ff*) that this law of force leads to elliptical, parabolic and hyperbolic orbits of the electron about the nucleus (or, more accurately, about the center of gravity of the system). The elliptical orbits correspond to a bound state of the system, whereas the other orbits correspond to states in which the electron is unbound. This is closely analogous to the motions of the planets and comets about the sun in the solar system, the law of force in the two cases being of similar form.*

Forces other than Coulombic ones also exist in atoms. The magnetic force between two moving charges is negligible compared with the Coulombic force (the ratio of the two is of the order of v^2/c^2 where v is the relative velocity and c is the velocity of light); so is the gravitational force between the elementary particles (it is of the order of 10^{-39} times the Coulombic force). On the other hand there exists a "radiation force" which is present whenever a charged particle is accelerated (see page 555). This force converts the kinetic energy of the particle into radiation, and therefore leads to energy dissipation. It should exist in the Rutherford model of the hydrogen atom since the electron is being continually accelerated towards the nucleus. Therefore the classical elliptical orbits of hydrogen should not be stable. The fact that normal hydrogen atoms do not continually emit radiation contradicts this conclusion. This was one of the great dilemmas raised by the Rutherford model of atomic structure.

b. The Bohr-Sommerfeld atom

The first step in the resolution of this situation was taken in 1912 by Bohr, who made the following assumptions:

(*i*) Atoms are able to exist only in certain states of constant energy.

(*ii*) Light is emitted and absorbed when the atom goes from one of these states to another. The frequency of the light is $\nu = \Delta E/h$, where ΔE is the difference in energy of the states and h is Planck's constant.

(*iii*) The angular momentum of the electron is quantized in multiples of $h/2\pi = \hbar$.

* Atoms containing more than one electron would behave very differently, however, from a solar system with the same number of planets, since two electrons repel one another, whereas a pair of planets will always attract. Furthermore, the gravitational force between two planets is very much smaller than that between the sun and a planet (the ratio of the planet-planet force to the sun-planet force for a given separation being 10^{-3} for a pair of Jupiters, 3×10^{-6} for a pair of Earths, 3×10^{-8} for a pair of Mercuries). The force between a pair of electrons, on the other hand, is relatively large compared with the force between the nucleus and an electron (the ratio is unity for hydrogen and 10^{-2} for the heaviest elements). Thus the effect of the electrons on each other's motion is much greater than that of the planets, and the sun dominates the motions of the planets to a much greater extent than the nucleus dominates the motion of the electrons. Perturbation methods of calculating planetary motions are therefore capable of far more precision than the same methods would be in an atom.

The third postulate was later generalized through the Wilson-Sommerfeld quantization rule. This rule states that for any cyclic coordinate in a system (that is, one periodically retraced by the system), the motion must be such that

$$\oint p\,dq = nh \tag{G-2}$$

where q is the coordinate, p is the momentum associated with it, n is an integer, and the integral is taken over one complete cycle of the periodic motion. This rule can be understood in terms of the de Broglie theory of the wave nature of matter. According to de Broglie, the momentum p and the wave length λ are related by $p = h/\lambda$ (*cf.* page 164). Therefore the Wilson-Sommerfeld rule merely requires that $\oint (dq/\lambda) = n$. That is, the distance travelled in any periodic orbit must be equal to a whole number of de Broglie wave lengths (this interpretation was not, of course, available to the original workers).

In the classical picture of the hydrogen atom, two of the coordinates of the electron retrace their values each time the electron goes about the proton in an elliptical orbit. These coordinates are the distance between the nucleus and the electron, and the angle through which a line drawn from the nucleus to the electron moves. It is found on applying the Wilson-Sommerfeld rule to these two coordinates (see *9*, pp. 39ff.) that the permitted orbits are ellipses whose semi-axes are

$$a = (k + l)^2\hbar^2/Zm_e e^2; \; b = (k + l)l\hbar^2/Zm_e e^2 \tag{G-3a}$$

giving an axial ratio $a/b = (k + l)/l$, where k and l are integers and m_e is the electronic mass. The energy is found to be (for a hydrogen atom or a hydrogen-like ion with an atomic number Z)

$$E = -Z^2R/n^2 \tag{G-3b}$$

where $n = k + l$ and R, the so-called *Rydberg constant*, is given by $R = m_e e^4/2\hbar^2$. The case $l = 0$ was excluded because it corresponds to motion of the electron along a straight line ($a/b = \infty$) passing through the nucleus, and this was felt to be unreasonable. We shall see, however, that the wave mechanical solution does include such a motion, so this restriction was not justified.

The energy levels (G-3) found in this way agree with those observed for hydrogen atoms and hydrogen-like ions. Furthermore, Sommerfeld was able to account in a remarkable way for the fine structure of the hydrogen spectrum by including relativistic corrections to the electronic mass. We now know, however, that the angular momenta assigned to the levels were partially in error and that to a certain extent the success of the Bohr-Sommerfeld theory was fortuitous. Nevertheless, the concept of quantized orbits will be most useful to us in interpreting the quantum mechanical results.

c. The quantum mechanical treatment of the hydrogen atom

The Schroedinger equation for the hydrogen atom and hydrogen-like ions is

$$-(\hbar^2/2m_e)\nabla^2\psi - (Ze^2/r)\psi = E\psi \tag{G-4}$$

where it is assumed that the nucleus does not move. (If this assumption is not made, it is found that the same equation is obtained, except that the electronic mass m_e is replaced by the reduced mass of the system, $\mu = m_eM/(m_e + M)$, M being the mass of the nucleus. Since $M \gg m_e$, μ is very close to m_e.)

If one rearranges the terms and converts to atomic units of distance and energy (see page 177), (G-4) takes the form

$$\nabla^2\psi + 2(E + Z/r)\psi = 0 \tag{G-5}$$

The occurrence of the distance r in this equation suggests that we use polar coordinates, giving

$$\frac{1}{r^2}\frac{\partial}{\partial r}\left(r^2\frac{\partial\psi}{\partial r}\right) + \frac{1}{r^2 \sin\theta}\frac{\partial}{\partial\theta}\left(\sin\theta\frac{\partial\psi}{\partial\theta}\right) + \cdot$$

$$+ \frac{1}{r^2\sin^2\theta}\frac{\partial^2\psi}{\partial\varphi^2} + 2\left(E + \frac{Z}{r}\right)\psi = 0 \tag{G-6}$$

On applying the method of separation of variables to this equation, assuming that

$$\psi(r, \theta, \varphi) = R(r)\,S(\theta, \varphi) \tag{G-7}$$

one readily obtains the equations

$$\frac{1}{\sin\theta}\frac{\partial}{\partial\theta}\left(\sin\theta\frac{\partial S}{\partial\theta}\right) + \frac{1}{\sin^2\theta}\frac{\partial^2 S}{\partial\varphi^2} + l(l+1)\,S = 0 \tag{G-8}$$

$$\frac{1}{r^2}\frac{d}{dr}\left(r^2\frac{dR}{dr}\right) + \left(2E + \frac{2Z}{r} - \frac{l(l+1)}{r^2}\right)R = 0 \tag{G-9}$$

where the separation constant has been written as $l(l + 1)$. We have seen, in discussing the normal modes of the ocean on a flooded planet (pages 85ff.), that Equation (G-8) has well behaved solutions only if $l = 0, 1, 2, \cdots$. Then $S(\theta, \varphi)$ is a surface spherical harmonic, which may be written in either the complex form

$$S(\theta, \varphi) = P_l^{|m|}(\cos \theta) \, e^{im\varphi} \tag{G-10}$$

where $m = 0, \pm 1, \pm 2, \cdots, \pm l$, or in the equivalent real forms

$$S(\theta, \varphi) = P_l^m(\cos \theta) \sin m\varphi \tag{G-10a}$$

and

$$S(\theta, \varphi) = P_l^m(\cos \theta) \cos m\varphi \tag{G-10b}$$

where m is zero or a positive integer no larger than l. Because of the relationship between $e^{im\varphi}$ and $\sin m\varphi$ and $\cos m\varphi$ (see page 21) the complex solutions (G-10) (with positive as well as negative values of m) merely represent different linear combinations of the real solutions Equations (G-10a) and (G-10b).

Let us now consider the radial equation (G-9). This may be simplified somewhat by making the substitutions

$$\rho = 2\sqrt{-2E}\, r \tag{G-11a}$$

$$\lambda = Z/\sqrt{-2E} \tag{G-11b}$$

$$R(r) = T(\rho) \tag{G-11c}$$

(Since we shall be concerned with the bound states of the hydrogen atom and since the zero of energy has been taken as the electron at an infinite distance from the nucleus, the energy E is a negative quantity. Thus the variable ρ and the constant λ are real quantities.) We then obtain

$$\frac{d^2T}{d\rho^2} + \frac{2}{\rho}\frac{dT}{d\rho} - \left[\frac{1}{4} - \frac{\lambda}{\rho} + \frac{l(l + 1)}{\rho^2}\right]T = 0 \tag{G-12}$$

It is important to determine the behavior of this equation for large values of ρ, where

$$d^2T/d\rho^2 - T/4 \cong 0 \tag{G-13a}$$

which has the solution

$$T = Ae^{-\rho/2} + Be^{\rho/2} \tag{G-13b}$$

A and B being arbitrary constants. It is obvious that we must set $B = 0$ in order to have solutions that remain finite at large distances between the electron and the nucleus. This suggests that in looking for the precise solution of (G-12) we set

$$T = K(\rho)e^{-\rho/2} \tag{G-14}$$

where $K(\rho)$ is a function whose form is to be determined but which must increase less rapidly than $e^{\rho/2}$ as ρ goes to infinity. Substituting (G-14) in (G-12) we find

$$\rho^2 \frac{d^2K}{d\rho^2} + \rho(2 - \rho) \frac{dK}{d\rho} + [(\lambda - 1)\rho - l(l + 1)] K = 0 \tag{G-15}$$

This equation must be valid for $\rho = 0$, which gives

$$l(l + 1)K(0) = 0 \tag{G-16}$$

Thus if $l \neq 0$ we must have $K(0) = 0$.

The form of $K(\rho)$ can be determined by the method of power series. Because of (G-16), however, we know that if $l \neq 0$ any power series for K must not contain a constant term (that is, a term in ρ^0). Therefore we write

$$K(\rho) = \rho^s \sum_{k=0}^{\infty} a_k \rho^k \tag{G-17}$$

where s is a positive number if $l \neq 0$. Substituting (G-17) into (G-15) we find

$$\sum_{k=0}^{\infty} \{[(s + k)(s + k - 1) + 2(s + k) - l(l + 1)]a_k + (\lambda - s - k)a_{k-1}\} \rho^k = 0 \tag{G-18}$$

This equation must be true for all values of ρ, which is possible only if the coefficient of each power of ρ vanishes. For the first coefficient, $k = 0$, we have (there being no coefficient a_{-1} by definition)

$$s(s - 1) + 2s - l(l + 1) = 0 \tag{G-19a}$$

or

$$s(s + 1) = l(l + 1) \tag{G-19b}$$

This means that either $s = l$ or $s = -(l + 1)$. The latter solution is not acceptable since it cannot lead to solutions consistent with (G-16). Therefore we must set $s = l$; we then obtain from (G-18) the recursion formula

$$a_{k+1} = \frac{l + k + 1 - \lambda}{(k + 1)(k + 2 + 2l)} a_k \tag{G-20}$$

From this relation we can find the values of all coefficients in the series in terms of the first coefficient a_0, which remains as an arbitrary constant. Thus the radial solutions have, in principle, been found.

We must, however, study the behavior of the function K at large values of ρ, in order to see if it gives us well behaved wave functions. As with the harmonic oscillator, the behavior of K for large values of ρ will be determined by the higher powers in the series. For large values of k the recursion formula (G-20) reduces to

$$a_{k+1} \cong a_k/k \tag{G-20a}$$

This means that the successive coefficients in $K(\rho)$ are related in the same way as the successive coefficients in the Taylor expansion of the function e^ρ, because

$$e^\rho = \sum_{k=0}^{\infty} \rho^k/k! \tag{G-20b}$$

Thus it appears that as ρ increases, $K(\rho)$ increases roughly in proportion to e^ρ, and the wave function, which is related to $K(\rho)$ through (G-14) and (G-7), has the form

$$\psi = K(\rho)e^{-\rho/2} \sim e\ e^{-\rho/2} = e^{\rho/2} \tag{G-20c}$$

for large values of ρ. Clearly this function is not well behaved. Therefore $K(\rho)$ cannot be an infinite series, and all coefficients must vanish beyond some value of k, say $k = n'$ where n' may be 0, 1, 2, $\cdots$. This can happen only if, when $k = n'$, the numerator on the right hand side of the recursion formula (G-20) vanishes; that is, if

$$\lambda = l + n' + 1 \tag{G-21}$$

Since λ is related to the energy, this means that the energy must be quantized. If we define a new integer

$$n = l + n' + 1 \tag{G-22}$$

and make use of (G-11b), we see that the energy must be given in atomic units by

$$E = -Z^2/2n^2 \tag{G-23}$$

or in ordinary units by

$$E = -m_e Z^2 e^4/\hbar^2 n^2 \tag{G-23a}$$

This is identical with the Bohr expression, Equation (G-3b), and of course is what is observed experimentally. Making use of (G-11a) we see that the variable ρ has the form

$$\rho = 2Zr/n \tag{G-23b}$$

If the energy of the system is positive (corresponding to unbound states) then by writing $\sigma = 2\sqrt{2E}r$ we obtain from (G-9) the differential equation

$$d^2R/d\sigma^2 + (2/\sigma)dR/d\sigma + [(1/4) - (\lambda/\sigma) + (l(l+1)/\sigma^2)]R = 0 \tag{G-24}$$

For large values of σ this has the approximate solution

$$R = A\, e^{i\sigma/2} + B\, e^{-i\sigma/2} \tag{G-24a}$$

which remains finite as $\sigma \to \infty$ for all values of the energy, and resembles the free particle solution (A-15b). The energy of the unbound electronic states is therefore not quantized, but can take on any value.

The polynomials $\sum_{k=0}^{n'} a_k\rho^k$ are related to functions that were studied in the nineteenth century by Laguerre. Let us define the function

$$L_r(\rho) = e^{\rho}d^r(\rho^r e^{-\rho})/d\rho^r \tag{G-25}$$

The function $L_r(\rho)$ is called the *Laguerre polynomial of degree r*. If this polynomial is now differentiated *s* times, the resulting polynomial,

$$L_r^s(\rho) = d^s L_r(\rho)/d\rho^s \tag{G-26}$$

is known as the *associated Laguerre polynomial of order s and degree r-s*. It can be shown that for any given set of values of the constants n and l,

$$K(\rho) = A\rho^l L_{n+l}^{2l+1}(\rho) \tag{G-27}$$

where A is an arbitrary constant.

Exercise. Prove (G-27). [Hint: Substitute (G-27) into (G-15) and make use of (G-26) to find the derivatives of the associated Laguerre polynomials.]

The wave functions of the hydrogen atom thus have the form

$$\psi(r, \theta, \varphi) = A\ \rho^l e^{-\rho/2} L_{n+l}^{2l+1}(\rho) P_l^{|m|}(\cos\theta) e^{im\varphi} \tag{G-28}$$

The constant A is conveniently chosen so that ψ is normalized; it then has the value

$$A = \left[\left(\frac{2Z}{n}\right)^3 \frac{(n-l-1)!}{2n\,[(n+l)!]^3} \frac{(2l+1)\,(l-|m|)!}{2(l+|m|)!} \frac{1}{2\pi}\right]^{1/2} \tag{G-28a}$$

Explicit expressions for the radial parts of some of the solutions are given in Table 6-2 and are plotted in Fig. 6-9. The angular functions have been described in Chapter 3 (page 92). Some of the complete normalized solutions are written out in Appendix III.

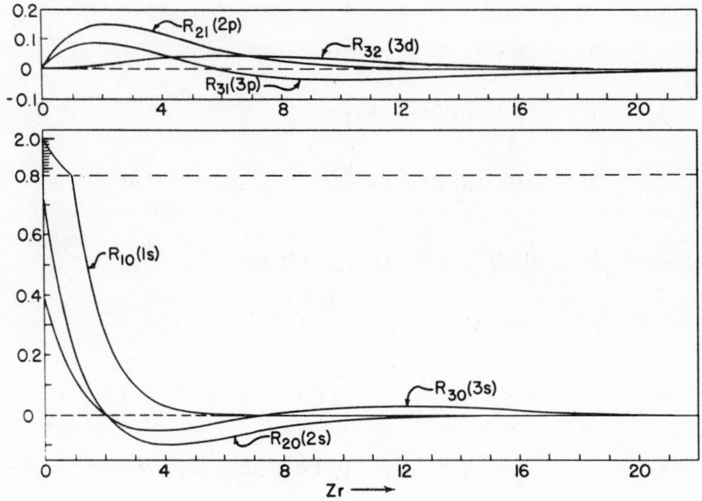

FIG. 6-9. Normalized radial parts of the wave functions of the hydrogen atom and hydrogen-like ions (atomic number = Z). Distances are measured in atomic units. The scale of the ordinates has been compressed above 0.8 so that the 1s-function will fit on the graph.

Evidently the energy of the hydrogen atom depends only on the quantum number n, which is called the *principle quantum number*. We see from the definition of n, Equation (G-22), that l (which will be called the *angular momentum quantum number*) may take any value from 0 up to $(n-1)$. Furthermore, for each value of l there are $2l+1$ possible independent solutions involving the φ coordinate. (That is, there are $2l+1$ values of m having values from $-l$ to $+l$, if we are using the complex solutions (G-10); for the real forms (G-10a, b), there are l solutions sin $m\varphi$, plus $l+1$ solutions cos $m\varphi$, where m can have values between 0 and l. For reasons that will be

TABLE 6-2

RADIAL PARTS OF THE HYDROGEN ATOM WAVE FUNCTIONS

$$R_{nl} = -\left[\frac{2Z^3}{n}\frac{(n-l-1)!}{2n\,[(n+l)!]^3}\right]^{1/2}\left(\frac{2Zr}{n}\right)^l L_{n+l}^{2l+1}(2Zr/n)\,e^{-Zr/n}$$

$$\left(\text{These functions are normalized so that } \int_{-\infty}^{\infty} R_{nl}{}^2 r^2 dr = 1\right)$$

n	l	R_{nl}
1	0	$2Z^{3/2}e^{-Zr}$
2	0	$(1/\sqrt{2})\,Z^{3/2}\,(1-\frac{1}{2}Zr)\,e^{-Zr/2}$
2	1	$(1/2\sqrt{6})\,Z^{5/2}\,r\,e^{-Zr/2}$
3	0	$(2/3\sqrt{3})\,Z^{3/2}\,(1-\frac{2}{3}Zr+\frac{2}{27}Z^2r^2)\,e^{-Zr/3}$
3	1	$(8/27\sqrt{6})\,Z^{3/2}\,(Zr-\frac{1}{6}Z^2r^2)\,e^{-Zr/3}$
3	2	$(4/81\sqrt{30})\,Z^{7/2}\,r^2\,e^{-Zr/3}$
4	0	$\frac{1}{4}Z^{3/2}\,(1-\frac{3}{4}Zr+\frac{1}{8}Z^2r^2-\frac{1}{192}Z^3r^3)\,e^{-Zr/4}$
4	1	$\frac{1}{16}\sqrt{\frac{5}{3}}\,Z^{3/2}\,(Zr-\frac{1}{4}Z^2r^2+\frac{1}{80}Z^3r^3)\,e^{-Z/r4}$
4	2	$(1/64\sqrt{5})\,Z^{3/2}\,(Z^2r^2-\frac{1}{12}Z^3r^3)\,e^{-Zr/4}$
4	3	$(1/768\sqrt{35})\,Z^{9/2}\,r^3\,e^{-Zr/4}$

given in Chapter 8, m is called the *magnetic quantum number*.) Thus each energy level of the hydrogen atom is degenerate (except for the lowest level, with $n = 1$). The degeneracy is given by

$$\sum_{l=0}^{n-1}(2l+1) = 2\sum_{l=0}^{n-1} l + n = 2\frac{(n-1)n}{2} + n = n^2 \tag{G-29}$$

There are four wave functions corresponding to $n = 2$; nine corresponding to $n = 3$, etc. (This result does not, however, take any account of the spin of the electron and of the nucleus; when these are included, the degeneracy is multiplied by a factor of $2(2I + 1)$, where I is the nuclear spin (see page 495).)

Since it is inconvenient to write out the expression (G-28) every time one wants to discuss a wave function of the hydrogen atom, a simple system of notation for these functions has been developed. If $l = 0, 1, 2, 3, 4, 5, \cdots$, then the state is called an $s, p, d, f, g, h, \cdots$ state. The principle quantum number, n, is indicated by writing it ahead of the l-symbol as $1s$, $2p$, $3s$, etc. If one wishes to use the complex form (G-10) for the φ-coordinate function, the value of m is indicated by means of a subscript after the l-symbol, as $1s_0$, $3d_{-2}$, $4p_1$, etc. If the φ function is written in the real form, as in (G-10a, b), then one uses subscripts x, y, z, xy, etc., in the manner described on page 94, giving symbols such as $2p_x$, $4d_{yz}$, etc.

d. Physical interpretations of the hydrogenic wave functions

As we shall see in Chapter 8, the wave functions that have just been derived for the hydrogen atom are eigenfunctions of the operator for the square of the angular momentum of the electron about an axis passing through the nucleus. The eigenvalue is $l(l + 1)$ atomic units. This fact suggests a very useful explicit picture of the motions of the electron in the hydrogen atom.

Suppose that the electron were actually moving about the nucleus in some definite orbit. According to classical mechanics, the angular momentum about the nucleus must be a constant, independent of the time. Furthermore, a centrifugal force must act on the electron. Elementary physics tells us that a particle of mass m moving with velocity v at the end of a rope of length r exerts a force $F = mv^2/r$ on the rope. The angular momentum of the particle is $M = mvr$. Thus the centrifugal force is related to the angular momentum by the equation

$$F = M^2/mr^3$$

The potential from which this force is derived is

$$V = -\int_{\infty}^{r} F\,dr = M^2/2mr^2$$

For the electron in the hydrogen atom, using atomic units, we must take $m = 1$ and $M^2 = l(l + 1)$, so that we would expect a term

$$V = l(l + 1)/2r^2$$

to occur in the expression for the potential energy of the hydrogen atom. Just such a term does, in fact, occur in Equation (G-9). Clearly, the radial motion of the electron is controlled not only by a Coulombic potential, $-Z/r$, but also by a centrifugal potential, $l(l+1)/2r^2$. These two potentials have opposite signs, since the Coulombic potential tends to bring the electron closer to the nucleus, whereas the centrifugal potential tends to keep the two particles apart.

As the distance between the electron and the nucleus is decreased, the centrifugal potential increases more rapidly than the Coulombic potential. Therefore we should expect that, other things being equal, an electron in a state of high angular momentum will tend to stay farther from the nucleus than an electron in a state of low angular momentum. This, in fact, is just what is found for the solutions (G-28). The radial dependence of these solutions is given by

$$R(r) \propto r^l \left(\sum_{k=0}^{n-l} b_k r^k \right) e^{-Zr/n}$$

where the coefficients b_k are related to the coefficients a_k discussed above. For small values of r, all terms but the constant term in the polynomial can be neglected, and the exponential is approximately unity. Therefore we may write for small r

$$R(r) \sim r^l$$

We see that the wave function of an s-state ($l = 0$) has a finite value at the nucleus, whereas all other types of states give wave functions that vanish at the nucleus. Furthermore, for a given small value of r, the function r^l is smaller, the larger the value of the exponent l, so that the electron tends to avoid the nucleus more and more as the value of l is increased. Thus we arrive at the important principle that *the higher the angular momentum of an electron, the less likely it is to be found close to the nucleus.* This principle is illustrated for the particular case of the normalized 5*s*, 5*p*, 5*d*, 5*f* and 5*g* functions of hydrogen in Fig. 6-10.

We have found from the description of the oscillations of the ocean on a flooded planet, that the number of nodal surfaces in the spherical harmonic $P_l^{|m|}(\cos\theta)e^{im\varphi}$ is equal to l. Similarly, it is found that the number of nodal surfaces in the radial part of the hydrogenic wave function (that is, in the associated Laguerre function $L_{n+l}^{2l+1}(\rho)$) is $(n-l)$ (this includes a nodal surface at $r = \infty$). Thus *the principal quantum number n is equal to the total number of nodal surfaces in the wave function.*

The number of nodes in the radial part of a wave function is a measure of the radial velocity of the electron, and the number of nodes in the angular part of the function is a measure of the tangential velocity. Among a group of states having the same principle quantum number (that is, the same total number of nodal surfaces), the *s*-state ($l = 0$) will always have the largest number of radial nodes. Since the angular momentum of the electron in this state is zero, the electronic orbits in *s*-states can be considered to be straight lines passing directly through the nucleus. It will be recalled that this kind of motion was arbitrarily excluded in the Bohr-Sommerfeld atom. There is, however, direct experimental support for it in the phenomenon of "*K*-capture" in nuclear reactions (see *3*, pp. 684ff).

In *p*-states ($l = 1$) some of the electronic motion is nonradial, but as the total energy increases in going from 2*p* to 3*p* to 4*p*, etc., the amplitude of the radial motion increases relative to that of the tangential motion. This can be

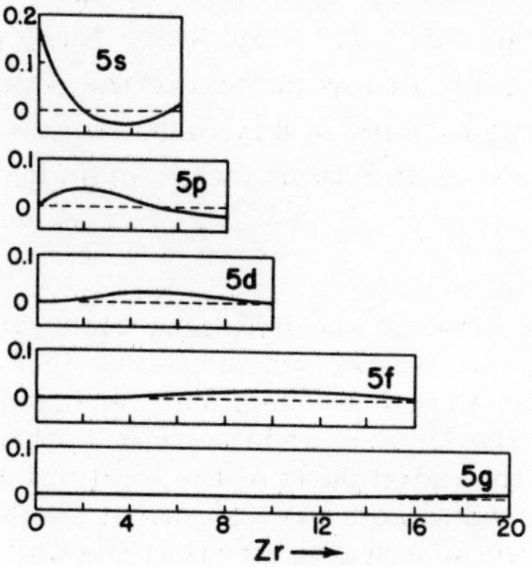

FIG. 6-10. Shapes of the normalized radial parts of the hydrogenic wave functions of principal quantum number five, in the vicinity of the nucleus, showing the tendency of the electron to avoid the nucleus to an increasing extent as its angular momentum is increased.

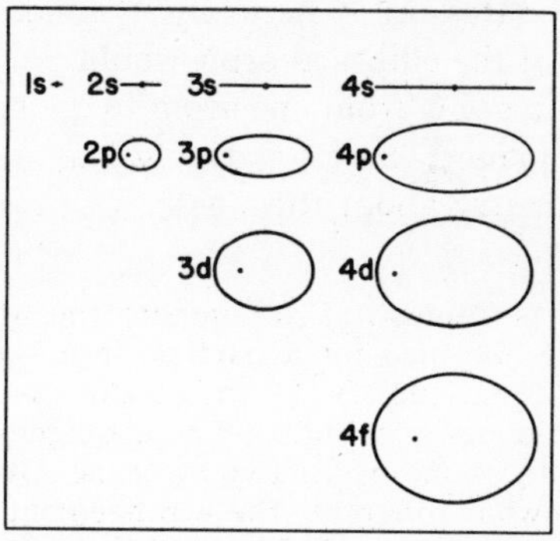

FIG. 6-11. Hypothetical classical orbits corresponding to the quantum mechanical states of hydrogen.

interpreted to mean that in *np*-states the electron moves in orbits that become more and more elliptical as *n* increases. Similar changes in ellipticity occur in states of still higher angular momentum.

Figure 6-11 shows an interpretation of the orbits of the electron in the hydrogen atom, which is based on the considerations outlined above. The student should be careful to realize, however, that these diagrams are idealizations not capable of experimental verification, because of the uncertainty principle (see Chapter 7). Nevertheless they are useful in helping the student retain a qualitative picture of the states of hydrogen and other atoms. The student should notice that these diagrams are quite different from the pictures of the hydrogen atom originally proposed by Bohr and by Sommerfeld. In particular, if Sommerfeld had not arbitrarily excluded the linear orbits, the lowest energy level of the hydrogen atom would be degenerate, consisting of an *s*-state and three *p*-states ($k = 1$, $l = 0$ and $k = 0$, $l = 1$, respectively in equations (G-3a)).

In relating the orbits of Fig. 6-11 to the hydrogenic wave functions, we must remember that the wave functions are related to the average behavior of a large number of hydrogen atoms. If we had many atoms in 1*s*-states we should expect that the directions of radial motion would be randomly distributed in the different atoms. It is for this reason that the wave functions of *s*-states are spherically symmetrical. Similarly, in a $3p_1$-state the long axis of the elliptical orbit would tend to be more or less normal to the *z*-axis, but in going from one atom to another it might point in any direction in the *x-y* plane. Therefore the square of the wave function shows cylindrical symmetry about the *z*-axis.

It is interesting to compare the wave functions of the hydrogen atom with those obtained for a particle in a spherical box. Inspection of these solutions (*cf.* Equations (B-13) and (G-28)) shows that they have identical dependence on the angles θ and φ. The arrangement of the angular nodal surface is therefore the same in the two systems. Although the radial parts of the solutions are somewhat different, the arrangement of the radial nodes is also rather similar. The outermost radial node of the particle in the spherical cavity is at the wall of the cavity, whereas in the hydrogen atom it is at infinity, but the inner radial nodes form concentric spherical surfaces in both cases. Since the electron increases its velocity as it approaches the nucleus (thereby decreasing its de Broglie wave length), the radial nodes near the center of the hydrogen atom tend to crowd together, whereas in the spherical cavity these surfaces are more evenly spaced. On the basis of this similarity in the arrangement of the nodal planes, we may label the states of the particle in a spherical cavity in the same way as their counterparts in the hydrogen atom are labeled. This has been done in Table 6-1, where all states with $l = 0$ have been called *s*-states, all states with $l = 1$ have been called *p*-states, etc., and a "principal quantum number" has been assigned on the basis of the total number of nodes.

In spite of the similarity in the wave functions of the two systems, the arrangement of the energy levels is very different. For one thing, the levels of a particle in the spherical cavity never converge to a limiting value. Even more striking is the complete absence of degeneracy between states having the same number of nodal surfaces (that is, the same "principal quantum number"). Thus if we arrange the levels of a spherical cavity in order of increasing energy, we obtain the succession $1s$, $2p$, $3d$, $2s$, $4f$, $3p$, $5g$, $4d$, $6h$, $3s$, $5f$, $7i$, $\cdots$, which is altogether different from the order obtained in the hydrogen atom. The physical reason is that in a spherical cavity the particle can have a longer de Broglie wave length (and hence a lower kinetic energy) by moving about close to the circumference of the cavity than by crossing back and forth on the shorter paths that pass close to the center. This stabilizes states of high angular momentum. In the hydrogen atom, on the other hand, the Coulombic attraction of the nucleus stabilizes orbits that pass close to the nucleus and we obtain the order of stability $1s$, $(2s, 2p)$, $(3s, 3p, 3d)$, etc. The tendency of energy levels in hydrogen to group together in this way, rather than in the order found for the spherical box, has chemical consequences of the utmost importance.

Exercises. *Electron moving around a partially shielded core.* In the alkali metal atoms (Li, Na, K, etc.) there is a highly charged nucleus surrounded by one or more tightly bound, compact, closed shells of electrons (the "core"). In addition there is a single, more loosely bound electron (the "valence electron"). Approximate values of the energy levels of the valence electron can be obtained by a slight modification of the theory just presented for the hydrogen atom.

The core may be regarded as a smeared-out, spherically symmetrical cloud of negative electricity whose charge is $(Z - 1)$ electronic charges, the nucleus having Z positive charges. The charge density in the electronic cloud of the core is negligible beyond a certain distance. If the valence electron is well outside of the core, the core electrons will exactly compensate for $(Z - 1)$ of the positive charges in the nucleus, so that the potential energy of the valence electron will vary with distance according to the relation $V = -1/r$ (in atomic units). If the valence electron should, however, penetrate into the core, it will be more strongly attracted by the nucleus. The exact dependence of the potential energy on the distance from the nucleus will depend on the shape of the core cloud, but we might expect that in general we could write

$$V = -\frac{1}{r} - \frac{A}{r^2} - \frac{B}{r^3} - \frac{C}{r^4} - \cdots$$

where A, B, C, $\cdots$ are constants. To a first approximation all but the first two terms of this expansion may be neglected.

(1) Show that when the Schroedinger equation is solved for the potential energy function

$$V = -\frac{1}{r} - \frac{A}{r^2}$$

one obtains wave functions of the form

$$\psi(r, \theta, \varphi) = S(\theta, \varphi) R'(r)$$

where $S(\theta, \varphi)$ is given by Equation (G-10) or (G-10a, b), and an angular momentum quantum number l is thereby introduced. The function $R'(r)$ is not, however, the same as the radial function in (G-28).

(2) Show that for bound states ($E < 0$) the energy is quantized with the values

$$E = -\frac{1}{2n^{*2}}$$

where

$$n^* = n' + l^* + 1$$

$$l^*(l^* + 1) = l(l + 1) - 2A$$

n' and l being either zero or positive integers analogous to n' and l in the hydrogen atom problem. In general, n^* and l^* are nonintegral, and for a given value of the principle quantum number, $n' + l + 1$, the energy varies as l is varied.

(3) Show that for small values of A and for a given value of $n' + l + 1$, the most stable state of the valence electron is that having the smallest value of l. That is, in lithium, for instance, the $2s$-state of the valence electron should be more stable than the $2p$-state. This is actually observed.

H. The Hydrogen Molecular Ion, H_2^+

A stable system composed of two protons and a single electron is known to exist in gas discharge tubes, and is called the hydrogen molecular ion. The potential energy function is

$$V = (e^2/R) - (e^2/r_A) - (e^2/r_B) \tag{H-1}$$

where R is the distance between the protons, and r_A and r_B are the distances between the electron and the two protons. Assuming that the two nuclei are stationary and using atomic units, the Schroedinger equation is

$$\nabla^2\psi + 2[E + (1/r_A) + (1/r_B) - (1/R)]\psi = 0 \tag{H-2}$$

where ∇^2 is the Laplacian operator containing the coordinates of the electron. (The justification of the neglect of the nuclear motions is found in the Born-Oppenheimer principle, which is discussed on page 533.) This partial differential equation can be separated if elliptical coordinates are used; these coordinates are denoted by (μ, ν, φ) and are defined as

$$\mu = (r_A + r_B)/R, \quad \nu = (r_A - r_B)/R \tag{H-3}$$

φ is the angle between the plane passing through all three particles and some fixed plane which passes through the two protons. The equation μ = a constant defines a series of ellipsoids of revolution whose foci are at the locations of the two protons, and the equation ν = a constant defines a series of hyperboloids of revolution whose foci are similarly located. It is found that in this coordinate system the Schroedinger equation takes the form (see *10*, p. 363 for the transformation of ∇^2)

$$\frac{\partial}{\partial \mu}\left[(\mu^2-1)\frac{\partial \psi}{\partial \mu}\right]+\frac{\partial}{\partial \nu}\left[(1-\nu^2)\frac{\partial \psi}{\partial \nu}\right]+\frac{\mu^2-\nu^2}{(\mu^2-1)(1-\nu^2)}\frac{\partial^2 \psi}{\partial \varphi^2}$$

$$-\frac{1}{2}\left[(\mu^2-\nu^2)(ER^2-R)-4\mu R\right]\psi=0 \qquad \text{(H-4)}$$

If we write

$$\psi(\mu,\nu,\varphi)=M(\mu)N(\nu)\Phi(\varphi) \qquad \text{(H-5)}$$

we obtain three ordinary differential equations,

$$d^2\Phi/d\varphi^2=-m^2\Phi \qquad \text{(H-6)}$$

$$\frac{d}{d\nu}\left[(1-\nu^2)\frac{dN}{d\nu}\right]-\left[A-\tfrac{1}{2}R^2E\nu^2+\frac{m^2}{1-\nu^2}\right]N=0 \qquad \text{(H-7)}$$

$$\frac{d}{d\mu}\left[(\mu^2-1)\frac{dM}{d\mu}\right]+\left[A+2R\mu-\tfrac{1}{4}R^2E\mu^2-\frac{m^2}{\mu^2-1}\right]M=0 \qquad \text{(H-8)}$$

where m and A are separation constants. Equation (H-6) is readily solved, giving

$$\Phi=e^{im\varphi} \qquad \text{(H-9)}$$

where $m=0, \pm 1, \pm 2, \cdots$. We see that Φ is an eigenfunction of the operator for angular momentum about the line joining the two protons, the eigenvalues being $m\hbar$ (*cf.* page 169).

Equations (H-7) and (H-8) are not so easily solved, and we shall not discuss them further at this point. Tables giving both the functions and the eigenvalues, E, are to be found in an article by Bates, Ledsham, and Stewart (*11*). Curves showing the energy as a function of internuclear distance for several of the electronic states of this system are given in Fig. 6-12. These curves will be discussed further in Chapters 11 and 12. Some of them obviously correspond to stable "molecular" systems.

I. The Morse Potential

Morse has pointed out that the potential energy curve of a chemical bond, relating the total energy to the interatomic distance r, ought to have the qualitative characteristics of the function

$$V=D(1-e^{-\beta(r-r_e)})^2 \qquad \text{(I-1)}$$

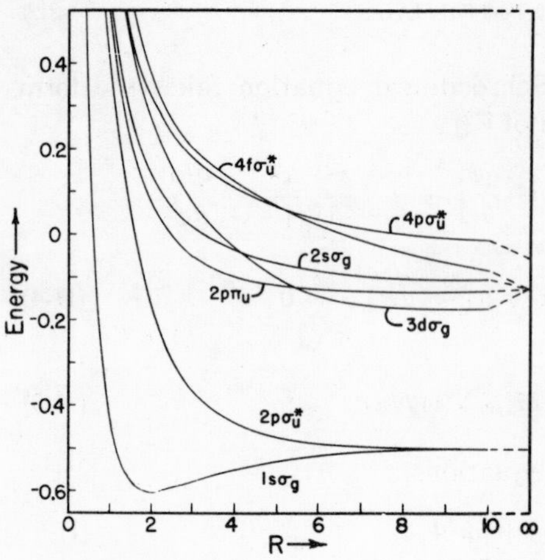

FIG. 6-12A. Total energies of some states of H_2^+ as a function of the internuclear distance Energy and distance in atomic units.

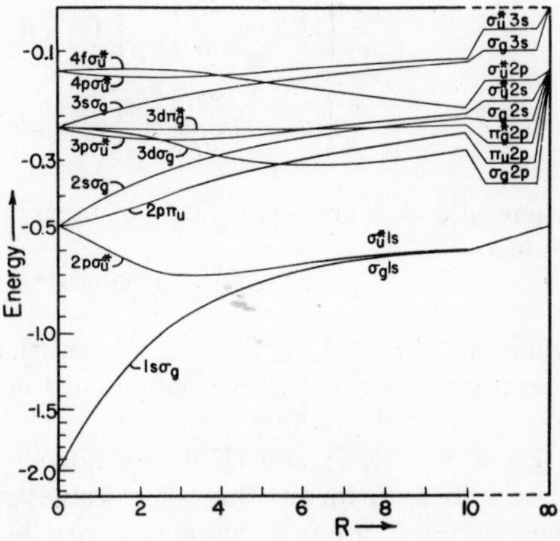

Fig. 6-12B Energies of states of H_2^+, not including the internuclear repulsion term, $1/R$. The scale of the ordinates has been distorted in order to clarify the correlations of states for $R = 0$ and $R = \infty$ The labelling of the states is discussed in Chapter 11 (pages 383*ff*).

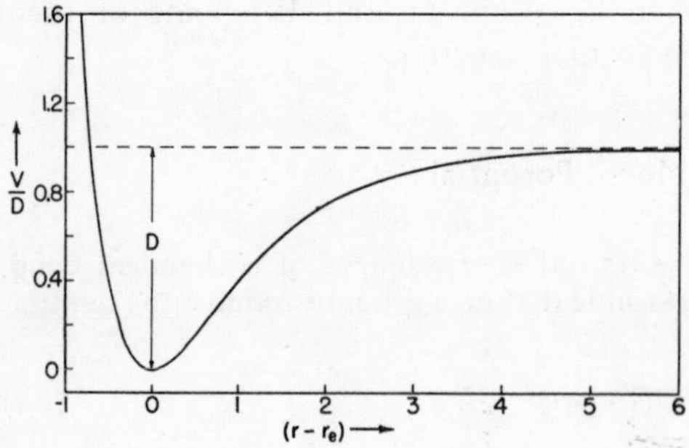

FIG. 6-13. The Morse potential, Equation (I-1), with $\beta = 1$·

where D is the dissociation energy of the bond, β is a constant related to the force constant of the bond, and r_e is the interatomic distance at the minimum of the curve (see Fig. 6-13). It is easy to see that as $r \to \infty$, V approaches a constant value D, and as r becomes small, V becomes very large (though not infinite).

Morse has shown that the Schroedinger equation for a diatomic molecule that has this potential function can be solved, and that the energy levels are

$$E = \beta \sqrt{\frac{D\hbar}{\pi\mu}} (n + \tfrac{1}{2}) - \frac{\hbar\beta^2}{4\pi\mu} (n + \tfrac{1}{2})^2 \qquad \text{(I-2)}$$

where μ is the reduced mass of the nuclei in the molecule and n is an integer. (As ter Haar has pointed out (*12*), (I-2) is not a precise expression, but merely a very close approximation.)

The observed vibrational energy levels of diatomic molecules often fit a formula of the form (I-2), and this is the basis of an important method of determining the dissociation energies of diatomic molecules. Poeschl and Teller (*13*) have shown, however, that other potentials will also give energy levels with the same functional dependence on a quantum number, n (see *14*, p. 101, for further discussion and references).

J. The Virial Theorem

The potential energy functions of many systems important in chemistry have the property that if all of the coordinates are multiplied by an arbitrary constant factor, λ, the potential energy is multiplied by λ^ρ, where ρ is a number typical of the system. Such a potential is said to be a homogeneous function of the coordinates of degree ρ. Consider for example a set of harmonic oscillators, for which

$$V = \tfrac{1}{2} \sum (k_{i1}x_i^2 + k_{i2}y_i^2 + k_{i3}z_i^2)$$

where x_i, y_i, and z_i are the displacements of the ith particle in the directions of the Cartesian coordinates. Multiplication of each variable, x_i, y_i, and z_i, by λ changes V by a factor λ^2. In this case, therefore, $\rho = 2$. Similarly, in a system made up of electrons and atomic nuclei between which the only forces are electrostatic, the potential energy depends on the reciprocals of the distances of the particles from each other. Therefore, multiplication of all of these distances by λ will mean that the potential energy V is multiplied by λ^{-1}. Thus $\rho = -1$ for the Coulombic potential.

Suppose that we have a quantum mechanical system containing N particles whose normalized wave function is $\psi(x_1, \cdots, z_N)$. The mean kinetic energy of this system is (in atomic units)

$$\overline{T} = -\frac{1}{2} \int \psi^* (\sum \nabla_i^2) \psi \, dx_1 \cdots dz_N \qquad \text{(J-1)}$$

and the mean potential energy is

$$\overline{V} = \int \psi^* \, V \, \psi \, dx_1 \cdots dz_N \qquad \text{(J-2)}$$

The virial theorem states that if V is a homogeneous function of the coordinates of degree ρ, then

$$\overline{T} = (\rho/2) \, \overline{V} \qquad \text{(J-3)}$$

We shall prove this important theorem by a method due to Fock. Let us write another wave function, $\psi'(x_1, \cdots, z_N)$, whose value for the configuration $(x_1, \cdots, z_N)$ is the same as that of the correct wave function when the system's configuration is $(\lambda x_1, \cdots, \lambda z_N)$. That is, we have merely changed the scale of the variables of the original wave function by a scale factor λ (the new wave function ψ' is of course not a correct wave function of the system). Then in order that the new wave function be normalized, we must have

$$\psi'(x_1, \cdots, z_N) = \lambda^{3N/2} \psi(\lambda x_1, \cdots, \lambda z_N) \qquad \text{(J-4)}$$

because if we let $x'_1 = \lambda x_1, \cdots, z'_N = \lambda z_N$, it must be true that

$$\int |\psi'(x_1, \cdots, z_N)|^2 \, dx_1 \cdots dz_N = 1 = \int |\psi(x'_1, \cdots, z'_N)|^2 \, dx'_1 \cdots dz'_N$$

$$= \lambda^{3N} \int |\psi(\lambda x_1, \cdots, \lambda z_N)|^2 \, dx_1 \cdots dz_N \qquad \text{(J-4a)}$$

If the system had the wave function ψ' rather than ψ, then its average kinetic energy would be

$$\overline{T'} = -\frac{1}{2} \int \psi'^*(x_1, \cdots, z_N) \, (\sum \nabla_i^2) \psi'(x_1, \cdots, z_N) dx_1 \cdots dz_N \qquad \text{(J-5)}$$

Let us replace the variables $x_1, \cdots, z_N$ in this integral by the variables $x'_1, \cdots, z'_N$ defined above. Then we must replace ∇_i^2 by $\lambda^2 \nabla'^2_i$ because

$\partial^2/\partial x_1^2 = \lambda^2\partial^2/\partial x'^2_1$, etc. Furthermore, $dx_1 = dx'_1/\lambda$, etc. Making use of Equation (J-4) we find

$$\overline{T}' = -\frac{1}{2}\lambda^2 \int \psi^*(x'_1, \cdots, z'_N)\left(\sum \nabla'^2_i\right)\psi(x'_1, \cdots, z'_N)dx'_1 \cdots dz'_N \quad \text{(J-6)}$$

The integral in (J-6) differs from that in (J-1) merely by having a prime on all of its variables, which can have no effect on the value of the integral. Therefore we have the result

$$\overline{T}' = \lambda^2 \overline{T} \quad \text{(J-7)}$$

In other words, shrinking a wave function by a factor λ increases the kinetic energy by λ^2.

Similarly the average potential energy of the wave function ψ' is

$$\overline{V}' = \int \psi'^*(x_1, \cdots, z_N)V(x_1, \cdots, z_N)\psi'(x_1, \cdots, z_N)dx_1 \cdots dz_N \quad \text{(J-8)}$$

Replacement of the variables in the homogeneous function V gives

$$V(x_1, \cdots, z_N) = \lambda^{-\rho}\, V(x'_1, \cdots, z'_N) \quad \text{(J-9)}$$

so that

$$\overline{V}' = \lambda^{-\rho} \int \psi^*(x'_1, \cdots, z'_N)V(x'_1, \cdots, z'_N)\psi(x'_1, \cdots, z'_N)dx'_1 \cdots dz'_N \quad \text{(J-10)}$$

or

$$\overline{V}' = \lambda^{-\rho}\, \overline{V} \quad \text{(J-11)}$$

According to the variation principle, the best wave function for this system of N particles is the one that minimizes the total energy of the system. Furthermore, this best wave function is our original wave function, ψ. That is, if we calculate the total energy corresponding to ψ'

$$E' = \overline{T}' + \overline{V}' \quad \text{(J-12)}$$

and then vary the parameter λ until E' is a minimum,

$$dE'/d\lambda = 0 \quad \text{(J-13)}$$

then this minimum must occur when $\lambda = 1$. Now

$$dE'/d\lambda = d(\lambda^2\overline{T} + \lambda^{-\rho}\overline{V})/d\lambda = 2\lambda\overline{T} - \rho\lambda^{-\rho-1}\overline{V} = 0 \quad \text{(J-14)}$$

so that, setting $\lambda = 1$, we obtain

$$\overline{T} = (\rho/2)\overline{V} \quad \text{(J-3)}$$

which is the virial theorem.

It is important to note that if our original wave function ψ is merely an approximate wave function for a system, rather than the correct one, then E' is minimized by a value of λ different from unity. We then obtain from (J-14)

$$2\lambda^2\overline{T} - \rho\lambda^\rho\overline{V} = 0 \qquad \text{(J-15)}$$

or

$$2\overline{T}' = \rho\overline{V}' \qquad \text{(J-16)}$$

Thus even an approximate wave function may always be made consistent with the virial theorem if one inserts a scale factor in front of all of the coordinates in the function and minimizes the energy with respect to this factor.

Exercises. (1) Verify the virial theorem for the ground state of the hydrogen atom, for which the normalized wave function and energy operations are

$$\psi = \sqrt{1/\pi}\, e^{-r}, \quad T = -\left(\frac{1}{2r^2}\right)\partial/\partial r\,(r^2\partial/\partial r), \quad V = -1/r,$$

and the volume element is $d\tau = 4\pi r^2 dr$.

(2) Show that for any system made up of charged particles that interact solely by electrostatic forces, $E = -\overline{T} = \overline{V}/2$.

(3) Show that for a system having any potential function, $V(x_1, \cdots, z_N)$, not necessarily homogeneous,

$$2\overline{T} = \int \psi^* \sum \left(x_i \frac{\partial V}{\partial x_i} + y_i \frac{\partial V}{\partial y_i} + z_i \frac{\partial V}{\partial z_i}\right) \psi\, dx_1 \cdots dz_N \qquad \text{(J-17)}$$

Hint: Make use of the relation

$$\frac{d}{d\lambda} = \sum \left(\frac{dx'_i}{d\lambda}\frac{\partial}{\partial x'_i} + \frac{dy'_i}{d\lambda}\frac{\partial}{\partial y'_i} + \frac{dz'_i}{d\lambda}\frac{\partial}{\partial z'_i}\right)$$

Note that the quantity $-\partial V/\partial x_i$ can be regarded as the component of force acting along the x_i-coordinate. Writing $F_{x_i} = -\partial V/\partial x_i$, we have the important result,

$$2\overline{T} = \overline{\sum (x_i F_{x_i} + y_i F_{y_i} + z_i F_{z_i})} \qquad \text{(J-18)}$$

The quantity $\sum (x_i F_{x_i} + y_i F_{y_i} + z_i F_{z_i})$ is known as the *virial* of a system.

Suppose that the nuclei at either end of a chemical bond are held in a fixed position by an external force that acts along the bond axis. If $E(R)$ is the energy of the bond when the nuclei are at a distance R from one another, then the external force necessary to fix the positions of the nuclei is $-dE/dR$. Using (J-18) we see that

$$2\overline{T} = -\overline{V} - R\, dE/dR \qquad \text{(J-19)}$$

where $\overline{T}$ is the average kinetic energy of the electrons in the bond (the nuclei, being held stationary, cannot contribute to $\overline{T}$), and $\overline{V}$ is the average potential due to all forces except the external force acting on the nuclei. $\overline{V}$ is clearly the mean potential energy arising from all of the electrostatic interactions of the electrons and nuclei with each other–that is, it is what we call the mean potential energy of the molecule. Since $E = \overline{T} + \overline{V}$, we have the interesting results,

$$\overline{T} = -E - R\,dE/dR \tag{J-20}$$

$$\overline{V} = 2E + R\,dE/dR \tag{J-21}$$

Thus it is possible to find the average kinetic and potential energies of the electrons in a bond from a knowledge of the way the bond energies change with bond distances.

Exercises. (1) For a pair of ions whose charges are Z_1 and Z_2, the energy is $E = E_0 + (Z_1Z_2/R)$, where R is the interionic distance and E_0 is the energy of the ions when $R = \infty$. How does the kinetic energy of the electrons in the ions vary with R?

(2) As will be shown in Chapter 13, the van der Waals forces between a pair of neutral atoms separated by a distance R leads to an energy $E = E_0 - (K/R^6)$, where K is a positive constant and E_0 is the total energy of the isolated atoms. Is the van der Waals attraction between atoms caused by a reduction in the kinetic energy of the electrons, or is it a result of a lowering of their potential energy?

(3) The repulsive forces ("Pauli forces", see pages 319ff.) between atoms when they are brought close together are probably fairly well described by the energy relation $E = A\exp(-kR)$ where A and k are positive constants and R is the interatomic distance. Do these repulsive forces result from changes in the kinetic energy of the electrons, or of their potential energy, or both?

REFERENCES CITED

1. E. Jahnke and F. Emde, "Tables of Functions," 2nd and 3rd eds., Teubner, Leipzig, 1933, 1938. (Reprinted by Dover, New York, 1943).
2. G. Gamow and C. L. Critchfield, "Theory of the Atomic Nucleus," Oxford U.P., New York, 1949.
3. J. M. Blatt and V. F. Weisskopf, "Theoretical Nuclear Physics," Wiley, New York, 1952.
4. H. Margenau and G. M. Murphy, "The Mathematics of Physics and Chemistry," Van Nostrand, New York, 1943.
5. K. S. Pitzer, "Quantum Chemistry," Prentice-Hall, New York, 1953.
6. G. W. King, R. M. Hainer, and P. C. Cross, *J. Chem. Phys.*, **11**, 27 (1943); **12**, 210 (1944); **15**, 820 (1947); **17**, 826 (1949).
7. E. U. Condon and P. M. Morse, "Quantum Mechanics," McGraw-Hill, New York, 1929.
8. P. M. Morse, *Phys. Rev.*, **35**, 1310 (1930).

9. L. Pauling and E. B. Wilson, "Introduction to Quantum Mechanics," McGraw-Hill, New York, 1935.
10. H. Eyring, J. Walter, and G. E. Kimball, "Quantum Chemistry," Wiley, New York, 1944.
11. D. R. Bates, K. Ledsham, and A. L. Stewart, *Phil. Trans. Roy. Soc.*, **A246**, 215 (1953).
12. D. ter Haar, *Phys. Rev.*, **70**, 222 (1946).
13. G. Poeschel and E. Teller, *Z. Physik.*, **83**, 143 (1933).
14. G. Herzberg, "Spectra of Diatomic Molecules," 2nd ed., Van Nostrand, New York, 1950.

GENERAL REFERENCES

A general summary of the soluble forms of the Schroedinger equation is given in P. M. Morse and H. Feshbach, "Methods of Theoretical Physics," pp. 655ff, McGraw-Hill, New York, 1953.

A detailed discussion of the solutions of Schroedinger's equation for one- and two-electron systems is given in the article by H. Bethe, in "Handbuch der Physik" (H. Geiger and K. Scheel, eds.), 2nd ed., Vol. 24, Part 1, Chapter 3. Springer, Berlin, 1933 (Reprinted by Edwards Brothers, Ann Arbor, Mich., 1943).

Chapter 7

The Uncertainty Relations

A. Limitations on the Simultaneous Measurement of Position and Momentum

According to Corollary VII of the laws of quantum mechanics (page 171), if we know the state function Ψ of a particle, it is impossible to predict precisely both the position and the linear momentum of the particle. This is in contradiction to classical mechanical ideas, according to which there is no reason why we should not know both quantities in any system to any degree of precision. It is worth examining this contradiction in some detail, since it gives a good insight into the nature of quantum mechanics and the limitations of classical mechanics.

For this purpose let us think of an experiment by which we might try to measure both the position and the momentum of a particle–say an electron (see Fig. 7-1). As a source of electrons we might have a hot filament.

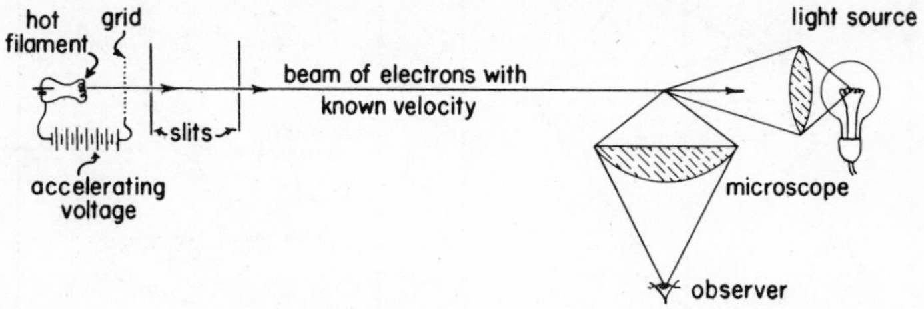

FIG. 7-1. Idealized apparatus for producing a beam of electrons of known velocity and for measuring their positions.

The electrons can be given a precisely known energy by accelerating them through a known potential difference, V, by means of a grid connected to the filament through a battery. An arrangement of slits will then give a well defined beam of electrons whose energy, and hence linear momentum, is known exactly. We have merely to find the exact positions of the electrons in this beam at any instant, and according to the laws of classical mechanics their positions at any future time can be predicted with absolute precision. In order to do this we set up a high-powered microscope with a suitable lighting arrangement. As the electrons collide with the photons from the light source (we must, of course, recognize the quantum nature of light), some of the photons bounce into the microscope and the observer sees flashes of light. He notes, in the field of view of the microscope, the positions at which these flashes occur, and in this way learns both the position and momentum of the particles at some instant of time. According to classical mechanics he should be able to predict their exact future motion. Nothing could be simpler; the quantum mechanical pessimism seems at first sight quite unwarranted.

There are, however, two fundamental limitations in such an experiment. First of all, the accuracy of the microscope in determining positions is limited by the laws of optics. Secondly, Compton showed that when light interacts with a particle, the particle suffers changes in momentum (the "Compton effect"). Let us examine these limitations a little more closely.

a. The optical limitation

Suppose that we observe a point source of light by means of a lens (see Fig. 7-2). If the source is at A, we will see at B an image in the form of a

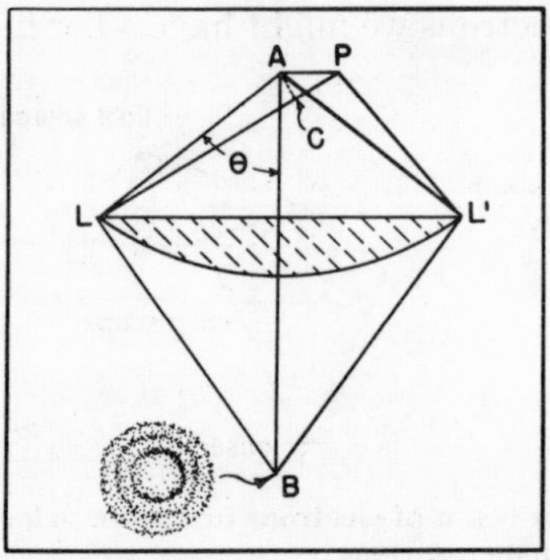

FIG. 7-2. Limitation on the measurement of the position of a particle due to diffraction.

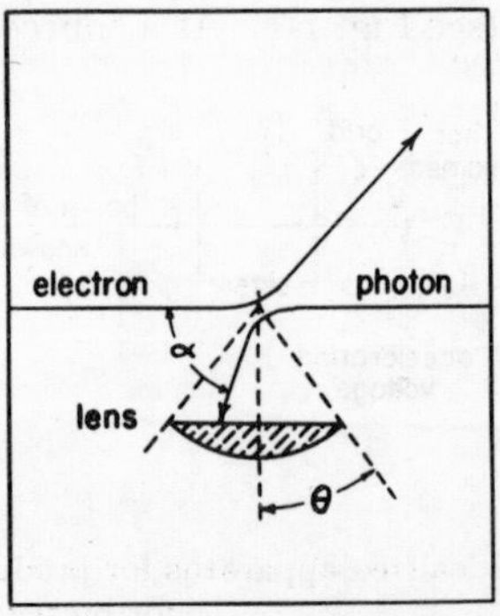

FIG. 7-3. Change in momentum of an electron resulting from collision with a photon (Compton effect).

bright disk surrounded by a group of bright rings, whose intensity falls off rapidly with distance from the center of the disk (i.e., a typical diffraction pattern). The source of light can be moved about through a small but finite distance, and the point B will still remain lighted. If, therefore, in our experiment we see a flash of light at B due to the collision of an electron with a photon, we cannot be sure that the electron is exactly at point A. The uncertainty in the position of the electron is roughly the distance through which a point source of light may be moved about while still giving at least partially constructive interference of the light waves at B. In general, if light from the source reaches opposite edges of the lens with a difference in path length of less than half a wave length, B will be brightly lighted. If in Fig. 7-2 the light source is at P, if C is the intersection of PL with a line through A normal to PL, and if the distance AP is small compared with AL, then $PL - AL \simeq PC \simeq AL - PL'$. If θ is half of the angular aperture of the lens, we have $PC \simeq PA \sin\theta$. Thus the difference in path length from P to the two opposite points L and L' on the edge of the lens is $PL - PL' \simeq 2PC \simeq 2PA \sin\theta$. The uncertainty in the location of the electron, δx, is approximately $2PA$ when this difference in path length is half a wave length, λ, of the light used in illuminating the electron. That is, $\lambda/2 \simeq PL - PL' \simeq 2PC \simeq 2PA \sin\theta \simeq \delta x \sin\theta$. Thus we find

$$\delta x \simeq \frac{\lambda}{2 \sin\theta} \tag{A-1}$$

b. The Compton limitation

According to Compton's experimental observations, a photon, on colliding with any material particle, acts as though it has a linear momentum of h/λ, where λ is the wave length of the light. Since momentum in any direction is conserved in a collision, and since we are concerned with a head-on collision (see Fig. 7-3), we have for the momenta in the direction of the beam

$$h/\lambda + p_x = (h/\lambda') \cos\alpha + p'_x$$

where p_x is the original momentum of the electron in the x-direction (across the field of the microscope), p'_x is the component of momentum in the x-direction after collision with the photon, α is the angle through which the photon is deflected by the collision, and λ' is the wave length of the photon after collision. It can be shown from the law of energy conservation that unless the electron is moving very rapidly, $\lambda = \lambda'$ very nearly. Thus

$$p'_x = p_x + (h/\lambda)(1 - \cos\alpha)$$

Now p'_x can be known accurately only if the angle α is known accurately. Unfortunately, the microscope objective collects all photons that have been scattered by angles between $\alpha_1 = 90° - \theta$ and $\alpha_2 = 90° + \theta$, and there is no way of telling what value of α between these limits belongs to the photon causing a flash of light in the field of view of the microscope. Therefore we can only know that p'_x falls between the limits

$$p'_{x1} = p_x + (h/\lambda)(1 - \cos \alpha_1) = p_x + (h/\lambda)(1 + \sin \theta)$$

and

$$p'_{x2} = p_x + (h/\lambda)(1 - \cos \alpha_2) = p_x + (h/\lambda)(1 - \sin \theta)$$

That is, there is a minimum uncertainty in the x-component of the momentum *after the observation* amounting to

$$\delta p_x \simeq p'_{x1} - p'_{x2}$$

$$= 2\frac{h}{\lambda}\sin \theta \tag{A-2}$$

Equations (A-1) and (A-2) are obviously working at cross purposes. If we try to improve our measurement of the electron's position by decreasing λ and increasing θ, we decrease our knowledge of the momentum after the observation; on the other hand, if we try to improve our knowledge of the electron's momentum after the observation by increasing λ and decreasing θ, we destroy the accuracy of the measurement of the electron's position. In fact, the precision in our knowledge of the position and of the momentum *after the observation* is subject to the condition

$$\delta p_x \delta x \simeq h \tag{A-3}$$

Similar relations must, of course, hold for the momentum component and the position coordinate corresponding to any direction in space. Recognizing the approximate nature of our arguments, this result is consistent with Corollary VII of the laws of quantum mechanics.

Equation (A-3) is one of the famous *Heisenberg uncertainty relations* and is evidently a consequence of the inevitable disturbance introduced in any system by any measurement of position and momentum. The relation conforms to the quantum mechanical laws, rather than to the ideas inherent in classical mechanics. No one has ever been able to think of an experimental arrangement by which the precision of measurement could be made to exceed that specified in (A-3). [For discussions of other types of experiments and the manner in which they are influenced by the uncertainty principle, see Heisenberg (*1*, Chapters II and III) and Bohr (*2*).]

It should be noted that the Heisenberg uncertainty principle does not necessarily imply that we cannot "know" the precise values of both the momentum and the position of a particle at times *prior* to making an observation on the particle. (Quotation marks have been placed around the word "know" for reasons given below.) In fact it is obvious that by means of the experimental set-up of Fig. 7-1, using light of very short wave length and a microscope of large aperture, the position of a particle can be ascertained to any desired precision. Since the momentum (and hence the velocity) just prior to the observation may also be known precisely, we should be able to compute with unlimited accuracy both the position and the momentum at any time before making the observation. In other words, there is no limitation in principle on our ability to calculate a precise path for a particle up to the moment of our last physical contact with it. Therefore the uncertainty principle does not at first sight seem to require any drastic revision of the old-fashioned concept of the motion of particles in definite paths, such as we all acquire through our everyday sense experiences and which we tend to justify by appealing to what we call "common sense."

At this point a most interesting philosophical point arises that spotlights an important difficulty in the relationship of science to "common sense." Suppose that a particle is given a precisely known momentum and that its position is accurately measured at some instant. Suppose further that we now calculate the precise path that the particle must have taken before its position was observed. How could we verify this calculation? It is obviously impossible to return to the past and make further measurements on the particle in the state in which it existed prior to the observation. Any measurements made after the observation of the position can have no value in checking the calculation, since the position measurement has thrown the particle off of the calculated path by an indeterminant amount. Thus as soon as we have gained enough information to compute a precise path for the particle we find that we are no longer able to verify the calculation. If science is concerned only with assertions that are capable of experimental verification, the calculation of such a path can hardly be of any scientific interest. (We can now see why the word "know" was placed in quotes in the sentence at the beginning of the last paragraph. If we say that we know something about a system in a scientific sense, we mean that we can make an assertion about the system that can be tested by an experimental measurement. Clearly, any "knowledge" that we may have about the exact path of a particle cannot be scientific knowledge in this sense.)

Thus it is possible to maintain that particles "really" do have definite positions and velocities and that it is merely the clumsiness of our instruments that makes it impossible to prove this. Such a thesis is, however, entirely metaphysical, since it cannot be either proven or contradicted by scientific means. Its chief justification seems to be that it allows us to believe literally what our gross senses tell us about the world about us. It also conforms with what the man in the street calls "common sense." Scientists know only too well, however, that our gross sensations are unreliable and that common sense is fallible. Most scientists therefore see little reason for clinging to a dogmatic position such as this. They are willing to accept the uncertainty principle as an expression of an inherent characteristic of the nature of things as we are fated to know it.

According to quantum mechanical Laws I and III, all that we may know about the future behavior of a system can be calculated from its state function. Since the precision of our knowledge of the future is limited, the state function must take these limitations into account. One of the most beautiful features of quantum mechanics is the way in which it automatically includes these limitations. The very act of ascertaining that some object in the world about us has a certain state function inevitably creates a disturbance that makes our knowledge of the properties of the object partially indeterminate. The lack of commutation of the position and momentum operators assures us that the quantum mechanical theory will never tell us more than these unavoidable disturbances make it possible for us to know.

B. Limitations on the Measurement of Energy in an Observation of Limited Duration

Suppose that we wish to know the energy of a free particle to a high degree of precision. Since $E = p^2/2m$, where p is the momentum and m is the mass, the error in the measurement of the energy is given by

$$\delta E = \frac{p}{m}\,\delta p$$

where δp is the error in the measurement of p. Now p/m, which is the velocity, may be measured by determining the time, Δt, required for the particle to go a distance Δx,

$$p/m = \Delta x/\Delta t$$

But Δx is the maximum uncertainty in the position of the particle during the measurement of p, so the minimum error in our knowledge of p after the measurement is

$$\delta p \simeq h/\Delta x$$

Thus the minimum uncertainty in our knowledge of E after the measurement is

$$\delta E \simeq \frac{\Delta x}{\Delta t}\frac{h}{\Delta x} = \frac{h}{\Delta t}$$

Since Δt represents the duration of the measurement, we can say that

$$\text{(Minimum uncertainty in energy)} \cdot \text{(Duration of measurement)} \simeq h \tag{B-1}$$

This is another of Heisenberg's uncertainty relations and is quite general.

Another illustration of (B-1) is to be found in the measurement of the energy of a photon. According to Planck's discovery, the energy of a photon is $E = h\nu$, so the error in the measurement of the energy is $\delta E = h\delta\nu$, where $\delta\nu$ is the error in the frequency measurement. The frequency may be measured by counting the number, n, of light waves passing a given point in time interval Δt, giving $\nu = n/\Delta t$. But the number n can be measured only to the nearest whole integer, so that ν will be uncertain by $1/\Delta t$. Thus we have $\delta E \simeq h/\Delta t$, and since Δt is the duration of the measurement, Equation (B-1) is again obtained.

If many systems are placed in identical states, and if they remain in that state on the average for a time τ (called the mean lifetime of the state), then it is obvious that on the average, measurements of the energy of the state can last no longer than time τ. Thus the minimum uncertainty in our knowledge of this energy is

$$\delta E \simeq h/\tau \tag{B-2}$$

C. Relationship of Zero-Point Energies to the Uncertainty Principle

The limitations on measurements prescribed by Heisenberg's uncertainty principles are automatically taken into account by quantum mechanics. As far as the momentum and the position are concerned, this is clearly shown in Corollary VII of the laws of quantum mechanics. It is interesting, however, to see how the uncertainty principle is taken into account in a particular example. For this purpose, let us consider a particle in a box of length a, for which the possible energy states are $E = n^2h^2/8ma^2$, where $n = 1, 2, 3, \cdots$. We have seen in Chapter 6 (Equations (B-7) and (B-8)) that the average value of the momentum is zero in any eigenstate, but that the square of the momentum is precisely $2mE = n^2h^2/4a^2$. Therefore the momentum itself has a value of either $+\ nh/2a$ or $-nh/2a$ and the uncertainty in the momentum is the difference between these, or

$$\delta p = nh/a$$

The uncertainty in the position of the particle is a, since we know only that the particle is in the box. Thus

$$\delta x = a$$

so that

$$\delta p\ \delta x = nh$$

This product has its lowest value in the ground state ($n = 1$), where

$$\delta p\ \delta x = h$$

Evidently our quantum mechanically derived knowledge of the properties of a particle in a box is consistent with the uncertainty principle. This would, however, not be the case if the lowest possible energy of the particle in the box were exactly zero (i.e., if the particle in the lowest energy state were stationary) since δp would then be zero. Thus we see that there must be some relationship between zero-point energies and the uncertainty principle.

Presumably *zero-point energies are a manifestation of the uncertainty principle*, reflecting the unavoidable disturbances that always occur even when a system is forced into its state of lowest energy. We must expect that any system, in which a particle is restricted to a limited region of space, will have a nonvanishing zero-point energy. This, of course, is actually the case with the hydrogen atom and the harmonic oscillator. On the other hand the zero-point energy of the rotator is actually zero. This is possible because in the ground state of the rotator we have no knowledge whatsoever of the orientation of the axis of rotation. The angular positions being completely unknown, the corresponding angular momenta can be precisely known. They can therefore, and do, assume the values zero.

Exercises. (1) Calculate $\overline{x^2}$ and $\overline{p^2}$ for the ground state of the harmonic oscillator, and show that their product is about what one would expect from the uncertainty principle.

(2) Calculate $\bar{r}$ and $\sqrt{\overline{p_r^2}}$ for the ground state of the hydrogen atom, and show that their product is about what one would expect from the uncertainty principle.

D. Zero-Point Energies and the Formation of Molecules

Consider the lowest energy states of a particle moving in two potential energy boxes of the shape shown in Fig. 7-4. Let the distance R be varied, while the length b of the boxes is held fixed. The solutions are just those of

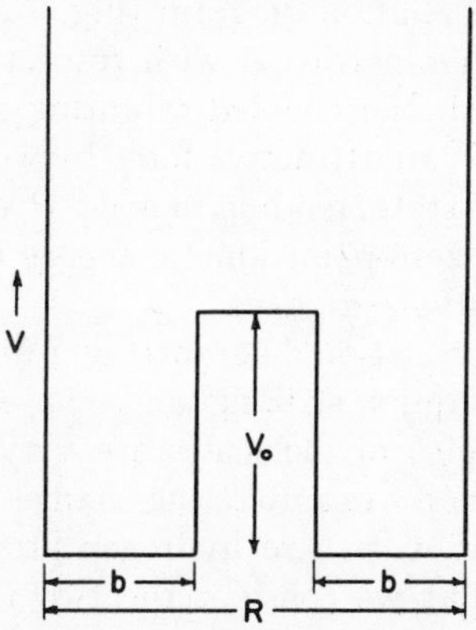

FIG. 7-4. Potential energy of a particle in two boxes separated by a barrier.

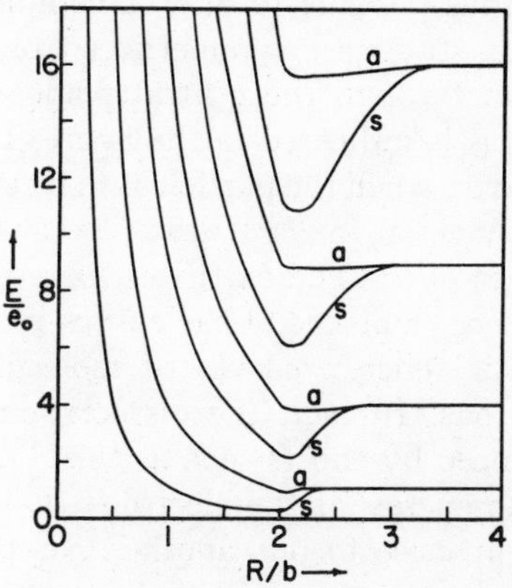

FIG. 7-5. Energy levels of a particle subject to the potential shown in Fig. 7-4.

two identical vibrating strings (page 146) coupled to one another by a string of large negative density. For very large values of R the energies are those of a particle in a single box, or slightly less than

$$E_\infty = n^2h^2/8mb^2 = n^2e_0$$

if the barrier V_0 is fairly high. Each level will be doubly degenerate. As R is decreased, nothing much happens to these energy levels until the central barrier is sufficiently thin to allow "leakage" of the particle from one box to the other to occur relatively frequently. The degeneracy will then be removed, half the states becoming more stable, the other half being not much affected. When $R = 2b$, the central barrier has zero thickness, the energies become precisely

$$n^2h^2/8m(2b)^2 = \frac{n^2}{4}e_0$$

and the eigenfunctions are those of a particle in a single box with infinite walls. For smaller values of R, the energies all increase with decreasing R according to the relation

$$E = n^2 \left(\frac{b}{R}\right)^2 e_0$$

It is thus possible to draw a complete correlation diagram (Fig. 7-5). The various states are symmetric (s) or antisymmetric (a) with respect to reflection through the central plane, as with the coupled vibrating strings, and this is indicated in the figure. There is an attractive force between the two boxes when the particle is in its ground state, tending to make R approximately equal to $2b$. It arises because the zero-point kinetic energy is 75% lower in one large box than it is in two half-sized boxes.

The resemblance of the curves in the correlation diagram (Fig. 7-5) to the potential energy curves of molecules is quite striking and not entirely fortuitous. In fact, the dissociation energies of molecules are very much influenced by effects such as this. Consider, for example, the changes in the kinetic energy of the electrons that result when two hydrogen atoms are brought close to one another. Suppose that we consider the two isolated atoms to be equivalent to two cubical boxes, each containing an electron. If the sides of the cubes have length B, the kinetic energy of the electron will be (*cf.* page 187)

$$T = 3h^2/8mB^2$$

The actual kinetic energy of an electron in the ground state of the hydrogen atom is (*cf.* Exercises (1) and (2) on page 232).

$$T = e^2/2a_0$$

Since $a_0 = \hbar^2/m_e e^2$, on equating the two kinetic energies we obtain $B = \sqrt{3}\pi a_0 = 2.88$Å. Thus, as far as the kinetic energy of the electron in a hydrogen atom is concerned, the atom may be regarded as a cube whose sides are 2.88Å. in length. If now the two hydrogen nuclei are brought together to a distance b, we may suppose that the electrons will have kinetic energies roughly equivalent to those of electrons in a rectangular box of dimensions $B \times B \times (B + b)$. The zero-point kinetic energy of the electrons in such a box will be less than that in the separated atoms by

$$\Delta E = 2\,\frac{h^2}{8m}\left[\frac{1}{B^2} - \frac{1}{(B+b)^2}\right]$$

In the hydrogen molecule the bond length is 0.74Å. Setting $b = 0.74$Å.

and $B = 2.88$Å. we find $\Delta E = 70$ kcal./mole. Since the observed energy required to dissociate the hydrogen molecule into atoms is 108 kcal./mole, we see that a "box-size" effect such as this must play some part in determining the stability of molecules.

In actual molecules, however, the above argument is oversimplified. It was shown in Chapter 6 (page 233), through an analysis based on the quantum mechanical virial theorem, that as two atoms are brought together the average kinetic energy of the electrons is given by

$$T = -E - R(\partial E/\partial R) \tag{D-1}$$

whereas the average potential energy is

$$V = 2E + R(\partial E/\partial R) \tag{D-2}$$

where E is the energy of the diatomic molecule when the nuclei are a distance R apart. These quantities are plotted for a typical molecule in Fig. 7-6. It is

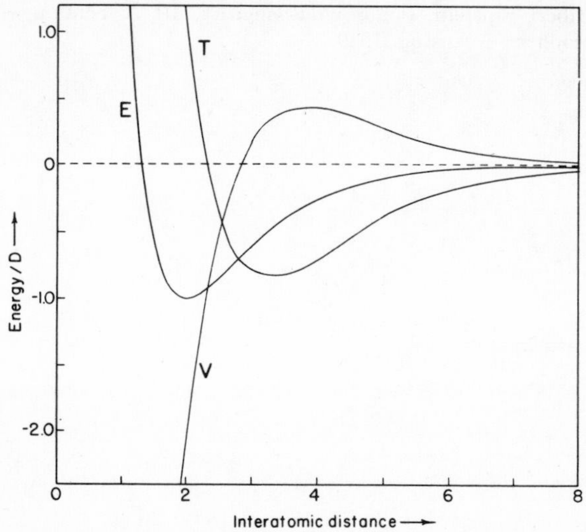

FIG. 7-6. Application of the virial theorem to the energy-*vs.*-distance curve of a chemical bond. The total energy of the bond is E, its potential energy is V, and the kinetic energy of the electrons is T. The total energy is assumed to vary with the bond length, R, according to the Morse function, $E = D(e^{-2\beta(R-R_0)} - 2e^{-\beta(R-R_0)})$ with $\beta = 1$ and $R_0 = 2$.

seen that the kinetic energy of the electrons does indeed at first decrease as the nuclei are brought together, indicating that more space becomes available to the electrons. But as the atoms are brought closer together, a point is reached at which, because of the strong simultaneous attraction of the two nuclei, the electrons begin to have *less* room in which to move about. This leads to an increase in their kinetic energy. At the equilibrium internuclear distance (where $\partial E/\partial R = 0$), the kinetic energy of the electrons must always be *greater* than that of the electrons in the separated atoms by an amount equal to the bond energy. Thus, although the "box-size" effect is responsible for the attraction that occurs at large interatomic distances, it causes repulsions at smaller separations. Under ordinary conditions the energy holding molecules together is therefore electrostatic potential energy.

REFERENCES CITED

1. W. Heisenberg, "The Physical Principles of Quantum Theory," U. of Chicago Press, Chicago, 1930. (Reprinted by Dover, New York).
2. N. Bohr, *in* "Albert Einstein, Philosopher-Scientist" (P. A. Schilpp, ed.), Library of Living Philosophers, Evanston, 1949.

Chapter 8

Angular Momentum in Quantum Mechanics

Angular momentum provides us with one of the most useful properties with which to classify atomic states and, less completely, molecular states. It is also a fairly simple quantity with which to work. In the next pages we shall outline the quantum mechanical procedure for dealing with this important property. We shall also mention experimental methods by which angular momenta in atoms and molecules may be studied.

According to the formulation of quantum mechanics that has been outlined in Chapter 5, we must first find the operators for angular momentum and then we must apply these operators to the state functions of atomic and molecular systems in accordance with the laws stated on page 159. The angular momentum operators are determined by finding the classical expressions for angular momentum in terms of the position coordinates and linear momenta, and then substituting the basic position and momentum operators ($x\cdot$ and $(\hbar/i)\partial/\partial x$, etc.) into these expressions.

A. The Angular Momentum of a Single Particle

a. Classical treatment

In Chapter 2, Equation (D-6), the angular momentum of a particle about a point 0 is defined as the vector

$$\mathbf{M} = m\mathbf{r} \times \mathbf{v}$$

where m is the mass of the particle, $\mathbf{r}$ is the vector drawn from the point 0 to the particle, $\mathbf{v}$ is the linear velocity of the particle, and $\mathbf{r} \times \mathbf{v}$ is the vector product defined by Equation (D-3) of Chapter 2. Since the linear momentum vector $\mathbf{p} = m\mathbf{v}$, we may also write

$$\mathbf{M} = \mathbf{r} \times \mathbf{p} \qquad \text{(A-1)}$$

If we set up a Cartesian coordinate system with the point 0 as the origin, the vectors in Equation (A-1) can be expressed in terms of the unit vectors $\boldsymbol{i}$, $\boldsymbol{j}$, and $\boldsymbol{k}$ along the x-, y-, and z-axes as follows

$$\boldsymbol{M} = M_x\boldsymbol{i} + M_y\boldsymbol{j} + M_z\boldsymbol{k} \tag{A-2}$$

$$\boldsymbol{r} = x\boldsymbol{i} + y\boldsymbol{j} + z\boldsymbol{k} \tag{A-3}$$

$$\boldsymbol{p} = p_x\boldsymbol{i} + p_y\boldsymbol{j} + p_z\boldsymbol{k} \tag{A-4}$$

Here (x, y, z) are the coordinates of the moving particle, (p_x, p_y, p_z) are the components of its linear momentum, and (M_x, M_y, M_z) are the components of angular momentum about the three axes. According to Equation (A-1) and the rules of vector multiplication, the components of angular momenta about the three axes are (*cf*. page 27)

$$M_x = yp_z - zp_y \tag{A-5a}$$

$$M_y = zp_x - xp_z \tag{A-5b}$$

$$M_z = xp_y - yp_x \tag{A-5c}$$

Furthermore, the square of the angular momentum (a scalar) is given by the scalar product of $\boldsymbol{M}$ with itself (*cf*. Equation (D-2) of Chapter 2)

$$\boldsymbol{M}\cdot\boldsymbol{M} = M^2 = M_x^2 + M_y^2 + M_z^2 \tag{A-6a}$$

The scalar magnitude of the angular momentum is, of course, simply the square root of (A-6a)

$$|\,\boldsymbol{M}\,| = \sqrt{M_x^2 + M_y^2 + M_z^2} \tag{A-6b}$$

b. The quantum mechanical operators for angular momenta

Expressions (A-5) and (A-6a) are in the form required for the construction of quantum mechanical operators. We find on replacing p_x by $(\hbar/i)\partial/\partial x$, x by x, etc., that the operators are

$$M_x = -i\hbar(y\partial/\partial z - z\partial/\partial y) \tag{A-7a}$$

$$M_y = -i\hbar(z\partial/\partial x - x\partial/\partial z) \tag{A-7b}$$

$$M_z = -i\hbar(x\partial/\partial y - y\partial/\partial x) \tag{A-7c}$$

$$M^2 = M_x^2 + M_y^2 + M_z^2 \tag{A-8}$$

The student can show that these operators are linear and Hermitian.

For many purpose it is more convenient to express these operators in polar coordinates

$$x = r \sin \theta \cos \varphi$$

$$y = r \sin \theta \sin \varphi$$

$$z = r \cos \theta$$

Applying this transformation to Equations (A-7) and (A-8), we find (the details are given in Eyring, Walter and Kimball, (*1*, page 40 and appendix III))

$$M_x = -i\hbar\,[-\sin \varphi\, \partial/\partial\theta - \cot \theta \cos \varphi\, \partial/\partial\varphi] \qquad \text{(A-9a)}$$

$$M_y = -i\hbar\,[\cos \varphi\, \partial/\partial\theta - \cot \theta \sin \varphi\, \partial/\partial\varphi] \qquad \text{(A-9b)}$$

$$M_z = -i\hbar\, \partial/\partial\varphi \qquad \text{(A-9c)}$$

$$M^2 = -\hbar^2 \left[\frac{1}{\sin \theta}\frac{\partial}{\partial\theta}\sin \theta\frac{\partial}{\partial\theta} + \frac{1}{\sin^2 \theta}\frac{\partial^2}{\partial\varphi^2}\right] \qquad \text{(A-10)}$$

c. The commutation of the angular momentum operators with each other

According to the principles of quantum mechanics it is only possible to know two properties precisely in the same system if their operators commute (Corollary VII, page 171). Which of the angular momentum operators commute? Using Equations (A-7) it is easy to show that

$$M_xM_y - M_yM_x = i\hbar M_z \qquad \text{(A-11a)}$$

$$M_yM_z - M_zM_y = i\hbar M_x \qquad \text{(A-11b)}$$

$$M_zM_x - M_xM_z = i\hbar M_y \qquad \text{(A-11c)}$$

The proof of (A-11a) is as follows

$$M_xM_y = i\hbar(y\partial/\partial z - z\partial/\partial y)\cdot i\hbar(z\partial/\partial x - x\partial/\partial z)$$

$$= -\hbar^2(y\partial/\partial x + yz\partial^2/\partial z\partial x - xy\partial^2/\partial z^2 - z^2\partial^2/\partial x\partial y + xz\partial^2/\partial y\partial z)$$

$$M_yM_x = -\hbar^2(yz\partial^2/\partial x\partial z - z^2\partial^2/\partial x\partial y - xy\partial^2/\partial z^2 + x\partial/\partial y + xz\partial^2/\partial y\partial z)$$

Subtracting these two expressions we obtain

$$M_xM_y - M_yM_x = -\hbar^2(y\partial/\partial x - x\partial/\partial y) = i\hbar M_z$$

Equations (A-11b) and (A-11c) are proven in analogous fashion. Evidently it is not possible to know more than one of the components of angular momentum precisely in any system. On the other hand, it is easily seen from (A-9c) and (A-10) that

$$M^2M_z - M_zM^2 = 0 \tag{A-12}$$

so that it should be possible to know both the z-component and the magnitude of the total angular momentum. Since there is nothing exceptional about the z-direction, the same must be true for any direction, so M^2 must also commute with M_x and M_y (this can, of course, be proven directly from (A-9a, b) and (A-10)).

Thus, although it is possible to prepare particles in states in which the (scalar) magnitude of the angular momentum and the component in any *one* direction are both known precisely, the magnitude and the components in *two* directions cannot be known precisely. This is in accordance with the uncertainty principle, for if we knew the total angular momentum and two of its components we could find the third component from Equation (A-6). This tells us the precise location of the axis of rotation (whose direction cosines are M_x/M, M_y/M, M_z/M). But if the axis of rotation is known, the linear momentum of the particle parallel to this axis must vanish, that is, the magnitude of the linear momentum parallel to the axis is precisely known. On the other hand, the uncertainty in the location of the particle in this direction will generally be finite. This, however, disagrees with the uncertainty principle, according to which the product of the uncertainty in linear momentum in any direction and the uncertainty in location in that direction can never vanish in any system whose state function is known.

Exercise. Does M_z commute with $\mathbf{M} = M_x\mathbf{i} + M_y\mathbf{j} + M_z\mathbf{k}$? How do you interpret this in terms of the uncertainty principle?

d. The eigenfunctions and eigenvalues of the angular momentum operators M^2 and M_z

It can be shown from the commutation rules (A-11) that the only functions that are simultaneously eigenfunctions of M^2 and M_z are the surface spherical harmonics in the complex form,

$$S_{lm}(\theta, \varphi) = P_l^{|m|}(\cos\theta)\, e^{im\varphi} \tag{A-13}$$

where l is a positive integer, m can be any integer between $-l$ and $+l$, and $P_l^{|m|}$ is the associated Legendre function of order $|m|$ and degree l. The argument used to show this is given by Eyring, Walter, and Kimball

(*1*, pp. 42-47) and by Condon and Shortley (*2*, pp. 46-53). For our purposes it will be sufficient to show by direct operation that S_{lm} is actually an eigenfunction of these operators. Thus

$$M_z S_{lm} = -i\hbar\,(\partial/\partial\varphi)\,P_l^{|m|}(\cos\theta)\,e^{im\varphi} = m\hbar S_{lm} \tag{A-14}$$

$$M^2 S_{lm} = -\hbar^2\left[\frac{1}{\sin\theta}\frac{\partial}{\partial\theta}\sin\theta\frac{\partial}{\partial\theta} + \frac{1}{\sin^2\theta}\frac{\partial^2}{\partial\varphi^2}\right] S_{lm}$$

$$= -\hbar^2\left[\frac{1}{\sin\theta}\frac{d}{d\theta}\sin\theta\frac{dP_l^{|m|}}{d\theta} - \frac{m^2}{\sin^2\theta}P_l^{|m|}\right] e^{im\varphi} \tag{A-15}$$

Referring to Chapter 3, we see from Equations (B-42, 60, and 62) that the quantity in brackets in (A-15) is equal to $-l(l+1)P_l^{|m|}$, so that

$$M^2 S_{lm} = l(l+1)\hbar^2 S_{lm} \tag{A-16}$$

Thus the only possible observable values of the angular momenta are $\sqrt{l(l+1)}\,\hbar$, and the only possible observable values of the component of angular momentum about a specified axis are $m\hbar$, where $-l \leq m \leq l$. It is interesting that the total angular momentum, $\sqrt{l(l+1)}\,\hbar$, always exceeds the maximum observable component of angular momentum, $l\hbar$. This is again a consequence of the uncertainty principle: if the total angular momentum were equal to its component in, say, the z-direction, we would know that the components in the x- and y-directions are precisely zero. We would therefore know all three components of angular momentum precisely, and we have already seen that this would lead to a contradiction of the uncertainty principle.

The solutions of the steady state Schroedinger equation for the hydrogen atom, for a free particle and for any other system in which the potential energy is spherically symmetric about some point can be written in the form

$$\psi = f(r)P_l^{|m|}(\cos\theta)\,e^{im\varphi}$$

where the form of $f(r)$ varies depending on the form of the spherically symmetric potential. These functions are clearly eigenfunctions of M^2 and M_z with eigenvalues $l(l+1)\hbar^2$ and $m\hbar$, respectively. They are also eigenfunctions of the Hamiltonian operator. Therefore they correspond to states for which we have precise knowledge of the values of three properties: the energy, the total angular momentum and the z-component of the angular momentum. According to Corollary VI (page 170), precise simultaneous knowledge of these properties is possible only if the operators H, M^2, and M_z commute with each other. The commutation of M^2 and M_z has been proven above; the commutation of H with M^2 and with M_z is most easily seen by expressing the energy classically in terms of the angular momentum

$$H = \frac{(\text{total angular momentum})^2}{2 \times \text{moment of inertia}} + \text{potential energy}$$

$$= (M^2/2I) + V(r)$$

where $V(r) = 0$ for the free particle and $-e^2/r$ for the hydrogen atom. To convert this to a quantum mechanical operator we merely write M^2 in operator form. The resulting Hamiltonian operator commutes with M^2 and M_z, since M^2 commutes with itself and with M_z, and since neither M^2 nor M_z have any effect on the variable r.

e. Effect of $(M_x + i\,M_y)$ and $(M_x - i\,M_y)$ on the angular momentum eigenfunctions

The operators $M_x + iM_y$ (henceforth denoted by μ) and $M_x - iM_y$ (henceforth denoted by ν) have an important effect on the surface spherical harmonics that we shall find very useful in understanding atomic spectra.* Referring to Equations (A-9a) and (A-9b), it is seen that

$$\mu = M_x + iM_y = \hbar e^{i\varphi} \ (\partial/\partial\theta \ + i \cot\theta\,\partial/\partial\varphi) \tag{A-17a}$$

$$\nu = M_x - iM_y = \hbar e^{-i\varphi} (-\partial/\partial\theta + i \cot\theta\,\partial/\partial\varphi) \tag{A-17b}$$

When these two functions operate on $e^{im\varphi}$ we obtain

$$\mu e^{im\varphi} = -m\hbar \cot\theta\; e^{i(m+1)\varphi} \tag{A-18a}$$

$$\nu e^{im\varphi} = -m\hbar \cot\theta\; e^{i(m-1)\varphi} \tag{A-18b}$$

Making use of Equation (B-77) of Chapter 3, setting $y = \cos\theta$ so that $\partial/\partial y = -(1/\sin\theta)\partial/\partial\theta$, we find that

$$\mu P_l^{|m|}(\cos\theta) = \hbar e^{i\varphi}\,[\,|m| \cot\theta\, P_l^{|m|}(\cos\theta) - P_l^{|m|+1}(\cos\theta)] \tag{A-19a}$$

Similarly

$$\nu P_l^{|m|}(\cos\theta) = -\hbar e^{-i\varphi}\,[\,|m| \cot\theta\, P_l^{|m|}(\cos\theta) - P_l^{|m|+1}(\cos\theta)] \tag{A-19b}$$

* The student who is not interested in working through some rather complicated calculations should skip Equations (A-17) through (A-22). The important results are given by (A-23) through (A-26) and the final paragraph of this section.

The effects of μ and ν on the surface spherical harmonics, $S_{lm}(\theta, \varphi)$, are thus found to be

$$\mu S_{lm} = e^{im\varphi}\, \mu P_l^{|m|} + P_l^{|m|}\, \mu e^{im\varphi}$$

$$= \hbar e^{i(m+1)\varphi}\, [-(m - |m|) \cot\theta\, P_l^{|m|} - P_l^{|m|+1}] \tag{A-20}$$

$$\nu S_{lm} = \hbar e^{i(m-1)\varphi}\, [-(m + |m|) \cot\theta\, P_l^{|m|} + P_l^{|m|+1}] \tag{A-21}$$

In Equations (A-20) and (A-21), m may be either positive or negative. If m is positive we have $|m| + 1 = |m + 1|$ and $m = |m|$ so that

$$m \geqq 0 \begin{cases} \mu S_{lm} = -\hbar P_l^{|m+1|}\, e^{i(m+1)\varphi} = -\hbar S_{l,m+1} & \text{(A-22a)} \\ \nu S_{lm} = -\hbar(2\,|m| \cot\theta\, P_l^{|m|} - P_l^{|m+1|})\, e^{i(m-1)\varphi} & \text{(A-22b)} \end{cases}$$

If, on the other hand, m is negative, $|m| + 1 = |m - 1|$ and $m = -|m|$ so that

$$m \leqq 0 \begin{cases} \mu S_{l,m} = \hbar(2\,|m| \cot\theta\, P_l^{|m|} - P_l^{|m+1|})\, e^{i(m+1)\varphi} & \text{(A-22c)} \\ \nu S_{l,m} = \hbar P_l^{|m-1|}\, e^{i(m-1)\varphi} = \hbar S_{l,m-1} & \text{(A-22d)} \end{cases}$$

Now according to Equation (B-74) of Chapter 3, setting $y = \cos\theta$,

$$2\,|m| \cot\theta\, P_l^{|m|} - P_l^{|m|+1} = (l + |m|)\,(l - |m| + 1)\, P_l^{|m|-1}$$

Thus (A-22b) becomes

$$m \geqq 0: \ \mu S_{l,m} = -\hbar(l + |m|)\,(l - |m + 1|)\, S_{l,m-1} \tag{A-22e}$$

and (A-22c) becomes

$$m \leqq 0: \ \mu S_{l,m} = \ \hbar(l + |m|)\,(l - |m - 1|)\, S_{l,m+1} \tag{A-22f}$$

If instead of using $S_{l,m}(\theta, \varphi)$ as defined by Equation (A-13), we use the functions

$$Y_{l,m}(\theta, \varphi) = a\left[\frac{2l + 1}{4\pi}\ \frac{(l - |m|)!}{(l + |m|)!}\right]^{1/2} S_{l,m}(\theta, \varphi) \tag{A-23}$$

where $a = (-1)^m$ if m is positive, and $a = 1$ if m is negative, then we find that the four Equations (A-22a, d, e, f) may be reduced to the two following expressions, regardless of the sign of m

$$\mu Y_{lm}(\theta, \varphi) = \hbar[l(l + 1) - m(m + 1)]^{1/2}\, Y_{l,m+1}(\theta, \varphi) \tag{A-24a}$$

$$\nu Y_{lm}(\theta, \varphi) = \hbar[l(l + 1) - m(m - 1)]^{1/2}\, Y_{l,m-1}(\theta, \varphi) \tag{A-24b}$$

The functions Y_{lm} are more convenient to use than S_{lm} because they are normalized in addition to being mutually orthogonal. That is

$$\int_0^{2\pi}\int_0^{\pi} Y_{lm}^* Y_{l'm'} \sin \theta d\theta d\varphi = \delta_{ll'}\delta_{mm'} \tag{A-25}$$

where δ_{ij} is the Kronecker delta. Equation (A-25) may be proven by means of (B-79) of Chapter 3 and Corollary VIII (page 172).

The surface spherical harmonics Y_{lm} are eigenfunctions of M^2 and M_z. According to Equations (A-24), *the functions* μY_{lm} and νY_{lm} *must also be eigenfunctions of* M^2 *and* M_z. Their eigenvalues are readily found from Equations (A-24) to be given by

$$M^2\mu Y_{lm} = l(l+1)\hbar^2\mu Y_{lm} \tag{A-26a}$$

$$M^2\nu Y_{lm} = l(l+1)\hbar^2\nu Y_{lm} \tag{A-26b}$$

$$M_z\mu Y_{lm} = (m+1)\hbar\mu Y_{lm} \qquad (m \neq l) \tag{A-26c}$$

$$M_z\nu Y_{lm} = (m-1)\hbar\nu Y_{lm} \qquad (m \neq -l) \tag{A-26d}$$

$$M_z\mu Y_{ll} = 0 \tag{A-26e}$$

$$M_z\nu Y_{l,-l} = 0 \tag{A-26f}$$

Thus the operators μ *and* ν *operating on a surface spherical harmonic have no effect on the total angular momentum, but* μ *increases the z-component by* $\hbar$, *while* ν *decreases the z-component by* $\hbar$. Although it has taken a certain amount of tedious algebra to prove it, this result, with Equations (A-24), will prove useful in interpreting the spectra of complex atoms.

f. Expression of M^2 in terms of μ, ν and M_z

The operator for M^2 is rather complicated, and it is often somewhat awkward to work with. The operators μ, ν, and M_z are, however, easy to use. It is therefore convenient to express M^2 in terms of these three operators. We know that

$$M^2 = M_x^2 + M_y^2 + M_z^2 \tag{A-27}$$

But

$$\begin{aligned} M_x^2 + M_y^2 &= (M_x + iM_y)(M_x - iM_y) + i(M_xM_y - M_yM_x) \\ &= \mu\nu + i(i\hbar M_z) \\ &= \mu\nu - \hbar M_z \end{aligned} \tag{A-28}$$

from Equation (A-11a). Similarly

$$M_x^2 + M_y^2 = \nu\mu + \hbar M_z \tag{A-29}$$

Thus we have the very useful relations

$$M^2 = \mu\nu - \hbar M_z + M_z^2 \tag{A-30a}$$

$$M^2 = \nu\mu + \hbar M_z + M_z^2 \tag{A-30b}$$

Exercise. Using Equations (A-30) and (A-24), show that $M^2 Y_{lm} = l(l+1)\,\hbar^2 Y_{lm}$.

B. The Angular Momentum of Systems Composed of Many Particles

a. Classical treatment

The classical total angular momentum, $\mathbf{M}_t$, of a group of particles about a point is a vector defined by the vector sum of the individual angular momenta, $\mathbf{M}_1, \mathbf{M}_2, \cdots$, of the particles about the given point

$$\mathbf{M}_t = \sum_{\substack{\text{all}\\ \text{particles}}} \mathbf{M}_n \tag{B-1}$$

(henceforth the subscript t will refer to the entire system, whereas the subscript n refers to the nth particle). $\mathbf{M}_t$ and $\mathbf{M}_n$ can be expressed in terms of their Cartesian components

$$\mathbf{M}_n = M_{xn}\mathbf{i} + M_{yn}\mathbf{j} + M_{zn}\mathbf{k} \tag{B-2a}$$

$$\mathbf{M}_t = M_{xt}\mathbf{i} + M_{yt}\mathbf{j} + M_{zt}\mathbf{k} \tag{B-2b}$$

so that from (B-1)

$$M_{xt} = \sum_{\substack{\text{all}\\ \text{particles}}} M_{xn}\,; \quad M_{yt} = \sum_{\substack{\text{all}\\ \text{particles}}} M_{yn}\,; \quad M_{zt} = \sum_{\substack{\text{all}\\ \text{particles}}} M_{zn} \tag{B-3}$$

where M_{xn} is the x-component of angular momentum of the nth particle, etc. The square of the total angular momentum, that is, the scalar product of $\mathbf{M}_t$ with itself, is

$$\mathbf{M}_t \cdot \mathbf{M}_t = M_t^2 = M_{xt}^2 + M_{yt}^2 + M_{zt}^2 \tag{B-4}$$

This expression, when written in terms of the angular momentum components of the individual particles, is complicated in that it contains many cross products between the angular momenta of different particles. Note especially that M_t^2 is *not* equal to $\sum M_n^2$.

b. The quantum mechanical operators and their commutation with each other

The operators for total angular momentum and its components are constructed by replacing M_{xn}, M_{yn}, and M_{zn} in (B-3) by the single particle operators discussed in Section A. For instance

$$M_{zt} = -i\hbar \sum_{\substack{\text{all} \\ \text{particles}}} \partial/\partial\varphi_n \tag{B-5}$$

where φ_n is the azimuthal angle of the nth particle. It is easy to show that the components M_{xt}, M_{yt}, and M_{zt} fail to commute

$$(M_{xt}\, M_{yt} - M_{yt}\, M_{xt}) = i\hbar M_{zt} \tag{B-6a}$$

$$(M_{yt}\, M_{zt} - M_{zt}\, M_{yt}) = i\hbar M_{xt} \tag{B-6b}$$

$$(M_{zt}\, M_{xt} - M_{xt}\, M_{zt}) = i\hbar M_{yt} \tag{B-6c}$$

It is also not hard to show that M_t^2 commutes with M_{xt}, M_{yt} and M_{zt}.

$$M_t^2\, M_{xt} = M_{xt}\, M_t^2\,;\quad M_t^2\, M_{yt} = M_{yt}\, M_t^2\,;\quad M_t^2\, M_{zt} = M_{zt}\, M_t^2 \tag{B-7}$$

It should therefore be possible to specify precisely the square of the total angular momentum as well as one (but no more than one) of its components in any system.

Exercises. (1) Show that all of the angular momentum operators of any one particle commute with those of any other particle. For instance $M_{x1}M_{y2} = M_{y2}M_{x1}$.

(2) Using the result of (1), prove Equations (B-6).

(3) Using the result of (2), prove Equations (B-7).

c. Commutation with molecular Hamiltonians

For molecules and atoms the Hamiltonian operator is, using atomic units (*cf.* Equation (E-1) of Chapter 5)

$$H = -\sum_{\substack{\text{all} \\ \text{particles}}} \frac{1}{2m_n} \nabla_n^2 + \sum_{\substack{\text{all pairs} \\ \text{of nuclei}}} \frac{Z_k Z_l}{r_{kl}} + \sum_{\substack{\text{all pairs} \\ \text{of electrons}}} \frac{1}{r_{ij}} - \sum_{\substack{\text{all electron-} \\ \text{nuclear pairs}}} \frac{Z_k}{r_{ik}} \tag{B-8}$$

where m_n is the mass of the nth particle (relative to an electronic mass), ∇_n^2 is the Laplacian operator of the nth particle, Z_k is the atomic number of the kth nucleus, and the r's are the distances between pairs of particles. We shall now show that M_{zt} (as defined in (B-5)) commutes with H. That is

$$M_{zt}H = HM_{zt} \tag{B-9}$$

Since the Laplacian operators involve no functions of the angles φ_n, but only a second derivative with respect to φ_n, all of the $\partial/\partial\varphi_n$ in M_{zt} commute with all of the ∇_n^2. Therefore the condition that M_{zt} commutes with H is that M_{zt} commutes with the terms $1/r_{mn}$ for every pair of particles m and n. Since r_{mn} depends only on the coordinates of the mth and n^{th} particles, we have

$$M_{zt}\frac{1}{r_{mn}} = \frac{1}{r_{mn}}M_{zt} + \frac{\hbar}{i}\left[\frac{\partial 1/r_{mn}}{\partial \varphi_m} + \frac{\partial 1/r_{mn}}{\partial \varphi_n}\right] \tag{B-10}$$

As can be seen from Fig. 8-1, r_{mn} depends only on the difference between the azimuthal angles, φ_m and φ_n, and not on the individual values of these angles. Setting $\Delta = \varphi_m - \varphi_n$, so that $\partial\Delta/\partial\varphi_m = -\partial\Delta/\partial\varphi_n$, we see that

$$\frac{\partial 1/r_{mn}}{\partial \varphi_m} + \frac{\partial 1/r_{mn}}{\partial \varphi_n} = \frac{d 1/r_{mn}}{d\Delta}\left[\frac{\partial\Delta}{\partial \varphi_m} + \frac{\partial\Delta}{\partial \varphi_n}\right] = 0 \tag{B-11}$$

Therefore the second term on the right in (B-10) vanishes, M_{zt} commutes with each of the terms $1/r_{mn}$, and we have demonstrated that for all molecular

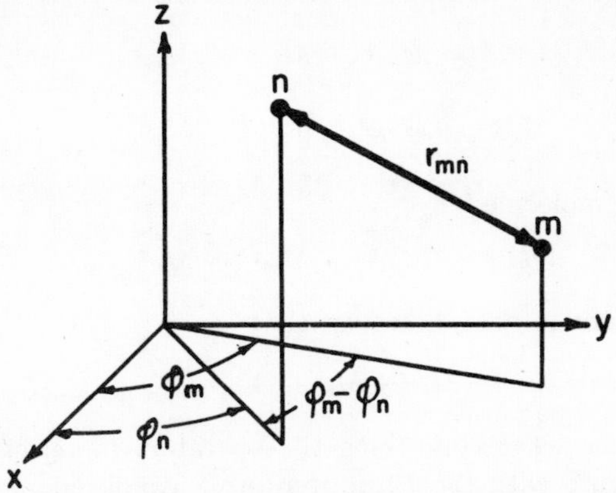

FIG. 8-1. Dependence of r_{mn} on the difference between the azimuthal angles φ_m and φ_n.

and atomic systems whose Hamiltonian is of the form (B-8), H commutes with M_{zt}. Since there is nothing exceptional about the z-axis, H must also commute with M_{xt} and M_{yt}, and hence also with M_t^2. *It is therefore possible to know simultaneously the energy and the magnitude of the total angular momentum and any one component of the total angular momentum of an atom or molecule to any desired degree of precision.* This statement is true only insofar as it is justifiable to neglect forces between particles that are not directed along the line joining the particles ("noncentral forces"). We shall see in Chapter 10 that under certain conditions (spin-orbit interactions) noncentral forces do become important, so this statement is no longer true.

It is interesting to see whether it is possible to separate the contributions of the electrons and the nuclei to the total angular momentum. To investigate this possibility, we set up new operators, M_{xe}, M_{ye}, M_{ze}, for the components of angular momentum of the electrons, and M_{xN}, M_{yN}, M_{zN}, for those of the nuclei. Thus

$$M_{xe} = \sum_{\substack{\text{all}\\\text{electrons}}} M_{xi}\,, \qquad M_{ye} = \sum_{\substack{\text{all}\\\text{electrons}}} M_{yi}\,, \qquad M_{ze} = \sum_{\substack{\text{all}\\\text{electrons}}} M_{zi} \tag{B-12a}$$

$$M_{xN} = \sum_{\substack{\text{all}\\\text{nuclei}}} M_{xk}\,, \qquad M_{yN} = \sum_{\substack{\text{all}\\\text{nuclei}}} M_{yk}\,, \qquad M_{zN} = \sum_{\substack{\text{all}\\\text{nuclei}}} M_{zk} \tag{B-12b}$$

$$M_{xt} = M_{xe} + M_{xN}\,, \; M_{yt} = M_{ye} + M_{yN}\,, \; M_{zt} = M_{ze} + M_{zN} \tag{B-12c}$$

where M_{xt}, M_{yt}, and M_{zt} are the total angular momentum operators discussed above. These components are combined in the usual way to obtain the squares of the angular momenta. For the electrons we have

$$M_e^2 = M_{xe}^2 + M_{ye}^2 + M_{ze}^2 \tag{B-13a}$$

for the nuclei

$$M_N^2 = M_{xN}^2 + M_{yN}^2 + M_{zN}^2 \tag{B-13b}$$

and for the entire molecule

$$M_t^2 = M_{xt}^2 + M_{yt}^2 + M_{zt}^2 \tag{B-13c}$$

It is most important to note, however, that

$$M_t^2 \neq M_e^2 + M_N^2 \tag{B-13d}$$

Let us see under what conditions the operators for electronic angular momentum commute with the Hamiltonian of a molecule or an atom. In particular we must consider the commutation of M_{ze} with $1/r_{ik}$, where r_{ik} is the distance between the ith electron and the kth nucleus. Clearly, out of all of the φ's of the electrons, r_{ik} depends only on φ_i. Therefore

$$M_{ze}\frac{1}{r_{ik}} = \frac{1}{r_{ik}}M_{ze} + \frac{\hbar}{i}\frac{\partial^1/r_{ik}}{\partial\varphi_i} \tag{B-14}$$

Evidently M_{ze} will commute with the Hamiltonian only if r_{ik} is independent of φ_i.This will be the case if the nucleus k lies on the z-axis. Thus if M_{ze} is to commute with H, all nuclei must lie on the z-azis.

The only molecules for which we can specify the electronic angular momentum precisely in a state of known energy are those in which all nuclei lie along a straight line (e.g., diatomic molecules, acetylene and CO_2). *Even in this case, only the component of electronic angular momentum about the linear axis of the molecule can be specified.*

On the other hand, *for atoms whose energies are known, it is possible to specify both the total electronic angular momentum about the nucleus and the component of this momentum along any one direction,* since in atoms the nucleus lies on the x-, y-, and z-axis, so that M_{xe}, M_{ye} and M_{ze}–and hence also M_e^2–all commute with H.

These results have a simple classical interpretation. As an electron moves about in a nonlinear molecule such as CH_4, it keeps running into atomic nuclei. Each encounter results in changes in size and direction of linear momenta, and hence also of angular momenta, of both the electrons and the nuclei. As a result there is a continual transfer of angular momentum to and from the nuclei; although the total angular momentum of the molecule is constant, it keeps shuttling back and forth between the nuclei and the electrons due to these collisions. In linear molecules, however, the component of motion of the electrons about the molecular axis (i.e., in the φ-direction) never results in collisions with nuclei. Therefore the electrons can maintain this part of their angular momentum indefinitely. Furthermore, in atoms the interactions between the electrons and the nucleus can never change the angular momentum of the electrons about the nucleus, since the force between the nucleus and the electrons is radial in direction and thus can have no effect on the angular motions.

Exercise. Is it possible to separate the electronic angular momentum in atoms having definite energies into the contributions of individual electrons? Justify your answer by a discussion of the commutation of the atomic Hamiltonian with the angular momentum operators of individual electrons. What classical interpretation can you give for this?

There is an interesting exception to these arguments in compounds of the rare earth elements. Some of the electrons (the "*f*-electrons") in the rare earth ions are so deeply buried inside the ionic core that they rarely run into surrounding nuclei with sufficient violence to change their angular momentum.* These electrons therefore act almost as if they were in isolated atoms.

* Since angular momentum occurs only in multiples of $\hbar$, collisions with surrounding atoms must be fairly violent in order to result in an exchange of angular momentum.

In the transition metals the "*d*-electrons" are also somewhat buried inside the ions, but not so deeply as are the *f*-electrons. Therefore, though somewhat unaffected by the surrounding atoms, the *d*-electrons act much less as if they were in isolated atoms than do the *f*-electrons. This has important effects on the magnetic properties, which we shall discuss in detail later (Chapter 10).

d. Useful expressions for the operator M_t^2

The operator M_t^2 is defined through Equations (B-3) and (B-4). Because of the occurrence of cross products in this expression, it is quite inconvenient for practical calculations. Now

$$M_{xt}^2 - M_{yt}^2 = (M_{xt} + iM_{yt})(M_{xt} - iM_{yt}) + i(M_{xt}M_{yt} - M_{yt}M_{xt})$$
$$= \mu_t\nu_t - \hbar M_{zt} \qquad \text{(B-15)}$$

from Equation (B-6a), where $\mu_t = M_{xt} + iM_{yt} = \sum \mu_n$ and $\nu_t = M_{xt} - iM_{yt} = \sum \nu_n$, and where μ_n and ν_n are the operators μ and ν for the nth particle. Thus we see that we can express M_t^2 as

$$M_t^2 = \mu_t\nu_t - \hbar M_{zt} + M_{zt}^2 \qquad \text{(B-16)}$$

It can be shown in a similar manner that an alternative expression for M_t^2 is

$$M_t^2 = \nu_t\mu_t + \hbar M_{zt} + M_{zt}^2 \qquad \text{(B-17)}$$

C. Spectroscopic Notation Based on Angular Momentum

It has become customary to label the energy states of atoms with letters depending upon the eigenvalue obtained when the wave function is operated on by M_e^2, the operator for the square of the angular momentum of the electrons moving about in their orbits around the nucleus (usually called *orbital angular momentum*). This eigenvalue invariably has the form $L(L + 1)\hbar^2$, where L is a positive integer. (That it must have this form is shown in Eyring, Walter, and Kimball (*1*, pp. 41-44).) The convention is that if

$L = 0, 1, 2, 3, 4, 5, 6, 7$, etc., then the state is denoted by

S, P, D, F, G, H, I, K, etc., respectively.

(The symbol J is omitted because it is used in another connection (A. state with a given L is $(2L + 1)$-fold degenerate because the component of angular momentum in any one direction can run from $-L\hbar$ to $+L\hbar$. The L-value of an atomic state can be determined experimentally by a study of the Zeeman effect (see page 266).

For linear molecules, we have shown that only the electronic angular momentum about the axis of the molecule can be precisely known for a molecule in a definite energy state. If this axis is called the z-axis, then the wave function can be an eigenfunction of M_{ze} with eigenvalues $\mathscr{M}\hbar$ where $\mathscr{M}$ is a positive or negative integer. This leads to the following terminology for the electronic states of linear molecules:

If $\lvert \mathscr{M} \rvert =$	0, 1, 2, 3, etc.,
the state is denoted by	Σ, Π, Δ, Φ, etc.

Electronic angular momentum does not contribute to the degeneracy of Σ-states, but since $\mathscr{M}$ can be positive or negative, Π-, Δ-, etc., states are always at least doubly degenerate. The $\mathscr{M}$-values of the states of linear molecules can be determined experimentally through the analysis of band spectra (page 393).

Additional features in spectroscopic notation depend on the spin of the electron. Their discussion will be deferred until this phenomenon has been considered in detail (see page 310).

D. Use of Magnetic Fields in Studying the Angular Momentum of Charged Particles

a. The magnetic moment associated with orbital motion

When an electric current I travels through a wire, it is accompanied by a circular magnetic field enclosing and centered on the wire (see Fig. 8-2).

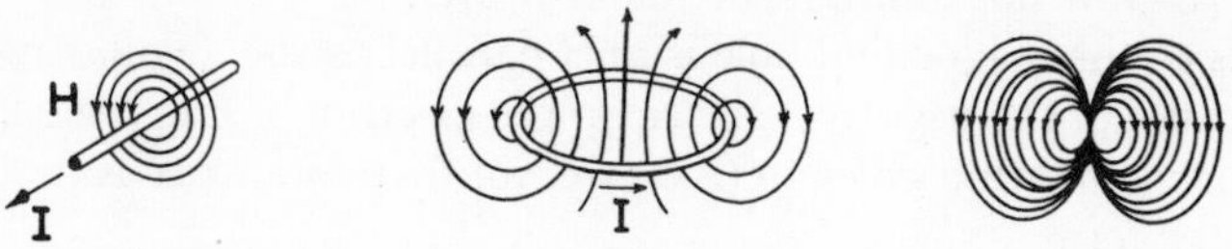

A B C

FIG. 8-2.
A. The magnetic field around a current element.
B. The magnetic field around a current ring.
C. The field around a magnetic dipole.

If the wire is bent into the form of a ring of radius a, the magnetic field takes on the same shape as that of a uniformly magnetized disc of the same area as the ring. (This law was first stated by Ampere. See Joos (*3*, p. 292).) At distances large compared with the size of the ring, the magnetic field is the same as that of a point dipole the magnitude of whose magnetic dipole moment is

$$|\boldsymbol{\mu}| = (\text{current})\cdot(\text{area of ring}) \tag{D-1}$$

(law of Biot and Savart; see *3*, p. 290). If the current is expressed in electrostatic units and the magnetic dipole moment is expressed in electromagnetic units, we have

$$|\boldsymbol{\mu}| = \pi I a^2/c \tag{D-2}$$

where c is the velocity of light in cm./sec.

An electron (charge $= e$) moving with linear velocity v in a circular orbit of radius a is equivalent to a circular current of this kind. Since the electron passes a given point in its orbit $v/2\pi a$ times per second, the total charge passing the point per second (i.e., the current) is

$$I = ev/2\pi a \tag{D-3}$$

Thus

$$|\boldsymbol{\mu}| = eva/2c \tag{D-4}$$

But the angular momentum of the electron is $m_e va = \sqrt{l(l+1)}\,\hbar$, where l is the angular momentum quantum number and m_e is the mass of the electron. Thus

$$|\boldsymbol{\mu}| = \sqrt{l(l+1)}\; e\hbar/2m_e c = \sqrt{l(l+1)}\,\beta_M \tag{D-5}$$

where

$$\beta_M = e\hbar/2m_e c \tag{D-6}$$

is the atomic unit of magnetic moment known as the *Bohr magneton*. Equation (D-5) is not limited to circular orbits.

The axis of the magnetic dipole is anti-parallel to the angular momentum vector because of the negative charge on the electron. Using the definition of the angular momentum given in (A-1) we can therefore write Equation (D-5) in the vector form

$$\boldsymbol{\mu} = (e/2m_e c)\,\mathbf{M} \tag{D-5a}$$

$$= (e/2m_e c)\,(\mathbf{r} \times \mathbf{p})$$

where e is negative. We have seen that the component of angular momentum along any chosen direction can take only the values $l\hbar$, $(l-1)\hbar$, $\cdots$, $-l\hbar$.

The angle θ between the axis of rotation and the chosen direction can therefore assume only the values (see Fig. 8-3)

$$\theta = \cos^{-1}(m/\sqrt{l(l+1)})$$

where m is an integer in the range $-l \leq m \leq l$. The component of the dipole moment in any chosen direction must therefore be

$$|\boldsymbol{\mu}| \cos \theta = m\beta_M \tag{D-7}$$

For this reason, the magnetic moment is said to be "space-quantized."

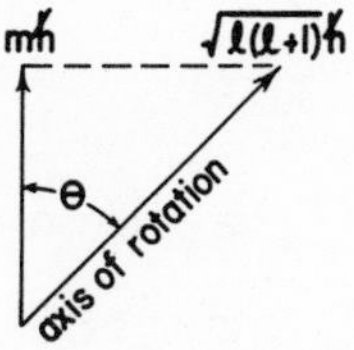

FIG. 8-3. Relation between the angular momentum and its component in a given direction.

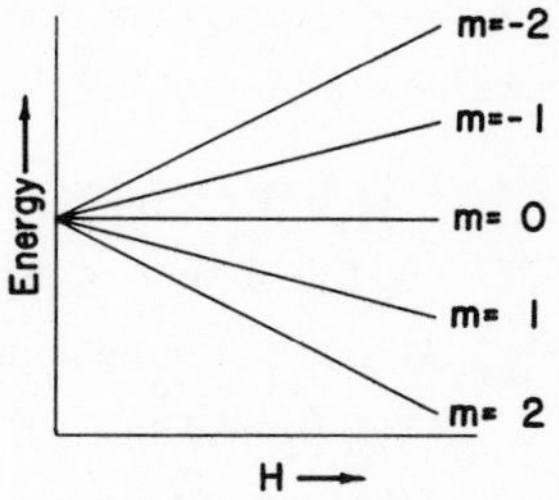

FIG. 8-4. Energy levels of a d-electron ($l = 2$) in a magnetic field $\boldsymbol{H}$.

b. Energies of electronic orbits in magnetic fields

According to classical electromagnetic theory, the energy of a magnetic dipole $\boldsymbol{\mu}$ in a magnetic field $\boldsymbol{H}$ is

$$E = -|\boldsymbol{\mu}||\boldsymbol{H}| \cos \theta = -\boldsymbol{\mu} \cdot \boldsymbol{H} \tag{D-8}$$

where θ is the angle between the dipolar axis and the field, $\boldsymbol{H}$. We have just seen, however, that for the magnetic dipole moment caused by an electron moving in an orbit

$$|\boldsymbol{\mu}| \cos \theta = m\beta_M$$

Thus the interaction energy of the orbital motion of an electron with a magnetic field is quantized, only the $(2l + 1)$ uniformly spaced values

$$E = -m\beta_M |\boldsymbol{H}| \quad (m = l, l-1, \cdots, -l+1, -l) \tag{D-9}$$

being possible (see Fig. 8-4). Because of its occurrence in this expression, the integer m is called the *magnetic quantum number*. It is, of course, the same as the quantum number for the z-component of angular momentum.

All of these arguments remain true when more than one electron is present, except that l and m must then be replaced by the total orbital angular momentum quantum number, L, and the total orbital magnetic quantum number, $\mathscr{M}$. The energy levels given by (D-9) are separated from one another by an amount $\beta_M \mid \boldsymbol{H} \mid$, which is independent of L and $\mathscr{M}$.

The Bohr magneton has the numerical value

$$\begin{aligned}\beta_M &= 0.9274 \times 10^{-20} \text{ ergs per molecule per gauss} \\ &= 1.34 \times 10^{-4} \text{ calories per mole per gauss.}\end{aligned}$$

In order to have some idea of the energies involved, we give the values in gauss of some typical magnetic fields:

One ampere turn	=	1.26
Earth's magnetic field	= *c.*	0.4
A good permanent magnet	= *c.*	10,000
Cyclotron field	= *c.*	20,000
Highest fields yet obtained (Kapitza)	= *c.*	500,000

The energy of a mole of Bohr magnetons in a field of 10,000 gauss is thus 5.6 joules, or 1.3 calories. This is small compared with the mean thermal energy at room temperature, $RT = 600$ cal./mole, so that such a field will be unable, by itself, to line up a very large fraction of the magnets.

It is interesting to compare this result with that for electrical dipoles in electric fields. The conventional unit for electric dipole moments is the Debye unit,

$$\begin{aligned}&1.000 \times 10^{-18} \text{ ergs per molecule per (e.s. volt/cm.)} \\ &= 0.465 \times 10^{-4} \text{ calories per mole per (practical volt/cm.)}\end{aligned}$$

Thus when a field of one practical volt per cm. is applied to a group of electric dipoles whose strength is one Debye unit, there is about one third as much orientation as when one gauss is applied to magnetic dipoles whose strength is one Bohr magneton. Furthermore, since there is no such thing as magnetic breakdown, it is possible to produce much higher magnetic field strengths than electrical ones. Therefore it is somewhat easier to orient magnetic dipoles than electrical dipoles, though without cooling it is generally quite difficult to obtain magnetic fields sufficiently strong to bring forth saturation effects. (Ferromagnetism is an exception here, since other factors cause the dipoles to line up in ferromagnetic materials.)

c. The study of atomic states through the effects of magnetic fields

Evidently magnetic fields completely remove the $(2L + 1)$-fold rotational degeneracy of the electronic orbital motions in atoms. Therefore, *in the absence of complicating effects due to electron spin* (into which we shall go later), *the total orbital angular momentum quantum number, L, of an atomic energy level can be found directly from the number of states* (*viz.*, $2L + 1$) *into which the level splits on applying a magnetic field.* (It is only fair to mention that, more often than not, electron spin does cause complicating effects here. It will not be difficult, however, to extend what we shall say below to include these effects.) This splitting can be studied experimentally in three ways – either (*i*) by paramagnetic resonance absorption, or (*ii*) by the effects of magnetic fields on optical spectra, or (*iii*) by the effects of inhomogeneous magnetic fields on atomic beams.

(*i*) Of these methods, the paramagnetic resonance absorption method is the most direct and also the most recently developed. If an atom possessing a magnetic moment due to orbital motions of its electrons is placed in a uniform magnetic field of 1000 gauss, a group of levels with an energy separation of 1000 β_M will be produced. This spacing is equivalent in energy to a photon whose wave length is 10.70 cm. – that is, in the microwave range. Thus if microwaves of wave length $10{,}700/|\boldsymbol{H}|$ cm. impinge on such an atom, transitions between the levels can be induced, and absorption of radiant energy will take place. This is known as *paramagnetic resonance absorption*. In actual practice, the absorption (which is very weak, and requires careful instrumentation in order to be detected) usually occurs at considerably different wave lengths than those just given; the discrepancy is due to the effects of electronic spin and will be discussed in Chapter 10. [The subject of paramagnetic resonance is reviewed by Bleaney and Stevens (*4*), by Gordy, Smith and Trambarulo (*5*) and by Wertz (*6*)].

(*ii*) If the principles described in Section b, above, were correct, the effect of a magnetic field on optical atomic spectra (that is, atomic spectra occurring in the optical and ultra-violet regions) should be quite simple. In Chapter 16 we shall find that these spectra result from transitions between different electronic states of an atom. In this transition the magnetic quantum number $\mathscr{M}$ changes by zero or by ± 1, except that a state with $\mathscr{M} = 0$ does not undergo a transition to another state with $\mathscr{M} = 0$. According to Section b, above, a level with a total angular momentum quantum number L will, in a magnetic field $\boldsymbol{H}$, split into $2L + 1$ levels, each having a different value of $\mathscr{M}$ and each separated from its nearest neighbor by the constant amount, $\beta_M |\boldsymbol{H}|$. Because of the selection rule $\Delta\mathscr{M} = 0$ or ± 1, this means that a given spectral line should split into no more than three components in a magnetic field (see Fig. 8-5). Such a splitting is in fact sometimes observed; it is known as the *normal Zeeman effect*. More often, spectral lines split into more than three lines in the presence of a magnetic field. This is known as the *anomalous Zeeman effect*. It occurs because, as we shall see in Chapters 9 and

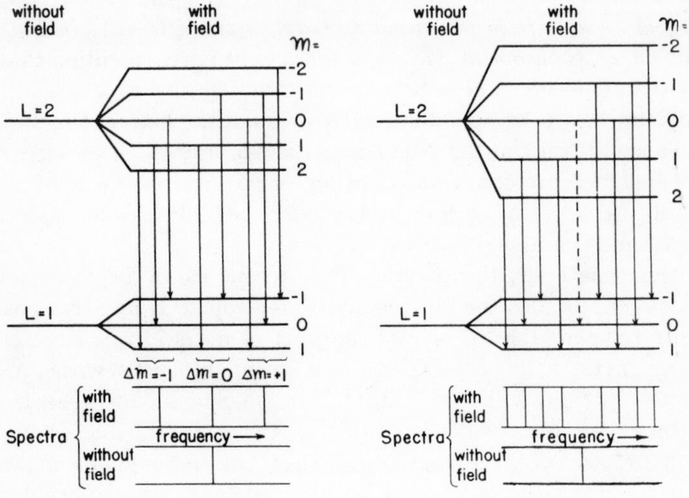

A B

FIG. 8-5.
A. The normal Zeeman effect for a transition from a state with $L = 2$ to one with $L = 1$. The spacings of the levels are the same in the two states in the presence of the magnetic field, so only three lines are observed, each being a superposition of the lines from several transitions.
B. The anomalous Zeeman effect for a transition from $L = 2$ to $L = 1$. The spacings of the energy levels differ in the two states, so eight lines are obtained. (Note that the transition $m = 0$ to $m = 0$, indicated by the dashed arrow, does not occur.)

10, electron spin can make the splitting of the energy levels in a magnetic field different in different atomic states. In such instances it is often possible to tell something about the L-values of the two states involved from the Zeeman effect. The interpretation depends on an understanding of electron spin, however, so we shall not go into this matter any further at this point (see Section D of Chapter 10).

Exercise. How many Zeeman components will be observed for a transition between two states if the spacings in the states are normal (i.e., $\beta_M|\boldsymbol{H}|$) and the orbital angular momentum quantum numbers, L, of the states are 0 and 0; 0 and 1; 0 and 2; 1 and 1?

(*iii*) The principle of the *atomic beam* (or *Stern-Gerlach*) *method* of studying magnetic moments is as follows. Consider a magnet in a uniform magnetic field. The force on the north pole tending to move the magnet in one direction is exactly balanced by the force on the south pole acting in the opposite direction. The only result is a torque; there will be no net force tending to cause translational motion of the entire magnet. On the other hand, if the magnetic field is not uniform the forces on the two poles will generally not be equal, and the magnet will accelerate in some direction. It is not hard to show that if there is a magnetic field in the z-direction, and if this field changes as one moves in the z-direction at a rate $d|\boldsymbol{H}|/dz$, then the net force on the magnet in the z-direction is

$$F_z = \mu_z d|\boldsymbol{H}|/dz \tag{D-10}$$

where μ_z is the z-component of the magnetic dipole. A varying field such as this is found just below the tip of the north pole of the magnet shown in cross section in Fig. 8-6A. Let us now send a beam of atoms past the north pole of such a magnet in an arrangement such as that shown in Fig. 8-6B. Let the

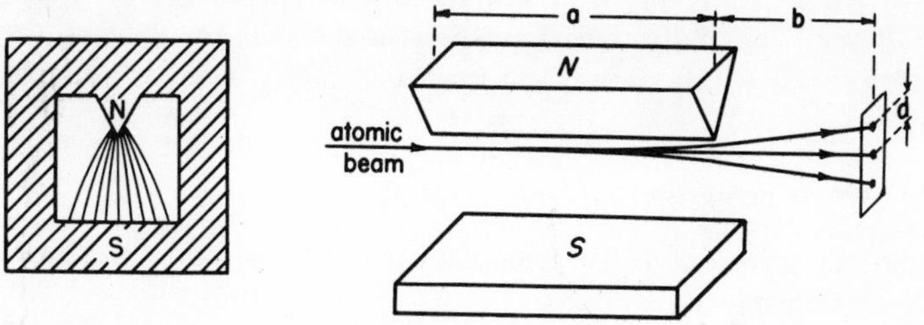

A B

FIG. 8-6. The Stern-Gerlach atomic beam method of studying magnetic moments. A. Cross section of the magnet used to produce an inhomogeneous magnetic field. B. Splitting of a beam of atoms having $L = 1$ into three beams by an inhomogeneous magnetic field.

atoms have a total orbital angular momentum quantum number L, and for the purposes of clarity let us disregard the effects of electron spin. Then as we have seen, only $2L + 1$ values of μ_z will be available to the atoms.

$$\mu_z = \mathscr{M}\beta_M \ (\mathscr{M} = L, L-1, \cdots, -L)$$

Each value of $\mathscr{M}$ will result in a different value of the force in the z-direction,

$$F_z = \mathscr{M}\beta_M d|\boldsymbol{H}|/dz \tag{D-11}$$

and hence, in a different acceleration of the atom in the z-direction. The original beam will therefore split into $2L + 1$ parts, each having a different value of $\mathcal{M}$. The number of beams can be counted by letting them fall on a photographic plate, or some other detection device, and L is thus determined. Furthermore, from the separation of the spots on the plate the value of β can be determined.

Exercise. Derive an expression for the distance, d, between the spots on a photographic plate in terms of the lengths a and b in Fig. 8-6B, β_M, the velocity v of the atoms, and the value of $d|\boldsymbol{H}|/dz$ for the magnetic field. (Assume that the angular deviation of the beam from its original direction is very small.)

In this experiment, the inhomogeneous magnetic field can be said to act as a "polarizer" which decomposes the beam of atoms into discrete beams having definite directional properties. To some extent it is similar to a Nicol prism, which splits a light beam into two beams having different directions of vibration.

This experiment was originally performed by Stern and Gerlach. It is not very often used any more in finding L, but the mere existence of such a method ought partially to satisfy the student that the quantum numbers L and $\mathcal{M}$ are something more than abstractions invented for his confusion.

d. The Larmor precession

From Equation (D-8) it is seen that a magnetic field exerts a torque on a dipole amounting to

$$\mathscr{T} = dE/d\theta = |\boldsymbol{\mu}|\,|\boldsymbol{H}| \sin\theta \tag{D-12}$$

the axis of the torque being perpendicular to both $\boldsymbol{\mu}$ and $\boldsymbol{H}$. A torque of this kind must act on atoms when they are placed in a magnetic field, and the atoms react to this torque in the same way as does a gyroscope or a child's top: the axis of the angular momentum of the atom slowly changes its direction, describing a cone about the direction of the field–a motion called *precession of the axis* (Fig. 8-7). According to classical mechanics (see Slater and Frank, 7, p. 118),

$$\mathscr{T} = |\mathbf{M}|\,\omega \sin\theta \tag{D-13}$$

where $|\mathbf{M}|$ is the magnitude of the angular momentum of the top about its own axis, ω is the precession rate (that is, the angular velocity of the axis of the top in its motion about the cone) and θ is the angle between the field and the axis of rotation of the top. (Equation (D-13) is true only if ω is much smaller than the angular velocity of the top about its own axis.) From (D-12) and (D-13)

$$|\boldsymbol{M}|\ \omega \sin\theta = |\boldsymbol{\mu}|\ |\boldsymbol{H}| \sin\theta \tag{D-14}$$

$$\omega = |\boldsymbol{\mu}|\ |\boldsymbol{H}|/M = \gamma |\boldsymbol{H}| \tag{D-15}$$

The quantity $\gamma = |\boldsymbol{\mu}| / |\boldsymbol{M}| = e/2m_e c$ is called the *gyromagnetic ratio for electronic orbital motion.* It has the value 8.8×10^6 gauss^{-1} sec.$^{-1}$

From (D-15) we see that

$$\omega = \beta_M\ |\boldsymbol{H}|/\hbar \tag{D-16}$$

The precession frequency (i.e., the number of complete precessional cycles per second) is

$$\nu_L = \omega/2\pi = \beta_M |\boldsymbol{H}|/h = 1.4 \times 10^6 |\boldsymbol{H}| \text{ sec.}^{-1} \tag{D-17}$$

when $\boldsymbol{H}$ is expressed in gauss; ν_L is known as the *Larmor frequency.* The gyroscope-like motion of the atom is known as the *Larmor precession,* after its discoverer. It is interesting to note that

$$h\nu_L = \beta_M |\boldsymbol{H}|$$

But $\beta_M |\boldsymbol{H}|$ is the level spacing in the normal Zeeman effect, so that ν_L is just the frequency separation of the three spectral lines in the normal Zeeman effect.

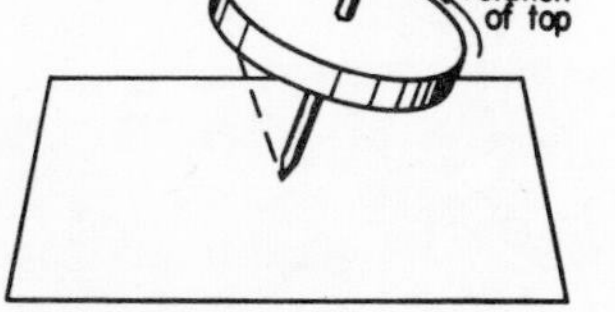

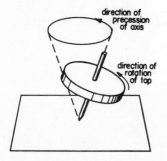

FIG. 8-7. Precession of a spinning top in the gravitational field of the earth.

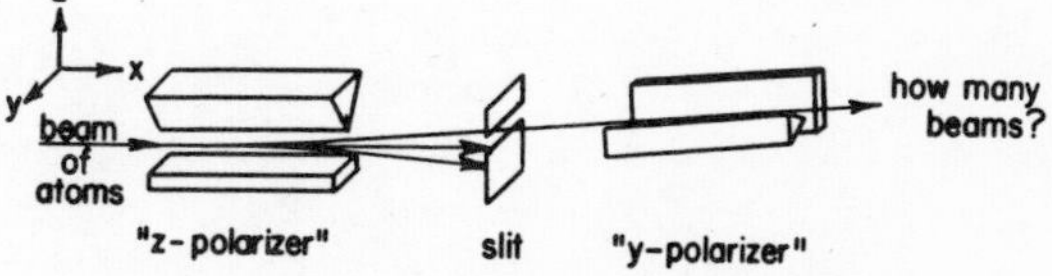

FIG. 8-8. Splitting of a beam of atoms with $L = 1$ by a series of inhomogeneous magnetic fields. See Exercise.

Exercise. A beam of atoms having $L = 1$ is sent in the x-direction through an inhomogeneous magnetic field ("polarizer") oriented in the z-direction (see Fig. 8-8).

The beam is thereby split into three beams having $\mathscr{M}_z = +1, 0$, and -1. By means of a slit, the beam for which $\mathscr{M}_z = +1$ is separated from the other beams and allowed to go through a second polarizer oriented in the y-direction. How many beams will emerge from the second polarizer? What does this mean in terms of the commutation of M_z and M_y? The beams from the second polarizer can be sent one at a time through a third polarizer oriented in the same way as the first one. How many beams will emerge in each case? What role does the Larmor precession play in the mechanistic explanation of this result?

REFERENCES CITED

1. H. Eyring, J. Walter, and G. E. Kimball, "Quantum Chemistry," Wiley, New York, 1944.
2. E. U. Condon and G. Shortley, "The Theory of Atomic Spectra," rev. ed., Cambridge U.P., New York, 1950.
3. G. Joos, "Theoretical Physics," Stechert, New York, 1934.
4. B. Bleaney and K. W. H. Stevens, *Repts. Progr. Phys.*, **16**, 108 (1953).
5. W. Gordy, W. V. Smith, and R. F. Trambarulo, "Microwave Spectroscopy," Wiley, New York, 1953.
6. J. E. Wertz, *Chem. Revs.*, **55**, 901 (1955).
7. J. C. Slater and N. H. Frank, "Mechanics," McGraw-Hill, New York, 1947.

GENERAL REFERENCE

E. Feenburg and G. E. Pake, "Notes on the Quantum Theory of Angular Momentum,' Addison-Wesley, Cambridge, Mass., 1953.

PART III

ATOMIC SYSTEMS

Chapter 9

Atomic Structure I
Hydrogen, Helium, and Electron Spin

Although chemists are almost always concerned with more complicated systems, an understanding of the properties of atoms must be the goal of the first stage of the application of quantum mechanics to chemistry. Some of the most important quantum chemical principles emerge at this stage, and much of the discussion of molecules depends on an understanding of the behavior of isolated atoms. Therefore it is necessary to devote a considerable space to atomic structure.

A. The Experimental Determination of the Energy Levels of Atoms

a. The analysis of spectra

In 1885 Balmer found that the wave lengths of the four visible lines of the hydrogen spectrum as measured by Ångstrom could be expressed very precisely by the formula

$$\lambda = 3645.6\, n^2/(n^2 - 4) \times 10^{-8} \text{ cm.} \tag{A-1}$$

where $n = 3, 4, 5,$ or 6. Rydberg turned this formula upside down to obtain an expression for the frequency

$$\tilde{\nu} = 1/\lambda = R\left[\frac{1}{4} - \frac{1}{n^2}\right] \tag{A-2}$$

In this expression the frequency is in units of wave numbers (i.e., the number of waves per centimeter) and R is a constant, known as the Rydberg constant. Its value, based on the most recent measurements of the hydrogen spectrum, is $R = 109678$ cm.$^{-1}$ Rydberg showed that the spectral lines of some other elements fall in series whose frequencies can be expressed fairly accurately by means of formulae of the type

$$\tilde{\nu} = R\left[\frac{1}{(n_1 + a_1)^2} - \frac{1}{(n_2 + a_2)^2}\right] \tag{A-3}$$

the constant R having the same value as in hydrogen. The quantities a_1 and a_2 are nonintegral constants that have definite values in a given series, whereas n_1 and n_2 are integers whose values are varied to give the different frequencies appearing in the series.

Although the occurrence of small integers and the constant R in Rydberg's formulae is interesting and significant, many spectra do not obey such simple laws. Another characteristic of the formulae has, however, been found to have a much more general importance. This is the fact that the formulae express the frequencies of spectral lines as differences between pairs of numbers

$$\tilde{\nu} = T_2 - T_1 \tag{A-4}$$

It was this feature that turned out to be the basic, universal, precise law of spectroscopic analysis. It is known as the *Rydberg-Ritz combination principle*: There exists for each element a set of numbers such that when differences are taken between different pairs of numbers, all of the observed spectral lines of the element are obtained. The numbers are called *terms*.

TABLE 9-1

APPLICATION OF THE RYDBERG-RITZ COMBINATION PRINCIPLE TO THE SPECTRUM OF MERCURY

Term Values (cm.$^{-1}$): $a = 5953.26$, $b = 10209.80$, $c = 21824.06$, $d = 40136.61$, $e = 44768.79$, $f = 46536.48$

Terms	Calculated Frequency (cm.$^{-1}$)	Calculated Wave Length (Å.)	Observed Wave Length* (Å.)
$d - a$	34183.35	2925.401	2925.406
$e - a$	38815.53	2576.288	2576.295
$f - a$	40583.22	2464.073	2464.068
$d - b$	29926.81	3341.485	3341.478
$e - b$	34558.99	2893.603	2893.595
$f - b$	36326.68	2752.798	2752.84
$d - c$	18312.55	5460.736	5460.740
$e - c$	22944.73	4358.299	4358.35
$f - c$	24712.42	4046.548	4046.561

* Data from Harrison (*1*)

A spectrum is said to have been completely analyzed when all of the terms required to account for the observed lines have been found.

In general any arbitrary set of n numbers can be derived from the differences between $n + 1$ other numbers. For instance, the four numbers 1, 5, 9, and 17 can be derived from the differences between no fewer than five other numbers (e.g., 0, 1, 5, 9, and 17). Therefore the Rydberg-Ritz principle is of no significance unless the number of terms required to express a spectrum is smaller than the number of lines in the spectrum plus one. In actual practice, the number of terms required is invariably much smaller than the number of lines, so that the combination principle results in a great simplification.

The Rydberg-Ritz principle is not limited to atoms, but is also valid for molecules, atomic nuclei, and any other system capable of emitting radiation.

As an example of the principle, we show in Table 9-1 how the frequencies of nine observed mercury lines can be calculated precisely from the differences between only six numbers. The agreement is within the experimental error.

Exercises. (1) Find four term numbers from which the frequencies of the four spectral lines of sodium having the following wave lengths can be found (units of 10^{-8} cm.)

$$\lambda = 11404.2$$
$$11382.4$$
$$5153.645$$
$$5149.090$$

(2) Find four more term numbers which, with the above four, give the frequencies of the following ten lines of sodium:

$\lambda =$ 5889.953	3302.988	22057
5895.923	3302.323	22084
6154.229	16378.0	
6160.760	16393.1	

(You now have eight numbers from which you can calculate the frequencies of 14 lines.)

(3) Check the term values you obtain against those given by Moore (*2*). (You will have to adjust your values by adding an appropriate constant to all of them.)

Before the application of quantum theory to spectroscopy by Bohr in 1912, there was a general feeling that the frequencies observed in atomic spectra were the frequencies of the normal modes of vibration of the electrons, which were by that time known to exist in atoms. The classical theories of radiation (*cf*. Chapter 15) required that any vibrating charged particle must radiate light of the same frequency as the vibration. Some interesting attempts at interpreting atomic spectra in terms of normal modes will be found in Lorentz (*3*), Ritz (*4*), Jeans (*5*), and Rayleigh (*6*). As we have seen, however, in our discussion of the classical theory of vibrating bodies (Chapters 3 and 4), normal mode frequencies are not ordinarily found by taking the differences between pairs of numbers from a relatively limited set. In fact, in spite of much effort, no one was ever able to conceive a plausible vibrating system whose normal mode frequencies could be expressed in this way. Another disturbing fact was the large number of spectral lines observed even for simple atoms such as hydrogen. This indicated that many different normal modes were available to the electrons in atoms – a conclusion offensive to one's faith in the ultimate simplicity of nature.

This and other failures of the classical vibration and radiation theories to account for the empirical laws of spectroscopy made Bohr's revolutionary ideas readily acceptable. Bohr assumed that atoms can exist only in "quantized" energy states, and that the energies are proportional to the term numbers. If the term numbers T are expressed in cm.$^{-1}$, the energies are given by

$$E = hcT \tag{A-5}$$

where h is Planck's constant and c is the velocity of light. Light is emitted or absorbed when an atom undergoes a transition from one energy level to another, the frequency being given (in cm.$^{-1}$) by

$$\tilde{\nu} = (E_i - E_j)/hc = T_i - T_j \tag{A-6}$$

The energy $hc\tilde{\nu}$ released or absorbed in the process is carried by the photon, which is emitted or absorbed during the transition.* In this way Bohr was able to account for the combination principle, and to give the term numbers a significance of the utmost importance in the understanding of the mechanics of atoms. The analysis of spectra became, through Bohr's work, one of the most important activities in physical science. It is the source of our most accurate knowledge of the energy states of atoms, molecules and nuclei. (Information about energy states may also be obtained by thermodynamic measurements and by studying collisions. These methods are used extensively in chemical problems and in nuclear physics, but for atoms, with which we are concerned here, the spectroscopic methods are all-important.)

* The assumption expressed by Equation (A-6) is a direct consequence of the laws of quantum mechanics, as we shall see in Chapter 16.

When the complete term system of an atom has been worked out, it is found that no spectral lines are observed corresponding to many of the pairs of terms. For instance, for the terms given in Table 9-1, no lines are observed whose frequencies are given by the differences between a, b, and c or between d, e, and f, even though some of these would occur in easily accessible portions of the spectrum (e.g., $c-a = 21824.06-5953.26 = 15870.80$ cm.$^{-1}$ = 6300.88 Å.). This is a result of certain rules, known as *selection rules*, which permit transitions to take place only between certain states of atoms. The most important of these rules depend upon the angular momenta of the states concerned. Transitions that violate the selection rules are said to be *forbidden*.

The origin of selection rules (which will be discussed in more detail in Chapter 16) can be made plausible by the following argument. The photon, which is the particle emitted by an atom when it undergoes a transition, has an angular momentum quantum of $\hbar$.* In quantum mechanics as in classical mechanics, the overall angular momentum vector must remain unchanged in any process. Therefore, transitions between states are possible only if the states have angular momenta that are able to compensate for the angular momentum removed by the photon. Evidently *the complete analysis of the spectrum of any system provides information on the angular momenta of atomic states*, since the angular momenta of term pairs that give observed spectral lines must be related in a definite way. Conversely, *if we know the angular momenta of a pair of atomic states, we should be able to say whether or not it is possible to have a transition between them resulting in emission or absorption of light.*

b. Energy level data

As a result of the careful study of the spectra of atoms, a vast amount of data is available on atomic energy levels. These data have been tabulated by Bacher and Goudsmit (*7*). A much more extensive tabulation is being made by Moore (*2*). A recent tabulation of the more important levels is given in Landolt-Bornstein (*8*).

The relationships of the energy levels of atoms are most effectively shown by means of energy level diagrams, in which the vertical scale represents

* Strictly speaking, this is only true of photons emitted by a dipolar transition mechanism (see Chapters 15 and 16). If the photons are emitted by a quadrupolar mechanism they carry an angular momentum of $2\hbar$ and the selection rule is correspondingly altered. Dipolar transitions are, however, by far the more usual source of photons.

the energy, states being indicated by short horizontal lines located at the appropriate energies. The horizontal dimension in the diagram is used to indicate other relationships between the states (e.g., their angular momenta). States between which optical transitions are possible are customarily connected by vertical lines on which the wave length of the corresponding spectral line is indicated. The intensity of the spectral line is sometimes indicated by drawing the vertical line heavy or light.

Some typical energy level diagrams are shown in Figs. 9-1 to 9-6; the factors that make them look as they do will be discussed in this and the next chapters. The notation appearing on the diagrams will also be discussed below, but the student will recognize that the capital letters, S, P, D, etc., indicate the orbital angular momentum of the electrons as discussed in Chapter 8 (page 260).

B. The Energy Levels, Wave Functions, and Spectrum of Hydrogen

We have seen in Chapter 6 that the Schroedinger equation for hydrogen can be solved exactly, and yields the energy levels

$$E = -1/2n^2 \tag{B-1}$$

where E is expressed in atomic units and n is an integer greater than zero. Each of these levels is n^2-fold degenerate because of the possibility of $(n-1)$ different values of the orbital angular momentum quantum number, l, and $(2l+1)$ values of the magnetic quantum number m for each value of l.

The observed energy levels of hydrogen are shown in Fig. 9-1. They represent almost the ultimate in simplicity, but are the starting point for the understanding of the energy level diagrams of all other elements. They are, of course, accounted for by Equation (B-1) and the solutions of the wave equation.

For reference the normalized, orthogonalized wave functions of the lowest hydrogen states are given in Appendix III. For generality the nuclear charge is assumed to be Z rather than one. All distances and energies are assumed to be in atomic units. These functions are all eigenfunctions of the angular momentum operators for the electron, M^2 and M_z, with eigenvalues $l(l+1)\hbar^2$ and $m\hbar$, respectively.

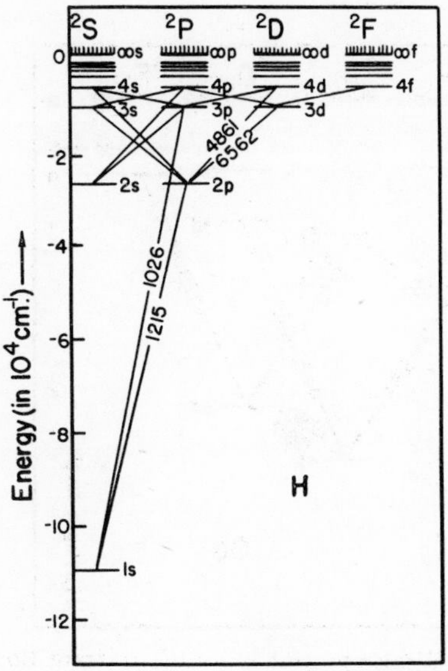

FIG. 9-1. Energy levels of the hydrogen atom. The diagonal lines join states responsible for spectral lines observed at the wave lengths indicated (in Å.). The 4861 Å. transition is responsible for the Fraunhofer *F* line in the sun's spectrum and the 6563 Å. transition is responsible for the Fraunhofer *C* line.

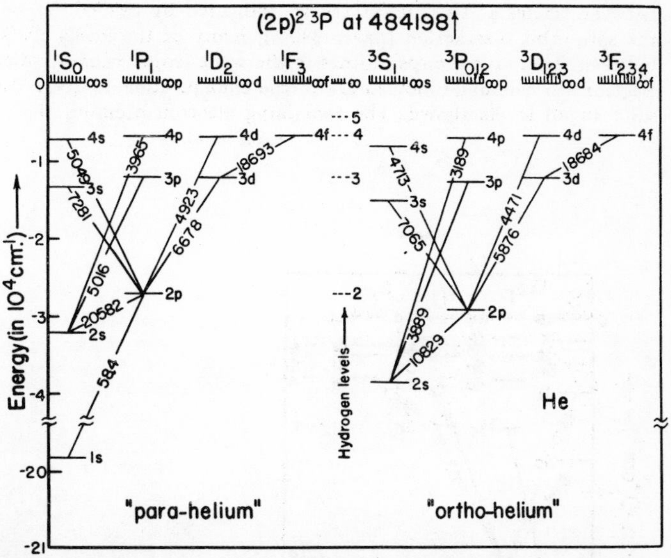

FIG. 9-2. Energy levels of the helium atom. Except for the $(2p)^2$-level, one of the two electrons is in the $1s$-orbital, the other electron being in the orbital indicated beside the level. The locations of the levels of the hydrogen atom are indicated in the center of the diagram to show the relationship between the excited states of He and those of H. Note that the scale of ordinates has been cut between 40,000 and 200,000 cm.$^{-1}$ in order to fit the ground state into the diagram without compressing the scale excessively.

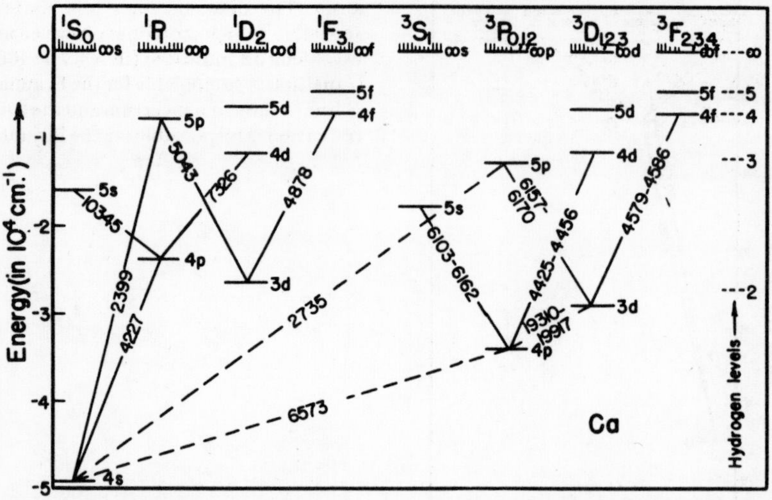

FIG. 9-3. Energy levels of the calcium atom. Weak spectral lines due to transitions between triplet and singlet states are indicated by dashed diagonal lines. Because of the spin-orbit interaction (pages 349*ff*.), many of the levels are split and transitions between them give groups of lines in the wave length ranges indicated. Of the twenty electrons in calcium, nineteen are in the configuration $1s^2 2s^2 2p^6 3s^2 3p^6 4s$, which is the same in all levels shown. The remaining electron occupies the orbital indicated for each level.

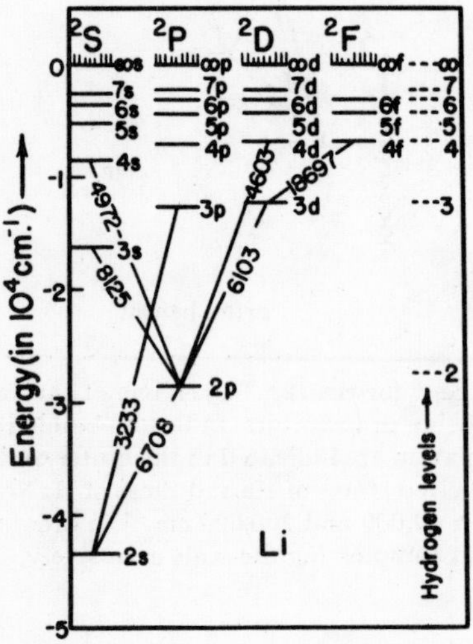

FIG. 9-4. Energy levels of the lithium atom. Two of the electrons are in the core, $1s^2$, and the third electron is in the orbital indicated.

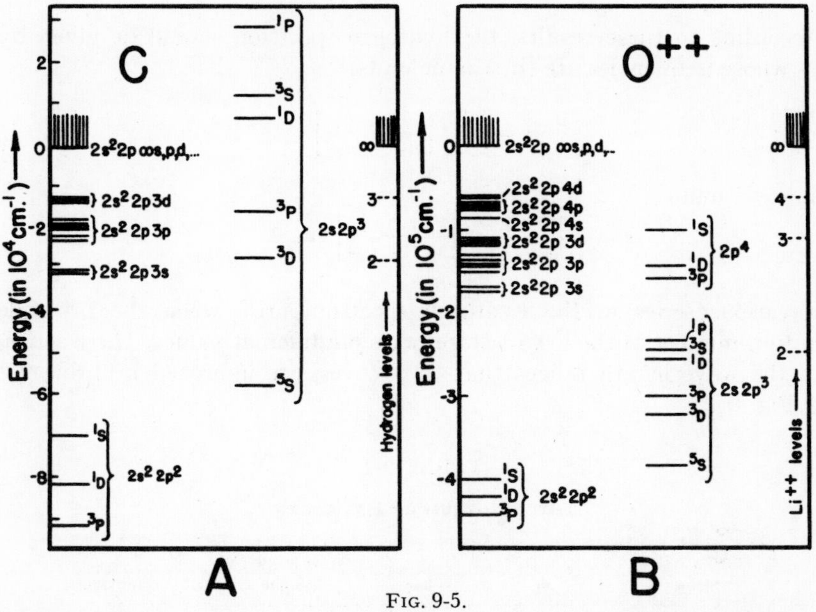

FIG. 9-5.

A. Energy levels of the carbon atom. Two electrons are in the core, $1s^2$, which is the same for all levels. The orbitals occupied by the remaining four electrons are shown for each level.

B. Energy levels of the doubly charged oxygen ion (O^{++}). This ion is isoelectronic with carbon.

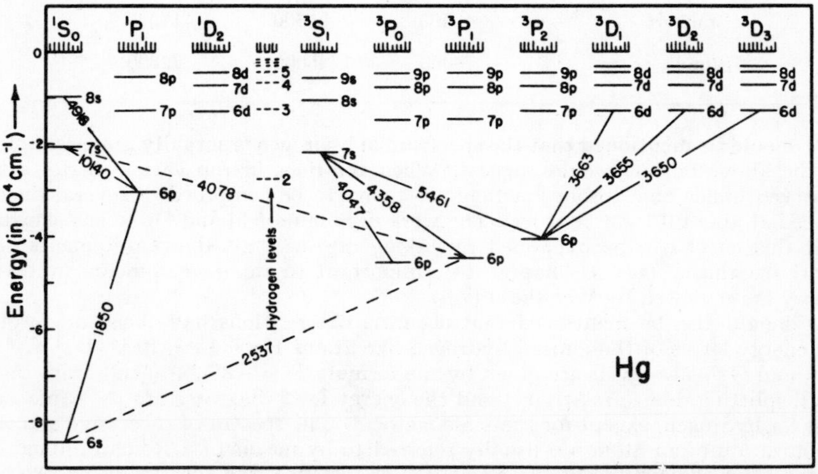

FIG. 9-6. Energy levels of the mercury atom. Of the eighty electrons in the atom, seventy-nine are in the configuration $1s^22s^22p^63s^23p^63d^{10}4s^24p^64d^{10}4f^{14}5s^25p^65d^{10}6s$, which is constant for all of the levels shown. The orbital occupied by the remaining electron is indicated for each level.

According to these results, the hydrogen spectrum should be given by lines whose frequencies are (in atomic units)

$$\nu = \frac{1}{2} Z^2 \left[\frac{1}{n_1^2} - \frac{1}{n_2^2}\right]$$

or in c.g.s. units

$$\tilde{\nu} = \frac{2\pi^2 m_e e^2}{h^3 c} Z^2 \left[\frac{1}{n_1^2} - \frac{1}{n_2^2}\right] \text{cm.}^{-1}$$

The various series in the hydrogen spectrum arise when the principle quantum number of the lowest state takes on different values. These series, with the wave length ranges that they cover, are indicated in Table 9-2 and Fig. 9-1.

TABLE 9-2

THE SPECTRUM OF HYDROGEN

Series	n_1	Wave length (Å.) 1st line	2nd line	Start of continuum
Lyman	1	1215.66	1025.8	911.7
Balmer	2	6562.79	4861.3	3647.0
Ritz-Paschen	3	18751.1	12818.1	8205.8
Brackett	4	40500.	26300.	14590.
Pfund	5	74000.	(46000)	22800.

It should be mentioned that the spectrum of hydrogen is actually not as simple as the above outline would suggest. When the lines in the various series are observed under high resolution they are found to be composed of several lines spaced at about 0.1 cm.$^{-1}$ or less. The work of Sommerfeld and Dirac has shown that this effect can be explained by taking into account the requirements of relativity theory (see *9*, Chapter 14); important recent developments in this theory are reviewed by Weisskopf (*10*).

It should also be mentioned that accurate observations have been made of the energy levels of the ionized hydrogen-like atoms He^{+1}, Li^{+2}, Be^{+3}, B^{+4}, C^{+5}, N^{+6}, and O^{+7}. The levels are given by the formula $E = -Z^2/2n^2$ (aside from the small splitting due to relativity) and the energy level diagrams are the same as that for hydrogen, except for the scale factor Z^2. The spectra of successive states of ionization of an atom are usually referred to by means of a Roman numeral placed after the symbol of the element. For instance, the spectra of Na, Na^+, Na^{++}, etc., are spoken of as the NaI, NaII, NaIII, etc., spectra.

C. The Energy Levels, Wave Functions, and Spectrum of Helium

The energy level diagram of helium is shown in Fig. 9-2. It is somewhat more complicated than the diagram of hydrogen, but we shall find that when it is completely understood, many of the important principles of atomic structure will have been clarified.

a. Formulation of the quantum mechanical problem

As our model for the helium atom we shall take a stationary nucleus around which move two electrons, identified by the numbers 1 and 2. By ascribing the charge $+ Ze$ to the nucleus, we can extend our results to helium-like ions, such as Li^{+1} and Be^{+2}. (The energy levels of ions up to F^{+8} have been studied experimentally. Though they closely resemble those of helium, they differ by more than merely a scale factor.)

The wave equation for this model is (using atomic units)

$$\nabla_1{}^2\psi + \nabla_2{}^2\psi + 2[E + (Z/r_1) + (Z/r_2) - (1/r_{12})]\psi = 0 \tag{C-1}$$

where $\nabla_1{}^2$ and $\nabla_2{}^2$ are the Laplacian operators for electrons no. 1 and no. 2, respectively; r_1 and r_2 are the distances of the respective electrons from the nucleus, and r_{12} is the distance of the electrons from each other. The wave function ψ will, of course, be a function involving the six coordinates required to locate the two electrons. If we use polar coordinates with the origin at the nucleus, the wave function will have the form,

$$\psi = \psi(r_1, \theta_1, \varphi_1, r_2, \theta_2, \varphi_2) \tag{C-2}$$

Equation (C-1) has never been solved by any direct method. The difficulty comes from the term $1/r_{12}$, representing the repulsion of the electrons; because of this term the method of separation of variables cannot be made to work in any coordinate system. One is therefore forced to rely on approximate methods, particularly the variation method and the perturbation method. Both methods have been applied to the problem, and each has its advantages.

The variation method is best adapted to finding accurate values of the energies of the one or two states of lowest energy. This approach has been exploited particularly by Hylleraas, who has been able to compute the energies of the ground states and the first excited states of helium and helium-like ions with a precision approaching and even exceeding that of the spectroscopic determinations. The variation approach to helium will be described on pages 288*ff*.

Although the perturbation method is not so well suited to giving precise results, it has the two important advantages of not being limited to low energy states and of providing an extremely useful conceptual basis for a general understanding of atomic structure. The procedure is simply to neglect the $1/r_{12}$ term in Equation (C-1), which is the source of all of the difficulty. The resulting partial differential equation,

$$\nabla_1^2\psi + \nabla_2^2\psi + 2[E + (Z/r_1) + (Z/r_2)]\psi = 0 \tag{C-3}$$

can be solved exactly. The $1/r_{12}$ term can then be introduced as a perturbation.

Writing

$$\psi = \chi_1(r_1, \theta_1, \varphi_1)\chi_2(r_2, \theta_2, \varphi_2) \tag{C-4}$$

we find from (C-3) in the usual way that

$$\nabla_1^2\chi_1 + 2[E_1 + (Z/r_1)]\chi_1 = 0 \tag{C-5a}$$

$$\nabla_2^2\chi_2 + 2[E_2 + (Z/r_2)]\chi_2 = 0 \tag{C-5b}$$

where

$$E_1 + E_2 = E \tag{C-6}$$

The solutions of (C-5) are the hydrogen wave functions (or, as they are more often called, "orbitals") given in Appendix III; E_1 and E_2 are given by

$$E_1 = -Z^2/2n_1^2$$

and

$$E_2 = -Z^2/2n_2^2 \tag{C-7}$$

so that, by neglecting the electronic repulsion, we obtain the "zero-order energies" of the helium atom,

$$E^0 = -\frac{Z^2}{2}[(1/n_1^2) + (1/n_2^2)] \tag{C-8}$$

The lowest energy of the helium atom is given by the state for which $n_1 = n_2 = 1$. Here both electrons are in $1s$-orbitals; a helium atom in such a state is said to have the "electronic configuration" $1s^2$. We can therefore write for the zero-order energy and wave function of the ground state

$$E^0(1s^2) = -Z^2 \tag{C-9}$$

$$\psi^0(1s^2) = (1s)_1\,(1s)_2 \tag{C-10}$$

where $(1s)_1$ signifies a $1s$ hydrogen wave function or orbital in which the coordinates of electron no. 1 appear, and similarly for $(1s)_2$

$$(1s)_1 = (Z^{3/2}/\sqrt{\pi})\ e^{-Zr_1}$$
$$(1s)_2 = (Z^{3/2}/\sqrt{\pi})\ e^{-Zr_2} \qquad \text{(C-11)}$$

This state is not degenerate.

The next zero-order energy level of helium is that for which $n_1 = 1$ and $n_2 = 2$, or $n_1 = 2$ and $n_2 = 1$. The zero-order energy is

$$E^0 = -5Z^2/8 \qquad \text{(C-12)}$$

This level is degenerate, there being eight wave functions having this energy:

$$\begin{aligned} \phi_1 &= (1s)_1(2s)_2 & \phi_5 &= (1s)_1(2p_0)_2 \\ \phi_2 &= (2s)_1(1s)_2 & \phi_6 &= (2p_0)_1(1s)_2 \\ \phi_3 &= (1s)_1(2p_{+1})_2 & \phi_7 &= (1s)_1(2p_{-1})_2 \\ \phi_4 &= (2p_{+1})_1(1s)_2 & \phi_8 &= (2p_{-1})_1(1s)_2 \end{aligned} \qquad \text{(C-13)}$$

These functions are normalized and orthogonal if we use the expressions in Appendix III.

We shall now apply the perturbation theory to these zero-order functions, carrying the calculation only to the first order in the energy and to the determination of the correct linear combinations of the zero-order wave functions for the first excited level.

b. The ground state

The lowest energy level being nondegenerate, the first order correction to the energy is given by (*cf.* Equation (D-14) of Chapter 5)

$$E' = \int \psi^{0*} H \psi^0 d\tau = \iint (1s_1)^2 (1s_2)^2 (1/r_{12}) d\tau_1 d\tau_2$$

$$= (Z^6/\pi^2) \iint \frac{e^{-2Zr_1} e^{-2Zr_2}}{r_{12}} d\tau_1 d\tau_2 \qquad \text{(C-14)}$$

The integrand has a simple physical interpretation. It represents the electrostatic energy of two charges, $e^{-2Zr_1}d\tau_1$ and $e^{-2Zr_2}d\tau_2$, at a distance r_{12} from one another. The integral itself therefore represents the mutual electrostatic energy of two overlapping, spherically symmetric clouds of electricity whose charge densities are both proportional to e^{-2Zr}.

This interpretation suggests a relatively simple method of evaluating the integral. It is a well-known result of the theory of electrostatics that the electrical potential inside a uniformly charged spherical shell is everywhere the same as the potential of the shell itself, whereas the potential outside the shell is the same as that resulting if the total charge of the shell were concentrated at the center of the shell (see *11*, pp. 58 and 60). Thus inside a shell of radius r_1, thickness dr_1, and charge density e^{-2Zr_1}, the potential due to the shell at a distance r from the origin is

$$\begin{aligned} V(r) &= \frac{1}{r_1} \times \text{(total charge on the shell)} \\ &= \frac{4\pi r_1^2\, e^{-2Zr_1}dr_1}{r_1} = 4\pi r_1\, e^{-2Zr_1}dr_1 \quad (r < r_1) \end{aligned} \tag{C-15}$$

while outside the shell the potential due to the shell is

$$\begin{aligned} V(r) &= \frac{1}{r} \times \text{(total charge on the shell)} \\ &= \frac{4\pi r_1^2\, e^{-2Zr_1}dr_1}{r} \qquad (r > r_1) \end{aligned} \tag{C-16}$$

The interaction energy of such a shell with an element of charge $e^{-2Zr_2}d\tau_2$ located at $r = r_2$ is

$$V(r_2)e^{-2Zr_2}d\tau_2$$

whereas the interaction energy of the entire cloud, e^{-2Zr_1}, with this element of charge is

$$\begin{aligned} d\varepsilon &= e^{-2Zr_2}\, d\tau_2 \left[\int_0^{r_2} \frac{4\pi r_1^2\, e^{-2Zr_1}}{r_2} dr_1 + \int_{r_2}^{\infty} 4\pi r_1 e^{-2Zr_1}\, dr_1 \right] \\ &= e^{-2Zr_2}\, d\tau_2 \left[\frac{\pi}{Z^3 r_2} \left(1 - e^{-2Zr_2}(Zr_2 + 1)\right) \right] \end{aligned} \tag{C-17}$$

Setting $d\tau_2 = 4\pi r_2^2 dr_2$, and integrating once again from $r_2 = 0$ to ∞, we obtain the total mutual energy of the two clouds, which is the integral in (C-14)

$$\int d\varepsilon = \iint \frac{e^{-2Zr_1} e^{-2Zr_2}}{r_{12}} d\tau_1 d\tau_2$$

$$= \frac{4\pi^2}{Z^3} \int_0^\infty [e^{-2Zr_2}r_2 - (Zr_2^2 + r_2)\, e^{-4Zr_2}]\, dr_2 = 5\pi^2/8Z^5 \qquad \text{(C-18)}$$

Substituting this result into (C-14), we find that the first order correction to the energy due to the electron repulsion is

$$E' = \frac{5Z}{8} \qquad \text{(C-19)}$$

and the energy of the ground state of a helium-like atom, corrected to the first order, is

$$E = E^0 + E' = -Z^2 + (5Z/8) \qquad \text{(C-20)}$$

This energy is relative to that of the completely ionized atom or ion (nucleus plus two electrons at infinity). If we wish to find the energy required to remove one electron, leaving the other electron in a 1*s* orbital, we must add to (C-20) the ionization energy of the second electron, $Z^2/2$. Thus to the first order, the ionization energy of the first electron is

$$E_{\text{ion}} = (Z^2/2) - (5Z/8) \qquad \text{(C-21)}$$

The observed ionization energies of the helium atom and of helium-like ions are given in Table 9-3, together with energies calculated according to the zero-order and first order approximations. The agreement is quite satisfactory. It would be even more impressive if the results were expressed as the total energy rather than as the ionization energy. In every instance the first order ionization energy is too low because it is based on a wave function in which the probability distribution of each electron is unaffected by the presence of the other electron in the atom. In reality, of course, the electrons try to keep out of each other's way to some extent, and the correct wave function would take this into account. This correction would appear, in fact, in the second order perturbation calculation. It tends to lower the energy of the state and hence to increase the ionization energy.

The stabilization of the system due to the tendency of the electrons to avoid one another is called the *correlation energy*. In the helium atom and the helium-like ions, the correlation energy may be defined as the difference

TABLE 9-3

IONIZATION ENERGIES OF THE GROUND STATES OF HELIUM AND HELIUM-LIKE IONS (ENERGY IN ATOMIC UNITS)

Atom or ion	Z	E_{ion} (zero-order)	E_{ion} (first order)	E_{ion} (observed)	Correlation energy
H^-	1	0.500	−0.125	0.026*	0.151
He	2	2.000	0.750	0.904	0.154
Li^+	3	4.500	2.625	2.780	0.155
Be^{+2}	4	8 000	5.500	5.655	0.155
B^{+3}	5	12.500	9.375	9.531	0.156
C^{+4}	6	18.000	14.250	14.409	0.159
N^{+5}	7	24.500	20.125	20.288	0.163
O^{+6}	8	32.000	27.000	27.169	0.169
F^{+7}	9	40.500	34.875	35.053	0.178

* Hyleraas' calculation.

between the observed energy and the first order energy that we have just calculated. Its values are given in Table 9-3. Evidently the correlation energy is not very dependent on the atomic number Z, so that its relative importance in helium-like ions decreases rapidly as the atomic number increases.

The next improvement in the computation of the energy of the ground state by the perturbation method would require the use of the second order perturbation terms. The calculations are rather awkward, however, and better values of the energy are more easily obtained by means of the variation method, which will now be considered. The method consists in guessing at the wave function, using physical intuition to establish its form, and allowing variable parameters to remain in the function, their numerical values being determined by means of the variation principle (see page 174).

The most obvious trial wave function to use is one having the same form as the zero-order wave function, (C-10, 11), but in which the exponent Z is considered as a variable parameter,

$$\phi = e^{-Z'r_1}e^{-Z'r_2} \tag{C-22}$$

When the energy is calculated by means of the relation

$$E_{\text{est}} = \int \phi H \phi \, d\tau \Big/ \int \phi^2 \, d\tau \tag{C-23}$$

where

$$H = -\tfrac{1}{2}(\nabla_1^2 + \nabla_2^2) - (Z/r_1) - (Z/r_2) + (1/r_{12}) \tag{C-23a}$$

it is found that the energy is minimized when $Z' = Z - 5/16$, and that the resulting estimated ionization energy is

$$E_{\text{ion; est}} = (Z^2/2) - (5Z/8) + (5/16)^2 \tag{C-24}$$

Evidently the trial wave function (C-22) gives a correlation energy of $(5/16)^2 = 0.098$, which is independent of Z, but is only about two-thirds of the observed value of 0.15 – 0.16.

The trial function (C-22) is important because it can be given a rather simple physical interpretation. The constant Z' can be said to represent an effective nuclear charge acting on each 1*s* electron, which is less by 5/16 of a unit than the true nuclear charge because the other 1*s* electron partially shields or "screens" it from the nucleus. The number 5/16 is called the "screening constant" of the 1*s* electron for another 1*s* electron. Because of the lowered effective charge on the nucleus the electrons have more space in which to move. They are therefore better able to keep out of each other's way – whence the origin of the correlation energy. But the screening constant also reduces the attraction between the electrons and the nucleus, so it is not necessarily the most effective way of producing the correlation.

Let us next consider the trial function

$$\phi = (1 + br_{12})\, e^{-Z'r_1}\, e^{-Z'r_2} \tag{C-25}$$

where both b and Z' are variable parameters, and r_{12} is the distance between the two electrons. The factor in parentheses is large when the electrons are far apart, and hence tends to keep the electrons out of each other's way. It does not necessarily do this at the expense of the attractive forces between the nucleus and the electrons, however, and therefore ought to give a much better approximation to the true wave function. It is found that for helium the optimum ionization energy obtainable with (C-25) is 0.891 atomic units, or only 0.013 units (1.4 %) below the observed value. The correlation energy is 0.141, which is only 8 % below the correct value. The values of the parameters that minimize the energy are $b = 0.364$ Bohr radii^{-1} and $Z' = 2 - 0.151$. Thus the screening constant in this case is only about half as large as the value obtained with trial function (C-22). The introduction of the r_{12} term is evidently a much more effective way of taking account of correlation than is the use of a screening constant alone.

Finally let us consider the trial function

$$\phi = [1 + br_{12} + c(r_1 - r_2)^2]\, e^{-Z'r_1}\, e^{Z'r_2} \tag{C-26}$$

where Z', b, and c are variable parameters. The term in $(r_1 - r_2)$ also helps to keep the electrons apart, since it favors configurations in which r_1 and r_2

are as different as possible. The optimum ionization energy obtainable for helium with (C-26) is 0.9024 atomic units, which is but 0.0012 atomic unit (0.1 %) below the observed value. The corresponding values of the parameters are $Z' = 2 - 0.184$, $b = 0.30$, $c = 0.13$.

Hylleraas (*12*) has applied the variation method to helium and helium-like ions using other trial wave functions of the type described above, but containing as many as fourteen variable parameters. In this way he was able to derive the following formula for the ionization energy (in atomic units) of a helium-like ion of atomic number Z

$$E_{\text{ion}} = \frac{M}{M+1}\left[(Z^2/2) - (5Z/8) + 0.15744 - \frac{0.00876}{Z} + \frac{0.00274}{Z^2}\right] \quad \text{(C-27)}$$

where M is the mass of the nucleus in atomic units (that is, the ratio of the nuclear mass to the electronic mass). This formula is regarded as more reliable than the experimentally determined values for values of Z greater than 2. In the case of helium, the ionization energy given by this formula is larger than the experimental value by 0.00006 atomic units, an apparent violation of the variation principle which probably arises because of neglect of relativistic and spin effects. In the case of the H^- ion, Hylleraas found a positive value for the ionization energy, demonstrating that this ion is a stable entity.

The calculations mentioned above are summarized in Table 9-4.

Exercise. Prove Equation (C-24) starting with Equations (C-22), (C-23) and (C-23a).

TABLE 9-4

CALCULATIONS OF THE IONIZATION ENERGY OF THE HELIUM ATOM

Approximation	Ionization energy (atomic units)	Per cent error	Correlation energy (atomic units)
Zero-order (electronic repulsions neglected)	2.000	120.	—
First order (Equation (C-21))	0.750	17.	0
Screening constant (Equation (C-22))	0.848	6.6	0.098
Screening constant plus r_{12} (Equation (C-25))	0.891	1.4	0.141
Screening constant plus r_{12} plus $(r_1 - r_2)^2$ (Equation (C-26))	0.9024	0.14	0.1524
Hylleraas' fourteen parameter function (Equation (C-27))	0.90374	−0.007	0.15374
Experimental	0.90368	—	0.15368

c. The first group of excited states

We have seen (Equation (C-13)) that in the zero-order approximation the first energy level above the ground state is eight-fold degenerate. In applying the perturbation theory we shall therefore have to set up a secular equation in the form of an eighth order determinant, requiring the evaluation of sixty-four integrals of the type

$$H_{ij} = \int \phi_i^* \, (1/r_{12}) \, \phi_j \, d\tau$$

It is, however, easy to show that most of these integrals vanish, and furthermore it is easy to find combinations of the zero-order wave functions that diagonalize the secular determinant. To do this we make use of the theorem (see page 144) that if two operators α and β commute and ϕ and ψ are eigenfunctions of β with eigenvalues b_1 and b_2, then

$$\int \psi \, \alpha \phi \, d\tau = 0 \quad \text{if} \quad b_1 \neq b_2$$

We have shown in Chapter 8 that the total electronic orbital angular momentum operators M_e^2 and M_{ze} commute with the atomic Hamiltonian operator, and in particular with the terms $1/r_{ij}$ for the electronic repulsion. Let us see if we can make use of this fact in reducing our secular equation. First we must see what effect M_e^2 and M_{ze} have on the ϕ's in (C-13). From the definition of these operators (pages 256-260) we have

$$M_{ze} = -i\hbar \, (\partial/\partial\varphi_1 + \partial/\partial\varphi_2) \tag{C-28a}$$

$$M_e^2 = (\mu_1 + \mu_2)\,(\nu_1 + \nu_2) - \hbar M_{ze} + M_{ze}^2 \tag{C-28b}$$

It is readily seen that all of our ϕ's are already eigenfunctions of M_{ze}. The eigenvalues of M_{ze} are given in Table 9-5.

TABLE 9-5

Function	ϕ_1	ϕ_2	ϕ_3	ϕ_4	ϕ_5	ϕ_6	ϕ_7	ϕ_8
Eigenvalues of M_{ze} (atomic units)	0	0	1	1	0	0	−1	−1
Eigenvalues of M_e^2 (atomic units)	0	0	2	2	2	2	2	2

The effect of M_e^2 on the functions is not so obvious. We shall need the following results, which come from Equations (A-26) of Chapter 8

$$\begin{aligned} \mu 1s &= 0;\ \mu 2p_{+1} = 0 \qquad \mu 2p_0 = \sqrt{2}\hbar\, 2p_{+1} \\ \nu 1s &= 0;\ \nu 2p_{+1} = \sqrt{2}\hbar\, 2p_0;\ \ \nu 2p_0 = \sqrt{2}\hbar\, 2p_{-1} \\ \mu 2p_{-1} &= \sqrt{2}\hbar\, 2p_0 \\ \nu 2p_{-1} &= 0 \end{aligned} \tag{C-29}$$

Then we find by substitution that

$$\begin{aligned} M_e^2\phi_1 &= 0 \\ M_e^2\phi_2 &= 0 \\ M_e^2\phi_3 &= (1s)_1\,[\mu_2\nu_2(2p_{+1})_2 - \hbar^2(2p_{+1})_2 + \hbar^2(2p_{+1})_2] \\ &= (1s)_1 2\hbar^2\,(2p_{+1})_2 = 2\hbar^2\phi_3 \\ M_e^2\phi_4 &= 2\hbar^2\phi_4 \\ M_e^2\phi_5 &= (1s)_1\,[\mu_2\nu_2(2p_0)_2] = 2\hbar^2\phi_5 \\ M_e^2\phi_6 &= 2\hbar^2\phi_6 \\ M_e^2\phi_7 &= (1s)_1\,[\mu_2\nu_2(2p_{-1})_2 - \hbar(-\hbar)\,(2p_{-1})_2 + (-\hbar)^2(2p_{-1})_2] \\ &= 2\hbar^2\phi_7 \\ M_e^2\phi_8 &= 2\hbar^2\phi_8 \end{aligned} \tag{C-30}$$

All of the ϕ's are therefore also eigenfunctions of M_e^2. The eigenvalues of M_e^2 are given in Table 9-5. They all have the form $L(L+1)$, with $L = 0$ for ϕ_1 and ϕ_2 and $L = 1$ for the six others. The number L is the total orbital angular momentum quantum number (see page 260). Therefore we shall expect to find two S-states and six P-states among this group.

Evidently all of the integrals H_{ij} vanish except those between ϕ_1 and ϕ_2, ϕ_3 and ϕ_4, ϕ_5 and ϕ_6, and ϕ_7 and ϕ_8. The secular equation therefore looks like this

$$\begin{vmatrix} * & * & 0 & 0 & 0 & 0 & 0 & 0 \\ * & * & 0 & 0 & 0 & 0 & 0 & 0 \\ 0 & 0 & * & * & 0 & 0 & 0 & 0 \\ 0 & 0 & * & * & 0 & 0 & 0 & 0 \\ 0 & 0 & 0 & 0 & * & * & 0 & 0 \\ 0 & 0 & 0 & 0 & * & * & 0 & 0 \\ 0 & 0 & 0 & 0 & 0 & 0 & * & * \\ 0 & 0 & 0 & 0 & 0 & 0 & * & * \end{vmatrix} = 0 \tag{C-31}$$

where all the elements vanish except those that are starred. That is, we have reduced our eighth order determinant to four second order determinants. These could be solved without much trouble, but it is easy to reduce them still further by means of the *permutation operator*, P_{12}. This operator, when applied to any function involving electrons no. 1 and no. 2, results in an interchange of the coordinates of the two electrons, so that in the function $P_{12}f(1, 2)$ electron no. 2 occurs where electron no. 1 was in $f(1, 2)$ and *vice versa*. This operator obviously has no effect on the Hamiltonian of the helium atom, since ∇_1^2 and ∇_2^2 are simply interchanged, as are the terms $-Z/r_1$ and $-Z/r_2$; and the electronic repulsion term, $1/r_{12}$, is unaffected because the distance between the electrons is not changed by merely interchanging the electrons. Therefore

$$P_{12}H = HP_{12} \tag{C-32}$$

We see, however, that our ϕ's are not eigenfunctions of P_{12}, since

$$\begin{aligned} &P_{12}\phi_1 = \phi_2; \quad P_{12}\phi_3 = \phi_4; \quad P_{12}\phi_5 = \phi_6; \quad P_{12}\phi_7 = \phi_8 \\ &P_{12}\phi_2 = \phi_1; \quad P_{12}\phi_4 = \phi_3; \quad P_{12}\phi_6 = \phi_5; \quad P_{12}\phi_8 = \phi_7 \end{aligned} \tag{C-33}$$

On the other hand, the following functions are normalized, orthogonal eigenfunctions of P_{12}:

$$\begin{aligned} \psi_1 &= (1/\sqrt{2})\,(\phi_1 + \phi_2); & \psi_3 &= (1/\sqrt{2})\,(\phi_3 + \phi_4) \\ \psi_2 &= (1/\sqrt{2})\,(\phi_1 - \phi_2)\;; & \psi_4 &= (1/\sqrt{2})\,(\phi_3 - \phi_4) \\ \psi_5 &= (1/\sqrt{2})\,(\phi_5 + \phi_6); & \psi_7 &= (1/\sqrt{2})\,(\phi_7 + \phi_8) \\ \psi_6 &= (1/\sqrt{2})\,(\phi_5 - \phi_6)\;; & \psi_8 &= (1/\sqrt{2})\,(\phi_7 - \phi_8) \end{aligned} \tag{C-34}$$

These new functions are also eigenfunctions of M_e^2 and M_{ze}. The eigenvalues are given in Table 9-6. (A function that changes sign on permutation is said to be *antisymmetric* with respect to the permutation; one which does not change sign is said to be *symmetric* with respect to the permutation.)

TABLE 9-6

Function		ψ_1	ψ_2	ψ_3	ψ_4	ψ_5	ψ_6	ψ_7	ψ_8
Eigenvalues of	M_{ze}	0	0	1	1	0	0	−1	−1
	M_e^2	0	0	2	2	2	2	2	2
	P_{12}	1	−1	1	−1	1	−1	1	−1

Since no two of these eight new functions have all three of their eigenvalues the same, all of the integrals between them must vanish

$$H_{ij} = \int \psi_i^*(1/r_{12})\psi_j \, d\tau = 0 \qquad (i \neq j) \tag{C-35}$$

Using the ψ's instead of the ϕ's as our zero-order wave functions, the secular equation is completely diagonal, the first order perturbations are given directly by

$$E'_i = H_{ii} = \int \psi_i^*(1/r_{12})\psi_i \, d\tau \tag{C-36}$$

and the ψ's are the correct zero-order wave functions.

Therefore the first excited states of helium have approximate wave functions of the form

$$\psi = (1/\sqrt{2})\,[(1s)_1\,(2s)_2 \pm (1s)_2\,(2s)_1] \qquad (S\text{-states})$$

$$\left.\begin{aligned} \psi &= (1/\sqrt{2})\,[(1s)_1\,(2p_{+1})_2 \pm (1s)_2\,(2p_{+1})_1] \\ \psi &= (1/\sqrt{2})\,[(1s)_1\,(2p_0)_2 \pm (1s)_2\,(2p_0)_1] \\ \psi &= (1/\sqrt{2})\,[(1s)_1\,(2p_{-1})_2 \pm (1s)_2\,(2p_{-1})_1] \end{aligned}\right\} \quad (P\text{-states}) \tag{C-37}$$

We see that, because of the $1/r_{12}$ term in the Hamiltonian, it is not possible to say which electron is in which orbital. Because of their electrostatic repulsions, the two electrons are continually exchanging their orbitals and their angular momenta. The total angular momentum, however, always remains constant in both magnitude and direction. (This conclusion must be modified when we come to consider spin-orbit interactions.) We can imagine that in a $1s2p$ P-state, first one electron is in the $2p$-orbital and has an angular momentum $\sqrt{1(1+1)}\hbar$. After several revolutions about the nucleus, this electron suffers a sufficiently violent encounter with the $1s$ electron to transfer to it its angular momentum, and the two electrons exchange orbitals.

It is possible to say a good deal about the relative energies of this group of excited states without going through the detailed evaluation of the integrals in (C-36). In the first place, any function whose eigenvalue for the permutation operator is -1 will give a small probability of finding the two electrons close to each other. For instance, consider the antisymmetric wave function

$$\psi_1 = (1/\sqrt{2})\,[(1s)_1\,(2s)_2 - (1s)_2\,(2s)_1]$$

If the two electrons are at the same point in space, then

$$(1s)_1\,(2s)_2 = (2s)_1\,(1s)_2$$

and ψ_1 vanishes. This is not the case with the symmetric wave function

$$\psi_2 = (1/\sqrt{2})\ [(1s)_1\,(2s)_2 + (1s)_2\,(2s)_1]$$

Since it is energetically unfavorable to allow two electrons to be close to one another, it is evident that the antisymmetric states should be more stable than the corresponding symmetric states.

The integrals in (C-36) have the form

$$E' = (\tfrac{1}{2}) \int [(1s)_1(2s)_2 \pm (1s)_2(2s)_1]^2/r_{12}\,d\tau = J_s \pm K_s \qquad \text{(C-38)}$$

where

$$J_s = \int \frac{1s_1^2 2s_2^2}{r_{12}}\,d\tau = \int \frac{1s_2^2 2s_1^2}{r_{12}}\,d\tau \qquad \text{(C-38a)}$$

$$K_s = \int \frac{1s_1 1s_2 2s_1 2s_2}{r_{12}}\,d\tau \qquad \text{(C-38b)}$$

For the states involving $2p$ electrons, we obtain in a similar fashion expressions of the type

$$E' = (\tfrac{1}{2}) \int [(1s)_1(2p_{+1})_2 \pm (1s)_2(2p_{+1})_1]^2/r_{12}\,d\tau = J_{p_{+1}} \pm K_{p_{+1}} \qquad \text{(C-39)}$$

where

$$J_{p_{+1}} = \int \frac{1s_1^2\,|\,2p_{+1}\,|_2^2}{r_{12}}\,d\tau \qquad \text{(C-39a)}$$

$$K_{p_{+1}} = \int \frac{1s_1 1s_2 (2p_{+1})_1^* (2p_{+1})_2}{r_{12}}\,d\tau \qquad \text{(C-39b)}$$

and similarly for states involving $2p_0$ and $2p_{-1}$. Thus we find

$$\begin{aligned}
E'_1 &= J_s + K_s \\
E'_2 &= J_s - K_s \\
E'_3 &= J_{p_{+1}} + (1/2)\,(K_{p_{+1}}^* + K_{p_{+1}}) \\
E'_4 &= J_{p_{+1}} - (1/2)\,(K_{p_{+1}}^* + K_{p_{+1}}) \\
E'_5 &= J_{p_0} + K_{p_0} \\
E'_6 &= J_{p_0} - K_{p_0} \\
E'_7 &= J_{p_{-1}} + (1/2)\,(K_{p_{-1}}^* + K_{p_{-1}}) \\
E'_8 &= J_{p_{-1}} - (1/2)\,(K_{p_{-1}}^* + K_{p_{-1}})
\end{aligned} \qquad \text{(C-40)}$$

But the $2p$ functions in $J_{p_{+1}}$, J_{p_0}, and $J_{p_{-1}}$ differ only in the orientation of the axis of rotation of the $2p$ electrons about the nucleus, and we should not expect a mere difference in orientation to have any effect on the energy. Therefore

$$J_{p_{+1}} = J_{p_0} = J_{p_{-1}} = J_p = \int \frac{1s_1{}^2 2p_{02}{}^2}{r_{12}}\, d\tau \tag{C-41a}$$

and for the same reason

$$\tfrac{1}{2}(K_{p_{+1}}{}^* + K_{p_{+1}}) = K_{p_0} = \tfrac{1}{2}(K_{p_{-1}}{}^* + K_{p_{-1}}) = K_p = \int \frac{1s_1 1s_2 2p_{01} 2p_{02}}{r_{12}}\, d\tau \tag{C-41b}$$

The eight states therefore group themselves into four energy levels, with

$$\left.\begin{aligned} E'_1 &= J_s + K_s \\ E'_2 &= J_s - K_s \end{aligned}\right\} S\text{-states}$$

$$\left.\begin{aligned} E'_3 = E'_5 = E'_7 &= J_p + K_p \\ E'_4 = E'_6 = E'_8 &= J_p - K_p \end{aligned}\right\} P\text{-states} \tag{C-42}$$

Evidently the P-states should occur in two triply degenerate levels. (As we shall see, however, this is not actually the case.)

The integrals J_s and J_p have a simple classical interpretation: J_s represents the electrostatic energy due to the mutual repulsion of a charge distribution $(1s)^2$ with a charge distribution $(2s)^2$; J_p represents the energy arising from the interaction of a $(1s)^2$ cloud of electricity with a $(2p)^2$ cloud. The two J-integrals are called *coulombic integrals*. They represent a classical correction to the zero-order energy that is very similar in origin to the first order correction to the ground state energy discussed earlier. Both J_s and J_p are positive.

The integrals K_s and K_p, on the other hand, have no analogy in classical electrostatics. They arise because the zero-order wave functions require that the electrons be able to exchange their places of residence among the different orbitals. For this reason the K-integrals are known as *exchange integrals*. The closest classical analogue to an effect of this kind occurs in the theory of vibration, where, as we have seen (page 145), the coupling of normal modes having the same frequency ("resonant normal modes") results in changes in normal mode frequencies that are mathematically of a similar origin. The splitting of energy levels caused by the exchange phenomenon is therefore often said to be caused by *resonance*. The sign of the exchange integrals is not obvious from inspection, since the $2s$ and $2p$ functions are in some regions negative and in others positive. But we know, from the argument given above (the effects on the energy of symmetry with respect to permutation) that ψ_2 is more stable than ψ_1, and that ψ_4 is more stable than ψ_3. Therefore both exchange integrals must be positive.

The coulombic integral J_s turns out to be smaller than J_p. The reason for this is also not hard to see. If the 1*s* cloud were concentrated entirely at the nucleus, the integral J_s would be exactly the same as J_p, because the coulombic integral would then be merely the mean potential energy of a hydrogen atom with the sign changed, and this is the same for a 2*p*-state as for a 2*s*-state.* But we have seen on page 286 that when an electron penetrates into a diffuse, spherically symmetrical charged cloud, there is a reduction in the interaction of the electron with the portion of the cloud that is farther from the center than the electron. Because of the centrifugal forces accompanying the angular motion of the 2*p* electron (page 222), it does not get as close to the nucleus as does the 2*s* electron. The electrostatic repulsion between the 2*s* and 1*s* electrons is therefore weaker than that between the 2*p* and 1*s* electrons, and J_s is smaller than J_p. Since both the J's are positive, this tends to make the *S*-states more stable than the *P*-states.

The same thing may be said in another way: If there were no penetration of the second electron into the 1*s* cloud, the 1*s* electron would act electrostatically as if it were concentrated entirely at the nucleus. The nucleus would then appear to have an atomic number of (Z-1) instead of Z. If there is penetration into the 1*s* cloud, however, the shielding is not so effective. *A 2s electron is therefore less effectively shielded and more firmly bound than is a 2p electron in the helium atom.*

This is a very general result of great significance in chemistry. *In atoms containing more than one electron, the binding of an electron in the orbitals ns, np, nd, nf, · · · , decreases as the angular momentum quantum number of the orbital increases.* It will be called the "penetration effect."†

Finally, what can we say about the relative magnitudes of the exchange integrals, K_s and K_p? The numerators of the integrands in K_s and K_p have the form

$$(f)_1(g)_1(f)_2(g)_2$$

* According to the virial theorem (page 230), in any system of charged particles the average kinetic energy, $\overline{T}$, and the average potential energy, $\overline{V}$, are related: $2\overline{T} = -\overline{V}$. Furthermore, the total energy is given by $E = -\overline{T} = \overline{V}/2$. Since E is the same for the 2*s* and 2*p*-states of the hydrogen atom, the average potential energy, $\overline{V}$, must be the same in the two states.

† The penetration effect is well illustrated by the model of an electron moving about a partially shielded core, discussed on page 225.

where f and g represent the orbitals between which the exchange takes place, and the subscripts refer to the electron whose coordinates are to be used in the orbital. It is obvious that if g is large only at places at which f is small, and if f is large only where g is small, the numerator will tend to be small everywhere. Only if both functions are large in the same region of space can the numerator, and hence the exchange integral, be large. We may state the general rule that *exchange integrals will tend to be large only if the orbitals concerned overlap effectively*. Since a 2*s* electron penetrates farther into the 1*s* cloud than does a 2*p* electron, K_s will be larger than K_p.

The expected relationships of the first group of excited helium levels are shown schematically in Fig. 9-7. The two *P*-levels should be above the two *S*-levels and closer together. Reference to Fig. 9-2 shows that the first four levels above the ground state in helium are indeed related in this way.

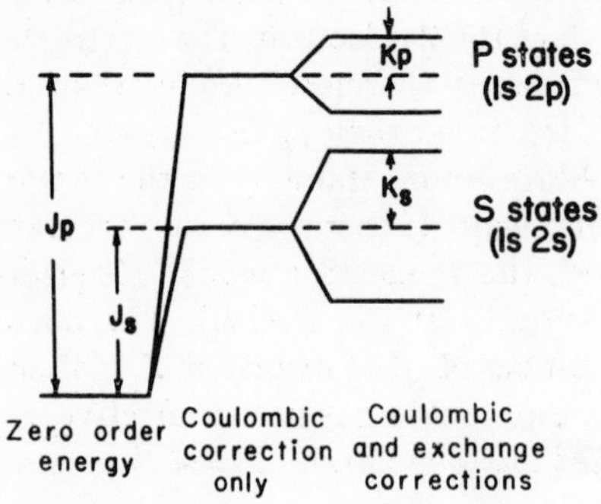

FIG. 9-7. Correlation diagram for the first group of excited levels of helium.

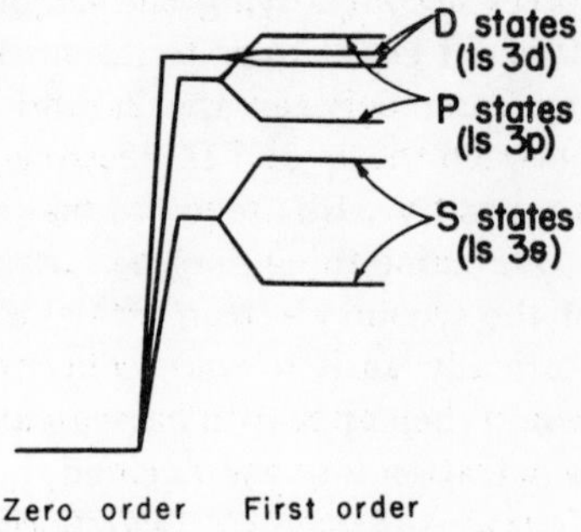

FIG. 9-8. Correlation diagram for the configurations 1*s*3*s*, 1*s*3*p*, and 1*s*3*d* of helium.

The quantitative agreement with the predictions of the first order perturbation theory is only fair. The integrals are found to have the following values (the method of evaluation of the integrals is outlined below)

$$\begin{aligned} J_s &= 17Z/81 &&= 0.2098Z \\ J_p &= 59Z/243 &&= 0.2428Z \\ K_s &= 16Z/729 &&= 0.02195Z \\ K_p &= 112Z/6561 &&= 0.01707Z \end{aligned} \qquad \text{(C-43)}$$

"Experimental values" of J_s, J_p, K_s, and K_p may be found as follows. Let E'_s and E''_s be the observed energies of the two S-levels, and E'_p and E''_p be those of the two P-levels. Then

$$\begin{aligned}(J_s)_{\text{exp}} &= (1/2)\,(E'_s + E''_s) - E_0 \\ (J_p)_{\text{exp}} &= (1/2)\,(E'_p + E''_p) - E_0 \\ (K_s)_{\text{exp}} &= (1/2)\,(E'_s - E''_s) \\ (K_p)_{\text{exp}} &= (1/2)\,(E'_p - E''_p)\end{aligned} \tag{C-44}$$

where E_0 is the zero-order energy, $-5Z^2/8$. The results are given in Table 9-7 for all known helium-like ions, along with the values calculated from Equations (C-43). The calculated values are invariably too large, the relative agreement being considerably better for the coulombic integrals than for the exchange integrals. The absolute error is almost the same for all of the ions, so the relative agreement improves as Z increases. (If the comparison had been made in terms of the actual energies of the states, the agreement would have seemed much better.)

TABLE 9-7

EXCHANGE AND COULOMBIC INTEGRALS IN HELIUM AND HELIUM-LIKE IONS (ALL ENERGIES IN ATOMIC UNITS)

		J_s		J_p		K_s		K_p	
Ion	Z	calc.	"expt."	calc.	"expt."	calc.	"expt."	calc.	"expt."
He	2	0.420	0.339	0.486	0.372	0.0439	0.0146	0.0341	0.0047
Li^{+1}	3	0.630	0.546	0.728	0.615	0.0659	0.0320	0.0512	0.0172
Be^{+2}	4			0.972	0.857			0.0683	0.0322
B^{+3}	5			1.214	1.100			0.0854	0.0481
C^{+4}	6			1.456	1.341			0.1024	0.0640
N^{+5}	7			1.700	1.584			0.1195	0.0809
O^{+6}	8			1.942	1.826			0.1366	0.0971
F^{+7}	9			2.186	2.068			0.1536	0.1137

The reason for the discrepancies lies, of course, in the higher order perturbations. It is not hard to see that these perturbations should improve the agreement if we look at the factors that have not been dealt with adequately in the first order energy calculation. Two of these factors are especially important, and they both work in the same direction. The first may be considered to arise from an error in the assumed shapes of the 2*s*- and 2*p*-orbitals that results from the neglect of the partial screening of the nucleus by the 1*s* electron Since the average distance between the nucleus and the 1*s* electron is considerably less than the average distance between the nucleus and the 2*s* or 2*p* electrons, the effective nuclear charge acting on the latter is probably almost one charge unit less than the actual nuclear charge. This means that the 2*s*- and 2*p*-orbitals must be somewhat less tightly bound (i.e., more spread out in space) than is indicated by the zero-order wave functions used in obtaining Equations (C-43). Presumably the higher order perturbation calculations would correct this by hybridizing 2*s* and 2*p* with other, less localized orbitals (especially, 3*s*, 4*s*, $\cdots$ and 3*p*, 4*p*, $\cdots$). As a result, the electrostatic repulsion between the 1*s* electron and the hybridized 2*s* and 2*p* electrons will be reduced. If the "experimental" values of J_s and J_p can be interpreted as the actual repulsion energy between the 1*s* electron and the 2*s* and 2*p* electrons, then the values of J_s and J_p calculated from (C-43) ought to be too high – as indeed they are. Furthermore, the spreading out of 2*s* and 2*p* should reduce the overlap of these orbitals with the 1*s* orbital, and if the "experimental" values of K_s and K_p can be said to be determined by the actual overlap, we can understand why the values obtained from (C-43) are too large.

The second factor handled inadequately by the first order theory is the correlation energy. In the evaluation of the coulombic energy, no account has been taken of the tendency of the electrons to keep out of each other's way. This tends to make the calculated values of J_s and J_p too large. On the other hand, in states of the type $(1s)_1\,(2s)_2 - (1s)_2\,(2s)_1$ the tendency of electrons to keep out of each other's way has probably been exaggerated, whereas in states of the type $(1s)_1\,(2s)_2 + (1s)_2\,(2s)_1$ the correlation has been underestimated. The difference in energy of these two states has thus been overestimated, and the calculated exchange integrals, K_s and K_p, are therefore too large.

The situation in which we now find ourselves is a typical one in quantum chemistry. The approximate theory gives a clear account of the qualitative features of the problem, but is not entirely adequate quantitatively. The reason for the disagreement is plain, and one can even account for its direction. But the improvement of the results by going to higher order perturbation theory would require extensive calculations, which hardly seem worth the effort.

A possible short cut method for the improvement of our calculations of the exchange and coulombic integrals suggests itself. Why not include variable parameters in the orbitals that take into account some of the effects mentioned above? For instance, we might use an "effective nuclear charge", $(Z - s)$, instead of the full nuclear charge Z in the 2*s*- and 2*p*-orbitals; this would serve the double purpose of delocalizing the orbital and of keeping the electrons out of each other's way to some extent. This method raises certain difficulties, however. Aside from the problem of deciding on an appropriate value for the "screening constant" s, the introduction of such a parameter destroys the orthogonality of our zero-order wave functions. In the process of patching up the first order theory in this way, we have created new weaknesses. *Evidently the very useful concepts of resonance and coulombic energies lose their meaning if we try to make them too precise*, since strictly speaking they are merely accompaniments of the first order perturbation theory. The real weakness in these proceedings, in fact, is the attempt to evaluate precise "experimental" coulombic and exchange energies.

On the other hand, there is no question that modified wave functions constructed in this way are much better than our zero-order wave functions. If they are used to evaluate the energy or other properties, much improvement will be found over the results of first order perturbation theory. Therefore this procedure of introducing screening constants and other variable parameters is generally used in spite of the above objections, the values of the parameters being chosen to fit a limited set of experimental results. The procedure essentially amounts to the construction of rational interpolation formulae on the framework of the first order perturbation theory. Its justification is that it gives better results. Its weakness lies in the fact that, lacking a sound theoretical basis (such as the variation principle or the perturbation theory), judgment is required in the evaluation of the significance of the results, and this judgment is not always easy to make.

With difficulties embarassing us in this way in a problem as simple as that of an atom containing only two electrons, our hopes should not be too high for any exact quantitative successes in the systematic application of quantum mechanics to even moderately complex systems. The true value of quantum mechanics lies in other directions. One of these is the development of concepts such as those of coulombic and exchange energies, which, as we have seen and as we shall see, can give semi-quantitative explanations of natural phenomena that are otherwise quite inexplicable. A second important direction is the development of reasonable interpolation formulae through which observed behavior can be interpreted quantitatively in terms of parameters of more or less definite physical significance. An example of this approach is the use of screening constants as parameters in constructing approximate atomic eigenfunctions. This will be discussed below.

d. The evaluation of coulombic and exchange integrals in atoms

Integrals of the type

$$\int (f)_1(g)_2/r_{12}\, d\tau$$

may be evaluated without much difficulty. Here $(f)_1$ and $(g)_2$ are functions of the coordinates of two electrons, and r_{12} is the distance between the two electrons. Use is made of the expansion

$$\frac{1}{r_{12}} = \sum_{l=0}^{\infty} \sum_{m=-l}^{l} \frac{(l-|m|)!}{(l+|m|)!} P_l^{|m|}(\cos\theta_1) P_l^{|m|}(\cos\theta_2) e^{im(\varphi_1-\varphi_2)} \frac{r_<^l}{r_>^{l+1}} \tag{C-45}$$

[proven in Eyring, Walter, and Kimball (*13*, page 369)], where r_1, θ_1, and φ_1 are the polar coordinates of electron no. 1 and r_2, θ_2, and φ_2 are those of electron no. 2, $r_<$ and $r_>$ are respectively the lesser and the greater of r_1 and r_2, and $P_l^{|m|}$ is an associated Legendre function. Ordinarily only one or two terms in the series fail to vanish on integration, since

$$\int_0^{2\pi}\int_0^{\pi} P_l^{|m|}(\cos\theta)\, e^{im\varphi} \sin\theta\, d\theta\, d\varphi = 0$$

for all l and m except for $l = m = 0$.

As an illustration of the use of this expansion we compute the value of J_p

$$J_p = \int \frac{1s_1{}^2 2p_0{}^2}{r_{12}}\, d\tau$$

$$= \frac{Z^8}{32\pi^2} \int\int\int\int\int\int \frac{e^{-2Zr_1}e^{-2Zr_2}r_2{}^2\cos^2\theta_2}{r_{12}} r_1{}^2 \sin\theta_1 r_2{}^2 \sin\theta_2 dr_1 dr_2 d\theta_1 d\theta_2 d\varphi_1 d\varphi_2$$

In the expansion of $1/r_{12}$, all terms but the one in which $l = m = 0$ vanish on integration over one or the other of the angular variables. Therefore

$$J_p = \frac{Z^8}{32\pi^2} I_1 \left[\int_0^{\infty} e^{-Zr_2} r_2{}^4 \left[\frac{1}{r_2} \int_0^{r_2} e^{-2Zr_1} r_1{}^2 dr_1 + \int_{r_2}^{\infty} e^{-2Zr_1} r_1 dr_1 \right] dr_2 \right]$$

where

$$I_1 = \int_0^{2\pi}\int_0^{2\pi}\int_0^{\pi}\int_0^{\pi} \cos^2\theta_2 \sin\theta_1 \sin\theta_2\, d\theta_1\, d\theta_2\, d\varphi_1\, d\varphi_2 = 16\pi^2/3$$

Substituting $x_1 = Zr_1$ and $x_2 = Zr_2$, we find

$$J_p = \frac{Z}{6}\int_0^\infty e^{-x_2}\left[x_2^3\int_0^{x_2} e^{-2x_1}x_1^2\,dx_1 + x_2^4\int_{x_2}^\infty e^{-2x_1}x_1\,dx_1\right]dx_2$$

Straightforward evaluation leads to the result, $J_p = 59Z/243$.

Exercises. (1) Complete the evaluation of J_p.
(2) Verify the expressions (C-43) for J_s, K_s, and K_p.

e. Other excited states

The procedure for calculating the energies of the rest of the excited states is a trivial extension of that outlined in the previous section and need not delay us for long. The next group of excited levels will involve the electronic

TABLE 9-8

COMPARISON OF ENERGY LEVELS OF HELIUM AND HYDROGEN
(ENERGIES IN CM.$^{-1}$)

Helium Level	Helium Energy	Hydrogen Level	Hydrogen Energy
$1s^2$	−198305	$1s$	−109719
$1s2s$	−38455 −32033		
		$2s$, $2p$	−27430
$1s2p$	−29234 −27176		
$1s3s$	−15074 −13446		
$1s3p$	−12746 −12101	$3s$, $3p$, $3d$	−12191
$1s3d$	−12209 −12206		
$1s4s$	−8013 −7371		
$1s4p$	−7093 −6818	$4s$, $4p$, $4d$, $4f$	−6858
$1s4d$	−6866 −6864		
$1s4f$	−6858 −6857		
$1s\infty s(= \text{He}^+)$	0	$\infty s(= \text{H}^+)$	0

configurations $1s\ 3s$, $1s\ 3p$, and $1s\ 3d$. Of these, the $1s\ 3s$ will be the most stable and the $1s\ 3d$ the least stable because of the decrease in penetration as the angular momentum increases. Each configuration will have two levels because of resonance, and the difference between the two levels will decrease in going from $1s\ 3s$ to $1s\ 3d$ because the overlap of the $1s$-orbital with the $3s$-, $3p$-, and $3d$-orbitals decreases in that order (see Fig. 9-8). The shielding by the $1s$-orbitals will be fairly effective, so the nucleus will act nearly as though it has a charge of $(Z - 1)$ instead of Z, and all of the energies will be close to those of a hydrogen-like atom of nuclear charge $(Z - 1)$.

In still higher levels in which one electron is in the $1s$-orbital, the energies of each set of levels will be even closer together and the resemblance to the hydrogen-like levels will be even more marked. This is shown in Table 9-8 and Fig. 9-2.

Levels in which both electrons are excited (e.g., those belonging to the configuration $(2p)^2$) are so high in energy that they are rarely observed experimentally. In helium, the most stable level of the $(2p)^2$ configuration has an energy far above that of the ground state of a singly ionized helium atom (see Fig. 9-2).

The general pattern of the energy level diagram of helium can in this way be accounted for. Unfortunately a number of important discrepancies remain. In the first place, the predicted degeneracies of the levels are in serious disagreement with what is observed. This discrepancy is shown in Table 9-9.

TABLE 9-9

DEGENERACIES OF THE HELIUM ENERGY LEVELS

Level		Predicted Degeneracy	Observed Degeneracy
$1s^2\ S$		1	1
$1s2s,\ S$	$1s2s + 2s1s$	1	1
	$1s2s - 2s1s$	1	3
$1s2p,\ P$	$1s2p + 2p1s$	3	3
	$1s2p - 2p1s$	3	9

In the second place, the Zeeman effect is not normal in many of the levels. Both of these discrepancies may be accounted for completely by introducing the concept of electron spin along with the Pauli principle. A third but minor discrepancy is a very small splitting of the antisymmetric P-level, which is accounted for by relativity and spin effects.

D. Electron Spin and the Pauli Principle

a. The experimental basis for the hypothesis of electron spin

Shortly before the discovery of quantum mechanics, Goudsmit and Uhlenbeck showed that many puzzling features of atomic spectra could be explained if it were assumed that the electron can spin about its own axis. This new property, called *spin angular momentum* (or more often simply *spin*), turned out to have some unusual characteristics. Goudsmit and Uhlenbeck had to assume that only two states of spin are possible. These two states differ only in that the component of spin angular momentum in any chosen direction is $+(1/2)\hbar$ in one state and $-(1/2)\hbar$ in the other. They also had to assume that the spin was accompanied by a magnetic dipole whose components in a chosen direction (e.g., the direction of a magnetic field) could have but two values, $\pm\ \beta_M$, where β_M is the Bohr magneton. This moment is twice $\pm\ (1/2)\beta_M$, the value that one would expect if spin angular momentum arose in the same way as orbital angular momentum.

The experimental evidence supporting these assumptions is now overwhelming, if somewhat indirect. The most direct evidence for it would, of course, come from sending a beam of electrons through the Stern-Gerlach apparatus (page 267). The beam would be expected to split into two parts if the electron spin hypothesis were correct. This simple experiment has never been successfully performed, however, because the splitting of the beam is inevitably obscured by the effects of the ordinary magnetic forces acting when a charge moves through a magnetic field (see *14*, p. 61). These extraneous forces can be eliminated by attaching the electron to a positive ion, so that the particle moving through the magnetic field is uncharged. If the electron in this neutral atom has no orbital angular momentum, a beam of these atoms ought to be split into two parts in the Stern-Gerlach experiment. This result was first found with a beam of silver atoms, but the effect has also been observed with copper, gold, and hydrogen atoms, in all of which there is good reason for believing that the electron has no orbital angular momentum in the ground state. In every instance the amount of the splitting indicated a magnetic dipole moment component of $\pm\ \beta_M$ in the direction of the magnetic field.

The assumption of electron spin also accounts for the anomalous Zeeman effect and for the so-called "fine structure" of many spectral lines, such as the sodium-D doublet. These matters will be discussed as we proceed. By giving the electron an extra degree of freedom the degeneracy of atomic energy levels will be increased; we shall find that this will help to remove the discrepancy noted in Table 9-9, where some of the observed levels of helium were found to have a higher degeneracy than was expected from the theory.

b. The quantum mechanical treatment of electron spin

According to the quantum mechanical rules given in Chapter 5, the procedure for incorporating the concept of spin into the framework of quantum mechanics is to find the classical analogue of the property, express it in terms of momenta and coordinates, and replace the momenta and coordinates by the appropriate operators. This should supply us with spin operators from which commutation rules and eigenfunctions can be derived in straightforward fashion, and then dealt with according to the quantum mechanical laws. Unfortunately, when we try to apply this program we find ourselves unable to begin, for *there is no classical analogue of electron spin.* The electron's spin angular momentum is certainly not the same as ordinary angular momentum, since only two eigenstates are ever observed and these involve half integral quantum numbers. We are therefore forced to proceed by postulating a formalism for dealing with spin without any appeal to classical analogues. This can be done in a number of ways, but for our purposes the following set of three postulates is adequate.

Postulate I – The operators for spin angular momentum commute and combine in the same way as those for ordinary angular momentum. Thus (*cf.* page 248) we can define S_x, S_y, and S_z as the operators for the components of the spin of a single electron in the x-, y-, and z-directions. The square of the spin angular momentum has the operator

$$S^2 = S_x{}^2 + S_y{}^2 + S_z{}^2 \tag{D-1}$$

The components fail to commute with each other

$$\begin{aligned} S_xS_y - S_yS_x &= i\hbar S_z \\ S_yS_z - S_zS_y &= i\hbar S_x \\ S_zS_x - S_xS_z &= i\hbar S_y \end{aligned} \tag{D-2}$$

They do commute with S^2, however

$$\begin{aligned} S_xS^2 - S^2S_x &= 0 \\ S_yS^2 - S^2S_y &= 0 \\ S_zS^2 - S^2S_z &= 0 \end{aligned} \tag{D-3}$$

Therefore it is possible to know the square of the spin angular momentum and any *one* of its components for an isolated electron.

If there is more than one electron, the considerations of Section B of Chapter 8 tell us that total electronic spin operators S_{xe}, S_{ye}, S_{ze}, and S_e^2 may be constructed from the operators of each of the electrons in the following way

$$S_{xe} = \sum S_{xj}; \quad S_{ye} = \sum S_{yj}; \quad S_{ze} = \sum S_{zj}$$

$$S_e^2 = S_{xe}^2 + S_{ye}^2 + S_{ze}^2 \qquad \text{(D-4)}$$

Postulate II – For a single electron there are only two eigenfunctions of S^2 and S_z. They will be called α and β. The eigenvalues are as follows

$$\begin{aligned} S_z\alpha &= (1/2)\hbar\alpha & S_z\beta &= -(1/2)\hbar\beta \\ S^2\alpha &= (1/2)\,[(1/2)+1]\,\hbar^2\alpha & S^2\beta &= (1/2)\,[(1/2)+1]\hbar^2\beta \end{aligned} \qquad \text{(D-5)}$$

By analogy with the operators $(M_x \pm iM_y)$ described on pages 252*ff*. we may set up the operators

$$\sigma = S_x + iS_y \qquad \text{and} \qquad \tau = S_x - iS_y \qquad \text{(D-6)}$$

which have the properties (*cf*. Equations (A-24) and (A-30) of Chapter 8)

$$\begin{aligned} \sigma\alpha &= 0 & \sigma\beta &= \hbar\alpha \\ \tau\alpha &= \hbar\beta & \tau\beta &= 0 \end{aligned} \qquad \text{(D-7)}$$

$$S^2 = \sigma\tau - \hbar S_z + S_z^2 = \tau\sigma + \hbar S_z + S_z^2 \qquad \text{(D-8)}$$

(These properties can be shown to follow from the commutation rules–see, for example, Rojansky (*9*, pages 479*ff*.).)

Exercises. (1) Show from Equations (D-7) that

$$\begin{aligned} S_x\alpha &= (\hbar/2)\beta & S_x\beta &= (\hbar/2)\alpha \\ S_y\alpha &= -(\hbar/2)\beta & S_y\beta &= (\hbar/2)\alpha \end{aligned} \qquad \text{(D-9)}$$

(2) Using (D-8) and $S_z\alpha = (\hbar/2)\alpha$, verify that $S^2\alpha = (1/2)[(1/2) + 1]\hbar^2\alpha$.

(3) Show that $S_z\sigma = \sigma(S_z + \hbar)$ and $S_z\tau = \tau(S_z - \hbar)$.

(4) From the result of the previous exercise, prove that $S_z\sigma\alpha = (3/2)\hbar\sigma\alpha$, so that either $\sigma\alpha$ is an eigenfunction of S_z with eigenvalue $3\hbar/2$, or $\sigma\alpha = 0$. Since Postulate II excludes the former alternative, this proves that $\sigma\alpha = 0$. Prove in a similar fashion that $\tau\beta = 0$. Also prove that $\sigma\beta$ and $\tau\alpha$ are eigenfunctions of S_z with eigenvalues $\hbar/2$ and $-\hbar/2$, respectively.

If there is more than one electron,

$$S_e^2 = \sigma_e \tau_e - \hbar S_{ze} + S_{ze}^2 = \tau_e \sigma_e + \hbar S_{ze} + S_{ze}^2 \qquad \text{(D-10)}$$

where $\sigma_e = \sum \sigma_j$; and $\tau_e = \sum \tau_j$, σ_j and τ_j being the operators for the jth electron defined in (D-6).

The essential characteristics of the eigenfunctions α and β will be clearer if, for the moment, we allow ourselves naively to think of the electron as truly a rotating sphere. In order to describe the orientation of such a sphere at any instant we would have to give the values of three coordinates–say, the Euler angles θ, ψ, and φ defined in Fig. 9-9. We might expect that α and β would be functions of these three angles. No one has ever been able to devise a method

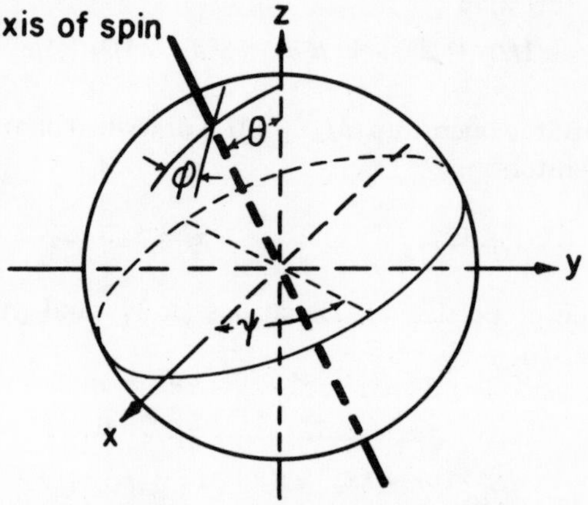

FIG. 9-9. Euler angles for defining the orientation of a rotating sphere. The angle between the z-axis and the axis of rotation is θ. The angle between the x-axis and the intersection of the equatorial plane with the $x-y$ plane is ψ. The angle between a reference plane passing through the axis of rotation and the plane determined by the z-axis and the axis of rotation is φ.

for measuring the angles ψ and φ for a single spinning electron. There is therefore little physical reason for trying to include these angles in α and β. On the other hand, the angle θ is precisely determined from the value of the total angular momentum, $\sqrt{(1/2)\,[(1/2)+1]}\hbar$, and its possible components in the z-direction, $\pm\,\hbar/2$. Thus

$$\theta = \cos^{-1}(\pm\hbar/2)/[\tfrac{1}{2}(\tfrac{1}{2}+1)\hbar^2]^{1/2} = \cos^{-1}(\pm\sqrt{3}/3) = 54.9^\circ \text{ and } 125.1^\circ$$

It is interesting that θ is greater than 45°, showing that the component of spin angular momentum perpendicular to the z-axis is greater than that along the z-axis. In fact, it has the value $(\sqrt{2}/2)\hbar = 0.707\hbar$.

Since $|\alpha|^2$ and $|\beta|^2$ are the probabilities of finding the electron in a given orientation, and since θ is precisely 54.9° if the electron is in state α and precisely 125.1° if it is in state β, it is clear that we can write

$$\alpha(\theta) = 0 \quad \text{when} \quad \theta \neq 54.9°$$

$$\beta(\theta) = 0 \quad \text{when} \quad \theta \neq 125.1°$$

$$\alpha(54.9) = 1$$

$$\beta(125.1) = 1$$

Instead of using θ as the variable, we might use the quantum number for the z-component of the spin, $\mathscr{S}_z$, writing

$$\alpha(\mathscr{S}_z) = 0, \text{ if } \mathscr{S}_z = -1/2 \text{ and } \alpha = 1, \text{ if } \mathscr{S}_z = 1/2$$

$$\beta(\mathscr{S}_z) = 0, \text{ if } \mathscr{S}_z = 1/2 \text{ and } \beta = 1, \text{ if } \mathscr{S}_z = -1/2$$

Thus α and β are really Kronecker delta functions (*cf.* page 115)

$$\alpha = \delta_{\mathscr{S}_z, \frac{1}{2}} \qquad \beta = \delta_{\mathscr{S}_z, -\frac{1}{2}} \tag{D-11}$$

These eigenfunctions are obviously orthogonal and normalized

$$\sum_{\mathscr{S}_z=-1/2}^{1/2} |\alpha|^2 = 1 \; ; \quad \sum_{\mathscr{S}_z=-1/2}^{1/2} |\beta|^2 = 1 \; ; \quad \sum_{\mathscr{S}_z=-1/2}^{1/2} \alpha^*\beta = 0 \tag{D-12}$$

Sums are used here instead of integrals because $\mathscr{S}_z$ can take on only two discrete values.

If there is more than one electron, the spin state is represented by products of the individual electron spin eigenfunctions. If both electrons are known to have their spins pointing in the positive z-direction ($\mathscr{S}_z = +1/2$), the spin wave function will be

$$\chi_a = \alpha_1\alpha_2$$

If electron no. 1 has $\mathscr{S}_z = +1/2$ and no. 2 has $\mathscr{S}_z = -1/2$, then we have

$$\chi_b = \alpha_1\beta_2$$

If electron no. 1 has $\mathscr{S}_z = +1/2$ but electron no. 2 has an equal probability of $\mathscr{S}_z = +1/2$ and $\mathscr{S}_z = -1/2$, then the spin wave function may be

$$\chi_c = (1/\sqrt{2})\alpha_1(\alpha_2 + \beta_2)$$

where the factor $1/\sqrt{2}$ is required for normalization because

$$\sum_{\mathscr{S}_{z_1}} \sum_{\mathscr{S}_{z_2}} \alpha_1^2(\alpha_2 + \beta_2)^2 = \sum \sum (\alpha_1^2\alpha_2^2 + \alpha_1^2\beta_2^2) = 2$$

If electrons no. 1 and no. 2 are known to have oppositely directed spin components, but with an equal chance that no. 1 has $\mathscr{S}_z = +1/2$ and $-1/2$, then we may have the spin function

$$\chi_d = (1/\sqrt{2})(\alpha_1\beta_2 + \alpha_2\beta_1)$$

With more than two electrons analogous spin wave functions can be constructed out of the spin eigenfunctions of the individual electrons.

Some of these wave functions are eigenfunctions of S_{ze} and of S_e^2, and some are not. If they are eigenfunctions, then it always happens that, writing the total eigenfunction as A, the eigenvalue equations take the form

$$S_{ze}A = \mathscr{S}_z\hbar A \qquad \text{(D-13)}$$

$$S_e^2 A = S(S+1)\hbar^2 A$$

where S is an integer if there is an even number of electrons and a half-integer if there is an odd number of electrons, and where $\mathscr{S}_z$ can have the values $-S, -S+1, \cdots, S-1, S$. S is known as the *total spin quantum number* of the group of electrons, or more concisely as the *total spin*.

The quantity $(2S + 1)$ is known as the *multiplicity* of a state. In the notation commonly used to describe atomic and molecular energy levels it is written as a superscript before the letter denoting the orbital angular momentum of the level. Thus one writes ^1S ("singlet S"), ^3P ("triplet P"), ^2D ("doublet D"), $^4\Sigma$ ("quartet sigma"), etc. In some cases the multiplicity is equal to the degeneracy of a level, but the two are often not equal and the terms should not be confused. The degeneracy of a level is always at least as great as the multiplicity, however.

Exercises. (1) Operate on χ_a, χ_b, χ_c, and χ_d with S_{ze} and S_e^2 using definitions (D-4) and the relations (D-5), D-6), (D-7), and (D-10). Which of them are eigenfunctions of these operators? In those cases where χ is an eigenfunction of both S_{ze} and S_e^2, are the eigenvalues of the form shown in (D-13)?

(2) Suppose that a system of three electrons has the spin wave function $\alpha_1\alpha_2\alpha_3$. Show that

$$S_{ze}\alpha_1\alpha_2\alpha_3 = (3/2)\hbar\alpha_1\alpha_2\alpha_3$$

$$S_e^2\alpha_1\alpha_2\alpha_3 = (3/2)\ [(3/2) + 1]\hbar^2\alpha_1\alpha_2\alpha_3$$

(3) Find a spin wave function for a system of four electrons in which any three electrons have $\mathscr{S}_z = 1/2$ and one has $\mathscr{S}_z = -1/2$ and which is an eigenfunction of both S_{ze} and S_e^2.

Postulate III – The spinning electron acts like a magnet the magnitude of whose dipole moment is

$$|\boldsymbol{\mu}| = 2 \times (\text{spin angular momentum}) \times e/2m_ec$$

$$= \sqrt{(1/2)\ [(1/2) + 1]}\ 2\beta_M \tag{D-14}$$

and whose direction is anti-parallel to the spin axis (β_M being the Bohr magneton defined on page 262). The only possible values of the component of this dipole moment in any direction are therefore

$$\mu_z = \pm \frac{\frac{1}{2}}{\sqrt{\frac{1}{2}(\frac{1}{2} + 1)}}\ |\boldsymbol{\mu}| = \pm\ \beta_M \tag{D-15}$$

The ratio of the spin magnetic moment of an electron to the spin angular momentum (that is, the gyromagnetic ratio for spin) is therefore twice as large as that for orbital motion of the electron, for which we found (Equation (D-5) of Chapter 8)

$$|\boldsymbol{\mu}| = (\text{orbital angular momentum}) \times e/2m_ec = \sqrt{l(l + 1)}\beta_M$$

This difference was shown by Dirac to be a consequence of the requirements of relativity theory. Recent work shows, however, that the factor is slightly greater than two (see *10*).

Because of the magnetic dipole moment of the spinning electron, the energy of an electron in an atom or molecule depends on its orbital angular momentum, since there is a magnetic field at the electron due to the apparent motion of the charged nucleus around the electron. This effect is responsible for the so-called "fine structure" of spectra (e.g., the existence of two sodium D lines rather than just one). It is very small in atoms of low atomic number, but becomes appreciable toward the bottom of the periodic table. The effect will be discussed in more detail in Section C of Chapter 10.

The "spin-orbit interaction" causes a small term containing the spin operator to appear in the Hamiltonian of atoms and molecules. Ordinarily this term is neglected. To this approximation the Hamiltonian is independent of the spin and all of the spin operators commute with it,

$$S_e^2H - HS_e^2 = 0 \; ; \qquad S_{ze}H - HS_{ze} = 0 \tag{D-16}$$

so that it is possible to know the energy, the square of the total electron spin angular momentum and its component in any one direction for any atom or molecule not containing nuclei of too great an atomic number.

c. Inclusion of electron spin in the helium wave function

The four possible spin states of a pair of electrons may be represented by the following spin wave functions

$$\begin{aligned} \chi_1 &= \alpha_1\alpha_2 \qquad & \alpha_3 &= \beta_1\alpha_2 \\ \chi_2 &= \alpha_1\beta_2 \qquad & \chi_4 &= \beta_1\beta_2 \end{aligned} \tag{D-17}$$

The first represents a pair of electrons in which both have their spin axes pointing upward (positive z-direction). The second represents electron no. 1 with its spin axis upward and electron no. 2 with its spin axis downward, etc. These functions are all eigenfunctions of S_{ze} but not all of them are eigenfunctions of S_e^2, as we can see from the following calculations

$$S_{ze}\chi_1 = (S_{z_1} + S_{z_2})\alpha_1\alpha_2 = \frac{\hbar}{2}\alpha_1\alpha_2 + \frac{\hbar}{2}\alpha_1\alpha_2 = \hbar\alpha_1\alpha_2$$

$$S_{ze}\chi_2 = (S_{z_1} + S_{z_2})\alpha_1\beta_2 = \frac{\hbar}{2}\alpha_1\beta_2 - \frac{\hbar}{2}\alpha_1\beta_2 = 0$$

$$\begin{aligned}
S_{ze}\chi_3 &= 0 \\
S_{ze}\chi_4 &= -\hbar\chi_4 \\
S_e^2\chi_1 &= (\tau_1 + \tau_2)(\sigma_1 + \sigma_2)\alpha_1\alpha_2 + \hbar S_{ze}\chi_1 + S_{ze}^2\chi_1 \\
&= 0 + \hbar^2\chi_1 + \hbar^2\chi_1 = 2\hbar^2\chi_1 \\
&= 1(1 + 1)\hbar^2\chi_1 \,. \\
S_e^2\chi_2 &= (\tau_1 + \tau_2)(\sigma_1 + \sigma_2)\alpha_1\beta_2 + \hbar S_{ze}\chi_2 + S_{ze}^2\chi_2 \\
&= (\tau_1 + \tau_2)\hbar\alpha_1\alpha_2 + 0 + 0 \\
&= \hbar^2(\alpha_1\beta_2 + \beta_1\alpha_2) = \hbar^2(\chi_2 + \chi_3) \\
S_e^2\chi_3 &= (\tau_1 + \tau_2)(\sigma_1 + \sigma_2)\beta_1\alpha_2 + \hbar S_{ze}\chi_3 + S_{ze}^2\chi_3 \\
&= (\tau_1 + \tau_2)\hbar\alpha_1\alpha_2 + 0 + 0 \\
&= \hbar^2(\alpha_1\beta_2 + \alpha_2\beta_1) = \hbar^2(\chi_2 + \chi_3) \\
S_e^2\chi_4 &= (\sigma_1 + \sigma_2)(\tau_1 + \tau_2)\beta_1\beta_2 - \hbar S_{ze}\chi_4 + S_{ze}^2\chi_4 \\
&= 0 + \hbar^2\chi_4 + \hbar^2\chi_4 \\
&= 1(1 + 1)\hbar^2\chi_4
\end{aligned} \tag{D-18}$$

Evidently χ_2 and χ_3 are not eigenfunctions of S_e^2, but it is readily seen that $(1/\sqrt{2})\,(\chi_2+\chi_3)$ and $(1/\sqrt{2})\,(\chi_2-\chi_3)$ are

$$S_e^2\,(1/\sqrt{2})\,(\chi_2+\chi_3) = (1/\sqrt{2})\hbar^2\,(\alpha_1\beta_2+\alpha_2\beta_1+\alpha_1\beta_2+\beta_1\alpha_2)$$

$$= 1(1+1)\hbar^2\,(1/\sqrt{2})\,(\chi_2+\chi_3) \tag{D-19}$$

$$S_e^2\,(1/\sqrt{2})\,(\chi_2-\chi_3) = (1/\sqrt{2})\hbar^2\,(\chi_2+\chi_3-\chi_2-\chi_3) = 0$$

Since it is convenient to have wave functions that are eigenfunctions of S_{ze} and S_e^2, we shall take the following as our four spin functions instead of χ_1, χ_2, χ_3, and χ_4

$$\text{Triplet states}\begin{cases} {}^3\xi_1 = \alpha_1\alpha_2 & \mathscr{S} = +1; \quad S = 1; \quad \mathscr{P} = +1 \\ {}^3\xi_0 = (1/\sqrt{2})\,(\alpha_1\beta_2+\alpha_2\beta_1) & \mathscr{S} = 0; \quad S = 1; \quad \mathscr{P} = +1 \\ {}^3\xi_{-1} = \beta_1\beta_2 & \mathscr{S} = -1; \quad S = 1; \quad \mathscr{P} = +1 \end{cases} \tag{D-20}$$

$$\text{Singlet state} \quad {}^1\xi_0 = (1/\sqrt{2})\,(\alpha_1\beta_2-\alpha_2\beta_1) \quad \mathscr{S} = 0; \quad S = 0; \quad \mathscr{P} = -1 \tag{D-21}$$

The quantum numbers for S_{ze}, S_e^2 and the permutation operator, P_{12}, are given alongside the functions. Since each has a different set of eigenvalues, these functions are orthogonal. They are also normalized.

The spin function ${}^3\xi_0 = (1/\sqrt{2})\,(\alpha_1\beta_2+\alpha_2\beta_1)$ is a rather surprising one in that it has a total spin quantum number $S = 1$ in spite of the fact that, according to the interpretation we have given, the two electrons have their z-components of spin pointing in opposite directions. The spin angular momentum in this case comes from the spin components perpendicular to the z-axis. We have seen that these components are larger than the z-components, having the value $(\sqrt{2}/2)\hbar$. If they are parallel they can give a net spin angular momentum perpendicular to the z-axis of $2\times[(\sqrt{2}/2)\hbar] = \sqrt{2}\hbar$. This, however, is exactly the spin angular momentum of the ${}^3\xi_0$ state $(\sqrt{1(1+1)}\hbar = \sqrt{2}\hbar)$. Therefore, we conclude that in the ${}^3\xi_0$state the spins are related as shown in Fig. 9-10A. Similarly, in the ${}^1\xi_0$ state the spins cancel completely, so they must be arranged as shown in Fig. 9-10B.

Exercise. Show that if a pair of electrons have a spin wave function

$$\xi = (1/\sqrt{2})\,(\alpha_1\beta_2 + e^{i\psi}\alpha_2\beta_1)$$

where ψ is a constant, then the average value of the square of the total spin angular momentum is

$$\overline{S_e^2} = \hbar^2\,(1+\cos\psi) = 2\hbar^2\cos^2(\psi/2)$$

Give a physical interpretation of the angle ψ. Evidently in systems containing more than one electron, something can be said about the relative values of the Eulerian angles ψ (*cf.* Fig. 9-9). Since the only observed values for S_e^2 are 0 and $2\hbar^2$, the only values of ψ that actually occur in nature must be $\psi = 0$ and π.

In setting up the zero-order wave functions for the treatment of the helium atom by the perturbation theory, each assignment of two electrons to two hydrogen-like orbitals can therefore be made in four different ways. Since the Hamiltonian does not depend on the spin except to a very slight degree which we shall neglect, this new factor will not have any effect on the calculated energy levels. It should, however, increase all of the predicted degeneracies in Table 9-9 by a factor of four.

The lowest energy level of helium should be made up of four states, three of them triplet states whose complete wave functions are

$$\phi_1 = (1s)_1\,(1s)_2\,{}^3\xi_1; \quad \phi_2 = (1s)_1\,(1s)_2\,{}^3\xi_0$$

$$\phi_3 = (1s)_1\,(1s)_2\,{}^3\xi_{-1}$$

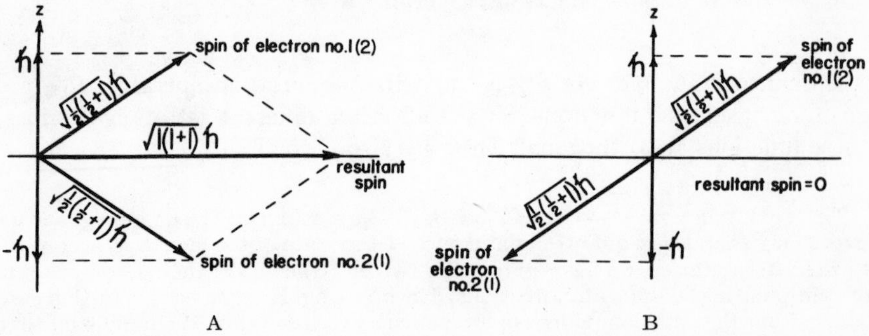

FIG. 9-10.
A. Relative orientations of the spin axes of two electrons having the spin wave function ${}^3\xi_0$.
B. Orientations of spin axes in the state ${}^1\xi_0$.

and one of them a singlet state whose complete wave function is

$$\phi_4 = (1s)_1\,(1s)_2\,{}^1\xi_0$$

Similarly there should be thirty-two states in the first group of excited levels. In Table 9-10 the wave functions belonging to all of these levels are represented, along with the eigenvalues for the permutation operator and

TABLE 9-10

WAVE FUNCTIONS, PERMUTATION EIGENVALUES, AND DEGENERACIES OF HELIUM ENERGY LEVELS

Type of Total Wave Function			Permutation Eigenvalues of			Degeneracies of		
Coordinate part	Spin part	Type of Level	Coordinate part	Spin part	Total Function	Coordinate part	Spin part	Total Function
$(1s)_1(1s)_2$	$^1\xi$	$1s^2\ ^1S$	+1	−1	−1	1	1	1
$(1s)_1(2s)_2 + (1s)_2(2s)_1$	$^1\xi$	$1s2s\ ^1S$	+1	−1	−1	1	1	1
$(1s)_1(2s)_2 - (1s)_2(2s)_1$	$^3\xi$	$1s2s\ ^3S$	−1	+1	−1	1	3	3
$(1s)_1(2p)_2 + (1s)_2(2p)_1$	$^1\xi$	$1s2p\ ^1P$	+1	−1	−1	3	1	3
$(1s)_1(2p)_2 - (1s)_2(2p)_1$	$^3\xi$	$1s2p\ ^3P$	−1	+1	−1	3	3	9
$(1s)_1(1s)_2$	$^3\xi$	$1s^2\ ^3S$	+1	+1	+1	1	3	3
$(1s)_1(2s)_2 + (1s)_2(2s)_1$	$^3\xi$	$1s2s\ ^3S$	+1	+1	+1	1	3	3
$(1s)_1(2s)_2 - (1s)_2(2s)_1$	$^1\xi$	$1s2s\ ^1S$	−1	−1	+1	1	1	1
$(1s)_1(2p)_2 + (1s)_2(2p)_1$	$^3\xi$	$1s2p\ ^3P$	+1	+1	+1	3	3	9
$(1s)_1(2p)_2 - (1s)_2(2p)_1$	$^1\xi$	$1s2p\ ^1P$	−1	−1	+1	3	1	3

the predicted degeneracies. The functions are grouped according to the permutation eigenvalue of the overall wave functions, the totally antisymmetric types of function being in the upper half of the table and the totally symmetric types in the lower half.

A striking result appears. In the last column of Table 9-10 the predicted degeneracies of the totally antisymmetric functions agree with the observed degeneracies given in Table 9-9. If we could, therefore, find some reason for discarding the totally symmetric states, our difficulty with the degeneracy would be resolved. This, in fact, is just what the Pauli principle tells us to do.

d. The Pauli principle

Evidently something is still missing from our system of quantum mechanical laws. Although we are able to account for the observed energy levels of the helium atom, we are unable to account for the degeneracies of these levels. Even more serious difficulties arise in the treatment of the energy levels of atoms beyond helium in the periodic table.

For instance, the application of perturbation theory to lithium leads one to expect an eight-fold degenerate ground level, with states of the types $(1s)^3\,{}^2S$ and $(1s)^3\,{}^4S$, the spin eigenfunctions of S_e^2 and S_{ze} for three electrons being

$$
\begin{aligned}
{}^4\xi_{3/2} &= \alpha_1\alpha_2\alpha_3 & {}^2\xi_{1/2} &= (1/\sqrt{3})\,(\alpha_1\alpha_2\beta_3 + \alpha_1\beta_2\alpha_3 - 2\beta_1\alpha_2\alpha_3)\\
{}^4\xi_{1/2} &= (1/\sqrt{3})\,(\alpha_1\alpha_2\beta_3 + \alpha_1\beta_2\alpha_3 + \beta_1\alpha_2\alpha_3) & {}^2\xi_{-1/2} &= (1/\sqrt{3})\,(\alpha_1\beta_2\beta_3 + \beta_1\alpha_2\beta_3 - 2\beta_1\beta_2\alpha_3)\\
{}^4\xi_{-1/2} &= (1/\sqrt{3})\,(\alpha_1\beta_2\beta_3 + \beta_1\alpha_2\beta_3 + \beta_1\beta_2\alpha_3) & {}^2\xi'_{1/2} &= (1/\sqrt{3})\,(\alpha_1\alpha_2\beta_3 - 2\alpha_1\beta_2\alpha_3 + \beta_1\alpha_2\alpha_3)\\
{}^4\xi_{-3/2} &= \beta_1\beta_2\beta_3 & {}^2\xi'_{-1/2} &= (1/\sqrt{3})\,(\alpha_1\beta_2\beta_3 - 2\beta_1\alpha_2\beta_3 + \beta_1\beta_2\alpha_3)
\end{aligned}
$$

The complete set of ground state wave functions of lithium would be obtained by multiplying each of these spin functions by the configurational function $(1s)_1(1s)_2(1s)_3$. For beryllium the ground level should be sixteen-fold degenerate and involve states of the types $(1s)^4\,{}^5S$, $(1s)^4\,{}^3S$, and $(1s)^4\,{}^1S$. None of these states is observed in either lithium or beryllium.

Exercise. Verify that the spin functions given above for three electrons are eigenfunctions of S_e^2 and S_{ze}, with eigenvalues as indicated.

Similarly, as we go further into the periodic table, we never find states having configurations involving more than two electrons in any s-orbital, or more than six electrons in any p-orbital, or more than $2(2l + 1)$ electrons in any group of orbitals with angular momentum quantum number l.

We notice that the non-spin parts of the wave functions, $(1s)^n$, are always symmetric with respect to all permutations of electrons, since they have the form

$$(1s)_1\,(1s)_2\,\cdots\,(1s)_n$$

We also notice that among the eight spin eigenfunctions listed above for the hypothetical $(1s)^3$-state of lithium only four are eigenfunctions of the permutation operator, and these four are symmetric with respect to all permutations. Antisymmetric spin functions do not occur at all. In fact it is easy to see that *for systems containing more than two electrons, no nonvanishing spin eigenfunctions can be constructed that are antisymmetric for all possible permutations of electrons.* For instance $\alpha_1\alpha_2\beta_3 - \alpha_1\beta_2\alpha_3$ is antisymmetric for P_{23}, the permutation of electrons 2 and 3, but it is not an eigenfunction for P_{13}. If the terms $-\beta_1\alpha_2\alpha_3 + \alpha_1\beta_2\alpha_3$ are added, we obtain an antisymmetric eigenfunction of P_{13} but we no longer have an eigenfunction of P_{23}.

Enough has been said to make it clear that all of our difficulties in accounting for the numbers of observed states will be removed if we accept the hypothesis that *total wave functions corresponding to systems occurring in nature must change their sign on permutation of any pair of electrons in the system.* This is known as the *Pauli principle* and should be considered as *the fourth of the fundamental laws of quantum mechanics.**

In the future we will save ourselves a lot of trouble if from the very beginning of all calculations involving more than one electron we use only antisymmetric wave functions. In most of our work we shall use wave functions composed of products of functions of the coordinates of individual electrons, just as we did with the helium atom. For such functions, there is an easy way of assuring antisymmetry. It is a universal property of determinants (page 47) that, if we interchange any two rows or columns, the sign of the determinant changes, whereas the numerical value is unaffected. Therefore if we set up wave functions in the form of determinants in which the permutation of a pair of electrons is equivalent to the interchange of a pair of rows or columns, we will have taken care of the Pauli principle automatically.

For instance, for helium in the ground state such a determinant would be

$$\psi_0 = \frac{1}{\sqrt{2}} \begin{vmatrix} (1s)_1\alpha_1 & (1s)_2\alpha_2 \\ (1s)_1\beta_1 & (1s)_2\beta_2 \end{vmatrix} = (1s)_1(1s)_2(\alpha_1\beta_2 - \beta_1\alpha_2)/\sqrt{2}$$

For the first group of excited states of helium we may write

$$\psi_1 = \frac{1}{\sqrt{2}} \begin{vmatrix} (1s)_1\alpha_1 & (1s)_2\alpha_2 \\ (2s)_1\alpha_1 & (2s)_2\alpha_2 \end{vmatrix}; \quad \psi_2 = \frac{1}{\sqrt{2}} \begin{vmatrix} (1s)_1\alpha_1 & (1s)_2\alpha_2 \\ (2s)_1\beta_1 & (2s)_2\beta_2 \end{vmatrix};$$

$$\psi_3 = \frac{1}{\sqrt{2}} \begin{vmatrix} (1s)_1\beta_1 & (1s)_2\beta_2 \\ (2s)_1\alpha_1 & (2s)_2\alpha_2 \end{vmatrix}; \quad \psi_4 = \frac{1}{\sqrt{2}} \begin{vmatrix} (1s)_1\beta_1 & (1s)_2\beta_2 \\ (2s)_1\beta_1 & (2s)_2\beta_2 \end{vmatrix}$$

It will be found that our previously derived helium wave functions (Table 9-10), when expressed in terms of these determinants, are given by

$$\text{Singlet function: } (\psi_2 - \psi_3)/\sqrt{2} \longrightarrow (1s)\,(2s)\ {}^1S$$

$$\text{Triplet functions: } \psi_1,\ (\psi_2 + \psi_3)/\sqrt{2},\ \psi_4 \longrightarrow (1s)\,(2s)\ {}^3S$$

* The Pauli principle holds not only for electrons, but also for protons and neutrons and for all atomic nuclei of odd atomic weight (that is, having an odd number of protons plus neutrons). On the other hand atomic nuclei of even atomic weight occur only in wave functions which remain unchanged when any pair of identical nuclei are interchanged. Electrons, protons and neutrons are said to be *antisymmetric* particles, while deuterons, alpha particles, etc., are *symmetric*.

We have found that an electron has four degrees of freedom, three having to do with its position in space, and the fourth arising from the possibility of different orientations of the spin axis. Therefore four quantum numbers are always required in specifying the state of an electron. For this purpose one usually uses the principle quantum number n, the angular momentum quantum number l, the magnetic quantum number m, and the spin quantum number s. In an atom containing N electrons, a state can be specified by giving N sets of these four quantum numbers. Let us call these sets

$$(n_1, l_1, m_1, s_1)\ (n_2, l_2, m_2, s_2) \cdots (n_N, l_N, m_N, s_N)$$

and let the orbital corresponding to any given set (n_i, l_i, m_i, s_i) be represented by $\phi(n_i, l_i, m_i, s_i)$. (For example the selection $n = 2, l = 1, m = 0, s = -1/2$ leads to the orbital $\phi(210 - \frac{1}{2})$, or $2p_0\beta$.) Then a wave function acceptable from the point of view of the Pauli principle is

$$\psi = \frac{1}{\sqrt{N!}} \begin{vmatrix} \phi(n_1l_1m_1s_1)_1 & \phi(n_2l_2m_2s_2)_1 & \cdots & \phi(n_Nl_Nm_Ns_N)_1 \\ \phi(n_1l_1m_1s_1)_2 & \phi(n_2l_2m_2s_2)_2 & \cdots & \phi(n_Nl_Nm_Ns_N)_2 \\ \vdots & \vdots & & \vdots \\ \phi(n_1l_1m_1s_1)_N & \phi(n_2l_2m_2s_2)_N & \cdots & \phi(n_Nl_Nm_Ns_N)_N \end{vmatrix} \tag{D-22}$$

Here the subscript after each ϕ represents the electron whose coordinates are to appear in the orbital. If the ϕ's are normalized and orthogonal, ψ will be normalized, since there are $N!$ terms when the determinant is evaluated, and on squaring and integrating each term contributes unity to the integral.

It is cumbersome to write out determinants in this way, so one usually indicates the determinant by writing merely its diagonal, sometimes even omitting the subscripts corresponding to the electrons. Thus (D-22) can be written as

$$\psi = (1/\sqrt{N!})\ |\ \phi(n_1l_1m_1s_1)_1\ \phi(n_2l_2m_2s_2)_2 \cdots \phi(n_Nl_Nm_Ns_N)_N\ | \tag{D-22a}$$

or as

$$\psi = (1/\sqrt{N!})\ |\ \phi(n_1l_1m_1s_1)\ \phi(n_2l_2m_2s_2) \cdots \phi(n_Nl_Nm_Ns_N)\ | \tag{D-22b}$$

Suppose that, using determinantal wave functions of this kind, we try to construct a wave function in which two sets of the four quantum numbers $(n_1l_1m_1s_1)$ are identical. The resulting determinant will have two identical columns, and such a determinant always vanishes (page 47). Therefore, we may state the general rule that *no system can exist in a state in which more than one electron has the same set of four quantum numbers*. This is the more familiar statement of the Pauli principle. It is not as general as the previous statement because it depends upon the idea of representing the state of an atom by products of individual electron orbitals, which, even though extremely useful, is only an approximation.

e. Pauli forces; Hund's multiplicity rule

We have seen that among the excited states of helium there occur states of two kinds: triplet states, in which the electron spins tend to be parallel, and singlet states, in which the electron spins cancel one another. Since triplet spin wave functions are symmetric with regard to electron permutations, the configurational part of the total triplet wave function must be antisymmetric; that is, of the type

$$^3\psi \sim 1s_1 2s_2 - 1s_2 2s_1$$

On the other hand the singlet spin function is antisymmetric so the singlet configurational wave function is of the type

$$^1\psi \sim 1s_1 2s_2 + 1s_2 2s_1$$

As we have seen, the correlation energy tends to make states of the first kind more stable, since electrons tend to keep out of each other's way in such states.

Evidently electrons in the same spin state avoid one another. We can say that such electrons act as though there is a repulsive "force" acting between them. This "force" may be called the "Pauli force" or "exchange force." The more nearly alike the quantum states of the two electrons, the greater is this "force", and if the states are identical, the "force" is infinite, since the two particles cannot co-exist at all.

Pauli forces can be said to affect the energy, but only in an indirect fashion. All of the chemically important terms in the Hamiltonians of atoms and molecules are electrostatic or electrodynamic in origin. Since Pauli forces are not electrostatic or electrodynamic in origin, they themselves cannot make any direct contribution to the energy; they act rather by controlling the movements of the electrons, which in turn affects the energy. Thus they determine the energy in much the same way that traffic lights in a busy city affect the number of automobile accidents. Strictly speaking, traffic lights do not prevent automobile accidents; only a two-foot layer of sponge rubber around every automobile, or some similar device, could do this. Traffic lights merely influence the flow of traffic so that the automobiles avoid one another. This decreases the number of collisions, and hence reduces the number of accidents. For conciseness, however, we ordinarily say that traffic lights do prevent accidents. It is in this sense that we may say that Pauli forces cause energies to be high or low.

Pauli forces are responsible for the apparent "solidity" of matter. We are told that most of the space in atoms is empty. Why, then, can we not push our hands through a piece of armor plate? The reason is that the electrons in our hands have the same spins and nearly the same velocities as some of the electrons in the armor plate. They therefore repel one another, and the armor plate feels "hard" to our touch. Clearly, if it were not for the Pauli principle, matter would have very different properties from what we know. In our everyday lives we therefore have a more direct experience of the Pauli principle than of most other laws of nature; a new-born baby becomes conscious of its consequences long before he finds it necessary to take account of the consequences of Newton's laws of motion.

Suppose that we have an atomic or molecular system having two states in which all of the electrons occupy the same orbitals, but which differ in their total spin quantum number. We have seen that the Pauli forces tend to keep electrons having the same spin out of each other's way, which increases the stability of the system. Therefore we may conclude that the most stable state of a group of electrons in a given set of orbitals should be that in which the greatest number of electrons have their spin components pointing in the same direction. Since the multiplicity of a state (that is, the value of $2S + 1$, where S is the total spin quantum number) is a direct measure of the number of parallel spins, we may assert as a general rule that, *other things being equal, the state of highest multiplicity will be the most stable*. This is known as *Hund's multiplicity rule*. It is illustrated by the fact that for the $1s2s$ configuration of helium the ^3S state is more stable than the ^1S state.

REFERENCES CITED

1. G. R. Harrison, "MIT Wave Length Tables," Wiley, New York, 1939.
2. C. E. Moore, "Atomic Energy Levels," Vols. 1 and 2, National Bureau of Standards, Washington, D.C., 1949, 1952.
3. H. A. Lorentz, "The Theory of Electrons," 2nd ed., Teubner, Leipzig, 1916. (Reprinted by Dover, New York, 1952).
4. W. Ritz, *Compt. rend.*, **144**, 634 (1907); **145**, 978 (1907).
5. J. Jeans, *Phil. Mag.*, [6] **11**, 421 (1901).
6. Lord Rayleigh, *Phil. Mag.*, [5] **44**, 356 (1897).
7. R. F. Bacher and S. Goudsmit, "Atomic Energy States," McGraw-Hill, New York, 1932.

8. H. Landolt and R. Börnstein, "Zahlenwerte und Funktionen" (A. Eucken ed.), 6th ed., Vol. 1, Part 1, Springer, Berlin, 1950.
9. V. Rojansky, "Introduction to Quantum Mechanics," Prentice-Hall, New York, 1938.
10. V. F. Weisskopf, *Revs. Mod. Phys.*, **21,** 305 (1949).
11. M. Abraham and R. Becker, "The Classical Theory of Electricity and Magnetism," Blackie and Son, London, 1937.
12. E. A. Hylleraas, *Skrifter Norske Videnskaps-Akad. Oslo, I. Mat.-Naturv. Kl.*, 1932, 5–141; *Z. Physik.*, **48,** 469 (1928); **54,** 347 (1929); **65,** 209 (1930).
13. H. Eyring, J. Walter, and G. E. Kimball, "Quantum Chemistry," Wiley, New York, 1944.
14. N. F. Mott and H. S. W. Massey, "The Theory of Atomic Collisions," 2nd ed. Oxford U.P., New York, 1949.

GENERAL REFERENCES

General discussions of atomic spectra:

G. Herzberg, "Atomic Spectra and Atomic Structure," Prentice-Hall, New York, 1937. (Reprinted by Dover, New York).

A. Sommerfeld, "Atomic Structure and Spectral Lines" (translated by H. L. Brose from 5th German ed. of 1931), Dutton, New York.

H. E. White, "Introduction to Atomic Spectra," McGraw-Hill, New York, 1934.

Detailed discussion of the quantum mechanical interpretation of atomic spectra:

E. U. Condon and G. Shortley, "The Theory of Atomic Spectra," Rev. ed., Cambridge U.P., New York, 1950.

Chapter 10

Atomic Structure II
Elements other than Hydrogen and Helium

A. The Energies of Orbitals in Elements beyond Helium; the Periodic System

a. The electronic configurations of atoms and ions

In calculating the energies of the atoms beyond helium by means of the perturbation theory, we may begin, as with helium, by neglecting the terms in the Hamiltonian that represent the electrostatic repulsions of the electrons. This leads to a partial differential equation of the type

$$\sum_{i=1}^{N} (1/2)\nabla_i^2\psi + \left[E + Z\sum_{i=1}^{N} 1/r_i\right]\psi = 0 \qquad \text{(A-1)}$$

where Z is the atomic number, N is the number of electrons in the atom ($N = Z$ for a neutral atom), ∇_i^2 is the Laplacian of the ith electron, and r_i is its distance from the nucleus. This equation can be solved by the method of separation of variables, writing

$$\psi = \prod_{i=1}^{N} \phi(n_i l_i m_i s_i)_i \qquad \text{(A-2)}$$

where n_i, l_i, m_i, and s_i are the four quantum numbers assigned to the ith electron and $\phi(n_i l_i m_i s_i)$ is the corresponding hydrogen-like orbital arising from a nuclear charge Z. Equation (A-2) gives a zero-order wave function of the atomic state. There are, of course, many different zero-order wave functions of this kind, including many which arise from permutations of electrons among the orbitals. The additional requirement imposed by the Pauli principle, however, necessitates the formation of determinants from those groups of zero-order functions that arise from mere permutation of electrons among a given set of orbitals. As a result, it is not possible to assign a particular electron to a particular orbital if more than one electron is present in the atom. Furthermore, the determinant vanishes if more than one electron is assigned the same set of four quantum numbers ($n_i l_i m_i s_i$).

The group of orbitals used in setting up the zero-order determinantal wave function of a state constitute what is called the *electron configuration* of the state. The configuration is indicated by writing the orbitals as shown in the examples given in Table 10-1; the values of the quantum numbers m and s are usually not indicated.

TABLE 10-1

SOME TYPICAL ATOMIC CONFIGURATIONS

Atom	Ground State	Typical Excited States
Li	$1s^22s$	$1s^22p$; $1s^23s$; $1s^23p$; $1s^23d$; $1s^24s$; $1s^24p$
Be	$1s^22s^2$	$1s^22s2p$; $1s^22s3s$; $1s^22s3p$; $1s^22s3d$; $1s^22s4s$; $1s^22s4p$
B	$1s^22s^22p$	$1s^22s2p^2$; $1s^22s2p3s$; $1s^22s^23s$
Ne	$1s^22s^22p^6$	$1s^22s^22p^53s$; $1s^22s^22p^53p$
Fe	$1s^22s^22p^63s^23p^63d^64s^2$	$1s^22s^22p^63s^23p^63d^54s^24p$

Most of the spectra ordinarily observed in the visible and ultraviolet are due to the excitation of one or, less often, two electrons in the least stable, outer orbitals of atoms and ions. The filled inner shells of orbitals of an atom are known as the *core*. Core electrons are only slightly affected by the transitions of outer electrons, but they can be excited by bombardment with high energy particles and by absorption of X-rays.

Since the core consists of more or less immutable filled shells, electronic configurations can be conveniently abbreviated by omitting the assignments of electrons to orbitals in the core. Thus the ground state of iron, $1s^22s^22p^63s^23p^63d^64s^2$, can be indicated by (core)$3d^64s^2$, or simply by $3d^64s^2$.

b. Classification of electronic repulsions in atoms

Three types of electronic repulsion are found when the perturbation theory is applied to the zero-order wave functions.

(*i*) *Interactions of core electrons with each other.* These interactions are relatively unaffected by the outer electrons. Therefore they are not of much importance in determining the optical spectrum of the chemical properties of the atom. Of course they have a major influence on the X-ray spectrum, which depends almost entirely on transitions involving core orbitals.

(*ii*) *Interactions of core electrons with outer electrons.* Since all of the spins in closed shells are paired, the total energy of interaction between the core electrons and the outer electrons cannot depend on the spin quantum numbers of the outer electrons. (For instance, compare the $1s-2s$ interactions in the $1s\alpha\ 1s\beta\ 2s\alpha$-configuration of Li with the same interactions in the $1s\alpha\ 1s\beta\ 2s\beta$-configuration. In reversing the spin of the $2s$-electron we merely replace a $1s\alpha-2s\alpha$ interaction by an identical $1s\beta-2s\beta$ interaction, and a $1s\beta-2s\alpha$ interaction by an identical $1s\alpha-2s\beta$ interaction. The two configurations therefore have the same energy.) Furthermore, since the core electrons move with speeds considerably different from those of the outer electrons, the Pauli forces between core and outer electrons will be small. Therefore the effect of the core on the outside electrons ought to be fairly close to the classical coulombic interaction between them, with penetration (that is, incomplete shielding) the predominant factor determining the energy. This will be discussed in more detail in Section c, below.

(*iii*) *Interactions of outer electrons with each other.* These interactions give rise to the so-called "multiplet structure," which will be discussed in Section B.

c. The effect of penetration on the energy of an orbital

The interactions of core electrons with outer electrons are best studied by means of the energy-level diagrams of atoms containing a single electron in addition to the closed shells (e.g., the alkali metals, Cu, Ag, Au, B, Al). If the shielding of the nucleus by the core were fully effective, the spectra of these atoms would be identical with that of hydrogen. The deviation from the hydrogen levels is therefore a direct measure of the effect of incomplete shielding or of penetration into the core.

The energy level diagram of lithium (Fig. 9-4) is typical. We see that all of the levels are slightly below the equivalent hydrogen levels. The *s*-levels show the greatest effect of penetration, the remaining levels showing decreasing effects as their angular momentum increases.

The effect of penetration can also be studied fairly well in atoms which in the ground state have two *s* electrons in their outermost orbitals (e.g., the alkaline earths, Zn, Cd, Hg). As one might expect, these elements have energy-level diagrams which closely resemble that of helium (compare Figs. 9-2, 3 and 6). Resonance effects complicate the picture here, but as we have seen, they may be eliminated to a first approximation by taking the average of the energies of the triplet and singlet states of a given configuration.

Penetration effects can also be isolated to a first approximation in certain other elements, but the interpretation is often confused by the existence of low-lying levels which have different electronic configurations, but the same total spin and angular momentum quantum numbers. As a result, resonance can occur between configurations, and the energies of the levels are shifted. Complications of this kind occur in some of the atoms mentioned above. In aluminium for instance, the rather stable $3s3p^2\,{}^2D$-level disturbs the positions of the $3s^2\,nd\,{}^2D$-levels, so that aluminum is not a good element for studying the effects of penetration on the energies of *d*-orbitals. A similar effect also occurs in the alkaline earth elements, where a relatively low-lying $(np)^2\,{}^1D$-level "perturbs" the $ms\,nd\,{}^1D$-level, so that it frequently lies below the corresponding $ms\,nd\,{}^3D$-level, in apparent contradiction to the predictions of the Hund multiplicity rule. These effects are called *series perturbations*, or *configuration interactions*. They are discussed in detail in White (*1*, Chapter 19) and Condon and Shortley (*2*, Chapter 15).

A survey of the energies of the various outer orbitals throughout the periodic system is shown in Fig. 10-1. The data from which this figure was constructed were obtained from the observed energy levels of appropriate atoms as outlined above. The zero of energy is the lowest state of the singly ionized atom. Figure 10-1 illustrates very clearly the penetration principle mentioned on page 297. In orbitals having a given principle quantum number, the order of decreasing stability is invariably *s*, *p*, *d*, *f*, and the differences in stability usually increase with increasing atomic number. The difference in stability of the *s*- and *d*-orbitals is so great that beyond nitrogen 4*s* becomes more stable than 3*d*. This, of course, is the reason for the late appearance of the transition metals in the periodic system. In fact, between magnesium and calcium the 3*d*-orbital is less stable even than 4*p*, so it is a little surprising that the transition elements do not occur even later in the system than they do. Once the 4*s* shell begins to fill, however, there is a very sudden increase in the stability of the 3*d*-orbital. This is because 4*s* and 3*d* overlap rather strongly, so that 3*d* is not well shielded from the nucleus by the 4*s* electrons. As a result the 3*d*-orbital is pulled in toward the nucleus and as more electrons are added they go into the *d*-shell. Since *d* electrons shield one another poorly, there is almost an autocatalytic effect on the stability; beyond copper the 3*d* electrons are once again more stable than the 4*s* electrons.

Similar abrupt drops in energy are observed with the 4*d*-, 5*d*- and 4*f*-orbitals, the drop with 4*f* being particularly sudden. In barium the 4*f*-level is just as high as it is in hydrogen. The single 5*d* electron in lanthanum evidently overlaps the 4*f*-orbital so effectively that penetration effects become very noticeable in lanthanum, and with the addition of more electrons 4*f* gains rapidly in stability. X-ray spectra show that beyond lead the 4*f*-level becomes even more stable than the 5*s*-level (see Fig. 10-4).

Exercises. (1) Why are closed shells so hard to break open? Is this a resonance effect, or is it entirely a penetration effect?

(2) The energy of 3*d* is almost unaffected by penetration before it begins to fill up, whereas the energies of 4*d* and 5*d* begin to drop long before they fill up. Suggest a reason for this. [Hint: Look at Fig. 7-9 on page 113 of White (*1*).]

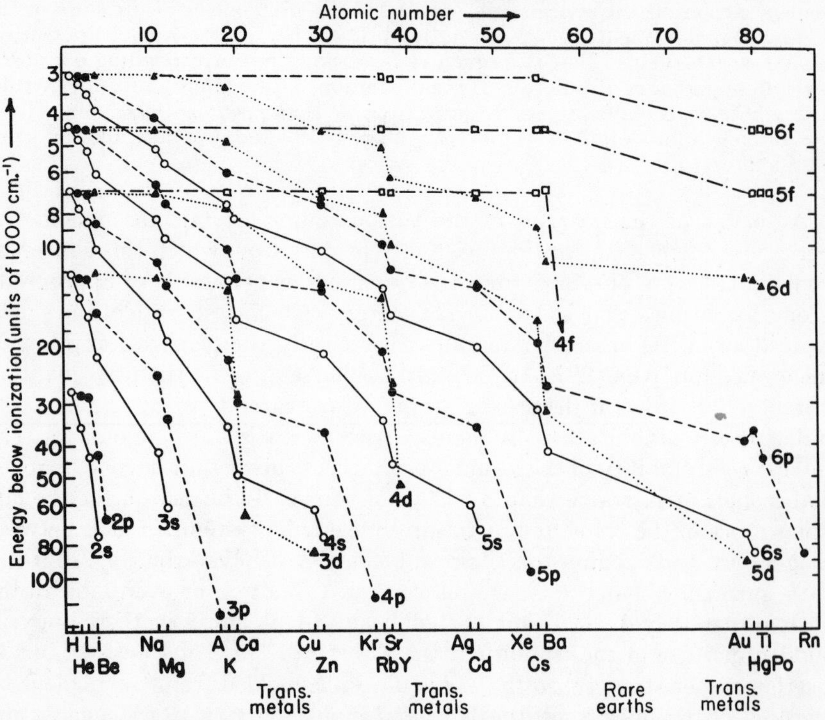

FIG. 10-1. Dependence of the energies of outer electrons on atomic number. The energies plotted are those required to remove electrons in the given orbitals from the neutral atoms, all other electrons in the atoms being in their lowest possible states. The atoms whose spectra have been used to obtain points on the curves are indicated along the abscissa. The scale of ordinates is logarithmic.

Energies of *s*-electrons ○———————○
Energies of *p*-electrons ●— — — — —●
Energies of *d*-electrons ▲· · · · · · · · · · · ▲
Energies of *f*-electrons □·—·—·—·—□

(3) Do you see any tendency for sudden drops in energy among the p-orbitals that might be accounted for in the same way that we have accounted for the breaks in the $3d$ and $4f$ curves?

(4) It is interesting that the sudden increase in the stability of the $4f$-orbital at lanthanum is accompanied by a decrease in energy of the higher f-levels. Suggest a reason for this. (It is also worth pointing out that the nf-levels of the elements around mercury have almost exactly the same energies as the $(n-1)f$-levels of the elements before cesium. There is a similar, but less striking, tendency for the d-levels to show this effect, too, after each transition series, and there is even some trace of this effect among the p-levels. Thus the np-level between calcium and zinc has nearly the same energy as the $(n\text{-}2)p$-level in hydrogen, the np-level between strontium and cadmium

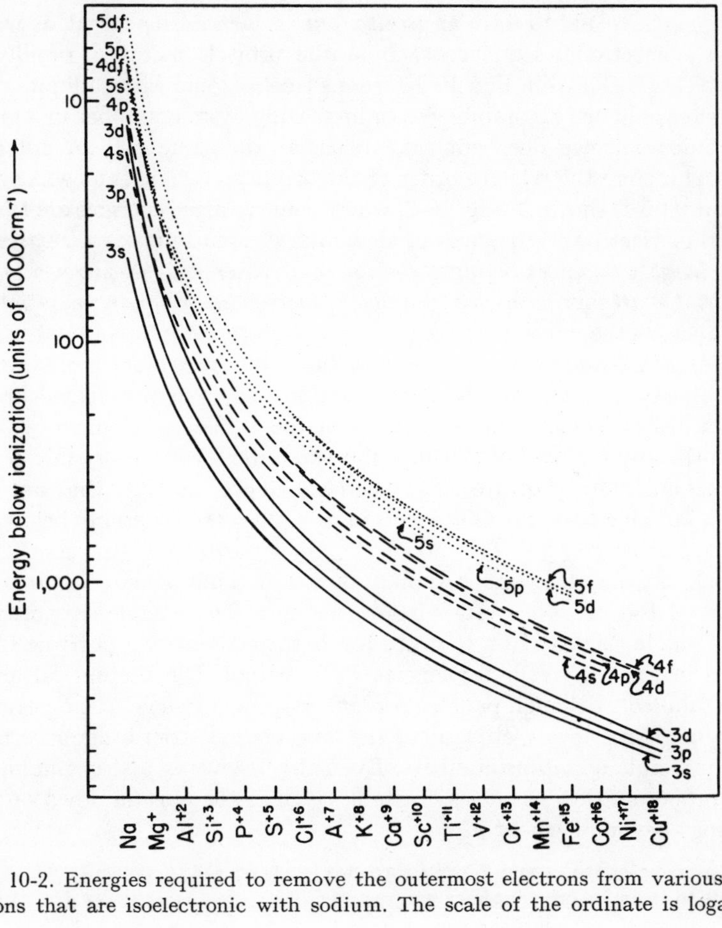

FIG. 10-2. Energies required to remove the outermost electrons from various orbitals in ions that are isoelectronic with sodium. The scale of the ordinate is logarithmic.

has about the same energy as the $(n-3)p$-level in hydrogen, and the lower np-levels between barium and mercury have roughly the same energies as the $(n-4)p$-levels in hydrogen. There seems to be no obvious reason for this interesting tendency.)

The data of Fig. 10-1 refer to the states of uncharged atoms. When atoms are subjected to the conditions of high voltage spark gaps, however, one or more electrons may be removed, producing singly or multiply charged ions, which can also exist in excited states. As one would expect, the energy level diagrams of such ions are qualitatively similar to those of the neutral atom containing the same number of electrons.

Systems having the same number of electrons are said to be *isoelectronic.* A series of atoms and ions such as H^-, He, Li^+, Be^{++}, etc., or C, N^+, O^{++}, F^{3+}, Ne^{4+}, etc., is said to form an *isoelectronic sequence.* In general, as we go down an isoelectronic sequence each of the orbitals increases rapidly in stability. This is shown in Fig. 10-2 for ions isoelectronic with sodium.

The increase in orbital stability with increasing atomic number in a given isoelectronic sequence does not take place at the same rate in different orbitals. Therefore the relative order of the orbitals changes somewhat with ionization. This is shown in Fig. 10-3, where some relative energies are traced in several isoelectronic sequences of elements. Evidently *if one increases the nuclear charge without changing the number of electrons, the energies of the orbitals tend to arrange themselves in order of their principle quantum numbers.* For instance, as the nuclear charge is increased in the sodium and potassium sequences, the $3d$- and $4d$-orbitals increase their stabilities much more rapidly than do the $4s$- and $5s$-orbitals. This effect is also clearly reflected in the ground states of isoelectronic sequences in the transition elements (Table 10-2). In the ground state of calcium, the two outer electrons are in the $4s^2$-configuration. In Sc^+ they are $4s3d$, whereas in Ti^{++} and V^{+++} they are $3d^2$. As shown in Table 10-2, the $4s3d$-levels in calcium lie considerably below the $3d^2$-levels, but on going to Ti^{++} and V^{+++}, their positions are reversed.

From X-ray spectra, one can find out something about the relative stabilities of the various orbitals inside the core. The energies required to extract a single electron fron the core levels throughout the periodic table are given in Fig. 10-4. The tendencies we have noted in the optical levels are here fulfilled. Although penetration effects persist (s-levels being always more stable than p-levels, etc.), all of the levels fall in strict order according to their principle quantum numbers. Even the $4f$-level is just beginning to find its proper station as nuclear instability rings the curtain down on the proceedings beyond element 92.*

* The splitting of the p- and d-levels into two levels on the right hand side of Fig. 10-4 is due to the spin–orbit interaction.

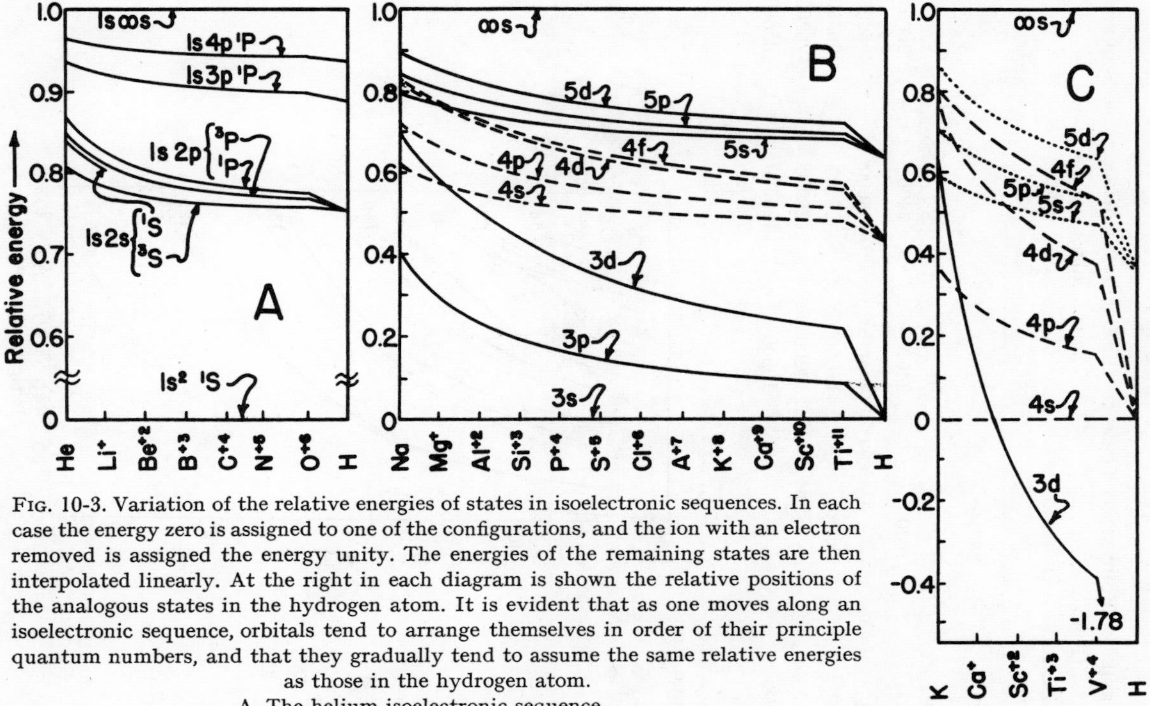

FIG. 10-3. Variation of the relative energies of states in isoelectronic sequences. In each case the energy zero is assigned to one of the configurations, and the ion with an electron removed is assigned the energy unity. The energies of the remaining states are then interpolated linearly. At the right in each diagram is shown the relative positions of the analogous states in the hydrogen atom. It is evident that as one moves along an isoelectronic sequence, orbitals tend to arrange themselves in order of their principle quantum numbers, and that they gradually tend to assume the same relative energies as those in the hydrogen atom.

A. The helium isoelectronic sequence.
B. The sodium isoelectronic sequence.
C. The potassium isoelectronic sequence.

TABLE 10-2

ENERGIES OF EQUIVALENT TERMS IN THE CALCIUM ISOELECTRONIC SEQUENCE (ENERGIES IN CM.$^{-1}$; $3s4d\ ^3D$ ARBITRARILY TAKEN AS ZERO)

Term Symbol	Ca	Sc^+	Ti^{++}	V^{+++}
$4s^2\ ^1S$	−20350*	+11600	—	—
$4s3d\ ^3D$	0	0*	0	0
^1D	1500	2400	3400	3600
$3d^2\ ^3F$	—	4800	−38000*	−96000*
^3P	28200	12000	−27700	−83300
^1G	—	14200	−23900	−78200
^1D	—	10800	−29800	−85000
^1S	—	—	−24200	—

* Ground state.

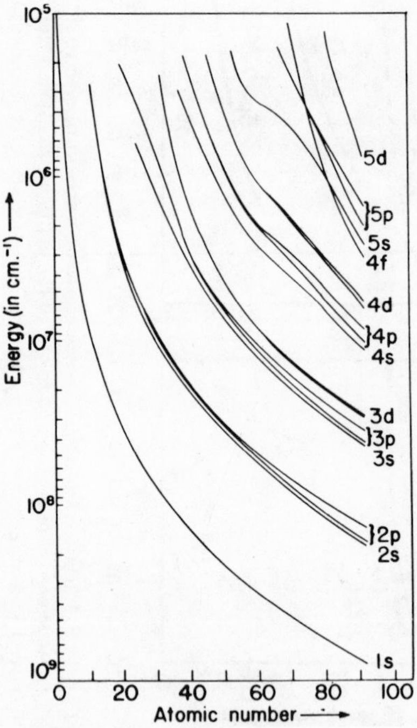

FIG. 10-4. Energies required to remove electrons from various core orbitals. (Data from M. Siegbahn, "Spektroskopie der Röntgenstrahlen," 2nd ed., Table 176. Springer, Berlin, 1931.)

Exercises. (1 An energy of 9.298×10^8 cm.$^{-1}$ is required to remove a 1*s* electron from a neutral uranium atom. Compare this with what you would expect for the removal of an electron from the U^{91+} ion and the U^{90+} ion on the basis of the theories presented earlier in this book for the hydrogen atom and the helium atom. Do the electrons in the 2*s*, 2*p*, etc., shells of uranium play much of a role in determining the energies of the innermost electrons?

(2) Can you explain the kinks in the 5*s* and 5*p* curves around atomic number 60–70, Fig. 10-4? [Hint: might the 5*s* and 5*p* orbits cross the 4*f* orbit? Could 4*f* have any shielding effect on them in spite of the fact that its energy is very much higher?]

Figures 10-1 and 10-4 can be joined together to give an over-all picture of the behavior of orbital energies throughout the periodic system. This is done schematically in Fig. 10-5, which may be regarded as a summary of the effects of penetration on orbital stability.

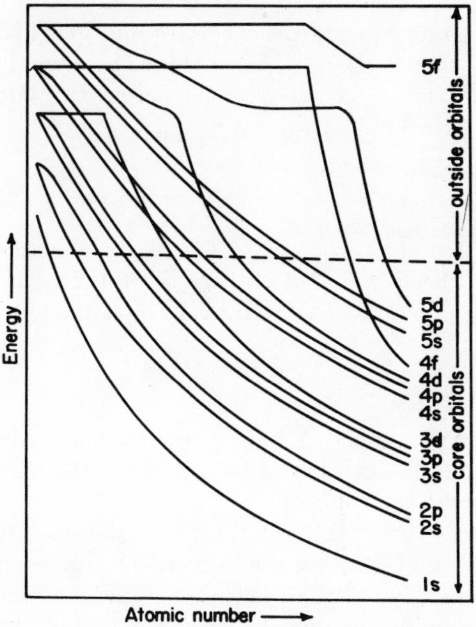

FIG. 10-5. Semi-schematic diagram of the variation of orbital energies with atomic number.

d. Limitations on the concept of an orbital energy

Orbital energies do not have an unequivocal meaning, so that we must be careful in using them. In the first place, the energy of an electron in an orbital depends on how many other electrons are present in the atom or ion, and on the orbitals they occupy. For instance, in beryllium, the energy of the $2s$-orbital in the configuration $2s^2 2p$ would not be the same as that of the same orbital in the configuration $2s2p^2$, since the shielding of a $2s$ electron by another $2s$ electron is not the same as the shielding of a $2s$ electron by a $2p$ electron. Similarly, the orbital energies in Fig. 10-2 correspond to ions in which only one electron occupies the shell in question, whereas Fig. 10-4 gives the energies required to remove a single electron from a completed shell. These energies are quite different.

In Figs. 10-1, 10-2, and 10-3, it is to be understood that the orbitals whose energies are given are always in the outside shells of the atom. In Fig. 10-4, on the other hand, the orbital is buried inside the core. Therefore, Fig. 10-5 is somewhat ambiguous.

If there is multiplet structure, a practical difficulty arises in defining an orbital energy, since the quantitative theory of multiplets is often confused by configuration interactions (see page 344). Therefore it is sometimes difficult to separate the effects of resonance from the purely Coulombic effects that determine orbital energies.

e. The Slater-Zener atomic orbitals

The perturbation theory is based on the use of hydrogen-like orbitals in the zero-order wave functions. These functions have the form

$$\psi_{nlm} = R_{nl}(r)\, e^{-Zr/n} Y_{lm}(\theta, \varphi)$$

where Z is the atomic number, Y_{lm} is a surface spherical harmonic, and $R_{nl}(r)$ is a polynomial in r containing $(n-l)$ terms, the highest power of r being r^{n-1}, regardless of the value of l.

In our discussion of the helium atom, it was mentioned that, because of the screening effects of orbitals on each other, the true wave functions are much less concentrated about the nucleus than would be expected from these hydrogen-like wave functions, in which all electronic repulsions are neglected. By replacing the true nuclear charge, Z, in the exponential term of the hydrogen-like orbitals with an effective nuclear charge, $(Z-s)$, the orbital can be made to spread out, away from the nucleus, and we ought to obtain a better representation of the true wave function. The constant s is known as the *screening constant*, and its value would be expected to vary from one orbital to another. It is, in effect, the average number of electrons spending their time in the region between the given orbital and the nucleus.

Using this approach Zener (*3*) and Slater (*4*) have set up simplified atomic orbitals that are widely used in making quantum chemical calculations. They showed that for many purposes only the term with the highest power of r in $R_{nl}(r)$ is of importance. Therefore they wrote their orbitals in the form

$$\psi = r^{n^*-1} e^{-(Z-s)r/n^*} Y_{lm}(\theta, \varphi)$$

where the constants n^* and s are chosen according to the following rules.

i. The value of n^* depends on the value of the true principle quantum number, n, of the orbital according to the following table:

n	=	1	2	3	4	5	6
n^*	=	1	2	3	3.7	4.0	4.2

ii. The screening constant, s, is found by arranging the orbitals in groups as follows:

$$(1s)(2s2p)(3s3p)(3d)(4s4p)(4d)(4f)(5s5p)(5d), \text{ etc.}$$

(that is, the s- and p-orbitals of given n are grouped together, but the d- and f-orbitals are separated). The value of s for a given orbital is then the sum of the following contributions:

a. Nothing from any electrons in groups outside that of the given orbital.

b. A contribution of 0.35 from each other electron in the same group, except that in the 1s-group the contribution is 0.30. (Compare this value for 1s electrons with the value $5/16 = 0.3125$ found by the variation method for helium and helium-like ions on page 289.)

c. For an s- or p-orbital, 0.85 from each electron whose total quantum number is less by one, and 1.00 from each electron further in.

d. For a d- or f-orbital, 1.00 from each electron inside the given group.

As an example, consider the neutral iron atom in the ground state, for which there are two 1s, eight 2s and 2p, eight 3s and 3p, six 3d, and two 4s electrons.

For a 1s-orbital, $s = 0.30$

2s or 2p-orbital, $s = 2 \times 0.85 + 7 \times 0.35 = 4.15$

3s or 3p-orbital, $s = 2 \times 1 + 8 \times 0.85 + 7 \times 0.35 = 10.25$

4s-orbital, $s = 2 \times 1 + 8 \times 1 + 14 \times 0.85 + 1 \times 0.35 = 23.25$

3d-orbital, $s = 2 \times 1 + 8 \times 1 + 8 \times 1 + 5 \times 0.35 = 19.75$

This system of selecting s and n^* was developed so as to give the best possible fit to the observed energies and sizes of atoms. The resulting orbitals, while much improved over the hydrogen-like orbitals and considerably simpler to work with because of the neglect of all but one term in the radial polynomial, do not reflect many of the finer points of Figs. 10-1 to 10-5. For instance, they do not lead to any difference in energy for *s*- and *p*-orbitals. Furthermore, any attempt to represent orbitals using screening constants is subject to the inadequacies mentioned on pages 289-290.

Improved Slater-type functions for the orbitals of the atoms in the first row of the periodic table are given by Duncanson and Coulson (*5*).

Boys (*6*) has proposed the use of orbitals based on error functions, $\exp(-ar^2)$, which have the advantage that they give more readily evaluated integrals. He has also perfected a very effective method for constructing atomic wave functions out of combinations of simple orbitals of the type $r^n e^{-nr} S_{lm}(\theta,\varphi)$ where m and n are integers and $S_{lm}(\theta,\varphi)$ is a surface spherical harmonic. The latter method is akin to the systematic introduction of configuration interactions into the wave function, and this more than compensates for the simplicity of the basic orbitals.

Exercise. Slater showed that the energy of an orbital in atomic units should be given approximately by $-[(Z - s)/n^*]^2$. Use this expression to calculate the energies of the 3*d*- and 4*f*-orbitals in neutral atoms throughout the periodic table. How do they agree with the observed behavior shown in Figs. 10-1 and 10-4? Can you suggest changes in the Slater-Zener system which would improve the agreement with these figures?

f. The Hartree self-consistent field method

Suppose that we write the wave function of an atom containing N electrons in the form of a product of one-electron orbitals,

$$\psi = \phi_1(1)\phi_2(2) \cdots \phi_N(N) \tag{A-3}$$

where $\phi_1, \phi_2, \cdots, \phi_N$ are the orbitals and $\phi_1(1)$ means that we put the coordinates of electron no. 1 in orbital ϕ_1, etc. Hartree has developed an interesting method of deducing the orbitals that will make the function in (A-3) give the best possible representation of the correct wave function of the atom.

The quantity $\phi_1^2(1)$ is the probability distribution of electron no. 1 in the atom, and similarly for $\phi_2^2(2)$, etc. If electron no. 1 is placed at a definite point in space, with coordinates x_1, y_1, z_1, then the average electrostatic interaction between electrons no. 1 and no. 2 is (in atomic units)

$$V_1(x_1, y_1, z_1) = \iiint \frac{\phi_2^2(2)}{r_{12}} dx_2 dy_2 dz_2 \tag{A-4}$$

where r_{12} is the distance between the electrons, and the integral is over all positions of electron no. 2. Similar contributions to the potential energy of electron no. 1 are made by each of the other electrons in the atom. In general, the total potential energy of the jth electron can be written

$$V_j\,(x_j, y_j, z_j) = -Z/r_j + \sum_{i \neq j} \iiint \frac{\phi_i^2(i)}{r_{ij}}\, dx_i dy_i dz_i \tag{A-5}$$

the first term on the right being the potential energy due to the nucleus, whose atomic number is Z. Similar equations can be written for the potential energies of the other electrons.

The wave function of an electron moving in the potential field (A-5) is given by the solution of the Schroedinger equation,

$$-(1/2)\,\nabla_j^2\phi_j + V_j\phi_j = E_j\phi_j \tag{A-6}$$

where E_j is the energy of the electron in orbital ϕ_j. Hartree's method consists in taking the solutions of (A-6) to form the wave function (A-3). It is clear, however, that Equation (A-6) cannot be solved until (A-5) has been evaluated, and we cannot evaluate (A-5) until we know the solutions ϕ_j of (A-6). This vicious circle may be broken by making a preliminary reasonable assumption of the forms of the functions ϕ_j in (A-5) (for instance, we might use Slater orbitals). The resulting potentials V_j may then be used to find improved functions ϕ_j by means of Equation (A-6). These improved functions may then be reintroduced into (A-5) to obtain an improved estimate of the potential function V_j. The process may be repeated until the functions obtained on successive cycles do not change appreciably. The resulting potential function V_j is then said to be a *self-consistent field* for electron j. The total energy of the atom or ion is

$$E = \sum_{j=1}^{N} E_j$$

where the E_j's are the eigenvalues of the Equations (A-6).

The method has been applied to numerous atoms and ions. In principle it may also be used to calculate the energies of molecular systems, but the calculations are more difficult here. For a review, see Hartree (*7*). A compilation of self-consistent field calculations will be found in Landolt and Bornstein (*8*, Vol. *1*, Part 1).

There are two serious shortcomings in a wave function of the form (A-3). In the first place, such a wave function takes no account of the tendency of electrons to keep out of each other's way (correlation effect, see page 287).

Thus, according to (A-3), the probability of finding electron no. 1 at a given point in space does not depend on the locations of the other electrons; if electrons nos. 2, 3, 4, etc. were already at the given point, this would have no effect on the probability of finding electron no. 1 at this point. In the correct wave function, however, we would expect a considerable reduction in the probability of finding an electron at a point that is already occupied by one or more electrons.

The second shortcoming of (A-3) is the neglect of the possibility of electronic exchange between orbitals – a phenomenon that may have an important effect on the energy, as we have seen in the discussion of the excited states of the helium atom (although it does not play a role in the ground state of helium). Fock has shown that Hartree's method may be modified to take into account the exchange between orbitals [see, for instance, Mott and Sneddon (*9*, pp. 122 and 135)]. This Hartree-Fock method increases the labor involved in arriving at the wave function, but the result is somewhat better than that obtained from (A-3).

A useful approximation to the electron density distribution in space (i.e., the square of the wave function) may be obtained by means of the Fermi-Thomas statistical method (see general references at the end of this chapter for details). The method can, in principle, be extended to molecules, but this is not easy. The best results are obtained with atoms having high atomic numbers.

B. Multiplet Structure

In the previous section we were concerned with the classical Coulombic interactions between the electron clouds represented by atomic orbitals. If there is degeneracy in the zero-order wave functions of an atomic configuration, however, exchange interactions make additional contributions to the energy. These interactions must be expected whenever atoms do not have closed shells. We have already encountered a simple example in the excited states of helium, where we found that each electronic configuration gives rise to states of two multiplicities – a singlet and a triplet. A similar behavior is found with nearly all partially filled shells. The resulting pattern of energy levels is known as the *multiplet structure* of an electronic configuration.

The detailed quantum mechanical theory of multiplet structure is rather complicated, though not inherently difficult. It is based on the same general procedure as that we used in the excited states of helium. The secular determinant is set up and, in order to bring it into diagonal form, use is made of the commutation of the spin and orbital angular momentum operators with the electronic repulsion terms of the Hamiltonian. We shall illustrate this procedure with the $1s^22s^22p^2$-configuration of the carbon atom. The method we shall use is due to Slater.

A much simpler, though qualitative, approach to the understanding of the principles of multiplet structure is provided by the so-called *vector model*. This will be discussed after we have worked through some of the detailed theory.

a. Detailed theory for the 2p²-configuration

The ground state of the carbon atom has the configuration $1s^22s^22p^2$. Fifteen zero-order wave functions can be written for this configuration. Since there are six electrons, each wave function should have the form of a sixth order determinant, as for example

$$\phi = (1/6!)^{1/2} \mid (1s\alpha)\ (1s\beta)\ (2s\alpha)\ (2s\beta)\ (2p_1\alpha)\ (2p_0\beta) \mid$$

But we have seen that the interactions of filled shells of electrons with the two outer p electrons is essentially Coulombic. Therefore they cannot give rise to any multiplet structure, which in this instance arises solely from the interactions of the two $2p$ electrons. As far as the relative spacing of the multiplet energies is concerned, the carbon atom can be treated as a two-electron problem, the Coulombic effect of the core electrons being taken into account indirectly by means of appropriate screening constants in the $2p$-orbitals.

The fifteen zero-order wave functions for the configuration $2p^2$ that are consistent with the Pauli principle are given in Table 10-3. These wave functions are to be subjected to the perturbation $1/r_{12}$ representing the electrostatic repulsion of the two electrons. Application of the perturbation theory in the usual way will thus lead to a fifteenth order determinant. We look to the operators M_e^2, M_{ze}, S_e^2, and S_{ze} for assistance in simplifying the determinant, since all of these operators commute with $1/r_{12}$.

Each of the fifteen functions is an eigenfunction of both M_{ze} and S_{ze}. For example,

$$M_{ze}\phi_1 = -i\hbar\,(\partial/\partial\varphi_1 + \partial/\partial\varphi_2)\,(1/\sqrt{2})\,[(2p_1\alpha)_1\,(2p_1\beta)_2 - (2p_1\beta)_1\,(2p_1\alpha)_2] = 2\hbar\phi_1$$

since $(\partial/\partial\varphi)\,(2p_1) = i(2p_1)$. The eigenvalues $\mathscr{M}_z$ and $\mathscr{S}_z$ of these operators are given alongside each function in Table 10-3. On the basis of these eigenvalues and the commutation theorem (page 144), we obtain a secular equation of the following form (columns 9 and 10 and rows 9 and 10 having been interchanged to give more symmetry to the determinant)

TABLE 10-3

ZERO-ORDER WAVE FUNCTIONS FOR THE CONFIGURATION $2p^2$

	$\mathscr{M}_z$	$\mathscr{S}_z$	L	S
$\phi_1 = (1/\sqrt{2}) \mid (2p_1\alpha)\,(2p_1\beta) \mid$	2	0	2	0
$\phi_2 = (1/\sqrt{2}) \mid (2p_1\alpha)\,(2p_0\alpha) \mid$	1	1	1	1
$\phi_3 = (1/\sqrt{2}) \mid (2p_1\alpha)\,(2p_0\beta) \mid$	1	0	*(1)	*(1)
$\phi_4 = (1/\sqrt{2}) \mid (2p_1\beta)\,(2p_0\alpha) \mid$	1	0	*(2)	*(0)
$\phi_5 = (1/\sqrt{2}) \mid (2p_1\beta)\,(2p_0\beta) \mid$	1	−1	1	1
$\phi_6 = (1/\sqrt{2}) \mid (2p_1\alpha)\,(2p_{-1}\alpha) \mid$	0	1	1	1
$\phi_7 = (1/\sqrt{2}) \mid (2p_1\alpha)\,(2p_{-1}\beta) \mid$	0	0	*(1)	*(1)
$\phi_8 = (1/\sqrt{2}) \mid (2p_1\beta)\,(2p_{-1}\alpha) \mid$	0	0	*(0)	*(0)
$\phi_9 = (1/\sqrt{2}) \mid (2p_1\beta)\,(2p_{-1}\beta) \mid$	0	−1	1	1
$\phi_{10} = (1/\sqrt{2}) \mid (2p_0\alpha)\,(2p_0\beta) \mid$	0	0	*(2)	*(0)
$\phi_{11} = (1/\sqrt{2}) \mid (2p_0\alpha)\,(2p_{-1}\alpha) \mid$	−1	1	1	1
$\phi_{12} = (1/\sqrt{2}) \mid (2p_0\alpha)\,(2p_{-1}\beta) \mid$	−1	0	*(1)	*(1)
$\phi_{13} = (1/\sqrt{2}) \mid (2p_0\beta)\,(2p_{-1}\alpha) \mid$	−1	0	*(2)	*(0)
$\phi_{14} = (1/\sqrt{2}) \mid (2p_0\beta)\,(2p_{-1}\beta) \mid$	−1	−1	1	1
$\phi_{15} = (1/\sqrt{2}) \mid (2p_{-1}\alpha)\,(2p_{-1}\beta) \mid$	−2	0	2	0

* Not eigenfunctions of M^2e or S^2e. Numbers in parentheses are the values for the linear combinations given in Table 10-4.

$$
\begin{vmatrix}
* & 0 & 0 & 0 & 0 & 0 & 0 & 0 & 0 & 0 & 0 & 0 & 0 & 0 & 0 \\
0 & * & 0 & 0 & 0 & 0 & 0 & 0 & 0 & 0 & 0 & 0 & 0 & 0 & 0 \\
0 & 0 & * & * & 0 & 0 & 0 & 0 & 0 & 0 & 0 & 0 & 0 & 0 & 0 \\
0 & 0 & * & * & 0 & 0 & 0 & 0 & 0 & 0 & 0 & 0 & 0 & 0 & 0 \\
0 & 0 & 0 & 0 & * & 0 & 0 & 0 & 0 & 0 & 0 & 0 & 0 & 0 & 0 \\
0 & 0 & 0 & 0 & 0 & * & 0 & 0 & 0 & 0 & 0 & 0 & 0 & 0 & 0 \\
0 & 0 & 0 & 0 & 0 & 0 & * & * & * & 0 & 0 & 0 & 0 & 0 & 0 \\
0 & 0 & 0 & 0 & 0 & 0 & * & * & * & 0 & 0 & 0 & 0 & 0 & 0 \\
0 & 0 & 0 & 0 & 0 & 0 & * & * & * & 0 & 0 & 0 & 0 & 0 & 0 \\
0 & 0 & 0 & 0 & 0 & 0 & 0 & 0 & 0 & * & 0 & 0 & 0 & 0 & 0 \\
0 & 0 & 0 & 0 & 0 & 0 & 0 & 0 & 0 & 0 & * & 0 & 0 & 0 & 0 \\
0 & 0 & 0 & 0 & 0 & 0 & 0 & 0 & 0 & 0 & 0 & * & * & 0 & 0 \\
0 & 0 & 0 & 0 & 0 & 0 & 0 & 0 & 0 & 0 & 0 & * & * & 0 & 0 \\
0 & 0 & 0 & 0 & 0 & 0 & 0 & 0 & 0 & 0 & 0 & 0 & 0 & * & 0 \\
0 & 0 & 0 & 0 & 0 & 0 & 0 & 0 & 0 & 0 & 0 & 0 & 0 & 0 & *
\end{vmatrix} = 0
$$

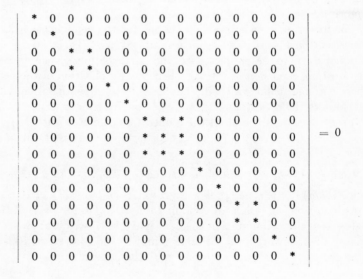

The secular equation is thus already almost in the diagonal form. The two remaining second order determinants and the single third order determinant could be solved without too much trouble, but it is more instructive to diagonalize the determinant further by means of M_e^2 and S_e^2.

Not all of the functions ϕ_1 to ϕ_{15} are eigenfunctions of M_e^2 and S_e^2. By straightforward operation in the usual way we find that (*cf.* Equations (A-24) of Chapter 8)

$$M_e^2\phi_1 = [(\nu_1 + \nu_2)(\mu_1 + \mu_2) + \hbar M_{ze} + M_{ze}^2](1/\sqrt{2})[(2p_1\alpha)_1(2p_1\beta)_2 - (2p_1\beta)_1(2p_1\alpha)_2]$$
$$= \hbar^2(2 + 4)\phi_1 = 2(2 + 1)\hbar^2\phi_1$$

$$M_e^2\phi_2 = [(\nu_1 + \nu_2)(\mu_1 + \mu_2) + \hbar M_{ze} + M_{ze}^2](1/\sqrt{2})[(2p_1\alpha)_1(2p_0\beta)_2 - (2p_0\alpha)_1(2p_1\alpha)_2]$$
$$= \sqrt{2}\hbar(\nu_1 + \nu_2)(1/\sqrt{2})[(2p_1\alpha)_1(2p_1\alpha)_2 - (2p_1\alpha)_1(2p_1\alpha)_2] + 2\hbar^2\phi_2$$
$$= 1(1 + 1)\hbar^2\phi_2$$

$$M_e^2\phi_3 = \sqrt{2}\hbar(\nu_1 + \nu_2)(1/\sqrt{2})[(2p_1\alpha)_1(2p_1\beta)_2 - (2p_1\beta)_1(2p_1\alpha)_2] + 2\hbar^2\phi_3$$
$$= 2\hbar^2(1/\sqrt{2})[(2p_0\alpha)_1(2p_1\beta)_2 + (2p_1\alpha)_1(2p_0\beta)_2 - (2p_0\beta)_1(2p_1\alpha)_2 - (2p_1\beta)_1(2p_0\alpha)_2] + 2\hbar^2\phi_3$$
$$= 2\hbar^2(\phi_3 - \phi_4) + 2\hbar^2\phi_3 = 2\hbar^2(2\phi_3 - \phi_4)$$

$$M_e^2\phi_4 = 2\hbar^2(2\phi_4 - \phi_3)$$

$$M_e^2\phi_5 = 1(1 + 1)\hbar^2\phi_5$$

$$M_e^2\phi_6 = \sqrt{2}\hbar(\nu_1 + \nu_2)(1/\sqrt{2})[(2p_1\alpha)_1(2p_0\alpha)_2 - (2p_0\alpha)_1(2p_1\alpha)_2]$$
$$= 2\hbar^2(1/\sqrt{2})[(2p_0\alpha)_1(2p_0\alpha)_2 + (2p_1\alpha)_1(2p_{-1}\alpha)_2 - (2p_{-1}\alpha)_1(2p_1\alpha)_2 - (2p_0\alpha)_1(2p_0\beta)_2]$$
$$= 1(1 + 1)\hbar^2\phi_6$$

$$M_e^2\phi_7 = 2\hbar^2(\phi_7 + \phi_{10})$$

$$M_e^2\phi_8 = 2\hbar^2(-\phi_8 + \phi_{10})$$

$$M_e^2\phi_9 = 1(1 + 1)\hbar^2\phi_9$$

$$M_e^2\phi_{10} = 2\hbar^2(\phi_7 - \phi_8 + \phi_{10})$$

$$M_e^2\phi_{11} = 1(1 + 1)\hbar^2\phi_{11}$$

$$M_e^2\phi_{12} = 2\hbar^2(2\phi_{12} - \phi_{13})$$

$$M_e^2\phi_{13} = 2\hbar^2(2\phi_{13} - \phi_{12})$$

$$M_e^2\phi_{14} = 1(1 + 1)\hbar^2\phi_{14}$$

$$M_e^2\phi_{15} = 2(2 + 1)\hbar^2\phi_{15}$$

$$S_e^2\phi_1 = 0\cdot\phi_1$$

$$S_e^2\phi_2 = [(\tau_1 + \tau_2)(\sigma_1 + \sigma_2) + \hbar S_{ze} + S_{ze}^2](1/\sqrt{2})[(2p_1\alpha)_1(2p_0\alpha)_2 - (2p_0\alpha)_1(2p_1\alpha)_2]$$
$$= 1(1+1)\hbar^2\phi_2$$
$$S_e^2\phi_3 = \hbar(\tau_1 + \tau_2)(1/\sqrt{2})[(2p_1\alpha)_1(2p_0\alpha)_2 - (2p_0\alpha)_1(2p_1\alpha)_2]$$
$$= \hbar^2(\phi_3 + \phi_4)$$
$$S_e^2\phi_4 = \hbar^2(\phi_3 + \phi_4)$$
$$S_e^2\phi_5 = 1(1+1)\hbar^2\phi_5$$
$$S_e^2\phi_6 = 1(1+1)\hbar^2\phi_6$$
$$S_e^2\phi_7 = \hbar^2(\phi_7 + \phi_8)$$
$$S_e^2\phi_8 = \hbar^2(\phi_7 + \phi_8)$$
$$S_e^2\phi_9 = 1(1+1)\hbar^2\phi_9$$
$$S_e^2\phi_{10} = 0\cdot\phi_{10}$$
$$S_e^2\phi_{11} = 1(1+1)\hbar^2\phi_{11}$$
$$S_e^2\phi_{12} = \hbar^2(\phi_{12} + \phi_{13})$$
$$S_e^2\phi_{13} = \hbar^2(\phi_{12} + \phi_{13})$$
$$S_e^2\phi_{14} = 1(1+1)\hbar^2\phi_{14}$$
$$S_e^2\phi_{15} = 0\cdot\phi_{15}$$

Thus of the fifteen zero-order functions, ϕ_3, ϕ_4, ϕ_7, ϕ_8, ϕ_{10}, ϕ_{12}, and ϕ_{13} are not eigenfunctions of both M_e^2 and S_e^2. By inspection one finds, however, that the linear combinations of these functions given in Table 10-4 are eigen-

TABLE 10-4

EIGENFUNCTIONS OF M^2 AND S^2 FOR THE CONFIGURATION $2p^2$

	$\mathscr{M}_z$	$\mathscr{S}_z$	L	S
$\phi'_3 = (1/\sqrt{2})(\phi_3 + \phi_4)$	1	0	1	1
$\phi'_4 = (1/\sqrt{2})(\phi_3 - \phi_4)$	1	0	2	0
$\phi'_7 = (1/\sqrt{2})(\phi_7 + \phi_8)$	0	0	1	1
$\phi'_8 = (1/\sqrt{3})(\phi_7 - \phi_8 - \phi_{10})$	0	0	0	0
$\phi'_{10} = (1/\sqrt{6})(\phi_7 - \phi_8 + 2\phi_{10})$	0	0	2	0
$\phi'_{12} = (1/\sqrt{2})(\phi_{12} + \phi_{13})$	−1	0	1	1
$\phi'_{13} = (1/\sqrt{2})(\phi_{12} - \phi_{13})$	−1	0	2	0

functions of these operators. [A systematic method of finding the proper linear combinations is given by Eyring, Walter, and Kimball (*10*, p. 138).] The values of L and S for these functions are introduced into Table 10-3 in parentheses. It is evident that if the primed ϕ's are used instead of the original ϕ's, the secular equation will be completely diagonalized, since no two ϕ's have the same set of the four eigenvalues $\mathcal{M}_z$, $\mathcal{S}_z$, L, and S. Therefore we now have the correct zero-order wave functions.

There are evidently five ^1D-states (that is, states with $L = 2$, $S = 0$; *viz.*, ϕ_1, ϕ'_4, ϕ'_{10}, ϕ'_{13}, ϕ_{15}), nine ^3P-states ($L = 1$, $S = 1$; ϕ_2, ϕ'_3, ϕ_5, ϕ_6, ϕ'_7, ϕ_9, ϕ_{11}, ϕ_{12}, ϕ_{14}), and one ^1S-state ($L = 0$, $S = 0$; ϕ'_8).

The five ^1D-states differ only in the orientations of their axis of total orbital angular momentum. We may therefore expect that all of these ^1D-states have the same energy. This is in fact the case, as we shall now show.

All of the ^1D wave functions can be converted into one another by the operation $\mu_e = \mu_1 + \mu_2$ or $\nu_e = \nu_1 + \nu_2$. For instance

$$(2\hbar)^{-1} \quad \mu_e\phi'_4 = \phi_1$$
$$(\sqrt{6}\hbar)^{-1} \mu_e\phi'_{10} = \phi'_4$$
$$(\sqrt{6}\hbar)^{-1} \mu_e\phi'_{13} = \phi'_{10}$$
$$(2\hbar)^{-1} \quad \mu_e\phi_{15} = \phi'_{13}$$

Exercises. (1) Prove these equations by direct operation.

(2) Prove that if $\phi(L, \mathcal{M})$ is an eigenfunction of M_e^2 and M_{ze} with eigenvalues $L(L+1)\hbar^2$ and $\mathcal{M}\hbar$, respectively, then $\mu_e\phi$ and $\nu_e\phi$ are eigenfunctions of M_e^2 with eigenvalues $L(L+1)\hbar^2$, and eigenfunctions of M_{ze} with eigenvalues $(\mathcal{M}+1)\hbar$ and $(\mathcal{M}-1)\hbar$, respectively. [Hint: Make use of the commutation rules for M_{xe}, M_{ye}, and M_{ze} noting that $M_{ze}(M_{xe} \pm iM_{ye}) = (M_{xe} \pm iM_{ye})(M_{ze} \pm \hbar)$.]

(3) Prove that

$$\phi(L, \mathcal{M}+1) = \{\hbar^2[L(L+1) - \mathcal{M}(\mathcal{M}+1)]\}^{-1/2} \mu_e\phi(L, \mathcal{M})$$

and

$$\phi(L, \mathcal{M}-1) = \{\hbar^2[L(L+1) - \mathcal{M}(\mathcal{M}-1)]\}^{-1/2} \nu_e\phi(L, \mathcal{M})$$

where the ϕ's are normalized eigenfunctions of M_e^2 and M_{ze}. [Hint: Find the normalizing factor for $\mu_e\phi$ and $\nu_e\phi$.]

(Similar relations hold, of course, for σ_e and τ_e. We may note in passing that operation by μ_e and ν_e in this way offers another systematic way of finding the linear combinations of zero-order wave functions which diagonalize the secular equation. Once one finds a single eigenfunction belonging to a given value of L and S, all of the other wave functions belonging to this L and S can be found simply by operating with μ_e, ν_e, σ_e, and τ_e.)

Using these expressions for the calculation of the first order perturbation of state ϕ_1, we find

$$E'_1 = \int \phi_1^* \frac{1}{r_{12}} \phi_1 d\tau = \frac{1}{4\hbar^2} \int (\mu_e^* \phi'_4{}^*) \frac{1}{r_{12}} (\mu_e \phi'_4) d\tau$$

$$= \frac{1}{4\hbar^2} \left[\int (M_{xe}{}^* \phi'_4{}^*) \frac{1}{r_{12}} (\mu_e \phi'_4) d\tau - i \int (M_{ye}{}^* \phi'_4{}^*) \frac{1}{r_{12}} (\mu_e \phi'_4) d\tau \right]$$

Since $M_{xe}{}^*$ and $M_{ye}{}^*$ are Hermitian, this gives

$$E'_1 = \frac{1}{4\hbar^2} \left[\int \phi'_4{}^* \frac{1}{r_{12}} (M_{xe} \mu_e \phi'_4) d\tau - i \int \phi'_4 \frac{1}{r_{12}} (M_{ye} \mu_e \phi'_4) d\tau \right]$$

$$= \frac{1}{4\hbar^2} \int \phi'_4{}^* \frac{1}{r_{12}} \nu_e \mu_e \phi'_4 d\tau = \frac{1}{4\hbar^2} \int \phi'_4{}^* \frac{1}{r_{12}} (M_e{}^2 - M_{ze}{}^2 - \hbar M_{ze}) \phi'_4 d\tau$$

$$= \frac{1}{4\hbar^2} \int \phi'_4{}^* \frac{1}{r_{12}} (2 \cdot 3 - 1 - 1) \phi'_4 d\tau = \int \phi'_4{}^* \frac{1}{r_{12}} \phi'_4 d\tau$$

Thus the perturbation energy for the state ϕ'_4 is the same as that for state ϕ_1. In the same way, all of the other ^1D-states can be shown to have the same energy as ϕ_1.

The nine ^3P-states differ only in the orientations of the axes of total spin and total orbital angular momentum, there being three possible orientations of each. Since we have included nothing in our Hamiltonian that depends on the spin, the three spin orientations should result in the same energy. Furthermore, the three orbital orientations should have the same energy for the same reason that all of the ^1D-states have the same energy. Therefore the nine ^3P-states form a single, nine-fold degenerate energy level.

Thus we conclude that the configuration $2p^2$ should give us three energy levels: a nine-fold degenerate ^3P-level, a five-fold degenerate ^1D-level, and a nondegenerate ^1S-level. The relative energies of these levels can be computed by evaluation of the integrals. This tedious procedure is performed in Eyring, Walter and Kimball (*10,* pages 143-151) and in Condon and Shortley (*2,* pages 197*ff.*). We shall content ourselves with a brief discussion of the results here.

Hund's multiplicity rule would lead us to expect that the ^3P-state, which is the state of highest multiplicity, would be the most stable. This is in fact observed (see Fig. 9-5). Thus the ground state of the carbon atom is $1s^2 2s^2 2p^2\ ^3P$. The next most stable level is the ^1D and the least stable state of the p^2-configuration is ^1S. This is a consequence of another rule proposed by Hund: *Among levels having the same electronic configuration and the same multiplicity, the most stable level is the one that has the largest orbital angular momentum* (*Hund's angular momentum rule*). The physical factor responsible for this rule is this: a large total orbital angular momentum is an indication that the electrons tend to move about the nucleus in the same direction. They are therefore better able to avoid collisions with one another. The same factor operates in a crowd of people moving about in a field. If the people move in random directions, they will make many contacts with each other, but if they form a parade and all march in the same direction, the number of contacts is much reduced.

Exercise. Expand the determinants in ϕ_1, ϕ'_3, and ϕ'_8 and show that the probability that two electrons are in the same place is much greater in ϕ_1 and ϕ'_8 than in ϕ'_3.

The detailed calculation of the energies of these states leads to the prediction that the ratio

$$R = [E(^1S) - E(^1D)] \,/\, [E(^1D) - E(^3P)]$$

should have the value 1.500, where $E(^1S)$, $E(^1D)$, and $E(^3P)$ are the energies of the three multiplet levels of p^2. This prediction is independent of the form of the radial part of the orbital used in the evaluation of the various integrals encountered. The observed values of R for these levels in a number of atoms

TABLE 10-5

RELATIVE SPACINGS OF THE ^1S, ^1D, AND ^3P LEVELS IN ATOMS HAVING THE CONFIGURATIONS np^2 AND np^4

Atom	R	Atom	R	Atom	R	Atom	R
C	1.128	P^{9+}	1.171	Sn	1.39	O	1.13
N^+	1.141	Si^{2+}	−3.93	La	18.43	F^+	1.155
O^{2+}	1.144	P^{3+}	28.5	Pb	0.63	Ne^{2+}	1.170
F^{3+}	1.145	Si	1.49	Be	−4.6	Na^{3+}	1.168
Ne^{4+}	1.149	P^+	1.50	B^+	7.4	Mg^{4+}	1.164
Na^{5+}	1.153	S^{2+}	1.51	C^{2+}	4.35	Al^{5+}	1.169
Mg^{6+}	~1.14	Cl^{3+}	1.49	N^{3+}	3.49	Si^{6+}	1.172
Al^{7+}	1.16	Ca	−0.014	O^{4+}	3.18	P^{7+}	1.181
Si^{8+}	1.162	Ge	1.50	F^{5+}	2.91	S^{8+}	1.190
						Cl^{9+}	~1.24

and ions are shown in Table 10-5. In all instances, except that of Ca, Si^{++}, and Be, the energies fall in the order ^3P, ^1D, ^1S, as predicted. Aside from this, the agreement is at best only fair; it shows that this method is usually able to make but rough quantitative predictions of the separations of multiplet levels.

Condon and Shortley (*2*, Chapter 7) discuss many further examples of the Slater method as applied to other configurations. The agreement is generally quite satisfactory in that the order of stability of the multiplet levels is as predicted, and even the relative spacings come out fairly well.

The most serious discrepancies of the Slater method can be accounted for by configuration interactions. For instance, in calcium there is a low-lying $4s4d\,^1D$-level which, since it has the same spin and orbital angular momentum, is able to resonate with the $4p^2\,^1D$-level. Since resonating levels "repel" one another (*cf.* page 143), the $4s4d\,^1D$-level is stabilized; it actually lies below the $4s4d\,^3D$-level, in contradiction to the Hund multiplicity rule. On the other hand, the energy of the $4p^2\,^1D$-level is raised, so that it has a higher energy than the $4p^2\,^1S$-level, in contradiction to the Hund angular momentum rule. Similarly, in the isoelectronic sequence from Be through F^{5+}, there is a low-lying $2s2p\,^3P$-level that raises the $2p^2\,^3P$-levels and causes the ratio R to be too large. Another factor leading to discrepancies is the spin-orbit interaction; this is important in atoms of high atomic number and probably accounts for the poor agreement found with lead.

Exercise. If we had included the core orbitals in our ϕ's, would we have obtained the same number of zero-order wave functions as are given in Table 10-3? Would they behave in the same way with respect to M_e^2, S_e^2, S_{ze} and M_{ze}? Are we justified in having neglected the core electrons?

b. A simple method for finding multiplet structures

A simple method is available for the prediction of the multiplets that can occur in a given configuration. It is based on the fact (not proven, but made plausible by the results of the previous section) that if a term with total angular momentum quantum number L occurs in a configuration, then states having all values of $\mathscr{M}_{ze}$ from $-L$ to $+L$ will be found. Similarly, if a state with a total spin quantum number S occurs, states having each value of $\mathscr{S}_{ze}$ from $-S$ to $+S$ will be found. It is always easy to write the zero-order determinantal wave functions that are the eigenfunctions of M_{ze} and S_{ze}. Therefore, all we have to do in order to find the multiplets of a configuration is to find the zero-order wave functions, find $\mathscr{M}_{ze}$ and $\mathscr{S}_{ze}$ for each, and determine the multiplet terms responsible for this array of values for $\mathscr{M}_{ze}$ and $\mathscr{S}_{ze}$.

Consider the configuration $2p^2$ once again. The fifteen functions of Table 10-3 have the eigenvalues of M_{ze} and S_{ze} shown in Table 10-6. The largest eigenvalue of M_{ze} is $\mathscr{M}_z = 2$, and for this there is only one function, having

$\mathscr{S}_z = 0$. This indicates that there must be a ^1D term level in this configuration. (If any ^3D-levels existed, there would have to be a wave function with $\mathscr{M}_z = 2$, $\mathscr{S}_z = 1$.) If a ^1D-level exists, there must be altogether five functions having $\mathscr{S}_z = 0$ and values of $\mathscr{M}_z$ running from 2 to -2. Obviously ϕ_{15} is one of these, with $\mathscr{M}_z = -2$. The function with $\mathscr{M}_z = 1$ must be one of the

TABLE 10-6

$\mathscr{M}_z$ \ $\mathscr{S}_z$	1	0	-1
2	–	$\underline{\phi_1}$	–
1	ϕ_2	$\phi_3\underline{\phi_4}$	ϕ_5
0	ϕ_6	$\phi_7\phi_8\underline{\phi_{10}}$	ϕ_9
-1	ϕ_{11}	$\phi_{12}\underline{\phi_{13}}$	ϕ_{14}
-2	–	$\underline{\phi_{15}}$	–

two linearly independent combinations of ϕ_3 and ϕ_4; $\mathscr{M}_z = 0$ must involve a combination of ϕ_7, ϕ_8, and ϕ_{10}; and $\mathscr{M}_z = -1$ must involve ϕ_{12} and ϕ_{13}. The ^1D-level thus "uses up" one function out of each value of $\mathscr{M}_z$ in the central column of Table 10-6. Five functions are therefore "earmarked" for ^1D by underlining them in the table. (This earmarking is done, of course, only as a bookkeeping device. The functions underlined in Table 10-6 need not be the actual wave functions of the ^1D-level.)

Among the remaining functions, the largest values of $\mathscr{M}_z$ and $\mathscr{S}_z$ are given by ϕ_2, for which $\mathscr{M}_z = 1$ and $\mathscr{S}_z = 1$. This indicates that a ^3P-level exists, which must account for nine states having $\mathscr{M}_z$ between 1 and -1 and $\mathscr{S}_z$ between 1 and -1. Thus we can earmark nine more functions, one for each entry inside the dotted square of Table 10-6. There remains a single unearmarked function having $\mathscr{M}_z = 0$ and $\mathscr{S}_z = 0$, which shows that a ^1S-state exists.

Let us apply this method to the configuration $2p3p$. The thirty-six zero-order wave functions are indicated in Table 10-7. (The symbol $(\bar{1}+0-)$ stands for the determinant $|(2p_{-1}\alpha)(3p_0\beta)|$, and similarly for the rest.) Inspection of this table following the principles outlined above, shows that $2p3p$ should give the levels ^3D (fifteen states, one from each entry of the entire table), ^1D (five states from the central vertical column), ^3P (nine states from the three middle rows), ^1P (three states from the central column), ^3S (three states from the central row), and ^1S (one state from the central entry of the table).

TABLE 10-7

$\mathscr{M}_z$ \ $\mathscr{S}_z$	1	0	−1
2	$(1+1+)$	$(1+1-)\ (1-1+)$	$(1-1-)$
1	$(1+0+)$	$(1+0-)\ (0+1-)$	$(1-0-)$
	$(0+1+)$	$(1-0+)\ (0-1+)$	$(0-1-)$
0	$(1+\bar{1}+)$	$(1+\bar{1}-)\ (\bar{1}+1-)$	$(1-\bar{1}-)$
	$(\bar{1}+1+)$	$(1-\bar{1}+)\ (\bar{1}-1+)$	$(\bar{1}-1-)$
	$(0+0+)$	$(0+0-)\ (0-0+)$	$(0-0-)$
−1	$(0+\bar{1}+)$	$(\bar{1}+0-)\ (0+\bar{1}-)$	$(\bar{1}-0-)$
	$(\bar{1}+0+)$	$(\bar{1}-0+)\ (0-\bar{1}+)$	$(0-\bar{1}-)$
−2	$(\bar{1}+\bar{1}+)$	$(\bar{1}+\bar{1}-)\ (\bar{1}-\bar{1}+)$	$(\bar{1}-\bar{1}-)$

The multiplet terms arising in this way from various common electron configurations are shown in Table 10-8.

c. The vector model

Although there is no logical necessity for doing so, it is helpful to most people to interpret the above procedure in terms of a relatively simple semi-classical picture known as the *vector model* of the atom. The total orbital angular momentum of an atom is considered as the vector sum of the quantized angular momenta of the individual electrons, the sum always being taken so that the resultant is also quantized. The same may be done with the spin angular momentum. The orbital and spin vector patterns that occur with two p electrons are shown in Fig. 10-6 (see page 348).

The various multiplet states of the atom can then be thought of as resulting from combination of these spin and orbital vector patterns. Where there are several electrons in the same shell ("equivalent electrons"), however, not all combinations are found. For instance, with the configuration p^2, no ^3D, ^3S, or ^1P-states are found. This is because of the Pauli principle, but, except perhaps for the ^3D-state, it is not at all easy to see in a direct

TABLE 10-8

MULTIPLET TERMS OF VARIOUS ELECTRON CONFIGURATIONS

a. Equivalent electrons

Configuration	Terms
s^2, p^6, and d^{10}	^1S
p and p^5	^2P
p^2 and p^4	$^3P, {}^1D, {}^1S$
p^3	$^4S, {}^2D, {}^2P$
d and d^9	^2D
d^2 and d^8	$^3F, {}^3P, {}^1G, {}^1D, {}^1S$
d^3 and d^7	$^4F, {}^4P, {}^2H, {}^2G, {}^2F, {}^2D, {}^2D, {}^2P$
d^4 and d^6	$^5D, {}^3H, {}^3G, {}^3F, {}^3F, {}^3D, {}^3P, {}^3P$ $^1I, {}^1G, {}^1G, {}^1F, {}^1D, {}^1D, {}^1S, {}^1S$
d^5	$^6S, {}^4G, {}^4F, {}^4D, {}^4P, {}^2I, {}^2H, {}^2G,$ $^2G, {}^2F, {}^2F, {}^2D, {}^2D, {}^2D, {}^2P, {}^2S$

b. Nonequivalent electrons

$s\,s$	$^1S, {}^3S$
$s\,p$	$^1P, {}^3P$
$s\,d$	$^1D, {}^3D$
$p\,p$	$^3D, {}^1D, {}^3P, {}^1P, {}^3S, {}^1S$
$p\,d$	$^3F, {}^1F, {}^3D, {}^1D, {}^3P, {}^1P$
$d\,d$	$^3G, {}^1G, {}^3F, {}^1F, {}^3D, {}^1D, {}^3P, {}^1P, {}^3S, {}^1S$
$s\,s\,s$	$^4S, {}^2S, {}^2S$
$s\,s\,p$	$^4P, {}^2P, {}^2P$
$s\,p\,p$	$^4D, {}^2D, {}^2D, {}^4P, {}^2P, {}^2P, {}^4S, {}^2S, {}^2S$
$s\,p\,d$	$^4F, {}^2F, {}^2F, {}^4D, {}^2D, {}^2D, {}^4P, {}^2P, {}^2P$

fashion just which multiplets the Pauli principle does not allow. The construction and manipulation of Table 10-6, however, makes this a simple matter to determine.

The electrostatic repulsion energy of the electrons will depend on the relative orientations of the orbital angular momenta of the individual electrons and not on the spatial orientation of their resultant. This is why the quantum number L is important in specifying atomic states while $\mathcal{M}_z$ is unimportant. Because of Pauli forces, the electrostatic energy will also depend on the relative orientations of the spins of the individual electrons, but not on the spatial orientation of their resultant. This, in turn, is why the quantum number S is important in specifying atomic states while $\mathscr{S}_z$ is unimportant.

The compounding of the angular momentum vectors is complicated by the fact that the vector corresponding to the quantum number l has a length $\sqrt{l(l+1)}$. Thus in adding $l_1 = 1$ to $l_2 = 1$ to give $L = 2$, we must combine two vectors of length $\sqrt{2}$ to obtain a vector whose length is only $\sqrt{6}$. The three vectors can therefore not be parallel. This is indicated in Fig. 10-6. The individual vectors l_1 and l_2 may be considered to be precessing about L, whose direction in space is constant. The torque causing the precession comes from the electrostatic repulsions of the two electron clouds.

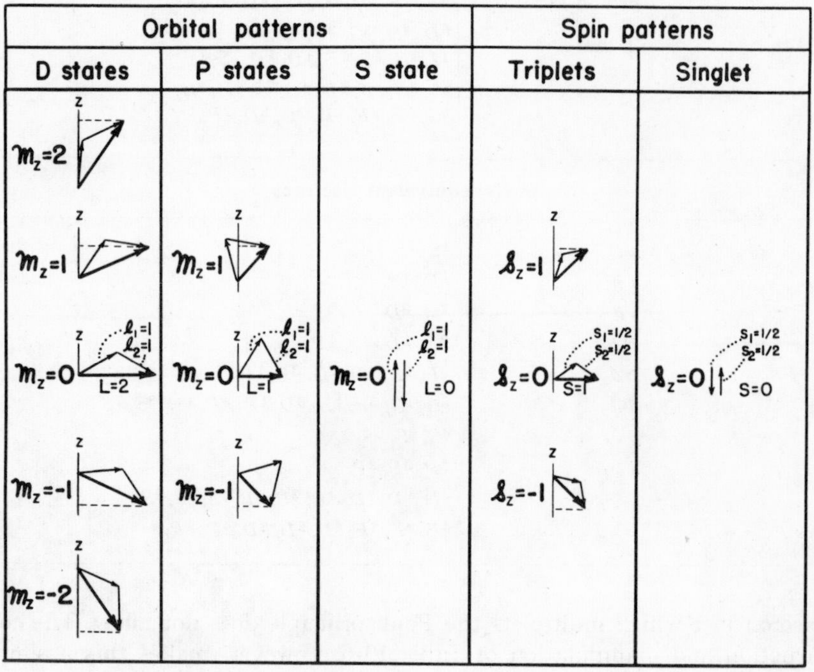

FIG. 10-6. Vector patterns for two p-electrons.

Some idea of the relative energies of the multiplet levels can often be obtained from Hund's rules, but there are frequently exceptions and ambiguities, especially among excited states, where the opportunity for configuration interaction is greatest. Hund's rules are, however, always reliable for predicting L and S for the ground state of any atom or ion.

Exercises. (1) Verify the multiplets arising from some of the configurations shown in Table 10-8.

(2) Show that closed shells always give only a ^1S-level.

(3) Show that a shell with n electrons missing always has the same multiplets as the same shell with n electrons (e.g., p^2 and p^4; d^3 and d^7).

(4) Do the multiplet levels of $2s2p^3$ of doubly ionized oxygen in Fig. 9-5 fall in the order expected from Hund's rules? What about the $2s^22p3p$-levels of carbon, which fall in the order ^1P (most stable), ^3D, ^3S, ^3P, ^1D, ^1S (least stable)?

C. Spin-Orbit Interaction (Fine Structure)

In addition to the electrostatic interaction between the electrons there is a magnetic interaction between the orbital motion of the electron and the electron's spin magnetic moment. We tend to think that the electron goes around the nucleus, but from the electron's point of view it is at the center of things, and the nucleus (with the core electrons) moves in an orbit around the electron. Since the nucleus and core bear a charge, the electron sees itself encircled by an electric circuit, just as if it were in the middle of a coil of wire carrying a current. This current causes a magnetic field, which, at the position of the electron, is perpendicular to the plane of the orbit. But the electron bears a magnetic moment parallel to its axis of spin, and its energy will vary depending on the relative directions of this spin magnetic moment and the magnetic field from the orbital motion (see Equation (D-8) on page 263). Therefore the energy of an atom must depend to some extent on the relative orientations of the spin axes and the orbital angular momentum axes of the electrons in the atom. This *spin-orbit interaction* splits most of the multiplet states of atoms into several closely spaced energy levels and is responsible for the so-called "fine structure" of atomic spectra.

For an alkali metal with a single unpaired electron, the spin can assume two orientations with respect to the magnetic field due to the orbital motion. We therefore expect that, except for S-levels (which have no orbital angular momentum), all of the levels of the alkali metals that we deduced on page 324 should occur as pairs having slightly different energies. This is actually observed. In fact, it is just this effect that causes the sodium D-line to be a doublet. (The D-line arises from the transition $3p\ ^2P \rightarrow 3s\ ^2S$.)

a. General theory of fine structure

In order to describe these states for atoms in general, a new property is introduced–the *total angular momentum*, usually represented by the symbol J. This property is the vector sum of the orbital and spin angular momenta, and its operators are defined accordingly.

For a single electron,	For more than one electron,	
$J_{xi} = M_{xi} + S_{xi}$	$J_{xe} = \sum J_{xi}$	
$J_{yi} = M_{yi} + S_{yi}$	$J_{ye} = \sum p_{yi}$	(C-1)
$J_{zi} = M_{zi} + S_{zi}$	$J_{ze} = \sum J_{zi}$	
$J_i^2 = J_{xi}^2 + J_{yi}^2 + J_{zi}^2$	$J_e^2 = J_{xe}^2 + J_{ye}^2 + J_{ze}^2$	

The commutation rules for these operators are easily seen to be analogous to those for orbital and spin angular momentum (Equations (A-11) of Chapter 8 and (D-2) of Chapter 9). It is therefore possible to have eigenstates both of J_e^2 and one component, say J_{ze}. As a further consequence of the commutation rules (see *2*, p. 46; or *11*, p. 479), the eigenvalues of J_e^2 must have the form $J(J+1)\hbar^2$, where J is an integer if there is an even number of electrons, and J is a half integer if there is an odd number of electrons. Eigenvalues of J_{ze} may have the values $\mathcal{J}_z = -J, -J+1, \cdots, J-1, J$. The quantum number J is called the *total angular momentum quantum number*.

We shall now derive an expression for the contribution to the Hamiltonian due to the spin-orbit interaction. According to the classical theory of electricity the magnetic field at an electron caused by the apparent motion of the nucleus with its core is (see *12*, p. 82)

$$\boldsymbol{H} = (\boldsymbol{E} \times \mathbf{v})/c \tag{C-2}$$

where $\boldsymbol{E}$ is the electric field at the electron, c is the velocity of light, and $\mathbf{v}$ is the velocity of the electron relative to the nucleus. For an atom, the field from the nucleus and core is always in a radial direction, being given by

$$\boldsymbol{E} = -(\partial V(r)/\partial r)\ (\mathbf{r}/r) \tag{C-3}$$

where $V(r)$ is the *electrostatic* potential due to the partially shielded nucleus and $\mathbf{r}$ is the radius vector from the nucleus. Furthermore, the angular momentum of an electron is given by

$$\mathbf{M} = m_e\,(\mathbf{r} \times \mathbf{v}) \tag{C-4}$$

Thus

$$\boldsymbol{H} = -\frac{1}{r}\frac{\partial V}{\partial r}\,\mathbf{M}/m_e c \tag{C-5}$$

where for quantum mechanical purposes we must regard $\mathbf{M}$ as an operator. We have seen in Chapter 9 (Equation (D-14)) that the spin magnetic moment $\boldsymbol{\mu}$ is related to the spin angular momentum $\mathbf{S}$ by

$$\boldsymbol{\mu} = -(e/m_e c)\,\mathbf{S} \tag{C-6}$$

where e is the electronic charge. The energy of a magnetic dipole in a magnetic field is (*cf.* Equation (D-8) of Chapter 8)

$$E = -\mathbf{H}\cdot\boldsymbol{\mu} = -\frac{e}{m_e^2 c^2}\frac{1}{r}\frac{\partial V}{dr}\mathbf{M}\cdot\mathbf{S} \tag{C-7}$$

Thomas and Frenkel showed that because of a relativity effect on the precession of the spin-axis in a magnetic field, the energy is actually only half this large. Therefore from (C-7) we obtain for the contribution of the spin-orbit interaction to the Hamiltonian of an atom,

$$H_{\text{sp}} = \sum_{\substack{\text{all}\\ \text{electrons}}} \xi(r_i)\mathbf{M}_i\cdot\mathbf{S}_i = \sum \xi_i(M_{xi}S_{xi} + M_{yi}S_{yi} + M_{zi}S_{zi}) \tag{C-8}$$

where

$$\xi(r_i) = -\frac{e}{2m_e^2c^2}\frac{1}{r_i}\frac{\partial V(r_i)}{\partial r_i} \tag{C-8a}$$

Here $\mathbf{M}_i$ and $\mathbf{S}_i$ are the orbital and spin operators for the ith electron, and r_i is the distance of the ith electron from the nucleus. This term must be added to the atomic Hamiltonian considered previously. We then have

$$H_{\text{total}} = H_0 + H_{\text{rep}} + H_{\text{sp}} \tag{C-9}$$

where H_0 is the operator given in Equation (A-1), for which exact eigenfunctions are available, and H_{rep} gives the electronic repulsions ($H_{\text{rep}} = \sum 1/r_{ij}$).

The usual methods of perturbation theory may be applied in evaluating the energy correction due to the spin-orbit interaction. The situation is complicated by the fact that H_{sp} does not commute with M_e^2, S_e^2, M_{ze}, or S_{ze}, although it does commute with J_e^2 and J_{ze} (see exercises below). Therefore, only the total angular momentum of an atomic level can be precisely specified. It is impossible to separate this total angular momentum into orbital and spin angular momenta because angular momentum is continually shuttled back and forth between the orbital and spin motions of the electrons through the action of the spin-orbit torque to be described below. The quantum numbers L, $\mathscr{M}$, S, and $\mathscr{S}$ are thus not really "good" quantum numbers; only J and $\mathscr{J}_z$ are accurately defined for a level. If the spin-orbit interaction is weak, however, L, $\mathscr{M}$, S, and $\mathscr{S}$ retain some of their validity as quantum numbers, as we shall see.

Exercises. (1) Show that $M_{ze}M_{xi}S_{xi} - M_{xi}S_{xi}M_{ze} = i\hbar M_{yi}S_{xi}$.

(2) Show that $S_{ze}M_{xi}S_{xi} - M_{xi}S_{xi}S_{ze} = i\hbar M_{xi}S_{yi}$.

(3) Show that M_{ze} and S_{ze} do not commute with $\mathbf{M}_i \cdot \mathbf{S}_i = (M_{xi}S_{xi} + M_{yi}S_{yi} + M_{zi}S_{zi})$.

(4) Show that J_{ze} does commute with $\mathbf{M}_i \cdot \mathbf{S}_i$.

(5) Show that M_e^2 and S_e^2 do not commute with $\mathbf{M}_i \cdot \mathbf{S}_i$, but that J_e^2 does.

The detailed procedure for the diagonalization of the secular equation where H_{sp} is included in the perturbation will be found in Eyring, Walter, and Kimball (*10*, pp. 151*ff*.) and in much greater detail in Condon and Shortley (*2*). Here we shall give an approximate, more intuitive treatment based on the vector model, which is adequate for many purposes.

The magnitude of the spin-orbit interaction in an atom depends on the average value of $(1/r)(\partial V/\partial r)$ for the electrons. For a bare nucleus, $V = Ze/r$, and $(1/r)(\partial V/\partial r) = -Ze/r^3$. For a nucleus surrounded by a core of electrons, however, the electrostatic potential changes much more rapidly with r because of the rapid change in the shielding by the core as one approaches the nucleus. The spin-orbit interaction therefore tends to be large in atoms toward the end of the periodic table and for orbits that penetrate into the core.

Equation (C-7) can be written

$$E_{\text{sp}} = -|\boldsymbol{H}|\,|\boldsymbol{\mu}|\cos\theta \tag{C-10}$$

where $|\boldsymbol{H}|$ is the numerical value of the magnetic field due to the orbital motion, and $|\boldsymbol{\mu}|$ is the numerical value of the spin magnetic moment of the electron. Thus there is a torque (*cf*. Equation (D-12) of Chapter 8)

$$\mathscr{T} = \partial E_{\text{sp}}/\partial\theta = |\boldsymbol{H}|\,|\boldsymbol{\mu}|\sin\theta \tag{C-11}$$

which tends to change the relative orientation of the axes of the orbital motion and the spin. Because of this torque, the spin axis and the orbital axis of each electron tend to precess about a common direction.

We have seen that as a result of the torque arising from the Pauli forces and from the electrostatic repulsions of the electrons, the axes of the orbital motions of the individual electrons precess about the axis of the total orbital angular momentum of all of the electrons, and the individual spin axes precess about the axis of the total spin of all of the electrons. Depending on the relative rates of these precessions and the spin-orbit precession, two extreme situations can be imagined. In one of these the spin-orbit torque is much smaller than the electrostatic and Pauli torques. This leads to the so-called *Russell-Saunders coupling* of the spin and orbital angular momenta of the individual electrons. In the other situation the spin-orbit torque is very large compared with the other torques. This leads to the *j* – *j* *coupling* of the angular momenta.

From what has been said above, we expect to find Russell-Saunders coupling in lighter atoms and in atoms having only a few outer electrons that do not penetrate each other's orbits very much. On the other hand, $j-j$ coupling should be important in heavy atoms and in atoms having nearly filled outer shells (such as excited rare gas and halogen atoms).

b. The case of *j*-*j* coupling

If the electrostatic interactions between electrons are neglected, the entire basis for our earlier discussion of multiplet structure (Section B) is lost and we must disregard the arguments leading to the quantum numbers L, S, $\mathscr{M}$, and $\mathscr{S}$. Instead, the spin and orbital angular momenta of each separate electron are first combined vectorially to give a total angular momentum for each electron. Since there are only two ways in which the spin can add to the orbital angular momentum (corresponding to the two possible orientations of the spin axis of an electron), no more than two different values of the total angular momentum are possible for an electron. Denoting the orbital and spin angular momentum quantum numbers of an electron by l (always an integer) and s (always one half), respectively, and the two possible values of the total angular momentum quantum numbers of the electron by j (always a half integer), these two states can be represented by the vector diagrams shown in Fig. 10-7.

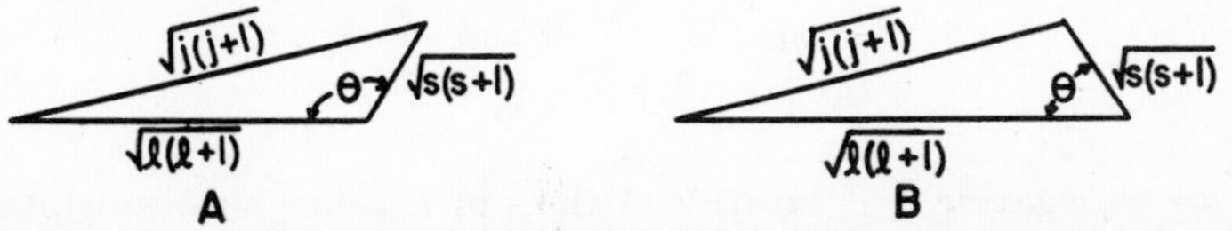

FIG. 10-7. Vector patterns for the angular momenta of single electrons. Because of the two possible orientations of the spin axis, each value of the orbital angular momentum quantum number, l, leads to two values of the total angular momentum quantum number, j, and two values of the angle θ. In one pattern (A), $j = l + \frac{1}{2}$ and θ is greater than a right angle. In the other pattern (B), $j = l - \frac{1}{2}$ and θ is smaller than a right angle.

For instance, a p electron ($l = 1$) can have $j = 3/2$ or $1/2$ (giving $\theta = 114°$ and $36°$, respectively), a d electron ($l = 2$) can have $j = 5/2$ or $3/2$ ($\theta = 118°$ and $45°$, and an f electron ($l = 3$) can have $j = 7/2$ or $5/2$ ($\theta = 120°$ and $48°$). Only one value of j is possible for s electrons ($j = 1/2$).

Each j-value can have $2j + 1$ states corresponding to the $2j + 1$ possible orientations of the axis of total angular momentum relative to some fixed axis.

The spin-orbit interaction energy of each electron is readily found from (C-8) as follows. The scalar product of the orbital and spin angular momenta of an electron is given by

$$\mathbf{M}\cdot\mathbf{S} = |\mathbf{M}|\ |\mathbf{S}| \cos(\pi - \theta) \tag{C-12}$$

$$= \sqrt{l(l+1)}\sqrt{s(s+1)}\hbar^2(-\cos\theta)$$

where θ is the angle shown in Fig. 10-7. By the law of cosines, from Fig. 10-7

$$-\cos\theta = \frac{j(j+1) - l(l+1) - s(s+1)}{2\sqrt{l(l+1)}\sqrt{s(s+1)}} \tag{C-13}$$

Thus the spin-orbit interaction energy for an individual electron is

$$E_{\text{sp}} = \xi\hbar^2\ \frac{j(j+1) - l(l+1) - s(s+1)}{2} \tag{C-14}$$

where

$$\xi = -(e/2m^2c^2)\left\langle\frac{1}{r}\frac{\partial V}{\partial r}\right\rangle \tag{C-14a}$$

and

$$\left\langle\frac{1}{r}\frac{\partial V}{\partial r}\right\rangle$$

is the average value of $(1/r)\,(\partial V/\partial r)$ for the orbital ϕ in which the electron occurs,

$$\left\langle\frac{1}{r}\frac{\partial V}{\partial r}\right\rangle = \int \phi^* \frac{1}{r}\frac{\partial V}{\partial r}\,\phi d\tau \tag{C-14b}$$

Since $\partial V/\partial r$ is negative, ξ is positive. Values of E_{sp} for various orbitals are given in Table 10-9. For a given value of l, the orbital having the smaller j is evidently the more stable.

If there are several electrons in the atom, their j-values combine vectorially to give the total angular momentum, $\sqrt{J(J+1)}\hbar$ where J is the total angular momentum quantum number. The manner in which they combine is best illustrated by means of an example. Consider a configuration in which there are two nonequivalent p electrons, $(mp\ np)$, for which the spin-orbit coupling is large. According to Table 10-9, each electron can have either $j = 1/2$ or $3/2$. The quantum numbers j_z for the components of j in some fixed direction are as follows:

$$\text{If } j = 1/2, \text{ then } j_z = 1/2 \text{ or } -1/2;$$

$$\text{If } j = 3/2, \text{ then } j_z = 3/2, 1/2, -1/2, \text{ or } -3/2.$$

TABLE 10-9

Type of Orbital	l	j	$E/\xi \hbar^2$
s	0	1/2	0
p	1	1/2	−1
		3/2	1/2
d	2	3/2	−3/2
		5/2	1

The possible values of J for the configuration $(mp\ np)$ are found by adding all possible combinations of these j_z-values, thus obtaining the values of $\mathcal{J}_z$, the z-component of J. This is done in Table 10-10. In this table the j-value of the mp electron is designated by j and that of the np electron is designated by j'. The pairs of numbers in parentheses at the top of each column, (3/2 3/2), etc., are the values of j and j' for that column. The numbers in brackets, [3/2 3/2], etc., refer to the values of j_z and j'_z (that is, the components of j and j' in the given direction).In the first column, where both mp and np have $j = 3/2$, there is one entry for $\mathcal{J}_z = 3$, so there must be a state of the atom with $J = 3$. This state accounts for one entry for each $\mathcal{J}_z$-value in this column. There is still one unused entry for $\mathcal{J}_z = 2$, so there must also be a state with $J = 2$. Continuing the argument in this way (compare pages 344*ff*.) we see that the configuration $(mp\ np)$ (3/2 3/2) must give states with $J = 3, 2, 1$, and 0. From Table 10-9 we find that the spin-orbit energy of an mp electron with $j = 3/2$ is $\xi_{mp}\hbar^2/2$, where ξ_{mp} is found from (C-14) using an mp-orbital for ϕ. Similarly an np electron with $j = 3/2$ has a spin-orbit interaction energy $\xi_{np}\hbar^2/2$, where ξ_{np} is evaluated for an np-orbital. If $n > m$, ξ_{np} should be smaller than ξ_{mp} since mp penetrates farther into the core than np, and $(\partial V/\partial r)$ increases rapidly as one approaches the nucleus.

TABLE 10-10

NONEQUIVALENT P ELECTRONS

$\mathcal{J}_z$	(j, j') $(\frac{3}{2}\ \frac{3}{2})$	$(\frac{3}{2}\ \frac{1}{2})$	$(\frac{1}{2}\ \frac{3}{2})$	$(\frac{1}{2}\ \frac{1}{2})$
3	$[\frac{3}{2}\ \frac{3}{2}]$	—	—	—
2	$[\frac{3}{2}\ \frac{1}{2}][\frac{1}{2}\ \frac{3}{2}]$	$[\frac{3}{2}\ \frac{1}{2}]$	$[\frac{1}{2}\ \frac{3}{2}]$	—
1	$[\frac{3}{2}\ -\frac{1}{2}][-\frac{1}{2}\ \frac{3}{2}]$ $[\frac{1}{2}\ \frac{1}{2}]$	$[\frac{3}{2}\ -\frac{1}{2}][\frac{1}{2}\ \frac{1}{2}]$	$[-\frac{1}{2}\ \frac{3}{2}][\frac{1}{2}\ \frac{1}{2}]$	$[\frac{1}{2}\ \frac{1}{2}]$
0	$[\frac{3}{2}\ -\frac{3}{2}][-\frac{3}{2}\ \frac{3}{2}]$ $[\frac{1}{2}\ -\frac{1}{2}][-\frac{1}{2}\ \frac{1}{2}]$	$[\frac{1}{2}\ -\frac{1}{2}][-\frac{1}{2}\ \frac{1}{2}]$	$[\frac{1}{2}\ -\frac{1}{2}][-\frac{1}{2}\ \frac{1}{2}]$	$[-\frac{1}{2}\ \frac{1}{2}][\frac{1}{2}\ -\frac{1}{2}]$
–1	$[-\frac{3}{2}\ \frac{1}{2}][\frac{1}{2}\ -\frac{3}{2}]$ $[-\frac{1}{2}\ -\frac{1}{2}]$	$[-\frac{3}{2}\ \frac{1}{2}][\frac{1}{2}\ -\frac{3}{2}]$	$[\frac{1}{2}\ -\frac{3}{2}][-\frac{1}{2}\ -\frac{1}{2}]$	$[-\frac{1}{2}\ -\frac{1}{2}]$
–2	$[-\frac{3}{2}\ -\frac{1}{2}][-\frac{1}{2}\ -\frac{3}{2}]$	$[-\frac{3}{2}\ -\frac{1}{2}]$	$[-\frac{1}{2}\ -\frac{3}{2}]$	—
–3	$[-\frac{3}{2}\ -\frac{3}{2}]$	—	—	—
J	3, 2, 1, 0	2,1	2,1	1,0
Energy	$\frac{\hbar^2}{2}\left(\xi_{mp} + \xi_{np}\right)$	$\frac{\hbar^2}{2}\left(\xi_{mp} - 2\xi_{np}\right)$	$\frac{\hbar^2}{2}\left(-2\xi_{mp} + \xi_{np}\right)$	$-\hbar^2(\xi_{mp} + \xi_{np})$

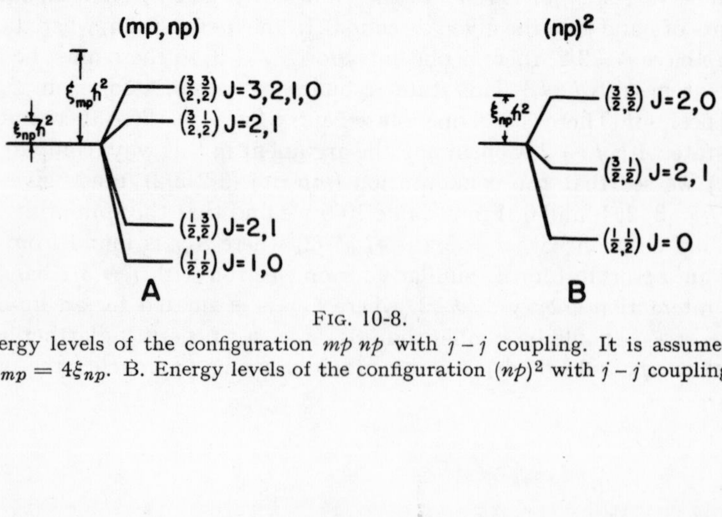

FIG. 10-8.
A. Energy levels of the configuration $mp\,np$ with $j - j$ coupling. It is assumed that $\xi_{mp} = 4\xi_{np}$. B. Energy levels of the configuration $(np)^2$ with $j - j$ coupling.

The total spin-orbit energy of the configuration $(mp\ np)$ $(3/2\ 3/2)$ is thus $(\xi_{mp} + \xi_{np})\hbar^2/2$. The possible J-values and energies of other configurations are given in Table 10-10. The arrangement of levels is indicated in Fig. 10-8.

If there are two equivalent electrons, as in the $(np)^2$ configuration, the Pauli principle must be taken into account and certain orientations of the j's are forbidden (e.g., j-values of (3/2 3/2) with both $j_z = 3/2$). Furthermore, some orientations are equivalent (e.g., j_z-values of (3/2 1/2) and (1/2 3/2) when the j-values are (3/2 3/2). Table 10-11 and Fig. 10-8 show the permitted states and their energies in this case.

TABLE 10-11

EQUIVALENT P ELECTRONS

J_z	(j, j') $(\frac{3}{2}\ \frac{3}{2})$	$(\frac{3}{2}\ \frac{1}{2})$	$(\frac{1}{2}\ \frac{1}{2})$
2	$[\frac{3}{2}\ \frac{1}{2}]$	$[\frac{3}{2}\ \frac{1}{2}]$	–
1	$[\frac{3}{2}\ -\frac{1}{2}]$	$[\frac{3}{2}\ -\frac{1}{2}][\frac{1}{2}\ \frac{1}{2}]$	–
0	$[\frac{3}{2}\ -\frac{3}{2}][\frac{1}{2}\ -\frac{1}{2}]$	$[\frac{1}{2}\ -\frac{1}{2}][-\frac{1}{2}\ \frac{1}{2}]$	$[\frac{1}{2}\ -\frac{1}{2}]$
1	$[-\frac{3}{2}\ \frac{1}{2}]$	$[-\frac{3}{2}\ \frac{1}{2}][-\frac{1}{2}\ -\frac{1}{2}]$	–
–2	$[-\frac{3}{2}\ -\frac{1}{2}]$	$[-\frac{3}{2}\ -\frac{1}{2}]$	–
J	2,0	2,1	0
Energy	$\hbar^2\,\xi_{np}$	$-\dfrac{\hbar^2}{2}\,\xi_{np}$	$-2\hbar^2\,\xi_{np}$

Exercises. (1) Find the energies and J-values of the states of the configurations sp, pd, d^2, and p^3, assuming $j-j$ coupling and neglecting electrostatic interactions between electrons.

(2) Show that for $(mp\ np)$, p^2, and the configurations in Exercise (1) $\sum_J (2J + 1)E_{sp} = 0$, where E_{sp} is the spin-orbit interaction energy of a level with total angular momentum quantum number J. This means that the mean energy of a configuration, weighted by $(2J + 1)$, is independent of the spin-orbit interaction. *This is a general rule.*

Electrostatic interactions will cause the energy levels derived above to split into various sublevels, depending on the value of J. The nature of this splitting will be better understood after we have considered Russell-Saunders coupling.

c. The case of Russell-Saunders coupling

If the spin-orbit interactions are small compared with the electrostatic repulsions between the electrons, it is permissible as a first approximation to use the concepts of the total orbital angular momentum L and the total spin angular momentum S deduced in discussing multiplet structure. These two angular momenta can then be considered nearly constant in direction and magnitude, since the rate of precession of each individual l and s about its resultant j is much less than the rates of precession of the l's about L and of the s's about S.

Using the vector model, it is an easy matter to deduce the possible values of the total quantum number J corresponding to give values of L and S. One merely adds L and S together vectorially in all possible ways that lead to half integral values of J (if there is an odd number of electrons present) or to integral values of J (if there is an even number of electrons present). The resulting J-value is written as a subscript to the right of the term symbol for the atomic level. For instance, with a ^3P-level we have the combinations shown in Fig. 10-9, leading to $^3P_2$, $^3P_1$, and $^3P_0$-levels. Such term symbols

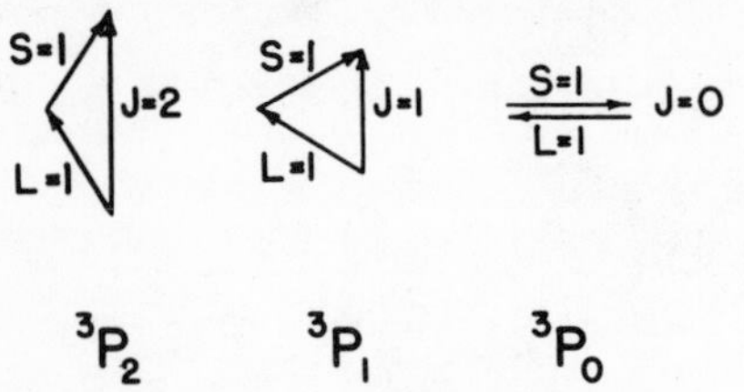

FIG. 10-9. Vector diagrams for J-values belonging to the ^3P-level (Russell-Saunders coupling).

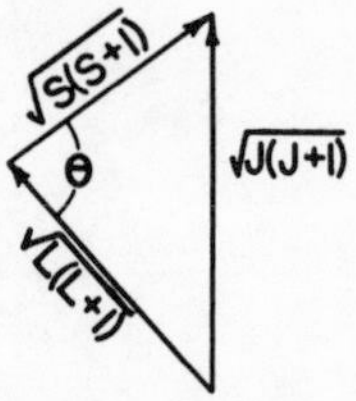

FIG. 10-10. Geometrical relationships between the spin, orbital, and total angular momentum vectors in Russell-Saunders coupling.

imply that one is basing the description of states on Russell-Saunders coupling although, as we shall see, these symbols can be uniquely correlated to states of $j - j$ coupling. Some further examples are shown in Table 10-12. In general J can take all positive values between $|L + S|$ and $|L - S|$. For each value of J, there are $2J + 1$ spatial orientations of the axis of total angular momentum, corresponding to the possible values of $\mathscr{J}_z$. Thus the energy level corresponding to each J-value is $(2J + 1)$-fold degenerate.

Exercise. Under what circumstances does the multiplicity of a level equal the number of fine structure components? Under what circumstances does the multiplicity equal the degeneracy?

TABLE 10-12

L	S	Fine Structure Components
0	0	$^1S_0$
1	0	$^1P_1$
0	1/2	$^2S_{1/2}$
1	1/2	$^2P_{3/2}$ $^2P_{1/2}$
2	1/2	$^2D_{5/2}$ $^2D_{3/2}$
1	1	$^3P_2$ $^3P_1$ $^3P_0$
2	0	$^1D_2$
2	1	$^3D_3$ $^3D_2$ $^3D_1$
2	3/2	$^4D_{7/2}$ $^4D_{5/2}$ $^4D_{3/2}$ $^4D_{1/2}$
1	2	$^5P_3$ $^5P_2$ $^5P_1$

Since each l precesses rapidly about L, only the component of each l that lies in the direction of L will fail to vanish when averaged over an interval of time. Similarly, only the component of each s lying in the direction of S will fail to vanish on the average. Therefore the spin-orbit energy in Russell-Saunders coupling is determined only by the interaction of the sum of the components of all the l's in the direction of L with the sum of the components of the s's in the direction of S. Of course the first sum is L itself, and the second sum is S. Thus the spin-orbit interaction energy is given by

$$E_{sp} = \xi_{av}\hbar^2 \mathbf{L}\cdot\mathbf{S} \tag{C-15}$$

where ξ_{av} is the average value of the quantity (C-8a) for the interacting electrons. Now

$$\mathbf{L}\cdot\mathbf{S} = -\sqrt{L(L+1)}\,\sqrt{S(S+1)}\,\cos\theta \tag{C-16}$$

where θ is the complement of the angle between $\mathbf{L}$ and $\mathbf{S}$. By the law of cosines, (see Fig. 10-10)

$$-\cos\theta = \frac{J(J+1) - L(L+1) - S(S+1)}{2\sqrt{L(L+1)}\,\sqrt{S(S+1)}} \tag{C-17}$$

giving

$$E_{sp} = \tfrac{1}{2}\xi_{av}\hbar^2[J(J+1) - L(L+1) - S(S+1)] \tag{C-18}$$

With Russell-Saunders coupling, the energy differences in the fine structure of a multiplet level depend on J in a very simple way. In a given multiplet level, the values of L and S are constant. Therefore the difference in energy of two successive fine structure levels will be given by

$$E_{sp}(J+1) - E_{sp}(J) = \tfrac{1}{2}\xi_{av}[(J+2)(J+1) - (J+1)J] \qquad \text{(C-19)}$$
$$= (J+1)\xi_{av}$$

This is known as the *Landé interval rule: The interval between two successive fine structure components (J and $J+1$) is proportional to $J+1$.* The rule is illustrated in Table 10-13 for the ^3P terms of various atoms. In this case, we should expect that if we define a ratio

$$T = [E(^3P_2) - E(^3P_1)] \,/\, [E(^3P_1) - E(^3P_0)]$$

then in the presence of Russell-Saunders coupling

$$T = 2$$

Deviations from this value of T are usually due to failure of the Russell-Saunders coupling model.* The deviations are largest in heavy atoms and in the rare gases, where spin-orbit interactions are large.

In Table 10-13, we observe that with some atoms the most stable level is that with the smallest J-value, whereas with others it is that with the largest J. Hund found empirically (and it was later shown to be a consequence of quantum mechanics; see Condon and Shortley (*2*, p. 300) and White (*1*, p. 256)) that *in configurations containing shells less than half full of electrons, the term having the lowest J-value is the most stable, whereas in those with shells more than half filled, the term having the largest J-value is the most stable.* This is the third (and last) of Hund's rules, which we shall call the *Hund fine structure rule.* The intervals in Table 10-13 all conform to this rule.

The actual separation of the fine structure components depends on the value of ξ_{av}, which in turn depends on the derivative of the potential about the nucleus, $\partial V/\partial r$. We should expect the greatest spin-orbit interactions with penetrating orbits and toward the end of the periodic table. Furthermore, the interaction should increase with nuclear charge in an isoelectronic sequence. These effects are clearly shown in Tables 10-13 and 10-14.

* With very light atoms, such as H and He, and with the innermost orbits of heavy atoms, relativity effects invalidate the assumptions underlying the Landé formula.

TABLE 10-13

TEST OF THE LANDÉ INTERVAL RULE FOR THE FINE STRUCTURE INTERVALS IN ^3P TERMS OF VARIOUS ATOMS (ENERGIES IN CM.$^{-1}$)

$$\Delta_{21} = E(^3P_2) - E(^3P_1) \quad \text{and} \quad \Delta_{10} = E(^3P_1) - E(^3P_0)$$

Atom	$C(2p^2)$	$Si(3p^2)$	$Ge(4p^2)$	$Sn(5p^2)$	$Pb(6p^2)$
Δ_{21}	27.1	146.2	853.	1736.	2831.
Δ_{10}	16.4	77.2	557.	1692.	7817.
T	1.65	1.90	1.53	1.025	0.362

Atom	$Be(2s2p)$	$Mg(3s3p)$	$Ca(4s4p)$	$Sr(5s5p)$	$Ba(6s6p)$
Δ_{21}	2.35	40.7	105.9	394.	878.
Δ_{10}	0.68	20.1	52.2	187.	371.
T	3.46	2.03	2.02	2.11	2.37

Atom			$Zn(4s4p)$	$Cd(5s5p)$	$Hg(6s6p)$
Δ_{21}			389.	1171.	4631.
Δ_{10}			190.	542.	1767.
T			2.05	2.16	2.62

Atom	$O(2p^4)$	$S(3p^4)$	$Se(4p^4)$	$Te(5p^4)$	
Δ_{21}	−158.5	−396.8	−1990.	−4751.	
Δ_{10}	−68.0	−176.8	−544.	44.	
T	2.33	2.25	3.66	−108.	

Atom	$He(1s2p)$	$Ne(2p^53s)$	$A(3p^54s)$	$Kr(4p^55s)$	$Xe(5p^56s)$	$Rn(6p^57s)$
Δ_{21}	−0.088	−417.5	−607.	−945.	−978.	−1369.
Δ_{10}	−0.99	−359.4	−803.	−4275.	−8252.	−11920.
T	0.089	1.16	0.757	0.221	0.118	0.115

Silicon Isoelectronic Sequence ($3p^2$)

Atom	Si	P^+	S^{2+}	Cl^{3+}	A^{4+}	K^{5+}	Ca^{6+}	Sc^{7+}
Δ_{21}	146.	304.	535.	850.	1267.	1793.	2443.	3230.
Δ_{10}	77.	167.	297.	491.	765.	1131.	1627.	2280.
T	1.90	1.82	1.80	1.73	1.66	1.59	1.51	1.42

TABLE 10-14

EFFECT OF ATOMIC NUMBER AND PENETRATION ON THE SPIN-ORBIT INTERACTION IN ONE-ELECTRON CONFIGURATIONS (ENERGIES IN CM.$^{-1}$)

$$\Delta = E(^2P_{3/2}) - E(^2P_{1/2})$$

a. Splitting of lowest ^2P-level

	Li(2*p*)	Na(3*p*)	K(4*p*)	Rb(5*p*)	Cs(6*p*)
Δ	0.44	17.2	57.7	238.	554.
	B(2*p*)	Al(3*p*)	Ga(4*p*)	In(5*p*)	Tl(6*p*)
Δ	16.	112.	826.	2213.	7793
	F(2p^5)	Cl(3p^5)	Br(4p^5)		
Δ	−404.	−881.	−3685.		
			Cu(4*p*)	Ag(5*p*)	Au(6*p*)
Δ			248.	921.	3816.

b. Splitting of the 6*p* ^2P-level in the alkali metals

	Na	K	Rb	Cs
Δ	1.25	8.41	77.5	554.

c. Effect of principal quantum number on the splitting of the ^2P-levels in sodium

Level	3*p*	4*p*	5*p*	6*p*	7*p*	8*p*	9*p*
Δ	17.20	5.63	2.52	1.25	0.74	0.47	0.47

d. Effect of orbital angular momentum (i.e., penetration) on splitting of levels of caesium

n	*np*	*nd*	*nf*
4	—	—	0.05
5	—	97.6	−0.16
6	554.	42.8	−0.11
7	181.	20.9	−0.07
8	81.	11.6	0.05

Exercises. (1) Why does Russell-Saunders coupling fail in the noble gases?

(2) Test the Landé interval rule on some of the multiplets of TiI and VI in Moore (*13*, Vol. 1, pp. 274, 292).

(3) Using Russell-Saunders coupling and Hund's three stability rules, predict the L, S, and J values of the ground states of the rare earth ion configurations $4f^n$, where n goes from 1 to 14. Compare your results with Table 10-16. [Hint: The case f^5, corresponding to Sm^{3+}, will be worked out as an example. According to Hund's first rule, the ground state should have the maximum possible multiplicity. Since there are five electrons and seven f-orbitals, all of the spins may be parallel, giving $S = 5/2$ or a sextet state. According to Hund's second rule, the orbital angular momentum should be as large as possible, consistent with the five parallel spins. Since only one electron can be placed in each orientation of the axis of rotation, this means that we can obtain the maximum z-component of total orbital angular momentum by using the f_{+3}, f_{+2}, f_{+1}, f_0, and f_{-1}-orbitals. Since $3 + 2 + 1 + 0 - 1 = 5$, this means that the largest L-value consistent with $S = 5/2$ is $L = 5$, so that the ground state is an H-state. According to Hund's third rule, the f-shell being less than half full, the lowest J-value gives the most stable state. Thus $J = 5 - 5/2 = 5/2$. We therefore conclude that the ground state of Sm^{3+} is $^6H_{5/2}$.]

d. Correlation of Russell-Saunders and j - j coupling

The Russell-Saunders and $j - j$ coupling schemes are two limiting ways of looking at atoms, and the actual coupling of angular momenta is always somewhere between these two extremes. A continuous sequence exists, in fact, and it is possible to make a unique correlation of the states in one coupling scheme with those in the other. The Russell-Saunders notation is therefore applicable even when $j - j$ coupling predominates and *vice versa*.

Consider the energy levels for the configuration sp according to the two methods of coupling. In Russell-Saunders coupling, this configuration gives $^1P_1$, $^3P_2$, $^3P_1$, and $^3P_0$-states, grouped into two energy levels–an upper singlet level and a lower triplet one. In $j - j$ coupling, one readily finds that there is a (1/2 3/2)-level with $J = 2$ and 1, and a more stable (1/2 1/2)-level with $J = 1$ and 0. Since J is a "good" quantum number regardless of the coupling, these levels must be related in such a way that the same J-value is preserved regardless of the amount of spin-orbit coupling. Furthermore, lines joining states with a given J can never cross (*cf.* page 142). And finally, for small amounts of spin-orbit coupling, the Landé interval rule must hold. These principles make it possible to relate uniquely the states in the two coupling schemes (see Fig. 10-11).

The correlation diagram drawn in this way for the configuration p^2 is shown in Fig. 10-12. Included in this figure is an indication of the observed relative spacings of the energy levels of some atoms having the p^2-configuration. The $^3P_0$-level is given an energy zero, and the $^1S_0$-level, an energy unity. The energies of the other levels are scaled to this interval. A straight line is drawn between $^3P_2$ and (3/2 1/2) $J = 2$. Each atom is located in the horizontal direction of the diagram so that the energy of its $^3P_2$-level falls on this line. The energies of the remaining levels of the atom are plotted vertically on a line through this point. It is seen that smooth curves can be drawn through the sets of points belonging to each value of J, showing that in actual atoms there must be a uniform sequence of coupling schemes intermediate between the two extremes.

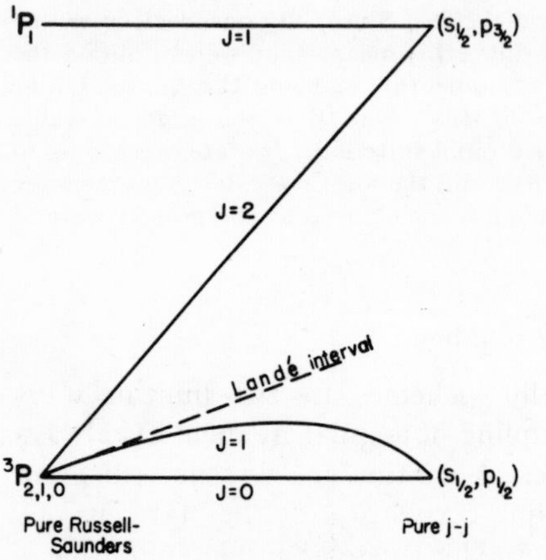

FIG. 10-11. Correlation of Russell-Saunders and $j-j$ coupling in the sp-configuration.

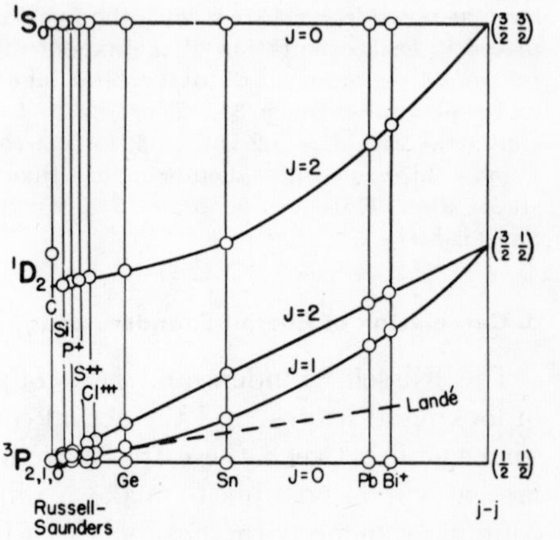

FIG. 10-12. Correlation of Russell-Saunders and $j-j$ coupling in the p^2-configuration.

These curves form a correlation diagram from which we can tell how closely these atoms approximate to one or the other of the two types of coupling. The tendency to show $j-j$ coupling as the atomic number increases is clearly shown. Thus CI and SiI show nearly pure Russell-Saunders coupling, whereas $j-j$ coupling predominates in PbI and BiII.

A similar semi-quantitative correlation diagram for the configuration p^5s is shown in Fig. 10-13. This configuration gives the same levels as sp but, because of the fact that it contains a shell more than half filled, the states with large values of j and J tend to be the most stable (Hund's fine structure rule). As a result, the correlation diagram is somewhat different from that for ps (Fig. 10-11). Various elements are placed in Fig. 10-13 in the same manner as in Fig. 10-12, a straight line between $^3P_0$ and (1/2 1/2) $J = 0$ being used for locating the atoms in a horizontal direction. It is evident that a smooth curve can be drawn between $^3P_1$ and (3/2 1/2) $J = 1$, thus showing once again that there is a gradual transition in the coupling of different elements. Figure 10-13 shows the tendency for $j-j$ coupling to predominate as the atomic number increases (Ne, A, Kr, Xe, and Rn approach the right hand side of the diagram in that order). It also shows that in a given element the $j - j$ coupling tendency increases as the s electron becomes more highly excited (NeI $2p^53s$ is closer to Russell-Saunders coupling, whereas NeI $2p^57s$ shows almost pure $j - j$ coupling).

A detailed quantitative treatment of intermediate coupling states will be found in Condon and Shortley (*2*, Chapter 11).

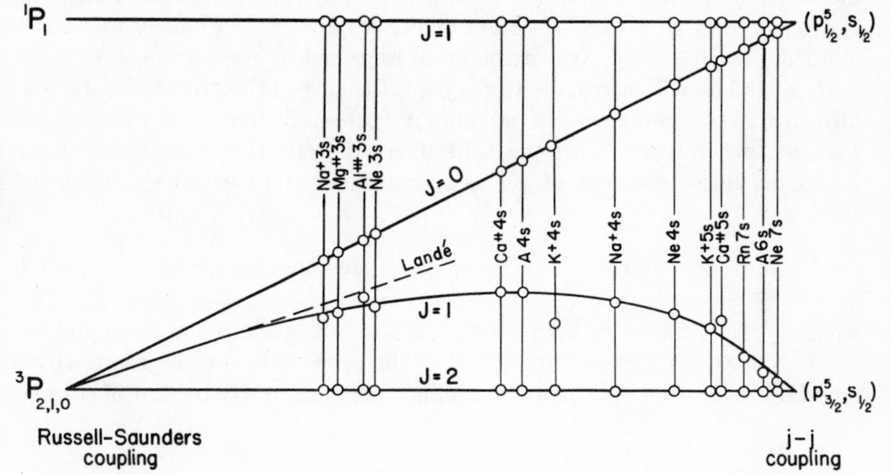

FIG. 10-13. Correlation of Russell-Saunders and $j - j$ coupling in the p^5s-configuration.

Exercises. (The data necessary for these exercises can be found in Bacher and Goudsmit (*14*), Moore (*13*) and Landolt-Börnstein (*8*).)

(1) Construct a correlation diagram for the configuration p^3 and locate the levels of N $2p^3$, P $3p^3$, As $4p^3$, Br^{++} $4p^3$, Sb $5p^3$, and Bi $6p^3$.

(2) Construct a correlation diagram for the configuration p^4 and locate the levels of O $2p^4$, S $3p^4$, Se $4p^4$, Kr^{++} $4p^4$, Te $5p^4$, I^+ $5p^4$, and Xe^{++} $5p^4$.

(3) Locate the levels of C $2p3s$, Si $3pns$ ($n = 4 - 8$), Ge $4p5s$, Sn $5p6s$, Ba $6s6p$, Hg $6snp$ ($n = 6 - 11$), Ti^+ $6snp$ ($n = 6,7$), and Ra $7s7p$ on Fig. 10-11.

D. The Magnetic Properties of Atoms

a. Magnetic dipole moments of atoms with Russell-Saunders coupling

The orbital motion of the electrons in an atom gives rise to a magnetic moment

$$|\mu_L| = -\beta_M \sqrt{L(L+1)} \tag{D-1}$$

in the direction of the axis of the orbit, where L is the total orbital angular momentum quantum number and β_M is the Bohr magneton. Similarly the electron spins contribute a magnetic dipole moment

$$|\boldsymbol{\mu}_S| = -2\beta_M \sqrt{S(S+1)} \tag{D-2}$$

in the direction of the axis of total spin, where S is the total spin angular momentum quantum number. Because of the spin-orbit interaction, $\boldsymbol{\mu}_L$ and $\boldsymbol{\mu}_S$ precess about the direction of the total angular momentum, whose quantum number is J. Now if the atom is placed in a weak magnetic field, both $\boldsymbol{\mu}_L$ and $\boldsymbol{\mu}_S$ will interact with the field. Because of the precession about J, however, the components of $\boldsymbol{\mu}_L$ and $\boldsymbol{\mu}_S$ perpendicular to J will, on the average, not interact with the field. Thus the effective magnitude of the magnetic dipole moment of a Russell-Saunders atom in a weak magnetic field is

$$|\boldsymbol{\mu}| = |\boldsymbol{\mu}_L| \cos(J, L) + |\boldsymbol{\mu}_S| \cos(J, S) \tag{D-3}$$

where (J, S) is the angle between the total spin angular momentum and the total angular momentum, and (J, L) is the angle between the total orbital angular momentum and the total angular momentum. By the law of cosines,

$$\cos(J, S) = [S(S+1) + J(J+1) - L(L+1)]/2\sqrt{S(S+1)\,J(J+1)}$$

and

$$\cos(J, L) = [L(L+1) + J(J+1) - S(S+1)]/2\sqrt{L(L+1)\,J(J+1)}$$

giving

$$|\boldsymbol{\mu}| = \left[1 + \frac{J(J+1) + S(S+1) - L(L+1)}{2J(J+1)}\right] \sqrt{J(J+1)}\,\beta_M$$

or

$$|\boldsymbol{\mu}| = g\sqrt{J(J+1)}\,\beta_M \tag{D-4}$$

where g is the so-called *Landé g-factor*,

$$g = 1 + \frac{J(J+1) + S(S+1) - L(L+1)}{2J(J+1)} \tag{D-5}$$

It is evident that g has the value unity for pure orbital motion ($S = 0$, $J = L$) and the value 2 for pure spin ($L = 0$, $J = S$).

b. Origin of the anomalous Zeeman effect

We now see the reason for the anomalous Zeeman effect described in Chapter 8. In a magnetic field, $\boldsymbol{H}$, the energy of an atom is (*cf.* page 263)

$$E = -\boldsymbol{\mu}\cdot\boldsymbol{H} \tag{D-6}$$
$$= -gJ\beta_M \mid \boldsymbol{H} \mid$$

Only if $g = 1$ (singlet levels), do we obtain the normal Zeeman splitting of the level predicted by Equation (D-9) of Chapter 8. For states of higher multiplicity, g may have any value between 0 and 2, and the anomalous Zeeman effect will be observed. It is evident that by a study of the spectra of atoms in magnetic fields we may determine the numerical values of the g-factors of states and in this way learn something of the values of J, S, and L for the states. Thus the study of the anomalous Zeeman effect can be of great assistance in analyzing spectra and in classifying terms.

c. The paramagnetism of the rare earth ions

The magnetic moment of an ion, atom, or molecule, as it exists in the solid, liquid, or gaseous state, may be found from the magnetic susceptibility, χ, through the relation [see, for example (*15,* p. 340) and (*16,* Chapters 2 and 4)]

$$\chi = N \mid \boldsymbol{\mu} \mid ^2/3kT$$

where N is the number of atoms per unit volume, k is Boltzmann's constant, and T is the absolute temperature. The magnetic moment may also be determined with great precision by direct measurement of the energy level spacing using paramagnetic resonance absorption of microwave radiation when the atom is placed in a magnetic field (*cf.* page 265).

In Table 10-15, the observed magnetic moments of the trivalent rare earth ions in aqueous solution and in crystalline salts are compared with the values calculated from Equation (D-4). The agreement is very good except for Sm and Eu, where there is thermal excitation of the ions to low-lying levels having different J-values, the fine structure splitting being particularly small for these two elements. Van Vleck and Frank (*16,* page 245) were able to show that when these higher levels are taken into account, the discrepancy is completely removed. This result is striking because it means that even in the solid and liquid states these ions show Russell-Saunders coupling.

TABLE 10-15

MAGNETIC MOMENTS OF THE TRIVALENT RARE EARTH IONS

Ion	Configuration	Ground State	J	L	S	g	$g\sqrt{J(J+1)} = \lvert\mu\rvert/\beta_M$	Observed $\lvert\mu\rvert/\beta_M$
La^{3+}	(core)	$^1S_0$	0	0	0	—	0.00	0
Ce^{3+}	$4f$	$^2F_{5/2}$	5/2	3	1/2	6/7	2.54	2.51
Pr^{3+}	$4f^2$	$^3H_4$	4	5	1	4/5	3.58	3.53
Nd^{3+}	$4f^3$	$^4I_{9/2}$	9/2	6	3/2	8/11	3.62	3.55
Pm^{3+}	$4f^4$	$^5I_4$	4	6	2	3/5	2.68	—
Sm^{3+}	$4f^5$	$^6H_{5/2}$	5/2	5	5/2	2/7	0.84	1.46
Eu^{3+}	$4f^6$	$^7F_0$	0	3	3	—	0.00	3.37
Gd^{3+}	$4f^7$	$^8S_{7/2}$	7/2	0	7/2	2	7.94	8.00
Tb^{3+}	$4f^8$	$^7F_6$	6	3	3	3/2	9.72	9.33
Dy^{3+}	$4f^9$	$^6H_{15/2}$	15/2	5	5/2	4/3	10.65	10.55
Ho^{3+}	$4f^{10}$	$^5I_8$	8	6	2	5/4	10.61	10.4
Er^{3+}	$4f^{11}$	$^4I_{15/2}$	15/2	6	3/2	6/5	9.60	9.5
Tm^{3+}	$4f^{12}$	$^3H_6$	6	5	1	7/6	7.56	7.35
Yb^{3+}	$4f^{13}$	$^2F_{7/2}$	7/2	3	1/2	8/7	4.53	4.5
Lu^{3+}	$4f^{14}$	$^1S_0$	0	0	0	—	0.00	0

d. The paramagnetism of the transition metal ions

In view of the success in the calculation of the paramagnetism of the rare earth ions, it is interesting to attempt a similar calculation for the transition metals. We shall restrict ourselves to the first series of transition elements, Sc through Cu. Because of their low atomic numbers, the spin-orbit interaction is much smaller in these ions than in the rare earths, the fine structure splitting of the multiplets being of the same order as the mean thermal energy, kT, at room temperature. This complicates the calculation considerably. The results of calculations based on three different sets of assumptions are shown in Table 10-16. In one calculation, it is assumed that the temperature is 0° K, so that all of the ions are in the lowest state of the fine structure components. In the second calculation, it is assumed that the temperature is very high, so that all of the states of different J of the multiplet are equally populated. Finally we give the results of a calculation by Laporte, who estimated the fine structure splitting in each ion and calculated the average magnetic moment at 293° K. On comparing the results of these calculations with the experimentally observed moments, it is found that the agreement is poor. Evidently the 3*d* electrons in the transition metal ions do not move about in the atom in the same way as do the 4*f* electrons in the rare earth ions. The reason for this behavior is that the 3*d* electrons are not so effectively buried inside the transition metal ion, so that they frequently collide with neighboring atoms and ions, thus changing

TABLE 10-16

MAGNETIC MOMENTS OF THE TRANSITION METAL IONS

[FROM VAN VLECK (*16*, P. 285)]

Ion	Configuration	Term symbol	Magnetic moment (in magnetons)				
			0° K.	∞° K.	Laporte	Spins only	Observed
$K^{+}\cdots V^{5+}$	(core)	^1S	0	0	0	0	0
Sc^{2+}	$3d$	^2D	1.55	3.00	2.57	1.73	—
Ti^{3+}	$3d$	^2D	1.55	3.00	2.18	1.73	—
V^{4+}	$3d$	^2D	1.55	3.00	1.78	1.73	1.75–1.79
Ti^{2+}	$3d^2$	^3F	1.63	4.47	3.36	2.83	—
V^{3+}	$3d^2$	^3F	1.63	4.47	2.73	2.83	2.76–2.85
V^{2+}	$3d^3$	^4F	0.77	5.20	3.60	3.87	3.81–3.86
Cr^{3+}	$3d^3$	^4F	0.77	5.20	2.97	3.87	3.68–3.86
Mn^{4+}	$3d^3$	^4F	0.77	5.20	2.47	3.87	4.00
Cr^{2+}	$3d^4$	^5D	0.00	5.48	4.25	4.90	4.80
Mn^{3+}	$3d^4$	^5D	0.00	5.48	3.80	4.90	5.05
Mn^{2+}	$3d^5$	^6S	5.92	5.92	5.92	5.92	5.2–5.96
Fe^{3+}	$3d^5$	^6S	5.92	5.92	5.92	5.92	5.4–6.0
Fe^{2+}	$3d^6$	^5D	6.70	5.48	6.54	4.90	5.0–5.5
Co^{2+}	$3d^7$	^4F	6.64	5.20	6.56	3.87	4.4–5.2
Ni^{2+}	$3d^8$	^3F	5.59	4.47	5.56	2.83	2.9–3.4
Cu^{2+}	$3d^9$	^2D	3.55	3.00	3.53	1.73	1.8–2.2
Zn^{2+}	$3d^{10}$	^1S	0	0	0	0	0

their orbital angular momenta. Stoner suggested that although such collisions should quench the orbital angular momentum, they should have no effect on the directions of the electronic spins, since the atoms, molecules and ions which surround the transition metal ions in aqueous solutions and in solid crystalline salts contain only closed electronic shells (e.g., water molecules in aqueous solution and in crystalline hydrates, and halide ions in the solid halides). Therefore one would expect that the paramagnetism in these substances should arise largely from the electron spins. That this is indeed the case is seen by comparing the values of $2\sqrt{S(S+1)}$ (next to last column in Table 10-16) with the observed magnetic moment in atomic units (last column in Table 10-16). The agreement is much better than that obtained in the calculations that assume unquenched orbital motions of the electrons.

Thus the $4f$ electrons in the rare earth ions act as if they were completely isolated from the atoms that surround the ions, whereas the $3d$ electrons in the transition elements are strongly affected by neighboring atoms and ions. We should therefore expect that in covalent inorganic complex compounds and in organic free radicals the orbital motions of the electrons should be completely quenched. In these substances the paramagnetism should depend entirely on the presence of unpaired electronic spins. Thus paramagnetism and paramagnetic resonance absorption offer ideal means of detecting and measuring the number of unpaired electrons in chemical compounds.

REFERENCES CITED

1. H. E. White, "Introduction to Atomic Spectra," McGraw-Hill, New York, 1934.
2. E. U. Condon and G. Shortley, "The Theory of Atomic Spectra," rev. ed., Cambridge U.P., New York, 1950.
3. C. Zener, *Phys. Rev.*, **36,** 51 (1930).
4. J. C. Slater, *Phys. Rev.*, **36,** 57 (1930).
5. W. E. Duncanson and C. A. Coulson, *Proc. Roy. Soc.* (*Edinburgh*), **62,** 37 (1944).
6. S. F. Boys, *Proc. Roy. Soc.*, **A200,** 542 (1950); **A201,** 125 (1950).
7. D. Hartree, *Repts. Prog. Phys.*, **11,** 113 (1946–7).
8. H. Landolt and R. Börnstein, "Zahlenwerte und Funktionen" (A. Eucken, ed.), 6th ed., Vol. 1, Part 1. Springer, Berlin, 1950.
9. N. F. Mott and I. N. Sneddon, "Wave Mechanics and Its Applications," Oxford U.P., New York, 1948.
10. H. Eyring, J. Walter, and G. E. Kimball, "Quantum Chemistry," Wiley, New York, 1944.
11. V. Rojansky, "Introductory Quantum Mechanics," Prentice-Hall, New York, 1938.
12. J. C. Slater and N. H. Frank, "Electromagnetism," McGraw-Hill, New York, 1947.
13. C. E. Moore, "Atomic Energy Levels," Vols. 1 and 2, National Bureau of Standards, Washington, D.C., 1949, 1952.

14. R. F. Bacher and S. Goudsmit, "Atomic Energy States," McGraw-Hill, New York, 1932.
15. J. Mayer and M. G. Mayer, "Statistical Mechanics," Wiley, New York, 1939.
16. J. H. Van Vleck, "The Theory of Electric and Magnetic Susceptibilities," Oxford U.P., New York, 1932.

GENERAL REFERENCES

On the Fermi-Thomas method:

H. Hellmann, "Einführung in die Quantenchemie," Chapter 1. Deuticke, Leipzig, 1937. (Reprinted by Edwards Brothers, Ann Arbor, Mich.).

P. Gombás, "Die statistische Theorie des Atoms," Springer, Vienna, 1949.

On multiplet and fine structure:

E. U. Condon and G. Shortley, reference *2*, above.

On magnetism:

J. H. Van Vleck, reference *16*, above.

PART IV

MOLECULAR SYSTEMS

Chapter 11

Molecules and the Chemical Bond I
The First Approximation

The application of quantum mechanics to problems of molecular structure and chemical binding has been remarkably successful within certain limits. Nearly all of the fundamental problems of chemistry can now be given plausible qualitative and in some instances even semi-quantitative explanations in quantum mechanical terms. We now have reasonably satisfactory answers to such basic questions as these: "What is responsible for the formation of a chemical bond between a pair of atoms?" "What is valence and why do elements have the valences they do?" "Why does carbon tetrachloride have a tetrahedral arrangement of chlorine atoms about the central carbon atom rather than some other arrangement, such as a square?" "Why are aromatic compounds so different in their chemical properties from other unsaturated compounds?" Furthermore, we known that classical mechanics could never have given satisfactory answers to these questions.

In nearly all of the early quantum mechanical discussions of molecular properties, the chief purpose was to show that quantum mechanics could furnish straightforward answers to chemical questions such as those mentioned above. Purely qualitative arguments were often all that was needed to demonstrate this. Frequently, however, approximate calculations were performed in order to make the case for quantum mechanics more quantitative. The results of these calculations usually correspond in a rough way to experimental observations, leaving little doubt of the validity of the quantum mechanical approach, though no one would claim that they proved any more than this.

As the quantum mechanical basis of chemistry became established, however, it was natural for chemists to hope and even to expect that these preliminary semi-quantitative and qualitative considerations would be followed by more precise calculations. There are many instances in which it is difficult experimentally to obtain information on the behavior of a chemical system, and if reliable quantum mechanical methods of calculation were available, they might be used to help solve some of these difficult problems. How useful it would be if one could calculate the exact manner in which hydrogen atoms react with oxygen molecules–a basic problem not at all easy to attack in the laboratory! Or, even more fundamental, to calculate the energy required for the dissociation of a nitrogen molecule into atoms, with a reliability of, say, five per cent.

Unfortunately quantum chemistry has not been very successful in solving quantitative problems of this type. True, there has been some progress in developing approximate interpolation methods by means of which certain properties of molecules can be deduced from the observed properties of other molecules (e.g., the correlations of resonance energies and the estimation of bond lengths in aromatic compounds). Even these methods, however, are based on rather severe approximations, which seem to have the habit of becoming seriously inadequate in some instances for no evident reason. What success these methods have had has only increased the disappointment at the failures of other more extensive efforts.

This chapter describes some of the successful, more rudimentary quantum chemical theories of molecular systems. In the next chapter we shall discuss the difficulties that have made progress with quantitative calculations so slow.

A. The Hydrogen Molecular Ion, H_2^+

a. The ground state of H_2^+; a primitive chemical bond

The system H_2^+, composed of two protons and an electron, has already been discussed in Chapter 6, where it was shown that, according to the Schroedinger equation, a stable molecule should exist. This molecule was in fact discovered many years ago by J. J. Thomson in cathode rays produced by the bombardment of ordinary hydrogen gas with electrons. From spectroscopic studies the ground state of H_2^+ is known to have an equilibrium internuclear distance of 1.060 Å. (2.00 Bohr radii) and a dissociation energy of 2.791 e.v. (0.1024 atomic units). Although the exact solution of the Schroedinger equation for this system agrees perfectly with these observed values, the solution is rather complex. It is therefore instructive to try to solve the Schroedinger equation for H_2^+ using approximate methods. These approximate methods have the advantage that they provide some valuable insight into the nature of chemical binding, which is not so readily obtained from the exact solution.

The Hamiltonian in atomic units is

$$H = -\frac{1}{2}\nabla^2 - \frac{1}{r_a} - \frac{1}{r_b} + \frac{1}{R} \qquad \text{(A-1)}$$

where ∇^2 is the Laplacian of the electron, r_a is the distance of the electron from proton a, r_b is the distance of the electron from proton b, and R is the distance between the two protons. The nuclei are assumed to be stationary, so their kinetic energy is disregarded and the term $(1/R)$ is a constant. The justification of this assumption is the so-called Born-Oppenheimer principle, which will be discussed in Chapter 14.

We shall proceed by making some reasonable guesses at the wave function for H_2^+, calculating the dependence of the energy on the internuclear separation, R, in each case. If the energy passes through a minimum at some value of R, the formation of a stable molecule is indicated.

As a first attempt at a solution we shall assume that the wave function is simply that of an electron localized in a 1s-orbital about one of the nuclei, say nucleus a. Then we have

$$\psi_1 = 1s_a = \pi^{-1/2} e^{-r_a} \tag{A-2}$$

The energy corresponding to this function is

$$E = \int 1s_a H\, 1s_a\, d\tau \Big/ \int (1s_a)^2\, d\tau \tag{A-3}$$

Since $1s_a$ is normalized, the denominator in (A-3) is equal to unity. The numerator may be broken up into three parts, giving

$$E = \int 1s_a \left(-\frac{1}{2}\nabla^2 - \frac{1}{r_a}\right) 1s_a\, d\tau + \int \frac{1s_a^2}{R}\, d\tau - \int \frac{1s_a^2}{r_b}\, d\tau \tag{A-4}$$

The first integral on the right in (A-4) is the energy of a hydrogen atom in a 1s-state, or $-1/2$ atomic unit. The second integral is $1/R$. The third integral is of the same type as one encountered in calculating the first order perturbation energy of the ground state of the helium atom (page 285). Using the arguments there presented, we find that this integral is equal to the electrostatic energy of a unit charge at a distance R from the center of an electron cloud whose charge density is $(1s)^2$. This energy is (*cf.* Equation (C-17) of Chapter 9)

$$\int \frac{1s_a^2}{r_b}\, d\tau = -\frac{1}{\pi R}\int_0^R e^{-2r_a} 4\pi r_a^2 dr_a - \frac{1}{\pi}\int_R^\infty \frac{e^{-2r_a}}{r_a} 4\pi r_a^2\, dr_a$$
$$= e^{-2R}\left[1 + \frac{1}{R}\right] - \frac{1}{R} \tag{A-5}$$

Thus the energy corresponding to the wave function (A-2) is

$$E = -\frac{1}{2} + \left(1 + \frac{1}{R}\right) e^{-2R} \tag{A-6}$$

This expression increases monotonically to infinity as R is decreased to zero, which is just what one would expect if a proton were to penetrate into a spherically symmetric negatively charged cloud without deforming it and if there were another proton in the center of the cloud. Therefore the wave function, $\psi_1 = 1s_a$, does not lead to the formation of a stable molecule and it must lack some essential features present in the correct wave function of the H_2^+ molecule.

Two of these essential features would be expected to be (1) the possibility that the electron can pass back and forth from one nucleus to the other, and (2) the possibility that the electron will tend to spend much of its time in the region between the two nuclei, where its potential energy is low. Both of these features may be taken into account by writing the normalized wave function

$$\psi_2 = \frac{1}{\sqrt{N}} (1s_a + 1s_b) \tag{A-7}$$

where $1s_a$ and $1s_b$ are hydrogenic orbitals centered respectively on protons a and b, and N is the normalization constant, so that

$$N = \int (1s_a + 1s_b)^2 d\tau = 2(1 + S) \tag{A-8}$$

where

$$S = \int 1s_a \, 1s_b \, d\tau \tag{A-9}$$

The so-called *overlap integral* S, vanishes when the internuclear distance is large. Under these circumstances

$$N = 2 \tag{A-10}$$

A function of the type (A-7) is called an LCAO function (for "Linear Combination of Atomic Orbitals"). Functions of this type are frequently used in quantum chemical calculations of the properties of molecules. This function places the electron with equal probability in the vicinity of each of the two nuclei and it also introduces some deformation into the atomic electron clouds by drawing the electron into the region between the two nuclei. Thus in the plane midway between the two protons we have $1s_a = 1s_b$ and if the two nuclei are far apart, so that Equation (A-10) is true, then $|\psi_2|^2 \simeq 2(1s_a)^2$. The electron density in the midplane of the molecule therefore has approximately double the value that would be produced in the same region by a simple $1s$-orbital of the type given in Equation (A-2).

The energy corresponding to the wave function ψ_2 is

$$E = \int (1s_a + 1s_b)\, H\, (1s_a + 1s_b)\, d\tau/2(1 + S) \tag{A-11}$$

This expression may be simplified by writing

$$\begin{aligned} H_{aa} &= \int 1s_a\, H\, 1s_a\, d\tau; \quad H_{ab} = \int 1s_a\, H\, 1s_b\, d\tau \\ H_{bb} &= \int 1s_b\, H\, 1s_b\, d\tau; \quad H_{ba} = \int 1s_b\, H\, 1s_a\, d\tau \end{aligned} \tag{A-12}$$

Furthermore, since the two nuclei are identical, we have $H_{aa} = H_{bb}$, and since H is Hermitian and real, $H_{ab} = H_{ba}$. Thus

$$E = (H_{aa} + H_{ab})/(1 + S) \tag{A-13}$$

The integral H_{aa} is identical with the integral in Equations (A-3) and (A-4). The integrals H_{ab} and S may be evaluated by transforming to elliptical coordinates (see page 226), defined by

$$\mu = \frac{r_a + r_b}{R}, \quad \nu = \frac{r_a - r_b}{R}$$

φ = azimuthal angle about the internuclear axis.

The volume element in this coordinate system is*

$$d\tau = \frac{R^3}{8} (\mu^2 - \nu^2)\, d\mu d\nu d\varphi$$

The values of μ range from 1 to ∞, those of ν from -1 to $+1$ and those of φ from 0 to 2π. Then.

$$S = \frac{R^3}{8\pi} \int_1^\infty \int_{-1}^1 e^{-R\mu} (\mu^2 - \nu^2)\, d\nu d\mu \int_0^{2\pi} d\varphi \tag{A-14}$$

$$= \frac{R^3}{4} \int_1^\infty e^{-R\mu} \left(2\mu^2 - \frac{2}{3}\right) d\mu$$

* This relationship is proven in Eyring, Walter, and Kimball (*1*, Appendix III) and in Margenau and Murphy (*2*, Section 5.7).

On integration by parts, it is easily shown that

$$\int_1^\infty \mu\, e^{-R\mu}\, d\mu = \frac{e^{-R}}{R^2}(R + 1) \tag{A-15}$$

and

$$\int_1^\infty \mu^2\, e^{-R\mu}\, d\mu = \frac{e^{-R}}{R} + \frac{2}{R^3}(R + 1)\, e^{-R} \tag{A-16}$$

so that

$$S = \left(\frac{1}{3}R^2 + R + 1\right) e^{-R} \tag{A-17}$$

For the integral H_{ab} we may write

$$H_{ab} = \int 1s_a \left(-\frac{1}{2}\nabla^2 - \frac{1}{r_b} + \frac{1}{R}\right) 1s_b\, d\tau - \int \frac{1s_a 1s_b}{r_a}\, d\tau \tag{A-18}$$

Since $1s_b$ is an eigenfunction of the hydrogen atom operator $-\frac{1}{2}\nabla^2 - 1/r_b$ with the eigenvalue $-1/2$, and since $1/R$ is a constant, the first integral in (A-18) gives $[(-1/2) + (1/R)]S$. The second integral may be evaluated by transforming to elliptical coordinates (or by the method described on page 286).

$$\int \frac{1s_a 1s_b}{r_a}\, d\tau = \frac{R^3}{8\pi}\int_1^\infty \int_{-1}^1 \frac{e^{-R\mu}}{\frac{1}{2}R(\mu - \nu)}(\mu^2 - \nu^2)\, d\nu\, d\mu \int_0^{2\pi} d\varphi$$

$$= \frac{R^2}{2}\int_1^\infty \int_{-1}^1 e^{-R\mu}(\mu + \nu)\, d\nu\, d\mu \tag{A-19}$$

$$= (R + 1)e^{-R}$$

Thus

$$H_{ab} = [(-1/2) + (1/R)]S - (R + 1)e^{-R} \tag{A-20}$$

Substituting these integrals into Equation (A-13) we find for the energy of the hydrogen molecular ion

$$E = -\frac{1}{2} + \frac{1}{R} + \frac{(R + 1)e^{-2R} - R(R + 1)e^{-R} - 1}{R\,[(1 + R + \frac{1}{3}R^2)e^{-R} + 1]} \tag{A-21}$$

This function is plotted in Fig. 11-1. If R is very large, it gives an energy of $-1/2$, which is the energy of an isolated hydrogen atom. As R decreases, the energy passes through a minimum at $R = 2.157$ Bohr radii (1.32 Å.) whose value is 0.0648 atomic units (1.76 e.v.) below the energy at $R = \infty$. The calculated equilibrium internuclear distance and the dissociation energy are comparable in order of magnitude with the experimentally observed values, 2.00 Bohr radii and 0.1024 atomic units, respectively. Thus we may say that the wave function ψ_2 is reasonably successful in reproducing the observed properties of the H_2^+ molecule.

By allowing the electron to move in the neighborhood of *both* nuclei, we have evidently reproduced the characteristic of the true wave function that is essential to the formation of a stable molecule. The possibility of moving from one nucleus to the other gives rise to the integral H_{ab} in the energy expression, and this integral is therefore known as the *exchange integral.* It is this integral that leads to a stable molecule in the calculation that we have just performed.

The decrease in the energy through exchange can, in a sense, be said to arise from a form of "resonance" between two structures: one structure (represented by the function $1s_a$) corresponds to an electron located on proton *a*, and the other structure (represented by $1s_b$) corresponds to an electron located on proton *b*. This resonance stabilization is mathematically equivalent to the mechanism through which the frequency of the fundamental normal mode is lowered when two vibrating systems are coupled (pages 145ff.). Thus we have here an elementary example of the stabilization of a molecule by resonance between several structures. A similar elementary example of stabilization by resonance has already been encountered in the excited states of the helium atom (page 296).

It is only fair to mention that the theory outlined above contains

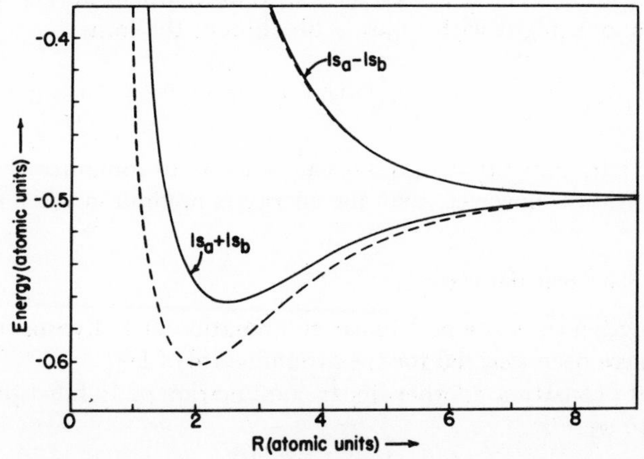

FIG. 11-1. Energy of the two lowest states of H_2^+ as a function of the distance between the protons. The solid curves are calculated from Equations (A-21) and (A-24). The dashed curves are the accurate solutions discussed in Chapter 6.

certain serious and basic inconsistencies. If we look at the wave function from a physical point of view, we are tempted to ascribe the stabilization to the fact that since $\psi_2 = 1s_a + 1s_b$ tends to move electrons into the region between the two nuclei, the potential energy ought to be lowered. Surprisingly enough, as we shall see in Chapter 12 (page 438), this is not the case at all: when the bond length has its equilibrium value, the function ψ_2 actually stabilizes the system by reducing the *kinetic* energy of the electrons. This, of course, is all wrong. We have shown in Chapter 7 (page 245) that according to the virial theorem the kinetic energy of the electrons in any molecule must be *increased* when the nuclei are at their equilibrium positions. Molecular stability always comes exclusively from a reduction in the *potential* energy of the electrons. Therefore the LCAO function ψ_2 has serious qualitative as well as quantitative shortcomings. These will be discussed in more detail in the next chapter. It turns out that the qualitative shortcomings with respect to the virial theorem are easily overcome, and that the true reason for the stability of the H_2^+ molecule–as well as of all other molecules containing covalent bonds–is the reduction in the potential energy resulting from the movement of electrons into the region between the nuclei at either end of the bond.

b. The excited states of H_2^+

The function ψ_2 (Equation (A-7)) is not necessarily the best LCAO function that we could have chosen for the ground state of the H_2^+ molecule. In particular, one might wish to use a function of the form

$$\psi_3 = \frac{1}{\sqrt{N'}}\,(1s_a + c\;1s_b) \tag{A-22}$$

where c is a parameter that may be varied so as to minimize the energy. It is readily shown, however, that the energy is minimized when c is equal to unity.

Exercise. Show that this is so.

Thus the function ψ_2 is the best linear combination of $1s$ hydrogen orbitals that could have been selected for the ground state of H_2^+.

It is easy to construct another linear combination of $1s$-functions that is orthogonal to ψ_2, *viz.*

$$\psi'_2 = (1s_a - 1s_b)/\sqrt{2(1-S)} \tag{A-23}$$

According to the principles of the variation method, ψ'_2 is an approximation to the wave function of an excited state of H_2^+. The energy obtained from this function is

$$E' = \frac{H_{aa} - H_{ab}}{1 - S} \tag{A-24}$$

Substitution into this expression of the previously calculated values of the integrals H_{aa}, H_{ab}, and S, gives the curve indicated in Fig. 11-1. Evidently it corresponds to a system that has the same energy as ψ_2 at large internuclear separations, but is unstable at all smaller separations. The exact calculation of the energies of the eigenstates of H_2^+ (Chapter 6) shows, in fact, that the first excited state is of just this kind; its wave function resembles ψ'_2 in that it has a node between the two nuclei and the energy *vs.* internuclear distance curve has no minimum. It is interesting that, as can be seen from Fig. 11-1, the energy calculated for this state of H_2^+ using the approximate LCAO function is remarkably close to the correct curve.

The physical reason for the instability of this state is easy to see. In a plane midway between the two protons we have $r_a = r_b$ and $1s_a = 1s_b$, so that $\psi'_2 = 0$. Thus the electron density in this plane vanishes; i.e., the electrons tend to stay away from the region between the two protons, where their potential energy is low. This, of course, tends to make the system unstable.

Although wave functions and energies of other excited states of H_2^+ are available from the exact solutions of the Schroedinger equation, it is much easier to visualize and classify these solutions by means of the method of the correlation diagram (*cf.* page 82). It is obvious that if the two protons are very far apart the states should have energies and wave functions reducing to those of the hydrogen atom, which are well known. (The energy levels, it will be recalled, are at $-1/2n^2$, where $n = 1, 2, \cdots$) Similarly, if the two protons are very close together and if their coulombic repulsion is disregarded, the system reduces to the He^+ ion, whose states are also well known (the energy levels being at $-2/n^2$). These two extremes are known as the *separated atoms* and the *united atom*, respectively. As the interproton distance is slowly varied from $R = \infty$ to $R = 0$, the states of the separated atoms should pass continuously into those of the united atom, in accordance with the adiabatic principle.

In making the correlation between these two extremes, it is helpful to use certain auxiliary operators that commute with the Hamiltonian of H_2^+ at all values of R. Four operators may be used for this purpose:

(*i*) M_φ, the angular momentum about the axis joining the protons–i.e., the axis about which the azimuthal angle φ is measured (see page 258 for a proof of the commutation of M_φ with H).

(*ii*) i, inversion of the electronic coordinates through the midpoint of the line joining the protons.

(*iii*) σ_d, reflection of the electronic coordinates in a plane normal to the molecular axis and midway between the two protons. (Operators i and σ_d have no effect on ∇^2, merely interchanging r_a and r_b with no net effect on H.)

(*iv*) σ_v, reflection in a plane passing through the two nuclei. (This operation has no effect on either ∇^2, r_a, or r_b.)

According to Corollary VI of the laws of quantum mechanics (page 170), it is possible to write eigenfunctions of the Hamiltonian that are at the same time eigenfunctions of all of these operators. This is easily done for both the united atom and the separated atoms; some examples are shown in Table 11-1.

TABLE 11-1

EIGENVALUES OF WAVE FUNCTIONS OF H_2^+

	Eigenvalues of					
Wave function	H (atomic units)	M_φ	i	σ_d	σ_v	Symbol for state
United atom						
$1s$	-2	0	1	1	1	$1s\sigma_g$
ns	$-2/n^2$	0	1	1	1	$ns\sigma_g$
np_0	$-2/n^2$	0	-1	-1	1	$np\sigma^*_u$
$np_{\pm 1}$	$-2/n^2$	± 1	-1	1	-1	$np\pi_u$
nd_0	$-2/n^2$	0	1	1	1	$nd\sigma_g$
$nd_{\pm 1}$	$-2/n^2$	± 1	1	-1	-1	$nd\pi^*_g$
$nd_{\pm 2}$	$-2/n^2$	± 2	1	1	1	$nd\delta_g$
Separated atoms						
$1s_a + 1s_b$	$-1/2$	0	1	1	1	$\sigma_g 1s$
$1s_a - 1s_b$	$-1/2$	0	-1	-1	1	$\sigma^*_u 1s$
$2s_a + 2s_b$	$-1/8$	0	1	1	1	$\sigma_g 2s$
$2s_a - 2s_b$	$-1/8$	0	-1	-1	1	$\sigma^*_u 2s$
$2p_{1a} + 2p_{1b}$	$-1/8$	1	-1	1	-1	$\pi_u 2p$
$2p_{1a} - 2p_{1b}$	$-1/8$	1	1	-1	-1	$\pi^*_g 2p$
$2p_{0a} + 2p_{0b}$	$-1/8$	0	-1	-1	1	$\sigma^*_u 2p$
$2p_{0a} - 2p_{0b}$	$-1/8$	0	1	1	1	$\sigma_g 2p$

A simple system of notation has been developed to denote these eigenvalues. If the eigenvalue for M_φ is 0, 1, 2, 3, $\cdots$, this is indicated by the small Greek letters, $\sigma, \pi, \delta, \phi, \cdots$, respectively. If the eigenvalue for inversion is $+1$, this is indicated by a subscript g (standing for the German "gerade," or "even"); if it is -1, the subscript u is used (German "ungerade", or "uneven"). If the eigenvalue for the reflection operator σ_d is -1, this is indicated by an asterisk. If the eigenvalue of the reflection operator σ_v is $+1$, a superscript $+$ is used; if it is -1, a superscript $-$ is used. Not all of these symbols are, however, always introduced, since some of them are redundant. For instance in the united atom symbols, it is really not necessary to use the g, u subscripts, since all s-states are g, all p-states are u, all d-states are g, etc. The eigenvalue for σ_v is also redundant, and for this reason it is not introduced into the terminology at this stage.

In the terminology based on the *separated atom point of view*, the Greek letter and its subscript (and asterisk, if called for) precede the symbol for the atomic orbital involved in the separated atoms; examples are shown in Table 11-1. The ground state of H_2^+ is $\sigma_g 1s$, and the repulsive state corresponding to the wave function ψ'_2 (Equation (A-23)) is $\sigma^*_u 1s$.

The translation from one terminology to the other is done by means of a correlation diagram, obtainable from the known exact solutions of the H_2^+ molecule (Fig. 6-12). It will be noted that in all instances the set of eigenvalues for M_φ, i, σ_d and σ_v must be the same for each correlated pair of united atom and separated atom states. Thus σ-states always correlate with σ-states, π-states with π-states, g-states with g-states, u-states with u-states, *-states with *-states, and so on.

It is obvious that those states in which electrons tend to be found in the region between the two nuclei should be more stable than those states in which the electron density is low in this region. This tendency parallels the behavior of the wave functions under the operation σ_d. If the wave function changes sign on reflection in the plane midway between the two nuclei, this plane must be a node of the function, and the electron density vanishes everywhere on this plane–a condition which we have seen leads to instability. Thus all orbitals with asterisks tend to give unstable molecules. For this reason they are called *antibonding orbitals*, whereas the unstarred orbitals are called *bond ng orbitals*. The antibonding orbitals of H_2^+ are thus $\sigma^*_u 1s$, $\sigma^*_u 2s$, $\pi^*_g 2p$, $\sigma^*_u 2p$, etc., and the bonding orbitals are $\sigma_g 1s$, $\sigma_g 2s$, $\sigma_g 2p$, $\pi_u 2p$, etc.

It will also be noted that the state $1s_a - 1s_b$ of the separated atoms goes into the $2p_0$-state of the united atom, and similarly $2p_{0a} + 2p_{0b}$ goes into $3d_0$. In these instances it is said that the electron is *promoted* (i.e., its principle quantum number is increased) by bringing the two nuclei together. Promotion is sometimes said to bring about instability; this, however, is not the case. The $3d\sigma_g$-state (which is promoted) forms a molecule that has a stronger bond than any other state of H_2^+ except the ground state, $1s\sigma_g$.

B. The Hydrogen Molecule, H_2

By adding an electron to H_2^+ we obtain the hydrogen molecule, which is the simplest molecular system familiar to chemists. The Hamiltonian for this molecule is (assuming as usual that the nuclei are stationary)

$$H = -(1/2)\nabla_1^2 - (1/2)\nabla_2^2 - (1/r_{1a}) - (1/r_{1b}) - (1/r_{2a}) - (1/r_{2b}) + (1/r_{12}) + (1/R) \tag{B-1}$$

where ∇_1^2 and ∇_2^2 are the Laplacian operators of the two electrons and the various distances, r_{1a}, r_{2b}, etc. are as indicated in Fig. 11-2. This Hamiltonian

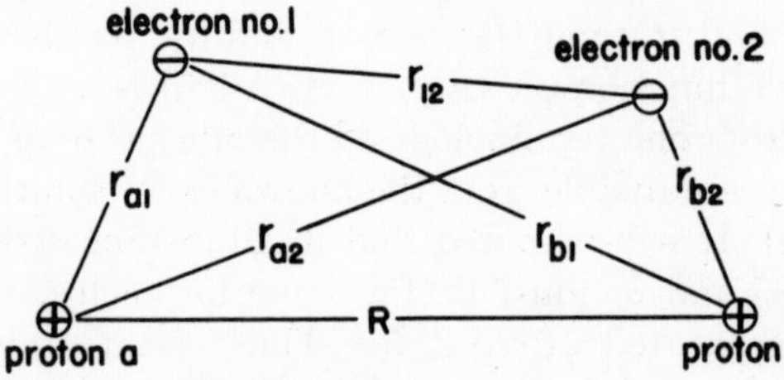

FIG. 11-2. Coordinates appearing in the Hamiltonian operator of the hydrogen molecule

leads to a considerably more complicated differential equation than that obtained with the H_2^+ molecule. The $1/r_{12}$ term, representing the mutual repulsion of the two electrons, makes it impossible to separate the variables in any coordinate system, and the equation is too difficult to solve by direct means. It is therefore necessary to rely completely on approximate methods of solving the Schroedinger equation for the hydrogen molecule, as well as for all molecules containing more than two electrons.

There are two especially important methods of setting up the first approximation to the wave function of the hydrogen molecule–the *molecular orbital* (*MO*) *method* and the *Heitler-London* or *valence bond* (*VB*) *method*. Both of these methods can be generalized for the study of more complicated molecules. It cannot be said that either method is "better" than the other, since each has its advantages and its disadvantages. Some types of problems are more easily understood in terms of one of the methods, others in terms of the other method. Sometimes both methods give the same answer, but often they do not. Therefore it is important that the student be acquainted with the advantages and shortcomings of both methods, as well as the types of problems to which each is best adapted.

a. The molecular orbital method for the ground state of hydrogen

It will be recalled that in constructing wave functions for atoms a most useful first approximation was obtained by forming products of hydrogen-like atomic orbitals, at first neglecting the repulsions between the electrons and then introducing them as perturbations. The orbitals selected determine the *configuration* of the atomic state, and the electronic repulsions determine the *multiplet structure.*

In the molecular orbital method a similar procedure is followed. If the term, $1/r_{12}$, in Equation (B-1) is omitted, the Schroedinger equation is separable in the coordinates of the two electrons, the two resulting equations being those of the H_2^+ molecular ion. To a first approximation, therefore, we can write the wave function of hydrogen as a product of H_2^+ orbitals. The same procedure may be followed in other diatomic molecules, suitably modified H_2^+ "molecular orbitals" playing a role analogous to that played by hydrogen-like "atomic orbitals" in atomic systems.

According to these arguments, the ground state of the hydrogen molecule would have the approximate wave function

$$\psi_1 = (1s\sigma_g)_1 \, (1s\sigma_g)_2 \, (\alpha_1\beta_2 - \alpha_2\beta_1) \tag{B-2}$$

A singlet spin function has been included here because the Pauli principle must be obeyed and the configurational part of the wave function is symmetric with respect to permutation of the two electrons. The ground state of hydrogen is, in fact, known to be a singlet, since hydrogen is a diamagnetic substance.

The excited states of hydrogen will be considered later in this chapter. Let us first discuss the energy of the ground state resulting from the wave function (B-1). It is easy to believe that this wave function will lead to a stable molecule because we known that each of the two occupied orbitals forms a strong bond in H_2^+. The electronic repulsions should weaken the bond somewhat, but certainly not enough to destroy it altogether.

Detailed calculations are made most easily by using the LCAO approximation to the molecular orbitals. This gives the approximate wave function (omitting the spin function since it has no direct effect on the energy)

$$\psi = [1s_a(1) + 1s_b(1)] \, [1s_a(2) + 1s_b(2)] \tag{B-3}$$

The energy can be found in the usual way from the expression

$$E = \int \psi H \psi \, d\tau \Big/ \int \psi^2 \, d\tau \tag{B-4}$$

Because of the $1/r_{12}$ term in the Hamiltonian, some of the integrations are not easy to perform, but in principle the calculation is straightforward and we shall merely give the result. Coulson (*3*) found that (B-3) leads to a stable molecule with an equilibrium internuclear distance of 0.850 Å. and an energy 2.681 e.v. less than the energy of two hydrogen atoms that are infinitely far apart. These values are to be compared with the observed values 0.740 Å. for the bond length and 4.75 e.v. for the dissociation energy. Although there is plenty of room for improvement, we may conclude that the molecular orbital method accounts for the stability of hydrogen in a qualitatively correct way.

b. The Heitler-London method for the ground state of hydrogen; the electron pair theory of valence

The wave function (B-3) has a rather serious and obvious shortcoming. It can be written in the form

$$\psi = 1s_a(1)1s_b(2) + 1s_a(2)1s_b(1) + 1s_a(1)1s_a(2) + 1s_b(1)1s_b(2) \qquad \text{(B-5)}$$

The third and fourth terms in this expression correspond to situations in which both electrons are close to the same proton – that is, to ionic structures $H_a^- + H_b^+$ and $H_a^+ + H_b^-$. Now if the two protons in H_2 are very far apart, we know that these ionic structures are not very stable. The ionization energy of the hydrogen atom is 0.50 atomic units while its electron affinity is only 0.026 atomic units (see Table 9-4). Thus the reaction $2H \rightarrow H^+ + H^-$ is highly endothermic. Yet the wave function (B-5) gives equal weights to the ionic structures and to the non-ionic structures, $1s_a(1)1s_b(2)$ and $1s_b(1)1s_a(2)$, which at large internuclear distances are equivalent to two neutral hydrogen atoms. This is highly unreasonable and leads one to suspect that an improvement might result if the last two terms in (B-5) were omitted altogether, giving the wave function

$$\psi = 1s_a(1)1s_b(2) + 1s_a(2)1s_b(1) \qquad \text{(B-6)}$$

This approximation to the wave function was originally suggested by Heitler and London (*4*). The reader can easily verify that it leads to an energy

$$E = \frac{J + K}{1 + S^2} \qquad \text{(B-7)}$$

where

$$J = \int 1s_a(1)1s_b(2)\ H\ 1s_a(1)1s_b(2)\ d\tau \qquad \text{(B-8)}$$

$$K = \int 1s_a(1)1s_b(2)\ H\ 1s_a(2)1s_b(1)\ d\tau \qquad \text{(B-9)}$$

and S is the overlap integral already encountered in the H_2^+ molecule (Equation (A-9)).

The integral J gives us the energy that we would have obtained if we had written the wave function of the hydrogen molecule simply as

$$\psi = 1s_a(1)1s_b(2) \tag{B-10}$$

that is, not allowing the electrons to exchange nuclei. It may be evaluated by methods similar to those discussed on pages 379*ff*. It is found that it leads to a very shallow energy minimum of about 0.4 e.v. at an internuclear separation of 1.0 Å. This is less than ten per cent of the observed dissociation energy of H_2, so the structure (B-10) obviously leaves out something essential to the stability of the hydrogen molecule.

The integral K is an exchange integral because its appearance in the energy expression depends on the possibility of exchange of the electrons between the nuclei. It is much more difficult to evaluate than J, and Heitler and London (*4*) were able only to estimate its value; its exact value was later computed by Sugiura (*5*). It is found that the exchange integral makes a large contribution to the stability of the hydrogen molecule. When it is included in (B-7) the energy passes through a minimum at 0.869 Å. giving a dissociation energy of 3.140 e.v. This is somewhat closer to the observed dissociation energy (4.75 e.v.) than the value determined from the molecular orbital method (2.681 e.v.). Evidently it is somewhat better to omit the ionic structures altogether rather than include them on an equal footing with the nonpolar structures. (Of course, it would be still better to add them with a multiplying parameter whose value could be chosen so as to minimize the energy. The results of such a procedure will be discussed in Chapter 12.)

It is sometimes argued that, because of the better value of the bond energy obtained with the Heitler-London method, this method is "better" than the molecular orbital method. This conclusion is, however, far from convincing. Both methods are quite far from the correct answer, and the mere fact that the one method happens to be closer to the truth in this particular instance does not mean that this will always be the case. It is unquestionably true that the molecular orbital method is seriously in error at large internuclear distances, but this error does not seem to be decisive in molecules such as hydrogen when the internuclear distance is near its equilibrium value.

Thus we have found once again that the inclusion of electron exchange in the wave function leads to a strong bond between a pair of atoms. We should, however, realize that there is nothing mysterious about the process of jumping back and forth between nuclei. Physically the Heitler-London wave function causes stability in part by increasing the density of electrons in the region between the two protons, where their potential energy is low. To a much greater extent, however, as we shall show later (page 438), the stability comes from a decrease in the kinetic energy of the electrons, and as we have mentioned above, this contradicts the virial theorem. Therefore the Heitler-London wave function has some serious quantitative shortcomings, just as does the MO function that we have used for both H_2^+ and H_2. These shortcomings will be discussed further in Chapter 12.

The success of the Heitler-London theory for hydrogen has led to an important generalization known as the *valence bond (VB) theory of bonding* (also called the *spin theory*, the *electron pair theory*, and the *Heitler-London-Slater-Pauling (HLSP) theory*). If two atoms, A and B, contain the orbitals a and b, then a stable bond involving a pair of electrons, 1 and 2, can be formed between these two atoms, its wave function being

$$\psi = a(1)b(2) + a(2)b(1) \tag{B-11}$$

This bond should be stabilized by an exchange integral in the same manner as was found in the Heitler-London theory of hydrogen.

Since the wave function (B-11) is symmetric with respect to permutation of the two electrons, the spins of the two electrons involved in the bond must be antiparallel. That is, the complete wave function for the bond must be written

$$\psi = [a(1)b(2) + a(2)b(1)]\,[\alpha(1)\beta(2) - \alpha(2)\beta(1)] \tag{B-12}$$

In general we should expect that a molecule will contain the maximum possible number of these bonds between the various pairs of singly occupied atomic orbitals that occur in the molecule. As a result, the spins of all of the electrons in the molecule will tend to be paired, and the multiplicity (that is, the quantity $2S + 1$, where S is the total spin quantum number) should have the smallest possible value. Furthermore, there should be a tendency for molecules to form in such a way that they contain an even number of electrons, since this makes it possible for all electrons to be paired, giving $S = 0$ and a multiplicity of unity. This, of course, is what is usually observed; most stable molecules that are held together by covalent bonds do contain an even number of electrons, and the few exceptional molecules that contain an odd number of electrons (for instance, NO_2 and triphenyl methyl) usually show a tendency to form dimers. Furthermore, in the great majority of molecules containing an even number of electrons, the ground state is a singlet. Therefore the electron pair theory is very attractive to chemists, who like the explicit justification that it gives for the lines that they have been drawing between the symbols for carbon atoms, hydrogen atoms, etc., since the time of Kekulé.

On the other hand there are some stubborn exceptions which raise doubts as to the universality of the electron pair theory. Outstanding among these is the fact that the ground state of the oxygen molecule is a triplet. There are also many other diatomic molecules whose ground states have higher multiplicities than would be expected on the basis of the electron pair theory (e.g., B_2, C_2, S_2, NH, OH^+, PH). True, none of these molecules is very familiar, yet all of them are entities to which the electron pair theory should apply if it is universally valid. It is one of the special merits of the molecular orbital theory that it is able to account for these exceptional cases in a rational manner.

c. The excited states of hydrogen; more about notation for electronic states

It is easy to show that if the wave function for the ground state of hydrogen is written

$$\psi = 1s_a(1)1s_b(2) + c1s_a(2)1s_b(1) \tag{B-13}$$

where c is a variable parameter, then the value $c = 1$ minimizes the energy. This, of course, was the choice made in writing (B-6). Moreover, the function

$$\psi = 1s_a(1)1s_b(2) - 1s_a(2)1s_b(1) \tag{B-14}$$

which is orthogonal to (B-6), must correspond to an excited state of hydrogen. The function (B-14) is antisymmetric with respect to permutation of the electrons, so it must be associated with a triplet spin function. It is easy to see that the energy resulting from (B-14) is

$$E = \frac{J - K}{1 - S^2} \tag{B-15}$$

Since K appears here with the opposite sign of that in the expression (B-7) for the energy of the ground state, this state does not lead to a stable molecule at all. A triplet "repulsive" state of this type is actually observed; it will be discussed below.

Wave functions of the Heitler-London type can be constructed for other excited states of hydrogen, but this method has not been widely used for this purpose, since the molecular orbital method is less awkward for qualitative discussions, and few quantitative calculations have been performed.

Let us now see what the molecular orbital method can tell us about the excited electronic states of hydrogen. Just as was found for the helium atom, a series of singlet and triplet states exists in which one electron has been excited into higher orbitals. For instance, the first excited configuration above the ground state should be $(1s\sigma_g)\ (2p\sigma^*_u)$, and from this configuration should arise two states (*cf.* the $1s2s$ configuration of helium, Table 9-10): a singlet state

$$\psi = [(1s\sigma_g)_1(2p\sigma^*_u)_2 + (1s\sigma_g)_2(2p\sigma^*_u)_1]\ (\alpha_1\beta_2 - \alpha_2\beta_1) \tag{B-16}$$

and a group of triplet states

$$\psi = [(1s\sigma_g)_1(2p\sigma^*_u)_2 - (1s\sigma_g)_2(2p\sigma^*_u)_1] \begin{cases} \alpha_1\alpha_2 \\ \alpha_1\beta_2 + \alpha_2\beta_1 \\ \beta_1\beta_2 \end{cases} \tag{B-17}$$

According to Hund's rule, the triplet state should have the lower energy of the two because of its more favorable correlation of the electrons. This, in fact, is the case. Furthermore, because of the presence of both bonding and antibonding electrons, neither of the two levels would be expected to show a very strong bond between the two hydrogen atoms. This is actually true of the triplet state, which does not even give a stable molecule.* The singlet state corresponding to (B-16) does, however, have a rather strong bond. The reason for this will be given presently.

In discussing the states of diatomic molecules it is inconvenient to have to write down the complete wave function as we have been doing in order to identify, describe and predict electronic states. It is worth interrupting the discussion of hydrogen at this point in order to outline a method of avoiding this difficulty. A similar problem arose in studying atoms, and we have seen that a useful system of notation has been developed for classifying the multiplets of atomic configurations. In atoms the system is based in part on the total angular momentum of the electrons about the nucleus–a property that can be measured experimentally. A similar system of notation has been developed for linear molecules, but here we may make use of only the component of total electronic angular momentum about the molecular axis; as was shown in Chapter 8, only this component can be known precisely

* It is important to note that the $2p\sigma^*_u$ antibonding orbital has a greater repulsive force than the attractive force of the $1s\sigma_g$ orbital. This is caused by the smaller denominator in the expression for its energy (see Equations (A-11) and (A-24), and note that $(1 + S)$ occurs in the denominator of the former, whereas $(1 - S)$ occurs in the latter, S being a positive quantity). Therefore in a competition between a $1s\sigma_g$ bonding electron and a $2p\sigma^*_u$ antibonding electron the repulsion of the latter should predominate, and no stable molecule should result.

because the other components continually change their values through collisions with the nuclei. The numerical value of this component of angular momentum can be determined experimentally through an analysis of the band spectrum, and we have mentioned in Chapter 8 that if its value in atomic units is 0, 1, 2, 3, $\cdots$, then the state is given the symbol Σ, Π, Δ, Φ, $\cdots$, respectively. The experimental methods used in assigning these term symbols to states are described by Herzberg (*6*). Unfortunately they are too complex to be given here.

If we restrict ourselves to molecules containing atoms of low atomic number, spin-orbit interactions may be neglected and the electronic spin becomes another property that is useful in identifying molecular states. The value of the spin in a state can also be determined experimentally (see Herzberg's book). Its value is indicated by writing the multiplicity, $(2S + 1)$, as a superscript before the Greek letter symbolizing the electronic angular momentum–in close analogy to the atomic system of nomenclature.

It is easy to deduce the possible values of the axial component of electronic angular momentum and the possible values of the spin that can arise from a given electronic configuration. We shall illustrate this by means of molecular orbital configurations involving two electrons. Further examples will be given later, and a much more complete discussion will be found in Herzberg's book (*6*, especially Chapter VI).

It is obvious that two σ-electrons can give only Σ-states. We can also readily see that if the two σ-electrons are in the same orbital (as in the $1s\sigma_g^2$ ground state configuration of hydrogen), the Pauli principle requires that the spins be antiparallel, so that only a $^1\Sigma$ ("singlet sigma") state is possible. If the electrons are in different orbitals (e.g., the configuration $1s\sigma_g 2p\sigma^*_u$ mentioned on page 392), then both a $^1\Sigma$-state and a $^3\Sigma$-state are possible. Similarly, two electrons in a $\sigma\pi$-configuration can give rise to a $^1\Pi$-state and a $^3\Pi$-state.

In deriving the states that may result from a $\pi\pi$-configuration we must remember that the symbol π really belongs to two orbitals; one, π_+, in which the electron moves in one direction about the molecular axis, and the other, π_-, in which it moves in the opposite direction. In the absence of electronic repulsions these two orbitals have exactly the same energy. Thus if we have the configuration $2p\pi\ 3p\pi$, the following possibilities may occur:

$$\left.\begin{array}{l}(a)\ 2p\pi_+ 3p\pi_+ \rightarrow \Delta \\ (b)\ 2p\pi_- 3p\pi_- \rightarrow \Delta\end{array}\right\} \text{(electrons moving in same direction; resultant angular momentum} = 2)$$

$$\left.\begin{array}{l}(c)\ 2p\pi_+ 3p\pi_- \rightarrow \Sigma \\ (d)\ 2p\pi_- 3p\pi_+ \rightarrow \Sigma\end{array}\right\} \text{(electrons moving in opposite directions; resultant angular momentum} = 0)$$

We should, however, be sufficiently aware of the resonance phenomenon to realize that (c) and (d) themselves are not suitable functions, but that we should use the linear combinations $(c + d)$ and $(c - d)$ instead. Each of the functions, (a), (b), $(c + d)$ and $(c - d)$ may occur with the electron spins either parallel or antiparallel. The two functions (a) and (b) will have the same energy, since they differ only in the direction of revolution of the electrons. The functions $(c + d)$ and $(c - d)$ will, however, have different energies. Thus the configuration $2p\pi\ 3p\pi$ will lead to one $^3\Delta$-level, one $^1\Delta$-level, two $^3\Sigma$-levels, and two $^1\Sigma$-levels.

For the configuration $(2p\pi)^2$, on the other hand, the combination $2p\pi_+\ 2p\pi_+$ can, because of the Pauli principle, exist only as a singlet. Furthermore, $2p\pi_+\ 2p\pi_-$ and $2p\pi_-\ 2p\pi_+$ are identical, and the linear combination equivalent to $(c - d)$ vanishes. Thus when the two π-orbitals are equivalent, the $\pi\pi$-configuration yields only a $^1\Delta$-level, a $^3\Sigma$-level, and a $^1\Sigma$-level.

Exercises. (1) Write out the complete wave functions (including the spin) for the configuration $\pi\pi'$ where the two orbitals are not equivalent. Hint: $^1\Delta$ is

$$[(\pi_+)_1(\pi'_+)_2 + (\pi_+)_2(\pi'_+)_1]\ [\alpha_1\beta_2 - \alpha_2\beta_1]$$

One of the $^3\Sigma$ is

$$[(\pi_+)_1(\pi'_-)_2 - (\pi_+)_2(\pi'_-)_1 + (\pi_-)_1(\pi'_+)_2 - (\pi_-)_2(\pi'_+)_1]\ [\alpha_1\alpha_2, \text{etc.}]$$

The other $^3\Sigma$ is

$$[(\pi_+)_1(\pi'_-)_2 - (\pi_+)_2(\pi'_-)_1 - (\pi_-)_1(\pi'_+)_2 + (\pi_-)_2(\pi'_+)_1]\ [\alpha_1\alpha_2, \text{etc.}]$$

(2) Write out the complete wave functions for the configuration π^2, where the two orbitals are equivalent.

The behavior of the wave functions with respect to the operations i and σ_v acting on the electronic coordinates is also useful in classifying states of symmetrical diatomic molecules. A subscript u is added to the Greek letter symbol if the wave function changes sign when all of the electrons are simultaneously inverted through the center of the molecule; if the wave function does not change sign, the subscript g is used. (Of course, if the two nuclei in the diatomic molecule are not the same, the wave functions will not be eigenfunctions of i, so the (g, u) characteristics cannot be used.) In the case of Σ-states a superscript $+$ or $-$ is added if the wave function does not or does change sign, respectively, when the electrons are subjected to the operation σ_v. The $(+, -)$ and (g, u) character of a state can be determined experimentally from the band spectrum (see *6*). It may also be predicted from the electronic configuration. Thus all σ-orbitals remain unchanged under the operation σ_v. A u-state can result only if an odd number of u-orbitals are occupied. The ground state of hydrogen can therefore be identified as $(1s\sigma_g)^2\ {}^1\Sigma^+{}_g$; the ${}^1\Sigma^+{}_g$-character of this state is, in fact confirmed by a study of the band spectrum. (The Heitler-London wave function (B-6) also clearly corresponds to a ${}^1\Sigma^+{}_g$-state.) The excited state (B-16) may be denoted by $1s\sigma_g 2p\sigma^*{}_u\ {}^1\Sigma^+{}_u$, and (B-17) may be denoted by $1s\sigma_g 2p\sigma^*{}_u\ {}^3\Sigma^+{}_u$. Similarly we may have the states $1s\sigma_g 2s\sigma_g\ {}^1\Sigma^+{}_g$ and ${}^3\Sigma^+{}_g$, and the states $1s\sigma_g 2p\pi_u\ {}^1\Pi_u$ and ${}^3\Pi_u$ (the (+, −) character of states other than Σ-states is not ordinarily indicated, for reasons mentioned in Exercise (1), below).

The (+ , −) character of the $\pi\pi$-configurations is not immediately obvious since π_+- and π_--orbitals are not eigenfunctions of σ_v. It is, however, easy to form eigenfunctions from them. Let us set up a cylindrical coordinate system (r, φ, z), where the z-axis passes through the two nuclei. This coordinate system is related to a Cartesian coordinate system by the equations

$$\cos\varphi = x/r$$

$$\sin\varphi = y/r$$

$$z = z$$

We have seen from the discussion of the $H_2{}^+$ molecule that the φ-dependence of the π_+- and π_--orbitals is given by the functions $e^{i\varphi}$ and $e^{-i\varphi}$, respectively. Then the linear combinations

$$\pi_+ + \pi_- = \pi_x$$

and

$$\pi_+ - \pi_- = \pi_y$$

are proportional to the coordinates x and y, respectively. They are therefore eigenfunctions for reflection in the yz-plane, π_x having the eigenvalue − 1, and π_y having the eigenvalue + 1. Using these new orbitals we obtain the following wave functions for the Σ-states of the π^2-configuration, where the two π-orbitals are equivalent

$$\pi_{x1}\pi_{x2} - \pi_{y1}\pi_{y2} \rightarrow {}^1\Sigma^+$$

$$\pi_{x1}\pi_{y2} - \pi_{x2}\pi_{y1} \rightarrow {}^3\Sigma^-$$

If the two π-orbitals are not equivalent we obtain

$$(\pi_{x1}\pi'_{x2} - \pi_{y1}\pi'_{y2}) + (\pi_{x2}\pi'_{x1} - \pi_{y2}\pi'_{y1}) \rightarrow {}^1\Sigma^+$$

$$(\pi_{x1}\pi'_{x2} - \pi_{y1}\pi'_{y2}) - (\pi_{x2}\pi'_{x1} - \pi_{y2}\pi'_{y1}) \rightarrow {}^3\Sigma^+$$

$$(\pi_{x1}\pi'_{y2} - \pi_{y1}\pi'_{x2}) + (\pi_{x2}\pi'_{y1} - \pi_{y2}\pi'_{x1}) \rightarrow {}^1\Sigma^-$$

$$(\pi_{x1}\pi'_{y2} - \pi_{y1}\pi'_{x2}) - (\pi_{x2}\pi'_{y1} - \pi_{y2}\pi'_{x1}) \rightarrow {}^3\Sigma^-$$

Exercises. (1) Show that these wave functions are the same as those obtained in the previous set of exercises. What states do the functions $\pi_{x1}\pi_{x2} + \pi_{y1}\pi_{y2}$ and $\pi_{x1}\pi_{y2} + \pi_{x2}\pi_{y1}$ lead to? Give their (+, −) characters. Note that these states have the same electron-cloud distributions and therefore have the same energy. The ${}^1\Sigma^+$- and ${}^1\Sigma^-$-

states of the π^2-configuration do not, however, have the same electron-cloud distributions, and therefore differ in energy. The same is true of the $^3\Sigma^+$- and $^3\Sigma^-$-states of the π^2-configuration. This is why the (+, −) character of Σ-states is important, whereas that of other types of states is not important.

(2) Show that molecules in which all orbitals are completely filled must be in $^1\Sigma^+_g$-states.

The characteristics of some of the low-lying excited states of hydrogen are shown in Table 11-2. The electronic configurations that can account for

TABLE 11-2

CHARACTERISTICS OF POTENTIAL ENERGY CURVES OF HYDROGEN

Electronic configuration	Energy of minimum† (cm.$^{-1}$) Singlet	Triplet	R_{min} (Å.) Singlet	Triplet	Nuclear vibration frequency (cm.$^{-1}$) Singlet	Triplet
$(1s\sigma_g)^2\ {}^1\Sigma^+_g$	124429	—	0.7417	—	4395	—
$1s\sigma_g 2p\sigma^*_u\ {}^{1,3}\Sigma^+_u$	32739	~68000‡	1.2927	unstable	1357	unstable
$1s\sigma_g 2p\pi_u\ {}^{1,3}\Pi_u$	24386	28685	1.0331	1.038	2443	2465
$1s\sigma_g 2s\sigma_g\ {}^{1,3}\Sigma^+_g$	24366	28491	1.012	0.9887	2589	2665
$1s\sigma_g 3d\sigma_g\ {}^{1,3}\Sigma^+_g$	11636	11652	1.085	1.83?	~2400	2266
$1s\sigma_g 3d\pi^*_g\ {}^{1,3}\Pi_g$	11364	11421	1.060		2265	2268
$1s\sigma_g 3d\delta_g\ {}^{1,3}\Delta_g$	11025	11231	1.077		2220	2265
$1s\sigma_g 3p\sigma^*_u\ {}^{1,3}\Sigma^+_u$		16652		1.107		2196
$1s\sigma_g 3p\pi_u\ {}^{1,3}\Pi_u$	10541	11659	1.034	1.05	2325	2372
$1s\sigma_g 3s\sigma_g\ {}^{1,3}\Sigma^+_g$	10540	11659	1.065	1.0496	2538	2395
$(2p\sigma^*_u)^2\ {}^1\Sigma^+_g$	20949	—	2.32	—	1000	—
$1s\sigma_g(H_2^+)\ {}^2\Sigma^+_g$	0		1.06_0		2297	

† Below the minimum of H_2^+.

‡ Energy at internuclear distance of 1.29 Å.; this state does not have a minimum in its potential energy curve.

these states are also indicated. Potential energy curves for some of these states are shown in Fig. 11-3.

All of the excited states shown in the table except the $(2p\sigma^*_u)^2\ {}^1\Sigma^+_g$ and $1s\sigma_g 2p\sigma^*_u\ {}^1\Sigma^+_u$ and $^3\Sigma^+_u$-states have potential energy minima at about 1.05 Å., which is close to the equilibrium internuclear distance in H_2^+. Furthermore, their curvatures at the minima are nearly the same, as shown by the close similarity of the nuclear vibration frequencies, and this frequency is nearly the same as that in H_2^+. This, of course, means that most of the binding in these molecules is caused by the $1s\sigma_g$ electron, as one would expect from the calculations of the energies of the H_2^+ orbitals, which show that this orbital gives by far the strongest bond. It will also be noted from Table 11-2 that the triplet state of a given configuration is always more stable than the singlet state, so that Hund's multiplicity rule evidently holds for molecules as it does for atoms.

It is interesting to investigate the reasons for the exceptional behavior of the $1s\sigma_g 2p\sigma^*_u\ {}^1\Sigma^+_u$ and ${}^3\Sigma^+_u$-states. Using the LCAO approximation for the molecular orbitals $1s\sigma_g$ ($= 1s_a + 1s_b$) and $2p\sigma^*_u$ ($= 1s_a - 1s_b$), we find that the two wave functions (B-16) and (B-17) take the form, for the singlet state

$$\psi = [1s_a(1)1s_a(2) - 1s_b(1)1s_b(2)]\,(\alpha_1\beta_2 - \alpha_2\beta_1)$$

and for the triplet state

$$\psi = [1s_a(1)1s_b(2) - 1s_a(2)1s_b(1)]\,[\alpha_1\alpha_2, \text{ etc.}]$$

The latter function is identical with the triplet state wave function derived from the Heitler-London calculations (Equation (B-14)), and its instability has already been accounted for. The singlet wave function is derived from

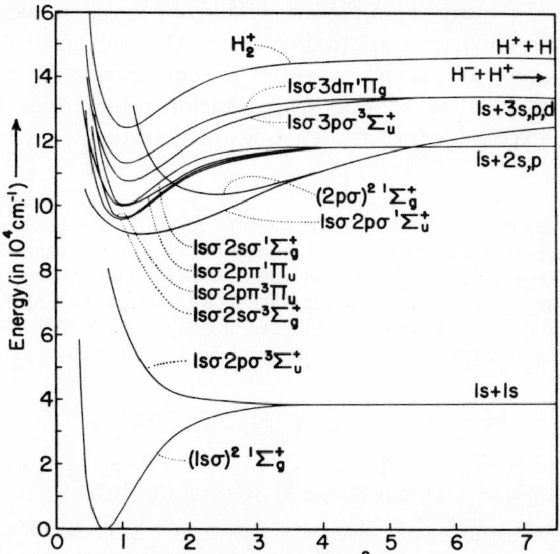

FIG. 11-3. Potential energy curves for various states of the hydrogen molecule. The atomic states from which the molecular states are derived are indicated on the right.

the ionic structures, $H_a^+ + H_b^-$ and $H_a^- + H_b^+$, in which both electrons are located on the same proton. A pair of ions would attract one another at large distances, whereas the repulsions from the antibonding orbital will become appreciable rather suddenly only at internuclear distances below 1.0 or 1.5 Å. This is apparently why the $1s\sigma_g 2p^*_u\ {}^1\Sigma^+_u$-state leads to a stable molecule in spite of the weak bonding resulting from the antibonding $2p\sigma^*_u$-orbital. It also accounts for the rather large bond length and the low value of the nuclear vibration frequency.

The "repulsive state," $1s\sigma_g 2p\sigma^*_u\ {}^3\Sigma^+_u$, should be able to resonate with other ${}^3\Sigma^+_u$-states, and thereby impart some of its instability to these states in the form of a somewhat elongated bond and lowered vibration frequency. This may account for the properties of the $1s\sigma_g 3p\sigma^*_u\ {}^3\Sigma^+_u$-state, whose bond length and frequency are slightly out of line with those of the other states. It is an example of "configuration interaction," which will be discussed in more detail in Chapter 12. A strong configuration interaction of the "ionic state," $1s\sigma_g 2p\sigma^*_u\ {}^1\Sigma^+_u$, with the $1s\sigma_g\sigma^*_u 2s\ {}^1\Sigma^+_u$ or $1s\sigma_g\sigma^*_u 2p\ {}^1\Sigma^+_u$-state may also take place at large internuclear distances, so that the singlet $1s\sigma_g 2p\sigma^*_u$-state may actually dissociate into two neutral hydrogen atoms rather than a positive and negative hydrogen ion (see *7*). It is evident that configuration interactions can cause some complications in the simple molecular orbital picture. This picture should, however, never be taken as more than a convenient first approximation to the true state of affairs, and such complications are to be expected as our knowledge become more complete. It will be recalled that configuration interactions are also important in the explanation of the energies of certain atomic states (page 325 and 344).

Exercise. The $(2p\sigma^*_u)^2\ {}^1\Sigma^+_g$ state of H_2 is formed from two antibonding orbitals yet it shows a minimum in its potential energy curve. Why is this? Why does the energy of this state change so slowly with interatomic distance at large distances (see Fig. 11-3)? [Hint: Consider the configuration interaction of $(2p\sigma^*_u)^2\ {}^1\Sigma^+_g$ with $(1s\sigma_g)^2\ {}^1\Sigma^+_g$ at very large interatomic distances, where the molecular orbitals $1s\sigma_g$ and $2p\sigma^*_u$ have the same energy. The zero-order wave functions of H_2 will then have the form

$$\psi_1 = (1s\sigma_g)_1\,(1s\sigma_g)_2 - (2p\sigma^*_u)_1\,(2p\sigma^*_u)_2$$

and

$$\psi_2 = (1s\sigma_g)_1\,(1s\sigma_g)_2 + (2p\sigma^*_u)_1\,(2p\sigma^*_u)_2$$

Write out these wave functions using LCAO molecular orbitals. To what states of hydrogen do they each correspond? As the nuclei are brought together the coefficients of the two configurations in each wave function become unequal, ψ_1 going predominantly to $(1s\sigma_g)^2$ and ψ_2 to $(2p\sigma^*_u)^2$.]

C. Other Diatomic Molecules

a. The system He_2

Let us consider next the states that may arise in a system containing two helium atoms. If all four electrons in this system are placed in the lowest available orbitals in accordance with the Pauli principle, we obtain the state $(1s\sigma_g)^2\,(2p\sigma^*_u)^2\,{}^1\Sigma^+_g$. This gives two bonding electrons and two antibonding electrons, and since the antibonding electrons are more potent than the bonding electrons (see footnote on page 392), a stable He_2 molecule in its ground state is not to be expected. Of course, aside from very weak atomic pairs stabilized by van der Waals forces, no such molecule is ever observed.

On the other hand, if one of the antibonding electrons is excited or removed altogether, the two remaining $1s\sigma_g$ bonding electrons should be able to cope with the antibonding tendencies of the one remaining $2p\sigma^*_u$ antibonding electron, so stable He_2 molecules should be expected. A large number of such He_2 molecules are in fact observed in gas discharge tubes. The properties of some of them are listed in Table 11-3. The equilibrium internuclear distance happens to be quite close to that found in the excited states of H_2, which makes it an easy matter to compare the energies of the various molecular orbitals of the two molecules under more or less equivalent conditions.

TABLE 11-3

CHARACTERISTICS OF THE POTENTIAL ENERGY CURVES OF THE EXCITED STATES OF He_2

Configuration $1s\sigma_g^2\ 2p\sigma^*_u$ plus	Energy of minimum† (cm.$^{-1}$)		Internuclear distance (in Å.)		Nuclear vibration frequency (cm.$^{-1}$)	
	Singlet	Triplet	Singlet	Triplet	Singlet	Triplet
$2p\sigma^*_u\ {}^1\Sigma^+_g$	‡	—	‡	—	‡	—
$2s\sigma_g\ {}^{1,3}\Sigma^+_u$	31958	34302	1.041	1.048	1791	1811
$2p\pi_u\ {}^{1,3}\Pi_g$	28423	29534	1.074	1.063		1698
$3s\sigma_g\ {}^{1,3}\Sigma^+_u$	13263	13910	1.076	1.071		1654
$3p\sigma^*_u\ {}^{1,3}\Sigma^+_g$		23412		1.097		1481
$3p\pi_u\ {}^{1,3}\Pi_g$	12481	12794	1.076	1.073	1651	1725
$3d\sigma_g\ {}^{1,3}\Sigma^+_u$	12581	12751	1.09	1.08		1549
$3d\pi^*_g\ {}^{1,3}\Pi_u$	12414	12550	1.09	1.08		1569
$3d\delta_g\ {}^{1,3}\Delta_u$	12063	12097	1.09	1.08		1636
$He_2^+\ {}^2\Sigma^+_u$	0		1.080		1627	

† Below the minimum of He_2^+.

‡ No minimum; molecule is unstable.

Comparison of Tables 11-2 and 11-3 shows that the energy of a given orbital in He_2 tends to be somewhat lower than that of the same orbital in H_2. This is particularly true of the $2s\sigma_g$-orbital, which in He_2 is considerably more stable than the $2p\pi_u$-orbital, whereas in H_2 the $2s\sigma_g$- and $2p\pi_u$-orbitals have about the same energy. The reason for this is easy to see. In Chapter 10 we have seen that in atoms *s*-orbitals tend to be more stable than *p*-orbitals because the *s*-orbitals penetrate farther into the core. The same penetration effect must be expected in He_2 in both the united atom and in the separated atoms. Since the $2s\sigma_g$ molecular orbital is related to 2*s* atomic orbitals in both the united atom and the separated atom, and since $2p\pi_u$ is a 2*p*-orbital in both of these extreme cases, one would expect the $2s\sigma_g$ molecular orbital to be more stable than the $2p\pi_u$ molecular orbital also at intermediate internuclear distances, as we have just found to be the case in He_2.

b. The correlation diagram for homonuclear diatomic molecules other than hydrogen

Because of these penetration effects the correlation of the united and separated atomic states of molecular orbitals is quite different in He_2, Li_2, C_2, etc., from what has been deduced for the H_2^+ molecule, where no penetration effects are present. In Fig. 11-4, a correlation diagram is shown which

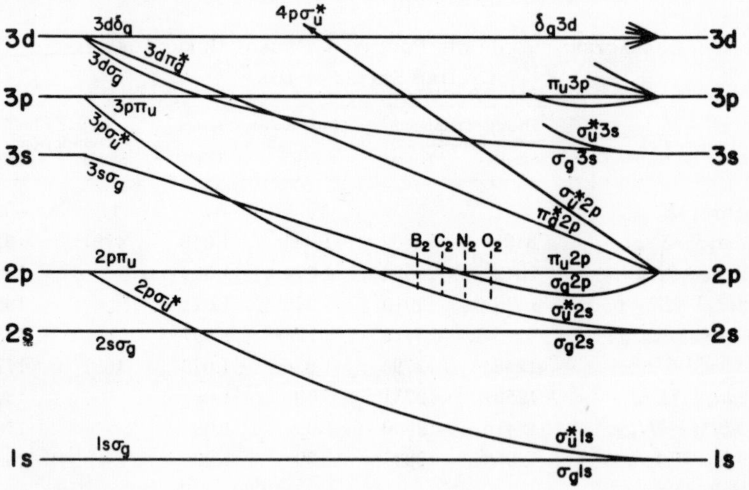

FIG. 11-4. Correlation of separated atom states (right side of diagram) and united atom states (left side of diagram) for molecular orbitals in diatomic molecules having identical nuclei.

takes these effects into account, and which is therefore applicable to diatomic molecules other than H_2 containing two identical atoms. (Such molecules are known as *homonuclear* diatomic molecules, to distinguish them from the *heteronuclear* diatomic molecules such as HCl, HeH, CO, etc.) The principles used in the construction of this diagram are as follows. The states of the united and separated atoms are arranged in order of their energies on the two sides of the diagram. The (g, u) character and the component of electronic angular momentum along the molecular axis are noted for each state. Lines are then drawn across the diagram connecting states with both the same (g, u) character and the same component of angular momentum. These lines must be drawn in such a way that no two lines of the same character cross (thus two σ_g-lines must not cross, though a σ_u- and a σ_g-line may cross). This noncrossing requirement is the result of the same considerations as those discussed in connection with the construction of correlation diagrams for vibrating systems (page 142).*

The abscissa of Fig. 11-4 can be regarded in a general way as a measure of the internuclear distance, being infinite on the right and zero on the left. There is, of course, no way of knowing exactly how the correlation lines should be drawn in the center of the diagram if some scale of internuclear distances is to be added to the abscissa. Among the states that correlate with the $2p$-level of the separated atoms, we should expect that $\sigma_g 2p$ will lead to the strongest bond when the internuclear distance is large, because it gives the largest concentration of electrons between the nuclei. Similarly $\sigma^*_u 2p$ ought to lead to the greatest repulsion. At large nuclear separations we should expect the correlation lines for $\pi_u 2p$ and $\pi^*_g 2p$ to fall between the $\sigma_g 2p$- and $\sigma^*_u 2p$-lines, with $\pi_u 2p$ being more stable than $\pi^*_g 2p$ because it is a bonding orbital. The correlation line starting at $\sigma_g 2p$ must, however, cross the $\pi_u 2p$-line at some intermediate internuclear distance, since $\sigma_g 2p$ is promoted to $3s\sigma_g$ in the united atom, whereas the $\pi_u 2p$-orbital becomes $2p\pi_u$ in the united atom. The position of this cross-over must be determined by reference to experimental observations, and in constructing Fig. 11-4 we have used the interpretations made by Mulliken (*7*), which will be described below.

Let us now use this correlation diagram to interpret the electronic states of the diatomic molecules formed from elements in the first row of the periodic table. Since the orbitals resulting from the $1s$-electrons will always contribute merely $(1s\sigma_g)^2 (2p\sigma^*_u)^2$ to the configuration, and since the

* Interestingly enough, the noncrossing rule does not hold in the H_2^+ correlation diagram, where the lines from the $2s\sigma_g$- and the $3d\sigma_g$-states cross. This appears to be an example of "accidental" degeneracy, which can be expected in this system alone.

overlap of the 1*s*-orbitals will be rather slight, so that they will continue to resemble the *K*-shells of the separated atoms, it is more realistic to indicate these four electrons by the symbols *KK*. They have little effect on the energy, other than to introduce a repulsion if the nuclei are brought quite close together. In discussing these molecules, it is also more realistic to use the separated atom terminology, since the nuclei are relatively far apart.

c. The system Li_2

The Li_2 molecule should have the ground state $KK\sigma_g 2s^2\,{}^1\Sigma^+{}_g$, and since the number of antibonding electrons is less than the number of bonding electrons, a stable molecule is to be expected. Such a molecule is indeed observed in lithium vapor. The bond strength is low (1.03 e.v.) and the internuclear distance is large (2.672 Å.), probably because the 2*s*-orbitals in the separated Li atoms are rather diffuse, so that the electrons cannot be very much concentrated in the bond between the two nuclei. On the other hand, the bond in Li_2 is much stronger than would be expected from the behavior of the $\sigma_g 2s$ orbital in H_2^+. The reason for this is that the incomplete screening of the nucleus by the two 1*s*-electrons of the Li atom pulls the 2*s*-cloud in toward the nucleus to a considerable extent. Therefore the 2*s*-cloud is much less diffuse in Li than it is in the H atom, and it is able to form a stronger bond. This effect is important because the bonds involving 2*s*- and 2*p*-electrons between atoms in the first row of the periodic table (Li through F) are very strong, and this would hardly be expected from the behavior of the same molecular orbitals in H_2^+.

d. The system Be_2

The ground state of Be_2 should be $KK\ (\sigma_g 2s)^2\ (\sigma^*{}_u\, 2s)^2\,{}^1\Sigma^+{}_g$. Since it contains two bonding and two antibonding electrons, we should not expect it to be stable. There is, in fact, no experimental evidence for a stable Be_2 molecule. Furthermore, the diatomic molecules formed by other elements of Group II in the periodic table either have very small dissociation energies (about 0.25 e.v. for Zn_2, 0.087 e.v. for Cd_2, 0.060 e.v. for Hg_2) or the molecules are not observed at all (as is the case with Ca_2, Sr_2, and Ba_2).

e. The system B_2

Two boron atoms contain six electrons in addition to the four *K*-electrons. The configuration of the ground state depends on the location of the cross-over point of the correlation lines of the $\sigma_g 2p$- and $\pi_u 2p$-orbitals. Thus we might expect either $KK(\sigma_g 2s)^2 (\sigma^*_u 2s)^2 (\sigma_g 2p)^2$, or $KK(\sigma_g 2s)^2 (\sigma^*_u 2s)^2 (\pi_u 2p)^2$ or $KK(\sigma_g 2s)^2 (\sigma^*_u 2s)^2 \sigma_g 2p \pi_u 2p$. In each of these configurations there are four bonding electrons and two antibonding electrons, so we should in any case expect the B_2 molecule to be stable. This is indeed found experimentally: the B_2 molecule has a dissociation energy of about 3.0 e.v. and an internuclear distance of 1.589 Å. It is found to be a $^3\Sigma^-{}_g$-state. This means that the configuration must be that containing two electrons in the $\pi_u 2p$-orbital, since neither of the other two configurations can give rise to a $^3\Sigma^-{}_g$-state. Furthermore the configuration $(\pi_u 2p)^2$ should give rise to $^1\Delta_g$-, $^1\Sigma^+{}_g$-, and $^3\Sigma^-{}_g$-states, and according to Hund's multiplicity rule, the triplet state should be the most stable of these states.

f. The system C_2

The ground state of the C_2 molecule (which exists in flames, the carbon arc, and gas discharges through organic vapors) is found to be $^3\Pi_u$, which indicates that its configuration is $KK(\sigma_g 2s)^2 (\sigma^*_u 2s)^2 (\pi_u 2p)^3 \sigma_g 2p$. Evidently the internuclear distance in C_2 is smaller than, but very close to, the cross-over point of the $\sigma_g 2p$ and $\pi_u 2p$ orbitals, but the molecule, instead of having a $(\pi_u 2p)^4\,{}^1\Sigma^+{}_g$-configuration for its ground state, finds it advantageous to place one electron in the $\sigma_g 2p$-orbital. In this way it can exist in a triplet state, thereby acquiring extra stability through the favorable electron correlation that is a consequence of increased multiplicity. An excited $^1\Sigma^+{}_g$-state is observed in the C_2 spectrum, but since transitions to the triplet system are forbidden, its position relative to the ground state is not known. The strong absorption and emission of C_2 vapor in the visible, known as the Swan bands, is caused by a transition to a $^3\Pi_g$-state that probably has the configuration $(\sigma g 2s)^2 (\sigma^*_u 2s) (\pi_u 2p)^3 (\sigma_g 2p)^2$.

g. The system N_2

The ground state of N_2 would be expected to be $KK(\sigma_g 2s)^2 (\sigma^*_u 2s)^2 (\pi_u 2p)^4 (\sigma_g 2p)^2\,{}^1\Sigma^+{}_g$, and the ground state is actually found to be a $^1\Sigma^+{}_g$. In N_2 all of the $2p$ bonding orbitals are filled and none of the $2p$ antibonding orbitals are occupied, so we should expect the molecule to be very strongly bonded, as of course it is (the dissociation energy is probably 9.756 e.v.). Since excitation must raise electrons to antibonding orbitals, which have much higher energies than the bonding orbitals, nitrogen begins to absorb strongly only in the far ultra-violet, around 1400 Å. (the so-called Lyman-Birge-

Hopfield bands). This strong absorption is caused by a transition to a $(\pi_u 2p)^4\,(\sigma_g 2p)\,(\pi^*{}_g 2p)\,{}^1\Pi_g$-state, but there is a lower lying $(\pi_u 2p)^3\,(\sigma_g 2p)^2\,\pi^*{}_g 2p\,{}^3\Sigma^+{}_u$-state whose transition to the ground state (forbidden by the selection rule $\Delta S = 0$) is responsible for the most prominent bands in the blue and violet light of the night sky (called the Vegard-Kaplan bands).

h. The system O_2

For the ground state of O_2 we should expect the configuration $KK(\sigma_g 2s)^2\,(\sigma^*{}_u 2s)^2\,(\sigma g 2p)^2\,(\pi_u 2p)^4\,(\pi^*{}_g 2p)^2$, for which the possible states are ${}^1\Delta_g$, ${}^1\Sigma^+{}_g$, and ${}^3\Sigma^-{}_g$. According to Hund's multiplicity rule, the most stable of these states should be ${}^3\Sigma^-{}_g$. This agrees with the experimental observations. It also explains unequivocally the well-known paramagnetism of oxygen gas. The other two states belonging to this configuration are also observed, the ${}^1\Delta_g$-state being at 0.98 e.v. and the ${}^1\Sigma^+{}_g$-state being at 1.63 e.v. above the ground state. (The fact that the ${}^1\Delta$-state is more stable than the ${}^1\Sigma$-state is consistent with Hund's angular momentum rule, which predicts the greatest stability for the state of highest orbital angular momentum, other things being equal.) Forbidden transitions to these levels from the ground state are responsible for very weak absorption bands in the infra-red, which are observable in the solar spectrum because of absorption in the earth's atmosphere. There are (aside from the KK-electrons) eight bonding electrons and four antibonding electrons in O_2, so we should expect the dissociation energy to be less in O_2 than in N_2. Its value is 5.080 e.v. The bond length is also longer than in N_2 (1.207 Å.). Because of this larger distance the $\pi_u 2p$-orbital is less stable than the $\sigma_g 2p$-orbital and the first excited configuration is $(\sigma_g 2p)^2\,(\pi_u 2p)^3\,(\pi^*{}_g 2p)^3$. The very strong absorption of oxygen below 1950 Å. is caused by a transition to a ${}^3\Sigma^-{}_u$-state of this configuration (this transition gives rise to the Schuman-Runge bands).

i. The system F_2

In fluorine the most stable configuration is $KK(\sigma_g 2s)^2\,(\sigma^*{}_u 2s)^2\,(\sigma_g 2p)^2\,(\pi_u 2p)^4\,(\pi^*{}_g 2p)^4$, which can give only a ${}^1\Sigma^+{}_g$-state, in agreement with experimental observations. There are eight bonding electrons and six anti-bonding electrons, so the dissociation energy is even less than that of oxygen, having a value of 1.6 e.v. The internuclear distance is 1.435 Å., which is also larger than that of oxygen, indicating a weaker bond.

j. The system Ne_2

Finally, in the first row of the periodic table we come to the molecule Ne_2, whose most stable configuration should be $KK(\sigma_g 2s)^2\,(\sigma^*_u 2s)^2\,(\sigma_g 2p)^2\,(\pi_u 2p)^4\,(\pi^*_g 2p)^4\,(\sigma^*_u 2p)^2$. Since there is an equal number of bonding and antibonding electrons, no stable molecule is to be expected, and this of course is actually the case.

k. Chemical bonds in diatomic molecules

Herzberg has pointed out that in the series of molecules just discussed, it is possible to define an effective number of bonds in such a way that it agrees with the chemist's concepts of the bonding in these substances. The rule is this: The number of bonds in a molecule is equal to one half the difference between the number of bonding electrons and the number of antibonding electrons. In Table 11-4 (modified from Herzberg, *6*, p. 343), we see that Ne_2 has no bond, F_2 a single bond, O_2 a double bond, and N_2 a triple bond, just as the chemist pictures them. The number of bonds also parallels, in a general way, the dissociation energy and the bond length. (A uniform relationship between these quantities would hardly be expected, because the different bonding and antibonding orbitals should not have the same effects on the bond strengths.)

It is interesting to note in Table 11-4 that when O_2^+ is formed from O_2 by removing a $\pi^*_g 2p$ *antibonding* electron, the dissociation energy is increased, whereas when N_2^+ is formed from N_2 by removal of a $\sigma_g 2p$ *bonding* electron, the dissociation energy is decreased. This, of course, is as expected.

l. Heteronuclear diatomic molecules

The discussion given above has been restricted to molecules in which the two nuclei are identical ("homonuclear diatomic molecules"). If this restriction is removed we should, strictly speaking, prepare a new correlation diagram in which the absence of degeneracy between the levels of the separated atoms is taken into account. If, however, the two nuclei are not too different, as for instance in carbon monoxide and nitric oxide, we may continue to use the homonuclear correlation diagram as a first approximation. In this way the properties of such molecules as CO, NO, CN, BO, and BN can be interpreted. If the two nuclei differ by too much, however, another approach is needed. This is the case with the hydrides LiH, BeH, BH, CH, NH, OH, HF, etc.; the reader is referred to Herzberg's book (*6*, pp. 339*ff*.) for discussions of these molecules.

If a diatomic molecule is formed from two atoms occurring in different rows of the periodic table, such as SiO, SrCl, etc., an approach similar to that used above can also be adopted. If, for example, the molecule is SiO,

TABLE 11-4

SUMMARY OF PROPERTIES OF GROUND STATES OF DIATOMIC MOLECULES†

Molecule	Electronic configuration	State	N_b = No. of bonding electrons	N_a = No. of antibonding electrons	$\frac{1}{2}(N_b - N_a)$ = No. of bonds	Dissociation energy (volts)	Bond length (Å.)	Force constant (dyne cm.$^{-1}$ × 10^5)
H_2^+	$(\sigma_g 1s)$	$^2\Sigma^+{}_g$	1	0	$\frac{1}{2}$	2.648	1.06	1.56
H_2	$(\sigma_g 1s)^2$	$^1\Sigma^+{}_g$	2	0	1	4.476	0.74	5.60
He_2^+	$(\sigma_g 1s)^2(\sigma^*{}_u 1s)$	$^2\Sigma^+{}_u$	2	1	$\frac{1}{2}$	(3.1)	1.08	3.13
He_2	$(\sigma_g 1s)^2(\sigma^*{}_u 1s)^2$ (= KK)	$^1\Sigma^+{}_g$	2	2	0	0	—	—
Li_2	$KK(\sigma_g 2s)^2$	$^1\Sigma^+{}_g$	2	0	1	1.03	2.67	0.25
Be_2	$KK(\sigma_g 2s)^2(\sigma^*{}_u 2s)^2$	$^1\Sigma^+{}_g$	2	2	0	—	—	—
B_2	$KK(\sigma_g 2s)^2(\sigma^*{}_u 2s)^2(\pi_u 2p)^2$	$^3\Sigma^-{}_g$	4	2	1	(3.0)	1.59	3.60
C_2	$KK(\sigma_g 2s)^2(\sigma^*{}_u 2s)^2(\pi_u 2p)^3(\sigma_g 2p)$	$^3\Pi_u$	6	2	2	(5.9)	1.31	9.55
N_2^+	$KK(\sigma_g 2s)^2(\sigma^*{}_u 2s)^2(\pi_u 2p)^4(\sigma_g 2p)$	$^2\Sigma^+{}_g$	7	2	$2\frac{1}{2}$	8.73	1.12	20.1
N_2	$KK(\sigma_g 2s)^2(\sigma^*{}_u 2s)^2(\pi_u 2p)^4(\sigma_g 2p)^2$	$^1\Sigma^+{}_g$	8	2	3	9.756	1.09	23.1
O_2^+	$KK(\sigma_g 2s)^2(\sigma^*{}_u 2s)^2(\pi_u 2p)^4(\sigma_g 2p)^2(\pi^*{}_g 2p)$	$^2\Pi_g$	8	3	$2\frac{1}{2}$	6.48	1.12	16.6
O_2	$KK(\sigma_g 2s)^2(\sigma^*{}_u 2s)^2(\pi_u 2p)^4(\sigma_g 2p)^2(\pi^*{}_g 2p)^2$	$^3\Sigma^-{}_g$	8	4	2	5.080	1.21	11.8
F_2	$KK(\sigma_g 2s)^2(\sigma^*{}_u 2s)^2(\pi_u 2p)^4(\sigma_g 2p)^2(\pi^*{}_g 2p)^4$	$^1\Sigma^+{}_g$	8	6	1	(1.6)	1.44	4.45
Ne_2	$KK(\sigma_g 2s)^2(\sigma^*{}_u 2s)^2(\pi_u 2p)^4(\sigma_g 2p)^2(\pi^*{}_g 2p)^4(\sigma^*{}_u 2p)^2$	$^1\Sigma^+{}_g$	8	8	0	0	—	—

† Revised from Herzberg (*6*)

we may construct molecular orbitals from linear combinations of the atomic orbitals of the outer shells of each atom, such as the σs-orbital $(a3s_{\mathrm{Si}} + b2s_{\mathrm{O}})$, the constants a and b being regarded as adjustable constants that allow for differences in the electron affinities of the two atoms. The orbitals will be generally similar to those we have described, and for this reason, Mulliken has suggested a generalized terminology for them. Thus for orbitals resembling

$$\sigma_g 2s, \ \sigma^*_u 2s, \ \sigma_g 2p, \ \pi_u 2p, \ \pi^*_g 2p, \ \sigma^*_u 2p$$

he uses the symbols

$$z\sigma, \quad y\sigma, \quad x\sigma, \quad w\pi, \quad v\pi, \quad u\sigma,$$

respectively.

Exercises. [The data required for the following exercises are to be found in Herzberg (*6*, Table 39), Rosen (*8*) or Landolt-Bornstein (*9*).]

(1) Discuss the ground states of the following groups of molecules on the basis of their experimentally observed term characteristics:

a. O_2, S_2, SO

b. CO, CS, CSe, SiO, SiSe, SiTe, GeO, GeS, GeSe, GeTe, SnO, SnS, SnSe, SnTe, PbO, PbS, PbSe, PbTe

c. N_2, P_2, As_2, Sb_2, Bi_2, PN, AsN, SbN

d. PO, NO, AsO, SbO

Do they show the resemblances expected in the light of the discussion given on page 405?

(2) Should the dissociation energy of NO^+ be greater than or less than the dissociation energy of NO? Check your answer by looking up the experimental values.

(3) The molecules BF, CO, and N_2 each contain fourteen electrons. Their ground states are all $^1\Sigma^+$ and the first excited singlet states are all $^1\Pi$. What orbital in $^1\Sigma^+$ does the excited electron probably come from, and what orbital in $^1\Pi$ does it occupy?

(4) On the basis of the experimentally observed spin and electronic angular momenta, give the probable configurations of the ground state and the first excited state of the following molecules, each of which contains thirteen electrons: BO, CN, CO^+ and N_2^+. Which of the orbitals, $\sigma_g 2p$ or $\pi_u 2p$, is the more stable in these molecules?

D. Directed Valence Bonds in H_2O and NH_3

One of the characteristic properties of molecules containing more than two atoms, such as the water molecule, ammonia, and methane, is their well-defined architecture. The two hydrogen atoms in H_2O are located at a definite distance from the oxygen atom and from each other. It is not difficult to see that this is a direct consequence of the principles that have been discussed earlier in this book.

In the first place, we must explain why water has the formula H_2O and not HO or H_3O. The ground state of the oxygen atom is $1s^2\,2s^2\,2p^4\,{}^3P$. Two of the $2p$-orbitals must therefore contain unpaired electrons which, according to the Heitler-London theory, can each form a bond with a hydrogen atom. Thus the oxygen atom has the power to form two bonds with hydrogen atoms, and no more than two.

These bonds will, of course, form in such a way that they have the maximum possible stability. We have seen that the factor making for stability in a bond is a high electron-cloud density in the region between the two nuclei at either end of the bond. This high density will be achieved if the atomic orbitals forming each bond (i.e., one of the oxygen orbitals and a hydrogen $1s$-orbital) overlap as effectively as possible. In more mathematical terms, we may say that increasing the amount of overlap tends to increase the value of the exchange integral, which is presumed to be a major factor in stabilizing a bond. The two partially filled orbitals on the oxygen atom can be concentrated much more effectively in space if the $2p_x$ and $2p_y$ forms are used rather than the complex forms, $2p_{+1}$ and $2p_{-1}$. The $2p_x$-orbital consists of two lobes protruding in opposite directions from the nucleus, and the $2p_y$-orbital has two identical lobes that point at right angles to the $2p_x$-lobes. The third $2p$-orbital, $2p_z$, contains two electrons, and thus cannot participate in the bonding. If, now, one hydrogen atom is placed in one of the lobes of the $2p_x$-orbital and the other hydrogen atom is placed in one of the lobes of the $2p_y$-orbital, two strong OH bonds will be formed with a 90° angle between them (see Fig. 11-5). Thus we should expect the H–O–H bond angle in water to be 90° and the two OH bonds to be indistinguishable. The bonds in water are actually found to be identical, but the bond angle turns out to be 104.5° instead of a right angle. At least part of this spreading of the bond angle can be blamed on the repulsive Pauli forces (see page 319) that act between the two hydrogen atoms, since these two atoms are not bonded together. Thus it appears that the major factor responsible for the shape of the water molecule is the geometry of the $2p$-functions. (It is interesting that we first encountered this geometrical property in Chapter 3 in studying the normal modes of the ocean.)

A similar argument may be applied to the ammonia molecule, NH_3. The ground state of the nitrogen atom is $1s^2\,2s^2\,2p^3\,{}^4S$. There are three singly occupied $2p$-orbitals, so nitrogen can form bonds with three hydrogen atoms. Again, we should expect the most favorable form of the orbitals for bonding to be $2p_x$, $2p_y$, and $2p_z$, and it is evident that these orbitals will tend to form three identical bonds at right angles to each other. The three bond lengths in NH_3 are found to be the same, but the H–N–H angles are 108° instead of 90°. If the molecule were planar, this angle would be 120°, so the observed angle is 12° away from the planar angle and 18° away from the predicted right angle. Although we can be fairly confident that we have hit upon the important factor responsible for the molecule's not being planar, the deviation from a right angle is embarassingly large. An additional factor that may be responsible for this deviation will be mentioned in the next section.

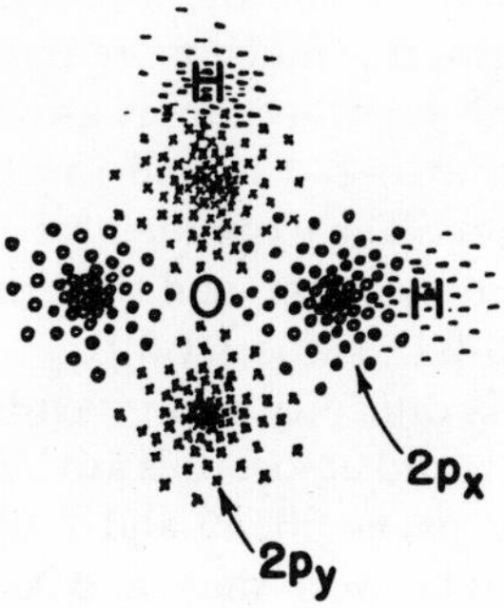

FIG. 11-5. Formation of the H_2O molecule from the overlap of the electron clouds about two hydrogen atoms with the $2p_x$- and $2p_y$-orbitals of oxygen.

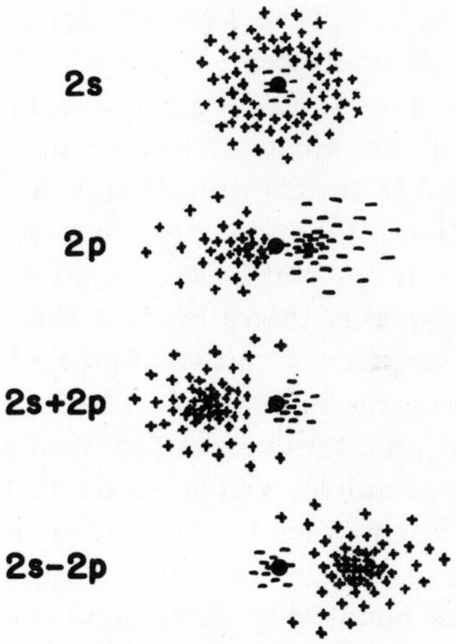

FIG. 11-6. Concentration of charge by the hybridization of $2s$- and $2p$-orbitals.

E. Hybridization and Directed Valence Bonds

The ground state of the carbon atom has the configuration $1s^2\,2s^2\,2p^2$, and its multiplet term is ^3P (see page 342). Being a triplet state, there are only two unpaired electrons, so one would think that carbon ought to be divalent. How, then, can we explain the occurrence of molecules such as CH_4 and CCl_4, in which *four* of the carbon electrons must be unpaired?

Evidently the compounds of carbon are derived from some electronic configuration that can have a multiplicity of five (that is, $S = 4 \times 1/2$, so that $2S + 1 = 5$). According to Fig. 9-5, the $1s^2\,2s\,2p^3$-configuration of carbon has a ^5S state with an energy 33735.2 cm.$^{-1}$ ($= 4.2$ e.v. $= 96$ kcal./mole) above the energy of the ^3P ground state. Since carbon is almost invariably tetravalent, it appears that the energy gained by forming four bonds in place of two is more than enough to supply the extra energy required to elevate one of the two $2s$-electrons in the ground state to a $2p$-orbital. The compounds of carbon must therefore be derived from atoms in this excited quintet state.

This explanation of the tetravalence of carbon is, however, incomplete because it fails to account for the observation that all four C–Cl bonds in CCl_4 are equivalent. If three of the valence electrons on carbon are in $2p$-orbitals, we should expect that three of the C–Cl bonds would make approximately 90° angles with one another. The fourth chlorine atom should be bonded to the $2s$-orbital; and since the $2s$-orbital is much more diffuse than the $2p$-orbital and cannot overlap the orbitals on other atoms as effectively as can a $2p$-orbital, we should expect the fourth C–Cl bond in CCl_4 to be much weaker than the other three C–Cl bonds. The experimental evidence, however, shows that the four C–Cl bonds are identical, and that the angles between these bonds are all tetrahedral ($109°28' = \cos^{-1}(1/3)$).

The quantum mechanical resolution of this difficulty was given by Slater and by Pauling, who pointed out that if the $2s$- and $2p$-orbitals are hybridized with each other (in the sense described on pages 66, 75 and 136), then a much higher electron-density can be produced between the C and X atoms in a C–X bond, so that the bond is much stronger. One of the possible hybrids of a $2s$-orbital with three $2p$-orbitals leads to tetrahedral bond angles and four completely equivalent bonds, as we shall show below.

a. Hybridization of an *s*-Orbital with a *p*-Orbital (*sp*-hybrids)*

The effect of hybridization in concentrating the electron cloud in the region of the bond can be seen most easily for an atom containing a $2s$ and a $2p_x$ valence electron. The $2p_x$-function is positive on one side of the nucleus and negative on the other side, whereas the $2s$-function has spherical symmetry (see Fig. 11-6). There is a radial node in the $2s$-function inside of which the orbital is negative and outside of which it is positive, but this

* The $2s$ and $2p$ orbitals in atoms such as carbon differ appreciably in energy, so one might think that they would not be able to hybridize to any great extent. The energy difference is, however, of the same order of magnitude as the perturbation energy resulting from the formation of a chemical bond. Therefore extensive hybridization is possible in spite of the lack of degeneracy (cf. page 136 and Eq. (D-16) on page 175).

nodal surface is quite close to the nucleus, and the amplitude of the orbital inside the nodal surface is small. Therefore the 2*s*-orbital is positive in most of the region involved in bonding. If we form a hybrid by taking the sum of the 2*s*- and $2p_x$-orbitals, the amplitudes tend to cancel on one side of the nucleus and add on the other side, giving an orbital with a lobe that is much more pronounced on one side of the atom than on the other side. Such an orbital should form a much stronger bond than either a 2*s*-orbital by itself or a 2*p*-orbital by itself. Similarly, by taking the *difference* between a 2*s*- and a $2p_x$-orbital, we may form a new orbital that points in the opposite direction to the orbital formed from the *sum* of 2*s* and $2p_x$. In the same way, improved bonding orbitals can be formed by combining 3*s*- with 3*p*-orbitals, 4*s*- with 4*p*-orbitals, etc.

By forming different mixtures of *s*-, p_x-, p_y-, and p_z- orbitals having the same principle quantum number, it is possible to obtain bonding hybrids that lead to bond angles having widely different values. For instance, out of the orbitals 2*s* and $2p_x$ we may obtain the hybrids

$$\begin{aligned} \psi_1 &= a_1 2s + b_1 2p_x \\ \psi_2 &= a_2 2s + b_2 2p_x \end{aligned} \tag{E-1}$$

These two orbitals will be normalized (assuming that 2*s* and $2p_x$ are themselves initially normalized) if

$$\int \psi_1^2\, d\tau = a_1^2 + b_1^2 = 1 \quad \text{and} \quad \int \psi_2^2\, d\tau = a_2^2 + b_2^2 = 1 \tag{E-2a}$$

and they will be orthogonal if

$$\int \psi_1 \psi_2\, d\tau = a_1 a_2 + b_1 b_2 = 0 \tag{E-2b}$$

In addition, they will be equivalent if

$$|a_1| = |a_2| \text{ and } |b_1| = |b_2| \tag{E-3}$$

which means that

$$a_1 = a_2 = b_1 = -b_2 = 1/\sqrt{2} \tag{E-3a}$$

or

$$\begin{aligned} \psi_1 &= (2s + 2p_x)/\sqrt{2} \\ \psi_2 &= (2s - 2p_x)/\sqrt{2} \end{aligned} \tag{E-4}$$

Bonding orbitals of this type appear to be present in the gaseous dihalides and dimethyl derivatives of zinc, cadmium and mercury, MX_2. The two M–X bonds in these molecules are known to be equivalent and to point in opposite directions. In forming these molecules, one of the two ns valence electrons of the ground state $(ns)^2\ {}^1S$-configuration of the neutral metal atom is evidently excited to the $ns\,np\ {}^3P$ state. On the approach of two halide atoms or two methyl radicals, this state hybridizes in the manner described above, giving a 180° angle between the two M–X bonds.

b. Hybridization of *s* with two *p*-orbitals (*sp*² hybrids)

A bond angle of 120° may be obtained by hybridizing two p-orbitals with the s-orbital in the following way. The three most general hybrids of these orbitals are

$$\begin{aligned} \psi_1 &= a_1 s + b_1 p_x + c_1 p_y \\ \psi_2 &= a_2 s + b_2 p_x + c_2 p_y \\ \psi_3 &= a_3 s + b_3 p_x + c_3 p_y \end{aligned} \tag{E-5}$$

These orbitals must be orthogonal and normalized, which leads to the six equations

$$a_i^2 + b_i^2 + c_i^2 = 1 \qquad (i = 1, 2, 3) \tag{E-6a}$$

$$a_i a_j + b_i b_j + c_i c_j = 0 \quad (i \neq j = 1, 2, 3) \tag{E-6b}$$

There are nine unknowns and six relationships, so we are free to choose three of the unknowns. If we take $a_1 = b_1/\sqrt{2} = 1/\sqrt{3}$ and $c_1 = 0$ we obtain the set of orbitals

$$\begin{aligned} \phi_1 &= (1/\sqrt{3})\,(s + \sqrt{2}p_x) \\ \phi_2 &= (1/\sqrt{3})\,[s - (\sqrt{2}/2)p_x + (\sqrt{6}/2)p_y] \\ \phi_3 &= (1/\sqrt{3})\,[s - (\sqrt{2}/2)p_x - (\sqrt{6}/2)p_y] \end{aligned} \tag{E-7}$$

Since the functions ϕ_2 and ϕ_3 may be obtained from ϕ_1 by rotation through $\pm$ 120° about the z-axis, these three hybrids must form three equivalent bonds with bond angles of 120° between them. This kind of hybridization evidently occurs in boron trimethyl, $B(CH_3)_3$, which is known to have the boron in the same plane as the three carbons, with the three boron–carbon bonds equivalent and making an angle of 120° with each other. The same is true of BF_3, BCl_3, and BBr_3.

A more general set of orthogonal and normalized hybrids obtainable from s, p_x, and p_y is

$$\chi_1 = (1/\sqrt{1+\gamma^2})\,(s + \gamma p_x)$$
$$\chi_2 = (1/\sqrt{2(1+\gamma^2)})\,(\gamma s - p_x + \sqrt{1+\gamma^2}p_y) \qquad \text{(E-8)}$$
$$\chi_3 = (1/\sqrt{2(1+\gamma^2)})\,(\gamma s - p_x - \sqrt{1+\gamma^2}p_y)$$

where γ is a parameter that can vary from zero to infinity. The case $\gamma = 0$ gives the three hybrids s, $(p_x + p_y)$, and $(p_x - p_y)$. These orbitals have the same shapes as the original orbitals, s, p_x, and p_y, but are rotated by 45° about the z-axis. The case $\gamma = \sqrt{2}$ gives the trigonal hybrid described above, with three equivalent bonds making 120° angles with one another. The case $\gamma = \infty$ gives the hybrids p_x, $(s + p_y)$, and $(s - p_y)$—that is, a pure p_x-orbital plus the pair of linear hybrids forming bonds at 180°, as described above. It is important to note that although χ_2 and χ_3 are always equivalent, being reflections of one another in the xz-plane, only the choice $\gamma = \sqrt{2}$ can give *three* equivalent orbitals. Furthermore, all bond angles between 90° and 180° are possible, but only the bond angle of 120° leads to three equivalent orbitals.

The value of γ appropriate to a given molecule can, in principle, be determined by minimizing the calculated total energy of the molecule with respect to γ. The calculations required in doing this are, however, too complex to be worthwhile at the present stage of development of quantum chemistry. Various other methods of estimating γ are discussed by Coulson (*10*, Chapter 8).

c. Hybridization of an *s*-orbital with three *p*-orbitals (sp^3 hybrids)

In order to find the possible hybrids of s with p_x, p_y, and p_z, we write

$$\psi_1 = a_1 s + b_1 p_x + c_1 p_y + d_1 p_z$$
$$\psi_2 = a_2 s + b_2 p_x + c_2 p_y + d_2 p_z \qquad \text{(E-9)}$$
$$\psi_3 = a_3 s + b_3 p_x + c_3 p_y + d_3 p_z$$
$$\psi_4 = a_4 s + b_4 p_x + c_4 p_y + d_4 p_z$$

The constants a_i, b_i, c_i, and d_i must be selected subject to the conditions of orthogonality and normalization of the functions. Thus

$$a_i^2 + b_i^2 + c_i^2 + d_i^2 = 1 \qquad (i = 1, 2, 3, 4) \qquad \text{(E-10a)}$$
$$a_i a_j + b_i b_j + c_i c_j + d_i d_j = 0 \qquad (i \neq j = 1, 2, 3, 4) \qquad \text{(E-10b)}$$

There are four normalization relations and six orthogonality relations, or a total of ten equations between sixteen unknowns. We can therefore choose six of the constants at will. Three of these six choices, however, merely result in different orientations of the hybrids in space, without affecting their shapes, so only three parameters need to be used to specify the hybridization.

Out of this infinite number of possible hybrids, one is particularly interesting. Let us make the choice $a_1 = b_1 = c_1 = d_1$. The orbital ϕ_1 resulting from this choice has a lobe that points away from the nucleus toward the (1, 1, 1) direction of a cube whose edges are parallel to the x, y, and z coordinate axes (see Fig. 11-7). It is easily verified that the normalization and orthogonalization conditions are satisfied by the following orbitals

$$\begin{aligned} \phi_1 &= (1/2)\,(s + p_x + p_y + p_z) \\ \phi_2 &= (1/2)\,(s + p_x - p_y - p_z) \\ \phi_3 &= (1/2)\,(s - p_x + p_y - p_z) \\ \phi_4 &= (1/2)\,(s - p_x - p_y + p_z) \end{aligned} \qquad \text{(E-11)}$$

These four orbitals point in the (1, 1, 1), (1, – 1, – 1), (– 1, 1, – 1) and (– 1, – 1, 1) directions of the unit cube. They are all equivalent and they

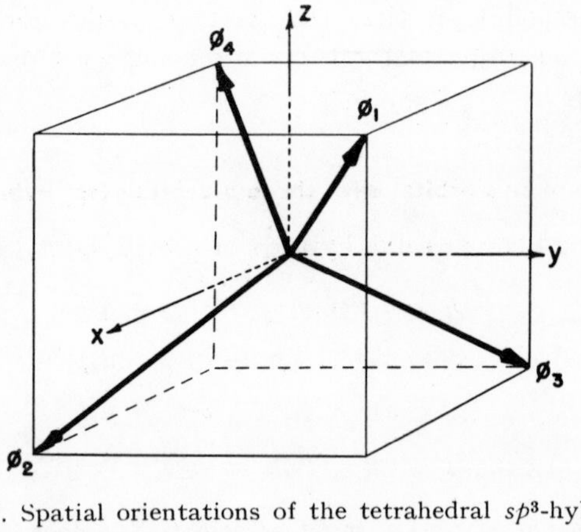

FIG. 11-7. Spatial orientations of the tetrahedral sp^3-hybrids.

make tetrahedral angles with one another. This arrangement of bonding hybrids is, of course, exactly the same as the stereochemical arrangement around the carbon atom in methane and other saturated organic molecules.

Exercise. Show that the most general sp^3 hybrids (aside from trivial differences in over-all orientation in space) are given by

$$\chi_1 = N_1 (s + \gamma p_x)$$

$$\chi_2 = N_2\left(s - \frac{1}{\gamma} p_x + \frac{\delta}{\gamma} p_y\right)$$

$$\chi_3 = N_3\left(s - \frac{1}{\gamma} p_x - \left(\frac{1+\gamma^2}{\gamma\delta}\right) p_y + \frac{\varepsilon}{\gamma} p_z\right)$$

$$\chi_4 = N_4\left(s - \frac{1}{\gamma} p_x - \left(\frac{1+\gamma^2}{\gamma\delta}\right) p_y - \frac{(1+\gamma^2)(1+\gamma^2+\delta^2)}{\gamma\varepsilon\delta^2} p_z\right) \qquad \text{(E-12)}$$

where γ, δ, and ε are independent variable parameters, and the N_i are normalization constants whose values are readily found (e.g., $N_1 = 1/\sqrt{1+\gamma^2}$). Show that the choice $\gamma = \sqrt{3}$, $\delta = 2\sqrt{2}$, $\varepsilon = 2\sqrt{3}$ leads to the tetrahedral orbitals having the same shapes as (E-11) but oriented differently in space. Describe the shapes of the hybrids obtained with the following sets of values of the parameters

a. $\gamma = 0, \quad \delta = 1, \quad \varepsilon = 0$

b. $\gamma = 1, \quad \delta = 0, \quad \varepsilon = 0$

c. $\gamma = 1, \quad \delta = 0, \quad \varepsilon = 1$

d. $\gamma = \sqrt{2}, \quad \delta = \sqrt{3}, \quad \varepsilon = 0$

e. $\gamma = \delta = \infty, \quad \varepsilon = 1$

If hybridization is important in methane, it is natural to ask if it might not also play a role in NH_3, H_2O, HF, and similar molecules. For instance, in H_2O we might express the electronic configuration of the oxygen atom in terms of the sp^3 hybrid functions, ϕ_1, ϕ_2, ϕ_3, and ϕ_4 of Equation (E-11), as $\phi_1^2\phi_2^2\phi_3\phi_4$, in which two of the tetrahedral orbitals are occupied by a single electron, and thus can form bonds with other atoms. If the parameters of the orbitals had appropriate values, the H–O–H bond angle would be 109° 28′. Since the observed angle is 104.5°, it would seem that the hybridization in H_2O is not quite tetrahedral, though it may not be far from tetrahedral. Similarly, the H–N–H bond angle of 108° indicates that the orbital occupied by the unshared electron pair is not quite equivalent to the other three orbitals on the nitrogen atom, presumably having slightly more *s* character.

We see that in going from CH_4 to NH_3 to H_2O the deviation from the tetrahedral angle becomes greater and greater. This undoubtedly reflects the increasing difference in stability of the 2*s*- and 2*p*-orbitals as the atomic number is increased (which is a consequence of the penetration effect discussed in Chapter 10). Thus an unshared pair of electrons would be expected to prefer an orbital with as much *s*-character as possible, since this allows them to get closer to the nucleus. This advantage is not so great for an orbital involved in bonding, since the bonding electron spends a smaller fraction of its time close to the nucleus of the hybridized atom.

The hybridization of orbitals is also very important in determining the stereochemistry of the transition metals, where mixtures of *s*-, *p*- and *d*-orbitals may take place. The variety in bond angles here is very great, but among the many possible arrangements, there is one that is particularly important. It involves six equivalent bonds directed towards the corners of a regular octahedron, and arises from the sp^3d^2-configuration. Since octahedral complexes of the transition metals are very common, this hybrid is probably especially favored. The square complex produced by sp^2d is also found. A more detailed discussion of these hybrids will be found in Pauling's book (*11*), and a general discussion of the possible arrangements of hybridized bonds is given in Eyring, Walter, and Kimball (*1*).

It is important to realize that the concept of hybridization is forced upon us from the moment we choose to describe the orbitals of a molecule in terms of hydrogen atomic wave functions. Hybridization is a means of concentrating electrons into the regions between the atoms at either end of a chemical bond without at the same time doing too much violence to the wave mechanical requirements of the Schroedinger equation. It is therefore merely a particularly convenient method of describing the distortion of the electron cloud in the vicinity of an atom that is bonded to other atoms. If we should choose other types of orbitals to describe bonds–and we shall see in Chapter 12 that orbitals exist which, though more awkward to use, are probably much better than orbitals based on hydrogen-like functions–the concept of hybridization would not have to be introduced. Furthermore, hybridization of the type described above gives only a partial account of the electron clouds in a bond. A really accurate description of the bonds in CH_4, for example, in terms of atomic orbitals would require the inclusion of $3s$, $3p$, $3d$, as well as higher orbitals on the carbon atom, and also $2p$ and higher orbitals on the hydrogen atom. We have no information on how well the $2s\ 2p^3$ hybrids of carbon by themselves reproduce the true wave function in the neighborhood of the carbon atoms in organic molecules, but there is every reason to believe that they lead to no more than a rather rough first approximation. This makes it very difficult to build a quantitative theory of hybridization that has any exact theoretical meaning. For instance, one might be tempted to characterize the difference in extent of hybridization in CH_4, NH_3, H_2O, and HF by means of a set of numbers such as γ, δ, and ε in Equations (E-12). It is questionable, however, whether these numbers could be given any precise meaning, though they might be useful in a rough, semiquantitative sense.

F. Multiple Bonds

Among the hybrids of the carbon atom in its sp^3-configuration given in (E-12), two are especially interesting. One of these hybrids ($\gamma = \sqrt{2}$, $\delta = \sqrt{3}$, $\varepsilon = 0$) makes use of the s atomic orbital with two of the p-orbitals to form three equivalent orbitals making an angle of 120° with each other. The third p-orbital is unhybridized. Something close to this mode of hybridization is believed to occur in ethylene ($H_2C{=}CH_2$) and other olefins. Two of the trigonal hybrids on each carbon atom are used in forming the C–H bonds of ethylene and the third trigonal hybrid forms a part of the double bond between the two carbon atoms. The rest of the double bond is formed from the two pure p-orbitals that remain on the two carbon atoms (see Fig. 11-8). Thus the two bonds involved in a double bond are quite

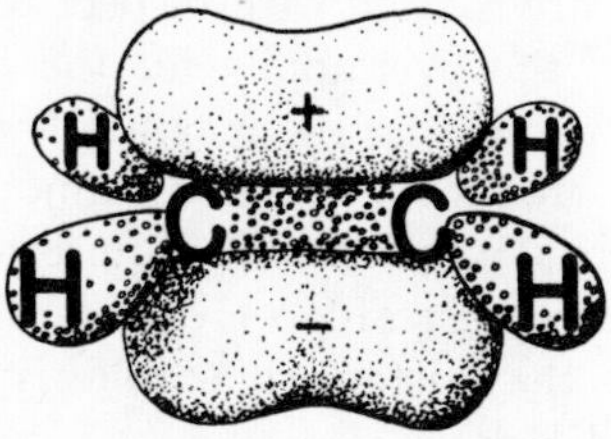

FIG. 11-8. Shapes of wave functions involved in the formation of the double bond in ethylene. The pi bond consists of the two "blobs" above and below the carbon atoms; it is formed from a linear combination of two $2p$-orbitals. The two "blobs" are identical, except that they have opposite signs. The sigma bond lies along the line joining the two carbon atoms.

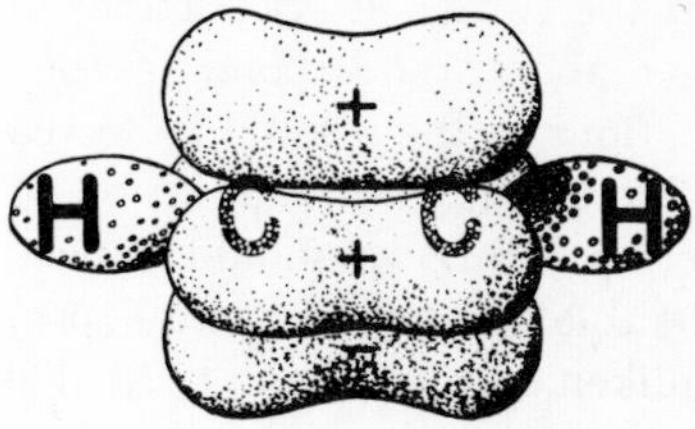

FIG. 11-9. Shapes of the wave functions involved in the triple bond of acetylene. There are two pi bonds, each similar to the pi bond in ethylene but lying in different planes. (These wave functions should not be confused with the shapes of the charge distributions, which are given by the sums of the squares of the wave functions. The charge distribution in acetylene has cylindrical symmetry about the bond, and does not show "blobs".)

different in character. One half of the double bond, formed from the pure p-orbitals, is called a *π-bond*, since if the molecule were linear (so that the electronic angular momentum about the C–C bond would be precisely determined), this bond would involve electrons having an angular momentum of one unit about the C–C axis. The second part of the double bond, formed from the overlap of one of the trigonal orbitals from each carbon atom, is called a *σ-bond*, because it is formed from electrons that would have no angular momentum about the C–C axis if the molecule were strictly linear.

The terms π-bond and σ-bond offer a useful and descriptive shorthand method of discussing the two components of a double bond, but care must be used not to confuse them with the true π-bonds and σ-bonds found in linear molecules.

The π-bond is considerably weaker than the σ-bond, and most of the reactions of double bonds probably involve chiefly the π-bonds.

Triple bonds between carbon atoms (as found, for instance, in acetylene, H–C≡C–H) are evidently formed from a pair of *sp* linear hybrids and two pairs of pure *p*-orbitals (the case $\gamma = 1$, $\delta = \varepsilon = 0$ in (E-12)). The remaining *sp* hybrid on each carbon atom goes into the C–H bond. This accounts for the fact that the four atoms in the acetylene molecule lie along a straight line. The triple bond thus consists of a σ-bond and two π-bonds (see Fig. 11-9).

If the two H–C–H portions of ethylene are rotated with respect to each other about the σ-bond in the C–C link, the overlapping of the two *p* functions in the π-bond is decreased, thus decreasing the magnitude of the exchange integral. A rotation of 90° will produce a minimum amount of overlap. Thus we should expect that such a rotation will require a considerable amount of energy (probably of the order of 2.5 e.v. according to Mulliken and Roothaan (*12*)). This explains the existence of stable *cis-trans*-isomers of molecules of the type $R_1R_2C{=}CR_3R_4$.

The pure carbon–carbon bond that occurs in a molecule such as ethane does not have any such large barrier to rotation, since the nearly cylindrical symmetry of the σ-bond results in practically no change in overlap when the two groups on either side of the bond are rotated. This accounts for the fact that no stable rotational isomers can be prepared from molecules of the type $R_1{-}CH_2{-}CH_2{-}R_2$. A small barrier (of approximately 3 kcal./mole) is, however, known to exist in ethane. Its origin is obscure.

G. Aromatic Compounds

One of the most puzzling facts of organic chemistry has always been the characteristic stability of aromatic compounds. According to their structural formulae these compounds are highly unsaturated, yet they show scarcely any of the properties typical of the double bond. Thus the typical reactions of a double bond are addition reactions, which lead to its replacement by single bonds, whereas the typical reactions of aromatic compounds are substitutions, which leave the double bond intact. The ability of quantum mechanics to explain this stability in a relatively straightforward manner has been one of its most impressive applications to chemistry.

The benzene ring is known to consist of a regular plane hexagon of carbon atoms each joined to a hydrogen atom by a single bond. According to the classical concepts of organic chemistry, the ring of carbon atoms consists of a set of alternating double and single bonds. Accepting this and the picture of the double bond presented in the previous section, the benzene molecule would be said to be held together in part by σ-bonds coplanar with the ring. Six of these σ-bonds are involved in the bonds between the hydrogen atoms and the carbon atoms. The remaining σ-bonds join the carbon atoms to each other. In addition there are six pure p-orbitals (or "π-orbitals") whose axes are normal to the plane of the ring, one being associated with each carbon atom. Since the π-orbitals are responsible for the instability and reactivity of ordinary double bonds, it would seem that the logical way to look for an explanation of the unusual characteristics of benzene would be to study the bonds formed between these π-orbitals.

This problem has been dealt with successfully by Pauling and Wheland (*13*) using the valence bond theory and by Hückel (*14*) using the molecular orbital theory. Both of these methods may be said to disregard the electrons in the σ-bonds and to consider the benzene molecule as a six-electron problem. (The effects of the σ-electrons may, however, be included indirectly through their effect on the screening and by various other means.)

a. The valence bond method for benzene

Let us indicate the six pure π atomic orbitals of benzene by the letters a, b, c, d, e, and f, in order of their arrangement about the ring. There are various ways in which bonds can be drawn between these orbitals, as for instance,

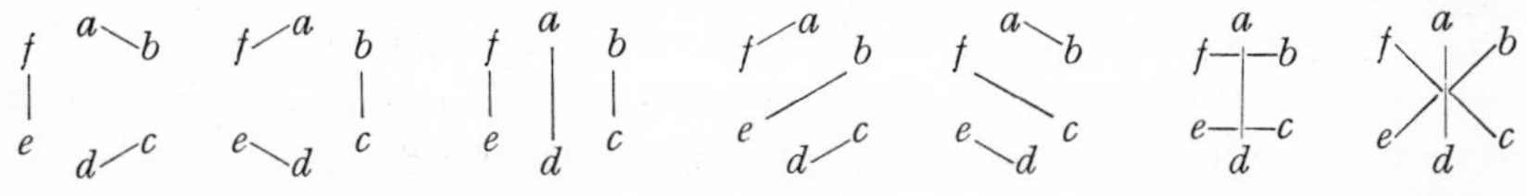

I II III IV V VI VII

In these diagrams the σ-bonds have been omitted for simplicity. Thus structure I actually corresponds to the formula

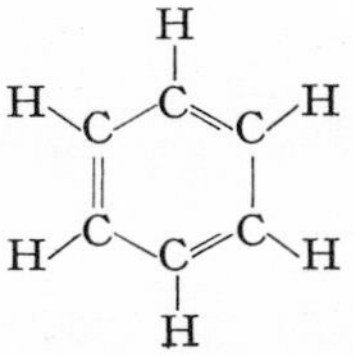

and structure VI corresponds to the structure

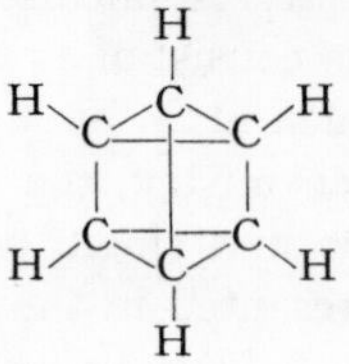

Structures I and II are those proposed by Kekulé; III, IV, and V were proposed by Dewar; VI was proposed by Ladenburg, and VII was proposed by Claus.

According to the VB method, each of the bonds in a molecule involves a pair of electrons having opposite spins. Therefore we may write wave functions corresponding to each of these bonding diagrams. These wave functions must, of course, be written in the form of determinants, in order to be sure that the Pauli principle is fulfilled. In order to save space we shall not write these determinants out, but will adopt the following symbolism. In Chapters 9 and 10 we have represented determinants by writing out their diagonal terms. Here we shall be even more concise and omit the symbols for the orbitals and the electrons. Thus the determinant

$$| a(1)\alpha(1) \; b(2)\beta(2) \; c(3)\alpha(3) \; d(4)\beta(4) \; e(5)\alpha(5) \; f(6)\beta(6) |$$

will be represented by

$$| \alpha\beta\alpha\beta\alpha\beta |$$

it being assumed that the orbitals are written in the order a, b, c, d, e, f in the successive rows of the determinant and the electrons in the order 1, 2, 3, 4, 5, 6 in successive columns. Thus if we write $| \alpha\alpha\alpha\beta\beta\beta |$, we mean that the electrons in orbitals a, b, and c have the spin function α, and that the electrons in orbitals d, e, and f have the spin function β.

The structure I, in which bonds occur between a and b, c and d, and e and f, will therefore have a wave function consisting of some linear combination of the eight determinants

$$\begin{aligned} D_1 &= | \alpha\beta\alpha\beta\alpha\beta | \qquad & D_5 &= | \beta\alpha\beta\alpha\alpha\beta | \\ D_2 &= | \beta\alpha\alpha\beta\alpha\beta | \qquad & D_6 &= | \beta\alpha\alpha\beta\beta\alpha | \\ D_3 &= | \alpha\beta\beta\alpha\alpha\beta | \qquad & D_7 &= | \alpha\beta\beta\alpha\beta\alpha | \\ D_4 &= | \alpha\beta\alpha\beta\beta\alpha | \qquad & D_8 &= | \beta\alpha\beta\alpha\beta\alpha | \end{aligned} \tag{G-1}$$

since these determinants and only these determinants give opposite spins to the orbitals that are bonded together in structure I.

The selection of the correct linear combination of these wave functions is accomplished in the following manner. Each of these determinants is an eigenfunction of the operator S_z for the component of the total electronic spin in an arbitrary fixed direction. Therefore any linear combination of these functions is also an eigenfunction of S_z. The eigenvalues are all zero.

Exercise. Prove this by writing out a few typical terms of one of the determinants and operating with S_z.

These determinants are not, however, eigenfunctions of S^2, the operator for the square of the total spin. Using Equation (D-10) of Chapter 9, it is readily shown that

$$
\begin{aligned}
S^2D_1 &= \hbar^2[3D_1 + D_2 + D_3 + D_4 + |\beta\beta\alpha\alpha\alpha\beta| + |\beta\beta\alpha\beta\alpha\alpha| + |\alpha\alpha\beta\beta\alpha\beta| \\
&\quad + |\alpha\beta\beta\beta\alpha\alpha| + |\alpha\alpha\alpha\beta\beta\beta| + |\alpha\beta\alpha\alpha\beta\beta|] \\
S^2D_2 &= \hbar^2[D_1 + 3D_2 + D_5 + D_6 + |\beta\beta\alpha\alpha\alpha\beta| + |\beta\beta\alpha\beta\alpha\alpha| + |\alpha\alpha\beta\beta\alpha\beta| \\
&\quad + |\beta\alpha\beta\beta\alpha\alpha| + |\alpha\alpha\alpha\beta\beta\beta| + |\beta\alpha\alpha\alpha\beta\beta|] \\
S^2D_3 &= \hbar^2[D_1 + 3D_3 + D_5 + D_7 + |\beta\beta\alpha\alpha\alpha\beta| + |\beta\beta\beta\alpha\alpha\alpha| + |\alpha\alpha\beta\beta\alpha\beta| \\
&\quad + |\alpha\beta\beta\beta\alpha\alpha| + |\alpha\alpha\beta\alpha\beta\beta| + |\alpha\beta\alpha\alpha\beta\beta|] \\
S^2D_4 &= \hbar^2[D_1 + 3D_4 + D_6 + D_7 + |\beta\beta\alpha\alpha\beta\alpha| + |\beta\beta\alpha\beta\alpha\alpha| + |\alpha\alpha\beta\beta\beta\alpha| \\
&\quad + |\alpha\beta\beta\beta\alpha\alpha| + |\alpha\alpha\alpha\beta\beta\beta| + |\alpha\beta\alpha\alpha\beta\beta|] \\
S^2D_5 &= \hbar^2[D_2 + D_3 + 3D_5 + D_8 + |\beta\beta\alpha\alpha\alpha\beta| + |\beta\beta\beta\alpha\alpha\alpha| + |\alpha\alpha\beta\beta\alpha\beta| \\
&\quad + |\beta\alpha\beta\beta\alpha\alpha| + |\alpha\alpha\beta\alpha\beta\beta| + |\beta\alpha\alpha\alpha\beta\beta|] \\
S^2D_6 &= \hbar^2[D_2 + D_4 + 3D_6 + D_8 + |\beta\beta\alpha\alpha\beta\alpha| + |\beta\beta\alpha\beta\alpha\alpha| + |\alpha\alpha\beta\beta\beta\alpha| \\
&\quad + |\beta\alpha\beta\beta\alpha\alpha| + |\alpha\alpha\alpha\beta\beta\beta| + |\beta\alpha\alpha\alpha\beta\beta|] \\
S^2D_7 &= \hbar^2[D_3 + D_4 + 3D_7 + D_8 + |\beta\beta\alpha\alpha\beta\alpha| + |\beta\beta\beta\alpha\alpha\alpha| + |\alpha\alpha\beta\beta\beta\alpha| \\
&\quad + |\alpha\beta\beta\beta\alpha\alpha| + |\alpha\alpha\beta\alpha\beta\beta| + |\alpha\beta\alpha\alpha\beta\beta|] \\
S^2D_8 &= \hbar^2[D_5 + D_6 + D_7 + 3D_8 + |\beta\beta\alpha\alpha\beta\alpha| + |\beta\beta\beta\alpha\alpha\alpha| + |\alpha\alpha\beta\beta\beta\alpha| \\
&\quad + |\beta\alpha\beta\beta\alpha\alpha| + |\alpha\alpha\beta\alpha\beta\beta| + |\beta\alpha\alpha\alpha\beta\beta|]
\end{aligned}
\tag{G-2}
$$

The student can verify from these equations that the linear combination

$$\psi_{\mathrm{I}} = D_1 - D_2 - D_3 - D_4 + D_5 + D_6 + D_7 - D_8 \tag{G-3}$$

is an eigenfunction of S^2 with the eigenvalue zero. Furthermore, no other linear combination of the determinants D_1 to D_8 can give this eigenvalue.

Since the ground state of benzene is known to be a singlet, this function must be the wave function corresponding to structure I.

In a similar fashion, a wave function can be deduced for each of the remaining structures, II to VII, for benzene. The procedure is not as tedious as it might at first seem since there is a simple rule by means of which wave functions corresponding to the different structures may be written down immediately without using the operator S^2 at all.

This rule is as follows: Given a structure in which chemical bonds are drawn between certain pairs of orbitals in the molecule. Write down all of the determinants that put opposite spins on each pair of atoms that are bonded in the structure. Choose any one of these determinants, and find out how many permutations of the α's and β's in each of the remaining determinants are required in order to bring them into the same order as that in the chosen determinant. If the number of permutations is even, the determinant occurs in the linear combination with the coefficient $+1$, and if the number of permutations is odd, the coefficient is -1. This rule is easily verified for the wave function ψ_{I}, since, for instance, D_2, D_3, and D_4 can be converted into D_1 by a single permutation of a pair of α's and β's; therefore D_2, D_3, and D_4 occur in ψ_{I} with a minus sign. The determinant D_8 requires three permutations, so it too occurs with a minus sign, whereas D_5, D_6, and D_7 can be converted into D_1 by an even number of permutations, so they occur in ψ_{I} with a plus sign.

Exercise. Write the eight determinants for structure III and find the wave function for this structure.

Thus it is not difficult to write down a wave function for each way of drawing bonds between the orbitals in any molecule. It was shown by Rumer, however, that not all of these structures are linearly independent. In particular, he showed that any structure in which bonds cross one another can always be expressed as a linear combination of other structures in which bonds do not cross. Thus it is necessary to consider only those structures in which bonds do not cross. These structures are called *canonical structures*. If N orbitals are arranged in a circle, there are

$$N!/[(\tfrac{1}{2}N)!\,(\tfrac{1}{2}N+1)!] \tag{G-4}$$

of these canonical structures, and in the case of benzene, where $N = 6$, there are five canonical structures. It is logical (though by no means necessary) to choose the two Kekulé structures and the three Dewar structures as the canonical structures of benzene. (If we had arbitrarily arranged the six π-orbitals in a different order around a circle, however, other sets of canonical structures could be used.)

If we wish, we may calculate the approximate energy of any one of these structures. For instance, the energy of structure I is

$$E_{\text{I}} = \int \psi_{\text{I}} H \psi_{\text{I}} d\tau \Big/ \int \psi_{\text{I}} \psi_{\text{I}} d\tau \qquad \text{(G-5)}$$

The most general wave function of benzene, however, would be a linear combination of all canonical structures,

$$\psi_{\text{Benzene}} = c_1\psi_{\text{I}} + c_2\psi_{\text{II}} + c_3\psi_{\text{III}} + c_4\psi_{\text{IV}} + c_5\psi_{\text{V}} \qquad \text{(G-6)}$$

the constants c_i being chosen in such a way that the energy of the system is minimized, in accordance with the variation method. Although there is nothing new in principle in this calculation the details are somewhat complicated and we shall give only the result [the complete calculation is performed in Eyring, Walter, and Kimball (*1*, Chapter 13) and in Coulson (*10*, Chapter 9)].

In order to simplify the calculation of the constants c_i and of the energy of the ground state of the benzene molecule, two important assumptions are made:

(1) The overlap integrals between any pair of different orbitals are neglected. That is, integrals of the type

$$\int a(1)b(2)d\tau \quad \text{and} \quad \int a(1)c(2)d\tau$$

are set equal to zero. The justification for this assumption is somewhat questionable, for these integrals are known to be by no means insignificant, but if the assumption is not made the calculation becomes much more difficult. (Overlap integrals of normalized $2p\pi$-functions at internuclear distances such as are found in benzene and other aromatic compounds amount to about 0.2, which is hardly negligible compared with unity.)

(2) Exchange integrals involving the interchange of two electrons not on adjacent carbon atoms are neglected, as are also exchange integrals involving the interchange of more than two electrons. This means that integrals of the type

$$\int a(1)b(2)c(3)d(4)e(5)f(6)Ha(2)b(1)c(4)d(3)e(5)f(6)\, d\tau$$

and

$$\int a(1)b(2)c(3)d(4)e(5)f(6)Ha(3)b(2)c(1)d(4)e(5)f(6)\, d\tau$$

are set equal to zero. The justification of this assumption is that the magnitudes of these integrals decreases very rapidly as the distance between the orbitals involved in the interchange is increased, as well as when simultaneous interchanges of more than two electrons are involved. Furthermore, the calculation is very much more difficult if these integrals are not neglected.

Making these assumptions, it is found that the energy is minimized when

$$c_1 = c_2 \qquad c_3 = c_4 = c_5 \tag{G-7}$$

and

$$c_3/c_1 = 0.4341$$

The energy is

$$E = Q + 2.6056J \tag{G-8}$$

where Q is the coulombic integral,

$$Q = \int a(1)b(2)c(3)d(4)e(5)f(6)Ha(1)b(2)c(3)d(4)e(5)f(6)\, d\tau \tag{G-9}$$

and J is the exchange integral

$$J = \int a(1)b(2)c(3)d(4)e(5)f(6)Ha(2)b(1)c(3)d(4)e(5)f(6)\, d\tau \tag{G-10}$$

For a single one of the two structures, I or II, in which the three double bonds are fixed (that is, assuming that, say, $c_1 = 1$ and all the other c_i in (G-6) are zero), the energy is

$$E_{\text{I}} = Q + 1.50J \tag{G-11}$$

Thus the energy resulting from the inclusion of all five canonical structures in the wave function of benzene is lowered by $(2.6056 - 1.50)J = 1.1056J$, as compared with the energy of a single one of the Kekulé structures. This stabilization can be said to result from "resonance" between the five valence bond structures of the canonical group, and the resonance energy is $1.1056J$. It is this resonance energy that is presumably responsible for the unusual stability of the highly unsaturated benzene molecule.

If the three Dewar structures are arbitrarily excluded from the benzene wave function, we obtain

$$\psi'_{\text{Benzene}} = \psi_{\text{I}} + \psi_{\text{II}} \tag{G-12}$$

and a resonance energy of $0.90J$. This is not much less than the resonance energy obtained when the Dewar structures are included, so the two Kekulé structures are responsible for the greater part of the stability of benzene. This might have been expected, because one of the three bonds in each Dewar structure is very elongated, so that these structures have a considerably higher energy than the Kekulé structures.

Thermal data indicate that the energy of benzene is about 40 kcal. per mole lower than that of the hypothetical molecule containing three double bonds in which no resonance is allowed to take place. This gives the exchange integral, J, a value of about 35 kcal. This is a reasonable value, since J would be expected to have a value considerably smaller than that of the dissociation energy of a normal single bond (about 80 kcal. for the C–C bond).

It should be noted that many other structures might have been included in the benzene wave functions. Thus there are polar structures such as

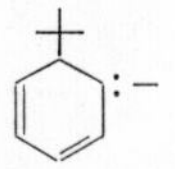

in which one electron is removed from orbital a and placed in orbital b. Structures of this type are probably even less important than the Dewar structures, so they are generally omitted. The very large number (there are thirty of them in benzene) makes their inclusion difficult, but it also makes the justification of their exclusion somewhat questionable.

b. The molecular orbital treatment of benzene

The MO method as applied to benzene is somewhat simpler than the VB method and seems to give equally satisfactory results. One begins by preparing a set of molecular orbitals from linear combinations of the six π-orbitals,

$$\phi = k_1a + k_2b + k_3c + k_4d + k_5e + k_6f \qquad \text{(G-13)}$$

where the k_i are constants whose values are to be determined, and a, b, c, d, e, and f are the six atomic π-orbitals located as shown on page 419. The variation method will give us six sets of constants, k_i, which define six approximate mutually orthogonal molecular orbitals. If we write H' for the Hamiltonian operator of a single electron moving about in the field of the

σ-bonded skeleton of the benzene ring, the procedure described on pages 125*ff*. leads to the simultaneous equations for the k_i

$$\begin{aligned}
(\alpha-E)k_1 + \beta k_2 + 0k_3 + 0k_4 + 0k_5 + \beta k_6 &= 0\\
\beta k_1 + (\alpha-E)k_2 + \beta k_3 + 0k_4 + 0k_5 + 0k_6 &= 0\\
0k_1 + \beta k_2 + (\alpha-E)k_3 + \beta k_4 + 0k_5 + 0k_6 &= 0\\
0k_1 + 0k_2 + \beta k_3 + (\alpha-E)k_4 + \beta k_5 + 0k_6 &= 0\\
0k_1 + 0k_2 + 0k_3 + \beta k_4 + (\alpha-E)k_5 + \beta k_6 &= 0\\
\beta k_1 + 0k_2 + 0k_3 + 0k_4 + \beta k_5 + (\alpha-E)k_6 &= 0
\end{aligned} \tag{G-14}$$

which in turn leads to the secular equation

$$\begin{vmatrix}
(\alpha-E) & \beta & 0 & 0 & 0 & \beta\\
\beta & (\alpha-E) & \beta & 0 & 0 & 0\\
0 & \beta & (\alpha-E) & \beta & 0 & 0\\
0 & 0 & \beta & (\alpha-E) & \beta & 0\\
0 & 0 & 0 & \beta & (\alpha-E) & \beta\\
\beta & 0 & 0 & 0 & \beta & (\alpha-E)
\end{vmatrix} = 0 \tag{G-15}$$

In setting up these equations, we have made the substitutions

$$\alpha = \int aH'a\,d\tau = \int bH'b\,d\tau = \int cH'c\,d\tau = \int dH'd\,d\tau =$$

$$= \int eH'e\,d\tau = \int fH'f\,d\tau$$

$$\beta = \int aH'b\,d\tau = \int bH'c\,d\tau = \int cH'd\,d\tau = \int dH'e\,d\tau$$

$$= \int eH'f\,d\tau = \int fH'a\,d\tau$$

and we have assumed that all overlap integrals such as

$$\int ab\,d\tau,\ \int ac\,d\tau,\ \int ad\,d\tau,$$

etc., are so small that they may be set equal to zero. We have also assumed that all exchange integrals between orbitals on nonadjacent carbon atoms, such as

$$\int aH'c\, d\tau \quad \text{and} \quad \int aH'd\, d\tau,$$

may be neglected. According to Mulliken and Rieke (*15*) and Wheland (*16*) the neglect of overlap integrals between adjacent atoms does not introduce any serious error in the form of the result.

The secular equation (G-15) is a cyclic determinant of the type discussed on page 48, and its roots are readily shown to be (*cf.* Exercise (3) on page 49)

$$E - \alpha = 2\beta\,(\cos \pi l/3) \quad (l = 1, 2, 3, 4, 5, 6)$$

$$= 2\beta, \beta, \beta, -\beta, -\beta, -2\beta$$

Since the exchange integral β is negative, the most stable orbital has the energy

$$E = \alpha + 2\beta$$

Substitution of this energy into the simultaneous equations (G-14) leads to the following relation between coefficients

$$k_1 = k_2 = k_3 = k_4 = k_5 = k_6$$

which gives for the most stable molecular orbital of benzene

$$\phi_1 = k_1(a + b + c + d + e + f)$$

Similarly, the next most stable root

$$E = \alpha + \beta$$

leads to the two linearly independent orbitals

$$\phi_2 = k_2(a + b - d - e)$$

$$\phi_3 = k_3(b + c - e - f)$$

while the energy

$$E = \alpha - \beta$$

is associated with the orbitals

$$\phi_4 = k_4(a - b + c - d)$$

$$\phi_5 = k_5(b - c + e - f)$$

and the least stable orbital, having energy

$$E = \alpha - 2\beta$$

has the wave function

$$f_6 = k_6(a - b + c - d + e - f)$$

In benzene six π-electrons are to be placed in these orbitals in accordance with the Pauli principle. In the ground state of the molecule only the most stable orbitals can be occupied; two electrons can be placed in ϕ_1, two in ϕ_2 and two in ϕ_3. To a first approximation, therefore, the energy of the ground state of the benzene is

$$2(\alpha + 2\beta) + 4(\alpha + \beta) = 6\alpha + 8\beta$$

This value of the energy is to be compared with the value that one would expect if the six electrons had been placed in three isolated double bonds whose length is the same as the carbon–carbon bond length in benzene. The energy of a molecular orbital formed from two π atomic orbitals would be $(\alpha + \beta)$, so the six electrons in a molecule containing three double bonds would have the energy $6(\alpha + \beta)$. The difference between this value and the value of the energy of benzene calculated above

$$(6\alpha + 8\beta) - (6\alpha + 6\beta) = 2\beta$$

represents the extra stability of a benzene ring as compared with a ring containing three normal double bonds. In this way the molecular orbital theory is able to account for the fact that benzene has an extra stability as compared with a molecule having three normal double bonds.

A number of approximations have been made in this calculation, of course, and it is really rather crude. In addition to the neglect of the overlap integrals and certain exchange integrals, no account has been taken of the energy resulting from the mutual repulsions of the six π-electrons; the energy of benzene was estimated by merely adding the energies of the occupied orbitals. We may expect, however, that some cancellation of errors will take place in the calculation, since the repulsions of the π-electrons are neglected in both the calculation of the energy of benzene and of the energy of three isolated double bonds.

Since the resonance stabilization of benzene is known to be about 40 kcal./mole, the integral β must have a value of about 20 kcal./mole. An independent check on this quantity is possible in the following way. The first excited state of benzene should have five of its six electrons located in

ϕ_1, ϕ_2, and ϕ_3, but the sixth electron should be excited from orbital ϕ_3 to the next higher orbital, ϕ_4. If we estimate the energy of this state by adding the energies of the occupied orbitals, this must result in an increase in the energy by 2β, or 40 kcal./mole. This excited state would be expected to give rise to an absorption band at about 7000 Å. A transition of this wave length is not, however, observed in benzene; the longest wave length at which reasonably strong electronic absorption takes place in benzene is at about 2700 Å., and if this transition is caused by excitation to the state mentioned above, it would require a value $2\beta = 106$ kcal./mole. Thus the spectrum of benzene is hardly consistent with the theory of resonance stabilization outlined above. This discrepancy, as well as other difficulties in the theory, will be discussed further in Chapter 12. It should be mentioned, however, that if the integral for the overlap of adjacent orbitals is included in the secular equation (G-16), the discrepancy between the value of β estimated from resonance stabilization and that obtained spectroscopically is reduced quite considerably (*14, 15*).

c. Results of the application of the valence bond and molecular orbital theories to other aromatic hydrocarbons

The two methods just described have been applied to the calculation of the resonance stabilization of many other aromatic hydrocarbons, and experimental data are available for comparison with the results of these calculations. In the VB method, the energies of the molecules are calculated in terms of the integrals Q and J, including different resonance forms analogous to the Kekulé, and in some cases also to the Dewar, formulae in benzene. This calculated value is compared with the energy calculated for the same molecule in one of its double bonded structures. The difference is called the resonance energy.

In the MO method, molecular orbitals are found for the π-electrons in the manner similar to that found for benzene. The most stable orbitals are then filled in accordance with the requirements of the Pauli principle, and the total energy of the ground state is found by adding the energies of all of the occupied orbitals. The result is expressed in terms of the coulombic and exchange integrals for a single electron, α and β. The difference between this energy and the energy of a molecule containing normal double bonds is found, and this difference is called the resonance energy.

In both of these methods, as employed in the calculations whose results are described here, it is assumed that the integrals Q, J, α, and β do not vary from one molecule to another. It is known, however, that there is a certain amount of variation in the bond lengths not only in different molecules, but even in different bonds of the same molecule. This variation should affect the values of these integrals.

Fig. 11-10 shows the excellent correlation between the resonance energies calculated by the two methods (using the values of J and β found from the benzene calculation) and the "observed" resonance energies found by comparing heats of combustion of the compounds with heats of combustion

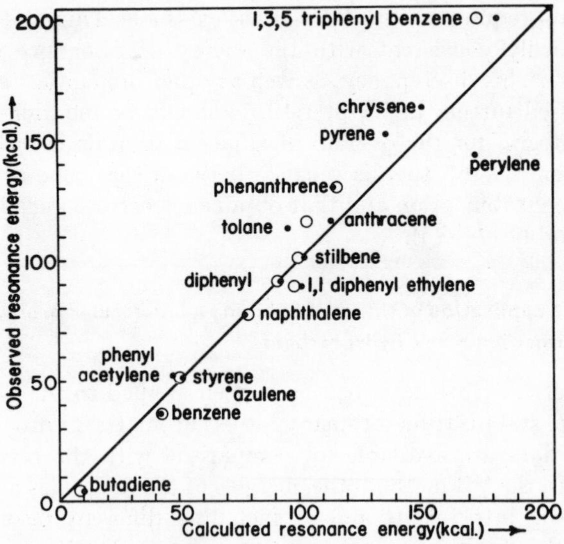

FIG. 11-10. Comparison of calculated and observed resonance energies of aromatic hydrocarbons using the valence bond (open circles) and molecular orbital (filled circles) methods. In the VB method the exchange integral J has been given the value 38.5 kcal. In the MO method the exchange integral β has been given the value 21 kcal. (Data from A. Pullman and B. Pullman, "Les théories électroniques de la chimie organique," pp. 226–227, Masson, Paris, 1952.)

expected from structures containing normal double bonds. We see that in spite of the rather crude assumptions that go into both theories, the observed stabilities of these compounds can be correlated very well by means of a single value of J or of β.

This is an excellent example of the use of an approximate quantum mechanical method to derive the properties of one molecule from the properties of another molecule. Although serious approximations are introduced into the calculation, the errors thereby introduced seem to be similar in magnitude in the different compounds, or else they compensate one another in some other way, for the results are very good. In spite of this apparent success, however, we know from the false prediction of the position of the absorption band of benzene that the procedure must contain serious quantitative errors of some kind. These errors will be discussed in the next chapter.

Exercise. Show that according to the MO theory that has just been used with benzene, the ground state of cyclobutadiene, C_4H_4, should be a triplet and the resonance energy should be zero. [Hint: Consider only the four π-electrons on the carbon atoms, denoting the $2p\pi$ atomic orbitals on these atoms by a, b, c, and d, in that order around the ring. Obtain and solve the secular equation corresponding to linear combinations of these four atomic orbitals,

$$\phi = k_1a + k_2b + k_3c + k_4d$$

Neglect exchange integrals between nonadjacent atoms (that is, those between a and c and between b and d). Put four electrons in the most stable orbitals, add up their energies and compare the result with the energy of the structure

HC – CH
‖ ‖
HC – CH

containing two normal double bonds.] It is interesting to note that according to the VB method the ground state of cyclobutadiene should be a singlet and it should be stabilized by resonance to almost the same extent as that found in benzene. Thus in this instance the MO and VB methods give very different answers. Unfortunately cyclobutadiene has never been synthesized, so it is not possible to compare these calculations with experimental observations.

REFERENCES CITED

1. H. Eyring, J. Walter, and G. E. Kimball, "Quantum Chemistry," Wiley, New York, 1944.
2. H. Margenau and G. M. Murphy, "The Mathematics of Physics and Chemistry," Van Nostrand, New York, 1943.
3. C. A. Coulson, *Trans. Faraday Soc.*, **33,** 1479 (1937).
4. W. Heitler and F. London, *Z. Physik.*, **44,** 455 (1927).
5. Y. Sugiura, *Z. Physik.*, **45,** 484 (1927).
6. G. Herzberg, "Spectra of Diatomic Molecules," 2nd ed., Van Nostrand, New York, 1950.
7. R. S. Mulliken, *Phys. Rev.*, **50,** 1029 (1939).
7a. R. S. Mulliken, *Revs. Mod. Phys.*, **4,** 1 (1932).

8. "Donnés spectroscopiques concernant les molecules diatomiques" (B. Rosen, ed.), Vol. 4 of "Tables de constantes et donnés numériques," Hermann et Cie., Paris, 1951.
9. H. Landolt and R. Börnstein, "Zahlenwerte und Funktionen" (A. Eucken, ed.), 6th ed., Vol. 1, Part 3, Springer, Berlin, 1951.
10. C. A. Coulson, "Valence," Oxford U.P., New York, 1952.
11. L. Pauling, "The Nature of the Chemical Bond," 2nd ed., Cornell U.P., Ithaca, 1940.
12. R. S. Mulliken and C. C. J. Roothaan, *Chem. Revs.*, **41,** 219 (1947).
13. L. Pauling and G. Wheland, *J. Chem. Phys.*, **1,** 362 (1933).
14. E. Hückel, *Z. Physik,* **70,** 204 (1931).
15. R. S. Mulliken and C. A. Rieke, *J. Am. Chem. Soc.*, **63,** 1770 (1941).
16. G. Wheland, *J. Am. Chem. Soc.* **63**, 2025 (1941).

GENERAL REFERENCES

References on band spectra of diatomic molecules and the nature of their excited states:

W. Weitzel, *in* Suppl. 1 of "Wien-Harms Handbuch der experimental Physik," M. Wien and G. Joos eds.), Akademische Verlags Ges., Leipzig, 1931.

R. S. Mulliken, *Revs. Mod. Phys.*, **4,** 1 (1932).

H. Sponer, "Molekülspektren," Springer, Berlin, 1936.

G. Herzberg, reference *6*.

A. G. Gaydon, "Dissociation Energies and Spectra of Diatomic Molecules," Wiley, New York, 1947. (Reprinted by Dover, New York.)

Extensive tables of data on the excited states of diatomic molecules will be found in Sponer, *ibid.*, Herzberg, *ibid.*, and Rosen, reference *8*.

The excited states of polyatomic molecules are discussed by H. Sponer and E. Teller, *Revs. Mod. Phys.*, **13,** 75 (1941); see also A. D. Walsh, *J. Chem. Soc.*, **1953,** 2260.

References on the quantum theory of valence:

J. H. Van Vleck and A. Sherman, *Revs. Mod. Phys.* **7**, 167 (1935).

L. Pauling, reference *11*.

Y. K. Syrkin and M. E. Dyatkina, "Structure of Molecules and the Chemical Bond' (translated by M. A. Partridge and D. O. Jordan), Butterworth's, London, 1950.

C. A. Coulson, reference *10*.

B. Pullman and A. Pullman, "Les théories électroniques de la chimie organique," Masson, Paris, 1952.

R. S. Mulliken, *J. chim. phys.*, **46,** 497, 675 (1949).

Excellent reviews of recent developments in the quantum theory of valence will be found in many of the volumes of "Annual Reviews of Physical Chemistry."

Chapter 12

Molecules and the Chemical Bond II
Difficulties in Developing Satisfactory Quantitative Theories

"The underlying physical laws necessary for the mathematical theory of a large part of physics and the whole of chemistry are thus completely known, and the difficulty is only that the exact application of these laws leads to equations much too complicated to be soluble." – Dirac (*1*).

"In so far as quantum mechanics is correct, chemical questions are problems in applied mathematics." – Eyring, Walter, and Kimball (*2*, p. *iii*).

The quotations given above lead to a most tantalizing situation. Chemistry still presents us with many unsolved problems of basic interest whose solutions are difficult to obtain by experimental means. On the other hand, these quotations claim that a theory exists that can resolve any chemical question. Unfortunately the claim is made "in principle" and not "in fact," and the distinction was never more important. Mathematical difficulties have rather effectively blocked the road between the basic equations of quantum mechanics and the complete and precise solutions of all but the simplest chemical problems. In this chapter we shall see what some of these difficulties are.

It is especially important that the nature of these difficulties be made clear to the student who is beginning to understand the principles of quantum chemistry. When he first realizes the implications of Schroedinger's equations, the student's natural reaction is to ask, "What are we waiting for? If the equations for solving all chemical problems are known, why has the machinery for solving them not been set in more rapid motion?" It seems to the student – and rightly so – that he has in his hands a locked chest full of treasures, and that all he need do is locate the key. Unfortunately there is every reason to believe that the key is very well hidden. Rather than look for the key, it is more profitable to look for cracks in the wall of the chest through which a few solutions might be pried out – usually with a certain amount of distortion.

It is wise to renounce at the outset any attempt at obtaining precise solutions of the Schroedinger equation for systems more complicated than the hydrogen molecular ion. We can keep ourselves fully and usefully occupied for the foreseeable future if we restrict ourselves to searching for improved methods of approximation. These methods must be good enough to permit quantum mechanical calculations that are useful in explaining and reliably predicting the chemical behavior of reasonably complex systems.

Much work has been done in the search for new methods of approximation and much work is still being done, but in recent years progress has been slow. It is usual to begin with first approximations such as those described in Chapter 11, and then to remove a few of the more serious assumptions. It often happens, however, that the results obtained in this way are in poorer agreement with the observations than were the original crude approximations. As one proceeds beyond the first approximation, the mathematical difficulties seem almost invariably to increase at a much greater rate than does the quality of the results of the calculations.

One general procedure in this search is to look for simple, flexible, tractable, and effective approximations to the wave functions of molecules. We shall call this approach *procedure A*. Once the rules for constructing good approximate wave functions have become clear, we can hope to make accurate *a priori* calculations of the properties of any molecular system. In spite of much effort, procedure A has not given much success with molecules more complex than H_2.

Another procedure is to develop methods by which the properties of an unknown system can be deduced from the observed properties of known systems, without necessarily obtaining explicit solutions of Schroedinger's equations. This approach (which we shall call *procedure B* and is illustrated by the treatments of benzene described in Chapter 11) has had somewhat greater success, but it too has its limitations. Unfortunately, it is not always easy to foresee what these limitations will be when procedure B is used to predict properties that cannot be checked by an independent experimental or theoretical method.

A. Criteria for the Reliability of Approximate Wave Functions

a. The insensitivity of the energy as a criterion

The energy is the most important property of molecules that chemists would like to know. It is therefore appropriate to test wave functions by comparing the energies calculated from them with the accurate values, if they are known. It is not hard to show, however, that this is a somewhat insensitive test. A wave function that gives a good account of itself by this criterion will usually be considerably less reliable for the calculation of other properties. [The discussion that follows is largely due to Eckart (*3*).]

Let the correct normalized wave function of the ground state of a system be ψ_0, and let ϕ be a normalized approximation to this function. The error in the trial wave function, ϕ, may be expressed in terms of the root mean square deviation, ε, defined by the relation

$$\varepsilon^2 = \int (\psi_0 - \phi)^2 \, d\tau \tag{A-1}$$

The function ψ_0 together with the wave function of the excited states of the system, $\psi_1, \psi_2, \cdots$, form a complete orthogonal and normalized set, so that we may write

$$\phi = \sum_{n=0}^{\infty} a_n \psi_n \tag{A-2}$$

We shall assume that all of the functions are real, so that

$$a_n = \int \psi_n \phi \, d\tau \tag{A-2a}$$

and since ϕ is normalized

$$\int \phi^2 \, d\tau = \sum a_n^2 = 1 \tag{A-2b}$$

Inserting (A-2) and (A-2a) into (A-1), we find

$$\varepsilon^2 = \int (\psi_0^2 + \phi^2 - 2\psi_0\phi) d\tau = 2(1 - a_0) \tag{A-3}$$

From equation (A-2b), however

$$a_0 = \sqrt{1 - \sum' a_n^2} \tag{A-4}$$

where the prime in the summation sign, $\sum'$, means that the term $n = 0$ is not included in the sum. If the function ϕ is a reasonably good approximation to ψ_0, the coefficient a_0 will be close to unity and $\sum' a_n^2$ will be small, so we may expand the square root in (A-4) by means of the binomial theorem and omit higher powers, giving

$$a_0 \cong 1 - \tfrac{1}{2} \sum' a_n^2$$

or from (A-3)

$$\varepsilon^2 \cong \sum' a_n^2 = 1 - a_0^2 \tag{A-5}$$

The estimate of the energy obtained from the approximate function ϕ is

$$E = \int \phi H \phi \, d\tau \tag{A-6}$$

If the true energy of the state ψ_n is W_n, then we have $H\psi_n = W_n\psi_n$, so that on substituting (A-2) into (A-6) we obtain

$$E = \int (\textstyle\sum a_n\psi_n)\,(\sum a_n W_n \psi_n) d\tau = a_0^2 W_0 + \sum' a_n^2 W_n \tag{A-7}$$

If W_1 is the energy of the first excited state, then it must be true that

$$\textstyle\sum' a_n^2 W_n \geqq W_1 \sum' a_n^2 = W_1(1 - a_0^2) \tag{A-8}$$

because by definition W_1 is less than all of the other W_n in the sum on the left, and all of the a_n^2 are positive quantities. Thus

$$E \geqq a_0^2 W_0 + W_1(1 - a_0^2)$$

The error in the estimated energy is

$$E - W_0 \geqq (W_1 - W_0)\,(1 - a_0^2) = (W_1 - W_0) \textstyle\sum' a_n^2 \tag{A-9}$$

and the root mean square deviation of the trial function from the correct function is

$$\varepsilon \leqq \sqrt{\frac{E - W_0}{W_1 - W_0}} \tag{A-10}$$

Thus the upper limit of the error in the approximate wave function is proportional to the *square root* of the error in the energy. If the error in the total energy is one per cent of the difference in energy between the ground state and the first excited state, then the wave function itself may be in error by as much as ten per cent.

Let us suppose that a trial wave function ϕ is used to estimate some property of the ground state other than the energy. Suppose further that the correct eigenfunctions $\psi_0, \psi_1, \ldots$ of the Hamiltonian are not eigenfunctions of the operator for this property. As an example, consider the estimation of the dipole moment of a molecule–the operator for the dipole moment being indicated by $\boldsymbol{m}$.

The estimated value of the dipole moment is

$$\boldsymbol{m}_{\text{est}} = \int \phi \boldsymbol{m} \phi \, d\tau = a_0{}^2 \int \psi_0 \boldsymbol{m} \psi_0 \, d\tau + 2a_0 \sum{}' a_n \int \psi_0 \boldsymbol{m} \psi_n \, d\tau$$

$$+ \sum{}' \sum{}' a_n a_m \int \psi_m \boldsymbol{m} \psi_n \, d\tau \tag{A-11}$$

$$= a_0{}^2 \boldsymbol{m}_{00} + 2a_0 \sum{}' a_n \boldsymbol{m}_{0n} + \sum{}' \sum{}' a_n a_m \boldsymbol{m}_{mn}$$

where $\boldsymbol{m}_{00}$ is the correct value of the dipole moment of the ground state. Thus the error in the estimated dipole moment is

$$\boldsymbol{m}_{\text{est}} - \boldsymbol{m}_{00} = (a_0{}^2 - 1)\boldsymbol{m}_{00} + 2a_0 \sum{}' a_n \boldsymbol{m}_{0n} + \sum{}' \sum{}' a_m a_n \boldsymbol{m}_{mn} \tag{A-12}$$

Since the ψ_n are not eigenfunctions of $\boldsymbol{m}$, the integrals $\boldsymbol{m}_{0n}$ do not necessarily vanish, and in fact some of them may be of the same order of magnitude as $\boldsymbol{m}_{00}$ itself. Furthermore, a_0 is presumed to be close to unity, so the error in the approximate dipole moment arises chiefly from the second term on the right in (A-12). Therefore the error in this case depends on the first powers of the coefficients a_n, instead of on the second powers, as was found for the energy in (A-9). An error of 10% in ϕ may be expected to result in an error of the same order of magnitude in the dipole moment, even though the same trial function might lead to an energy that was correct within 1 % of the energy difference between the ground state and the first excited state. A similar enhanced sensitivity to error will obviously be present in the estimation of any property of a system if the states of the system are not eigenstates of the operator belonging to the property.

b. The virial theorem as a criterion

In Chapter 6, we found that if a system is made up of electric charges between which only electrostatic forces are significant, the virial theorem must be obeyed. That is

$$E = -\overline{T} = \overline{V}/2$$

where E is the total energy, $\overline{T}$ is the average kinetic energy, and $\overline{V}$ is the average potential energy. It is therefore always an easy matter to evaluate the kinetic and potential energies of a molecule or an atom from its observed energy. The wave functions of atomic and molecular systems are never eigenfunctions of either the operator T or the operator V. Therefore, from the arguments presented in the previous section, these two properties provide a more sensitive test of the reliability of wave functions than does the total energy.

Let us apply this test to some of the molecular wave functions that were discussed in Chapter 11. For the hydrogen molecular ion, H_2^+, Hirschfelder and Kincaid (*4*) found that the approximate LCAO wave function in Equation (A-7) of Chapter 11 gives

$$E = -15.30 \text{ e.v.}$$

$$-\overline{T} = -10.37 \text{ e.v.}$$

$$\overline{V}/2 = -12.84 \text{ e.v.}$$

at the equilibrium internuclear separation. These three quantities are not equal, so the virial theorem is apparently violated by this wave function. It is interesting to note that the energy of an isolated hydrogen atom in a 1*s*-state is −13.6 e.v., so the kinetic energy of the electron in this state is 13.6 e.v. This is considerably larger than the value (10.37 e.v.) found for the kinetic energy in H_2^+ with the LCAO trial wave function mentioned above. According to this function, a major part of the stability of the H_2^+ molecule is therefore caused by a decrease in the kinetic energy of the electron. This is, however, completely inconsistent with the virial theorem; on page 246 we have seen that the kinetic energy of the electrons must always be greater when the two atoms forming a bond are at their equilibrium distance than when they are infinitely far apart.

Similarly the Heitler-London wave function for hydrogen (Equation (B-6) of Chapter 11) gives (*4*)

$$E = -30.20 \text{ e.v.}$$

$$-\overline{T} = -22.79 \text{ e.v.}$$

$$\overline{V}/2 = -26.49 \text{ e.v.}$$

which is again inconsistent with the virial theorem. Furthermore the kinetic energy of two electrons in two hydrogen atoms infinitely far apart is $2 \times 13.6 = 27.2$ e.v., which is 4.4 e.v. greater than the value obtained from the Heitler-London wave function. Thus the Heitler-London wave function also wrongly derives nearly all of the stability of the H_2 molecule from a decrease in the kinetic energy of the electrons.

In Chapter 6 (page 232) it was shown that trial wave functions that satisfy the virial theorem are easy to construct. One merely inserts a scale factor in front of all coordinates that appear in the trial wave function and minimizes the energy with respect to this scale factor. The scale factor that minimizes the energy will automatically give kinetic and potential energies consistent with the virial theorem.

Thus the LCAO wave function of the H_2^+ molecular ion and the Heitler-London wave function of H_2 can be made consistent with the virial theorem by writing the 1*s* hydrogen atom orbitals appearing in these functions in the form

$$1s = \sqrt{\frac{Z^3}{\pi}}\, e^{-Zr}$$

where Z is a parameter that must be varied so as to minimize the energy. The results of the use of such a parameter in these molecules will be described later in this chapter.

It is evident that in using procedure A, a rather poor trial wave function can give a fairly good estimate of the total energy, and even the more stringent requirements imposed by the virial theorem can be satisfied by merely changing the scale of the coordinates in the trial wave function. Therefore, if one is trying to construct approximate wave functions that will give reliable estimates of properties other than the energy, it is necessary to test these wave functions with other properties than the energy. Among the properties that might be used for this purpose are the following: force constants of chemical bonds, bond lengths, dipole moments, diamagnetism, polarizability, and the interactions of nuclear magnetic moments and nuclear quadrupole moments (pages 484ff.) with the magnetic and electric fields due to the valence electrons. Unfortunately, it is not always feasible to make use of these properties, so the energy remains the most frequently used property for checking the validity of wave functions. One should not be surprised, however, if a wave function built up with an eye on the energy should turn out to be less satisfactory for interpreting other properties.

It should also be pointed out that chemists are usually not interested in the *total* energy of a system, but rather in the *change* in energy when, for example, two atoms are brought together to form a bond. This change in energy may amount to but a small fraction of the total energy of the system, so that even if the total energy is known with good accuracy, the percentage error in the energy change may be quite large.

B. Some Problems in Constructing Accurate Molecular Wave Functions

In using procedure A (page 434) to explain and predict the properties of a molecule, the quantum chemist looks for adequate approximations to the molecule's wave function. Once a good wave function has been developed, of course, all of the molecule's properties can be computed (Law III, Chapter 5). The procedure is largely intuitive. The theoretician squeezes, gouges, and distorts his trial wave functions in much the same way that a sculptor molds a statue out of clay, seeking the shape that corresponds to reality.

In Chapter 11 we have seen illustrations of two general methods for obtaining molecular wave functions–the molecular orbital (MO) method and the valence bond (VB) method. Both of these methods construct their wave functions from products of independent orbitals, each occupied by a different electron, the wave function then being antisymmetrized by taking the sums and differences of the various permutations of the electrons among the orbitals. In the VB method the orbitals are atomic orbitals, while in the MO method they are molecular orbitals.

Two problems arise in following this procedure. The first is the problem of obtaining orbitals that have the optimum shape and are at the same time simple enough in mathematical form to make calculations feasible. The second problem is that of dealing adequately with the repulsions between electrons. We have encountered the same problems in our discussion of atomic systems, but in molecular systems they are even more difficult to solve. The orbital shape problem in molecules is most clearly illustrated by reviewing the studies that have been made on the wave function of the hydrogen molecular ion, H_2^+. The difficulties in dealing with electronic repulsions appear when we consider the wave function of the hydrogen molecule H_2. We shall now discuss these two systems in some detail in order to illuminate some of the basic difficulties that confront the quantum mechanical approach to molecular systems using procedure A.

a. The hydrogen molecular ion and the problem of obtaining correct orbital shapes

The exact wave functions of the ground state and many of the excited states of H_2^+ are available, so that we know exactly what the true wave functions look like [Chapter 6 and Bates, Ledsham, and Stewart (*5*)]. These exact functions are, however, too complex to be useful in most practical calculations, such as the application of the MO method to diatomic molecules. There is therefore good reason to look for functions that resemble the correct ones, but have a simpler analytical form. Since most of our information concerns the wave function of the ground state of H_2^+, we shall limit our discussion to this state.

In Chapter 11 we found that a simple LCAO function constructed from 1*s* hydrogen atomic orbitals

$$\psi = (1/\sqrt{N})\,(e^{-r_a} + e^{-r_b}) \tag{B-1}$$

leads to a potential energy curve that resembles the actual curve of the molecule. The calculated bond energy is, however, 1.76 e.v. as compared with the observed value of 2.79 e.v.–that is, too small by almost 40 per cent. Furthermore, the calculated equilibrium bond length, 1.32 Å., differs considerably from the observed value of 1.06 Å. This discrepancy is not surprising in view of the crudeness of the LCAO function. Figure 12-1 shows two cross sections of the correct, normalized ground state wave function of H_2^+ (as obtained from the exact solutions of Equation (H-2) of Chapter 6), along with cross sections of approximate normalized wave functions that have been proposed by various workers. It is obvious that whereas the simple LCAO function of Equation (B-1) resembles the correct wave function in its general shape, it is too small in the region between the two protons and too large elsewhere. We have also seen that the wave function (B-1) is not consistent with the virial theorem.

The simplest improvement is the introduction of a scale factor, Z, into the LCAO function, giving

$$\psi = (1/\sqrt{N})\,(e^{-Zr_a} + e^{-Zr_b}) \tag{B-2}$$

The value of Z is selected so as to minimize the energy. This wave function was used by Finkelstein and Horowitz (*6*) who found a bond energy of 2.25 e.v. and an equilibrium bond length within one or two hundredths of an Ångstrom unit of the correct value. The discrepancy between the calculated and the correct bond energy is in this way cut in half (though there is still an error of 19 per cent). Furthermore the virial theorem is now satisfied. As seen from Fig. 12-1, the Finkelstein-Horowitz wave function is somewhat closer to the correct wave function than was the simple LCAO function of (B-1), but it still gives too low an electron density between the nuclei.

The value of Z that corresponds to the stable H_2^+ molecule is 1.228. In (B-1), Z is given the value unity. The Finkelstein-Horowitz function,

(B-2), thus tends to draw the electron closer to the nuclei, and in so doing it shifts the electron cloud to some extent into the region between the nuclei. At other internuclear distances the optimum value of Z has different values. If the internuclear distance is zero, we have a helium ion in its ground state, for which the correct wave function is a 1s-orbital corresponding to a

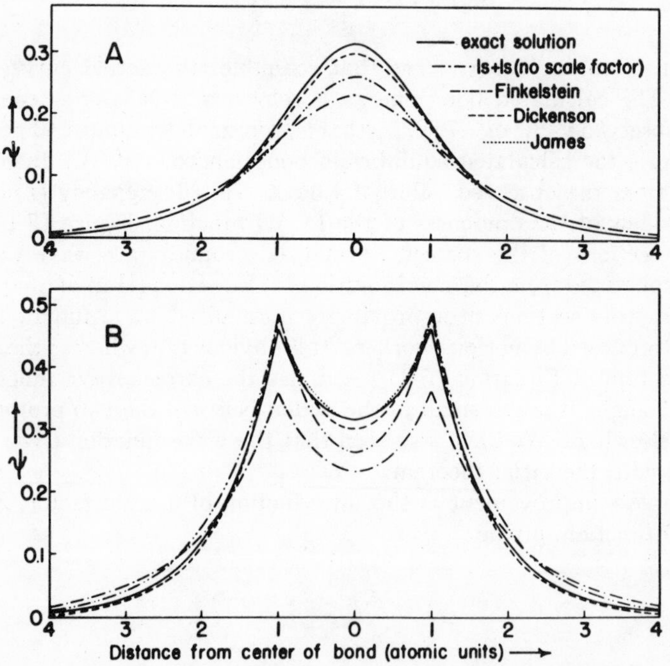

FIG. 12-1. Comparison of the exact wave function of the ground state of the H_2^+ molecule-ion with various approximate wave functions. All of the wave functions have been normalized. The wave function of James is indistinguishable from the exact solution on the scale of this drawing at nearly all points.

A. Values of the wave functions along a line normal to the H-H bond and passing through the midpoint of the bond.

B. Values of the wave functions along a line passing through the two protons. The protons are located at the two peaks.

nuclear charge of 2. Clearly in this case $Z = 2$. On the other hand, if the two nuclei are infinitely far apart the wave function must reduce to that of a normal hydrogen atom, with $Z = 1$. Thus the scale factor, Z, may be regarded as an effective nuclear charge whose value increases, as the internuclear distance is decreased, because of the electrical attractions that the nuclei exert on the electron.

A further improvement of the LCAO form results on replacing the pure $1s$ atomic orbitals of (B-2) with hybridized atomic orbitals of the type $(1s + c2p_z)$, where c is a variable parameter (to be determined by minimizing the energy) and $2p_z$ is a hydrogen-like $2p$ atomic orbital whose axis lies along the axis of the molecule. Since $2p_z$ is positive on one side of its origin and negative on the other side, it can raise the electron density in the region between the two nuclei, where the electron evidently wants to be, and it can decrease the density at the two ends of the molecule, where the electron does not want to be (*cf.* discussion of $s-p$ hybrids on page 410). The $2p_z$ function can be said to make possible the mutual polarization of the two atomic orbitals in the LCAO expression.

Dickenson (*7*) has calculated the energy corresponding to the hybridized LCAO function,

$$\psi = (1/\sqrt{N})\,(e^{-Zr_a} + cr_a \cos\theta_a e^{-\varepsilon Zr_a} + e^{-Zr_b} + cr_b \cos\theta_b e^{-\varepsilon Zr_b}) \qquad \text{(B-3)}$$

involving three parameters, Z, c, and ε, where θ_a and θ_b are the angles between the internuclear axis and the lines joining the electron respectively to nucleus a and nucleus b. He found that the energy was minimized when the internuclear distance was 1.06 Å., $\varepsilon = 1.15$, $Z = 1.247$, and $c = 0.145$. The resulting dissociation energy was 2.652 e.v.–only 5 per cent below the correct value. If it was assumed that $\varepsilon = 1$, the energy was minimized when $Z = 1.2537$ and $c = 0.1605$, and the resulting dissociation energy and bond length are 2.637 e.v. and 1.05 Å., respectively. (The more recent calculations of Pritchard and Skinner (*8*) using $Z = 1$ and $\varepsilon = 0.5$ and varying only c are inconsistent with Dickenson's results, since they gave a dissociation energy of 2.706 e.v., which is greater than the maximum value found by Dickenson on varying all three parameters, Z, ε, and c.) The shape of the Dickenson wave function with the optimum values of the three parameters is shown in Fig. 12-1. As expected, it moves the electron cloud into the region between the nuclei and away from the ends of the molecule, but the agreement with the exact wave function is still far from perfect.

Another method of shifting electrons into the region between the nuclei is to "detach" the $1s$-orbitals from their protons, moving their centers toward each other by a distance x along the nuclear axis (see Fig. 12-2) to give the function resembling (B-2)

$$\psi = (1/\sqrt{N})\,(e^{-Zr'_a} + e^{-Zr'_b}) \qquad \text{(B-4)}$$

except that r'_a and r'_b are measured from the displaced centers of the orbitals and not from the two protons. This type of function was first suggested by Gurnee and Magee (*9*) for the hydrogen molecule and it has been applied to H_2^+ by Hurley (*10*), who calculated from it a dissociation energy of 2.66 e.v. This is less than the observed value by only 4 per cent. The calculated bond length was found to be 1.05 Å. The effective nuclear

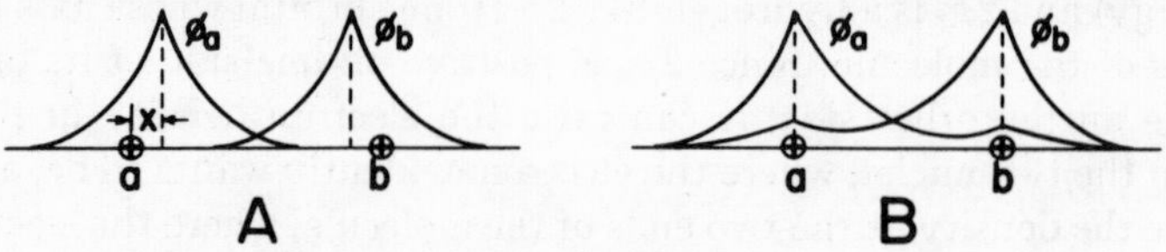

FIG. 12-2.
A. Orbitals used by Gurnee and Magee for H_2.
B. Orbitals used by Coulson and Fisher for H_2.

charge Z is 1.23 and the optimum displacement x of the orbitals from the nuclei is 0.061 Å. (that is, about 6 per cent of the bond length). Thus, although the Gurnee-Magee wave function contains only two variable parameters (the scale factor Z and the orbital displacement x), it gives a more accurate value of the energy than the Dickenson wave function, which contains three parameters. Therefore, it might be regarded as the most satisfactory of all LCAO functions now available. Unfortunately, the movement of the orbitals from the nuclei complicates some of the integrations that are necessary if this type of function is to be generally applied to chemical bonds. Furthermore, this type of function is certainly not accurate in the immediate vicinity of the two nuclei, so that it would be very bad for calculating nuclear quadrupole coupling constants (page 489), for instance.

All of the approximate wave functions of H_2^+ that have been discussed so far were constructed by taking linear combinations of atomic orbitals associated with the two nuclei involved in the bond. The results that have been obtained give a preliminary idea of the adequacy of atomic orbitals in molecular problems, and especially, therefore, of the valence bond method, which is based on atomic orbitals. Evidently polarization (or hybridization) is necessary in order to obtain adequate atomic orbitals for bonds, since the necessary concentration of electrons between nuclei cannot be achieved in any other way.

If we are willing to give up the LCAO form, even more accurate wave functions of very simple form may be written for the H^+_2 molecule. Consider the function

$$\psi = (1/\sqrt{N}) \exp\left[-Z(r_a + r_b)\right] \qquad \text{(B-5)}$$

The quantity $(r_a + r_b)$ has its minimum value along the straight line segment drawn between the two nuclei, and it is constant at all points along this line. Therefore the function (B-5) is constant and has its maximum value along the axis of the bond; obviously it overdoes the concentration of electrons in the bond. Furthermore it does not lead to the correct wave function in the limit of large internuclear distances. Nevertheless James (*11*) found that the energy calculated from this function passes through a minimum at an internuclear distance of 1.06 Å. and is 2.16 e.v. below the energy of an isolated hydrogen atom. Thus the simple function (B-5) is almost as good as the Finkelstein-Horowitz LCAO function if the nuclei are not too far apart.

The function (B-5) can easily be improved by introducing a term that reduces the electron density symmetrically about the midpoint of the bond. This may be accomplished by writing

$$\psi = (1/\sqrt{N})\, e^{-Z(r_a+r_b)}\, [1 + a(r_a - r_b)^2] \tag{B-6}$$

James found that when $Z = 1.35$ and $a = 0.448$, a binding energy of 2.772 e.v. is obtained, which is within one per cent of the observed value. In the curves shown in Fig. 12-1 this wave function is almost indistinguishable from the correct function. Equally good results were obtained by Guillemin and Zener (*12*) with the function

$$\psi = (1/\sqrt{N})\, e^{-Z(r_a+r_b)} \cosh a(r_a - r_b) \tag{B-7}$$

the hyperbolic cosine term having the same effect as the factor $[1 + a(r_a - r_b)^2]$ in the James function.

It appears that the best and most simple approximations to the wave functions of H_2^+ are obtained if the LCAO form is not used.

b. The hydrogen molecule and the problem of electronic repulsions

(*i*) *The Molecular Orbital Approach.* We have seen (page 387) that if simple LCAO $1s\sigma$ molecular orbitals of the form of Equation (B-1) are used to construct an MO wave function of hydrogen

$$\psi = 1s\sigma_1\, 1s\sigma_2 \tag{B-8}$$

a dissociation energy of 2.681 e.v. is obtained for the H–H bond. This is more than 40 per cent below the observed value of 4.745 e.v. Furthermore, the equilibrium bond length is 0.11 Å. too long and the virial theorem is violated. Since we have just found that the LCAO function employed in this calculation is a rather poor approximation in the case of H_2^+, and since improved molecular orbitals are available, we should hope to be able to increase the accuracy of the MO method considerably by using these improved orbitals.

The effects of various improvements in the form of the molecular orbitals $1s\sigma$, are summarized in Table 12-1, which is based on the work of Coulson (*13*). The most simple improvement is made by the introduction of a scale

TABLE 12-1

RESULTS OF MOLECULAR ORBITAL METHOD FOR HYDROGEN

$\psi = (1s\sigma)_1(1s\sigma)_2$

Molecular orbital used for $1s\sigma$ †	Bond energy (e.v.)	Optimum values of parameters	Reference
$e^{-r_a} + e^{-r_b}$	2.681	—	‡
$e^{-Zr_a} + e^{-Zr_b}$	3.470	$Z =$ 1.197	‡
$e^{-ZR\mu}(1 + a\nu^2)$ (James' orbital)	3.096	$Z =$ 0.535 $a =$ 0.23342 $R =$ 1.40 Bohr radii	‡
$e^{-ZR\mu}(1 + a\nu^2 + b\mu)$	3.535	$Z =$ 0.535 $a =$ 0.21948 $b =$ −0.079575 $R =$ 1.40	‡
$e^{-ZR\mu}(1 + a\nu^2 + b\mu + c\mu^2)$	c. 3.540	—	‡
$e^{-ZR\mu}(1 + a\nu^2 + b\mu + c\mu^2 + d\mu\nu^2)$	3.603	$Z =$ 0.535 $a =$ 0.2787 $b =$ −0.12863 $c =$ 0.012503 $d =$ −0.039589 $R =$ 1.40	‡
"Best molecular orbital"	c. 3.63	—	‡
$(e^{-r_a} + e^{-r_b})$ plus configuration interaction with $(2p\sigma^*_u)^2$	3.21	—	§
$(e^{-Zr_a} + e^{-Zr_b})$ plus configuration interaction with $(2p\sigma^*_u)^2$	4.00	$Z =$ 1.193	§

† $\mu = (r_a + r_b)/R$, $\nu = (r_a - r_b)/R$, $R =$ internuclear distance $= 0.74$ Å. in all but first two calculations

‡ Coulson (*13*)

§ Weinbaum (*14*)

factor into the LCAO $1s\sigma$ molecular orbital in the manner of Finkelstein and Horowitz. The resulting bond energy is too low by 27 per cent, though the equilibrium bond length is too short by only 0.008 Å. Of course the introduction of a scale factor assures us that we have satisfied the virial theorem. The improvement in energy is, however, only moderate–much less than that achieved by the similar use of a scale factor in H_2^+.

The next obvious improvement is to utilize the wave function that James found gave such good results for H_2^+. Surprisingly enough, it is found that in H_2, the James orbital gives a bond energy farther from the observed value than the energy obtained with the Finkelstein-Horowitz orbital. The James orbital may, however, be further modified by adding to it terms involving powers of $(r_a + r_b)$. These terms are found to enter in such a way as to raise the electron density between the two nuclei. On the basis of the improvements that result from them, Coulson estimated that the best bond energy that can be obtained using molecular orbitals of the type found in the ground state of H_2^+ is 3.63 e.v.–which is still 23 per cent below the observed value.

Evidently a wave function of the form of Equation (B-8) cannot give a very good representation of the ground state of hydrogen. The reason is that a molecular orbital function of this type is not at all effective in keeping the two electrons out of each other's way because it forces both electrons to spend about half of their time in the vicinity of the same nucleus. Such a wave function is therefore unable to deal adequately with the correlation energy (page 388).

There is, however, a means of improving the situation while still maintaining the molecular orbital approach. This consists in taking account of configuration interaction. We have seen in Chapter 10 (pages 335 and 344), that resonance can take place in atoms between atomic states having different electronic configurations but the same spin and orbital angular momentum. Similarly in the hydrogen molecule the configuration $(2p\sigma^*_u)^2$ can give rise to a $^1\Sigma^+_g$ wave function, which can resonate with the $(1s\sigma_g)^2\ ^1\Sigma^+_g$ wave function. Because of this resonance, we may write an improved wave function for the ground state of hydrogen in the form

$$\psi = 1s\sigma_g(1)\, 1s\sigma_g(2) + a2p\sigma^*_u(1)\, 2p\sigma_u^*(2) \tag{B-9}$$

where a is a parameter that is to be varied so as to minimize the energy. Writing the $1s\sigma_g$ and $2p\sigma^*_u$ orbitals in the LCAO form, we obtain

$$\begin{aligned}\psi &= [1s_a(1) + 1s_b(1)]\,[1s_a(2) + 1s_b(2)] + a[1s_a(1) - 1s_b(1)]\,[1s_a(2) - 1s_b(2)]\\ &= (1 + a)\,[1s_a(1)\,1s_a(2) + 1s_b(1)\,1s_b(2)]\\ &\quad + (1 - a)\,[1s_a(1)\,1s_b(2) + 1s_a(2)\,1s_b(1)]\end{aligned} \tag{B-10}$$

It is evident that if the constant a is set equal to unity, this function corresponds to the pure ionic structures H^+H^- and H^-H^+, whereas if we set $a = -1$, we obtain the Heitler-London wave function, Equation (B-6) of Chapter 11. Thus the LCAO molecular orbital wave function which includes configuration interaction with the $(2p\sigma^*_u)^2\ {}^1\Sigma^+_g$ state is exactly equivalent to a Heitler-London function to which ionic wave functions have been added.

Weinbaum (*14*) has calculated the energy of H_2 using a function of the form of Equation (B-10). When no scale factor Z was used in the 1s-orbitals, a dissociation energy of 3.21 e.v. and an equilibrium bond length of 0.884 Å. were obtained when $a = -0.727$. If the scale factor Z is introduced into the 1s-orbitals and the energy is minimized with respect to both Z and a, a dissociation enrgy of 4.00 e.v. and a bond length of 0.748 Å. are obtained, with $Z = 1.193$ and $a = -0.591$. We see that a marked improvement in the molecular orbital wave function results when interaction with the $(2p\sigma^*_u)^2$ configuration is introduced in this way; the dissociation energy obtained with crude LCAO 1s molecular orbitals is considerably better than that obtained with even the best $1s\sigma_g$-orbitals in the absence of configuration interaction. The reason for this improvement is, of course, the fact that we have reduced the emphasis given to the ionic structures by the molecular orbital wave function for the $(1s\sigma_g)^2$-configuration. In other words, configuration interaction in this example provides an effective means of dealing with the correlation energy.

The use of configuration interactions is a general method of improving the approximate wave functions of all molecules. If we were to include configuration interactions with the approximate wave functions of all of the excited states of a molecule, this would be equivalent to the expansion of the true wave function in terms of a complete set of functions. In principle, therefore, if one has a set of rough approximations to the wave functions of all of the states of a molecule (such as a set based on LCAO molecular orbitals employing hydrogen-atomic orbitals) it should be possible to write the correct wave function for any one of the states of the molecule by including configuration interactions of the rough wave function for the state with the rough approximations to the wave functions of a sufficient number of other states. In practice, of course, this procedure will be feasible only if the number of interacting configurations required to attain the desired precision is not too large. The improvement resulting from the inclusion of the single interaction with the $(2p\sigma^*_u)^2\ {}^1\Sigma^+_g$-state is promising. A further significant improvement resulting from interaction with the $(2p\pi_u)^2\ {}^1\Sigma^+_g$-state will be mentioned below.

(*ii*) *The Valence Bond Approach.* In the Heitler-London method for the hydrogen molecule, we write the wave function in the form

$$\psi = \phi_a(1)\,\phi_b(2) + \phi_a(2)\,\phi_b(1) \qquad \text{(B-11)}$$

where ϕ_a and ϕ_b are atomic orbitals associated with the two nuclei. This type of wave function has the initial advantage over the MO method that it discourages the simultaneous presence of both electrons close to the same nucleus, and thus gives a better account of the correlation energy. Of course the actual molecule does presumably have a certain amount of "piling up" of electrons on the same nucleus, but the calculations indicate that if one has to make a choice between introducing ionic structures or omitting them, it is much better to leave them out altogether than to put them on equal footing with the nonpolar structures, as is done in the MO method.

In the original Heitler-London treatment, the atomic orbitals ϕ were written as simple 1*s* hydrogen wave functions,

$$\phi_a = e^{-r_a} \qquad \text{(B-12)}$$

The resultant dissociation energy, though in somewhat better agreement with the observed value than was the first MO approximation, is still too small by 33 per cent. Furthermore, the equilibrium bond length is too long, and the kinetic energy and the potential energy are inconsistent with the virial theorem.

The results of various improvements in the Heitler-London atomic orbitals are summarized in Table 12-2. The simplest of these improvements is the introduction of a scale factor Z into the exponent of the 1*s*-function, giving

$$\phi_a = e^{-Zr_a} \qquad \text{(B-13)}$$

Wang (*15*) found that the dissociation energy is thereby brought within 20 per cent of the correct value, the equilibrium bond length being too large by only 0.003 Å. Of course, the virial theorem is now automatically satisfied.

The study of the H_2^+ molecule has shown the importance of allowing the electrons in a bond to move into the region between the two nuclei–that is, of allowing for the mutual polarization of the atoms involved in the bond. Therefore it is natural to expect further improvement in the wave function by using atomic orbitals that draw the electrons into the internuclear region. This can be done in a variety of ways. Rosen (*16*) hybridized the 1*s*-orbital with a p_z-orbital in a manner similar to that used by Dickenson for H_2^+ (page 443), writing

$$\phi_a = e^{-Zr_a} + aZr_a \cos\theta_a e^{-Zr_a} \qquad \text{(B-14)}$$

TABLE 12-2

RESULTS OF VALENCE BOND AND OTHER METHODS FOR HYDROGEN

Name of wave function	Type of orbital used	Number of parameters	Calculated dissociation energy (e.v.)	Values of parameters (all distances measured in atomic units)
Heitler-London	e^{-r_a}	0	3.14	—
Wang (Scale factor)	e^{-Zr_a}	1	3.76	$Z = 1.166$
Rosen (Polarization)	$e^{-r_a}(1 + ar_a \cos \theta_a)$	1	3.34	$a = 0.095$
Rosen (Polarization plus scale factor)	$e^{-Zr_a}(1 + aZr_a \cos \theta_a)$	2	4.02	$Z = 1.17$ $aZ = 0.123$
Gurnee-Magee (Polarization plus scale factor)	$e^{-Zr'_a}$ (orbital displaced a distance x along bond)	2	4.16	$Z = 1.17$ $x = 0.07$
Inui (Scale factor polarization)	$e^{-Zr_a-Z'r_b}$	2	4.04	$Z = 1.081$ $Z' = 0.131$
Mueller-Eyring (Polarization plus scale factor; "semi-localized")	$(e^{-Zr_a-Z'r_b} + \lambda e^{-Z'r_a}e^{-Zr_b})$	3	4.20	$Z = 1.150$ $Z' = 0.042$ $\lambda = 0.138$
Weinbaum (Ionic plus Heitler-London)	(B-10)	1	3.21	
Weinbaum (Ionic plus scale factor)	(B-10)	2	4.00	$Z = 1.193$
Weinbaum (Ionic plus scale factor plus Rosen)	(B-10)	3	4.10	$Z = 1.190$ $a = 0.07$
Hirschfelder-Linnett (Polarization)	(B-21)	2	3.38	$Z = 1.000$; $\alpha = -0.04341$ $\beta = -0.07035$; $\gamma = 0$

Name of wave function	Type of orbital used	Number of parameters	Calculated dissociation energy (e.v.)	Values of parameters (all distances measured in atomic units)
Hirschfelder-Linnett (Ionic plus polarization)	(B-21)	3	3.52	$Z = 1.000$; $\alpha = -0.05444$ $\beta = -0.11031$ $\gamma = 0.2450$
Hirschfelder-Linnett (Scale factor plus polarization)	(B-21)	3	3.92	$Z = 1.146$; $\alpha = -0.04624$ $\beta = -0.04496$ $\gamma = 0$
Hirschfelder-Linnett (Scale factor plus ionic plus polarization)	(B-21)	4	4.25	$Z = 1.195$; $\alpha = -0.05735$ $\beta = -0.06944$ $\gamma = 0.3334$
Lennard-Jones-Pople (Scale factor plus ionic plus azimuthal correlation)	(B-22)	3	4.16	$Z = 1.19$ $a = -0.054$ $\gamma = 0.273$
Frost-Braunstein (Scale factor plus r_{12} term; "correlated molecular orbital")	(B-23)	2	4.10	$Z = 1.285$ $a = 0.28$
James-Coolidge	(B-24)	—	c. 4.29	Estimated max. dissociation energy in absence of r_{12} terms
James-Coolidge	(B-24)	5	4.5289	—
James-Coolidge	(B-24)	11	4.7043	—
James-Coolidge	(B-24)	13	4.7198	—
Observed	—	—	4.7451	—

When Rosen set the scale factor in his orbitals equal to unity, he obtained a dissociation energy of 3.34 e.v., which is only 0.20 e.v. better than the Heitler-London value. Thus the hybridization is of itself not as effective as the scale factor in improving the wave function. When the energy was minimized with respect to both a scale factor and the coefficient of the p_z-function, the dissociation energy was found to be 4.02 e.v., which is 0.26 e.v. better than the value obtained by use of the scale factor alone.

An interesting and somewhat more effective way of introducing polarization into the atomic orbitals has been suggested by Gurnee and Magee (*9*), who merely shifted the center of each 1*s*-orbital a distance x along the bond away from the proton with which it is associated (Fig. 12-2). Thus they write for the orbital on nucleus a,

$$\phi_a = e^{-Zr'_a} \tag{B-15}$$

where r'_a is the distance of the electron, not from the nucleus as in (B-13), but from the displaced point, along the bond away from nucleus a and toward nucleus b. When the energy was minimized with respect to both the scale factor Z and the distance x, a dissociation energy of 4.16 e.v. was obtained. The improvement resulting from the Gurnee-Magee method of polarization is thus considerably better than that obtained by the Rosen method.

Inui (*17*) has made use of still another artifice for polarizing the atomic orbitals of the Heitler-London function. He has written the atomic orbital on nucleus a in the form

$$\phi_a = e^{-Zr_a}\, e^{-Z'r_b} \tag{B-16}$$

where Z and Z' are variable parameters. The optimum value of Z' turns out to be rather small compared with Z. The introduction of the factor $e^{-Z'r_b}$ tends to increase the magnitude of the orbital belonging to nucleus a, when the electron is in the vicinity of nucleus b, since $e^{-Z'r_b}$ then becomes large. That is, this factor tends to move the electron into the regions between the two nuclei, where r_a and r_b are both small, and away from the other side of nucleus a, where r_a may still be small but r_b is larger. Inui's orbital gives a dissociation energy of 4.04 e.v., which is slightly better than Rosen's result.

Mueller and Eyring (*18*) have combined the Inui type of polarization with another interesting method of polarizing the atomic orbitals. Following Coulson and Fischer (*19*) they begin by writing the atomic orbital on atom a in the form

$$\phi_a = 1s_a + \lambda 1s_b \tag{B-17}$$

where λ is a variable parameter. This orbital polarizes the atom by adding a "pip" to it in the vicinity of the other atom in the bond (see Fig. 11-2). Presumably the pip will be small (value of λ small) when the two atoms are far apart, but as the atoms are brought together the pip will become larger (increase in the value of λ). On introducing elliptical coordinates, $\mu = (r_a + r_b)/R$ and $\nu = (r_a - r_b)/R$ and a scale factor, ζ, the orbital (B-17) takes the form

$$\phi_a = e^{-\zeta R(\mu+\nu)/2} + \lambda e^{-\zeta R(\mu-\nu)/2} \tag{B-18}$$

In order to obtain more flexibility, a new parameter, ζ', is introduced into this function, such that

$$\phi_a = e^{-\zeta R\mu/2} e^{-\zeta' R\nu/2} + \lambda e^{-\zeta' R\mu/2} e^{\zeta R\nu/2} \tag{B-19}$$

This can be written in the equivalent form

$$\phi_a = e^{-Zr_a} e^{-Z'r_b} + \lambda e^{-Zr_b} e^{Z'r_b} \tag{B-20}$$

where $Z = (\zeta + \zeta')/2$ and $Z' = (\zeta - \zeta')/2$. The resulting orbital resembles Inui's, except that it also contains a pip at the second nucleus. When the energy was minimized with respect to the three parameters Z, Z', and λ, a dissociation energy of 4.20 e.v. was found. This is the best value yet reported for a valence bond function of the form (B-11). The Mueller-Eyring orbitals automatically introduce a certain amount of ionic structure into the molecule. They also bear an obvious resemblance to LCAO molecular orbitals. For this reason they are called "semi-localized orbitals." Since the Inui orbital (B-16) is similar to the molecular orbital (B-5), it too can be regarded as an orbital of the semi-localized type.

(iii) Other Wave Functions Not Explicitly Involving Interelectronic Distance. We have already mentioned Weinbaum's improvement of the Heitler-London and MO wave functions by the manipulation of terms corresponding to ionic structures. Weinbaum (*14*) has also studied the improvement of the Rosen wave function resulting from the addition of ionic terms. He finds that they result in less than 2 per cent improvement in the dissociation energy, his final value being still 13 per cent below the experimental value.

Hirschfelder and Linnett (*20*) have used the four-parameter function

$$\begin{aligned}\psi = {} & 1s_a(1)1s_b(2)\,[1 + \alpha Z^2(x_{a1}x_{b2} + y_{a1}y_{b2}) + \beta Z^2(z_{a1}z_{b2})] \\ & + 1s_a(2)1s_b(1)\,[1 + \alpha Z^2(x_{a2}x_{b1} + y_{a2}y_{b1}) + \beta Z^2(z_{a2}z_{b1})] \\ & + \gamma[1s_a(1(\ 1s_a(2) + 1s_b(1)\ 1s_b(2)]\end{aligned} \tag{B-21}$$

where a scale factor Z is included in the $1s$-orbitals and x_{a1}, y_{a1}, z_{a1}, are the coordinates of electron no. 1 in a Cartesian coordinate system whose origin is at nucleus a and whose z-axis lies along the bond, and similarly for $x_{b1}, \cdots, z_{b2}$. Terms such as $x_{a1}x_{b2}$ and $y_{a1}y_{b2}$ take account of the mutual polarization of the two atoms perpendicular to the molecular axis, while the terms such as $z_{a1}z_{b2}$ take some account of the polarization along the bond. We shall find in Chapter 13 that terms of these types are responsible for van der Waals forces between molecules. These forces are predominant at large interatomic distances, so that we should expect the function (B-21) to be especially accurate when the two nuclei are far apart. The maximum dissociation energies calculated when the energy is minimized with respect to different combinations of the four parameters in (B-21) are shown in Table 12-2. The best value is 4.25 e.v., still 10 per cent below the observed value.

(*iv*) *The Importance of Including r_{12} Terms in Providing for Correlation.* In all of the calculations of the bond energy of hydrogen that have been described so far in this chapter, a discrepancy of more than 10 per cent remains even when relatively complex functions are used. The reason for this persistent discrepancy seems to be that none of the functions take adequate account of the tendency of the electrons to avoid each other (that is, the correlation effect). True, some of these functions correlate the electrons better than others, the MO functions being particularly bad in this respect. But even the best functions of the valence bond type give a relatively high probability of finding the two electrons at the same point in space.

The importance of this factor can be seen in several ways. Let us consider first the wave function used by Lennard-Jones and Pople (*21*),

$$\begin{aligned}\psi = {} & 1s_a(1)\, 1s_b(2)\, [1 + aZ^2 r_{a1} r_{b2} \sin\theta_{a1} \sin\theta_{b2} \cos(\varphi_1 - \varphi_2)] \\ & + 1s_a(2)\, 1s_b(1)\, [1 + aZ^2 r_{a2} r_{b1} \sin\theta_{a2} \sin\theta_{b1} \cos(\varphi_2 - \varphi_1)] \\ & + \gamma[1s_a(1)\, 1s_a(2) + 1s_b(1)\, 1s_b(2)] \qquad \text{(B-22)}\end{aligned}$$

The angle θ_{a1} is measured between the molecular axis and the line drawn from nucleus a to electron no. 1, and the remaining angles θ_{a2}, θ_{b1}, and θ_{b2} are defined in a similar way. The angles φ_1 and φ_2 are the azimuthal angles of the respective electrons about the molecular axis. If a scale factor is included in the $1s$-orbitals, the function (B-22) has three parameters: Z, a, and γ. The function does not contain a polarization term of the Rosen type, but the factor $\sin\theta_{a1} \sin\theta_{b2} \cos(\varphi_1 - \varphi_2)$ and its counterpart in the fourth term of (B-22) operate in such a way as to tend to keep the electrons apart.

This can be seen from the following observations. It is found that the energy is minimized when the constant α has a negative value. Therefore the terms containing α tend to make the wave function larger when $\cos(\varphi_1 - \varphi_2)$ is negative – that is, when φ_1 and φ_2 differ by $180° \pm 90°$. This happens whenever the two electrons are on opposite sides of the molecular axis. Furthermore, if either of the electrons comes close to the molecular axis, the $\sin\theta$ terms belonging to it vanish, and the terms involving α in Equation (B-22) disappear. Thus the electrons tend to stay on opposite sides of the axis and they tend to keep away from the axis. The effect on the wave function of the terms involving α is referred to as "azimuthal correlation."

A dissociation energy of 4.16 e.v. is obtained when the parameters in (B-22) are optimized. This result is as good as any obtained from the three-parameter functions that take account of polarization parallel to the bond. Thus the inclusion of azimuthal correlation appears to be about as effective as the inclusion of orbitals polarized along the bond.

The function (B-22) is another example of the effectiveness of configuration interaction in improving a molecular wave function. Making use of the trigonometric identity

$$\cos(\varphi_1 - \varphi_2) = \cos\varphi_1 \cos\varphi_2 + \sin\varphi_1 \sin\varphi_2$$

we can write the second and fourth terms in (B-22) as

$$e^{-Z(r_{a1}+r_{b2})} r_{a1} r_{b2} \sin\theta_{a1} \sin\theta_{b2} \cos(\varphi_1 - \varphi_2)$$

$$+ e^{-Z(r_{a2}+r_{b1})} r_{a2} r_{b1} \sin\theta_{a2} \sin\theta_{b1} \cos(\varphi_2 - \varphi_1)$$

$$= (2p_x)_{a1}(2p_x)_{b2} + (2p_x)_{a2}(2p_x)_{b1} + (2p_y)_{a1}(2p_y)_{b2}$$

$$+ (2p_y)_{a2}(2p_y)_{b1} \qquad \text{(B-22a)}$$

where $2p_x$ and $2p_y$ are hydrogen atom orbitals in the real form. But the right-hand side of (B-22a) is merely the Heitler-London form of the wave function of the $2p\pi_u^2\,{}^1\Sigma^+_g$ state of hydrogen. Thus we see again that inclusion of configuration interactions can help to give some electronic correlation (*cf.* page 447).

An even more striking example of the importance of correlation is the result found by Frost and Braunstein (*22*), using the modified LCAOMO function

$$\psi = [1s_a(1) + 1s_b(1)]\,[1s_a(2) + 1s_b(2)]\,[1 + ar_{12}] \qquad \text{(B-23)}$$

where r_{12} is the distance between the two electrons and a is a parameter. If a scale factor Z is included in the 1s-orbitals, this function contains only two parameters and makes no provision for orbital polarization. Yet on minimizing the energy with respect to Z and a, a dissociation energy of 4.11 e.v. is obtained; the optimum value of the constant a is positive, so that the term $[1 + a\, r_{12}]$ favors situations in which the electrons are far apart. This calculated energy is better than the energies obtained from any other two-parameter wave function except the Gurnee-Magee function. Again it appears that the importance of the correlation is equal to that of polarization.

The clearest indication of the importance of correlation, however, is to be found in the work of James and Coolidge (*23*). These workers used wave functions expressed in terms of powers of the elliptical coordinates of the two electrons, μ_1, ν_1, μ_2, and ν_2 (defined in Table 12-1), and the interelectronic distance, r_{12}

$$\psi = e^{-\delta(\mu_1+\mu_2)} \sum_{\substack{m,n,j,\\ k,p,}} C_{mnjkp} \, [\mu_1{}^m \mu_2{}^n \nu_1{}^j \nu_2{}^k + \mu_1{}^n \mu_2{}^m \nu_1{}^k \nu_2{}^j] \, r_{12}{}^p \tag{B-24}$$

where the exponents m, n, j, k, and p are integers, δ is set equal to 0.75, the coefficients C_{mnjkp} are variable parameters, and the sum is taken over various sets of values of the exponents. James and Coolidge studied the effect on the energy of the inclusion of different numbers and types of terms in the sum in (B-24). They found that if they set $p = 0$ in each of the terms (so that ψ did not depend explicitly on r_{12}), then regardless of how many terms were included in the sum, it was not possible to obtain a dissociation energy greater than about 4.29 e.v. This is 9.5 per cent below the observed value. On the other hand they obtained a dissociation energy of 4.5289 e.v. using an expression containing only five terms, one of which involved $p \neq 0$. An eleven-term expression in which three of the terms involved r_{12} gave a dissociation energy of 4.7043 e.v. and a thirteen-term expression containing five terms involving r_{12} gave a dissociation energy of 4.7198 e.v. This last result is smaller by 0.025 e.v. (about 0.5 per cent) than the observed value, 4.745 e.v.

Thus, through the calculations of James and Coolidge, we have obtained a wave function of hydrogen that must be quite close to the true function. These calculations demonstrate the necessity of including terms involving r_{12} (either directly or indirectly through configuration interactions) if reasonably precise wave functions are to be constructed for molecular bonds.

c. Problems encountered in more complex molecules

A rather detailed discussion of the wave functions of H_2^+ and H_2 has been given in order to show that even in the simplest molecules, reasonably good wave functions are not at all easy to obtain. The problem of obtaining good orbital shapes is apparently especially difficult in the hydrogen molecule, in spite of the fact that several excellent and simple orbitals are available for H_2^+. In addition, the problem of electron correlation in the hydrogen molecule is not easily dealt with.

When we turn to more complex molecules such as the homonuclear diatomic molecules of the first period in the periodic table (Li_2, O_2, N_2, F_2, etc.), the same difficulties are found, but they occur in a magnified form. They become particularly evident if one tries to take care of the correlation energy by introducing interelectronic distances into the wave functions in the manner of James and Coolidge or Frost and Braunstein, since the number of electrons in the system is increased. Thus, whereas H_2 has only a single pair of interacting electrons, so that only one interelectronic distance need be introduced, Li_2, with six electrons, has fifteen interelectronic distances, and O_2, with sixteen electrons, has 120 interelectronic distances. It is obvious that the Frost-Braunstein and James-Coolidge approaches, which were successful in the hydrogen molecule, cannot be used here, since the computational problems associated with these large numbers of electron pairs are staggering. Furthermore, there is every reason to believe that the complexity in the shape of the accurate wave functions, which increased markedly in going from H_2^+ to H_2 would increase to a corresponding degree in going from H_2 to molecules with more than two electrons.

Therefore a different approach has generally been made in treating these more complex systems. Instead of trying to find the total energy of the system *ab initio,* one tries to find approximate wave functions from which one can calculate the difference in energy of two molecules, or of a molecule in different electronic states. For instance, if we want to find the energy change in the reaction

$$\text{benzene} + 3H_2 \rightarrow \text{cyclohexane}$$

it should not be necessary to calculate precisely the total energy of benzene and the total energy of cyclohexane. If suitable approximate wave functions were employed in calculating these energies, one might hope that the errors introduced by the approximations would be similar in the two calculations, so that they would cancel when the difference in energy of the two systems is evaluated. Therefore it might be possible to find differences in energy quite accurately with procedure A using wave functions that are not particularly good.

Unfortunately there is usually no way of being sure that this cancellation of errors will take place without making the hopelessly difficult complete calculation of the total energy. Therefore this procedure is fraught with dangers and deceptions. One proceeds by making approximations that one hopes will be valid. The validity of the approximations cannot be assessed in practice without making use of the facts that are being explained. Even if the facts seem to justify the assumptions, one cannot be sure to what extent this is a matter of luck and to what extent one is justified in extending the calculations to systems for which experimental data are not available. The range of application of these theories is therefore rather limited. One loses confidence in them soon after passing beyond the frontiers of what is already known and into the region in which useful new quantitative information might be uncovered. (Of course, there is no question of the usefulness of many of these theories in their *qualitative* aspect; but it is the *quantitative* aspect that is under discussion here.)

We shall now mention briefly some of the more specific difficulties which, in addition to those already discussed for H_2^+ and H_2, arise in more complicated molecules.

(1) The effects of the electrons in inner shells are difficult to take into account in any reasonably simple yet accurate way. Although these shells are usually taken to be chemically inert, they exert powerful forces – of both coulombic and exchange types – on the valence electrons. Therefore it is essential that they be properly dealt with.

(2) The orbitals that one is required to use in these molecules are more complex in shape than the simple, nodeless 1s-type orbitals arising in the ground state of hydrogen. The number of variable parameters required in order to optimize their shapes is even greater than was found necessary in H_2^+ and H_2.

(3) When the total energy is not calculated, the variational method of determining the numerical values of parameters by minimizing the energy is not applicable. The choice of the values of the parameters must therefore be made on the basis of other information, usually derived from experimental observations. But the information that we have at our disposal is limited. Therefore only a few parameters can be introduced. This means that we are forced to use wave functions with relatively little flexibility in their shape, which in turn seriously impairs the accuracy attainable with the functions. Thus in considering the very problems that require the use of more parameters, we are forced to use fewer parameters.

(4) Attempts are often made to simplify the calculations by considering the wave functions of only a few of the electrons present in the molecule, the influence of the other electrons being taken into account by selecting suitable parameters (e.g., screening constants). The numerical values of these parameters cannot be determined by the variational method, so we are once again thrown back upon the limited amount of available experimental data for the estimation of their values. This makes it difficult to deal adequately with the effects of electrons not explicitly included in the wave function.

(5) The orbitals required are more complex than those occurring in H_2. The evaluation of integrals involving these orbitals is therefore more difficult. (This difficulty is often avoided by simply neglecting the offending integrals!)

(6) There is some hope that many of these difficulties may be overcome by the use of configuration interaction because, as we have seen, inclusion of a sufficient number of configurations in the wave function ought always to result in a good wave function. Of course, the practical question here is that of how many configurations need be included. At the present time this approach is not unpromising [see, for instance, the calculation of the energy of the ground state of oxygen by Meckler (*24*)].

(7) There is also some hope of obtaining accurate energies by means of numerical methods made possible by high-speed computing machines.

In order to illustrate some of these points, we shall now consider an important quantum chemical problem on which a considerable amount of computational effort has been expended.

C. The Quantitative Treatment of Aromatic Hydrocarbons

The aromatic hydrocarbons, and especially benzene, have been favorite subjects for theoretical discussion since the early days of quantum mechanics. The chief reason for this popularity was the initial success that Hückel, Pauling, and Wheland and others had in explaining the stability, dipole moments, reactivities, and other properties of these substances and their derivatives. The symmetry of many of these molecules and the apparent possibility of treating the π-electrons separately from the other electrons were also attractive simplifying features. If a reasonably good quantitative quantum mechanical theory is possible for any group of chemical compounds, the aromatic compounds would be expected to produce such a theory.

Three kinds of experimental information are especially useful in developing a quantitative theory of these compounds. They are resonance energies, bond lengths, and absorption spectra. We shall now give a brief survey of the difficulties that have been encountered in using these three properties to develop good wave functions for aromatic compounds.

a. The resonance energy

The resonance energy of a molecule may be defined as the difference between the actual energy of the molecule at 0° K. and the energy of a hypothetical molecule having an electronic structure equivalent to the most stable valence bond structure that the organic chemist can draw for the substance. Thus in benzene, the resonance energy is the difference between the true energy of benzene and one of its Kekulé forms. It is immediately obvious that resonance energies cannot be precisely determined from experimental observations without careful specification of the hypothetical reference molecule. Unfortunately there is at the present time no unequivocal way of determining the energy of this hypothetical molecule, and different ways of defining it give somewhat different numerical values of the resonance energy.

For instance, in benzene it is not obvious whether the double bond energy should be taken from the value for ethylene, cyclohexene, one of the *n*-hexenes, or some other olefin. The energies of the double bonds in these molecules differ because of the interaction of the double bond with the groups that are attached to it. Therefore, in order to obtain a reliable "experimental" energy for the hypothetical Kekulé form, these interactions must be completely understood. The theory of these interactions is as yet rather primitive and hardly quantitative.

Furthermore, the carbon–carbon distances in benzene are not the same as those in the olefins, so a correction (known as the "compression energy") must be applied. This involves a knowledge of the force constants of the C–C and C = C bonds. Since these force constants depend to some extent on the groups attached to the bonded atoms, we are again faced with the necessity of choosing the appropriate model compound, and of correcting our choice to conform with the conditions that exist in the hypothetical Kekulé form. Once again, the incompleteness of our theoretical understanding–this time of the factors affecting force constants–makes it impossible to calculate the precise value of the compression energy. Therefore it is not possible to give an exact value for the resonance energy of benzene at the present time [for a more detailed discussion of these matters, see (*25–28*)].

The uncertainty in the resonance energy that results from these incompletely understood theoretical factors amounts to several kilocalories–that is, something of the order of 10 per cent of the resonance energy itself. Since the approximate theories of Hückel and of Pauling and Wheland already give values of the resonance energy that agree almost within this range of uncertainty, it appears that the resonance energy does not give a particularly sensitive criterion with which to test the wave functions of the ground states of aromatic hydrocarbons. The excellent results obtained with these theories are all the more remarkable because, as we shall see, they are known to contain questionable approximations. For instance, the overlap integrals and exchange integrals between nonadjacent atoms are known to be fairly large, yet they are neglected in both the Hückel theory and the Pauling and Wheland theory. Therefore these theories might better be considered to be empirical as far as their quantitative aspects are concerned, their success being the result of a fortunate cancellation of errors.

b. Bond lengths

The recent work of Robertson and his students (*29*) has made available reliable values for the distances between carbon atoms in many aromatic hydrocarbons. That these bond lengths should be different in different parts of the molecule is easily shown from the valence bond theory of naphthalene. Considering only the three Kekulé forms

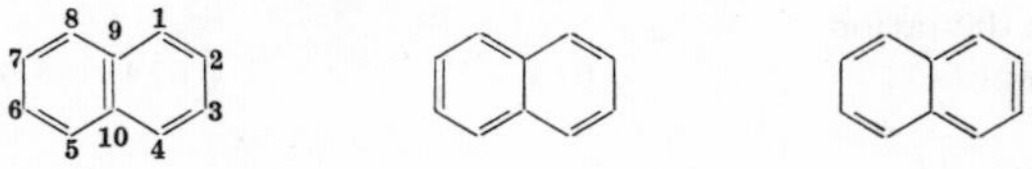

we see that the bond between atoms 1 and 2 is a double bond in two out of the three structures. Therefore we may say that this bond has a "bond order" of $(2 + 2 + 1)/3 = 5/3$. Similarly, the order of the bond between atoms 2 and 3 is $(1 + 1 + 2)/3 = 4/3$. Since double bonds are shorter than single bonds, we are not surprised to find that the 1–2 bond is the shortest bond in the molecule (see Table 12-3).

There are more rigorous ways of defining the bond order. In the VB theory the order of a bond was defined by Pauling, Brockway, and Beach (*30*) as the weighted average number of bonds between the atoms in the bond for all valence bond structures, the weighting factor being the square of the coefficient of the structure in the total molecular wave function. That is, if

TABLE 12-3

CALCULATION OF BOND LENGTHS IN AROMATIC HYDROCARBONS

Hydrocarbon	Bond	VB[a] (Kekulé only)	VB[a] (Kekulé + excited)	MO[b]	SCMO[b]	Observed
naphthalene						
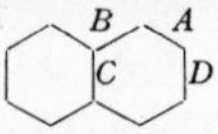	*A*	1.374	1.382	1.377	1.364	1.365[c]
	B	1.415	1.424	1.410	1.424	1.425
	C	1.415	1.402	1.417	1.396	1.393
	D	1.415	1.417	1.400	1.418	1.404
root mean square deviation from observed lengths		0.011	0.014	0.015	0.007	—
anthracene						
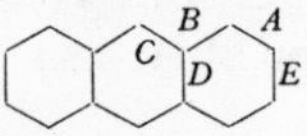	*A*	1.365	—	1.375	1.357	1.371[d]
	B	1.425	—	1.414	1.438	1.424
	C	1.393	—	1.400	1.395	1.396
	D	1.425	—	1.424	1.412	1.436
	E	1.425	—	1.404	1.431	1.408
root mean square deviation from observed lengths		0.008		0.008	0.017	—
coronene						
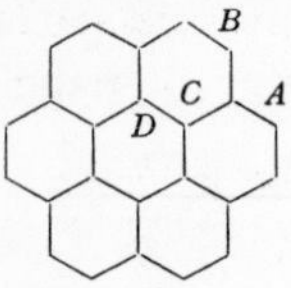	*A*	1.418	—	1.411	1.428	1.415[e]
	B	1.370	—	1.372	1.360	1.385
	C	1.405	—	1.411	1.391	1.430
	D	1.418	—	1.415	1.429	1.430
root mean square deviation from observed lenghts		0.014		0.014	0.024	—
ovalene						
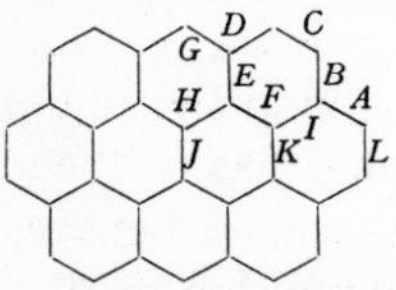	*A*	1.405[f]	—	1.410	1.413	1.407[g]
	B	1.433	—	1.417	1.443	1.416
	C	1.360	—	1.370	1.353	1.355
	D	1.433	—	1.419	1.445	1.438
	E	1.418	—	1.419	1.408	1.432
	F	1.418	—	1.416	1.437	1.431
	G	1.393	—	1.400	1.394	1.396
	H	1.405	—	1.413	1.404	1.427
	I	1.405	—	1.414	1.394	1.413
	J	1.433	—	1.422	1.445	1.428
	K	1.418	—	1.417	1.419	1.410
	L	1.382	—	1.377	1.371	1.386
root mean square deviation from observed lengths		0.011	—	0.012	0.016	—

the wave functions for the different valence bond structures of a molecule are $\chi_1, \chi_2, \cdots$, and if the wave function of the molecule is (*cf.* page 423)

$$\psi = \sum_{\substack{\text{all}\\ \text{structures, } i}} c_i \chi_i$$

then the order of the bond between adjacent atoms r and s is

$$p_{rs} = 1 + \sum_i c_i^2 \, n_{rsi}$$

where $n_{rsi} = 0$ if there is a single bond between the two atoms in the structure χ_i, and $n_{rsi} = 1$ if there is a double bond between the atoms in this structure. In the LCAOMO theory we recall that the jth molecular orbital has the form

$$\psi_j = \sum_{\substack{\text{all}\\ \text{atoms, } r}} d_{jr} \phi_r$$

where ϕ_r is the $2p\pi$ atomic orbital of the rth atom and the coefficients d_{jr} are found from the secular determinant and simultaneous equations of the type described for benzene on page 426. In the LCAOMO theory, the bond order for a pair of adjacent atoms, r and s, was defined by Coulson (*39*) as

$$p_{rs} = 1 + 2 \sum_j d_{jr} d_{js}$$

where the sum is taken over all occupied orbitals and the coefficients d_{jr} and d_{js} are assumed to be real.

The relationship of the bond order to the bond length may also be obtained in several ways. Pauling, Brockway, and Beach (*30*) drew a smooth curve through a plot of bond order *vs.* bond length for the C–C single bond in diamond (bond order = 1), benzene (bond order = 3/2), graphite (bond

---

[a] Recalculated by the method of Pauling, Brockway, and Beach (*30*), using a slightly different bond order *vs.* bond length curve.

[b] From Pritchard and Sumner (*31*).

[c] Data of Abrahams, Robertson, and White (*32*), recalculated by Ahmed and Cruickshank (*33*).

[d] Data of Sinclair, Robertson, and Mathieson (*34*), recalculated by Ahmed and Cruickshank (*35*).

[e] Data of Robertson and White (*36*).

[f] Using the Kekulé structures given by Robertson (*37*).

[g] Data of Donaldson and Robertson (*38*).

order = 4/3), and ethylene (bond order = 2). Coulson (*39*) derived a relationship that gives a curve similar to that of Pauling, Brockway, and Beach,

$$x = s - \frac{s - d}{1 + k\left(\frac{2 - p}{p - 1}\right)}$$

where x is the bond length, p is its order, s is the single bond length, d is the double bond length, and k is an empirical constant.

Typical results of these considerations are shown in Table 12-3. In the MO calculations cited in this table, the coefficients d_{jr} were calculated by neglecting all overlap integrals, but exchange integrals between non-adjacent atoms were included. The SCMO ("self-consistent molecular orbital") calculation refers to a more elaborate procedure that takes into account the interactions of π-electrons in determining the coefficients appearing in the orbitals, in a manner analogous to the Hartree-Fock method for atoms (see page 336). In the VB treatment of naphthalene, the results of two calculations are shown: one in which only the three Kekulé structures were taken into account with equal weights, and the other in which all 42 non-ionic canonical structures were included with the weights found by Sherman (*35*). The VB calculations for the remaining hydrocarbons include only the Kekulé structures assuming equal weights for all. From the results in the table, both VB and MO theories appear to give equally good accounts of the bond lengths. The results with other hydrocarbons show that neither theory is reliable for prediction of bond lengths within ±0.01 Å. on the average, and sometimes greater deviations are found.

The Pullmans (*40*) have compared the contributions to the VB wave function of anthracene made by the Kekulé structures such as

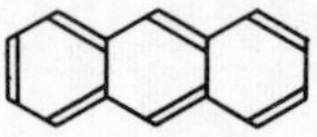

and by the Dewar-like or "excited" structures such as

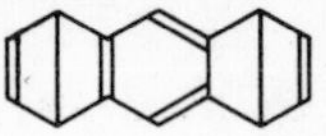 and 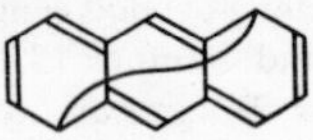

There are altogether only four Kekulé structures and 198 excited structures in anthracene. Although each one of the excited structures is considerably less stable than any one of the Kekulé structures, the large number of them makes the total weight (as measured by the sum of the squares of their coefficients in the wave function) nine times as great as the total weight of the Kekulé structures. The success of the Pauling-Brockway-Beach theory with anthracene is particularly significant in the light of this finding: an entirely inadequate wave function can evidently give reasonably good predictions of the bond lengths. Even in naphthalene, where the total weight of the three Kekulé structures is only slightly greater than the total weight of the excited structures, the Kekulé structures alone give just as good agreement as do the MO wave function and the more complete VB wave function including the excited structures. A rather poor wave function again seems to be able to give adequate agreement with the experimental results.

The experimental measurements of bond lengths in aromatic hydrocarbons are probably reliable to only about $\pm$ 0.01 Å. in the most favorable instances. Since the relatively crude theories described above already account for the bond lengths almost within this range of error, this property, like resonance, does not seem to be a particularly sensitive test of molecular wave functions.

c. The spectrum of benzene

In discussing the application of the variation method to the calculation of the frequencies of the normal modes of vibrating bodies, it was found that better agreement is usually obtained in computing the frequency of the fundamental mode than in computing the overtone frequencies (see pages 124 and 127). It is not surprising, therefore, that the inadequacies of the quantum mechanical theories of aromatic compounds become strikingly apparent when they are used to interpret quantitatively the spectrum of benzene, where excited states must be considered.

(*i*) *The Observed Absorption Spectrum of Benzene.* Benzene shows evidence of four electronic absorption bands at wave lengths above 1600 Å. All four of these bands are believed to result from transitions of electrons to excited molecular orbitals involving predominantly the $2p\pi$-orbitals on the carbon atoms (much of the absorption immediately below 1600 Å. probably results from excitation of the $2p\pi$-electrons to excited states such as $3s\sigma$, $3p\sigma$, $3p\pi$, $4s\sigma$, etc.). The four observed bands above 1600 Å. consist of (1) an extremely weak absorption in the region from 3000 to 3500 Å., (2) a weak absorption between 2200 and 2700 Å., (3) a moderately strong absorption around 2000 Å., and (4) a very strong absorption around 1750 Å. This means that there are four excited levels with energies at approximately 3.8 e.v., 4.9 e.v.,

6.2 e.v., and 7.0 e.v. above the ground state. Of course we cannot know from the spectrum alone what the wave functions of these levels look like, but as we shall see presently, some insight into the shapes of the wave functions can be gained from the intensity of the absorption and from the analysis of its vibrational structure.

(*ii*) *Use of Symmetry Properties to Describe Orbitals and Wave Functions of Benzene.* The carbon and hydrogen nuclei in the benzene molecule are known to be located at the vertices of concentric regular plane hexagons. This means that the Hamiltonian operator for the electrons in the benzene molecule is unaffected by all of the symmetry operations characteristic of a regular hexagon. Therefore the Hamiltonian of the electrons must commute with all of the symmetry operators of a hexagon. According to the discussion following Corollary VI (page 170), this means that it must be possible to set up the molecular orbitals and wave functions of benzene in such a way that they are eigenfunctions of the symmetry operators.

The eigenvalues of these symmetry operations provide us with a convenient means of labeling and describing the molecular orbitals and the electronic states of benzene. We shall now describe a system of notation based on these eigenvalues that is widely used to classify the states of molecules having hexagonal symmetry. (Similar systems of notation have been developed for use with molecules showing other types of symmetry. They are described in Eyring, Walter, and Kimball (*2*, Chapter 10 and Appendix VII).)

Let us investigate the effects of the following symmetry operations on the orbitals and wave functions of benzene (see Fig. 12-3A):

C_2 = rotation by 180° about the symmetry axis normal to the benzene ring

C_3 = rotation by 120° about the symmetry axis normal to the ring

$C'_{2a}, C'_{2b}, C'_{2c}$ = rotation by 180° about the three two-fold axes that lie in the plane of the ring and pass through opposite pairs of vertices of the hexagon

i = inversion through the center of the ring

The various combinations of eigenvalues of the operators C_2, C_3, and any of the three C'_2 operators are indicated by the letters and subscripts shown in Table 12-4. The eigenvalue for the inversion operation is indicated by adding the subscript g if the eigenvalue is $+1$ and the subscript u if the eigenvalue is – 1. In referring to individual molecular orbitals, lower case letters are used, whereas capital letters are used for describing the symmetry of the total wave function of the molecule. The spin eigenvalue for the total wave function is indicated in the usual way by inserting the multiplicity, $(2S + 1)$, as a superscript ahead of the letter symbol.

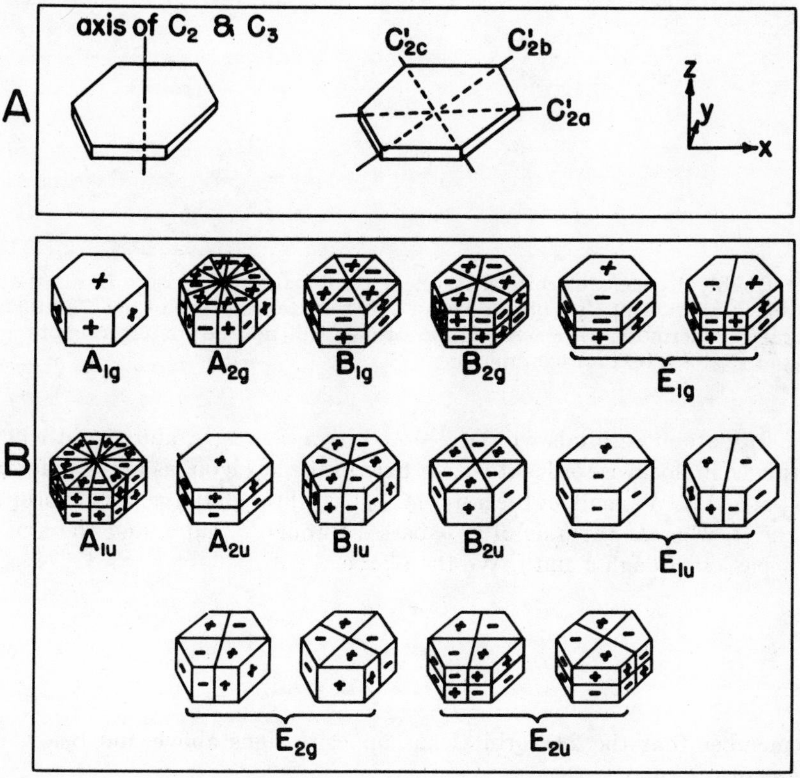

FIG. 12-3.
A. Symmetry axes of the regular hexagon.
B. General types of eigenfunctions of the symmetry operators of a regular hexagon.

Figure 12-3B shows different arrangements of nodal planes reflecting the twelve possible symmetry types among the wave functions and orbitals of benzene.

Let us deduce the symmetry types of the molecular orbitals that were derived on page 427 for the π-electrons on the carbon atoms of benzene.

TABLE 12-4

SYMMETRY TYPES ARISING FROM THE SYMMETRY OPERATIONS OF A REGULAR HEXAGON

Symbols		Eigenvalues		
Molecular Orbital	Wave Function	C_2	C_3	C'_{2a}, C'_{2b}, or C'_{2c}
a_1	A_1	1	1	1
a_2	A_2	1	1	–1
b_1	B_1	–1	1	1
b_2	B_2	–1	1	–1
e_1	E_1	–1	*	*
e_2	E_2	1	*	*

* Doubly degenerate orbitals. A linear combination of these orbitals that is an eigenfunction of C_3 will not be an eigenfunction of any of the C'_2, and *vice versa*. Furthermore, an eigenfunction of C'_{2a} will not be an eigenfunction of C'_{2b} or C'_{2c}. See text for examples.

The $2p\pi$ atomic orbitals will be denoted by a, b, c, d, e, and f, and will be arranged in clockwise order at the vertices of the hexagon (as in the diagrams on page 419). We shall let the axis of C'_{2a} pass through the centers of orbitals a and d, whereas the axis of C'_{2b} passes through b and e, and the axis of C'_{2c} passes through c and f. We then have

$$C_2a = d$$

$$C_3a = c$$

$$ia = -d$$

(remember that the $2p\pi$-orbital has opposite signs above and below the plane of the ring)

$$C'_{2a}a = -a$$

$$C'_{2b}a = -c$$

$$C'_{2c}a = -e$$

Similar transformations are found for other atomic orbitals.

In Chapter 11 we found that the most stable molecular orbital that can be formed from these atomic orbitals is

$$\phi_1 = a + b + c + d + e + f$$

It is easy to see that ϕ_1 is an eigenfunction of the symmetry operations of the hexagon

$$C_2\phi_1 = C_2(a + b + c + d + e + f)$$
$$= d + e + f + a + b + c = \phi_1$$
$$C_3\phi_1 = c + d + e + f + a + b = \phi_1$$
$$C'_{2a}\phi_1 = -a - f - e - d - c - b = -\phi_1$$
$$C'_{2b}\phi_1 = -c - b - a - f - e - d = -\phi_1$$
$$C'_{2c}\phi_1 = -e - d - c - b - a - f = -\phi_1$$
$$i\phi_1 = -d - e - f - a - b - c = -\phi_1$$

On referring to Table 12-4, we see that the set of eigenvalues, $+1$ for C_2 and C_3 and -1 for C'_{2a}, C'_{2b}, and C'_{2c} is denoted by the symbol a_2. The eigenvalue -1 for i is indicated by adding a subscript u. Thus we may characterize the orbital ϕ_1 as an a_{2u}-orbital.

Next in energy above ϕ_1 is a degenerate pair of orbitals,

$$\phi_2 = a + b - d - e$$
$$\phi_3 = b + c - e - f$$

We find that these two orbitals are both eigenfunctions of C_2 and i

$$C_2\phi_2 = -\phi_2; \qquad C_2\phi_3 = -\phi_3$$
$$i\phi_2 = \phi_2; \qquad i\phi_3 = \phi_3$$

For the operations C_3 and C'_{2a}, however, various linear combinations of ϕ_2 and ϕ_3 must be taken, in order to obtain eigenfunctions. Making use of the relations

$$C_3\phi_2 = c + d - f - a = \phi_3 - \phi_2$$
$$C_3\phi_3 = d + e - a - b = -\phi_2$$
$$C'_{2a}\phi_2 = -a - f + d + c = \phi_3 - \phi_2$$
$$C'_{2a}\phi_3 = -f - e + c + b = \phi_3$$

the student can readily verify the following results

$$C_3(\phi_2 - e^{-i\pi/6}\phi_3) = -e^{i\pi/6}(\phi_2 - e^{-i\pi/6}\phi_3)$$
$$C_3(\phi_2 - e^{i\pi/6}\phi_3) = -e^{-i\pi/6}(\phi_2 - e^{i\pi/6}\phi_3)$$
$$C'_{2a}\phi_3 = \phi_3$$
$$C'_{2a}(2\phi_2 - \phi_3) = -(2\phi_2 - \phi_3)$$

Evidently eigenfunctions can be constructed for these two operators by combining ϕ_2 and ϕ_3, but different combinations are required for different operators. It is therefore clear that ϕ_2 and ϕ_3 must be characterized by the symbol e_{1g}.

Exercise. Find the linear combinations of ϕ_2 and ϕ_3 that are eigenfunctions of C'_{2b}. Also find linear combinations that are eigenfunctions of C'_{2c}. [Hint: Determine the results of the operation of C'_{2b} on ϕ_2 and ϕ_3. Then find the values of the constants α, β and γ that give

$$C'_{2b}(\alpha\phi_2 + \beta\phi_3) = \gamma(\alpha\phi_2 + \beta\phi_3)$$

Repeat the procedure for C'_{2c}.]

The student can readily show that the degenerate orbitals ϕ_4 and ϕ_5 are of the type e_{2u}, and ϕ_6 is of the type b_{2g}.

Next let us determine the symmetry type of the ground state of benzene. If we put two electrons into the lowest orbital, ϕ_1, we obtain the determinantal wave function

$$\psi_1 = | \phi_1{}^+(1)\, \phi_1{}^-(2) |$$

where the superscripts + and – refer to the spin states of the electrons in the orbitals, and the numbers in the parentheses refer to the electron whose coordinates are to be inserted in the orbital. It is easy to see that this wave function has the eigenvalue + 1 for all of the symmetry operators of the hexagon. For instance

$$i\psi_1 = | i\phi_1{}^+(1)\, i\phi_1{}^-(2) | = | -\phi_1{}^+(1) - \phi_1{}^-(2) | = \psi_1$$

and similarly for the rest of the operators. Therefore ψ_1 is of the type ${}^1A_{1g}$ (a capital letter is used here because ψ_1 is a wave function containing more than one electron; the superscript 1 refers to the fact that ψ_1 is a singlet).

Similarly, if we place four electrons in the degenerate pair of orbitals ϕ_2 and ϕ_3, we obtain the wave function

$$\psi_2 = | \phi_2{}^+(1)\, \phi_2{}^-(2)\, \phi_3{}^+(3)\, \phi_3{}^-(4) |$$

It is easy to see that this function is an eigenfunction of C_2 and i, with eigenvalues + 1 for both operators. It is also found that ψ_2 is an eigenfunction of C_3, C'_{2a}, C'_{2b}, and C'_{2c} with eigenvalues of + 1 for each. For instance

$$\begin{aligned} C_3\psi_2 = {} & | [\phi_3{}^+(1) - \phi_2{}^+(1)]\, [\phi_3{}^-(2) - \phi_2{}^-(2)]\, [-\phi_2{}^+(3)]\, [-\phi_2{}^-(4)] | \\ = {} & | \phi_3{}^+(1)\, \phi_3{}^-(2)\, \phi_2{}^+(3)\, \phi_2{}^-(4) | - | \phi_2{}^+(1)\, \phi_3{}^-(2)\, \phi_2{}^+(3)\, \phi_2{}^-(4) | \\ & - | \phi_3{}^+(1)\, \phi_2{}^-(2)\, \phi_2{}^+(3)\, \phi_2{}^-(4) | + | \phi_2{}^+(1)\, \phi_2{}^-(2)\, \phi_2{}^+(3)\, \phi_2{}^-(4) | \\ = {} & | \phi_3{}^+(1)\, \phi_3{}^-(2)\, \phi_2{}^+(3)\, \phi_2{}^-(4) | = \psi_2 \end{aligned}$$

(these transformations make use of properties (3) and (4) of determinants mentioned on page 47). Similarly

$$\begin{aligned} C'_{2a}\psi_2 &= | [\phi_3^+(1) - \phi_2^+(1)] [\phi_3^-(2) - \phi_2^-(2)] \phi_3^+(3) \phi_3^-(4) | \\ &= | \phi_3^+(1) \phi_3^-(2)\phi_3^+(3) \phi_3^-(4) | - | \phi_2^+(1)\phi_3^-(2) \phi_3^+(3) \phi_3^-(4) | \\ &\quad - | \phi_3^+(1) \phi_2^-(2) \phi_3^+(3) \phi_3^-(4) | + | \phi_2^+(1) \phi_2^-(2)\phi_3^+(3) \phi_3^-(4) | \\ &= | \phi_2^+(1) \phi_2^-(2) \phi_3^+(3) \phi_3^-(4) | = \psi_2 \end{aligned}$$

Evidently the function ψ_2 has the character ${}^1A_{1g}$.

It should be obvious from these examples that if the molecular orbitals of an electronic configuration are completely filled with electrons, the corresponding wave function must have the character ${}^1A_{1g}$. In the ground state of benzene, we have the configuration $\phi_1{}^2\phi_2{}^2\phi_3{}^2$. Since the three molecular orbitals are all filled in this configuration, the resulting wave function

$$\psi = | \phi_1^+(1) \phi_1^-(2) \phi_2^+(3) \phi_2^-(4) \phi_3^+(5) \phi_3^-(6) |$$

has the character ${}^1A_{1g}$.

Suppose that we set up a Cartesian coordinate system with its origin at the center of the benzene ring, its z-axis normal to the ring, and the x- and y-axes in the plane of the ring (see Fig. 12-4). Then we find that

$$\begin{array}{lll} C_2x = -x & C_2y = -y & C_2z = z \\ C_3x = -\tfrac{1}{2}x + \tfrac{1}{2}\sqrt{3}y & C_3y = -\tfrac{1}{2}y - \tfrac{1}{2}\sqrt{3}x & C_3z = z \\ ix = -x & iy = -y & iz = -z \end{array}$$

$$\left.\begin{array}{l} C'_{2a} \\ C'_{2b} \\ C'_{2c} \end{array}\right\} x = \begin{array}{c} x \cos 2\alpha \\ + y \sin 2\alpha \end{array} \qquad \left.\begin{array}{l} C'_{2a} \\ C'_{2b} \\ C'_{2c} \end{array}\right\} y = \begin{array}{c} -y \cos 2\alpha \\ + x \sin 2\alpha \end{array} \qquad \left.\begin{array}{l} C'_{2a} \\ C'_{2b} \\ C'_{2c} \end{array}\right\} z = -z$$

where α is the angle between the x-axis and the axis of C'_2. It is evident that the coordinates x and y are of the symmetry type E_{1u}, and z is of the symmetry type A_{2u}.

(*iii*) *Effect of Symmetry Properties on the Intensity of Spectral Absorption.* In Chapter 16 (page 654), we shall show that strong light absorption is possible in a transition between two states, Ψ_A and Ψ_B, only if at least one of the integrals

$$\int \Psi_A{}^*(\textstyle\sum x_i) \Psi_B \, d\tau, \quad \int \Psi_A{}^*(\sum y_i) \Psi_B \, d\tau, \quad \text{or} \quad \int \Psi_A{}^*(\sum z_i) \Psi_B \, d\tau$$

fails to vanish (x_i, y_i, and z_i are the Cartesian coordinates of the ith electron, and the sums are taken over all electrons). Because of the theorem mentioned on page 140, an integral of this type can fail to vanish only if its integrand has eigenvalues of + 1 for all of the symmetry operations belonging to the system; that is, for benzene, the integrand as a whole must be of the type A_{1g}. This means that if Ψ_A is the ground state of benzene (which is of the type $^1A_{1g}$), then the excited state Ψ_B must have the same character as one of the Cartesian coordinates x, y, or z. Furthermore, to the extent that spin–orbit coupling can be neglected (and this is a good approximation for the light atoms of which benzene is composed) all of these integrals will vanish unless Ψ_A and Ψ_B have the same spin. The coordinates x and y have the character E_{1u}, and z has the character A_{2u}; therefore strong absorption from the ground state of benzene is possible only if the upper state has the character $^1A_{2u}$ or $^1E_{1u}$. We may conclude that the very strong 1750 Å. absorption band of benzene involves either a $^1A_{2u}$ or a $^1E_{1u}$ upper state, whereas the other three observed absorption bands do not involve upper states of either of these two types.

The analysis of the observed vibrational structure of the 2500 Å. absorption band of benzene shows that the upper state must have a $^1B_{2u}$ electronic wave function (*41, 42*). Similar analysis of the very weak 3300 Å. band shows that it must have either a $^3B_{2u}$ or a $^3B_{1u}$ upper electronic state (*43*). Because of the fairly high intensity of the absorption, the level at 6.2 e.v. probably does not belong to a triplet state.

Thus we have the following experimental facts to explain:

(1) There is a $^3B_{2u}$- or $^3B_{1u}$-level of benzene at 3.8 e.v.

(2) There is a $^1B_{2u}$-level at 4.9 e.v.

(3) There is a level at 6.2 e.v. that is neither $^1A_{2u}$ nor $^1E_{1u}$, and is also probably not a triplet.*

(4) There is a level at 7.0 e.v. that is either $^1A_{2u}$ or $^1E_{1u}$.

(*iv*) *Theoretical Treatment of the Excited States of Benzene.* Many attempts have been made to calculate the positions of the excited states of benzene using various methods of approximation. The results of seven of these attempts are shown in Fig. 12-4, together with the positions of the observed levels. Three sets of these calculations were made with the VB method,

* Dunn and Ingold (*Nature*, **176**, 65 (1955)) have shown from a study of the vibrational structure of the 2000 Å. absorption band of benzene that the 6.2 e.v. level is probably E_{2g} and definitely not B_{1u}. On the other hand, Albrecht and Simpson (*J. Chem. Phys.*, **23**, 1480 (1955); *J. Am. Chem. Soc.*, **77**, 4454 (1955)) have concluded from the spectroscopic behavior of several para-disubstituted benzenes that the 6.2 e.v. level of benzene has a B_{1u} symmetry.

whereas the remaining four were made with the MO method. All of the calculations agree in giving the lowest singlet excited state a $^1B_{2u}$-character, which is in agreement with experiment. They also agree in making the lowest triplet level a $^3B_{1u}$-state, which is consistent with the experimental

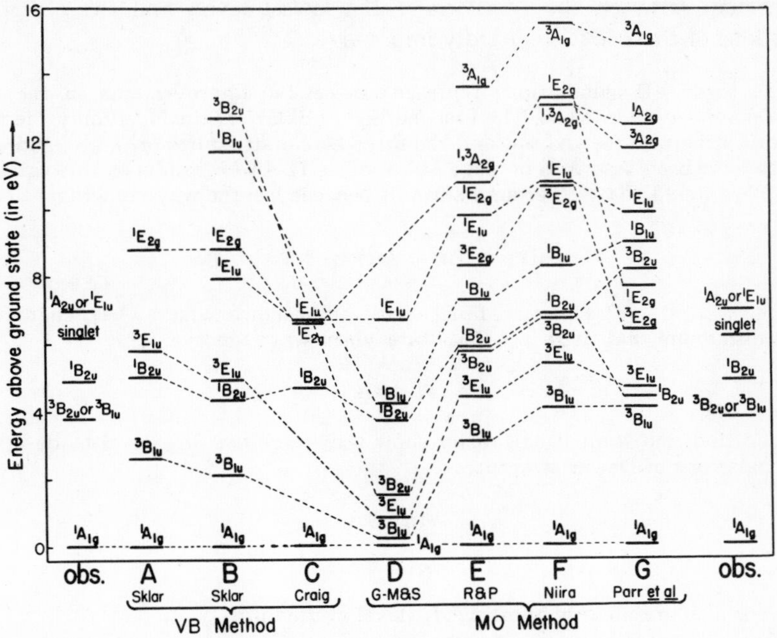

FIG. 12-4. Results of various calculations of the energies of excited states of benzene Dashed lines are drawn joining levels having the same symmetry types and spins in the different calculations.

Valence bond method:

A. Sklar (*44*). Nonpolar structures only.
B. Sklar (*44*). Nonpolar plus 12 polar structures.
C. Craig (*45*). Nonpolar plus 24 polar structures.

Molecular orbital method:

D. Goeppert-Mayer and Sklar (*46*). ASMO, with many of the integrals neglected (as corrected by Parr, Crawford, and Craig (*47*)).
E. Roothaan and Parr (*48*). ASMO, no integrals neglected.
F. Niira (*49*). ASMO with $\sigma - \pi$ interactions included.
G. Parr, Craig, and Ross (*50*). ASMO with configuration interactions.

observations. The lowest triplet state is always more stable than the lowest singlet state, which is again in agreement with the observations. Aside from this satisfactory qualitative agreement in the relative positions of the lowest levels, however, the results of the different calculations are rather discordant. Indeed, the most striking feature of Fig. 12-4 is the drastic manner in which the relative energies of many of the states change according to the different methods of calculation. Furthermore, the quantitative agreement between the positions of the lowest levels and the observed positions of these levels is hardly impressive.

The three VB calculations represent successive improvements in the approximate wave functions. The first method of Sklar (*44*) included only the two Kekulé structures (ψ_I and ψ_{II}) and the three Dewar structures (ψ_{III}, ψ_{IV}, and ψ_V), which have been described on page 419 (see Fig. 12-4A for results of this calculation). We found that the ground state of benzene has the wave function

$$\psi = c_1(\psi_I + \psi_{II}) + c_2(\psi_{III} + \psi_{IV} + \psi_V)$$

where $|c_1|^2 > |c_2|^2$. This wave function is readily shown to have a $^1A_{1g}$-character. It is also found that there is a $^1B_{2u}$-state whose wave function is*

$$\psi_I - \psi_{II}$$

In addition, the student can easily show that there is a $^1E_{2g}$-level made up of combinations of Dewar structures

$$\psi_{III} - \psi_{IV}$$

and

$$\psi_{IV} - \psi_V$$

There is also presumably another $^1A_{1g}$-level of the form

$$c'_1(\psi_I + \psi_{II}) + c'_2(\psi_{III} + \psi_{IV} + \psi_V)$$

where

$$|c'_2|^2 > |c'_1|^2$$

The five wave functions just described lead to the only possible singlet states that can be derived from Kekulé and Dewar structures. Since none of these wave functions is of the type $^1E_{1u}$ or $^1A_{2u}$, they cannot account for the strong absorption band at 1750 Å. It is only when polar structures of the type

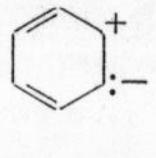

* By inspection of the valence bond structures, we find that, for instance, $C_2\psi_I = \psi_{II}$, $C_2\psi_{II} = \psi_I$, $C_3\psi_I = \psi_I$, $C_3\psi_{II} = \psi_{II}$, $C'_{2a}\psi_I = \psi_{II}$, $C'_{2a}\psi_{II} = \psi_I$, $i\psi_I = \psi_{II}$, $i\psi_{II} = \psi_I$. From these results, we see that the linear combination $(\psi_I + \psi_{II})$ is A_{1g}, and the linear combination $(\psi_I - \psi_{II})$ is B_{2u}. Since ψ_I and ψ_{II} are singlets, both of these linear combinations must also be singlets.

are introduced into the wave function that a ${}^1E_{1u}$-state appears. Sklar made an additional calculation in which he included the twelve such polar structures that can be derived from the Kekulé forms by converting a double bond into an ionic bond. He made use of empirical data to estimate the values of the integrals that appeared in his secular equations. The results of this calculation are shown in Fig. 12-4B. Sklar also investigated the triplet levels of benzene, and found one very low-lying ${}^3B_{1u}$-level. Craig (*45*) extended Sklar's calculations by including the twelve polar structures derived from the Dewar structures, such as

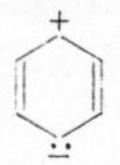

in addition to the twelve polar forms derived from the Kekulé structures. The values of the integrals appearing in Craig's calculation were estimated from the observed spectrum of ethylene, which depends on the same integrals. It is seen from Fig. 12-4C that the observed 7.0 e.v. and 4.9 e.v. levels lie very close to the ${}^1E_{1u}$- and ${}^1B_{2u}$-levels found by Craig. The identity of the 6.2 e.v. level is not, however, made clear by these calculations.

Somewhat more elaborate calculations have been made using the MO approximation. Goeppert-Mayer and Sklar (*46*) took the Pauli principle into account and set up wave functions in determinantal form (so-called *antisymmetrized molecular orbitals*, usually abbreviated as ASMO). Thus they wrote, for the ground state of benzene, the determinant

$$|\,\phi_1^+(1)\,\phi_1^-(2)\,\phi_2^+(3)\,\phi_2^-(4)\,\phi_3^+(5)\,\phi_3^-(6)\,|$$

The superscript + here means that a spin function α is to be assigned to the given column in the determinant, and a superscript – indicates the spin function β. For the excited states they formed suitable linear combinations of the eight wave functions obtained by exciting one of the electrons in the ϕ_2 or ϕ_3 orbitals to the next higher ϕ_4 or ϕ_5 orbitals, such as

$$|\,\phi_1^+(1)\,\phi_1^-(2)\,\phi_2^+(3)\,\phi_2^-(4)\,\phi_3^+(5)\,\phi_4^-(6)\,|$$

$$|\,\phi_1^+(1)\,\phi_1^-(2)\,\phi_2^+(3)\,\phi_2^-(4)\,\phi_3^+(5)\,\phi_5^-(6)\,|$$

$$|\,\phi_1^+(1)\,\phi_1^-(2)\,\phi_2^+(3)\,\phi_4^-(4)\,\phi_3^+(5)\,\phi_3^-(6)\,|$$

$$|\,\phi_1^+(1)\,\phi_1^-(2)\,\phi_2^+(3)\,\phi_5^-(4)\,\phi_3^+(5)\,\phi_3^-(6)\,|$$

These linear combinations produce states of the types ${}^3B_{1u}$, ${}^1B_{1u}$, ${}^3B_{2u}$, ${}^1B_{2u}$, ${}^3E_{1u}$, and ${}^1E_{1u}$. The energies of the states were computed by evaluation of the integrals appearing in the secular equation, a screening constant in the $2p\pi$ atomic orbital being the only empirical constant appearing in the calculation. Many of the exchange integrals involving nonadjacent atoms were neglected, however. The numerical results of this calculation are shown in Fig. 12-4D (these results include corrections of numerical errors later reported by Parr and Crawford (*47*)). The agreement with the observed positions of the singlet levels is fairly good, the permitted transition ${}^1A_{1g} \rightarrow {}^1E_{1u}$ occurring at a shorter wave length than the two forbidden transitions, ${}^1A_{1g} \rightarrow {}^1B_{2u}$ and ${}^1A_{1g} \rightarrow {}^1B_{1u}$. The triplet levels, however, are much too low.

Roothaan and Parr (*48*) extended these calculations by including all of the integrals. The result (see Fig. 12-4E) was a very considerable upward shift of all of the excited levels, with especially favorable movements of the triplet levels. The calculated energies of the excited singlet levels all tend to be considerably higher than the observed energies of these states, however. An important result of this calculation was the demonstration of the importance of exchange integrals between nonadjacent atoms. These integrals are often neglected in approximate calculations, but they can apparently have considerable effects on the energy, especially in excited states.

Niira (*49*) has elaborated on the Roothaan-Parr calculation by taking account of the mutual interactions of the σ and π electrons of the benzene ring. He found a rather marked increase in the energies of all of the excited states relative to the ground state (see Fig. 12-4F). The inclusion of this interaction therefore puts all of the singlet–singlet transitions much farther into the ultraviolet than is actually observed. Here we have an example of a discouraging result that often occurs in quantum chemical calculations. After a great many numerical calculations, passably good agreement between the theory and the experimental observations is obtained. At this point an "improvement" is introduced into the calculations, taking into account a factor that had previously been omitted. As a result of this "improvement", the agreement between the theory and the observations becomes much worse. This, of course, means that other important factors remain to be taken into account in the theory, and that the apparently good agreement obtained in the first instance must have been a matter of luck.

The most extensive calculation of the energy levels of the benzene molecule has been made by Parr, Craig, and Ross (*50*). These workers included, in their wave functions, configurations in which a single electron is excited to the molecular orbital ϕ_6. They included also other configurations not previously considered, in which two electrons are simultaneously excited to the orbitals ϕ_4, ϕ_5, and ϕ_6. These new configurations produce many states having the same symmetry and the same spin as the wave functions used previously by Goeppert-Mayer and Sklar. They can therefore resonate freely with these states, thus altering their energies. This is another example of configuration interaction, which has already been encountered in atoms (pages 325 and 344) and in molecules (page 447). It might have been anticipated that, because of the rather high energies of these new configurations, they would not have much effect on the lower-lying levels; one might at least have hoped that their effects would be similar for all levels. Neither expectation turns out to be true. Large movements of the levels result when configuration interactions are included in the wave function, and there are even shifts in the relative positions of some of the levels (see Fig. 12-4G).

In spite of the impressive amount of labor that Parr, Craig, and Ross expended on their calculation, it cannot be said with any assurance that their energy levels are necessarily close to the correct energy levels. For instance, their calculations of the singlet energy levels do not give a very close correspondence to the observed singlet levels. Much more important, however, is the fact that several additional physical effects were not taken into account in these calculations. For instance, the σ-π interaction has not been studied simultaneously with the configuration interaction. Although Niira's calculation indicated that in the absence of configuration interaction the σ-π interaction merely shifted all of the excited level supward by about the same amount, it is by no means obvious that this will also be the case in the presence of configuration interaction. Furthermore, all of the calculations that have been made on the excited states of benzene have the serious shortcoming that they are based on simple Slater $2p\pi$ atomic orbitals. The π-bonds that result from such atomic orbitals are, however, undoubtedly much less concentrated about the C–C σ-bond than are the actual π-bonds in the benzene molecule. In other words, polarization effects are inadequately taken into account when molecular orbitals are constructed from linear combinations of Slater $2p$ atomic orbitals. We have found that polarization effects make important contributions to the energies of H_2^+ and H_2; there is every reason to believe that they would be important here. Moreover, there is no reason to believe that these polarization effects would be similar in all of the excited states; their inclusion would not necessarily shift all levels in the same direction by about the same amount.

Thus extensive alterations of the relative positions of the excited levels could well result from the inclusion of σ-π interactions and polarization terms. Unfortunately the computations that are required in order to decide on the importance of these factors are too lengthy and complex to be practical at the present time.

The theoretical account of the spectrum of benzene is therefore only partially satisfactory. There seems to be little doubt that the lowest excited singlet state is of the $^1B_{2u}$-type, that a triplet state should lie somewhat below it, and that there is a $^1E_{1u}$-level at some distance above it. Other details of the energy level diagram are, however, still not at all clear, and it is doubtful if calculations capable of resolving these uncertainties are feasible at the present time.

A similar situation exists in the interpretation of the excited states of other conjugated molecules. Quite apart from the question of the importance of σ–π interactions and polarization in the π-bond, the necessity of including configuration interactions has evoked the following comment by Coulson, Craig, and Jacobs (*51*): "There appear to be no simple rules for determining in advance which configurations will interact most strongly with one another. Attempts to restrict the extent of this interaction, such as by a better choice of basic molecular orbitals, are not really satisfactory. In particular, no reliable estimate of the intensity can be made without such interaction. This is a melancholy situation, suggesting very strongly that progress in more complicated molecules is not possible in this direction, but must be looked for in some entirely new technique."

An important lesson may be learned from this work on the ground state and the excited states of aromatic hydrocarbons. We see that if a given set of assumptions happens to lead to a convincing account of some observed phenomenon, it is by no means true that the same set of assumptions will be reliable for predicting other phenomena, even though the predicted phenomena may seem to be quite closely related to the observed phenomena with which the assumptions were tested. Errors that happen to cancel in one calculation may not cancel at all in another calculation, and it is difficult to predict when and why this will be so. It is therefore important to keep a very close rein on quantum chemical calculations by constant reference to experimental observations. Unfortunately experimental data suitable for this purpose are hard to obtain for complex molecules. For this reason there is now a tendency to return to simpler systems, such as diatomic molecules and even atoms (for which there is a vast quantity of experimental information) in order to test quantum chemical methods of calculating molecular properties by means of procedure A.

D. The Quantitative Comparison of Chemical Bonds between Different Pairs of Atoms

a. The properties of chemical bonds

Much of chemistry is ultimately concerned with the characteristics of chemical bonds, so it is important that chemists know not only the general reasons for the formation of chemical bonds, but also why the properties of the bonds between different pairs of atoms differ in the way that they do. It is pertinent to begin a discussion of this question by listing the most important measurable properties of chemical bonds, since it is these properties that a quantitative theory of the chemical bond must explain. The most important are the following:

(1) The bond length.

(2) The bond energy, or dissociation energy.

(3) The force constant.

(4) The dipole moment.

(5) The change in dipole moment with the bond length (which may be determined from the intensity of the infrared absorption).

The study of the interactions of atomic nuclei with the electrons in molecules has recently provided two new properties that are strongly influenced by chemical bonding:

(6) The quadrupole coupling constant.

(7) The chemical shift of the interaction of the nuclear magnetic moment with external magnetic fields.

The last two properties will be described in detail on pp. 484ff.

In addition to the seven bond properties listed above, chemists are interested in what is called the "reactivity" of bonds. This property probably has something to do with the ease with which the electron distribution in a bond changes its shape on the approach of reagent molecules. Unfortunately it is not easy to measure quantitatively, so we shall not consider it further here, despite its great importance in determining the rates of chemical reactions.

If the quantum mechanical theory of chemical bonds is at all adequate, it ought to be able to account for all of the differences that exist between bonds. In the next few pages we shall consider some of the efforts that have been made in this direction, and we shall also point out some of the inherent difficulties that the theory is sure to encounter. Before doing this, however, it should be mentioned that in many instances it is difficult to define some of these properties. For instance, except in diatomic molecules the dissociation energy of a bond is strongly affected by the atoms adjacent to the bond. Thus in speaking of the energy of the C–C bond in ethane, it is necessary to be quite explicit in specifying the state of the methyl groups after the bond is broken. A thermochemical measurement of the heat of dissociation of ethane into two methyl radicals would give the energy required to form methyl radicals in their most stable state, which is believed to be planar. But for theoretical purposes it might be more convenient to discuss the energy required to rupture the C–C bond without changing the C–H bond lengths and bond angles in the two methyl groups that are formed. These two energies are probably very different. Similarly, the force constant of a bond in a polyatomic molecule is determined by the analysis of the normal mode vibrations of the molecule, but it is found that (except, of course, in diatomic molecules) there are by no means negligible interactions between the deformations of different bonds, which complicate the evaluation of the force constant of a given bond. The dipole moment of a bond in a polyatomic molecule is subject to polarization effects and other interactions involving other bonds that may be quite difficult to separate from the "inherent" dipole moment of the bond.

Thus even before we try to develop a theory of the variation in chemical bonds between different atoms, we encounter difficulties in establishing an unequivocal set of facts that have to be explained.

b. The electronegativity of atoms and the polar character (ionicity) of bonds

One striking characteristic of bond strengths is the fact that for any pair of atoms, A and B, the $A-B$ single bond is usually stronger than the average strengths of the $A-A$ and $B-B$ single bonds. Thus the reaction

$$A-A + B-B \rightarrow 2\,A-B$$

is almost always exothermic. This is illustrated in Table 12-5. (A few exceptions to this rule are also given in this table. In addition, the table shows that a similar relation seems to be valid for the double bonds between carbon and oxygen, but not for the triple bonds between nitrogen and carbon.)

Pauling explained this observation in the following way (*52*). We have seen (page 448) that in writing the wave function for the bond in the hydrogen molecule an improvement in the calculated energy was obtained when ionic structures, H^+H^- and H^-H^+, were included along with the Heitler-London structures. A similar improvement can be expected if ionic structures are included in the wave function for the bond between any two atoms. The improvement will in general be significant, however, only if one of the ionic structures has a relatively low energy. Now the energy of the structure A^+B^- is determined predominantly by the difference between the ionization potential of the A atom and the electron affinity of the B atom: if A has a small ionization energy and B has a large electron affinity, then the structure A^+B^- will be relatively stable, and it should make a large contribution to the wave function of the $A-B$ bond. Consider, then,

TABLE 12-5

TYPICAL HEATS OF REACTIONS IN WHICH HOMONUCLEAR BONDS ARE REPLACED BY HETERONUCLEAR BONDS

Reaction		ΔH_{298} (kcal.)
$H_3C–CH_3(g) + H_2(g)$	$\longrightarrow 2CH_4$	-15.54
$\genfrac{}{}{0pt}{}{C}{C}{>}C{<}\genfrac{}{}{0pt}{}{C}{C}$ (diamond) $+ 2H_2(g)$	$\longrightarrow CH_4(g)$	-17.44
$HO–OH(g) + H_2(g)$	$\longrightarrow 2H_2O(g)$	-83.77
$H_2N–NH_2(aq) + H_2(g)$	$\longrightarrow 2NH_3(aq)$	-46.80
$H_2N–NH_2(aq) + H_3C–CH_3(g)$	$\longrightarrow 2H_3C–NH_2(aq)$	-23.90
$H_3C–CH_3(g) + HO–OH(g)$	$\longrightarrow 2H_3C–OH(g)$	-44.09
$H_3C–CH_3(g) + Cl–Cl(g)$	$\longrightarrow 2H_3C–Cl(g)$	-27.8
$H_2N–NH_2(aq) + HO–OH(aq)$	$\longrightarrow 2H_2N–OH(aq)$	-5.9
$H–H(g) + F–F(g)$	$\longrightarrow 2H–F(g)$	-128.4
$H–H(g) + Cl–Cl(g)$	$\longrightarrow 2H–Cl(g)$	-44.12
$F–F(g) + Cl–Cl(g)$	$\longrightarrow 2F–Cl(g)$	-26.6
$H–H(g) + Br–Br(g)$	$\longrightarrow 2H–Br(g)$	-17.32
$Br–Br(g) + Cl–Cl(g)$	$\longrightarrow 2Br–Cl(g)$	$+7.02$
$H_2C{=}CH_2(g) + O{=}O(g)$	$\longrightarrow 2H_2C{=}O(g)$	-67.9
$HC{\equiv}CH(g) + N{\equiv}N(g)$	$\longrightarrow 2HC{\equiv}N(g)$	$+8.2$

the stabilities of the ionic structures of the homonuclear bonds, $A–A$ and $B–B$. It is invariably true that atoms with small ionization energies have small electron affinities, and that atoms with large electron affinities have large ionization energies; this is true because the same factor that makes it easy to remove an electron from an atom also leads to a small increase in energy when an electron is added to the atom. Therefore the ionic structures A^+A^- and B^+B^- are invariably high in energy. As a result, these structures make only small contributions to the wave functions of the $A–A$ and $B–B$ bonds and do not lead to much stabilization of the system. On the other hand, if A has a small ionization energy and B has a large electron affinity, or *vice versa*, one of the two ionic structures for the $A–B$ bond will make an important contribution to the wave function and to the stability of the bond. The $A–B$ bond can in this way acquire extra stability as compared with the $A–A$ and $B–B$ bonds.

In general we should expect that the extra stability contributed in this way by ionic-covalent resonance will be greatest for bonds in which one atom has a small ionization energy and the other has a large electron affinity. Inspection of Table 12-5 will show that this is indeed the case.

Pauling went on to develop an empirical quantitative scheme for dealing with the extra stability conferred on bonds by resonance involving ionic structures. He showed that numbers, called *electronegativities*, could be assigned to atoms, such that if x_A and x_B are the electronegativities of atoms A and B, respectively, then the dissociation energy of the $A-B$ single bond is given by the expression

$$w_{AB} = \tfrac{1}{2}(w_{AA} + w_{BB}) + (x_A - x_B)^2 \qquad \text{(D-1)}$$

where w_{AA} and w_{BB} are the dissociation energies of the $A-A$ and $B-B$ bonds.* Figure 12-5 shows the values of the electronegativities of various

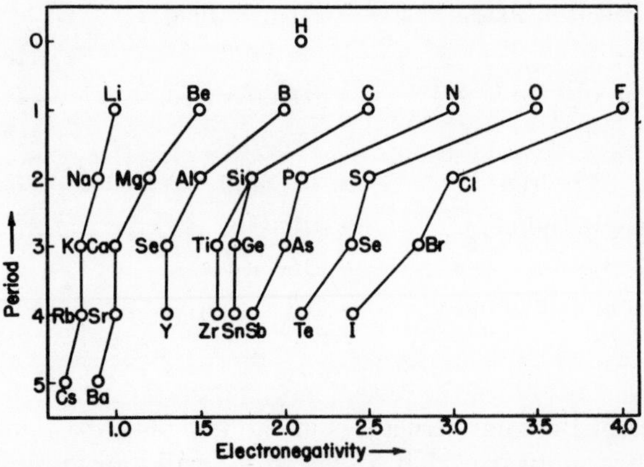

FIG. 12-5. Electronegativities of atoms as related to their positions in the periodic table (after Pauling (*52*)).

* Pauling actually found that better results are obtained using the relation

$$w_{AB} = \sqrt{w_{AA}w_{BB}} + (x_A - x_B)^2 \qquad \text{(D-2)}$$

but this formulation is less convenient to use than that given in (D-1). The quantity

$$w_{AB} - \tfrac{1}{2}(w_{AA} + w_{BB})$$

is equal to the heat of the reaction

$$\tfrac{1}{2}A_2 + \tfrac{1}{2}B_2 \rightarrow AB$$

which is obtainable directly from thermochemical data without necessitating the determination of the individual bond energies, w_{AA}, w_{BB}, and w_{AB}. Equation (D-2), on the other hand, requires the knowledge of all three of these quantities, and these are often not available.

atoms. In accordance with general usage, the numerical values of the electronegativities are such that, when substituted into (D-1), the energy difference

$$w_{AB} - \tfrac{1}{2}(w_{AA} + w_{BB})$$

is given in electron volts.

As one would expect, there is a distinct parallelism between the electronegativity scale and the ionization potentials and electron affinities of atoms. The alkali metals (which have small ionization energies and small electron affinities) have small electronegativities, and the halogens (which have large ionization energies and large electron affinities) have large electronegativities. There is, furthermore, a close relationship between the position of an atom in the periodic table and its location in Fig. 12-5. This relationship arises from the fact that as a shell is filled with electrons, they tend to be bound more and more firmly (*cf.* the penetration effect, discussed on pages 324*ff.*). Mulliken (*53*) has discussed this in some detail and has shown that the electronegativity of an atom should be proportional to the sum of its ionization potential and its electron affinity.

A further piece of evidence for Pauling's theory of the role of ionic structures and electronegativities in determining bond properties is the fact that bonds between atoms having widely different electronegativities tend to have large dipole moments. This is expected if these bonds have a great deal of ionic character. As we shall mention below, however, the relationship between the dipole moment of a bond and its ionic character is complicated by hybridization.

The development of the electronegativity scale is an excellent example of the use of procedure B (p. 434) to increase our understanding of chemically important phenomena.

c. Other concepts required in describing bonds

The concept of ionic character is of great importance in understanding the properties of chemical bonds. It is, however, by no means the only factor that must be reckoned with. For instance, the bond between a hydrogen atom and a carbon atom is quite different in methane, ethylene, and acetylene, as we know from the different C–H bond lengths and different chemical reactivities of the hydrogen in these compounds. Thus the nature of the hybridization on each atom of a bond may have an influence on the bond between the atoms. (This effect may arise at least in part from the fact that *s*-electrons are more firmly bound than *p*-electrons, so that the electronegativity of an $s-p$ hybridized orbital will tend to increase as the amount of s-character is increased.) Furthermore, resonance structures involving the rest of the molecule can affect the bond between a pair of atoms. Thus the carbon–carbon bond in benzene is different from the carbon–carbon bond in ethylene. Similarly, the bond between carbon and chlorine is affected by the other atoms bonded to the carbon atom, being quite different in methyl chloride and in vinyl chloride because of the possibility of the structure

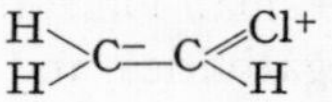

in the latter compound.

In short, the bond between a pair of atoms is influenced in some degree by the other atoms that are joined to them, and according to the theories of chemical binding described in the last chapter, we shall need the concepts of hybridization and of resonance between single bonds and double bonds in order to understand completely the properties of bonds between different pairs of atoms.

d. Nuclear quadrupole coupling

It was shown on page 95 that a distribution of charges in a closed region gives rise to an electrostatic potential outside the region that can be expressed in the form

$$V(r,\theta,\varphi) = Z/r + (1/r^2)\,(m_x p_x + m_y p_y + m_z p_z) + (1/r^3)\,(Q_{z^2} d_{z^2} + Q_{xz} d_{xz} + Q_{yz} d_{yz} + Q_{xy} d_{xy} + Q_{x^2-y^2} d_{x^2-y^2}) + \cdots \qquad \text{(D-3)}$$

The polar coordinates r, θ, and φ are measured with respect to some fixed set of axes inside the charged region, Z is the net electrical charge in the region, p_x, p_y, p_z, d_{z2}, d_{xz}, d_{yz}, d_{xy}, and d_{x2-y2} are the familiar surface spherical harmonics, and the m's and Q's are the components of dipole moment and quadrupole moment of the distribution.

Let us consider the electrostatic potential in the vicinity of an atomic nucleus in terms of this expression. Here the total net charge, Z, is merely the atomic number, if the charge is expressed in atomic units. It can be shown (*54*, p. 24) that it is impossible for an atomic nucleus in a stationary state to have a permanent electric dipole moment, so the second term in (D-3) must vanish for all stable nuclei. On the other hand, if the spin quantum number of a nucleus has the value $I \geqq 1$, it is found that the nucleus can have a quadrupole moment (*54*, p. 26).

The quadrupole moment can interact with the electrostatic fields arising from the electrons in a molecule in such a way that the energy of the molecule will depend on the orientation of the spin axis of the nucleus relative to the molecular framework. This interaction is known as *nuclear quadrupole coupling*. It is important because it is sensitive to the shape of the electronic charge distribution in molecules. Since it is presumed that electronic distributions can be described in terms of hybridization, resonance and electronegativities, one might hope that measurements of nuclear quadrupole coupling would be helpful in making quantitative statements about these basic concepts of chemical bonding. In the next few pages we shall see how such statements may be formulated.

(i) Nuclear Quadrupole Moments. The origin of a nuclear quadrupole moment can be understood if instead of supposing that the nucleus is spherical, as we might at first imagine, we think of it as a charged ellipsoid. Because of the nuclear spin, this ellipsoid must act as a figure of revolution with the axis of cylindrical symmetry lying along the axis of spin. In that case, placing the origin of the polar coordinate system, r, θ, φ, at the center of the nucleus and measuring the colatitude θ from the axis of spin (which thus becomes the z-axis), all of the components of the quadrupole moment vanish except for Q_{z^2}, which has the value (*54*, p. 26)

$$Q = \int \rho(r, \theta)\,(3 \cos^2 \theta - 1) r^2\, d\tau \qquad \text{(D-4)}$$

where $\rho(r,\theta)$ represents the cylindrically symmetrical charge density inside the nucleus. If the charge density is constant at all points within the ellipsoid and zero elsewhere, then integration of (D-4) leads to the result

$$Q = \tfrac{2}{5} Ze(c^2 - a^2) = \tfrac{8}{5} ZeR^2\, \varepsilon \qquad \text{(D-5)}$$

where $2c$ is the length of the ellipsoid along the axis of revolution, $2a$ is the width of the ellipsoid in the dimension perpendicular to this axis, ε is the ellipticity ($\varepsilon = (c-a)/(c+a)$), R is the mean radius of the ellipse ($R = (c+a)/2$), Z is the atomic number, and e is the electronic charge (sign taken as positive).

The numerical values of Q are known for many nuclei from the interpretation of the hyperfine structure of atomic spectra. It is found that among the chemically most interesting nuclei, the following have quadrupole moments: D, B^{10}, B^{11}, N^{14}, O^{17}, S^{33}, S^{35}, Cl^{35}, Cl^{36}, Cl^{37}, Mn^{55}, Cu^{63}, Cu^{65}, Ge^{73}, As^{75}, Se^{79}, Br^{79}, Br^{81}, Sb^{121}, Sb^{123}, I^{127}, I^{129}, and Bi^{209}. The largest known quadrupole moment, that of Lu^{176}, corresponds to an ellipticity of only about 15 per cent. Both positive and negative values of Q are observed, so the nuclear ellipsoid may be either prolate (cigar shaped, with $\varepsilon > 0$) or oblate (disc shaped, with $\varepsilon < 0$).

(*ii*) *The Interaction of Quadrupoles with Electric Fields.* The interaction energy of a nucleus with the electrons in a molecule can be found by replacing the nucleus with a point charge, whose charge is equal to the net charge of the nucleus (first term on the right in (D-3)) plus a linear array of two equal and oppositely directed dipoles (third term in (D-3)). In the prolate case, this array has the form

$$+ - \quad - +$$

and in the oblate case the arrangement would be

$$- + \quad + -$$

The energy of a single dipole in a uniform electric field $\boldsymbol{E}$ is

$$E = \boldsymbol{m} \cdot \boldsymbol{E} = |\boldsymbol{m}|\,|\boldsymbol{E}| \cos \theta \tag{D-6}$$

where $\boldsymbol{m}$ is the dipole moment and θ is the angle between the axis of the dipole and the direction of the field, as shown in Fig. 12-6A. Therefore when

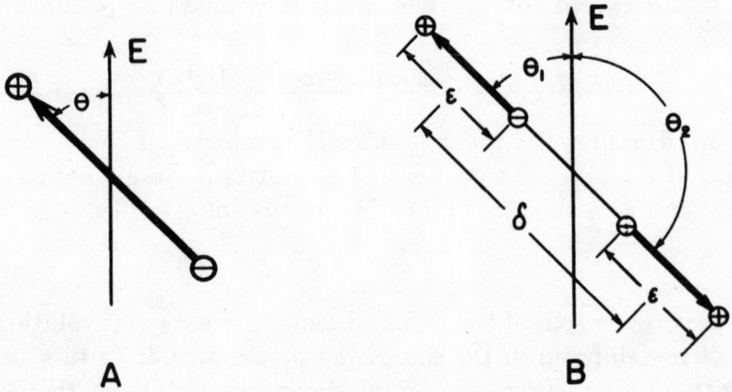

FIG. 12-6.
A. Dipole in an electric field.
B. Quadrupole in an electric field.

we place an array consisting of two identical, colinear, and opposed dipoles in a uniform electric field as in Fig. 12-6B, the energy is

$$\begin{aligned} E &= |\boldsymbol{m}|\,|\boldsymbol{E}| \cos \theta_1 + |\boldsymbol{m}|\,|\boldsymbol{E}| \cos \theta_2 \\ &= |\boldsymbol{m}|\,|\boldsymbol{E}|\,[\cos \theta_1 + \cos (\pi - \theta_1)] = 0 \end{aligned}$$

Thus the energy is independent of the orientation of the quadrupole.

On the other hand, suppose that the quadrupole is placed in a non-uniform electric field. For simplicity let us assume that the field has cylindrical symmetry about the z-direction, thus making the x- and y-directions equivalent. (This assumption will be valid for such cases as the Cl nucleus in Cl_2, HCl, and CH_3Cl, all of which are symmetrical-top molecules. It should also be nearly true of the Cl nucleus in CH_2Cl_2, since the C–Cl bond is probably almost cylindrically symmetrical.) We shall also assume that the variation of the components of the electric field with displacement is linear. (This assumption will be valid over the small distances between different portions of a nucleus.) Thus we may write for the electric field

$$\mathbf{E} = E_x\mathbf{i} + E_y\mathbf{j} + E_z\mathbf{k} = \mathbf{E}_0 + x(\partial E_x/\partial x)\mathbf{i} + y(\partial E_y/\partial y)\mathbf{j} + z(\partial E_z/\partial z)\mathbf{k}$$

where $\mathbf{E}_0$ is the field at the center of the nucleus (where $x = y = z = 0$). Since $\mathbf{E}_0$ has no effect on the energy of the quadrupole, it can be disregarded. The components of the electric field are related to the electrostatic potential V, by the equations

$$E_x = -\partial V/\partial x; \qquad E_y = -\partial V/\partial y; \qquad E_z = -\partial V/\partial z$$

and Poisson's equation (page 95) states that

$$\nabla^2 V = 0$$

Therefore

$$\partial E_x/\partial x + \partial E_y/\partial y + \partial E_z/\partial z = 0$$

Because of the equivalence of the x- and y-directions, we may write

$$\partial E_x/\partial x = \partial E_y/\partial y$$

so that from Poisson's equation

$$\partial E_x/\partial x = -\tfrac{1}{2}\partial E_z/\partial z$$

Thus the electric field is given by

$$\mathbf{E} = \tfrac{1}{2}(\partial^2 V/\partial z^2)\,(x\mathbf{i} + y\mathbf{j} - 2z\mathbf{k})$$

When the quadrupole is placed in such a field, the energy of dipole no. 1 is given by

$$E_1 = \mathbf{m}_1 \cdot \mathbf{E} = \tfrac{1}{2}(\partial^2 V/\partial z^2)\,(xm_x + ym_y - 2zm_z)$$

where x, y, and z are the coordinates of the dipole and m_x, m_y, and m_z are the components of the moment in the corresponding directions. Let the two dipoles in the quadrupole be separated by a distance δ. Introducing polar coordinates, with $x = (\delta/2) \sin\theta \cos\varphi$, $m_x = |\,\boldsymbol{m}\,| \sin\theta \cos\varphi$, and similarly for y, z, m_y, and m_z, we find for the energy of dipole no. 1

$$E_1 = \frac{|\,\boldsymbol{m}\,|\,\delta}{4}\,(\partial^2 V/\partial z^2)\,[\sin^2\theta(\cos^2\varphi + \sin^2\varphi) - 2\cos^2\theta]$$

$$= -\frac{|\,\boldsymbol{m}\,|\,\delta}{4}\,(\partial^2 V/\partial z^2)\,(3\cos^2\theta - 1)$$

Since $\cos\theta$ appears in this expression raised to an even power, and the value of $\cos\theta$ for dipole no. 2 is the same as that for dipole no. 1 except for its sign, the energy of dipole no. 2 must be identical with the energy of dipole no. 1. Therefore the total energy of the quadrupole is

$$E = -\frac{|\,\boldsymbol{m}\,|\,\delta}{2}\,(\partial^2 V/\partial z^2)\,(3\cos^2\theta - 1) \tag{D-7}$$

The quadrupole moment of an array of charges such as this is defined as

$$Q = \sum e_i(3z_i^2 - r_i^2) \tag{D-8}$$

where e_i is the charge on the ith particle and z_i and r_i are its coordinates. For the array $+ - - +$ we find, since $z_i = r_i$,

$$Q = e\sum 2z_i^2 = 4e\,[-(\tfrac{1}{2}\delta - \tfrac{1}{2}\varepsilon)^2 + (\tfrac{1}{2}\delta + \tfrac{1}{2}\varepsilon)^2] = 4e\varepsilon\delta = 4\,|\,\boldsymbol{m}\,|\,\delta \tag{D-9}$$

where ε is the separation of the positive and negative charges in each dipole and δ is the distance between the two dipoles (see Fig. 12-6B). Substituting in (D-7) we find for the total energy

$$E = -\tfrac{1}{8}\,Q\,(\partial^2 V/\partial z^2)\,(3\cos^2\theta - 1) \tag{D-10}$$

Evidently the energy of a quadrupole in a nonuniform electric field depends on its orientation in space.

If a molecule contains a nucleus with a quadrupole moment, the energy of the molecule must depend on the orientation of the nuclear spin axis relative to any nonuniform electric fields in the molecule. If the spin quantum number of the nucleus is I, only $(2I + 1)$ values of $\cos\theta$ are possible so $(2I + 1)$ energy levels will result. The separation of the energy levels in typical molecules is found to correspond to radiation of frequencies of the order to 10 to 1000 megacycles (wave lengths of 30 meters to 30 centimeters), which is in the radiofrequency range. Transitions between these energy levels can be induced by subjecting substances to radio waves having these frequencies, and in this way it is possible to measure the separations of the levels. The numerical value of $Q(\partial^2 V/\partial z^2)$ can be determined by analysis of this splitting. This quantity is called the *nuclear quadrupole coupling constant.* The numerical value of $(\partial^2 V/\partial z^2)$ can be obtained from the nuclear quadrupole coupling constant if the nuclear quadrupole moment Q is known. Even if Q is not known, however, the relative magnitudes of $(\partial^2 V/\partial z^2)$ at a given nucleus as it exists in different molecules can be compared by comparing the nuclear quadrupole coupling constants, because the magnitude of Q for the nucleus is the same in all molecules in which the nucleus occurs.

The magnitude of $(\partial^2 V/\partial z^2)$ is determined by the shape of the wave function in the vicinity of the nucleus. This quantity therefore provides us with a new method of getting information about the wave functions of molecules. It is convenient to give this important molecular property the symbol q

$$q = (\partial^2 V/\partial z^2) \tag{D-11}$$

According to the assumptions on which the present treatment rests, q is related to the curvature of the electrostatic potential at the nucleus along the axis of cylindrical symmetry. (If such an axis does not exist, a more complicated treatment is necessary (see *55*, pp. 139, 257, 286).) Let us now consider the relationship between q and the shape of the wave function of a molecule.

(*iii*) *Interpretation of q in Terms of the Theory of Chemical Bonding.* A small increment of charge de, at a distance r from a point, makes a contribution to the electrostatic potential at that point amounting to

$$dV = de/r$$

Thus the contribution to q at the point is

$$dq = de\frac{\partial^2(1/r)}{\partial z^2} = de\frac{\partial^2}{\partial z^2}(x^2 + y^2 + z^2)^{-1/2} = de\frac{3z^2 - r^2}{r^5}$$

$$= de\frac{3\cos^2\theta - 1}{r^3}$$

Let N electrons be placed in the orbitals $\psi_1, \psi_2, \cdots$. Then the increment of charge in the volume element $d\tau$ is $de = -e \sum |\psi_i|^2 d\tau$, where ψ_i is the orbital occupied by the ith electron and e is the electronic charge (taken as a positive quantity). The resulting value of q at a particular nucleus is

$$q = -e \sum_{\substack{\text{all} \\ \text{electrons}}} \int |\psi_i|^2 \, [(3\cos^2\theta_i - 1)/r_i^3] \, d\tau \tag{D-12}$$

where θ_i, φ_i, and r_i are the polar coordinates of the ith electron relative to axes whose origin is at the nucleus and in which the direction $\theta = 0$ lies along the axis of symmetry.

If $\sum |\psi_i|^2$ is spherically symmetrical, then it is possible to integrate this expression directly over the polar angle θ_i,

$$q = -\int_0^{\pi} (3\cos^2\theta - 1)\sin\theta \, d\theta \int_0^{2\pi} d\varphi \int_0^{\infty} e(\sum |\psi_i|^2 / r_i^3) r_i^2 \, dr_i \tag{D-13}$$

But

$$\int_0^{\pi} \cos^2\theta \sin\theta \, d\theta = 2/3$$

and

$$\int_0^{\pi} \sin\theta \, d\theta = 2$$

so the first integral in (D-13) vanishes. Thus we see that no spherically symmetrical distribution of charges can make any contribution to q. Since completed valence shells are to a first approximation spherically symmetrical in molecules, this means that the magnitude of q is determined (to this approximation) by the valence electrons alone. It was pointed out by Townes and Dailey (*56*) that this dependence of the nuclear quadrupole coupling on the wave function of the valence electrons makes it potentially valuable in giving information on the nature of the bonding in molecules.*

Let us consider the contributions to q made by electrons in the important orbitals involved in chemical bonding. Since all s-orbitals are spherically

* The spherical symmetry of filled shells can be removed by polarization effects due to the interaction of the filled shells with the electric fields from neighboring atoms. The asymmetries produced in this way are, however, believed to be very small. (See (*55*) for detailed discussion.)

symmetrical, s-electrons can make no contribution to q. For a $2p_x$-electron in the field of a nucleus with effective nuclear charge Z, we have

$$\psi_{2p_x} = \sqrt{\frac{Z^5}{32\pi}}\, re^{-Zr/2} \sin\theta \cos\varphi$$

$$q_{2p_x} = -e\frac{Z^5}{32\pi}\int_0^{2\pi}\int_0^{\pi}\int_0^{\infty} r^2e^{-Zr}\sin^2\theta\cos^2\varphi\left[\frac{3\cos^2\theta - 1}{r^3}\right] r^2 \sin\theta\, dr\, d\theta\, d\varphi$$

$$= -e\frac{Z^5}{32\pi}\int_0^{\infty} re^{-Zr}\, dr \int_0^{2\pi}\cos^2\varphi\, d\varphi \int_0^{\pi}\sin^3\theta(3\cos^2\theta - 1)\, d\theta$$

$$= Z^3e/60 \qquad \text{(D-14a)}$$

Because of the equivalence of the x- and y-directions, $2p_x$- and $2p_y$-electrons must give the same contribution to q in an axially symmetric molecule having the z-axis as the axis of symmetry. For a $2p_z$-electron, on the other hand, we have

$$q_{2p_z} = -e\frac{Z^5}{32\pi}\int_0^{2\pi}\int_0^{\pi}\int_0^{\infty} r^2e^{-Zr}\cos^2\theta\,\frac{3\cos^2\theta - 1}{r^3}\, r^2\sin\theta\, dr\, d\theta\, d\varphi$$

$$= -eZ^3/30 \qquad \text{(D-14b)}$$

Thus we find that

$$q_{2p_z} = -2q_{2p_x} = -2q_{2p_y} \qquad \text{(D-15a)}$$

This important result can be proven more directly in the following way for any np-electron. By symmetry $q_{np_x} = q_{np_y}$. Furthermore, $q_{np_x} + q_{np_y} + q_{np_z} = 0$ because an atom containing one np_x-electron, one np_y-electron, and one np_z-electron would have a spherically symmetrical charge distribution. This leads directly to the result

$$q_{np_z} = -2q_{np_x} = -2q_{np_y} \qquad \text{(D-15b)}$$

Because of the slight penetration of d-orbits into regions close to the nucleus, it is found that the contribution to q by d-electrons is small compared with the contribution of p-electrons in the same quantum shell. (The contribution of an nd-electron is of the order of 10 to 15 per cent of the contribution of an np-electron.) Therefore the value of q at a nucleus is determined essentially by the number of electrons in p-orbitals in the atom to which the nucleus belongs.

If the number of np_x-, np_y-, and np_z-electrons in an atom is N_x, N_y, and N_z, respectively, then the q-value of the atom is

$$q = [N_z - \tfrac{1}{2}(N_x + N_y)]q_{np_z} \tag{D-16}$$

The quantity $[\frac{1}{2}(N_x + N_y) - N_z]$ is called the *number of unbalanced p-electrons* in a shell. This number is affected by the hybridization and by the ionic character of the bonds between an atom and its neighbors. Consider for example the chlorine nucleus. In atomic chlorine the quantity qQ for the isotope Cl^{35} has the value -109.74 megacycles, as determined from the hyperfine structure of the chlorine atomic spectrum. Since the free chlorine atom has the configuration $1s^2 2s^2 2p^6 3s^2 3p^5$, we need consider the effect of only the $3p$-electrons and take $N_z = 1$, $N_x = N_y = 2$, giving $q = -q_{3p_z}$. Thus the energy $qQ = -109.74$ mc. corresponds to a single unpaired p-electron in the chlorine atom. It is found that qQ in solid Cl_2 is within one per cent of the value for the free atom. This indicates that the orbitals used in forming the Cl–Cl bond must be nearly pure p in character. On the other hand, qQ for the Cl^{35} nucleus in the diatomic NaCl molecule in sodium chloride vapor is very close to zero. This indicates that all of the p-orbitals in the chlorine atom must be occupied in the NaCl molecule, so that $N_x = N_y = N_z = 2$ and $q = 0$. That is, the bond in NaCl must be almost completely ionic, with practically no covalent character at all.

In general, we can write the wave function of the bond formed by a chlorine (or other halide) atom with another atom, A, in the following way:

$$\psi = \sqrt{1 - \beta - \gamma}\,\psi_{A-\mathrm{Cl}} + \sqrt{\beta}\,\psi_{A^+\mathrm{Cl}^-} + \sqrt{\gamma}\,\psi_{A^-\mathrm{Cl}^+}$$

where the first term represents a pure covalent structure, and the second and third terms refer to the two possible ionic structures of the bond (the parameters β and γ are the fractions of the different ionic structures present). We should expect that as β increases qQ should decrease, with $qQ = 0$ when $\beta = 1$ and $\gamma = 0$. On the other hand, qQ should increase as γ increases because the structure $A^-\,Cl^+$ gives $N_x + N_y = 4$ and $N_z = 0$, so that there are two unbalanced p-electrons in the structure $A^-\,Cl^+$ as compared with only one for the pure covalent structure A–Cl. In addition, the orbitals on the chlorine atom that are used in forming the covalent A–Cl bond may be hybridized with the $3s$-orbital, so that the chlorine p_z bonding orbital is replaced by

$$\phi = \sqrt{a}\,s + \sqrt{1-a}\,p_z$$

and the unshared pair of electrons that were formerly in the 3s-orbital have to be placed in the orbital

$$\phi' = \sqrt{1-a}\, s - \sqrt{a}\, p_z$$

a being the fraction of s-character in the hybridized bond. Since there is one electron in ϕ and there are two electrons in ϕ', we have

$$N_z = (1 - a) + 2a = 1 + a$$

and the number of unbalanced electrons is

$$N_z - \tfrac{1}{2}(N_x + N_y) = 1 + a - \tfrac{1}{2}(2 + 2) = -(1 - a)$$

Therefore, as the amount of hybridization represented by the parameter a increases, qQ must decrease. (It should be noted that if there is any hybridization involving d-orbitals, a further decrease in qQ will result.)

Taking the two ionic structures and the s–p hybridization into account, we may write

$$q \simeq (1 - a)(1 - \beta + \gamma)\, q_{p_z} \qquad \text{(D-17a)}$$

Because of the fact that the effective nuclear charge Z in (D-14a) and (D-14b) will be smaller in the neutral chlorine atom than in the Cl^+ ion, however, q_{p_z} should be larger for the structure $A^- Cl^+$ than for the covalent structure A–Cl. The variation of the parameter γ should therefore produce a somewhat larger effect than that given in (D-17a). This may be taken into account by writing

$$q = (1 - a)(1 - \beta + k\gamma)\, q_{p_z} \qquad \text{(D-17b)}$$

where k is a constant greater than unity. Townes and Dailey (*56*) estimate that $k = 1.5$.

The application of Equation (D-17b) can be illustrated by the interpretation of the observed quadrupole coupling constants of Cl^{35} given in Table 12-6. The fact that qQ for the chlorine nucleus in FCl is larger than qQ for Cl_2 indicates that the former molecule must have some Cl^+F^- structure (effect of γ in (D-17b)); this might be expected from the fact that fluorine is more electronegative than chlorine. The smaller values of qQ for BrCl and ICl presumably indicate the importance of ionic structures with negatively charged chlorines, Br^+Cl^- and I^+Cl^- (effect of β in (D-17b)). This is in line

TABLE 12-6

QUADRUPOLE COUPLING CONSTANTS OF VARIOUS CHLORIDES AND BROMIDES

Substance	Value of qQ (megacycles) Gaseous state	Solid state	Electronegativity of atom attached to halide
For the Cl^{35} nucleus in:			
Cl (atomic)	−109.74	—	—
Cl_2	—	−108.5	3.15
FCl	−146.0	−141.4	3.90
BrCl	−103.6	—	2.95
ICl	−82.5	—	2.65
ClCN	−83.2	—	2.60
CH_3Cl	−74.77*	−68.40	2.60
SiH_3Cl	−40.0	—	1.90
GeH_3Cl	−46.	—	1.90
H–C≡C–Cl	−79.67	—	2.60
CH_2Cl_2	−78.	−72.47	2.60
$CHCl_3$	—	−76.5	2.60
CCl_4	—	−81.85	2.60
For the Br^{79} nucleus in:			
Br (atomic)	769.62	—	—
Br_2	—	765	2.95
FBr	1089.	—	3.90
ClBr	876.8	—	3.15
BrCN	686.06	—	2.60
CH_3Br	577.15*	528.90	2.60
SiH_3Br	336.	—	1.90
GeH_3Br	380.	—	1.90
CH_3–C≡C–Br	647.	—	2.60
CF_3Br	619.	—	2.60

Data from Gordy, Smith, and Trambarulo (*55*) except for those marked with asterisk, which are from Kraitchman and Dailey (*57*). Electronegativities are those given by Huggins (*58*).

with the smaller electronegativities of Br and I; to what extent it may also be caused by $s-p$ hybridization (effect of α in (D-17b)) is not clear, but it is hard to believe that, in view of the apparent absence of hybridization in Cl_2, there should be much of it in the other interhalogen compounds. On the other hand, methyl chloride gives a somewhat smaller value of qQ than ICl, in spite of the fact that iodine and carbon have about the same electronegativities. Therefore some additional factor such as hybridization must be operating here. There is a gradual increase in qQ for Cl^{35} in the series CH_xCl_{4-x} as more chlorines are introduced. This presumably reflects an increase in the electronegativity of the carbon atom induced by the chlorine atoms; as more chlorines are bonded to the carbon, they would be expected to draw electrons away from it, increasing its effective nuclear charge and thus its electronegativity.

Exercise. The values of qQ for the Br^{79} nucleus in some bromides are given in Table 12-6. Discuss the variation of qQ in terms similar to those just outlined for the chlorides.

It is interesting to consider also the case of NH_3 and other nitrogen compounds, since the quadrupole coupling constant throws some light on the state of hybridization of the nitrogen orbitals. If the three valence electrons of the nitrogen atom were all in pure p-orbitals,

$$N_x = N_y = N_z = 1 \quad \text{and} \quad [\tfrac{1}{2}(N_x + N_y) - N_z] = 0$$

In this case we would have $q = 0$ and there would be no nuclear quadrupole coupling in NH_3 and similar nitrogen compounds. It is found, however, that the quadrupole coupling does not vanish in NH_3, thus proving that the bonding orbitals of the nitrogen are indeed hybridized. Unfortunately the quadrupole moment of the N^{14} nucleus is not known, so the actual extent of hybridization cannot be determined. It is interesting that qQ for N^{14} is about the same in HCN, BrCN, and CH_3CN as it is in NH_3, but that in CH_3NC ($CH_3 - N^+ \equiv C^-$) and in the central nitrogen atom in N_2O (principal structures: $N^- = N^+ = O$ and $N \equiv N^+ - O^-$), where the nitrogen atom is ionic, qQ is much smaller.

Although quadrupole coupling has been studied in many other types of molecules, the discussion given above should be sufficient to illustrate the kind of information that can be obtained, as well as some of the uncertainties. For further discussion see Gordy, Smith, and Trambarulo (*55*, pp. 279*ff*.) and Townes and Dailey (*56*).

e. Chemical shifts in nuclear magnetic resonance

Many nuclei have magnetic dipole moments. These moments are directed along the axis of spin of the nucleus, and if the nucleus is placed in a magnetic field, the energy will depend on the relative orientations of the spin axis and the magnetic field. If the nuclear spin quantum number is I, the spin axis can assume $(2I + 1)$ orientations relative to any fixed direction in space. Therefore the nucleus in a magnetic field can have $(2I + 1)$ energy levels. If the magnetic moment of the nucleus is $\boldsymbol{\mu}$ and the magnetic field at the nucleus is $\boldsymbol{H}'$, the maximum energy is $|\boldsymbol{\mu}|\,|\boldsymbol{H}'|$ (when $\boldsymbol{\mu}$ and $\boldsymbol{H}'$ point in opposite directions) and the minimum energy is $-|\boldsymbol{\mu}|\,|\boldsymbol{H}'|$ (when $\boldsymbol{\mu}$ and $\boldsymbol{H}'$ are in the same direction). Thus the total spread in energy is $2|\boldsymbol{\mu}|\,|\boldsymbol{H}'|$. Since there are $2I$ intervals between the $(2I + 1)$ levels, and since the levels are equally spaced, the energy difference between successive levels is

$$E = |\boldsymbol{\mu}|\,|\boldsymbol{H}'|\,/\,I$$

The nuclear magnetic resonance phenomenon consists of the absorption of radiation due to transitions between these levels. For magnetic fields of the order of a few thousand gauss, the radiation lies in the radiofrequency range. The frequencies involved, and therefore the separation of the levels, can be measured with great precision.

Because of the precessional motions of the electrons in their orbits induced by the applied magnetic field (see page 268), the field $\boldsymbol{H}'$ acting on the nuclei is not quite the same as the field $\boldsymbol{H}$ in the free space outside of the atoms. This change in the magnetic field due to electronic motions is called *internal diamagnetic shielding*. For isolated atoms Lamb (*59*) has shown that $\boldsymbol{H}$ and $\boldsymbol{H}'$ are related by

$$\frac{|\boldsymbol{H}'|}{|\boldsymbol{H}|} = 1 - \frac{e}{3m_ec^2}V$$

where V is the electrostatic potential at the nucleus produced by the electrons in the atom, e and m_e being the electronic charge and mass, and c being the velocity of light. For molecules, Ramsey (*60*) has shown that the internal diamagnetic shielding depends not only on the potential V at the nucleus, but also on another more complex term, called the second order paramagnetic term, which is difficult to evaluate or interpret.

When the nuclear magnetic resonance of a given nucleus is studied in different molecules, it is found that there are slight differences in the ratio $|\boldsymbol{H}'|\,/\,|\boldsymbol{H}|$ which must reflect changes in the electron distribution in going from one molecule to another. These variations in $|\boldsymbol{H}'|\,/\,|\boldsymbol{H}|$ are called *chemical shifts*. If it were not for the second order paramagnetism, the chemical shifts would be directly related to the potential V, and might be discussed in a manner analogous to that used in interpreting the magnitude of $\partial^2V/\partial z^2$ as deduced from nuclear quadrupole coupling. Because of the second order paramagnetism, however, it is necessary to use a more empirical approach.

This has been done especially by Gutowsky and his students (*61*), who have found, for instance, that there is a direct relation between the chemical shift of the magnetic shielding of the F^{19} nucleus in a compound and the electronegativity of the atom to which the fluorine is bonded. They have also observed a parallelism between the chemical shifts of F^{19} in substituted fluorobenzenes and the ortho–para or meta directing powers of the substituents.

Although the interpretation of these chemical shifts is less straightforward than that of the nuclear quadrupole coupling, this property should give useful additional semiquantitative information which will help in the elucidation of the nature of chemical binding.

f. Significance of the interpretation of bond properties in terms of elementary concepts such as electronegativities, hybridization, and resonance

We have seen some examples of the interpretation of several properties of chemical bonds in terms of the concepts of hybridization and resonance involving ionic and covalent bond structures. Other bond properties can be discussed in similar terms, and reasonable pictures can be deduced from many of them (see Pauling (*52*) and Cottrell and Sutton (*62*) for discussions of bond lengths and Walsh (*63*) for a discussion of force constants).*

The question now arises, whether all of the different pictures of the chemical bond, each derived from the study of a different property, are consistent with each other. We must ask whether it is possible to write down a simple recipe for a chemical bond, that accounts simultaneously for all bond properties. Such a recipe would involve the assignment of a set of, say, three or four numbers that prescribe:

(a) The difference in the electronegativities of the atoms in the bond.

(b) The nature of the hybridization present in the atoms at the two ends of the bond.

* The dipole moment is unfortunately rather difficult to interpret in terms of these simple concepts. The ionic character of a bond has an important influence, but the unsymmetrical arrangements of electron clouds that result from hybridization also make significant contributions to the dipole moment. For instance, an electron in an *s–p* hybrid on a hydrogen atom can give the hydrogen atom a large dipole moment because the center of gravity of the electron cloud does not coincide with the nucleus. This matter is discussed further by Coulson (*64*, pp. 128, 207). There also seem to be serious limitations in the use of these simple bond concepts in the interpretation of another bond property mentioned on page 479–*viz.*, the rate of change of the dipole moment with bond length (see Hornig and McKean (*65*)).

(c) The ratio of the coefficients of the ionic and covalent structures. (This may require more than one number, because in some instances more than one ionic or covalent structure might be important.)

There is no doubt that a quantitative understanding of all of the properties of chemical bonds would require taking at least these factors into account; the question is, do these factors alone provide an adequate quantitative basis on which procedure B may be used to construct a general theory of chemical bonds?

The answer to this question is not yet known. The task of gathering the data needed for the analysis is still not completed, and the interpretation of the data from a unified point of view has barely begun. One thing, however, is fairly certain: the best that we can expect in the way of a quantitative theory must be essentially empirical, and it will use quantum mechanical manipulations as no more than a framework for the interpretation of the observations. In order to understand some of the implications of this situation, let us suppose that an empirical scheme of numbers could be devised with which all of the properties of bonds could be satisfactorily correlated using procedure B. What would such a scheme mean in terms of a precise quantum mechanical theory of the chemical bond–i.e., a theory based entirely on procedure A?

For the sake of simplicity we may assume that the correct wave function for a single bond between two atoms, A and B, may be represented by a function of the coordinates of only two electrons. We shall denote this correct wave function by $\psi(1, 2)$, where the numbers 1 and 2 refer to the two electrons in the bond. Now let us try to approximate the function $\psi(1, 2)$ by means of a function, $\phi(1, 2)$, which represents the bond by means of orbitals located on the two atoms

$$\phi(1,2) = A(1)\,B(2) + A(2)\,B(1) + aA'(1)\,A'(2) + bB'(1)\,B'(2)$$

where A and A' are atomic orbitals centered on atom A, B and B' are orbitals centered on atom B, and a and b are constants. The first two terms in ϕ represent a covalent structure, and the last two terms represent ionic structures. Although the orbitals A and A' presumably resemble each other, and B and B' also resemble each other, somewhat more flexibility is possible if the orbitals appearing in the ionic structures are allowed to differ from those appearing in the covalent structures.

The function $\phi(1, 2)$ has the most general form that still maintains the concepts of ionic and covalent structures for the bond. Although this wave function appears to be based on the VB theory of bonding, we know from the discussion on page 447 that it is also derivable from the MO theory if configuration interactions are included.

The function $\phi(1, 2)$ has a great deal of flexibility. The coefficients a and b are related to the ionic character of the bond, and may be varied to give the best fit to the correct function $\psi(1, 2)$. Furthermore, the form of the orbitals A, A', B, and B' may be varied within wide limits. Perhaps the most convenient approach is to write these orbitals as sums of hydrogen-like atomic orbitals

$$A = \sum_n (\alpha_{An}(ns) + \beta_{An}(np) + \gamma_{An}(nd) + \cdots)$$

where n is the principle quantum number of the orbital and $s, p, d, \cdots$ have their usual meaning. Similar expressions could be written for A', B, and B'. The parameters α_{An}, β_{An}, γ_{An}, $\cdots$, a, and b might then be determined by minimizing the integral

$$\int (\psi - \phi)^2 \, d\tau$$

if ψ is known, or by minimizing the energy integral,

$$\int \phi H \phi \, d\tau,$$

if ψ is not known. The coefficients α_{An}, β_{An}, γ_{An}, $\cdots$ define the hybridization of orbital A, except that allowance has been made for different kinds of hybridization in the ionic and covalent structures. Furthermore, we have allowed for the possibility of including among the hybrids a great many functions not ordinarily employed in discussing hybridization. (The adequacy of the usual discussions of hybridization depends on the relative values of α_{An}, β_{An}, γ_{An}, $\cdots$, belonging to the lowest possible value of n that the valence electrons can have, as compared with the coefficients of functions having higher values of the principle quantum number. If the coefficients belonging to the smallest n's are large and all of the other coefficients are small, then the usual discussions could be said to give a good approximation to the true function.)

In this way we might be able to establish in principle a rigorous basis for the concepts of ionicity and hybridization. There are, however, some serious questions as to the adequacy and significance of such a procedure:

(1) The procedure will have significance only if it is possible to construct a simple function of the form of $\phi(1,2)$ that really resembles the true function, $\psi(1, 2)$. From the calculations that have been performed on the hydrogen molecular ion and the hydrogen molecule, however, we can be

fairly certain that the resemblance between any simple form of the function ϕ and the true function ψ can at best be only limited. Since we know from the hydrogen molecule that a simple function of the type ϕ cannot deal adequately with the correlation energy, it is even possible that the resemblance might be rather poor.

(2) The choice of the set of hydrogen-like orbitals (ns), (np), (nd), $\cdots$, that are used as the basis of the expansion of the atomic orbitals A, A', B, and B' is inevitably somewhat arbitrary. For instance, there is some leeway in the selection of the effective nuclear charge in the hydrogenic orbitals. The numerical values of the parameters will depend very much on this initial choice of orbitals, and since there is no obvious way of deciding which choice is to be preferred, the hybridization and ionic character that are obtained must be somewhat ambiguous. The extent of the ambiguity cannot be estimated without making the detailed calculations, and this has not been done. It would be surprising, however, if the ionic character of a bond and the coefficients of the hybrids on A and B could be defined by means of such a calculation with an uncertainty of less than ten per cent or so.

In the light of these considerations, it seems rather pointless to worry about whether, for instance, the bond in the NaCl molecule is 90 per cent or 100 per cent ionic. It might be possible to set up an empirical scheme for interpreting certain sets of data, which from internal consistency will clearly distinguish between 90 per cent and 100 per cent ionic character. Others chemes could probably be set up, however, which would lead to other values of the ionicity, and the choice between the schemes would be largely subjective.

The description of chemical bonds in terms of hybridization, ionicity and resonance therefore resembles in many ways the following situation: A blind man is given a large pile of shoes and is asked to sort it out according to the length and width of the shoe and whether the shoe is for the left or right foot. He proceeds by picking up a number of stones in a field, which he tries to fit into each shoe. According to the stone that gives the best fit, he is able to separate the large pile into several smaller ones, each of which he can describe as "corresponding" to one of the stones. The resulting classification of the shoes would probably be useful to the owner of the shoes. Nevertheless, the blind man cannot be expected to accomplish a complete classification of the shoes. It would be possible to improve the classification if a more careful selection of the stones were made in the first place. At best, however, certain aspects of the style, such as the color, type of sole, and decoration, are bound to be missed by such a procedure.

REFERENCES CITED

1. P. A. M. Dirac, *Proc. Roy. Soc.*, **A123,** 714 (1929).
2. H. Eyring, J. Walter, and G. E. Kimball, "Quantum Chemistry," Wiley, New York, 1944.
3. C. Eckart, *Phys. Rev.*, **30,** 891 (1930).
4. J. O. Hirschfelder and J. Kincaid, *Phys. Rev.*, **52,** 658 (1937).
5. D. R. Bates, K. Ledsham, and A. L. Stewart, *Phil. Trans. Roy. Soc.*, **A246,** 215 (1953).
6. B. N. Finkelstein and G. E. Horowitz, *Z. Physik*, **48,** 118 (1928).
7. B. N. Dickenson, *J. Chem. Phys.*, **1,** 317 (1933).
8. H. O. Pritchard and H. A. Skinner, *J. Chem. Soc.*, **1951,** 945.
9. E. F. Gurnee and J. L. Magee, *J. Chem. Phys.*, **18,** 142 (1950).
10. A. C. Hurley, *Proc. Roy. Soc.*, **A226,** 179 (1954).
11. H. M. James, *J. Chem. Phys.*, **3,** 9 (1935).
12. V. Guillemin and C. Zener, *Proc. Natl. Acad. Sci. (U.S.)*, **15,** 314 (1929).
13. C. A. Coulson, *Trans. Faraday Soc.*, **33,** 1479 (1937); *Proc. Cambridge Phil. Soc.*, **34,** 204 (1938).
14. S. Weinbaum, *J. Chem. Phys.*, **1,** 593 (1933).
15. S. C. Wang, *Phys. Rev.*, **31,** 579 (1928).
16. N. Rosen, *Phys. Rev.*, **38,** 2099 (1931).
17. T. Inui, *Proc. Phys. Math. Soc. Japan*, **20,** 770 (1938); **23,** 992 (1941).
18. C. R. Mueller and H. Eyring, *J. Chem. Phys.*, **19,** 1495 (1951).
19. C. A. Coulson and I. Fischer, *Phil. Mag.* 7, **40,** 386 (1949).
20. J. O. Hirschfelder and J. W. Linnett, *J. Chem. Phys.*, **18,** 130 (1950).
21. J. Lennard-Jones and J. A. Pople, *Proc. Roy. Soc.*, **A 210,** 190 (1951).
22. A. A. Frost and J. Braunstein, *J. Chem. Phys.*, **19,** 1133 (1951).
23. H. M. James and A. S. Coolidge, *J. Chem. Phys.*, **1,** 825 (1933).
24. A. Meckler, *J. Chem. Phys.*, **21,** 1750 (1953).
25. T. L. Cottrell and L. E. Sutton, *J. Chem. Phys.*, **15,** 685 (1947).
26. D. F. Hornig, *J. Am. Chem. Soc.*, **72,** 5772 (1950).
27. R. S. Mulliken and R. G. Parr, *J. Chem. Phys.*, **19,** 1271 (1951).
28. G. Glockler, *J. Chem. Phys.*, **21,** 1249 (1953).
29. J. M. Robertson, "Organic Crystals and Molecules," Cornell U.P., Ithaca, 1953.
30. L. Pauling, L. O. Brockway, and J. Y. Beach, *J. Am. Chem. Soc.*, **57,** 2705 (1935).
31. H. O. Pritchard and F. H. Sumner, *Proc. Roy. Soc.*, **A226,** 128 (1954).
32. S. C. Abrahams, J. M. Robertson, and J. G. White, *Acta Cryst.*, **2,** 238 (1949).
33. F. R. Ahmed and D. W. J. Cruickshank, *Acta Cryst.*, **5,** 852 (1952).
34. V. C. Sinclair, J. M. Robertson, and A. M. Mathieson, *Acta Cryst.*, **3,** 251 (1950).
35. J. Sherman, *J. Chem. Phys.*, **2,** 488 (1934).
36. J. M. Robertson and J. G. White, *J. Chem. Soc.*, **1945,** 607.
37. J. M. Robertson, *Proc. Roy. Soc.*, **A 207,** 101 (1951).
38. D. M. Donaldson and J. M. Robertson, *Proc. Roy. Soc.*, **A 220,** 157 (1953).
39. C. A. Coulson, *Proc. Roy. Soc.*, **A 169,** 413 (1939); **A 207,** 91 (1951).
40. A. Pullman and B. Pullman, *Ann. chim.* (*Paris*), **2,** 5 (1947).
41. H. Sponer, G. Nordheim, A. L. Sklar, and E. Teller, *J. Chem. Phys.*, **7,** 207 (1939).
42. F. M. Garforth, C. K. Ingold, and H. G. Poole, *J. Chem. Soc.*, **1948,** 406.
43. H. Shull, *J. Chem. Phys.*, **17,** 295 (1949).
44. A. L. Sklar, *J. Chem. Phys.*, **5,** 669 (1937).

45. D. P. Craig, *Proc. Roy. Soc.*, **A 200,** 401 (1950).
46. M. Goeppert-Mayer and A. L. Sklar, *J. Chem. Phys.*, **6,** 645 (1938).
47. R. G. Parr and B. Crawford, *J. Chem. Phys.*, **16,** 1049 (1948); **17,** 726 (1949).
48. C. C. J. Roothaan and R. G. Parr, *J. Chem. Phys.*, **17,** 1001 (1949).
49. K. Niira, *J. Chem. Phys.*, **20,** 1498 (1952); *J. Phys. Soc. Japan*, **8,** 630 (1953).
50. R. G. Parr, D. P. Craig, and I. G. Ross, *J. Chem. Phys.*, **18,** 1561 (1950).
51. C. A. Coulson, D. P. Craig, and J. Jacobs, *Proc. Roy. Soc.*, **A 206,** 297 (1951).
52. L. Pauling, "Nature of the Chemical Bond," 2nd ed., Cornell U.P., Ithaca, 1940.
53. R. S. Mulliken, *J. Chem. Phys.*, **2,** 782 (1934); **3,** 573 (1935).
54. J. M. Blatt and V. F. Weisskopf, "Theoretical Nuclear Physics," Wiley, New York, 1952.
55. W. Gordy, W. V. Smith, and R. F. Trambarulo, "Microwave Spectroscopy," Wiley, New York, 1953.
56a. C. H. Townes and B. P. Dailey. *J. Chem. Phys.*, **17,** 782 (1949).
56b. B. P. Dailey, *Ann. Rev. Phys. Chem.*, **4,** 436 (1953).
56c. B. P. Dailey and C. H. Townes, *J. Chem. Phys.*, **23,** 118 (1955).
57. J. Kraitchman and B. P. Dailey, *J. Chem. Phys.*, **22,** 1477 (1954).
58. M. L. Huggins, *J. Am. Chem. Soc.*, **75,** 4123 (1953).
59. W. E. Lamb, *Phys. Rev.*, **60,** 817 (1941).
60. N. F. Ramsey, *Phys. Rev.*, **78,** 699 (1950).
61. H. S. Gutowsky, *Ann. Rev. Phys. Chem.*, **5,** 343 (1954).
62. T. L. Cottrell and L. E. Sutton, *Quart. Revs. (London)*, **2,** 260 (1948).
63. A. D. Walsh, *Proc. Roy. Soc.*, **A 207,** 22 (1951).
64. C. A. Coulson, "Valence," Oxford U.P., New York, 1952.
65. D. F. Hornig and D. C. McKean, *J. Phys. Chem.*, **59,** 1133 (1955).

GENERAL REFERENCES

Bibliographies of tables of integrals important in the quantum mechanical calculations of molecular energies will be found in:

C. C. J. Roothaan. *J. Chem. Phys.*, **19,** 1445 (1951).
K. Ruedenberg, *J. Chem. Phys.*, **19,** 1459 (1951).
K. Ruedenberg, C. C. J. Roothaan, and W. Jaunzemis, *J. Chem. Phys.*, **24,** 219 (1956).

A compilation of papers on bond properties will be found in:

Prcc. Roy. Soc., **A 207,** 1–136 (1951).

An extensive discussion of the concept of electronegativity is given by:

H. O. Pritchard and H. A. Skinner, *Chem. Revs.*, **55,** 745 (1955).

Chapter 13

Van der Waals Forces

In Chapters 11 and 12 we have discussed the origin of the strong attractive forces between atoms that are responsible for the stability of the chemical entities known as molecules. We are all aware, however, that these are not the only attractive forces that can act between atoms and molecules. Thus the atoms in a rare gas such as neon, helium, or argon cannot form a chemical bond with each other. Nevertheless they form liquid and solid phases with the evolution of heat, showing that energy is released when the atoms are brought together, and that an attractive force must exist. Forces such as these, which become evident when chemical bonding is not possible, are known as van der Waals forces.

A. General Discussion of Intermolecular Forces Not Involving Chemical Bonds

Van der Waals forces have two outstanding characteristics: (1) They are much weaker than the forces in chemical bonds. Thus the energy required to separate the two chlorine atoms in Cl_2 is 57.2 kcal. per mole, whereas the energy required to remove an argon atom from its neighbors on the surface of crystalline argon is only 1.88 kcal. per mole. (2) In many cases, van der Waals forces are additive and cannot be "saturated." That is, the van der Waals force between two atoms, *A* and *B*, is not very much affected if other atoms are brought into the vicinity of *A* and *B*. As a result, the energy of, say, three neon atoms, Ne_a, Ne_b, and Ne_c, in a given configuration can be calculated by adding the energy that would be obtained if only Ne_a and Ne_b were present, plus the energy of Ne_b and Ne_c by themselves, plus the energy of Ne_c and Ne_a by themselves. This is very different from the behavior of chemical forces. For instance, if there are no other hydrogen atoms in the vicinity, the interaction of two hydrogen atoms, H_a and H_b, is not at all the same as the interaction that occurs when H_b is already chemically bonded to another hydrogen atom, H_c.

Several explanations for van der Waals forces have been proposed. In 1921 Keesom pointed out that if two molecules bear permanent electric dipole moments $\boldsymbol{m}_1$ and $\boldsymbol{m}_2$ and if they undergo thermal motions, they will on the average assume orientations leading to attraction. If $\mathbf{R}$ is the vector joining the two dipoles, the potential energy of a pair of dipoles is (*cf.* Exercise on page 507)

$$V = \frac{1}{|\mathbf{R}|^3}\left[\boldsymbol{m}_1 \cdot \boldsymbol{m}_2 - 3\frac{(\boldsymbol{m}_1 \cdot \mathbf{R})(\boldsymbol{m}_2 \cdot \mathbf{R})}{|\mathbf{R}|^2}\right] \tag{A-1}$$

Since the thermal motions are subject to the Boltzmann distribution, so that an arrangement whose energy is V has the probability $\exp(-V/kT)$, orientations with low energy are favored. On averaging the quantity $V\exp(-V/kT)$ over all orientations, and assuming that $kT >> V$, it is found that the average potential energy of a pair of dipoles whose separation is R is approximately

$$\overline{V} = -\frac{2|\boldsymbol{m}_1|^2 \ \boldsymbol{m}_2|^2}{3kTR^6} \tag{A-2}$$

This force evidently decreases in importance as the temperature increases. It is known as the *dipole orientation force*, or the *Keesom force*. It is difficult to see how this force could show the property of additivity, especially if the potential V is not much smaller than kT.

In 1920 Debye pointed out that if one molecule has a permanent dipole moment $\boldsymbol{m}$, and a second molecule has a polarizability α, then the first molecule will induce in the second molecule a dipole oriented in such a direction that the induced and permanent dipoles attract each other. If the interaction energy is averaged for all orientations of the permanent dipole, it is found that

$$\overline{V} = -\alpha\,|\boldsymbol{m}|^2/R^6 \tag{A-3}$$

If both molecules are polarizable, with polarizabilities α_1 and α_2, and if both have permanent dipole moments $\boldsymbol{m}_1$ and $\boldsymbol{m}_2$, then the mean potential energy is

$$\overline{V} = -(\alpha_2\,|\boldsymbol{m}_1|^2 + \alpha_1\,|\boldsymbol{m}_2|^2)/R^6 \tag{A-4}$$

The resulting force is independent of the temperature. It is referred to as the *induction force* or the *Debye force*. These forces are also not additive.

The Keesom and Debye forces depend on the presence of a dipole moment in at least one of each interacting pair of molecules. The molecules of many gases, however, do not have permanent dipole moments, yet they show van der Waals forces that are not much smaller than those found between molecules that do bear permanent moments. For instance, the heat of sublimation of HCl (dipole moment 1.03×10^{-18} e.s.u.) is 4850 calories per mole, and the heat of sublimation of xenon (which has no permanent dipole moment) is 3850 calories per gram atom. Although it is possible to explain these interactions qualitatively by assuming that the atoms possess quadrupole moments, a much more plausible explanation was given by London in 1930.

According to the quantum mechanical theory of atomic and molecular structure, the electrons in an atom are in continual motion even when the atom is in its ground state. As a result of these "zero point motions," all atoms possess rapidly fluctuating dipole moments. The electric field from the temporary dipole moment in one atom can induce a dipole moment in any neighboring atom, and as in the Debye induction effect, the relative orientations of the temporary moment and the moment that it induces in the other atom always lead to an attractive force between the two atoms. In other words, when two atoms are brought together, their electrons tend to move in such a way that the energy is lowered. London was able to show that the forces arising in this manner are large enough to account for the observed van der Waals forces in gases whose molecules do not have permanent dipole moments. He was also able to show that in many cases, even though permanent dipole moments are present, these forces are larger than the Debye and Keesom forces. These forces are found to be additive.

The forces described by London are usually referred to as *dispersion forces* or *London forces.*

B. The Interaction of a Pair of Dipoles

Consider a pair of positive charges e_a and e_b at a distance R apart. Let each positive charge have next to it an equal negative charge. Then the potential energy is

$$V = e_a e_b \left[\frac{1}{R} + \frac{1}{r_{12}} - \frac{1}{r_{1b}} - \frac{1}{r_{2a}} \right] - \frac{e_a^2}{r_{1a}} - \frac{e_b^2}{r_{2b}} \qquad \text{(B-1)}$$

where the r's are defined in Fig. 13-1. The first four terms in this expression represent the mutual interaction of two dipoles, and it is convenient to

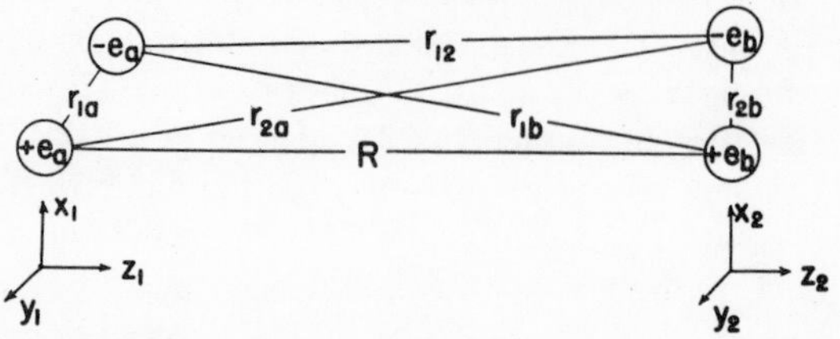

FIG. 13-1. Distances involved in the interaction of two dipoles.

derive an approximate expression for this interaction when R is large compared with r_{1a} and r_{2b}. We shall write

$$V_d = e_a e_b \left[\frac{1}{R} + \frac{1}{r_{12}} - \frac{1}{r_{1b}} - \frac{1}{r_{2a}} \right] \tag{B-2}$$

where V_d is this dipole interaction.

Let us set up two Cartesian coordinate systems whose z-axes run in the direction from e_a toward e_b and whose x- and y-axes are respectively parallel. Then if x_1, y_1, z_1 are the coordinates of charge no. 1 with e_a as the origin and x_2, y_2, z_2 are the coordinates of charge no. 2 with e_b as the origin, we see that

$$\begin{aligned} 1/r_{12} &= [(R + z_2 - z_1)^2 + (x_1 - x_2)^2 + (y_1 - y_2)^2]^{-1/2} \\ 1/r_{1b} &= [(R - z_1)^2 + x_1^2 + y_1^2]^{-1/2} \\ 1/r_{2a} &= [(R + z_2)^2 + x_2^2 + y_2^2]^{-1/2} \end{aligned}$$

Since it is assumed that R is much greater than any of the coordinates $x_1, x_2, y_1, \cdots$, each of these expressions can be expanded using the binomial theorem

$$(1 + x)^{-1/2} = 1 - \tfrac{1}{2}x + \tfrac{3}{8}x^2 - \cdots$$

giving

$$\frac{1}{r_{12}} = \frac{1}{R}\left[1 - \frac{z_2 - z_1}{R} - \frac{1}{2}\frac{(x_1 - x_2)^2 + (y_1 - y_2)^2 + (z_1 - z_2)^2}{R^2} + \frac{3}{2}\frac{(z_1 - z_2)^2}{R^2} + \text{terms in } R^{-3} + \cdots\right]$$

$$\frac{1}{r_{1b}} = \frac{1}{R}\left[1 + \frac{z_1}{R} - \frac{1}{2}\frac{x_1^2 + y_1^2 + z_1^2}{R^2} + \frac{3}{2}\frac{z_1^2}{R^2} + \text{terms in } R^{-3} + \cdots\right]$$

$$\frac{1}{r_{2a}} = \frac{1}{R}\left[1 - \frac{z_2}{R} - \frac{1}{2}\frac{x_2^2 + y_2^2 + z_2^2}{R^2} + \frac{3}{2}\frac{z_2^2}{R^2} + \text{terms in } R^{-3} + \cdots\right]$$

Substituting into (B-2) we find that

$$V_d = e_a e_b (x_1 x_2 + y_1 y_2 - 2z_1 z_2)/R^3 + (\text{terms in } R^{-4}) + \cdots \qquad \text{(B-3)}$$

Exercise. Show that Equation (A-1) is equivalent to the first term in (B-3).

C. The London Forces between a Pair of Oscillating Dipoles

Suppose that the two oppositely charged particles in a dipole interact with each other according to a harmonic oscillator potential

$$V = kr^2/2 = \tfrac{1}{2}k(x^2 + y^2 + z^2) \qquad \text{(C-1)}$$

where r is the distance between the two charges, and x, y, and z are the Cartesian coordinates of the negative charge relative to the positive charge. If two of these dipoles are separated by a distance R, the potential energy is (neglecting terms involving powers of $1/R$ greater than three)

$$V = \frac{1}{2}k(x_1^2 + y_1^2 + z_1^2 + x_2^2 + y_2^2 + z_2^2) + \frac{e^2}{R^3}(x_1x_2 + y_1y_2 - 2z_1z_2) \qquad \text{(C-2)}$$

where e is the charge on each end of the dipoles. We must now solve the Schroedinger equation for this potential. If we make the substitutions

$$\begin{aligned} X_+ &= (x_1 + x_2)/\sqrt{2} \qquad & X_- &= (x_1 - x_2)/\sqrt{2} \\ Y_+ &= (y_1 + y_2)/\sqrt{2} \qquad & Y_- &= (y_1 - y_2)/\sqrt{2} \\ Z_+ &= (z_1 + z_2)/\sqrt{2} \qquad & Z_- &= (z_1 - z_2)/\sqrt{2} \end{aligned} \qquad \text{(C-3)}$$

then the potential energy takes the form

$$\begin{aligned} V = {} & \frac{1}{2}\left(k + \frac{e^2}{R^3}\right)(X_+^2 + Y_+^2) + \frac{1}{2}\left(k - \frac{e^2}{R^3}\right)(X_-^2 + Y_-^2) \\ & + \frac{1}{2}\left(k - \frac{2e^2}{R^3}\right)Z_+^2 + \frac{1}{2}\left(k + \frac{2e^2}{R^3}\right)Z_-^2 \end{aligned} \qquad \text{(C-4)}$$

The Laplacian operator for the two particles is

$$\nabla_1^2 + \nabla_2^2 = \frac{\partial^2}{\partial X_+^2} + \frac{\partial^2}{\partial Y_+^2} + \frac{\partial^2}{\partial Z_+^2} + \frac{\partial^2}{\partial X_-^2} + \frac{\partial^2}{\partial Y_-^2} + \frac{\partial^2}{\partial Z_-^2} \qquad \text{(C-5)}$$

It is seen that if the new coordinates are used the Schroedinger equation for the system is completely separable into six equations: a pair of the form

$$-\frac{\hbar^2}{2m}\frac{\partial^2\psi}{\partial X_{\pm}^2}+\frac{1}{2}\left(k\pm\frac{e^2}{R^3}\right)X_{\pm}^2\psi=E_{\pm}\psi \tag{C-6a}$$

a similar pair of equations for Y_+ and Y_-, and a pair

$$-\frac{\hbar^2}{2m}\frac{\partial^2\psi'}{\partial Z_{\pm}^2}+\frac{1}{2}\left(k\mp\frac{2e^2}{R^3}\right)Z_{\pm}^2\psi'=E'_{\pm}\psi' \tag{C-6b}$$

The lowest eigenvalues of the pair of equations involving $X_{\pm}$ and $Y_{\pm}$ are (*cf.* pages 202*ff.*)

$$E_{\pm}=\frac{1}{2}\hbar\sqrt{\left(k\pm\frac{e^2}{R^3}\right)/m} \tag{C-7a}$$

and for the pair of equations involving $Z_{\pm}$, the lowest eigenvalues are

$$E'_{\pm}=\frac{1}{2}\hbar\sqrt{\left(k\mp\frac{2e^2}{R^3}\right)/m} \tag{C-7b}$$

Thus the lowest energy of the system of two interacting dipoles is

$$E=\frac{1}{2}\hbar\sqrt{\frac{k}{m}}\left[2\left(\sqrt{1+e^2/kR^3}+\sqrt{1-e^2/kR^3}\right)+\sqrt{1+2e^2/kR^3}+\sqrt{1-2e^2/kR^3}\right] \tag{C-8}$$

If $e^2/kR^3 \ll 1$ the square roots may be expanded by the binomial theorem

$$\sqrt{1+x}=1+\tfrac{1}{2}x-\tfrac{1}{8}x^2+\cdots$$

Setting $\nu_0=(\sqrt{k/m})/2\pi$, which is the natural frequency of the oscillators when they are far apart, we find that the linear terms in the binomial expansion of (C-8) cancel, and the energy of the ground state is

$$E=\frac{1}{2}h\nu_0\left[6-2\left(2\frac{e^4}{8k^2R^6}\right)-2\frac{4e^4}{8k^2R^6}\right]=3h\nu_0-\frac{3}{4}h\nu_0\frac{e^4}{k^2R^6} \tag{C-9}$$

Thus the energy of the two oscillators decreases as they are brought together, showing that they attract one another.

It is convenient to express this result in a somewhat different way, in order to compare it with the results of calculations to be made later in this chapter.

Let a single oscillator be placed in an electric field $\boldsymbol{E}$. Then a force $e\boldsymbol{E}$ acts on the two charges in the dipole, tending to separate them. The force resisting this displacement is the Hooke's law force, kx, where k is the constant appearing in Equation (C-1). Equating these two forces, we find for the displacement in the field $\boldsymbol{E}$,

$$\boldsymbol{r} = (e/k)\boldsymbol{E}$$

The dipole moment induced in this way is

$$\boldsymbol{m} = e\boldsymbol{r} = (e^2/k)\boldsymbol{E}$$

The polarizability α of an atom is defined as the ratio of the induced moment to the field, so we see that the force-constant k is directly related to the polarizability,

$$\alpha = e^2/k$$

Using this result in (C-9), we find that the energy of interaction of the two oscillators is

$$E'' = -\tfrac{3}{4}h\nu_0\alpha^2/R^6 \tag{C-10}$$

It should be noted that if one or more of the six normal modes of the two dipoles is not in its ground state, then the linear terms in the binomial expansion of (C-8) do not cancel, and the interaction energy is inversely proportional to the *cube* of R rather than to the sixth power. Furthermore, depending on which mode is excited, the interaction can lead to either attraction or repulsion.

D. London Forces between Two Hydrogen Atoms

Suppose that a pair of hydrogen atoms in the ground state is separated by a large distance, so that the perturbation is given reasonably well by the first term of V_d in Equation (B-3). This means that the interatomic distance R is much greater than the dimensions of the region in which the 1*s* wave function is large. Therefore the 1*s* wave functions of the two atoms do not overlap appreciably, and the discussion of the energy using the concept of an exchange energy, as in Chapters 11 and 12, does not lead to any interaction.

The problem is best considered from the point of view of perturbation theory. The zero-order wave functions of the system are

$$\psi_{mn} = \phi_{am}(1)\,\phi_{bn}(2) \tag{D-1}$$

where the letters a and b refer to the two nuclei, the numbers 1 and 2 refer to the two electrons, and m and n indicate the quantum state. The zero-order energies in atomic units are

$$E_{mn} = -\frac{1}{2}\left[\frac{1}{m^2} + \frac{1}{n^2}\right] \tag{D-2}$$

For the ground state the zero-order wave function is

$$\psi_{11} = 1s_a(1)\, 1s_b(2) \tag{D-3}$$

and the zero-order energy is $E_{11} = -1$. From Equation (D-14) of Chapter 5 the first order perturbation energy is

$$\begin{aligned} E' &= \int \psi_{11}{}^* V_d \psi_{11} d\tau \\ &= \frac{e^2}{R^3} \iint 1s_a(1) 1s_b(2) [x_1x_2 + y_1y_2 - 2z_1z_2] 1s_a(1) 1s_b(2) d\tau_1 d\tau_2 \\ &= \frac{e^2}{R^3} [(\bar{x})^2 + (\bar{y})^2 - 2(\bar{z})^2] \end{aligned} \tag{D-4}$$

where $\bar{x}$, $\bar{y}$, and $\bar{z}$ are the mean values of the coordinates of the electron in the ground state of a hydrogen atom, the nucleus being taken as the origin. Since the $1s$-state is spherically symmetric, the mean values of all of the coordinates vanish, and $E' = 0$. Thus it is necessary to go to second order perturbation theory in order to find the interaction of a pair of hydrogen atoms. From Equation (D-15) of Chapter 5, the second order energy for the ground state in atomic units is

$$\begin{aligned} E'' &= \sum_{m,n\neq 1} \frac{H'_{11,mn} H'_{mn,11}}{E_{11} - E_{mn}} \\ &= \frac{1}{R^6} \sum \frac{\left| \int \psi_{11}{}^* [x_1x_2 + y_1y_2 - 2z_1z_2]\, \psi_{mn} d\tau \right|^2}{-1 + \frac{1}{2}[(1/m^2) + (1/n^2)]} \\ &= -\frac{1}{R^6} \sum \frac{2m^2n^2 [\,|x_{1m}|^2 |x_{1n}|^2 + |y_{1m}|^2 |y_{1n}|^2 + 4 |z_{1m}|^2 |z_{1n}|^2]}{2m^2n^2 - (m^2 + n^2)} \end{aligned} \tag{D-5}$$

The sum was evaluated numerically by Eisenschitz and London, who found that it is approximately 6.47 if the distances are measured in atomic units. More careful treatments by Slater and Kirkwood and by Pauling and Beach lead to a value of 6.499 for this coefficient. Thus we have for the van der Waals interaction of two hydrogen atoms at large distances

$$E'' = -6.499/R^6 \tag{D-6}$$

the energy being measured in atomic units.

An approximate evaluation of the sum in (D-5) that gives an answer close to the correct one is easily performed in the following way. If either m or n is equal to unity, the corresponding term in the sum vanishes for the same reason that the first order perturbation vanishes ($x_{11} = \bar{x} = 0$, and similarly for y_{11} and z_{11}). Thus the minimum value of m and n in the sum is 2. But if $m = n = 2$, then $2m^2n^2/(2m^2n^2 - m^2 - n^2) = 4/3$, which is not far from unity. Larger values of m and n give values of this factor even closer to unity. The factor can therefore be neglected to a first approximation. Then

$$E'' \cong -\frac{1}{R^6}\,[(\sum |x_{1m}|^2)^2 + (\sum |y_{1m}|^2)^2 + 4(\sum |z_{1m}|^2)^2]$$

But it can be shown that (see Equation (A-38) of Chapter 16)

$$\sum |x_{1m}|^2 = \sum x_{1m}x_{m1} = (x^2)_{11} = \int 1s\, x^2\, 1s\, d\tau$$

$$\sum |y_{1m}|^2 = \sum y_{1m}y_{m1} = (y^2)_{11} = \int 1s\, y^2\, 1s\, d\tau$$

$$\sum |z_{1m}|^2 = \sum z_{1m}z_{m1} = (z^2)_{11} = \int 1s\, z^2\, 1s\, d\tau$$

Furthermore, since the $1s$ state is spherically symmetric and $x^2 + y^2 + z^2 = r^2$

$$(x^2)_{11} = (y^2)_{11} = (z^2)_{11} = \tfrac{1}{3}(r^2)_{11}$$

Thus

$$E'' \cong -\frac{1}{R^6}\,\frac{2}{3}\,[(r^2)_{11}]^2$$

Since the normalized $1s$ function is

$$1s = \frac{1}{\sqrt{\pi}}\, e^{-r}$$

and

$$(r^2)_{11} = (1/\pi) \iiint e^{-r}r^2 e^{-r}r^2 \sin\theta \, dr \, d\theta \, d\varphi$$

$$= 4 \int_0^\infty r^4 e^{-2r} \, dr$$

$$= 3$$

we find

$$E'' = -6/R^6 \tag{D-7}$$

which is within 10 per cent of (D-6).

E. London Forces between More Complex Atoms and Molecules

For a pair of atoms or molecules, A and B, each containing many electrons, the mutual interaction energy at large distances is

$$V = \frac{e^2}{R^3}[X_a X_b + Y_a Y_b - 2Z_a Z_b] \tag{E-1}$$

where X_a is the sum of the x-coordinates of all of the electrons in molecule A and similarly for X_b, Y_a, etc. The first order perturbation energy is

$$E' = \frac{e^2}{R^3}(\bar{X}_a \bar{X}_b + \bar{Y}_a \bar{Y}_b - 2\bar{Z}_a \bar{Z}_b) \tag{E-2}$$

where $\bar{X}_a$, etc., are the mean values of the sums of the coordinates in the respective molecules. Since the x-component of the dipole moment of molecule A is

$$m_{xa} = e\bar{X}_a$$

and similarly for the remaining coordinates of the two molecules, (E-2) may be written

$$E' = \frac{1}{R^3}(m_{xa} m_{xb} + m_{ya} m_{yb} - 2m_{za} m_{zb}) \tag{E-3}$$

which is equivalent to (A-1). This expression therefore merely gives the interaction of two permanent dipoles. It leads to forces of the Keesom type, and if there is no permanent dipole moment in the molecules, it must vanish.

Turning to the second order perturbation we find

$$E'' = \frac{e^4}{R^6} \sum_{m,n\neq 0} \frac{|X_{am0}X_{bn0} + Y_{am0}Y_{bn0} - 2Z_{am0}Z_{bn0}|^2}{E_{a0} + E_{b0} - E_{am} - E_{bn}} \tag{E-4}$$

where $$X_{am0} = \int \psi_{am}{}^* X_a \psi_{a0} d\tau,$$

ψ_{am} and ψ_{a0} being wave functions of molecule A, and similarly for the remaining quantities in the numerator of (E-4). If the energy is averaged for all orientations of the molecules, cross products involving XY, YZ, and XZ, vanish, and we obtain

$$E'' = \frac{e^4}{R^6} \sum_{m,n\neq 0} \frac{|X_{am0}|^2 |X_{bn0}|^2 + |Y_{am0}|^2 |Y_{bn0}|^2 + 4 |Z_{am0}|^2 |Z_{bn0}|^2}{E_{a0} + E_{b0} - E_{am} - E_{bn}} \tag{E-5}$$

If the molecules are spherically symmetric

$$|X_{am0}|^2 = |Y_{am0}|^2 = |Z_{am0}|^2 = \tfrac{1}{3} |R_{am0}|^2 \tag{E-6a}$$

$$|X_{bn0}|^2 = |Y_{bn0}|^2 = |Z_{bn0}|^2 = \tfrac{1}{3} |R_{bn0}|^2 \tag{E-6b}$$

so that

$$E'' = \frac{2}{3}\frac{e^4}{R^6} \sum_{m,n\neq 0} \frac{|R_{am0}|^2 |R_{bn0}|^2}{E_{a0} + E_{b0} - E_{am} - E_{bn}} \tag{E-7}$$

Equations (E-5) and (E-7) are London's general formulae for the dispersion forces. They show that the energy should vary as the inverse sixth power of the interatomic distance. Since the energies of the ground states, E_{a0} and E_{b0}, are always more negative than those of the excited states, E_{am} and E_{bn}, the interaction energy must be negative, indicating an attractive force if both interacting atoms are in their ground states.

The reason for the additivity of the dispersion forces is easy to understand. Suppose that the interaction of three atoms is being considered. According to the approach being used here, the energy would be made up of the sum of three pairs of direct perturbations of the atoms on each other (which is the effect given by London's equations, above), plus the perturbation of these perturbations due to the presence of the third atom. Since the perturbations themselves are small, the perturbations of the perturbations are negligible.

London transformed his equations into a more useful approximate form in the following way. In Chapter 16 we shall find that the polarizability of a molecule exposed to electromagnetic radiation of frequency ν is given by the "dispersion equation" (see Equation (A-19b) of Chapter 16)

$$a = \frac{2e^2}{3h} \sum \frac{\nu_{j0} \mid R_{j0} \mid^2}{\nu_{j0}^2 - \nu^2} \tag{E-8}$$

where $\nu_{j0} = (E_j - E_0)/h$ is the frequency corresponding to the transition from state 0 to state j. The polarizability in a static field ($\nu = 0$) is therefore

$$a = \frac{2e^2}{3} \sum \frac{\mid R_{j0} \mid^2}{E_j - E_0} \tag{E-9}$$

The experimental investigation of the frequency dependence of the polarizability of many substances reveals that the most important terms in (E-8) and (E-9) arise from states whose energies $E_j - E_0$ are not very different from the ionization energy, I, of the atom or molecule. Therefore we may write

$$a \cong \frac{2e^2}{3I} \sum \mid R_{j0} \mid^2 \tag{E-10}$$

For similar reasons the most important terms in the sum in (E-7) must arise from states for which $E_{am} + E_{bn} - E_{ao} - E_{b0}$ is close to the sum of the ionization energies of the two interacting molecules, $I_a + I_b$. Therefore we may write for the interaction energy

$$E'' \cong - \frac{2e^4}{3R^6} \frac{1}{I_a + I_b} \sum \mid R_{am0} \mid^2 \mid R_{bn0} \mid^2 \tag{E-11}$$

Combining (E-10) and (E-11)

$$E'' \cong - \frac{3I_aI_b}{2(I_a + I_b)} \frac{a_a a_b}{R^6} \tag{E-12}$$

If the two molecules are identical

$$E'' \cong -\tfrac{3}{4} I \alpha^2/R^6 \tag{E-13}$$

Equations (E-12) and (E-13) may be used to obtain a fairly reliable estimate of the magnitude of the dispersion forces between a pair of atoms or simple molecules.

The reason for calling these forces dispersion forces is now evident. The same factors that determine the dispersion of the polarizability (and therefore, as we shall see on pages 646*ff*., the dispersion of the index of refraction) also influence the attractive forces between molecules. We see that dispersion forces tend to be large between molecules of large polarizability. This accounts for the increase in interatomic attraction (as reflected by the increase in boiling point) as one goes through the series He, Ne, A, Kr, Xe, or through the series CH_3F, CH_3Cl, CH_3Br, CH_3I, since the polarizability of the atoms in a given group of the periodic table always increases with increasing atomic weight.

It is interesting to note that if we identify $h\nu_0$ in Equation (C-10) with I in Equation (E-13), these two equations become identical.

London made an interesting comparison of the orientation, induction, and dispersion forces between different pairs of simple molecules and atoms, using (E-13) to estimate the dispersion forces. The results of his calculations are shown in Table 13-1. It is found that only in strongly polar molecules of low polarizability, such as H_2O, is the dispersion force of secondary importance in comparison with the orientation force.

In all of the formulae in this chapter that relate interatomic forces with interatomic distance, the distance is raised to the inverse sixth power. The student should realize, however, that this term is merely the first term of an infinite series involving higher powers of the distance. Thus Margenau (*1*) included the terms involving higher powers of $1/R$ in (B-3), the expression for the interaction energy of two dipoles. Using an approach similar to that employed in deriving the approximate expression (D-7), he found that the interaction of two hydrogen atoms is given by

$$E'' \cong -(6/R^6)\,[1 + (22.5/R^2) + (236/R^4) + \cdots] \tag{E-14}$$

Margenau (*1*) also estimated that the interaction energy between a pair of helium atoms is (in atomic units)

$$E'' = -(1.6/R^6)\,[1 + (7.9/R^2) + (30/R^4) + \cdots] \tag{E-15}$$

TABLE 13-1

COMPARISON OF THE DIFFERENT TYPES OF VAN DER WAALS FORCES IN VARIOUS MOLECULES (FROM LONDON (*2*) AND MARGENAU (*1*)).

Substance	Polarizability (cm.^{3} × 10^{24})	Dipole moment (e.s.u. × 10^{18})	Ionization energy (electron volts)	Interaction Energy* (ergs × 10^{12})		
				Orientation effect (*1*)	Induction effect (*1*)	Dispersion effect (*2*)
H	0.667	0	13.6	0	0	6.10
He	0.205	0	24.5	0	0	1.49
Ne	0.39	0	25.7	0	0	7.97
A	1.63	0	17.5	0	0	69.5
Kr	2.46	0	14.7	0	0	129.
Xe	4.0	0	12.2	0	0	273.
H_2	0.81	0	14.5	0	0	11.4
N_2	1.74	0	15.8	0	0	57.2
O_2	1.57	0	13.6	0	0	39.8
Cl_2	4.60	0	12.7	0	0	321.
CO	1.99	0.1	14.3	0.003	0.057	67.5
HCl	2.63	1.03	13.4	18.6	5.4	111.
HBr	3.58	0.78	12.1	6.2	4.05	185.
HI	5.4	0.38	10.5	0.35	1.68	370.
NH_3	2.24	1.5	11.7	84.	10.	70.
H_2O	1.48	1.84	18	190.	10.	47.

* Coefficent K of $E = -K/R^6$ when R is in Å. and E is in ergs. Most of the values of K for the dispersion effect were obtained by Margenau from a modified formula obtained from (C-10).

(Note that because of their small polarizability, helium atoms attract one another much less strongly than do a pair of hydrogen atoms at the same distance.) At the interatomic distance observed in liquid helium, R is about 5.5 atomic units. At this distance the ratio of the second term in (E-15) to the first is 0.26 and the ratio of the third term to the first is 0.18. Therefore the series does not converge very rapidly when the atoms are this close together.

The calculation of the interaction of two molecules requires an entirely different approach when the molecules come sufficiently close to one another for their electron clouds to overlap appreciably. Under these conditions the first term in Equation (B-3) is inadequate for describing the interaction. Even more important, however, is the necessity of taking the Pauli principle and electron exchange into account. If there are electrons on one of the molecules that have about the same velocity as some of the electrons on the other molecule, but if their spins are not in the same direction, then a chemical bond may be formed. If the electron spins do point in the same direction, strong repulsive forces arise. These are the Pauli exclusion forces discussed on page 319.

REFERENCES CITED

1. H. Margenau, *Revs. Mod. Phys.*, **11**, 1 (1939).
2. F. London, *Trans. Faraday Soc.*, **33**, 8 (1937).

GENERAL REFERENCES

The two references cited above are also excellent general references on the subject of van der Waals forces.

PART V

SYSTEMS IN NON-STATIONARY STATES

Chapter 14

Time-Dependent Processes

In principle, the solutions of the steady-state Schroedinger equation,

$$H\psi = E\psi$$

tell us everything we can ever hope to know about atoms and molecules in states of definite energy. If we could somehow surmount the mathematical difficulties preventing the exact solution of this equation for most chemical systems, we would presumably be able to account precisely for a large portion of the facts of chemistry now known or remaining to be discovered. Certain important phenomena cannot, however, be handled in this way. For instance, the steady-state Schroedinger equation does not tell us how molecules can go from one energy state to another, nor does it say anything about the behavior of systems not in definite energy states. Questions of this kind arise when light or other radiation shines on or is emitted by matter, as well as in chemical reactions in which a change of electronic state occurs ("nonadiabatic" reactions). For the solution of these problems, we must use the more general time-dependent Schroedinger equation

$$H\Psi = -(\hbar/i)\,\partial\Psi/\partial t$$

A. The Behavior of Localized Clusters of Free Particles

Few problems other than those of the steady state are ordinarily attacked by seeking a closed solution of the time-dependent Schroedinger equation. There is, however, a simple and instructive example of this procedure, which we shall discuss before considering the more usual type of treatment.

Suppose that we try to locate a particle of mass m so that its x-coordinate has a value as close as possible to $x = x_0$ at some instant. What will happen if we then release the particle? Of course, because of experimental difficulties, the particle can never be placed precisely at a given point. Its initial location is therefore best described in terms of a probability function, $P(x, 0)$, and its location at any later time can be represented by the probability function $P(x, t)$. In the Schroedinger formulation, this function is directly related to the state function, $\Psi(x, t)$, of the particle

$$P(x,t) = |\Psi(x,t)|^2 \tag{A-1}$$

The problem, stated in quantum mechanical language, is to find $\Psi(x, t)$ if the initial wave function, $\Psi(x, 0)$, is known.

Another interpretation of $\Psi(x, 0)$ is that it represents a cluster or "packet" of free particles distributed along the x-axis in accordance with the initial experimental uncertainty. How will this cluster behave as time passes?

Without solving any equations at all, we can immediately guess something of the general nature of the solution. We know from the uncertainty principle that, once the location of a particle is specified within a range Δx, its momentum and hence its velocity becomes indefinite by an amount

$$\Delta v_x = \Delta p_x/m \simeq h/(m\Delta x)$$

Therefore all particles in the cluster cannot have the same velocity in the x-direction and the cluster must spread as time goes by. *According to quantum mechanics, if we place a particle at some point we cannot expect to find it precisely at that point later, even in the absence of any intervening disturbance.* This, of course, seems to be at variance with our general experience, for objects are often placed in definite places and found years later, apparently in the same place. Before we condemn quantum mechanics on these grounds, however, its conclusions must be expressed in quantitative terms, to see under what conditions the effect is sufficiently large to be observed.

In doing this, it is natural to assume that the initial distribution of the packet is a Gauss error function

$$P(x,0) = (a\sqrt{\pi})^{-1} \exp(-(x - x_0)^2/a^2) \tag{A-2}$$

corresponding to the state function

$$\Psi(x,0) = (a\sqrt{\pi})^{-\frac{1}{2}} \exp(-(x - x_0)^2/2a^2) \tag{A-3}$$

The length, a, is the standard deviation of x from its mean value, x_0, and is a measure of how precisely the particle's initial location is known. Since the eigenfunction of the ground state of a harmonic oscillator is a Gauss error function, we could obtain this initial distribution by attaching the particle to a spring and cooling the entire system to absolute zero, so that it would be in its lowest energy state, the particle being released from the spring at time $t = 0$.

Exercise. What would the force constant of the spring have to be if you wanted to locate a one-gram weight in this way with a standard deviation $a = 10^{-8}$ cm. from some definite point?

The state function (A-3) is not an eigenfunction of the Hamiltonian of a free particle, so Ψ must change its shape with time. According to the time-dependent Schroedinger equation

$$\partial\Psi/\partial t = -(i/\hbar)H\Psi = i\left[\frac{\hbar}{2m}\frac{\partial^2}{\partial x^2} - \frac{V}{\hbar}\right]\Psi \tag{A-4}$$

and for the free particle, $V = 0$ everywhere. The solution of this differential equation which gives (A-3) at $t = 0$ is (as can be verified by direct substitution)

$$\Psi(x,t) = (ab\sqrt{\pi})^{-\frac{1}{2}} \exp(-(x - x_0)^2/2a^2b) \tag{A-5a}$$

where

$$b = 1 + (i\hbar t/ma^2) \tag{A-5b}$$

This gives for the probability function at time t

$$P(x,t) = \left[\frac{1}{a\sqrt{\pi}(1 + \hbar^2t^2/m^2a^4)^{1/2}}\right] \exp\left[-\frac{(x - x_0)^2}{a^2(1 + \hbar^2t^2/m^2a^4)}\right] \tag{A-6}$$

The distribution therefore remains at all times Gaussian, but spreads out more and more, as expected from our qualitative discussion. The distribution spreads rapidly at times greater than $t = ma^2/\hbar$.

Exercises. (1) Suppose that an object weighing one gram is placed near a point, so that the standard deviation of its distance from the point (as measured by a in Equation (A-2) is 1 Å. How long will it take for the uncertainty to double? How long if $a = 1$ cm.? How long if the particle is an electron?

(2) If beads weighing 0.1 gram were placed along a line on a smooth table in the Great Pyramid in 2600 B.C. with a Gaussian distribution in their distance from the line having $a = 0.001$ cm., how much would the distribution parameter have changed by the year 1950 A.D. if no earthquakes or other disturbances had taken place in the intervening period? Does this result contradict the general experience of mankind with macroscopic bodies? How long would it take for the variance to double if the "beads" were electrons?

(3) The wave packet belonging to a group of free particles is known to have the form $\Psi_0(x)$ at $t = 0$. This packet can be described in terms of the eigenfunctions of a free particle, $\exp(ikx)$, by means of the Fourier integral (see page 33)

$$\Psi_0(x) = \int_{-\infty}^{\infty} A(k) \exp(ikx)\, dk \tag{A-7}$$

where

$$A(k) = (1/2\pi) \int_{-\infty}^{\infty} \Psi_0(x) \exp(-ikx)\, dx \tag{A-8}$$

Show that at any other time the wave packet will have the form

$$\Psi(x, t) = \int_{-\infty}^{\infty} A(k) \exp\left(ikx - \frac{i\hbar k^2 t}{2m}\right) dk \tag{A-9}$$

Obtain the solution given by Equation (A-5) from Equation (A-8) and (A-9), using Equation (A-3) for $\Psi_0(x)$.

(4) Equation (A-7) represents a Fourier analysis of the cluster $\Psi_0(x)$ in terms of the wave functions e^{ikx} of the free particles having a momentum $P_x = k\hbar$ and a de Broglie wavelength $\lambda = 2\pi/k$; the "coefficients" $A(k)$ give the contribution of each wavelength to $\Psi_0(x)$. What does $|\Psi_0(x)|^2$ look like for a packet made up of $A(k) = 1$ for values of k lying between k_1 and k_2, and $A(k) = 0$ elsewhere? How would the shape of such a packet change with time?

B. Perturbation Theory for Time-Dependent Processes

a. General theory

One often encounters the following type of problem in chemistry. An atom or molecule is known to be in a certain electronic state. The system is then disturbed (i.e., "perturbed"), say by shining light on it or by bringing a second atom or molecule close to it for some time. When the disturbance is removed, the system will not necessarily be found in its original state. One then wants to know which states may be occupied and the probability of finding the system in each state. The absorption and emission of light and the important related problem of the dependence of color on chemical constitution are examples that involve these questions. We shall now outline the quantum mechanical method of dealing with them.

Let $\phi_0(x)$ be the wave function of the system in its original state, having a definite energy E_0. Then $\phi_0(x)$ is an eigenfunction of the energy operator of the unperturbed system. This operator will be called H_0 and the perturbation will be denoted by H', H_0 being independent of the time, whereas H' may depend on the time. At any time t after applying the perturbation, the state function of the system, $\Psi(x, t)$, can be expressed in the usual way in terms of the complete set of steady state eigenfunctions, ϕ_k, of H_0

$$\Psi(x,t) = \sum a_k(t)\, \phi_k(x) \tag{B-1}$$

If the perturbation is suddenly removed at time τ, then the probability of finding the system in the state k is the square of the absolute value of the coefficient $a_k(\tau)$.

This interpretation is made plausible by calculating the average energy of the system after removing the perturbation (so that the energy operator is again H_0). If E_k is the energy of the unperturbed system in the kth eigenstate of H_0, and if the probability of occurring in the kth state is $|a_k|^2$, then the average energy immediately after removing the perturbation should be (cf. Equation (E-2a) on page 28)

$$E_{\text{av}} = \sum |a_k|^2 E_k \left(\sum |a_k|^2\right)^{-1} \tag{B-2}$$

The average-value theorem of quantum mechanics, on the other hand, gives the identical result

$$\begin{aligned} E_{\text{av}} &= \int \Psi^* H_0 \Psi\, dx \left(\int \Psi\Psi^*\, dx\right)^{-1} \\ &= \sum_{j,k} a_k^* a_j \int \phi_k^* H_0 \phi_j dx \left(\sum_{j,k} a_k^* a_j \int \phi_k^* \phi_j\, dx\right)^{-1} \\ &= \sum |a_k|^2 E_k \left(\sum |a_k|^2\right)^{-1} \end{aligned} \tag{B-2a}$$

thus supporting the proposed interpretation. (The cross products $a_k^* a_j$ in (B-2a) disappear because of the orthogonality of the ϕ_k.)

A similar result will be found on calculating the average value of any property for whose operator the ϕ_k are eigenfunctions, so that each state has a definite value of the property. On the other hand, properties for which the ϕ_k are not eigenfunctions do not give such a simple result. If α is the operator for such a property, then the integral

$$\int \phi_k^* \alpha\, \phi_j dx$$

usually does not vanish for two different states, and cross products of the a_k appear in the numerator of (B-2a). The states of the system do not contribute independently to properties of this latter kind.

Evidently the perturbation H' can be said to induce transitions from the initial state to other states of the unperturbed system; the squares of the coefficients $|a_k|^2$ give the probabilities of these transitions. *The problem of evaluating the effects of time-dependent perturbations is to find out how these coefficients change with time.* This is readily determined by introducing (B-1) into the time-dependent Schroedinger equation†

$$-(\hbar/i)\partial\Psi/\partial t = (H_0 + H')\Psi \tag{B-3}$$

$$-(\hbar/i)\sum(da_k/dt)\phi_k = \sum a_k H_0 \phi_k + \sum a_k H' \phi_k \tag{B-4}$$

Multiplying both sides by ϕ_j^* and integrating over all space we obtain

$$-(\hbar/i)da_j/dt = a_j E_j + \sum_k a_k H'_{jk} \tag{B-5}$$

where

$$H'_{jk} = \int \phi_j^* H' \phi_k \, d\tau \tag{B-6}$$

When repeated for all eigenfunctions ϕ_j, this leads to a set of linear first order differential equations which can be somewhat simplified by defining new coefficients b_j through the relation

$$a_j = b_j(t) \exp[-i(E_j + H'_{jj})t/\hbar] \tag{B-7}$$

and new energies

$$E_{kj} = (E_k + H'_{kk}) - (E_j + H'_{jj}) \tag{B-8}$$

giving

$$db_j/dt = (-i/\hbar) \sum_{k \neq j} b_k H'_{jk} \exp(-iE_{kj}t/\hbar) \tag{B-9}$$

If the system is known to be in the definite eigenstate ϕ_0 initially, we can say that at $t = 0$, $b_0 = 1$ and all of the other b's are zero. As long as all b's can be neglected in comparison with b_0, the only important term in the

† In obtaining Equation (B-4) it is assumed that the operator H′ commutes with the coefficients $a_k(t)$. This will be true for all of the perturbations with which we shall be concerned in the rest of this book.

summation in Equation (B-9) is the term having $k = 0$. We can therefore write

$$b_0 = 1 \; ; \; b_j = (-i/\hbar) \int_0^t H'_{jj} \exp(-iE_{00}t/\hbar)\, dt \qquad (j \neq 0) \quad \text{(B-10a)}$$

$$a_0 = \exp(-i(E_0 + H'_{00})t/\hbar)$$
$$a_j = (-i/\hbar) \exp[-i(E_j + H'_{jj})t/\hbar] \int_0^t H'_{j0} \exp(-iE_{0j}t/\hbar)dt \qquad (j \neq 0) \quad \text{(B-10b)}$$

$$|a_0|^2 = 1$$
$$|a_j|^2 = |b_j|^2 = (1/\hbar^2) \left| \int_0^t H'_{j0} \exp(-iE_{0j}t/\hbar)dt \right|^2 \qquad (j \neq 0) \quad \text{(B-11)}$$

This result is a first approximation. It is physically equivalent to taking account of only those transitions to the excited state ϕ_j that occur directly from the initial state ϕ_0. As time passes, however, and other states begin to acquire populations, transitions from these states to ϕ_j must also be taken into account (see Fig. 14-1). This leads to a second approximation, which

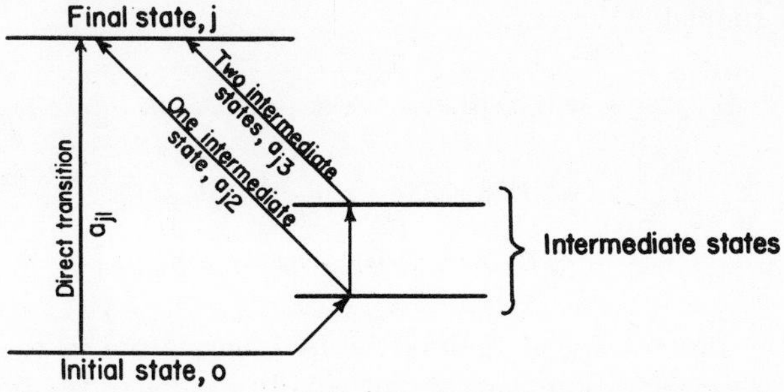

FIG. 14-1. Role of intermediate states in the time-dependent perturbation theory.

may be computed in the following way. Let us denote the a_j given by (B-10b) as a_{j1} and write

$$a_j = a_{j1} + a_{j2} \qquad \text{(B-12)}$$

where a_{j1} is the contribution to a_j due to direct transitions from the initial state to state j, and a_{j2} is the contribution of indirect transitions involving only one other intermediate state. It is found that

$$(-\hbar/i)\, da_{02}/dt = \sum a_{k1} H'_{0k} \qquad \text{(B-12a)}$$

$$(-\hbar/i)\, da_{j2}/dt = a_{j1}(E_j + H'_{jj}) + \sum_{\substack{k \neq 0 \\ k \neq j}} a_{k1} H'_{jk} \qquad (j \neq 0) \quad \text{(B-12b)}$$

These equations are in a readily soluble form.

Exercise. Prove Equations (B-12a) and (B-12b).

This procedure can be extended if necessary to take account of third and higher approximations resulting from transitions involving two or more intervening states. The relationship of this procedure to the perturbation theory for steady states considered in Chapter 5 (page 175) will become clearer after we consider a few examples.

b. Sudden application of a constant perturbation

Suppose that the perturbation, once applied, is independent of the time, such as a uniform electric field $\boldsymbol{E}$ suddenly applied to a hydrogen atom. Then if the position of the nucleus is fixed

$$H' = -|\boldsymbol{E}|ex \tag{B-13}$$

where e is the charge on the electron and x is its position coordinate in the direction of the field. We find

$$b_j = (i/\hbar)\,|\boldsymbol{E}|ex_{j0}\int_0^t \exp\,(-iE_{0j}t/\hbar)\,dt \tag{B-14a}$$

$$= -(|\boldsymbol{E}|ex_{j0}/E_{0j})\,[\exp(-\,iE_{0j}t/\hbar) - 1]$$

$$|\,a_j\,|^2 = (2|\boldsymbol{E}|^2e^2\,|\,x_{j0}\,|^2/E_{0j}{}^2)\,[1 - \cos E_{0j}t/\hbar] \tag{B-14b}$$

Evidently the probability of finding the hydrogen atom in the jth excited state under these conditions varies sinusoidally from zero to a maximum of $4|\boldsymbol{E}|^2e^2\,|x_{j0}\,|^2/E_{0j}{}^2$ with frequency E_{0j}/h.

If $|\boldsymbol{E}|ex_{j0}$ is small compared with $(E_j - E_0)$, then $|\,a_j\,|^2$ will always be small and the above will remain a good approximation for a long time. The initial state may, however, be degenerate, in which case some of the E_{j0} will be zero. Equation (B-14) then gives

$$b_j = (i/\hbar)\,|\boldsymbol{E}|ex_{j0}t \tag{B-15a}$$

$$|\,a_j\,|^2 = (|\boldsymbol{E}|^2e^2\,|\,x_{j0}\,|^2/\hbar^2)t^2 \tag{B-15b}$$

and the probability of a transition will increase continuously instead of oscillating. Furthermore, the probability of a transition does not increase linearly with the time, but rather as the square of the time. That is, the longer a system is subject to a constant perturbation after it has been established in a definite state, the greater is its rate of transition to another state of equal energy, if one is available. This is a quite different law of transition from the more familiar first-order law found in radioactive decay and in unimolecular reactions, where the transition probability is independent of the time.

c. Gradual application of a perturbation. The adiabatic principle

Let an electric field $\boldsymbol{E}_0$ be established in the vicinity of a hydrogen atom by increasing the field at a constant rate, $d\boldsymbol{E}/dt = \alpha \boldsymbol{E}_0$, over a time interval τ such that $\alpha\tau = 1$, α being the rate of application of the perturbation. Then, while the field increases, the perturbation operator is

$$H' = -|\boldsymbol{E}_0|ex\alpha t \tag{B-16}$$

and at the end of this interval

$$b_j = \frac{i}{\hbar}\,|\boldsymbol{E}_0|\,ex_{j0}\alpha \int_0^{1/\alpha} te^{-iE_{0j}t/\hbar}\,dt \tag{B-17a}$$

$$|a_j|^2 = \frac{|\boldsymbol{E}_0|^2e^2\,|x_{j0}|^2}{E_{0j}{}^2}\left\{1 - 2\,\frac{\hbar\alpha}{E_{0j}}\sin\frac{E_{0j}}{\hbar\alpha} + 2\,\frac{\hbar^2\alpha^2}{E_{0j}{}^2}\left[1 - \cos\frac{E_{0j}}{\hbar\alpha}\right]\right\} \tag{B-17b}$$

If the field is applied very slowly, so that α is very small, and if $E_{0j} \neq 0$, the second and third terms on the right in Equation (B-17b) can be neglected and we have at the end of the build-up period

$$|a_j|^2 = \frac{|\boldsymbol{E}_0|^2e^2\,|x_{j0}|^2}{E_{0j}{}^2} \qquad (t = 1/\alpha) \tag{B-17c}$$

Suppose that subsequent to the very long build-up period (i.e., for $t > 1/\alpha$) the electric field is held constant at $\boldsymbol{E}_0$. Then if $\boldsymbol{E}_0$ is small, so that the a_j's are all small and we can use Equation (B-10a), we find

$$\begin{aligned} b_j(t) &= b_j(1/\alpha) + \frac{i}{\hbar}\,|\boldsymbol{E}_0|ex_{j0}\int_{1/\alpha}^{t} e^{-iE_{0j}t/\hbar}\,dt \\ &= -\frac{|\boldsymbol{E}_0|ex_{j0}}{E_{0j}}\,e^{-iE_{0j}t/\hbar} \end{aligned}$$

$$|a_j|^2 = \frac{|\boldsymbol{E}_0|^2e^2\,|x_{j0}|^2}{E_{0j}{}^2} \qquad (t > 1/\alpha) \tag{B-17d}$$

where $b_j(1/a)$ is the value of b_j given by Equation (B-17a). Equation (B-17d) is the same as Equation (B-17c). Thus, if the perturbation is slowly applied, the probability $|a_j|^2$ of finding a system in a given state on sudden removal of the perturbation is a constant independent both of the time and of the rate of application of the perturbation, provided, however, that the energy of this state is different from that of the initial state (that is, provided $E_{0j} \neq 0$).

If we deal with states having the same energy as the initial state, so that $E_{0j} = 0$, then as long as the degeneracy remains

$$b_j = \frac{i}{\hbar} |\mathbf{E}_0| e x_{j0} a \int_0^{1/a} t\, dt = \frac{i}{2\hbar} \frac{|\mathbf{E}_0| e x_{j0}}{a}$$

$$|a_j|^2 = \frac{|\mathbf{E}_0|^2 e^2 |x_{j0}|^2}{4\hbar^2 a^2} \qquad (t = 1/a) \qquad \text{(B-17e)}$$

The probability of a transition now appears to depend on the rate of application of the perturbation. But if the two states are to have the same energy throughout the application of the perturbation, the first order corrections to their energies, H'_{jj} and H'_{00}, must be the same. It can be shown that if this is the case, then $H'_{j0} = |\mathbf{E}_0|^2 e^2 |x_{j0}|^2 = 0$, the right hand side of Equation (B-17e) vanishes, and no transition occurs at all. If the degeneracy is removed by the perturbation and $H'_{j0} \neq 0$, the situation is more complicated.

Exercise. Work out the transition probability for a slowly applied perturbation when

$$E_{j0}\,(= H'_{jj} - H'_{00}) = at\Delta H$$

that is, where the degeneracy is removed by the perturbation and the splitting of the energy levels increases linearly with the time. Here a has the same meaning as in Equation (B-16) and ΔH is the total final splitting.

It is interesting to note that if the initial state is nondegenerate and if the perturbation has been applied slowly, the wave function in the presence of the perturbation is

$$\Psi(x, t) = \left[\phi_0(x) - \sum_{j \neq 0} \frac{|\mathbf{E}_0| e x_{j0}}{E_{0j}} \phi_j(x)\right] e^{-iE_{00}t/\hbar} \qquad \text{(B-18)}$$

This wave function is, however, merely the first order approximation to the solution of the *steady state* Schroedinger equation for the *perturbed* system, expressed in terms of the unperturbed wave functions according to the first order perturbation theory

$$(H_0 + H')\phi = E\phi$$

$$\phi = \phi_0 + \sum \left(\frac{H'_{j0}}{E_0 - E_j}\right)\phi_j \tag{B-19}$$

$$E = E_0 + H'_{00}$$

where $H'_{j0} = -|\mathbf{E}_0|\, ex_{j0}$

If the second approximation (Equation (B-12)) is carried through for this case, one finds that $\Psi(x, t)$ has just the form given by the ordinary second order perturbation theory, with

$$\Psi(x,t) = \phi(x)e^{-iEt/\hbar}$$

$$\phi(x) = \phi_0 + \sum_{j\neq 0} \frac{H'_{j0}}{E_0 - E_j}\phi_j + \sum_{k\neq 0}\left[\sum_{j\neq 0} \frac{H'_{jk}H'_{k0}}{(E_0 - E_k)(E_0 - E_j)} - \frac{H'_{00}H'_{k0}}{(E_0 - E_k)^2}\right]\phi_k$$

$$E = E_0 + H'_{00} + \sum_{j\neq 0} \frac{H'_{0j}H'_{j0}}{E_0 - E_j} \tag{B-20}$$

This leads to an interesting interpretation of the ordinary perturbation theory (as used with the steady-state Schroedinger equation). This theory appears to be valid only for perturbations that are very slowly applied. As the perturbation is applied it evidently causes transitions between the states of the unperturbed system in the sense that sudden removal of the perturbation may leave the system in different states from that of the original system. Thus the first order perturbation theory gives us the results of direct transitions from the initial state, second order perturbation theory gives the result of transitions in which one other state intervenes, and so on for higher order perturbations.

A further important conclusion can be made from this analysis. In Equation (B-1) we have based our description of the perturbed system on the eigenfunctions of the unperturbed system. Using this frame of reference the perturbation appears to "induce" transitions between the states of the *unperturbed* system (that is, the eigenstates of the Hamiltonian H_0). Suppose, however, that we choose to describe the perturbed state function, $\Psi(x, t)$, in terms of the steady state eigenfunctions of the *perturbed* Hamiltonian, $H_0 + H'$. Since the wave function ϕ in Equations (B-19) and (B-20) is, at least to a second approximation, a steady state eigenfunction of $H_0 + H'$ (expanded in terms of the unperturbed eigenfunctions), it would appear that a slowly applied perturbation does not induce transitions between eigenstates of the *perturbed* system. In other words, if a system is placed originally in a definite nondegenerate state and then a perturbation is slowly built up, the system will always be found in a particular eigenstate of the perturbed Hamiltonian. If it is possible to go from one system to another by applying a continuous perturbation, then each eigenstate of the one system can be related to a definite eigenstate of the other. This is the *adiabatic principle of Ehrenfest: a system will always remain in a definite quantum level if its surroundings are changed sufficiently slowly.* It is analogous to the principle of the continuity of normal modes in the theory of vibrations. The proof given here is valid only to the approximation of second order perturbation theory. A more general proof will be found in Tolman (*1*, pp. 412–4).

Exercises. (1) It is possible to convert a hydrogen atom into an isotropic harmonic oscillator by applying to the hydrogen atom (having the Hamiltonian

$$H_0 = -\frac{\hbar^2}{2m_e}\nabla^2 - \frac{e^2}{r}$$

the perturbation

$$H' = \lambda\left(\frac{e^2}{r} + \tfrac{1}{2}kr^2\right)$$

where λ varies from zero (corresponding to the hydrogen atom) to unity (corresponding to the oscillator). It should therefore be possible to correlate the states of the hydrogen atom with those of the isotropic oscillator. Draw a schematic diagram in which λ is plotted against the energy showing the correspondence of the two systems (see Exercise 2 on page 207 for the states of the oscillator and Section C of Chapter 6 for the states of the hydrogen atom).

(2) Do the same for the states of a hydrogen atom and those of an electron in (a) a cubical box and (b) a spherical box.

d. Condition for the validity of the adiabatic principle. The Born-Oppenheimer principle. Adequacy of the steady-state formulation in chemical kinetics

Let us now investigate the conditions for the validity of the adiabatic principle: how rapidly can one apply a perturbation without forcing transitions to new states? Referring back to Equation (B-17b) we see that if for all possible states the rate of perturbation, a, fulfills the condition

$$a \ll E_{0j}/\hbar \qquad \text{(B-21)}$$

then the perturbed state will have the form of Equation (B-18), and the system will remain in an eigenstate. Equation (B-21) is therefore the condition for the validity of the adiabatic principle.

The potential energy curves of molecules are always calculated assuming that the nuclei are stationary. Can we be sure that when the nuclei are allowed to move, as they do in molecular vibrations, transitions to new electronic states will not result? It is easy to show from (B-21) and the known properties of molecules that such transitions will not be very likely: *the nuclear motions in ordinary molecular vibrations are so slow that they do not affect the electronic states of molecules.* This is the *Born-Oppenheimer principle.*

Exercise. Show that the Born-Oppenheimer principle is plausible using the following considerations. The nuclei in a molecule invariably oscillate. (If the molecule is at 0° K., the oscillations will result from the zero-point energy; if the molecule is not at 0° K. they may also arise from the thermal energy.) These oscillations must disturb the motions of the electrons in the molecule. Thus the nuclear motions are equivalent to a variable perturbation applied at the rate of α times per second, where α is the frequency of the nuclear oscillations. The energy differences between electronic states, E_{0j}, are typically of the order of several electron volts, and most nuclear vibrations have frequencies lying in the infra-red below 2000 cm.$^{-1}$ = 6×10^{13} sec.$^{-1}$. Find and compare the orders of magnitude of α and $E_{0j}/\hbar$.

Because of the Born-Oppenheimer principle, the changes in the energies of the electrons in molecules that are brought about by nuclear motions are essentially adiabatic. In most problems involving nuclear motions, the behavior of the electrons can therefore be handled by means of the Schroedinger method for steady states. This includes the important problem of chemical reactions, which differ from molecular vibrations only in the relatively drastic movements of nuclei that they entail. Strange as it might otherwise seem, the quantum mechanical study of the rates of chemical reactions therefore rarely necessitates the use of the time-dependent Schroedinger method in spite of the fact that it is concerned with a process occurring in time. As was first pointed out by London (*2*), the electronic motions involved in chemical reactions are adiabatic and the nuclear motions are very nearly classical. In computing the rate of a reaction, one has merely to determine the potential energies of all configurations of the atoms involved (that is, find the so-called "potential energy surface" of the reaction) assuming that the nuclei are stationary in each configuration. The actual motions of the nuclei then follow the rules of classical mechanics, subject to the potential energy due to the electrons given by the adiabatic surface. Eyring has been particularly active in pursuing this program (*3*).

C. Resonance and the Rate of Electronic Tautomerism

Suppose that to a first approximation a system can exist in two states having the same energy, E_0, represented by the wave functions $\phi_a \exp(-iE_0t/\hbar)$ and $\phi_b \exp(-iE_0t/\hbar)$. An example might be the H_2^+ molecular ion at moderate internuclear distances for which an electron can occur in a 1*s*-orbital on either of the two nuclei. Another example might be the benzene molecule with its two Kekulé structures. Then we know that to a second and closer approximation the steady-state wave functions of the system are resonance hybrids of the two forms

$$\Psi_1 = (\phi_a + \phi_b) \exp[-i(E_0 - \delta)t/\hbar] \tag{C-1a}$$

$$\Psi_2 = (\phi_a - \phi_b) \exp[-i(E_0 + \delta)t/\hbar] \tag{C-1b}$$

where δ is the resonance energy. If H is the Hamiltonian of the system, then we know that approximately

$$H\Psi_1 \cong (E_0 - \delta)\Psi_1 \tag{C-2a}$$

$$H\Psi_2 \cong (E_0 + \delta)\Psi_2 \tag{C-2b}$$

It should always be possible, at least in principle, to set up systems of this kind so that at some instant they are known to be in a particular one of the two forms, ϕ_a or ϕ_b. For instance, using some kind of "atomic forceps" whose construction we need not specify, a proton could be held in one's left hand and a hydrogen atom in one's right hand, and the two particles could then be brought instantaneously to within a few Ångstrom units of one another to form an H_2^+ molecule in which it is definitely known that the electron is initially on the right hand proton. What will happen subsequently?

We first find a wave function having the form ϕ_a at $t = 0$ that obeys the time dependent Schroedinger equation. It is easy to see that, to the extent that Equations (C-1) and (C-2) are valid, these conditions are fulfilled by the function

$$\Psi = \tfrac{1}{2}(\Psi_1 + \Psi_2) \tag{C-3}$$

Exercise. Show that this wave function satisfies the time-dependent Schroedinger equation if (C-1) and (C-2) are true.

Equation (C-3) can be written in the form

$$\Psi = [\phi_a \cos(\delta t/\hbar) + i\phi_b \sin(\delta t/\hbar)] \exp(-iE_0t/\hbar) \tag{C-4}$$

Taking the square of the absolute value, in order to find the probability distribution of the electron, we find

$$| \Psi |^2 = | \phi_a |^2 \cos^2(\delta t/\hbar) + | \phi_b |^2 \sin^2(\delta t/\hbar) \tag{C-5}$$

Evidently if the system is prepared initially in one of its two resonance forms, it will subsequently oscillate back and forth between the two forms with a frequency $2\delta/h$. In this situation we can imagine that resonance leads to an "electronic tautomerism" between the two structures, each structure having a lifetime of $h/4\delta$. It is important to realize, however, that this tautomerism can occur only if the system is prepared initially in a single one of the structures and that a system prepared in this way is not in an eigenstate. Except in certain special circumstances (one of which will be described in the next section), molecules are not prepared in a single resonance structure. Therefore it is quite incorrect to say that molecules in their ground state (or in any other eigenstate) actually tautomerize in this way; ordinary benzene, for instance, cannot in any sense be said to tautomerize between the two Kekulé structures.

The situation is analogous to the motion of a pair of coupled oscillators, which has been described in Exercise 4 on page 149. Each oscillator is analogous to a structure, the normal modes are analogous to eigenstates, and the coupling is analogous to the resonance energy. If one of the oscillators is set in motion, this does not constitute a normal mode; the energy will subsequently shuttle back and forth ("tautomerize") between the oscillators at a rate proportional to the strength of the coupling. If the system is in one of its normal modes, however, both oscillators will move with a constant amplitude and there will be no "tautomerism" of energy from one oscillator to the other.

Resonance forms are therefore in a sense mathematical abstractions, which are not usually observed in nature, but which are useful in the description of atomic and molecular systems. This does not mean, however, that the idea of resonance does not have scientific stature; indeed, all of science is constructed of fictions of this sort—that is, of concepts one or more steps removed from direct sensual contact with the outside world.

A certain amount of care must also be used in speaking of the properties of individual resonance forms. Suppose that we tried to measure the energy E_0 of the system of which we have been speaking while it was in one of its resonating forms, say ϕ_a. The measurement would have to be completed in less time than $\Delta t = h/4\delta$, since this is the lifetime of the form. Because of the Heisenberg principle, this introduces an uncertainty into the energy equal to $\Delta E \approx h/\Delta t = 4\delta$. But the resonance hybrids themselves differ in energy by only 2δ. Therefore, within the limits of experimental error, the individual resonating forms would be found to have the same energy as the resonance hybrids. Of course it is always possible to calculate an energy E_0 corresponding to the particular form ϕ_a by evaluating the integral

$$E_0 = \int \phi_a{}^* H\phi_a \, dx \tag{C-6}$$

This energy, however, cannot be measured directly; it is again a conceptual (though entirely respectable) aid to the understanding of the energetics of atomic and molecular systems.

D. Nonadiabatic Transitions

a. The "crossing" of potential energy surfaces

Let us compare the interaction of a sodium atom and a chlorine atom with that of a sodium ion and a chlorine ion. The ionization potential of sodium is 5.14 volts and the electron affinity of chlorine is 3.72 volts. Therefore if the two atoms are infinitely far apart, the system Na(g) + Cl(g) is more stable by 5.14 – 3.72 = 1.42 electron volts than the system $Na^+(g)$ + $Cl^-(g)$. On the other hand, if instead of being infinitely far apart the two ions are a distance r from each other, they will be stabilized by the Coulombic energy, e^2/r. If r is measured in Ångstrom units this energy is $14.42/r$ electron volts. Below 10.15 Å. the Coulombic stabilization of the ions becomes greater than 1.42 volts. Since the un-ionized atoms probably do not interact appreciably until they are less than five or six Ångstroms apart (when they begin to attract due to the formation of a covalent bond), the potential energy curves of the two systems must cross. At distances less than about 10 Å. the ionized state must be more stable than the un-ionized one. At very small interionic distances the core electron orbitals begin to overlap, causing a repulsion, so that the energy of the ion pair passes through a minimum. The equilibrium interionic distance in crystalline NaCl being 2.81 Å., we should expect the minimum in energy to occur at approximately this distance for the ion pair. These relationships are indicated in Fig. 14-2.

Because of the possibility of resonance between the ionic and covalent structures, however, the curves in Fig. 14-2 do not represent the true potential energy curves of the NaCl molecule. In order to obtain a better approximation to the true curves we first write wave functions for the covalent and ionic states as follows

$$\chi_{\text{ion}} = \phi_{\text{Cl}}(1)\,\phi_{\text{Cl}}(2)\,(\alpha_1\beta_2 - \alpha_2\beta_1) \tag{D-1a}$$

$$\chi_{\text{cov}} = [\phi_{\text{Na}}(1)\,\phi_{\text{Cl}}(2) + \phi_{\text{Na}}(2)\,\phi_{\text{Cl}}(1)]\,(\alpha_1\beta_2 - \alpha_2\beta_1) \tag{D-1b}$$

The functions ϕ_{Na} and ϕ_{Cl} are 3*s*- and 3*p*-orbitals on Na and Cl, respectively, while α and β are electron spin functions and the numbers 1 and 2 refer to the two electrons involved in the bonding. Figure 14-2 gives the potential energy curves corresponding to the two structures, χ_{ion} and χ_{cov}. The true wave function of the NaCl molecule is a hybrid of the above two

$$\psi = a\chi_{\text{ion}} + b\chi_{\text{cov}} \tag{D-2}$$

the coefficients a and b being functions of the interatomic distance, whose values, along with that of the corrected energy, may be found by means of the Rayleigh-Ritz method. Without going into the details of this method, we can foresee that it will result in the following behavior. The potential energy curves of the two true molecular states derived from χ_{ion} and χ_{cov}

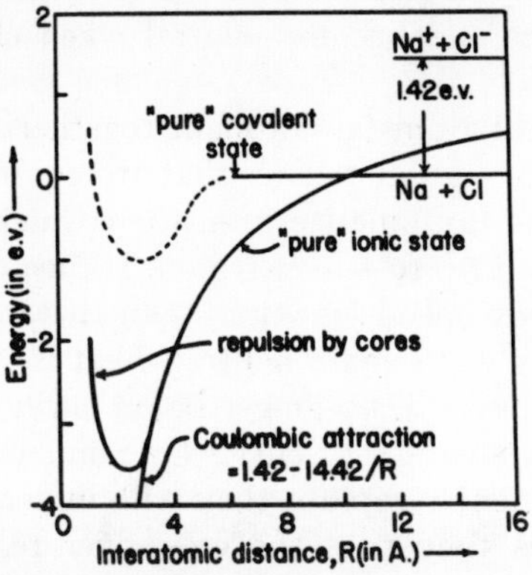

FIG. 14-2. Potential energy curves for the hypothetical "pure" ionic and covalent states of the NaCl molecule.

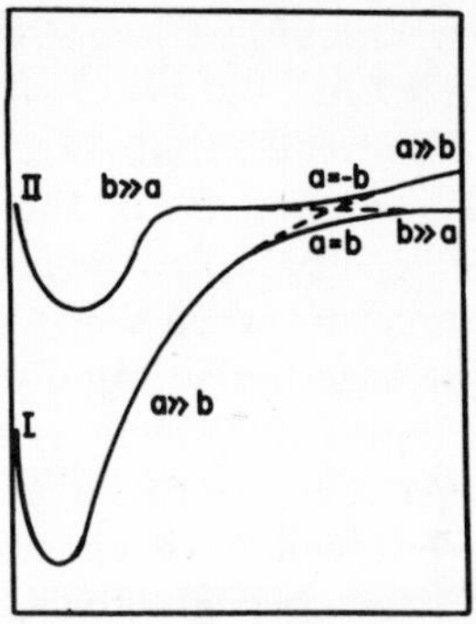

FIG. 14-3. True (adiabatic) potential energy curves of the NaCl molecule. The dashed curves are those of Fig. 14-2. The magnitude of the separation of curves I and II near the "cross-over" point is greatly exaggerated.

will fail to cross, giving curves of the form shown in Fig. 14-3. For the more stable state (I) the NaCl molecule will be essentially covalent ($a \ll b$)

at large internuclear distances and ionic ($b \ll a$) at small ones, whereas for the less stable state (II) the reverse is true. At the "cross-over point" in Fig. 14-2 the one state has $a = b$, giving the wave function

$$\psi = a(\chi_{\text{ion}} + \chi_{\text{cov}}) \tag{D-3a}$$

and the energy $E = E_0 + e$, whereas the other state has $a = -b$ giving

$$\psi = a(\chi_{\text{ion}} - \chi_{\text{cov}}) \tag{D-3b}$$

and the energy $E = E_0 - e$, where E_0 is the energy of the two "pure" structures at the cross-over point and e is the resonance energy between the two forms at this internuclear distance.

Because of the adiabatic principle we can say that curve I of Fig. 14-3 gives the potential energy if a sodium atom and a chlorine atom are *slowly* brought together; the two atoms will remain nearly neutral until they are about 10 Å. apart, and they become essentially charged ions at smaller separations. Curve II, on the other hand, gives the energy obtained when a sodium ion and a chlorine ion are *slowly* brought together; here the ions retain their charges at large separations and become neutral when closer than about 10 Å.

Suppose, however, that the two neutral atoms are brought together very rapidly. The arguments of the previous section tell us that it requires a finite time, $\tau = h/4e$, for the electrons to tautomerize from the covalent to the ionic structure when the system is in the cross-over region. If the atoms are brought together so rapidly that they spend less time than this in the cross-over region, it is obvious that they will remain neutral. That is, after they have passed the cross-over point they will no longer be on curve I of Fig. 14-3, but will have undergone a transition to curve II. Similarly, if Na^+ and Cl^- are brought together sufficiently rapidly they will undergo a transition from curve II to curve I as they pass the cross-over region. Furthermore if the NaCl molecule in its ground state is rapidly dissociated, it will split into ions rather than into neutral atoms.

Processes of this type are called *nonadiabatic transitions*, because they result from conditions incompatible with the adiabatic principle. We shall find that these transitions are fairly common. For instance, they must occur whenever light is converted into heat without either inducing a photochemical reaction or causing fluorescence. The phosphorescence of organic molecules is believed to involve a nonadiabatic transition from a singlet excited state to a triplet state which then emits light (see page 700). The cis-trans isomerization of a double bond may in some instances also involve such a transition (*4*). Nonadiabatic transitions must be looked for whenever the potential energy surfaces of different electronic states of a molecule intersect.

b. The probability of a nonadiabatic transition

It is quite easy to make an estimate of the probability of a nonadiabatic transition in terms of the configuration of the potential energy curves at the cross-over point and the velocity of the system past the point.

The essential features of a typical cross-over region are shown (in a magnified view) in Fig. 14-4. The straight lines 1 and 2 represent portions

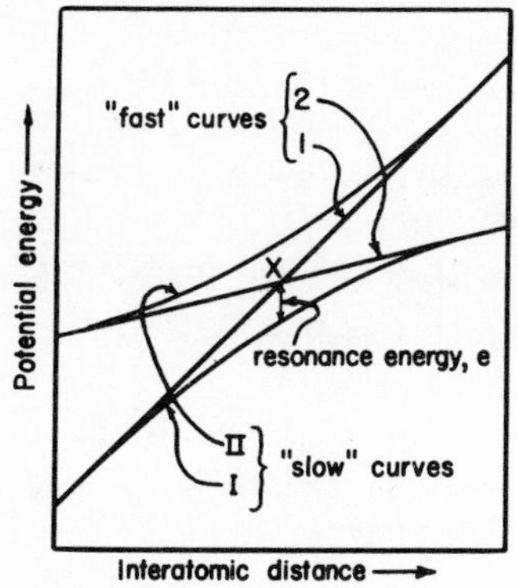

FIG. 14-4. Magnified view of the cross-over region.

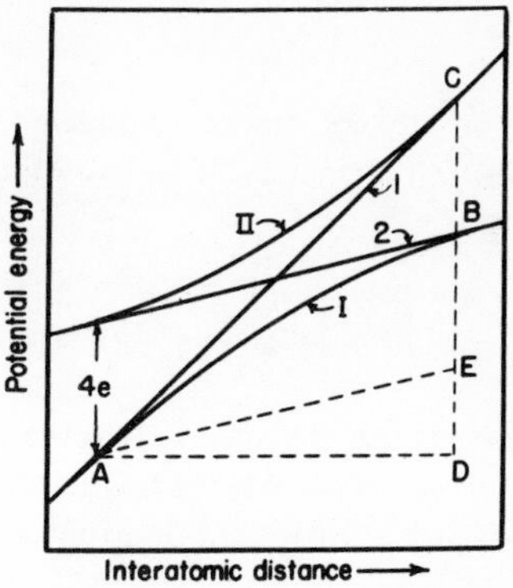

FIG. 14-5. Construction for the approximate calculation of the probability of a nonadiabatic transition.

of the potential energy curves of two states of a molecule when no resonance is permitted between them. They may be called the "fast" curves since they give the potential energy of the system when the atoms move very rapidly. Their intersection is at X. The curved lines I and II are the potential energy curves of the same system when complete resonance is permitted. They may be called the "slow" or adiabatic curves, since they give the potential energy when the atoms move slowly. The minimum vertical separation of the two slow curves is at X, and amounts to twice the resonance energy e.

Suppose that a molecule moves along curve I, its interatomic distance changing with velocity v. What is the probability that in passing through the cross-over region it will undergo a transition to curve II?

The time required for the system to tautomerize from state 1 to state 2 is, as we have seen, $\tau = h/4e$. Let Δt be the effective time spent in the cross-over region. Then if τ is much greater than Δt, the probability that a transition from curve I to curve II will not occur is approximately $\Delta t/\tau$. The probability, P, of a nonadiabatic transition is therefore

$$P \cong 1 - (\Delta t/\tau) = 1 - (4e\Delta t/h) \tag{D-4}$$

In order to be able to proceed, we must find a means of estimating Δt. This depends on the shape of the cross-over region, which is ordinarily determined in turn solely by the energies of the fast curves (which we shall call E_1 and E_2) and the resonance energy e. Straightforward application of the variation principle to the wave function

$$\psi = a\phi_1 + b\phi_2 \tag{D-5}$$

(where ϕ_1 and ϕ_2 are orthogonal wave functions of the two "pure" structures and a and b are variable parameters) gives for the energy of the slow curves

$$E_{\rm I} = \tfrac{1}{2}(E_1 + E_2) + \tfrac{1}{2}(E_1 - E_2)\ [1 + (4e^2/(E_1 - E_2)^2)]^{\frac{1}{2}} \tag{D-6a}$$

$$E_{\rm II} = \tfrac{1}{2}(E_1 + E_2) - \tfrac{1}{2}(E_1 - E_2)\ [1 + (4e^2/(E_1 - E_2)^2)]_{\frac{1}{2}} \tag{D-6b}$$

where E_1 is the energy of structure ϕ_1 ($E_1 = \int \phi_1 H \phi_1 dx$), E_2 is the energy of structure ϕ_2 ($E_2 = \int \phi_2 H \phi_2 dx$), and e is the exchange integral ($e = \int \phi_1 H \phi_2 dx = \int \phi_2 H \phi_1 dx$). From these equations we find that if

$$E_1 - E_2 \geqq 4e \tag{D-7}$$

then $(E_1 - E_2)/(E_{\rm I} - E_{\rm II}) \geqq 0.89$. That is, if the difference between the fast curves for a given interatomic distance is more than double the difference between the slow curves at the cross-over point, then the fast curves and the slow curves differ by less than about 10 % of the interval between them. We shall use the condition (D-7) as a somewhat arbitrary, but not unreasonable, definition of the limits of the cross-over region, to be used in estimating Δt.

In Fig. 14-5 the points A and B correspond to the equality sign in condition (D-7); to the left of A and to the right of B the slow curves in effect merge with the fast ones. The line AE is parallel to curve 2; AD is horizontal and CD is vertical. The time interval Δt which we seek is the time required for the system to move the distance AD; thus

$$\Delta t = AD/v \tag{D-8}$$

where v is the rate of change of the interatomic distance. Substituting in (D-4), we find

$$P \cong 1 - (AD \cdot 4e/hv) \tag{D-9}$$

The distance AD is determined geometrically by the resonance energy and the inclinations of the two fast curves. Let s_1 and s_2 be the slopes of curves 1 and 2 ($s_1 = dE_1/dr$ and $s_2 = dE_2/dr$, where r is the interatomic distance). Then

$$\begin{aligned} AD &= CD/s_1 \\ CD &= DE + EB + BC \\ DE &= s_2 AD \\ EB &= BC = 4e \end{aligned} \tag{D-10}$$

giving finally

$$AD = \frac{8e}{|s_1 - s_2|} \tag{D-10a}$$

Therefore the nonadiabatic transition probability is

$$P \cong 1 - \frac{32e^2}{hv\,|s_1 - s_2|} \tag{D-11}$$

This result is, of course, valid only if P is close to unity. A more exact calculation by Landau (*5*) and Zener (*6*) gives

$$P = \exp(-4\pi^2 e^2/hv\,|s_1 - s_2|) \tag{D-12}$$

which, for values of P close to unity, is in good agreement with our rough value

$$P \cong 1 - \frac{4\pi^2 e^2}{hv\,|s_1 - s_2|} = 1 - \frac{39.5e^2}{hv\,|s_1 - s_2|} \tag{D-12a}$$

Nonadiabatic transitions are evidently favored by small resonance energies, high velocities, and large differences in the slopes of intersecting potential energy curves.

Exercises. (1) Estimate the probability of the formation of an NaCl molecule in its lowest state by the collision of a neutral sodium atom with a chlorine atom. The resonance energy at the cross-over is probably of the order of 5×10^{-5} volts. Assume that the velocity of the nuclei at the cross-over point is the mean thermal velocity at 300° K.

(2) Estimate the positions of the cross-over points (if any exist) of the ionic and covalent states of some of the alkali and hydrogen halides using the information in Table 14-1. Assume that the potential energy curves of the covalent states are horizontal

TABLE 14-1

	Ionization energy (e.v.)	Covalent radius (Å.)		Electron affinity (e.v.)	Covalent radius (Å.)
H	13.595	0.29	F	4.13	0.64
Li	5.39	1.30	Cl	3.72	0.99
Na	5.138	1.60	Br	3.49	1.14
K	4.340	1.95	I	3.14	1.33
Rb	4.176	2.08	H	0.7157	0.29
Cs	3.893	2.20			

for interatomic distances greater than double the sum of the covalent radii listed in this table. Which of the halides would be expected to form most readily by the collision of the free atoms?

c. The effect of nonadiabatic transitions on vibrational energy levels. Predissociation

If two potential energy curves of a molecule intersect, the vibrational energy levels of the molecule are disturbed. Suppose that two curves, *AOB* and *COD* in Fig. 14-6, cross at a point *O*, and suppose that we prepare a

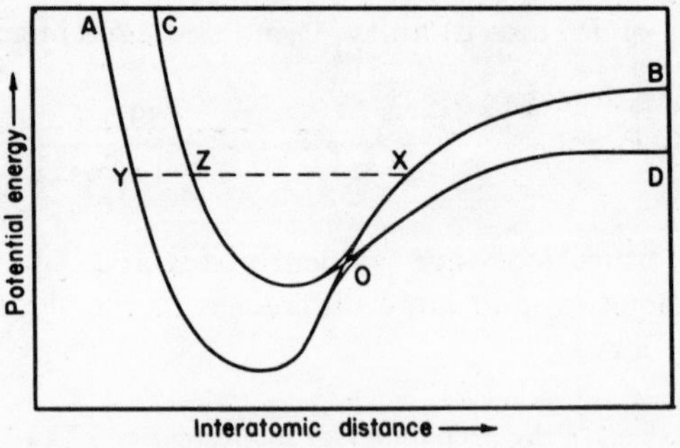

FIG. 14-6. Influence of the crossing of potential energy curves on the vibrational behavior of a molecule.

molecule so that its interatomic distance and potential energy correspond to the point X on OB. When the molecule is released it will begin to oscillate. Whether the endpoint of the first oscillation is at Y or at Z depends, however, on the probability of a nonadiabatic transition at O. If the probability of this transition is high, the oscillations will tend to take place along AOB. If it is low, the nuclei will move along COB. Since these two curves have very different shapes, the spacings of the vibrational levels must depend on the transition probability at O.

We have seen in the previous section that the transition probability increases as the velocity at the cross-over point increases. The higher the point X lies above the point O, the more rapidly the system will move past O, and the more probable it will be that the system will continue along the curve AOB. If, therefore, a system is initially located on OB and if it has an energy considerably greater than the energy of the cross-over point, then the energy levels of the system will be those that would be obtained from curve AOB alone (as if the curve COD were not present). Furthermore, if the energy of a system lies below the cross-over point, a transition from one curve to the other is improbable, and it becomes even more improbable as the energy is lowered.

Thus we should expect that the energy levels most strongly affected by the crossing of two potential energy curves are those whose energies are most nearly the same as the energy of the cross-over point. Fig. 14-7 shows the typical effects on the energy levels when two potential energy curves cross. In the left hand side of the figure are shown the locations that the energy levels of AOB and COD would have if the two curves were completely

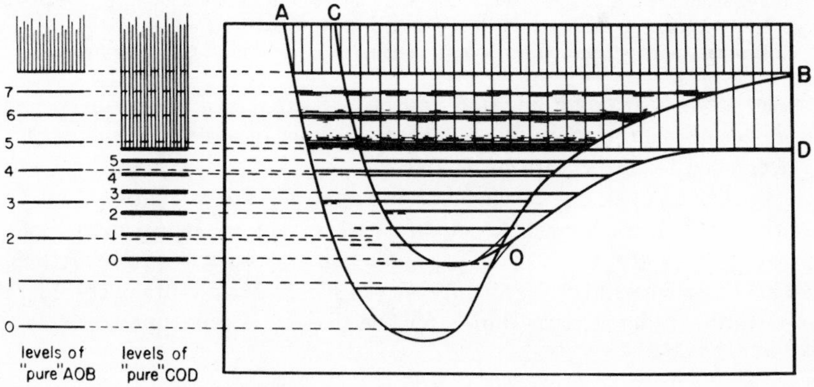

FIG. 14-7. Perturbation of vibrational levels by the crossing of potential energy curves.

independent. To the right are shown the actual positions of the levels. Except in the immediate vicinity of O, the levels occur at nearly the same positions as those shown on the left. The levels near O are shifted, however, and it may not be possible to say which level belongs to which potential energy curve.

Above the dissociation limit of COD there is a continuum of levels, superimposed on which are the discrete levels of AOB. There is, however, always a possibility that a transition from AOB to the continuum of COD may occur, from which there is no possibility of a return. The vibrational states of AOB that lie in the continuum of COD will then have only a limited lifetime, τ (where τ is equal to the reciprocal of the product of the vibration frequency and the probability that a nonadiabatic transition will not occur). According to the uncertainty principle, the vibrational energy will be indefinite by an amount $\Delta E \simeq h/\tau$. Therefore these vibrational levels of AOB will not be sharp, but will be "fuzzed out" over a frequency range $\Delta\nu = \Delta E/h \simeq 1/\tau$. The fuzziness will be less, the higher the level lies above the cross-over. It will also set in sharply with the first level lying above the dissociation limit of COD. As a result, the spectral bands corresponding to transitions to different vibrational levels of AOB will look like Fig. 14-8. Emission lines

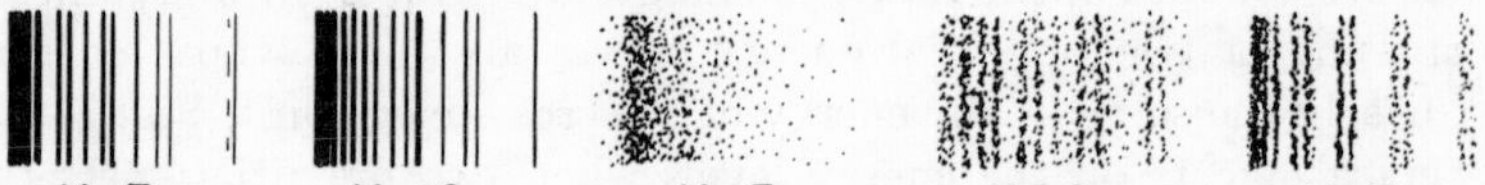

FIG. 14-8. Appearance of rotation–vibration bands in the spectrum due to transitions from some state x to various vibrational levels of state AOB in Fig. 14-7. The bands resulting from transitions to levels 3 and 4 are sharp, but those belonging to levels 5, 6, and 7 are diffuse, because these levels are mixed with the continuum of COD. The diffuseness is most marked in level 5 because it is closest to the cross-over point.

caused by transitions from AOB will be less intense, and may even vanish for levels above the cross-over, since the number of molecules in these levels is decreased by the transition to COD.

This fuzzing of energy levels involving nonadiabatic transitions between a continuum and a stable state is called *predissociation*. It was discovered by Henri in 1923 and explained by Bonhoeffer and Farkas and by Kronig in 1927. The fuzzing and change in intensity of spectral bands are of considerable importance in determining limits for the dissociation energies of molecules (7, pp. 405–450).

Exercise. The vibration frequency of the ground state of NaCl is 380 cm.$^{-1}$ Using the transition probability calculated in Exercise 1, page 541, estimate the extent of the fuzzing of the higher levels of the ground state due to the intersection of the curves for the covalent and ionic states. Should predissociation be noticeable in NaCl?

REFERENCES CITED

1. R. C. Tolman, "The Principles of Statistical Mechanics," Oxford U.P., New York, 1938.
2. F. London, *Z. Elektrochem.*, **35**, 552 (1929); also Sommerfeld Festschrift, "Probleme der modernen Physik," p. 104, Hirzel, Leipzig, 1928.
3. S. Glasstone, K. J. Laidler, and H. Eyring, "The Theory of Rate Processes," McGraw-Hill, New York, 1941.
4. J. L. Magee, W. Shand, and H. Eyring, *J. Am. Chem. Soc.*, **63**, 677 (1941).
5. L. Landau, *Physik. Z. Sowjetunion*, **2**, 46 (1932).
6. C. Zener, *Proc. Roy. Soc.*, **A 137**, 696 (1932); **A 140**, 660 (1933).
7. G. Herzberg, "Spectra of Diatomic Molecules," 2nd ed., Van Nostrand, New York, 1950.

GENERAL REFERENCE

D. Bohm, "Quantum Theory," Chapters 18–20, Prentice-Hall, New York, 1951. Contains discussions of sudden and slowly applied perturbations and the relationship between time-dependent and steady-state perturbation theory.

Chapter 15

The Interactions of Matter with Light I
The Classical Electron Theory of Optics

When a beam of light strikes a piece of matter, many changes can take place, both in the light and in the matter. These changes are of great interest to the chemist because they provide the basis for many of his most important research tools (e.g., the spectroscope, the refractometer, the colorimeter, the polarimeter, light scattering, x-ray diffraction, and the photoactivation of chemical reactions). It is therefore advisable that the chemist understand in some detail what happens when matter and radiation interact. Briefly, some of the more important changes that may occur are these:

The direction of the incident beam can be changed by *reflection* and *refraction.*

The beam can be decomposed into other beams by *diffraction, double refraction,* and *scattering.*

If scattering occurs, the scattered light may have the same wave length as the incident beam (*Rayleigh scattering* or *coherent scattering*) or the scattered light may have a different wave length (*Raman scattering* or *incoherent scattering*).

If the incident beam is plane polarized, the plane of polarization may be rotated on passing through the substance (*optical rotation*) and the degree of polarization may be reduced (*depolarization*).

The light beam can disappear, the energy of the light reappearing as heat or sometimes in other forms (*absorption*). The extent of the absorption may depend on the orientation of the plane of polarization in the incident beam (*dichroism*). Absorption may be followed, after a time interval, by re-emission of light that usually has a different color than the incident light (*fluorescence* if the time interval is short and *phosphorescence* if it is long). Light absorption may also produce chemically reactive substances (*photoactivation* and *photochemical reaction*).

Finally, matter can be made to *emit* light if it is properly "excited"—for instance by heat or by bombardment with high velocity electrons, as in a spark gap.

Theories are available that account for these and other interactions in terms of the atomic composition of matter. None of these theories are completely satisfactory, but in this chapter and in the next chapter we shall outline an approach that is adequate for essentially all of the interactions a chemist is likely to encounter.* In this chapter we shall outline the classical approach known as the *electron theory of optics*. This theory was one of the major developments in physics toward the end of the nineteenth century. Much of the theory was worked out by the Dutch physicist, H. A. Lorentz. After 1900 the theory ran into a series of fundamental difficulties, which played a predominant part in the establishment of quantum theory. (Lorentz' book (*1*) is an interesting and detailed account of the theory that describes the difficulties as they appeared to the pre-quantum theorist.) Although the classical electron theory has proven to be inadequate, it has one very useful attribute: the correct quantum mechanical theory gives many of the same equations for optical properties as did the electron theory. Thus in many instances one has merely to take the old classical equations and reinterpret some of the quantities that appear in them. This will be shown in the next chapter.

The justification for describing the classical electron theory in some detail is clear. The classical theory gives an understandable physical explanation of the optical properties of matter. This explanation is much easier for the average student to grasp in his first encounter with optics than is the much more mathematical quantum theory of radiation. Furthermore, even though the classical theory is wrong, it can be corrected by a mere reinterpretation of some of the basic classical concepts.

A. The Basic Assumptions in the Classical Electron Theory of Optics

Around the year 1900 it seemed that the final concepts were at hand out of which a complete classical theory of the atom could be constructed. J. J. Thomson had discovered that electrons are present in all matter and the

* The interactions for which this approach fails are chiefly those involving radiation of very short wave length, especially where the Compton effect comes into play and where matter is converted into radiation, and *vice versa*.

movements of electrons, controlled by Newton's laws of motion, were expected to account for all phenomena that depend on the internal dynamics of atoms. It was supposed that atoms consist of a massive diffuse cloud of positive charge in which the electrons were embedded in such a way that they could oscillate in simple harmonic motion. Even though this idea had ultimately to be discarded, it is interesting and instructive to look at some of its consequences, particularly as they concern the interactions of radiation with matter.

Our discussion of this theory will be based on a set of simple assumptions, which we shall list together with brief justifications in each instance.

Assumption 1. A charge e in an electric field $\boldsymbol{E}$ is acted on by a force given by

$$\mathbf{F} = e\boldsymbol{E} \tag{A-1}$$

in the direction of the field. This can be taken as a definition of an electric field.

Assumption 2. Light consists of oscillating electric and magnetic fields. The directions of both fields are perpendicular to the direction in which the light is propagated. At a time t the electric field associated with a beam of light moving through a vacuum in the positive z-direction is given by

$$\boldsymbol{E} = \boldsymbol{E}_0 \cos \omega(t - z/c) \tag{A-2}$$

where $\boldsymbol{E}_0$ is a vector whose size and direction are constant in both time and space, c is the velocity of light, $\omega = 2\pi\nu$, and ν is the number of oscillations per second as the light wave passes a fixed point. The constant ω is called the "circular frequency" and ν is called the "ordinary frequency." We shall use the term "frequency" for both of these quantities; the context will always reveal which quantity is referred to.

The magnetic field, $\boldsymbol{H}$, at any point in a light wave is equal in magnitude to the electric field, $\boldsymbol{E}$, provided that $\boldsymbol{E}$ is expressed in electrostatic units (e.s.u.) and $\boldsymbol{H}$ is expressed in electromagnetic units (e.m.u.).* Thus

$$|\boldsymbol{H}| = |\boldsymbol{E}| = |\boldsymbol{E}_0| \cos \omega(t - z/c) \tag{A-3}$$

* Practical electrical and magnetic units are related to electrostatic units (e.s.u.) and electromagnetic units (e.m.u.) in the following way.

Electric field strength	1 e.s.u.	= 300 volts/cm.
	1 e.m.u.	= 10^{-8} volts/cm.
Magnetic field strength	1 e.s.u.	= 3.34×10^{-11} gauss
	1 e.m.u.	= 1 gauss
Quantity of electricity	1 e.s.u.	= 3.34×10^{-10} coulomb
	1 e.m.u.	= 10 coulombs

The two fields, $\boldsymbol{E}$ and $\boldsymbol{H}$, differ, however, in their directions, being perpendicular to each other and to the direction of propagation, such that $\boldsymbol{E}$, $\boldsymbol{H}$, $\mathbf{z}$ form a right handed system (*cf.* page 26), as shown in Fig. 15-1. That is, light is a transverse wave.

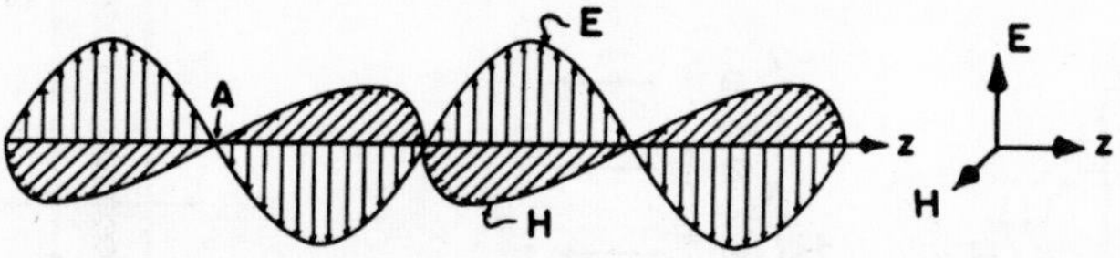

FIG. 15-1. The structure of a light wave moving toward the right. Note that when the electric field, $\boldsymbol{E}$, is pointing toward the top of the page, the magnetic field, $\boldsymbol{H}$, points out of the plane of the paper toward the reader.

A beam of light having a finite cross section consists of a bundle of parallel rays each having the structure shown in Fig. 15-1. For a given value of the coordinate z, the magnitudes of $\boldsymbol{E}$ and $\boldsymbol{H}$ are identical for all rays. That is, $\boldsymbol{E}$ and $\boldsymbol{H}$ are constant in each plane normal to the z-direction within the light beam. Such a beam of light is said to have a *plane wave front.*

The justification of Equations (A-2) and (A-3) and of Fig. 15-1 is to be found in Maxwell's equations, which rest in turn essentially on the laws of electromagnetic induction. Although Equations (A-2) and (A-3) and Fig. 15-1 are all that we shall need to know about light for our present purposes, it is of some interest to give a qualitative discussion of the nature of the relationship between the laws of induction and the structure of the light wave. (For a more quantitative discussion, see *2*, Chapter 10; *3*, Chapters 16 and 17; and *4*, pp. 397*ff*.).

Consider the situation at point A in the light wave in Fig. 15-1. An observer located at this point would find that the electric field increases in the vertical direction with time, and he might note that it is similar to the changing field that would be produced if either a flow of positive charges took place in the upward vertical direction or a flow of negative charges took place in the downward vertical direction. Such a flow of charge corresponds in either case to a flow of positive current in the upward direction and, according to the well-known electromagnetic laws, would be encircled by a magnetic field as shown in Fig. 15-2A. Similar vertical current elements would exist in the light rays next to point A in a plane perpendicular to the direction of propagation. As seen in Fig. 15-2B, however, the portions of these circular magnetic fields that are parallel to the direction of propagation will tend to cancel each other, whereas the components normal to the direction of propagation tend to reinforce. The net result is a magnetic field normal to both $\mathbf{z}$ and $\boldsymbol{E}$ at all points in the wave front passing through A, as shown in Fig. 15-2C.

The currents that we have assumed as generators of the magnetic field are known as *polarization currents*. It is easy to see how they might arise in the presence of matter through the displacements of electrons. In order to account for the propagation of light through a vacuum, Maxwell had to assume that these currents also exist in free space. This seemed to Maxwell to require the existence of a polarizable medium pervading all of space, which he termed the "ether." The nature of this ether has provided theoretical physics with one of its most fruitful and difficult problems.

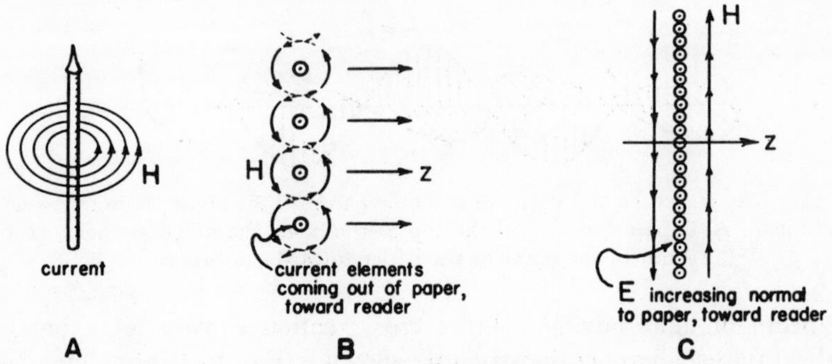

FIG. 15-2. Qualitative explanation of the origin of the fields in a light wave.
A. Magnetic field about a current element.
B. Magnetic field about a row of parallel current elements.
C. Magnetic field on either side of a layer in which the electric field increases in a direction parallel to the layer. This is the situation in the plane normal to z passing through point A in Fig. 15-1.

A stationary observer at point A in Fig. 15-1 would note that the magnetic field also changes with time. According to the laws of magnetic induction, this should result in an electric field that tends to induce an electric current in any conductor encircling the changing magnetic field. By arguments similar to those given above, we see that the changing magnetic field can account for the distribution of the electric fields in the light wave shown in Fig. 15-1. Thus a light wave can be said to arise from the interactions of the changing magnetic and electric fields associated with the flow of the polarization current. The electric field pattern sustains the magnetic field pattern and *vice versa*, in accordance with the laws of electromagnetic induction.

Assumption 3. If the electric field at a point in a vacuum is $\boldsymbol{E}$ (measured in e.s.u.) and the magnetic field is $\boldsymbol{H}$ (measured in e.m.u.), then the energy density, in ergs per cubic centimeter, stored in the field at that point is

$$W = \frac{1}{8\pi}(|\boldsymbol{E}|^2 + |\boldsymbol{H}|^2) \tag{A-4}$$

This very general result can be proven from Maxwell's equations (see M. Abraham and R. Becker (*2*, page 145)). In a light wave, since $|\boldsymbol{E}| = |\boldsymbol{H}|$, the energy density is

$$W = |\boldsymbol{E}|^2/4\pi \tag{A-4a}$$

Assumption 4. When a moving electric charge changes its velocity, it emits electromagnetic radiation. Simple arguments (depending ultimately on Maxwell's equations; see Compton and Allison (*5*, Appendix II)) show that if an observer is located at a distance r from a charge that is accelerated at a rate a, he will notice a pulse of radiation in which the magnitude of the electric field is

$$|\boldsymbol{E}| = (ea \sin \theta)/rc^2 \qquad \text{(A-5)}$$

where e is the magnitude of the charge, c is the velocity of light, and θ is the angle between the direction of acceleration and the vector $\boldsymbol{r}$ drawn from the charge to the observer. The electric field $\boldsymbol{E}$ is normal to $\boldsymbol{r}$ and in the same plane as $\boldsymbol{r}$ and the direction of acceleration. Since it takes time for light to move from one point to another, this expression gives the field at a distance r due to an acceleration that occurred at a time r/c previously.

The physical basis of Equation (A-5) is not difficult to understand. If a charged particle moves with constant velocity, it carries with it its lines of force, which are straight. If the velocity of the particle is increased, the lines of force also tend to move with the new velocity, but because of the finite rate of propagation of electric fields the adjustment of the lines of force to the change in velocity will take place only out to a distance $r = ct$, where t is the time since the acceleration began and c is the velocity of light. Beyond a spherical surface of radius ct the lines of force continue to move with the original velocity of the particle. Now the lines of force which leave the particle after it has been speeded up must be continuous with those that left it before the acceleration. As can be seen from Fig. 15-3, this is possible only if the lines are "kinked," and if the kinks move outward with the speed of light along the lines of force. These kinks therefore constitute a pulse of radiation. We see from Fig. 15-3 that the kinks tend to be small ahead of and behind the charge and are largest in directions lateral to the direction of acceleration. This is the origin of the sin θ term in (A-5). The kinks are also sharper (giving a larger pulse, and hence larger values of $\boldsymbol{E}$) the greater the acceleration, and the fields will be greater, the larger the charge, whence the terms e and a in the numerator of (A-5). The factor r in the denominator arises because of the spreading of the energy of the pulse over larger and larger areas at greater distances from the charge; the area varies as the square of r, so the energy density varies as the inverse square of r and the field strength (*cf.* Equation (A-4a)) must vary inversely with r. (It is interesting that the electrostatic field from a charge varies inversely with the *square* of the distance; thus the radiation field decreases much less rapidly with distance than does the electrostatic field.)

Assumption 4a. If a charge e moves harmonically along the x-axis so that its displacement is

$$x = x_0 \cos \omega t$$

where ω is the circular frequency, then the acceleration of the charge is

$$a = d^2x/dt^2 = -\omega^2 x_0 \cos \omega t$$

It will, according to Equation (A-5), emit light in which the magnitude of the electric field strength at a distance r from the charge is*

$$|\boldsymbol{E}| = -\frac{e x_0 \omega^2}{r c^2} \sin \theta \cos \omega(t - r/c) \qquad \text{(A-5a)}$$

If a stationary charge e of opposite sign is present at $x = 0$, we have a so-called *Hertzian dipole oscillator*, which emits radiation having the same

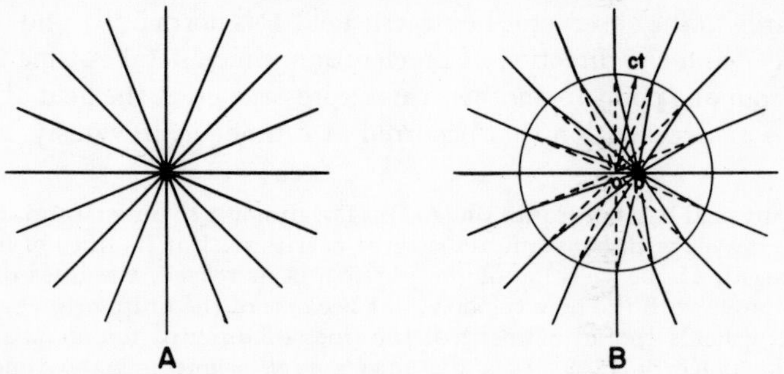

FIG. 15-3. The origin of radiation from accelerated electric charges.
A. Lines of force around a charge moving with constant velocity. The lines are straight and move with the charge.
B. The charge is subjected to a sudden change in velocity at time $t = 0$. If the velocity had not been changed the charge would have been located at point o at time t. Because of the finite velocity of propagation of electrical disturbances, the lines of force beyond a sphere of radius ct will still point toward o, c being the velocity of light. Inside the sphere the lines of force point toward the actual location of the charge, p. Therefore the lines of force are kinked at the surface of the sphere. This kink, moving outward with the velocity of light, is a pulse of electromagnetic radiation.

field strength as (A-5a) at large distances from the dipole. The dipole moment of such an oscillator is

$$m = ex = m_0 \cos \omega t$$

* Strictly speaking, no minus sign should appear in (A-5a) because the magnitude of a vector should be a positive quantity. In this chapter and the next one, however, we shall want to allow for the fact that a vector which is parallel to a given line, may point toward either of the two ends of the line (*i.e.*, the *sense* of the vector must be specified). The sign in equations such as (A-5a) should be taken to specify the sense of the vector, which may change in both space and time. Note that in Equation (A-5) the sense of the vector $\boldsymbol{E}$ is such that when $\theta = 90°$, $\boldsymbol{E}$ points in the same direction as the acceleration if the charge is positive.

where $m_0 = ex_0$, which is the amplitude of the variation of the dipole moment. Thus we can write for the magnitude of the electric field strength in the radiation from a Hertzian dipole

$$|\boldsymbol{E}| = -(m_0\omega^2/rc^2)\sin\theta\cos\omega(t - r/c) \tag{A-5b}$$

This equation is valid only at distances much larger than the wave length of the radiation. Close to the dipole other electromagnetic effects become important, and in particular, at distances small compared with the wave length, the electrostatic field of the dipole becomes predominant and has to be added to (A-5b). This electrostatic field varies inversely with the cube of the distance and therefore dies out much more rapidly than the radiation field. Furthermore, its angular dependence and direction are entirely different from the angular dependence and direction of the radiation field.

The rate at which energy is emitted by a charge undergoing a *constant* acceleration a is the total amount of electromagnetic energy contained in a sphere of radius $r = c$. From Equations (A-4a) and (A-5), this is

$$\mathscr{W} = \int_0^c \int_0^\pi \int_0^{2\pi} (|\boldsymbol{E}|^2/4\pi) r^2 \sin\theta \, d\varphi d\theta dr$$

$$= \frac{e^2a^2}{4\pi c^4}\int_0^c dr \int_0^\pi \sin^3\theta d\theta \int_0^{2\pi} d\varphi = \frac{e^2a^2}{4\pi c^4}\cdot c\cdot\frac{4}{3}\cdot 2\pi = \frac{2e^2a^2}{3c^3} \tag{A-6}$$

The mean rate at which a Hertzian dipole emits radiation is therefore

$$\mathscr{W} = \frac{2e^2\overline{a^2}}{3c^3}$$

where $\overline{a^2}$ is the mean square acceleration in the dipole. This is

$$\overline{a^2} = \int a^2 dt \Big/ \int dt$$

both integrals being over one cycle. Thus

$$\overline{a^2} = \omega^4 x_0^2 \int_0^{2\pi/\omega} \cos^2\omega t \, dt \Big/ \int_0^{2\pi/\omega} dt$$

$$= \omega^4 x_0^2/2$$

so that the mean rate of emission is

$$\mathscr{W} = e^2 x_0^2 \omega^4/3c^3 \tag{A-7}$$

Introducing the wave length associated with the frequency ω

$$\lambda = 2\pi c/\omega$$

we obtain

$$\mathscr{W} = 16\pi^4 e^2 x_0^2 c/3\lambda^4 \tag{A-7a}$$

According to this result, we see that if a group of oscillators have the same amplitude, x_0, but different frequencies, they will emit radiation in inverse proportion to the fourth power of the wave length of the radiation. The reason for this high power is that for harmonic motion the acceleration is proportional to the square of the frequency, and the intensity of the radiation is proportional to the square of the acceleration. We shall see that this simple fact explains the blue color of the sky (page 586).

Assumption 4b. It is possible to show from Maxwell's equations that a magnetic dipole whose moment varies sinusoidally with the time will emit radiation in which the direction of the magnetic field is in the same plane as the direction of the dipole. The field strengths in this radiation have exactly the same magnitudes as those from an oscillating electric dipole of equal amplitude. That is, the amplitude m_0 in (A-5b) can represent either a magnetic or an electric dipole moment, the units of measurement being respectively e.m.u. and e.s.u.

Thus we can have electric dipole radiation and magnetic dipole radiation. Later in this chapter we shall also consider a third type of radiation—that produced by quadrupoles. Of these three types, however, the electric dipole radiation is usually the most important by far. In most of the discussion in this and the next chapter, therefore, the term "dipole" refers to an electric dipole, and the effects of quadrupoles will not be considered.

The four assumptions discussed above are perfectly general and are not restricted to the electron theory of matter. The properties of the electron are introduced in the following assumption.

Assumption 5. The electron is a particle that moves according to the laws of Newtonian mechanics. Its motion is therefore determined by the forces acting upon it. These forces consist of the following:

(*i*) An inertial force, $\mathbf{f}_1 = -m_e \boldsymbol{a}$, where m_e is the mass of the electron and $\boldsymbol{a}$ is its acceleration.

(*ii*) The restoring force responsible for the harmonic motion the electron is assumed to undergo in the atom, $\mathbf{f}_2 = -k\mathbf{x}$, where $\mathbf{x}$ is the displacement of the electron from its equilibrium position and k is a constant. The origin of this force was never made clear in the classical electron theory.

(*iii*) The force on the electron due to whatever electric fields are present, $\mathbf{f}_3 = e\boldsymbol{E}$, where e is the charge on the electron.

(*iv*) The force on the electron due to any magnetic field that might be present, $\mathbf{f}_4 = e(\mathbf{v} \times \boldsymbol{H})/c$, where $\mathbf{v}$ and $\boldsymbol{H}$ are the velocity and magnetic field strength (both vectors) and $(\mathbf{v} \times \boldsymbol{H})$ represents the vector product (page 26). In a light wave, since $|\boldsymbol{E}| = |\boldsymbol{H}|$, this force will be negligible compared with $\mathbf{f}_3$ as long as the velocity of the electron is small compared with the velocity of light ($|\mathbf{v}| \ll c$). This is always true in chemical systems. We therefore conclude that when light interacts with ordinary matter, it is generally the electric field of the wave that does most of the interacting, rather than the magnetic field. There are, however, some exceptions (e.g., in the phenomenon of optical rotation, where the interactions with both magnetic and electric fields in the light wave are important).

(*v*) Since a moving electron may dissipate energy in the form of radiation, there must be a force associated with the emission of radiation. This is called the *radiation damping force*. It is convenient for later applications to assume that this force is proportional to the velocity, $\mathbf{f}_5 = -\mu\mathbf{v}$. The constant μ is called the radiation damping coefficient, and its value can readily be calculated for an electron undergoing harmonic motion. The rate of energy dissipation as radiation is $-(\mathbf{f}_5 \cdot \mathbf{v}) = \mu |\mathbf{v}|^2$. The energy radiated per second by the oscillating electron is (taking $\mathbf{x} = \mathbf{x}_0 \cos\omega t$ and $\mathbf{v} = d\mathbf{x}/dt$)

$$\mathscr{W} = (\text{number of cycles per second}) \times (\text{energy radiated per cycle})$$

$$= \nu \int_0^{1/\nu} |\mathbf{f}_5|\,|\mathbf{v}|\, dt = \nu\mu x_0^2 \omega^2 \int_0^{1/\nu} \sin^2 \omega t\, dt$$

$$= \nu\mu x_0^2 \omega \int_0^{2\pi} \sin^2 x\, dx = \mu x_0^2\, \omega^2/2$$

But according to Equation (A-7)

$$\mathscr{W} = e^2 x_0^2 \omega^4/3c^3$$

Thus the radiation damping coefficient is

$$\mu = 2e^2 \omega^2/3c^3 \tag{A-8}$$

(*vi*) It is also assumed that the electron can dissipate energy in forms other than light—particularly as heat. The forces responsible for this are called frictional forces, or "damping forces" because they tend to "damp out" the electronic oscillations. In analogy with ordinary viscous forces, they are assumed to be proportional to the velocity of the electron

$$\mathbf{f}_6 = -\eta \mathbf{v}$$

where η is a constant analogous to a viscosity. The physical origin of these forces was presumed to lie in a sort of "viscous drag" on the electron as it moved through whatever made up the non-electronic portions of the atom. We shall find that collisions between atoms can have the same effects as $\mathbf{f}_6$.

(*vii*) If there are other electrons in the atom, they will exert forces on a given electron according to Coulomb's law

$$\mathbf{f}_7 = e^2 \sum (\mathbf{r}_i / |\mathbf{r}_i|^3)$$

where $\mathbf{r}_i$ is the distance from the ith electron (considered as a vector) and the sum is a vector sum over all of the other electrons in the atom.

B. The Emission of Light by Excited Atoms and Molecules

a. Emission by a single electron

Suppose that we have an atom containing only one electron, or alternatively, an atom in which the electrons exert no forces on one another. Then $\mathbf{f}_7 = 0$, and in the absence of light the equation of motion of an electron is, according to the classical theory,

$$\mathbf{f}_1 + \mathbf{f}_2 + \mathbf{f}_5 + \mathbf{f}_6 = 0 \tag{B-1}$$

Assuming that the motion takes place only in the x-direction, this gives

$$m_e d^2x/dt^2 + (\eta + \mu)dx/dt + kx = 0 \tag{B-2}$$

The solution of this differential equation is, as can be verified by substitution

$$x = x_0 e^{-\omega' t} \cos \omega_1 t \tag{B-3}$$

where

$$\omega' = (\eta + \mu)/2m_e \tag{B-3a}$$

$$\omega_1^2 = \omega_0^2 - \omega'^2 \tag{B-3b}$$

$$\omega_0^2 = k/m_e \tag{B-3c}$$

Equation (B-3) represents the damped vibrations of an oscillator whose natural frequency in the absence of dissipative forces is $\omega_0/2\pi$ (see Fig. 15-4). Because of the accelerations that it undergoes, the electron must emit radiation in accordance with Equation (A-5). The acceleration can be computed by differentiating (B-3) twice, but the result can be simplified

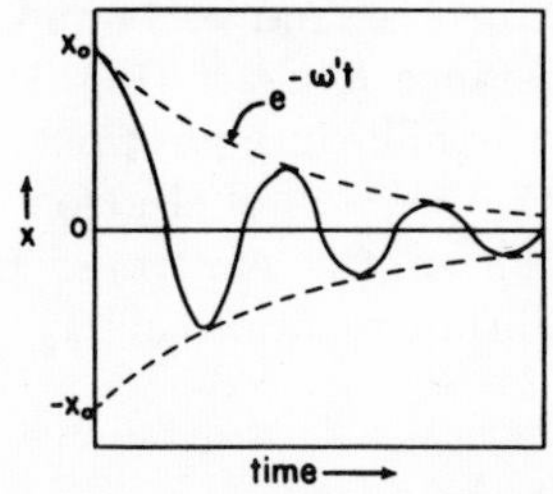

FIG. 15-4. Damped oscillations of an elastically bound electron. In the classical atom the damping would be much less rapid than is shown here.

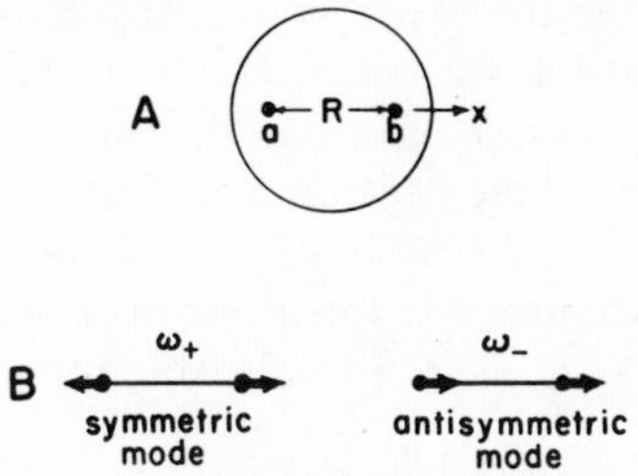

FIG. 15-5. Normal modes of interacting electrons.

because of the fact that ω' is always much smaller than the oscillation frequency ω_1. Therefore the magnitude of the electric field in the radiation is very nearly

$$|\boldsymbol{E}| = |\boldsymbol{E}_0| e^{-\omega' t} \cos \omega_1 t \tag{B-4}$$

where $|\boldsymbol{E}_0| = e^2\omega_1^2 x_0^2 \sin\theta / rc^2$.

Thus the electron emits a damped train of waves of the same form as the displacement shown in Fig. 15-4. According to the classical electron theory, this is the mechanism of the emission of visible and ultraviolet light by atoms. The electron is set in motion either by heat (through collisions with other molecules) or by bombardment with other electrons (as in a gas discharge or a spark). It then moves according to (B-4), emitting a continuous train of waves whose frequency is close to the natural frequency ω_0 (but always slightly smaller) and whose amplitude decreases exponentially with the time.

b. Emission by coupled electrons

If an atom contains N electrons all moving independently of each other, and if the force constants of the electrons differ from each other and also differ for displacements in different directions, the atom can emit as many as $3N$ different frequencies.

If the forces between the electrons, represented by $\mathbf{f}_7$ in Assumption 5, cannot be neglected, the electrons will not move independently, but must be discussed in terms of their $3N$ normal modes of vibration. For instance, consider an atom containing a pair of electrons, a and b, bound with identical force constants, k, to two sites a distance R apart (Fig. 15-5A). Let the x-axis lie along the line joining the two equilibrium positions. Then the forces on electron a will be influenced by the position of electron b and *vice versa* because of the Coulombic repulsion of the electrons. If the displacements of the electrons from their equilibrium positions are x_a, y_a, z_a and x_b, y_b, z_b, the Coulombic repulsion is

$$|\mathbf{f}_7| = e^2/[(R + x_a - x_b)^2 + (y_a - y_b)^2 + (z_a - z_b)^2]$$

Assuming that the displacements are small, so that their squares can be neglected in comparison with their first powers, this expression can be approximated by

$$|\mathbf{f}_7| \simeq (e^2/R^2)\left[1 - \frac{x_a - x_b}{R}\right] \tag{B-5}$$

This force is directed along the x-axis, so that only the vibrations in this direction will be affected; the vibrations in the y and z directions will take place independently as if the two electrons were on separate atoms. Neglecting the contribution of e^2/R^2 to $\mathbf{f}_7$ since it is independent of the displacement, and neglecting the damping forces, we can write the equations of motion of the two electrons as follows

$$m_e\ddot{x}_a = -k'x_a + (e^2/R^3)x_b \tag{B-6a}$$

$$m_e\ddot{x}_b = (e^2/R^3)x_a - k'x_b \tag{B-6b}$$

where $k' = k + e^2/R^3$. For the normal modes we must look for solutions of the form

$$x_a = x_{a0}\sin\omega t \tag{B-7a}$$

$$x_b = x_{b0}\sin\omega t \tag{B-7b}$$

Substitution into Equations (B-6) results in the relations

$$-\omega^2 m_e x_{a0} = -k'x_{a0} + (e^2/R^3)x_{b0} \tag{B-8a}$$

$$-\omega^2 m_e x_{b0} = (e^2/R^3)x_{a0} - k'x_{b0} \tag{B-8b}$$

which leads in the usual manner to the secular equation

$$\begin{vmatrix} (-k' + \omega^2 m_e) & e^2/R^3 \\ e^2/R^3 & (-k' + \omega^2 m_e) \end{vmatrix} = 0 \tag{B-9}$$

having the roots

$$\omega^2 = (k'/m_e) \pm (e^2/m_e R^3) \tag{B-10}$$

The upper sign leads to the result

$$\omega_+^2 = (k/m_e) + (2e^2/m_e R^3) \tag{B-10a}$$

$$x_{a0} = -x_{b0} \tag{B-11a}$$

In this normal mode the two electrons move with equal amplitudes and opposite phases (see Fig. 15-5B); we shall refer to it as the *symmetric mode.* The other normal mode (lower sign in Equation (B-10)) gives

$$\omega_-^2 = k/m_e \tag{B-10b}$$

$$x_{a0} = x_{b0} \tag{B-11b}$$

In this mode the two electrons have equal amplitudes and the same phase (see Fig. 15-5B); it will be called the *antisymmetric mode.*

It is obvious that the emission from the two electrons in the symmetric mode will tend to cancel, whereas that in the antisymmetric mode will give constructive interference. The frequency ω_- will thus be emitted strongly, whereas ω_+ will be weak. (The intensity of emission of the symmetric mode will be discussed further on page 611; it is an example of so-called quadrupole emission.) The interaction of the electrons has thus removed part of the degeneracy of the normal modes and has altered the intensity of the emission.

In general the coordinate x in Equation (B-3) can be taken as the displacement of a given electron in some normal mode. A similar equation is obtained for each electron in each normal mode. The electric field in the radiation from the entire atom is then simply the resultant of the fields due to all of the electrons in all of the normal modes that happen to be excited. In some normal modes, radiation from different electrons may interfere destructively, so that there will be little or no net emission; in other normal modes, there may be constructive interference with enhanced emission intensity. These normal modes can be called "inactive" and "active," respectively. It is clear that the radiation from active modes should have the form of a damped oscillation, as in Equation (B-4).

The physicists responsible for this theory were greatly puzzled by the fact that the emission spectra of practically all known atoms seem to contain a very large number of frequencies. This is true even of hydrogen, the simplest known atom. It would seem to require that all atoms have an almost unlimited number of normal modes, which is hardly conceivable. We now know that this paradox results from fundamental shortcomings of the theory (*cf.* Chapter 16).

c. Emission by molecules

Similar considerations can be used to explain the emission of light by molecules. If a molecule contains a permanent dipole moment, its over-all rotation will cause the component of the moment in any particular direction to vary sinusoidally with time. The rotating molecule can thus act as a Hertzian oscillator and will emit radiation having the same frequency as the frequency of rotation. For molecules at ordinary temperatures these rotation frequencies correspond to the far infra-red and microwave regions of the spectrum, where we know that there is strong absorption, and hence presumably also emission, by molecules with permanent dipoles. According to the classical theory, however, a molecule can rotate at any frequency, so the emission ought to be continuous over a range of frequencies. This is not observed because, as we now know, rotational motion is quantized. At the time that the classical electron theory was developed, however, these regions of the spectrum had not been investigated with sufficient resolution to show the discontinuous nature of the pure rotational spectra of molecules.

Molecules can not only rotate, but they can also undergo skeletal harmonic vibrations with frequencies equal to the normal mode frequencies of the molecule. If these vibrations result in the movement of the charged ends of the permanent dipoles present in the molecule, they may also cause emission of spectroscopic lines having frequencies equal to the normal mode frequencies. This is the classical explanation of the origin of the infra-red spectra of molecules, lying between approximately 10,000 Å. and 300,000 Å. (1μ to 30μ).

In addition to these over-all molecular motions, we should expect electronic vibrations in molecules, similar to those just postulated in atoms. Molecules should therefore give rise to electronic line spectra in the visible and ultraviolet. This is in fact observed, but there is a complication in that electronic motions are "coupled" to the rotations and vibrations of the molecule. In order to show this, let us for simplicity consider the behavior of a diatomic molecule in which the electrons are undergoing harmonic motion. We should expect that if the electrons shifted their positions in the bond of diatomic molecule, the strength of the bond would be changed slightly and the atoms at either end of the bond would tend to assume new equilibrium positions. The result is that the motions of the electrons are influenced to some extent by the motions of the atoms. The intensity of the emission by the vibrating electrons will therefore vary as the atoms move; if the two atoms themselves are vibrating, the intensity of the emission by the electrons will fluctuate (or be "modulated") with the much lower frequency of the atomic oscillations. The situation is very similar to that in ordinary "amplitude modulated" radio broadcasting, where a "carrier wave" of high frequency has superimposed on it a message having a much lower frequency (see Fig. 15-6).

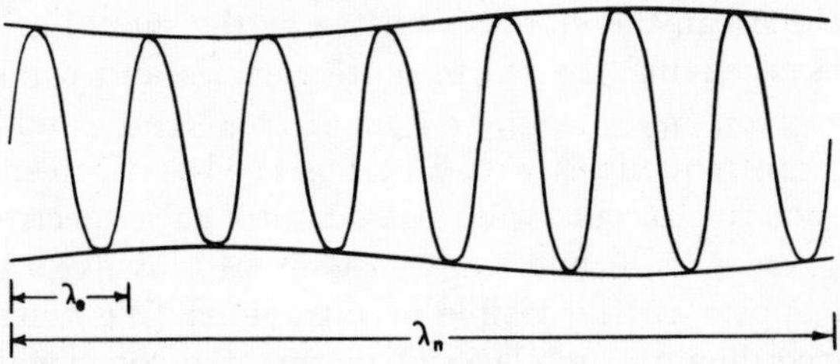

FIG. 15-6. Modulation of light emitted by electrons as a result of nuclear motions in a molecule. The wave length corresponding to the frequency of nuclear motion is λ_n, and that corresponding to electronic motion is λ_e.

Furthermore, the intensity of the electronic radiation from a molecule depends on the orientation of the molecule with respect to the observer, because of the sin θ term in Equation (A-5). The orientation changes as the molecule rotates, so the intensity of the observed electronic emission will also be modulated by the frequency of the molecular rotation. As a result of all of these factors, a rotating and vibrating diatomic molecule containing a single electron ought to produce a radiation field that varies with time in the following manner

$$
\begin{aligned}
|\boldsymbol{E}| = {} & k_1 \sin \omega_1 t + k_2 \sin \omega_2 t + k_3 \sin \omega_3 t + k_{12} \sin \omega_1 t \sin \omega_2 t \\
& + k_{13} \sin \omega_1 t \sin \omega_3 t + k_{23} \sin \omega_2 t \sin \omega_3 t \\
& + k_{123} \sin \omega_1 t \sin \omega_2 t \sin \omega_3 t
\end{aligned}
$$

Here the k's are constants, ω_1 is the frequency of the electron, ω_2 is the frequency of the atomic (infrared) vibration, and ω_3 is the rotational frequency (far infrared). The first three terms represent pure electronic, vibrational, and rotational spectra. The next four terms represent the mutual modulations of these motions.

Now trigonometry shows that, for instance

$$\sin \omega_1 t \sin \omega_2 t = [\cos (\omega_1 - \omega_2)t - \cos (\omega_1 + \omega_2)t]/2$$

and

$$\sin \omega_1 t \sin \omega_2 t \sin \omega_3 t = [\sin (\omega_1 - \omega_2 + \omega_3)t + \sin (\omega_1 - \omega_2 - \omega_3)t$$
$$- \sin (\omega_1 + \omega_2 + \omega_3)t - \sin (\omega_1 + \omega_2 - \omega_3)t]/4$$

The radiation field $\boldsymbol{E}$ can therefore be written as the sum of individual sine and cosine terms involving the thirteen frequencies and combinations of frequencies, ω_1, ω_2, ω_3, $|\omega_1 \pm \omega_2|$, $|\omega_1 \pm \omega_3|$, $|\omega_2 \pm \omega_3|$, and $|\omega_1 \pm \omega_2 \pm \omega_3|$. These thirteen frequencies would be observed as thirteen lines if the radiation from such a molecule were passed through a spectroscope; only ω_1, $|\omega_1 \pm \omega_2|$, $|\omega_1 \pm \omega_3|$, and $|\omega_1 \pm \omega_2 \pm \omega_3|$ involve the electronic motion and will appear in the visible or ultraviolet. We must remember, however, that according to the classical picture the rotational frequency, ω_3, can assume any value in a molecule. Therefore the visible and ultraviolet emission of a diatomic molecule caused by a single mode of electronic vibration would be expected to cause continuous *regions* of emission, in contrast to the relatively sharp *lines* produced by atoms. This is more or less what is observed; diatomic molecules produce so-called "band spectra" containing very many more lines than the spectra of atoms, and if the resolution of the spectroscope is not high the bands will seem to be continuous regions.

Exercises. (1) Calculate the average rotational frequency of the HCl molecule (interatomic distance 1.27 Å.) and of the ICl molecule (interatomic distance 2.32 Å.) at room temperature. (Assume that the average rotational energy is kT with $k = 1.38 \times 10^{-16}$ erg degree^{-1}molecule^{-1}.) Show that these frequencies lie in the far infrared and microwave regions of the spectrum.

(2) Sketch the emission spectrum that you would expect from a single excited electronic normal mode of a diatomic molecule on the basis of the classical considerations presented above. Assume that the molecule has been heated to some definite temperature, and take account of the fact that, because of the Boltzmann expression, $\exp(-\varepsilon/kT)$, only a limited range of rotational frequencies will be found. Assume that the electronic frequency is 30,000 cm.$^{-1}$, the atomic vibration frequency is 1000 cm.$^{-1}$, and the rotational frequency covers the approximate range 1 to 10 cm.$^{-1}$ with a broad maximum around 7 cm.$^{-1}$.

C. The Widths and Shapes of Spectral Lines

a. Preliminary discussion in terms of line widths

As we have seen on pages 36 and 241, no wave train that lasts for a finite time can be strictly monochromatic. If the train consists of n waves and takes τ seconds to pass by an observer, its frequency will be given by $\nu = n/\tau$, but it is not possible to measure n with much more precision than the nearest whole integer, so the frequency must be indefinite to the extent $\Delta\nu \simeq \Delta n/\tau \simeq 1/\tau$. Therefore the damped radiation emitted by a vibrating electron does not produce a perfectly sharp line when passed through a spectroscope, but it is spread out over a frequency range of the order of $\Delta\nu$ corresponding to a range of wave lengths of the order of $\Delta\lambda = (c/\nu^2)\Delta\nu$. The quantities $\Delta\nu$ and $\Delta\lambda$ are thus rough measures of the "width" of a spectral line. We shall now show how these quantities may be estimated. In Section *b* we shall give a more precise discussion of the shapes of spectral lines.

A convenient measure of the length of the wave train from an atom or molecule in which electrons oscillate according to Equation (B-3) is the quantity

$$Q = \omega_1/2\pi\omega' \qquad \text{(C-1)}$$

This quantity is evidently equal to the number of oscillations that occur before the amplitude has dropped to $1/e^{\text{th}}$ of its initial value. We can say that the "duration" of the wave train is therefore of the order of magnitude

$$\tau \cong Q \times (\text{duration of one cycle}) = 2\pi Q/\omega_1 = 1/\omega' \qquad \text{(C-2)}$$

An upper limit to τ (lower limit to $\Delta\nu$) is set by the radiation damping, which must be present even if no other dissipative forces exist. This limit is given by the expression

$$\tau_{\max} \cong 1/\Delta\nu_{\min} \cong 2\,m_e/\mu = 3m_e c^3/e^2\omega_1^2$$

where μ is the radiation damping coefficient defined in Equation (A-8). The corresponding upper limit to Q is

$$Q_{\max} = 2m_e\omega_1/2\pi\mu = 3m_e c^3/2\pi e^2\omega_1$$

For visible light (say wave length $= 6000$ Å., or $\omega_1 = 3 \times 10^{15}$ sec.$^{-1}$) we obtain

$$Q_{\max} = 16{,}000{,}000$$

$$\tau_{\max} \cong 3 \times 10^{-8} \text{ sec.}$$

$$\Delta\nu_{\min} \cong 3 \times 10^{7} \text{ sec.}^{-1}$$

The minimum uncertainty in the wave length of the emitted light is of the order of

$$\Delta\lambda_{\min} = 4\pi^2 e^2/3m_e c^2 = 0.00037 \text{ Å}. \qquad \text{(C-3)}$$

which is a constant for all regions of the spectrum. Accordingly this should be the lower limit to the sharpness of the spectral line that originates from the vibration of a *single isolated electron*.*

The observed widths of spectral lines from atoms and molecules are usually greater than the value given in (C-3). Three factors contribute to the broadening. In the first place, because of the Doppler effect, thermal motions of the radiating particles along the line of vision change the wave lengths of the emitted light as seen by a stationary observer. Since some of the atoms are moving toward the observer and some are moving away, and since there is a continuous range of velocities, even perfectly monochromatic radiators moving in this way would appear to emit a range of frequencies. This is known as the "Doppler broadening" of a spectral line.

The Doppler shift in frequency due to motion of a radiating body toward or away from the observer is

$$\Delta\nu = \nu_1 v/c \quad \text{or} \quad \Delta\lambda = v/\nu_1$$

where v is the velocity of the emitter, c is the velocity of light, and ν_1 is the frequency that would be observed if the body were not moving. In a gas at a temperature T, the velocity v can be related to the mean thermal energy in a given direction, ($v = \sqrt{2RT/M}$, where M is the molecular weight of the emitter). For a molecular weight of 20 at 300° K. and light of wave length 6000 Å. this gives

$$v = 35{,}000 \text{ cm./sec.}$$

$$\Delta\lambda = 0.0012 \text{ Å}.$$

which is several times larger than the limiting width due to radiation damping.

Doppler broadening is most serious in atoms of low atomic weight. It can be reduced by cooling the radiating particles so that they do not move so fast, and also by collimating the particles through a series of slits into a narrow beam in which the velocities normal to the beam are very low.

* If several electrons are coupled together, however, so that there is constructive or destructive interference between the fields produced by different electrons, then the radiation damping can be, respectively, greater or smaller than the values given by these relations.

The second factor affecting the line width is molecular collision ("collision broadening"). If in each collision between atoms or molecules the vibrational state of the electrons is changed, the emitted wave train cannot on the average be any longer than the average time between collisions, Δt. The kinetic theory of gases shows that

$$\Delta t = 1/(\sqrt{2}\, vn\sigma) \tag{C-4}$$

where v is the mean velocity, n is the number of molecules per cubic centimeter and σ is the effective cross-sectional area for collision. As a result, the frequency and wave length of the emitted line are indefinite by approximately

$$\Delta \nu \cong \sqrt{2}\, vn\sigma, \quad \text{or} \quad \Delta \lambda \cong \sqrt{2}\, vn\sigma \lambda^2/c \tag{C-5}$$

This phenomenon has been used to obtain information about the collision cross sections of radiating atoms (see *6*).

The third factor affecting the breadth of spectral lines is the interaction of atoms or molecules when they are brought close together ("interaction broadening"). Under these conditions, the electrons of one particle are influenced by the electrons in neighboring particles, and as a result the vibration frequencies are altered. The magnitude of the frequency shift depends on the distance between the particles, being large when they are close together. These interactions become noticeable in gases at high pressures and in liquids, where the atoms or molecules are not very far apart on the average. Furthermore, the interatomic distances in gases and liquids, and hence the magnitude of the interaction, varies considerably from one pair of particles to another because of the randomness of the structure of gases and liquids. As a result the observed spectrum is a superposition of emission spectra whose natural frequencies cover a range of values and the relatively narrow emission "lines" of an isolated atom or molecule appear to be spread out over continuous bands of frequencies. Broadening from this source can also occur in gases at low pressures when there is a tendency for the atoms to undergo slightly "sticky" collisions because of the formation of weak "quasi-molecules" in which the interatomic distance is not sharply defined.

Interaction broadening and collision broadening in gases can be reduced by lowering the pressure. These types of broadening are particularly important in liquids, and are responsible for the relatively broad absorption regions almost universally observed with colored substances in solution.

b. The shapes of spectral lines

It is not difficult to obtain an expression giving the shape of the spectral line emitted by an oscillator that moves according to Equation (B-3). By the shape of a spectral line is meant the distribution of the intensity in the line with respect to frequency when it is observed through a spectrograph. Spectrographs break down polychromatic radiation into its monochromatic components, and as we have seen in Chapter 2, Section F, these components can be regarded as Fourier components in a Fourier expansion of the electric field, $\boldsymbol{E}(t)$, emitted by the atom. Since a continuous range of frequencies is involved, a Fourier integral expansion must be utilized (page 33). Thus we may write

$$|\,\boldsymbol{E}(t)\,| = \int_{-\infty}^{\infty} A(k)e^{ikt}dk \tag{C-6}$$

where $\boldsymbol{E}(t)$ is given by Equation (B-4) when $t>0$, and $\boldsymbol{E}=0$ for $t<0$. Here $A(k)dk$ is the amplitude of the total electric field strength associated with all components having circular frequencies between k and $k+dk$. Since the intensity of radiation (that is, the energy density of the radiation) is proportional to the square of the field strength, according to Equation (A-4a), the intensity of the radiation in this frequency range must be proportional to $|\,A(k)\,|^2$. Thus $|\,A(k)\,|^2$ gives the shape of the spectral line directly.

According to the theory of Fourier transforms, $A(k)$ is given by

$$A(k) = (1/2\pi)\int_{-\infty}^{\infty} |\,\boldsymbol{E}(t)\,|\; e^{-ikt}dt$$

Using the relation

$$\cos\omega_1 t = (e^{i\omega_1 t} + e^{-i\omega_1 t})/2$$

and remembering that $\boldsymbol{E}(t) = 0$ for $t<0$, we find

$$A(k) = (|\,\boldsymbol{E}_0\,|/4\pi)\int_0^{\infty}\left\{e^{-[i(k-\omega_1)+\omega']t} + e^{-[i(k+\omega_1)+\omega']t}\right\}dt$$

$$= (|\,\boldsymbol{E}_0\,|/4\pi)\left[\frac{1}{\omega'+i(k-\omega_1)} + \frac{1}{\omega'+i(k+\omega_1)}\right] \tag{C-7}$$

If we restrict ourselves to frequencies very close to ω_1, so that $(k-\omega_1)$ is of the same order as ω', and recall that $\omega' \ll \omega_1$, the second term in the brackets in (C-7) can be neglected and we obtain for the shape of the spectral line

$$|A(k)|^2 = (\mathbf{E}_0^2/16\pi^2)/[(k-\omega_1)^2 + \omega'^2] \qquad \text{(C-8)}$$

Replacing the circular frequencies, ω_1 and k, by ordinary frequencies, ν_1 and ν, we find

$$\text{Intensity} = \frac{\gamma/2\pi}{(\nu-\nu_1)^2 + (\gamma/2)^2} \qquad \text{(C-9)}$$

where

$$\gamma = \omega'/\pi = (\eta + \mu)/2\pi m_e$$

and the numerator is chosen so that the integrated intensity under the curve is unity. This function has the appearance shown in Fig. 15-7. It is

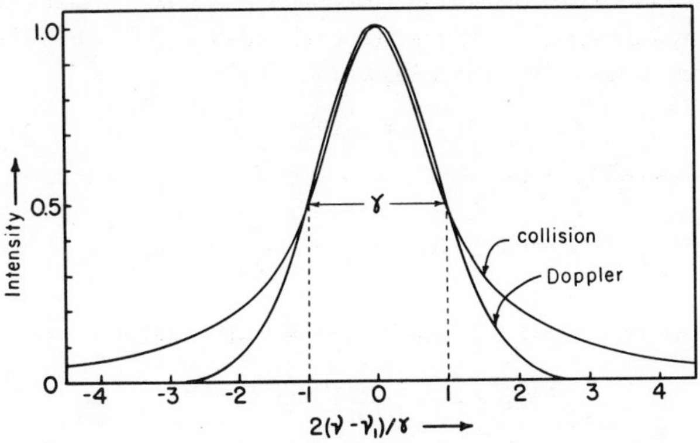

FIG. 15-7. Line shapes due to collision and radiation broadening and to Doppler broadening. Both curves have the same half-width, γ.

sometimes called the *Lorentz formula*. The intensity is a maximum at ν_1 and falls to half of its maximum value when the frequency is larger or smaller than ν_1 by $\gamma/2$. Therefore γ is called the "half width" of the line.*

If dissipation is caused by radiation damping alone, then

$$\gamma = \gamma_0 = \mu/2\pi m_e = 4\pi e^2 \nu_1^2/3m_e c^3 \qquad \text{(C-10)}$$

where γ_0 is called the "natural half width" of a spectral line.

* The term "half width" is sometimes applied to the quantity $\gamma/2$. Strictly speaking, γ is the "width at half maximum intensity" and $\gamma/2$ is "half the width at half maximum intensity."

Lorentz showed (*1*, page 37) that when line broadening is caused by collisions as well as by radiation damping, the same line shape as Equation (C-9) is obtained, except that the half width now has the value

$$\gamma = \sqrt{2}\sigma v n/\pi + \gamma_0 \tag{C-11}$$

where σ, v, and n are the same quantities as those appearing in (C-4). Collisions therefore act in the same way on the electron as a viscous drag for which the drag coefficient is $\eta = 2\sqrt{2}\sigma v m_e n$.

The line shape due to Doppler broadening is easily derived in the following way. The frequency observed by a stationary observer when a radiator of frequency ν_1 approaches with velocity v is

$$\nu = \nu_1(1 + (v/c)) \tag{C-12}$$

The Boltzmann principle states that the number of atoms having a velocity component v in a given direction is proportional to $\exp(-Mv^2/2RT)$, where M is the atomic weight. But from Equation (C-12)

$$v^2 = c^2(\nu - \nu_1)^2/\nu_1^2$$

Hence the line shape is given by

$$\text{Intensity} = K \exp\left[-\left(\frac{\nu - \nu_1}{\gamma_D/2}\right)^2 \ln 2\right] \tag{C-13}$$

where K is a constant and γ_D is the "Doppler half width,"

$$\gamma_D = 2\nu_1 \sqrt{2RT \ln 2/Mc^2} \tag{C-13a}$$

This shape is that of a Gauss error curve, and as seen from Fig. 15-7 the intensity drops off much more rapidly at large values of $(\nu - \nu_1)$ than it does in the Lorentz formula, Equation (C-9). Therefore, even in the presence of Doppler broadening that has a half width several times that due to radiation damping and collisions, it is possible to observe the effects of the latter in the "wings" of the line.

D. The Response of Bound Electrons to Light

a. The motion of an isotropically bound electron

The equation of motion of an elastically bound electron that is exposed to a beam of monochromatic light is the same as Equation (B-1), except for the additional force, $\boldsymbol{f}_3$, arising from the interaction of the electron with the incident light

$$\boldsymbol{f}_1 + \boldsymbol{f}_2 + \boldsymbol{f}_3 + \boldsymbol{f}_5 + \boldsymbol{f}_6 = 0 \quad \text{(D-1)}$$

The magnetic force, $\boldsymbol{f}_4$, is also present, but as has been explained previously, it can be neglected because it is very small compared with $\boldsymbol{f}_3$. If the atom is small compared with the wave length of the light, the spatial variation of the electric field in the light wave (Equation (A-2)) can be disregarded giving

$$m_e d^2x/dt^2 + (\eta + \mu)\, dx/dt + kx = e\,|\boldsymbol{E}_0|\, \sin \omega t \quad \text{(D-1a)}$$

where ω is the circular frequency of the light, x is the displacement of the electron in the direction of $\boldsymbol{E}_0$, the field of the light wave, and k is the force constant in the x-direction.* The electron would be expected to respond to the light by oscillating with it at the same frequency, though perhaps with a different phase. After some time a steady state will have been reached in which the amplitude of the electron's oscillations becomes constant. We can then say that

$$x = x'_0 \sin \omega(t + \delta) = x_0(\sin \omega t + \beta \cos \omega t) \quad \text{(D-2)}$$

where x_0 is the amplitude of the component of the oscillation that moves in phase with the light wave, and βx_0 is the amplitude of the component that is 90° out of phase. Substituting (D-2) into (D-1a), we find that

$$\{[-m_e\omega^2 - \beta(\eta + \mu)\omega + k]x_0 - e|\boldsymbol{E}_0|\} \sin \omega t$$
$$+ \{[-m_e\omega^2\beta + (\eta + \mu)\omega + \beta k]x_0\} \cos \omega t = 0 \quad \text{(D-3)}$$

This equation must be true for all values of the time t, which can be the case only if the coefficients of sin ωt and cos ωt vanish. Thus, replacing k/m_e by ω_0^2, $(\eta + \mu)/2m_e$ by ω', and $\omega_0/2\pi\omega'$ by Q, we obtain

$$\beta = \frac{(\eta + \mu)\,\omega/m_e}{\omega^2 - \omega_0^2} = \frac{2\,\omega\,\omega'}{\omega^2 - \omega_0^2} = \frac{1}{\pi Q}\;\frac{\omega\,\omega_0}{\omega^2 - \omega_0^2} \quad \text{(D-4a)}$$

$$x_0 = \frac{e}{m_e}\left[\frac{\omega_0^2 - \omega^2}{(\omega_0^2 - \omega^2)^2 + 4\omega'^2\omega^2}\right]|\boldsymbol{E}_0| = \frac{e}{k}\left[\frac{(\omega_0^2 - \omega^2)\,\omega_0^2}{(\omega_0^2 - \omega^2)^2 + \omega_0^2\omega^2/\pi^2Q^2}\right]|\boldsymbol{E}_0| \quad \text{(D-4b)}$$

* For the moment we shall assume that the force constant k is the same for displacements of the electron in any direction (isotropic binding). This means that the solution of (D-1a) will be independent of the orientation of the atom relative to the field in the light. Later, however, we shall consider systems in which k differs in different directions (see pp. 608 *ff*.).

$$\beta x_0 = -\frac{e}{m_e}\left[\frac{2\omega'\omega}{(\omega_0^2-\omega^2)^2+4\omega'^2\omega^2}\right]|\boldsymbol{E}_0| = -\frac{e}{k}\left[\frac{1}{\pi Q}\,\frac{\omega\omega_0^3}{(\omega_0^2-\omega^2)^2+\omega_0^2\omega^2/\pi^2Q^2}\right]|\boldsymbol{E}_0| \tag{D-4c}$$

Equations (D-4a, b, and c) describe the steady-state motions of a bound electron in a light wave. Since the positive charge in the environment is assumed to be stationary, the movement of the electron can be said to be responsible for an oscillating induced dipole moment amounting to

$$m = ex \tag{D-5}$$

This induced moment has two components: one component oscillates in phase with the incident light and has the amplitude

$$m_0 = ex_0 \tag{D-5a}$$

and the other component oscillates 90° out of phase with the light wave and has the amplitude

$$m'_0 = -\beta e x_0 \tag{D-5b}$$

The polarizability of an atom or molecule is defined as the induced dipole moment divided by the field strength responsible for the induced moment. We may therefore define an in-phase polarizability of the electron

$$a = m_0/|\boldsymbol{E}_0| = (e^2/k)\left[\frac{(\omega_0^2-\omega^2)\omega_0^2}{(\omega_0^2-\omega^2)^2+4\omega'^2\omega^2}\right] \tag{D-6a}$$

and an out-of-phase polarizability of the electron

$$a' = m'_0/|\boldsymbol{E}_0| = (e^2/\pi k Q)\left[\frac{\omega\omega_0^3}{(\omega_0^2-\omega^2)^2+4\omega'^2\omega^2}\right] \tag{D-6b}$$

$$= \frac{\omega\omega_0}{\omega_0^2-\omega^2}\,\frac{1}{\pi Q}\,a$$

We have shown that the quantity ω' determines the contribution of dissipative forces to spectral line widths, and for atoms it is found to be very much smaller than ω_0. Therefore, except for light waves having frequencies very close to the natural frequency of the electron, ω_0, we may neglect the term $4\omega'^2\omega^2$ in the denominators of (D-6a) and (D-6b). Using ordinary frequencies, ν and ν_0, instead of circular frequencies, ω and ω_0, we then obtain, for the in-phase polarizability of the electron, the expressions

$$a = \frac{e^2}{k}\,\frac{\nu_0^2}{\nu_0^2-\nu^2} = \frac{e^2}{k}\,\frac{\lambda^2}{\lambda^2-\lambda_0^2} = \frac{e^2}{4\pi^2 m_e}\,\frac{1}{\nu_0^2-\nu^2} \tag{D-6c}$$

where $|\nu - \nu_0| >> \omega'/2\pi$, and λ and λ_0 are the wave lengths corresponding to ν and ν_0, respectively. Furthermore, Q is generally a very large number (of the order of 10^6), so that when $|\nu - \nu_0| >> \omega'/2\pi$, we have for the out-of-phase polarizability,

$$\alpha' = 0 \tag{D-6d}$$

Figure 15-8 shows the variation of the in-phase and out-of-phase polarizabilities of the electron with the frequency of the incident light. In order to show more clearly the behavior in the immediate vicinity of ν_0, Q has been

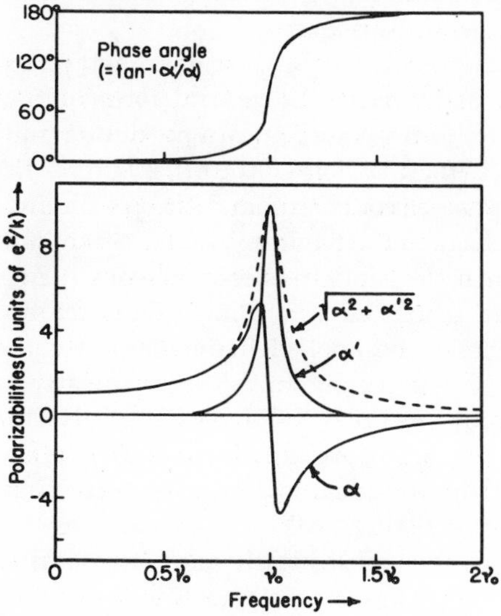

FIG. 15-8. Frequency variation of the polarizability and phase for a system in which $\pi Q = 10$.

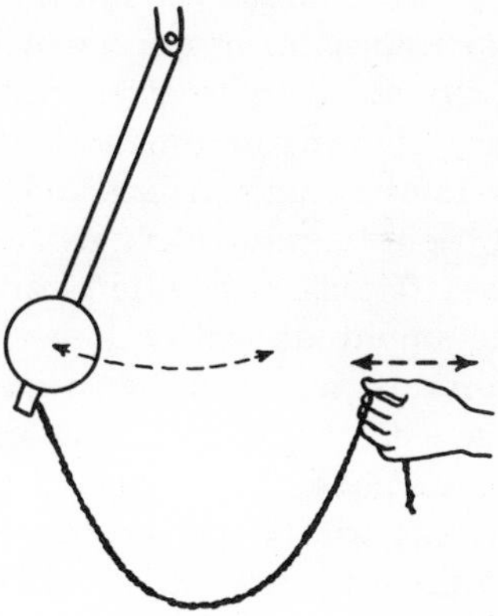

FIG. 15-9. Model illustrating the behavior of an oscillating system when subjected to a periodically applied force.

given the unusually low value of $10/\pi$. Except in the immediate vicinity of ν_0, Equations (D-6c) and (D-6d) dominate the behavior. At very low frequencies, $\alpha \cong e^2/k$, which is the static polarizability as measured in a constant field. As the frequency is raised, α increases, going through a maximum at $\nu = \nu_0(1 - (1/\pi Q))^{1/2}$. At still higher frequencies α decreases, going through zero at $\nu = \nu_0$ and attaining negative values. It reaches a minimum at $\nu = \nu_0(1 + (1/\pi Q))^{1/2}$ and then approaches zero assymptotically from below. The out-of-phase polarizability merely goes through a sharp and very large maximum at $\nu = \nu_0$. Thus if the frequency of the radiation is much less than the natural frequency of the electron, the electron oscillates in phase with the light. If the radiation has a much higher frequency than the electron's natural frequency, the electron moves 180° out of phase with the light. Only when the frequency of the light is close to the natural frequency of the electron does the phase differ from 0° or 180°.

It should be emphasized that the value of Q assumed in Fig. 15-8 is very much smaller than values typically found for electronic oscillations in atoms and molecules. Therefore the curve for α' should be much more sharply peaked and the maximum and minimum in α should occur in a much narrower frequency range than is shown in this figure.

This type of response is always found when oscillating systems are subjected to periodic forces. The student can easily verify the general form of the relationships mentioned above in the following way: Set up a pendulum and attach a string to its lower end (Fig. 15-9). Take hold of the other end of the string, allowing it to hang loosely between your hand and the pendulum. Now move your hand back and forth at a definite frequency, ν. The pendulum will begin to swing back and forth with the same frequency as your hand, soon attaining a definite steady-state amplitude and phase. This steady-state amplitude will be larger, the closer the applied frequency is to the natural frequency of the pendulum, ν_0. The pendulum will move nearly in phase with your hand if ν is less than ν_0, and it will move with nearly the opposite phase if ν is larger than ν_0. If ν and ν_0 are nearly equal, the phase difference will be intermediate and if the frictional losses in the pendulum suspension are small the amplitude of oscillation will become very large.

If an atom or molecule contains a number of elastically and isotropically bound electrons that exert no forces on each other, we should expect that as long as the frequency of the incident light is not close to any of the natural frequencies, ν_i, of any of the electrons, the total polarizability of the system will be given by the sum of the polarizabilities of all of the electrons; that is

$$\alpha = (e^2/4\pi^2 m_e) \sum_{\substack{\text{all} \\ \text{electrons, } i}} 1/(\nu_i^2 - \nu^2) \qquad \text{(D-7)}$$

where each electron contributes one term to the sum. Figure 15-10 shows a plot of polarizability *vs.* frequency of the radiation for a hypothetical atom containing three independent electrons whose natural frequencies are in the ratio $\nu_1 : \nu_2 : \nu_3 = 1 : \sqrt{2} : \sqrt{3}$.

According to (D-7), it should be possible to predict the frequency dependence of the polarizability of an atom or molecule from a knowledge of the natural frequencies ν_i of the electrons in the atom or molecule, since each electron makes a contribution $(e^2/4\pi^2 m_e)\,[1/(\nu_i^2 - \nu^2)]$ to the polarizability, and e and m_e are fundamental constants which are the same for all electrons. Experimental measurements of the frequency dependence of the polarizability show, however, that although (D-7) gives the correct functional form for the frequency dependence outside of the absorption bands of substances, the contributions of most natural frequencies are numerically smaller than predicted. Agreement with the experimental observations is obtained if factors f_i are introduced such that instead of (D-7) we have

$$\alpha = (e^2/4\pi^2 m_e) \sum f_i/(\nu_i^2 - \nu^2) \qquad \text{(D-8)}$$

The constants f_i are called *oscillator strengths* (also sometimes the *effective number of electrons*). In the next section we shall show how oscillator strengths different from unity may arise in the classical theory, and in Chapter 16 we shall give a quantum mechanical interpretation that usually leads to oscillator strengths less than unity.

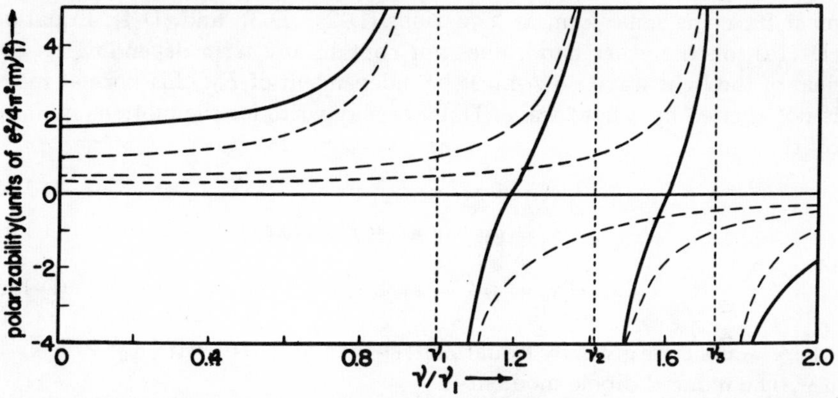

FIG. 15-10. Dependence of the polarizability on the frequency for a hypothetical atom containing three electrons whose natural frequencies are in the ratio $\nu_1:\nu_2:\nu_3 = 1:\sqrt{2}:\sqrt{3}$. The dashed curves give the contributions of the individual electrons. The solid curve is the sum of these dashed curves. The effect of damping is neglected.

b. Response of systems containing interacting electrons

The motions of interacting electrons are best discussed in terms of their normal modes of vibration. Consider, for instance, the system already described on page 558, consisting of an atom containing two electrons.

Assume that the line joining these two electrons is parallel to the electric field in the incident light. Then the equations of motion are (*cf.* Equations (B-6))

$$m_e\ddot{x}_a + (\eta + \mu)\,\dot{x}_a + k'x_a - (e^2/R^3)x_b = e\,|\,\boldsymbol{E}_0\,|\sin\omega t \tag{D-9a}$$

$$m_e\ddot{x}_b + (\eta + \mu)\,\dot{x}_b + k'x_b - (e^2/R^3)x_a = e\,|\,\boldsymbol{E}_0\,|\sin\omega t \tag{D-9b}$$

Taking the sum and difference of these two equations and setting

$$\begin{aligned} q_1 &= x_a - x_b \qquad & \omega_1{}^2 &= \frac{k}{m_e} + \frac{2e^2}{R^3 m_e} \\ q_2 &= x_a + x_b \qquad & \omega_2{}^2 &= \frac{k}{m_e} \end{aligned} \tag{D-10}$$

we find

$$\ddot{q}_1 + [(\eta + \mu)/m_e]\,\dot{q}_1 + \omega_1{}^2 q_1 = 0 \tag{D-11a}$$

$$\ddot{q}_2 + [(\eta + \mu)/m_e]\,\dot{q}_2 + \omega_2{}^2 q_2 = 2e\,|\,\boldsymbol{E}_0\,|\sin\omega t \tag{D-11b}$$

Equation (D-11b) has the same form as Equation (D-1a), so its solutions must have the same form as Equations (D-2), (D-3), and (D-4). Equation (D-11a), on the other hand, does not contain any term depending on the field of the light wave, so q_1 must be independent of $\boldsymbol{E}_0$; this normal mode is not excited by a light wave. Thus we may write for the motions induced by the light

$$q_1 = 0 \tag{D-12a}$$

$$q_2 = q_{20}\,(\sin\omega t + \beta\cos\omega t) \tag{D-12b}$$

or

$$x_a = x_b = q_2/2 = x_0(\sin\omega t + \beta\cos\omega t) \tag{D-12c}$$

where x_0 and β are given by Equations (D-4a) and (D-4b), with $\omega_0{}^2 = k/m_e = \omega_2{}^2$. The induced dipole moment is

$$m = e(x_a + x_b) = 2ex_0\,(\sin\omega t + \beta\cos\omega t) \tag{D-13}$$

If the frequency of the incident light differs appreciably from ω_2 we thus find for the in-phase polarizability

$$\alpha = \frac{e^2}{m_e}\left[\frac{2}{\omega_2{}^2 - \omega^2} + \frac{0}{\omega_1{}^2 - \omega^2}\right] \tag{D-14}$$

Therefore we may say that each normal mode of the system makes a contribution to the in-phase polarizability; the contribution of the antisymmetric mode is double that of a single isolated electron (first term in Equation (D-14)), and the symmetric mode makes the contribution zero (second term in Equation (D-14)). The same factors appear in the expressions for the out-of-phase polarizability, the antisymmetric mode giving twice as large a value of α' as would be obtained for a single electron, whereas the symmetric mode gives $\alpha' = 0$.

Generalizing this result, we may expect that the contributions of the electronic motions to the in-phase polarizability of an atom or molecule will result in an expression of the form

$$\alpha = \frac{e^2}{4\pi^2 m_e} \sum_{\substack{\text{all normal} \\ \text{modes}}} \frac{f_i}{\nu_i{}^2 - \nu^2} \tag{D-15}$$

where the dimensionless constants f_i are the *oscillator strengths* mentioned in the previous section, and ν_i is the frequency of the ith normal mode. The oscillator strengths are characteristics of the normal modes and, according to the classical point of view, they can be regarded as an effective number of electrons active in the respective normal modes. In the simple example described above, the oscillator strength of the antisymmetric mode is two, and that of the symmetric mode is zero. If the electrons in a normal mode tend to move in phase, the oscillator strength of that mode will be large. If they tend to move out of phase, it will be small or even zero.

Just as we must introduce the oscillator strength, f_i, into the expression for the contribution of a normal mode to the in-phase polarizability, so we must introduce the same factor into the expression for the out-of-phase polarizability associated with each normal frequency. Thus, when the light frequency is close to one of the normal mode frequencies, ω_i, we shall write Equation (D-6b) in the form

$$\alpha'_i = \frac{f_i e^2}{\pi m_e Q} \frac{\omega\,\omega_i}{(\omega_i{}^2 - \omega^2)^2 + (\omega_i\,\omega/\pi Q)^2} \tag{D-16}$$

where k has been replaced by $m_e\omega_i{}^2$, and Q has the value corresponding to a single isolated electron.

An oscillator strength different from unity also occurs if electrons are bound anisotropically. For example, consider an atom in which a single electron is constrained to move along a straight line with a characteristic frequency ν_0. If this line makes an angle θ with the direction of the electric field of the incident light, the component of force, $\mathbf{f}_3$, along the path must be written

$$e\,|\,\boldsymbol{E}_0\,|\cos\theta\,\sin 2\pi\nu t$$

and Equations (D-4b) and (D-4c) will contain an additional factor, cos θ, on the right. The component of the induced dipole moment in the direction of the applied field also contains a factor of cos θ, so that the amplitude of the oscillating dipole moment component is

$$m_0 = ex_0 \cos \theta$$
$$= \alpha_0 \cos^2 \theta \, | \boldsymbol{E}_0 |$$

where α_0 is the polarizability when $\theta = 0$. Suppose that the atom is free to orient at random in all directions relative to the direction of $\boldsymbol{E}_0$. The components of the induced moment perpendicular to $\boldsymbol{E}_0$ will cancel, so that the average induced moment lies in the direction of $\boldsymbol{E}_0$. Since the average value of $\cos^2 \theta$ is one-third, the mean in-phase polarizability will have the value

$$\alpha = \frac{e^2}{4\pi^2 m_e} \frac{1/3}{\nu_0^2 - \nu^2}$$

The same factor of one-third also appears in the expression for α'. Thus the oscillator strength is one-third for a randomly oriented linear oscillator.

If we assign oscillator strengths to normal modes in the manner suggested above, it is necessary to reinterpret the oscillator strength of an isolated, isotropically bound electron. Such an electron will have, of course, three independent normal modes, which consist of vibrations in three mutually perpendicular directions. These three normal modes have the same frequency. They therefore contribute three identical terms to Equation (D-15). In order that (D-15) give the same result as (D-6c), which was obtained for the identical system, it is necessary to assign values $f_i = 1/3$ to each of these normal modes.

It is evident that if an atom contains N uncoupled electrons, each capable of three modes of oscillation, then it must be true that

$$\sum_{\substack{\text{all} \\ \text{modes}}} f_i = N \tag{D-17}$$

This is known as the *Kuhn-Thomas sum rule*. We shall see that it is true also in the quantum mechanical modification of the electron theory of optics, in spite of the fact that the oscillator strengths are given an entirely different interpretation from the one just described.

At very high frequencies, when ν is much greater than all of the normal mode frequencies ν_i, Equation (D-15) takes the form

$$\alpha \cong (e^2/4\pi^2 m_e) \sum (- f_i/\nu^2) = - Ne^2/4\pi^2 m_e \nu^2 \tag{D-18}$$

This expression is, in fact, valid for x-rays passing through matter, and was used by J. J. Thomson to calculate the number of electrons in atoms. It is interesting to note that matter has a negative polarizability for x-radiation.

If an oscillation involving several coupled electrons has an oscillator strength f, the damping constant, ω', appearing in Equation (B-3) will be given by

$$\omega' = 3f \times \text{(damping factor due to each normal mode of a single electron)}$$

If radiation damping is the only source of dissipation, this means that

$$\omega' = fe^2\omega_0^2/m_ec^3 \tag{D-19}$$

Exercises. (1) Show that the maximum out-of-phase polarizability of an electron is larger than the static polarizability by a factor πQ.

(2) Show that for large values of Q the maximum and minimum in-phase polarizabilities are $\pm\,\pi Q/2$ times the static polarizability.

(3) Show that the curve of the half width of the out-of-phase polarizability *vs.* ordinary frequency satisfies $\nu_0/\pi Q = 4\pi\omega'$. Compare this with the half width of the emission line of the same electron.

(4) Show that $x_0^2\,(1 + \beta^2) = [\alpha'/\omega\,(\eta + \mu)]\,|\mathbf{E}_0|^2$ (D-20)

(5) Discuss the oscillator strengths of the normal modes of an atom containing three electrons that lie long a straight line, the equilibrium positions of the two end electrons being equidistant from the equilibrium position of the central electron.

c. The absorption of light by bound electrons

The term $(\eta + \mu)dx/dt$ in Equation (D-1a) represents a force that dissipates energy in the form of scattered radiation and heat. It must therefore be responsible for the phenomenon of *light absorption*.

The rate of energy dissipation is the scalar product of the dissipative force multiplied by the velocity of the electron

$$\text{Rate of dissipation} = (\mathbf{f}_5 + \mathbf{f}_6)\cdot d\mathbf{x}/dt = (\eta + \mu)\,(dx/dt)^2 \tag{D-21}$$

The mean rate of dissipation per second is the dissipation per cycle multiplied by the number of cycles per second

$$\mathscr{W} = \nu \int_0^{1/\nu} (\eta + \mu)\,(dx/dt)^2\,dt \tag{D-22}$$

Using Equations (D-2) and (D-20) to evaluate dx/dt and writing $z = \omega t$, we obtain

$$\mathscr{W} = 2\pi(\eta + \mu)\nu^2 x_0^2 \int_0^{2\pi} (\cos z - \beta \sin z)^2\, dz \qquad \text{(D-23)}$$

$$= 2\pi^2(\eta + \mu)\nu^2 x_0^2 (1 + \beta^2)$$

$$= \tfrac{1}{2}\omega\alpha' \, |\boldsymbol{E}_0|^2$$

If there are n atoms in each cubic centimeter, the rate of energy dissipation per cubic centimeter is

$$n\mathscr{W} = \tfrac{1}{2}n\,\omega\alpha' \, |\boldsymbol{E}_0|^2 \qquad \text{(D-24)}$$

Exercise. The oscillating electrons are equivalent to a current with a current density of

$$i = ne\, dx/dt$$

The electrons move in the direction of the electric field $\boldsymbol{E}$ of the light wave, and they therefore consume power at the rate $|\boldsymbol{E}|i$. The net power consumed per second per cubic centimeter is

$$\nu \int_0^{1/\nu} |\boldsymbol{E}|i\, dt$$

Show that this gives the same result as Equation (D-23).

The tendency of a substance to absorb light is measured experimentally in terms of the *molar extinction coefficient*, ε (also sometimes called the *molar absorptivity* and the *molar absorbancy index*) which is defined by the equation

$$\varepsilon = -(1/M)\, d(\log_{10} I)/dx = -1/(2.303MI)\, dI/dx \qquad \text{(D-25)}$$

where M is the number of moles of the substance per liter, $I(x)$ is the intensity of the light at point x, and dI is the decrease in the intensity that results when the light passes through a layer of substance of thickness dx. When Equation (D-24) is integrated, one obtains the familiar expression

$$I = I_0\, 10^{-\varepsilon M x} \qquad \text{(D-25a)}$$

where I_0 is the intensity of the light falling on the surface of the absorbing medium, and I is the intensity remaining after the light has travelled a distance x through the medium. Equations (D-25) and (D-25a) are, of course, merely the mathematical expressions of Beer's law and Lambert's law of light absorption.

The molar extinction coefficient defined in this way must be directly related to the energy dissipation, $\mathscr{W}$, given in Equation (D-24). When a beam of light whose cross-section is one square centimeter passes through a thickness dx of a medium containing n absorbing atoms per cubic centimeter, the energy dissipated in unit time is

$$dI = -n\mathscr{W}\,dx \tag{D-26}$$

On comparing this result with (D-24) we see that

$$\varepsilon = n\mathscr{W}/2.303MI \tag{D-27}$$

Now the intensity, I, of the incident light is defined as the radiant energy passing through an area of one square centimeter in one second. It is given by the velocity of light multiplied by the radiant energy contained in one cubic centimeter. From Equations (A-4a) and (A-2), the radiation energy in one cubic centimeter is

$$\begin{aligned}(1/4\pi)\int_0^{1\text{ cm}} |\boldsymbol{E}|^2 dx &= (|\boldsymbol{E}_0|^2/4\pi)\int_0^{1\text{ cm}} \sin^2\omega(t - x/c)\,dx \\ &= |\boldsymbol{E}_0|^2/8\pi \end{aligned} \tag{D-28}$$

Thus we find that

$$I = c|\boldsymbol{E}_0|^2/8\pi \tag{D-29}$$

Furthermore the number of molecules per cubic centimeter is

$$n = MN/1000 \tag{D-30}$$

where N is Avogadro's number. Combining Equations (D-24), (D-27), (D-29), and (D-30) we find

$$\varepsilon = (4\pi N/2303c)\,\omega\alpha' \tag{D-31}$$

Thus the out-of-phase polarizability, α', is directly related to the molar extinction coefficient, which is commonly used to express the ability of matter to absorb light.

According to Equation (D-31), the shape of an absorption band (that is, the dependence of the extinction coefficient on the frequency) should be essentially the same as the curve shown in Fig. 15-8 relating α' and the frequency, the width of the band being determined by the magnitude of the dissipation factors η and μ. This, however, is almost never the case. In any sample of matter electrons in different atoms never have exactly the same natural vibration frequencies, ν_i. One reason for this is the Doppler effect, which causes a difference in the relative values of the light frequency, ν, and the natural frequencies, ν_i, depending on the direction of motion of the atom relative to the light wave. A much more important cause of variation, however, is the fact that because of thermal fluctuations the environments of all atoms are not the same. This is especially true of atoms in the liquid and solid states. Since the environment affects the values of the normal mode frequencies, the values of ν_i will be different in different atoms. The observed absorption band in a liquid or solid is therefore a superposition of many narrow bands, each having a shape determined by α' in Equation (D-16), but centered on different values of ν_i. Consequently, the over-all shape of the band depends on the interactions between the environment and the electrons.

These environmental effects are most conveniently taken into account in the following way. Suppose that in an isolated atom the electrons have a normal mode whose frequency is ν_{i0}. When the atoms are placed in a solution or in a solid, this normal mode frequency will take on other values because of the interactions with neighboring molecules. Let us define a function $\phi(\nu_i)$ such that $\phi(\nu_i)d\nu_i$ is the fraction of the atoms present in the solution in which the frequency of this i^{th} normal mode lies between ν_i and $\nu_i + d\nu_i$. Then, using Equation (D-16) for the out-of-phase polarizability and replacing the circular frequencies by ordinary frequencies, we find from (D-31) that the molecular extinction coefficient at a particular frequency ν in the absorption band arising from the ith normal mode is

$$\begin{aligned}\varepsilon_i &= \frac{2N}{2303\,c}\frac{e^2}{m_e}\int_0^\infty \frac{f_i\nu^2\nu_i\phi(\nu_i)d\nu_i}{\pi Q[(\nu_i^2-\nu^2)^2+(\nu\nu_i/\pi Q)^2]} \\ &= \frac{Ne^2}{2303\,cm_e}\int_0^\infty \frac{f_i\nu^2\phi(\nu_i)\,d\nu_i^2}{\pi Q\,[(\nu_i^2-\nu^2)^2+(\nu\nu_i/\pi Q)^2]}\end{aligned} \tag{D-32}$$

Since Q is generally much larger than unity, the denominator in the integrand is very large except when ν_i is very close to ν. The entire contribution to the integral therefore comes from the immediate vicinity of $\nu_i = \nu$, so the numerator, which varies relatively slowly with ν_i, can be treated as a constant and can be taken out of the integral, and the quantity $\nu\nu_i$ in the denominator can be replaced by ν^2. Furthermore,

$$\int (x^2 + b^2)^{-1}dx = (1/b)\tan^{-1}(x/b)$$

Therefore, setting $x = \nu_i^2 - \nu^2$ inside the integral, we obtain

$$\varepsilon_i = \frac{Ne^2}{2303\ cm_e} \frac{\nu^2 f_i \phi(\nu)}{\pi Q} \int_{-\infty}^{\infty} \frac{dx}{x^2 + (\nu^2/\pi Q)^2}$$

$$= \frac{\pi Ne^2}{2303\ cm_e} f_i \phi(\nu) \quad \text{(D-33)}$$

Under these circumstances, the shape of the absorption band is evidently determined by ϕ rather than by Equation (D-16).

Equation (D-33) provides us with an experimental method of finding the oscillator strength of the electronic vibrations responsible for a given absorption band. Suppose that the molar extinction coefficient of an absorption band is plotted against the frequency, and the area under the curve is measured. According to (D-33) this area is

$$\int \varepsilon_i d\nu = \frac{\pi Ne^2}{2303\ cm_e} f_i \int \phi(\nu) d\nu \quad \text{(D-34)}$$

But from the definition of ϕ, we have

$$\int \phi(\nu) d\nu = 1 \quad \text{(D-35)}$$

Therefore

$$f_i = \frac{2303\ m_e c}{\pi Ne^2} \int \varepsilon_i d\nu \quad \text{(D-36)}$$

Thus the oscillator strength of a normal mode is directly proportional to the area under its absorption curve. If the frequency is expressed in cm.$^{-1}$ (indicated by the symbol $\tilde{\nu}$), Equation (D-36) gives*

$$f_i = 4.319 \times 10^{-9} \int \varepsilon_i d\tilde{\nu} \quad \text{(D-36a)}$$

* In this derivation, a small internal field correction has been omitted. If it is included, (D-36) becomes

$$f_i = \frac{2303\ m_e c}{\pi Ne^2} \frac{9n_0}{(n_0^2 + 2)^2} \int \varepsilon_i d^{\nu} \quad \text{(D-36b)}$$

where n_0 is the refractive index of the medium in which the absorbing molecule is dissolved [see page 603 and (7)]. Since the index of refraction of most solvents is between 1.3 and 1.6, the correction factor usually lies between 0.69 and 0.86.

The numerical values of the oscillator strengths of absorption bands found in this way will be discussed in detail in the next chapter, but it is interesting to give a few typical values at this point. The absorption bands of the rare earth ions in the visible part of the spectrum give oscillator strengths between 10^{-6} and 10^{-8}. The bands responsible for the colors of the transition metal cations (Cu^{++}, Cr^{+++}, Fe^{+++}, etc.) have oscillator strengths of the order of 10^{-4}. The intense purple of permanganate ion is caused by a band with an oscillator strength of only 0.03. Among the familiar colored substances of everyday life, only the organic dyestuffs give oscillator strengths approaching unity. Values greater than unity are exceptional, and values much greater than unity are never found. This fact is not easy to explain in terms of the classical electron theory, but is readily dealt with by the quantum mechanical theory. Further discussion will therefore be deferred to the next chapter.

The absorption bands of most organic and inorganic molecules in solutions have half-widths of the order of 5000 cm.$^{-1}$ If we use the approximation

$$\int \varepsilon d\tilde{\nu} \simeq 5000\ \varepsilon_{\max}$$

where $\varepsilon_{\max}$ is the maximum value of the extinction coefficient of the band, then we may estimate an upper limit to the molar extinction coefficient of typical colored substances. Since oscillator strengths greater than unity are unusual, we see from (D-36a) that values of $\varepsilon_{\max}$ much greater than about 5×10^4 are not to be expected for substances in condensed phases.

In Section Db of this chapter we have seen that the contribution of each normal mode to the in-phase polarizability is proportional to the oscillator strength of the mode. We now see that the oscillator strength also determines the intensity of the absorption of light. *Thus there is a close relationship between the contribution of a band to the polarizability and the intensity of the absorption in the band; weak bands lead to small contributions to the polarizability and strong bands give large contributions.* Evidently the absorption bands responsible for the colors of most substances other than organic dyes give very small contributions to the polarizability.

In the next chapter, we shall need to know the total rate of energy dissipation by a normal mode when an absorbing atom is exposed to "white" light—that is, light that contains all of the frequencies belonging to a given absorption band of the atom. Light of this kind may be described by means of a density function, $\varrho(\nu)$, where $\varrho(\nu)d\nu$ is the energy per unit volume belonging to frequencies lying in the range between ν and $\nu + d\nu$. We shall be particularly interested in what happens when $\varrho(\nu)$ is constant for the frequencies in some absorption band of a substance. From Equation (D-28), the mean energy density in a light wave is $|\boldsymbol{E}_0|^2/8\pi$, where $|\boldsymbol{E}_0|$ is the amplitude of the electric field in the light wave. Thus the total amplitude of the electric field, $|\boldsymbol{E}_0|$, that arises from the frequencies lying between ν and $\nu + d\nu$ can be found from the relation

$$|\boldsymbol{E}_0|^2/8\pi = \varrho(\nu)d\nu \tag{D-37}$$

The energy absorbed per cubic centimeter by the ith normal mode under these conditions is, from (D-24) and (D-16)

$$\begin{aligned} n\mathscr{W} &= \tfrac{1}{2}n \int_0^\infty 8\pi\omega\alpha'\varrho(\nu)d\nu \\ &= 2n\varrho(\nu)\frac{fe^2}{m_e}\int_0^\infty \frac{\omega^2\omega_i\,d\omega}{\pi Q\,[(\omega_i^2 - \omega^2)^2 + (\omega_i\omega/\pi Q)^2]} \end{aligned} \tag{D-38}$$

The integral is evaluated by noting that the integrand is significantly different from zero only when ν is very close to ν_i because of the large value of Q. Using the same procedure as that employed in obtaining (D-33) from (D-32), we find for the energy absorbed in unit time per cubic centimeter

$$n\mathscr{W} = n\frac{fe^2}{m_e}\varrho(\nu)\frac{\nu_i^2}{\pi Q}\int_0^\infty \frac{d\nu^2}{(\nu_i^2 - \nu^2)^2 + (\nu_i^2/\pi Q)^2} = \pi n\frac{fe^2}{m_e}\varrho(\nu) \tag{D-39}$$

It is important to note that this result is independent of the nature of the dissipative forces. The integrated absorption evidently depends only on the magnitude of the oscillator strength. The dissipative forces (that is, the forces giving rise to the coefficients η and μ and to the Q-factor of a vibration) can at most affect the shape of a band, and not the total amount of the absorption.

Exercises. (1) Show that Equation (D-36) is true even if there is no broadening of the absorption line by interactions with neighboring molecules. [Hint: Substitute (D-16) into (D-31), and integrate over the light frequencies, ν.]

(2) What is the amplitude of the electric field at a point one meter from a 100-watt light bulb, if 10% of the power supplied is converted into light? [Answer: 0.052 millivolts per centimeter.]

(3) The cross-section of a molecule for an absorption process may be defined as the equivalent opaque area that the molecule seems to present to an incident beam of radiation. Thus if there are N molecules per cubic centimeter and if each has a cross-section σ, then a layer of thickness dx would "black out" a fraction $N\sigma dx$ of the area of the layer. As a result of passing through the layer, an incident beam of intensity I would be decreased in intensity by the fractional amount $dI/I = -N\sigma dx$. Comparing this with Equation (D-25), calculate the effective cross-section of a molecule whose molar extinction coefficient is 5×10^4 (corresponding approximately to an oscillator strength of unity). [Answer: $\sigma = 2 \times 10^{-16}$ cm.^{2}]

E. The Scattering of Light and Some of Its Consequences

In Section D of this chapter we have found that when an atom is exposed to light, its electrons undergo oscillations that have the same frequency as the incident light. According to Equation (A-5a), however, any oscillating charge must emit light whose frequency is the same as the frequency of the oscillation. Therefore, when light shines on an atom, the atom must itself emit light of the same frequency as the incident light. The atom is said to *scatter* the incident light. The light that is scattered in this way is responsible for many important optical phenomena, such as refraction, x-ray diffraction, and optical rotatory power.

a. Scattering by an isolated particle

The field strength in the light radiated from an oscillating dipole is given by Equation (A-5b)

$$|\boldsymbol{E}_s| = -(m_0 \omega^2/rc^2) \sin\theta \cos\omega(t - r/c) \tag{A-5b}$$

where m_0 is the amplitude of the oscillation of the dipole moment. If an atom is placed in a beam of polarized light whose frequency does not coincide with any of the absorption frequencies of the atom, we may replace m_0 in (A-5b) by $\alpha|\boldsymbol{E}_0|$, where α is the in-phase polarizability of the atom and $|\boldsymbol{E}_0|$ is the amplitude of the electric field in the incident light. We shall assume that $\boldsymbol{m}_0$ lies in the same direction as $\boldsymbol{E}_0$ (this means that the polarizability of the atom is the same in all directions, an assumption which will be modified in Section h, below). Then the amplitude of the electric field in the scattered light is

$$|\boldsymbol{E}_s| = -\alpha|\boldsymbol{E}_0|(\omega^2/rc^2) \sin\theta \cos\omega(t - r/c) \tag{E-1}$$

where θ is the angle between the direction of polarization in the incident light (that is, the direction of $\boldsymbol{E}_0$) and a line drawn from the atom to the observer.

The energy in the light scattered in one second within a solid angle $d\Omega$ in a given direction is

$$\begin{aligned} I_s &= \int_{r=0}^{c} (|\boldsymbol{E}_s|^2/4\pi) r^2 \, dr \, d\Omega \\ &= (\alpha^2 \omega^4/c^4) I_0 \sin^2\theta \, d\Omega \end{aligned} \tag{E-2}$$

where $I_0 = c|\boldsymbol{E}_0|^2/8\pi$ is the intensity of the incident light (see Equation (D-29)). The total energy scattered in all directions is

$$I_{st} = \int I_s \, d\Omega = \int_0^{2\pi} \int_0^{\pi} I_s \sin\theta \, d\theta \, d\varphi$$

$$= 2\pi(\alpha^2\omega^4/c^4)I_0 \int_0^{\pi} \sin^3\theta \, d\theta$$

$$= \frac{8\pi\alpha^2\omega^4}{3c^4} I_0 \qquad \text{(E-3)}$$

Exercise. Show that if the incident light is not polarized, the energy of the light scattered into the solid angle $d\Omega$ in a direction making an angle χ with the direction of the incident beam is

$$I_s = (\alpha^2\omega^4/2c^4)\,(1 + \cos^2\chi) I_0 d\Omega \qquad \text{(E-4)}$$

Also show that the total energy scattered in all directions in this case is the same as Equation (E-3).

b. Scattering by more than one particle

If more than one atom is present in the light beam, the scattered fields from all of the atoms must first be added and the resultant must then be squared in order to find the intensity of the scattered light. Suppose, for instance, that two atoms are separated by a distance small compared with the wave length of the incident light. Then the electric fields in the scattered light will have the same phase and will add, giving a resultant field strength double that obtained from a single atom, while the scattered intensity will be four times that obtained from a single atom. If, on the other hand, the two atoms are farther apart, and are so located that the scattered beams arrive at the observer exactly out of phase (see Fig. 15-11), then the observer will see no scattered light at all. Thus the intensity of scattering from a group of atoms may depend very strongly on the geometrical arrangement of the atoms with respect to one another, as well as on the position of the observer. We can see that the calculation of the intensity of the scattered light in different directions from an array of atoms may be a rather complicated problem. A few general remarks can, however, be made about the properties of the scattered light in certain situations.

(*i*) The light scattered in the forward direction by any number and arrangement of atoms always gives constructive interference. We shall see that this forward-scattered beam is responsible for the phenomenon of refraction.

(*ii*) The intensity of the scattered light varies as the fourth power of the frequency. Blue light is therefore scattered to a much greater degree by isolated atoms than is red light, and if the incident beam consists of white light, the scattered light will appear bluish. This, of course, is the reason for the blue color of the sky, as was first pointed out by Lord Rayleigh.

(*iii*) Suppose that a great many atoms are arranged in a perfectly uniform fashion to give a large crystal, and that the wave length of an incident light beam is long compared with the distance between the atoms (see Fig. 15–12). Consider the light scattered in the direction OP by the atoms in the thin layer $ABCD$ of Fig. 15-12. The phase of the light scattered by each of the atoms in this layer is different, so it is obvious that when the scattered light from the layer is gathered to a focus by the eye of the observer, nearly complete destructive interference will take place. (There may be a small net contribution by the atoms closer to the crystal surface than one wave length, but in a large crystal this will be negligible.) The same is true

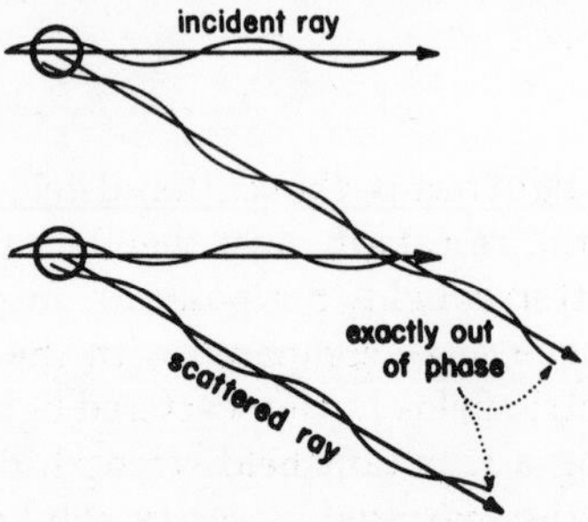

FIG. 15-11. Scattering of light by a pair of particles. For the scattering direction shown, the scattered rays from the two particles are exactly out of phase, so the net intensity scattered in this direction is zero.

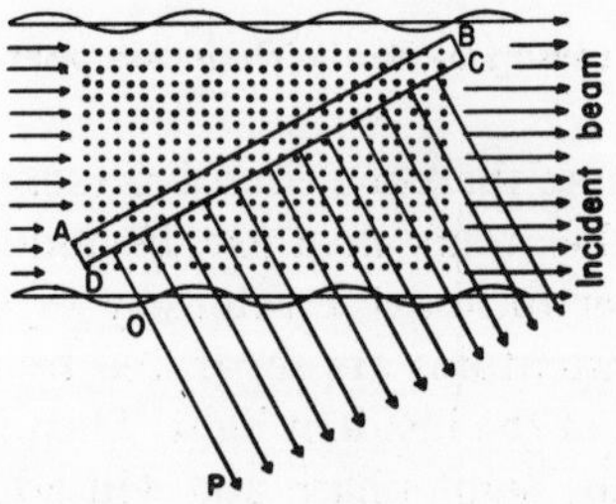

FIG. 15-12. Scattering by a crystal. The wave length of the light is long compared with the interatomic distance.

of the light scattered from any other layer parallel to $ABCD$. Therefore we see that there is practically no net sidewise scattering from a large, homogeneous specimen of matter in which the interatomic spacing is small compared with the wave length of the incident light. This is the reason for the weakness of the scattering in the non-forward direction by a perfectly homogeneous crystal.

(*iv*) If atoms are randomly arranged in space, as in a dilute gas, the intensity of the scattered light is proportional to the number of atoms present. The general tenor of the proof of this statement is shown by means of the following argument (for a more mathematical proof, see the exercise below). Consider the average intensity scattered by a pair of atoms that can move freely throughout space. If one atom is held fixed and the other is moved about, the relative phases of the light scattered by the two atoms in any direction other than the forward direction will change continuously. Oversimplifying the argument somewhat, we can say that destructive interference occurs half of the time on the average, and constructive interference occurs the rest of the time. If I_0 is the intensity of the light scattered by a single atom, we have seen that when there is constructive interference the scattered intensity is $4I_0$, and when there is destructive interference the scattered intensity is zero. The mean intensity scattered by a pair of atoms is thus $\frac{1}{2}(4I_0 + 0) = 2I_0$, or just double that from a single atom. If four atoms are present the same argument can be applied to each pair of atoms, so that on the average four atoms scatter four times as much light as a single atom, and similarly for larger numbers of atoms. Thus we can expect that the intensity of the light scattered by unit volume of a gas will be given by

$$I_{st} = (8\pi N\alpha^2\omega^4/3c^4)I_0 \tag{E-5}$$

where N is the number of molecules in unit volume.

Exercises. (1) Suppose that N identical molecules are placed in a beam of light. The light scattered in a given direction by the ith atom has the field strength $\mathbf{K}\sin(\omega t + \delta_i)$, where $\mathbf{K}$ is the same for all of the atoms and δ_i is a phase factor depending only on the location of the ith atom in the incident beam. Then the total field strength in the scattered beam is

$$\boldsymbol{E} = \mathbf{K}\sum_{i=1}^{N}\sin(\omega t + \delta_i)$$

and the intensity is

$$I = |\boldsymbol{E}|^2/4\pi = (|\mathbf{K}|^2/4\pi)\,[\textstyle\sum\sin(\omega t + \delta_i)]^2$$

If the atoms are free to move about independently of one another, the average intensity of the scattered light will be given by

$$I_{av} = \frac{\dfrac{|\mathbf{K}|^2}{4\pi}\displaystyle\int_0^{2\pi}\cdots\int_0^{2\pi}[\textstyle\sum\sin(\omega t + \delta_i)]^2 d\delta_1\cdots d\delta_N}{\displaystyle\int_0^{2\pi}\cdots\int_0^{2\pi}d\delta_1\cdots d\delta_N}$$

Show that
$$I_{av} = NI_1$$

where I_1 is the average intensity of the light scattered by a single isolated atom

$$I_1 = \frac{|\mathbf{K}|^2}{4\pi} \int_0^{2\pi} \sin^2(\omega t + \delta)\, d\delta \Big/ \int_0^{2\pi} d\delta$$

(2) When light of wave length 4358 Å. passes through a layer of liquid carbon tetrachloride one centimeter thick, the fraction of the incident light scattered is about 2.6×10^{-4} (see *8*). Compare this with the fraction of the light that would be scattered by a layer of carbon tetrachloride vapor containing the same number of molecules per unit area. (The density of liquid CCl_4 is 1.46 gm./cc. and the polarizability of the CCl_4 molecule is 10.5×10^{-24} cc.)

c. Scattering by density fluctuations in pure liquids and solids

We have seen that if a solid or liquid is perfectly homogeneous no scattering of radiation with long wave lengths is to be expected. Because of the thermal motions of molecules that occur in all matter, however, small and random local variations in the numbers of molecules per unit volume take place throughout any body. It is easy to see that these fluctuations result in scattering of light.

Consider the scattering produced by a sample of matter whose volume is one cubic centimeter. Let us divide the sample mentally into cells of equal volume whose dimensions are small compared with the wave length of the incident light (the exact size of the cells need not be further specified, as we shall see). If we were to count the number of molecules in each of these cells at a given instant, we would find that this number would be different in different cells, because of thermal fluctuations. Therefore the cells have randomly different polarizabilities, and they should scatter light to some extent independently of each other, the intensity being determined by the fluctuation in the polarizability. Now the polarizability of a volume V of a substance is given by (see *3*, p. 267, together with the definition of the polarizability in Equation (D-6a))

$$a = \frac{\varepsilon - 1}{4\pi} V \qquad \text{(E-6a)}$$

where ε is the dielectric constant of the substance. The fluctuation of the polarizability is thus

$$\Delta a = \frac{\Delta \varepsilon}{4\pi} V \qquad \text{(E-6b)}$$

where $\Delta\varepsilon$ is the fluctuation in the local dielectric constant resulting from the fluctuation in the number of molecules in the volume V.

If a particular cell had exactly the same polarizability as all of the other cells, no scattering would be observed. Because of the fluctuation $\Delta\varepsilon$ in the dielectric constant, however, the cell will cause observable scattering. The energy scattered into a solid angle $d\Omega$ is (assuming the incident light to be unpolarized, as in Equation (E-4))

$$i = [(\Delta\alpha)^2\omega^4/2c^4]\,(1 + \cos^2\chi)\,I_0\,d\Omega$$

$$= \frac{\pi^2}{2\lambda^4}\,(1 + \cos^2\chi)I_0\,d\Omega(\Delta\varepsilon)^2V^2 \qquad \text{(E-7a)}$$

Since the fluctuations in each cell are independent of those in the other cells, the phases of the scattered light from the different cells will be random, so the scattered intensity I_s from the entire unit volume of the sample in a given direction is the total number of elements, $1/V$, multiplied by the average intensity, $\bar{i}$, scattered by a single element

$$I_s = \frac{1}{V}\,\bar{i} = \frac{\pi^2}{2\lambda^4}\,(1 + \cos^2\chi)I_0 d\Omega\,\overline{\Delta\varepsilon^2}\,V \qquad \text{(E-7b)}$$

where $\overline{\Delta\varepsilon^2}$ is the mean square fluctuation in the dielectric constant in one of the elements.

In order to evaluate $\overline{\Delta\varepsilon^2}$ we must turn to statistical mechanics, which tells us that the probability of any state that is a fluctuation from the mean is proportional to

$$Ne^{-\Delta E/kT}$$

where N is the number of ways the state can be established and ΔE is the difference of the energy in the state from the mean value (this, of course, is simply the Boltzmann relation). We may write

$$\Delta S = k\ln N \quad \text{or} \quad N = e^{\Delta S/k}$$

and

$$\Delta A = \Delta E - T\Delta S$$

where ΔS and ΔA are the changes in the entropy and the Helmholtz free energy caused by the fluctuation. Thus we can say that the probability of a fluctuation is proportional to

$$e^{-\Delta A/kT} \qquad \text{(E-8)}$$

In order to calculate the magnitude of ΔA for the fluctuations with which we are concerned we shall make use of the fact that the dielectric constant in each cell could equally well be varied by applying pressure. Thus a pressure P acting on one of the cells would change its volume by ΔV and its dielectric constant by

$$\Delta\varepsilon = \frac{d\varepsilon}{dV}\Delta V = -\frac{\varrho}{V}\frac{d\varepsilon}{d\varrho}\Delta V \tag{E-9}$$

where ϱ is the density of the substance in the cell. The change in Helmholtz free energy that accompanies this change in pressure is the negative reversible work performed,

$$\Delta A = -\int_0^{\Delta V} P\, d(\Delta V) \tag{E-10}$$

But P and ΔV are related through the compressibility β, defined by

$$\beta = -\frac{1}{V}\left(\frac{\partial V}{\partial P}\right)_T$$

giving

$$P = -\Delta V/\beta V$$

Thus

$$\Delta A = (\Delta V)^2/2V\beta \tag{E-11}$$

$$= kTa(\Delta\varepsilon)^2$$

where

$$a = \frac{1}{2kT\beta}\frac{V}{\varrho^2}\left(\frac{d\varrho}{d\varepsilon}\right)^2$$

The mean square fluctuation in the dielectric constant is therefore

$$\overline{\Delta\varepsilon^2} = \frac{\int_{-\infty}^{\infty}(\Delta\varepsilon)^2 e^{-\Delta A/kT} d\Delta\varepsilon}{\int_{-\infty}^{\infty} e^{-\Delta A/kT} d\Delta\varepsilon}$$

$$= \frac{\int_{-\infty}^{\infty} x^2 e^{-ax^2} dx}{\int_{-\infty}^{\infty} e^{-ax^2} dx} \tag{E-12}$$

$$= \frac{1}{2a} = \left(\frac{d\varepsilon}{d\varrho}\right)^2 \frac{kT\beta\varrho^2}{V}$$

Substituting in (E-6b), we find for the scattered energy in a given solid angle

$$I_s = \frac{\pi^2}{2\lambda^4}(1 + \cos^2 \chi) I_0 d\Omega \left(\frac{d\varepsilon}{d\varrho}\right)^2 kT\,\varrho^2\beta \qquad \text{(E-13a)}$$

Integrating over all directions (*cf.* Equation (E-3))

$$I_{st} = \frac{8\pi^2}{3\lambda^4} kT\varrho^2\beta \left(\frac{d\varepsilon}{d\varrho}\right)^2 I_0 \qquad \text{(E-13b)}$$

The value of $d\varepsilon/d\varrho$ may be found from the variation of the index of refraction and density with the temperature, since $n^2 = \varepsilon$, so that

$$d\varepsilon/d\varrho = dn^2/d\varrho = (dn^2/dT)/(d\varrho/dT)$$

It can also be found from any of the various empirical or theoretical relations between the index of refraction and the density, such as the Lorentz-Lorenz equation

$$\frac{n^2 - 1}{n^2 + 2} = K\varrho$$

or the Eykman equation

$$\frac{n^2 - 1}{n - 0.4} = K'\varrho$$

or the Gladstone-Dale equation

$$n - 1 = K''\varrho$$

K, K', and K'' being constants for a given substance.

It is interesting that the cell volume V does not appear in the result, Equations (E-13); this is why its magnitude could be left arbitrary. It is merely necessary that V be sufficiently large for the fluctuations in neighboring cells to be considered independent and that it be sufficiently small for its dimensions to be small compared with the wave length of the incident light, thus assuring that all parts of a given cell scatter with the same phase.

It will be noted that according to Equations (E-13), the intensity of the scattering is proportional to the compressibility, β. Einstein and Smoluchowski pointed out that this is responsible for the unusually strong opalescence observed at the critical point of gases, where p–V isotherms are very nearly horizontal, so that dV/dp and, therefore, β are very large.

Exercises. (1) The refractive indices of water at 20° C. and 30° C. are 1.33299 and 1.33192, respectively, and the densities at these two temperatures are 0.998203 and 0.995646 gm./cc. The compressibility of water is 50×10^{-6} atm.$^{-1}$. Find the ratio of the total light scattered by one cubic centimeter of water to the intensity of the incident beam (assume $\lambda = 6000$ Å.).

(2) In a dilute gas the dielectric constant, ε, is given by $(\varepsilon - 1) = 4\pi N a$, where N is the number of molecules per unit volume and a is the polarizability of a molecule. Show that this relation, with (E-13b), gives (E-3a).

(3) Show that, assuming the Lorentz-Lorenz equation, one obtains

$$I_{st}/I_0 = (8\pi^2/27\lambda^4)kT\beta\,(n^2 - 1)^2\,(n^2 + 2)^2$$

Use this result to calculate the scattering by water at 20° C. and compare your answer with that found in Exercise 1.

d. Concentration fluctuations and scattering from solutions

If a solution were perfectly homogeneous (having exactly the same concentration at all points), it would not scatter light at all, except in the forward direction. Because of the atomic nature of matter and because of statistical fluctuations, however, the concentration of a dissolved substance in a solution is not everywhere exactly the same. Since these concentration fluctuations are statistical in origin, they are randomly distributed in space, and we should expect light to be scattered by them in all directions.

The intensity of the scattering by concentration fluctuations may be found by the procedure used in the previous section. A unit volume of the solution is divided into small equal cells that scatter according to Equation (E-7a), the scattering by the entire unit volume of solution being given by (E-7b). The quantity $\overline{\Delta\varepsilon^2}$ is the mean square fluctuation of the dielectric constant in a cell, which results from concentration fluctuations in the cell.

If C is the concentration of solute in grams per cubic centimeter and if ΔN molecules are transferred from the bulk of the solution into the cell, the dielectric constant inside the cell will change by

$$\Delta\varepsilon = (d\varepsilon/dN)\Delta N = (d\varepsilon/dC)\,(dC/dN)\,\Delta N$$

If the mass of a molecule in grams is m, and the volume of the cell is V, then

$$(dC/dN) = m/V \tag{E-14}$$

and the mean square fluctuation in the dielectric constant is

$$\overline{\Delta\varepsilon^2} = \left(\frac{d\varepsilon}{dC}\right)^2 \frac{m^2}{V^2} \int_{-\infty}^{\infty} (\Delta N)^2\, e^{-\Delta A/kT} d\Delta N \Big/ \int_{-\infty}^{\infty} e^{-\Delta A/kT} d\Delta N\,. \tag{E-15}$$

where ΔA is the change in the free energy of the entire system resulting from the transfer of ΔN molecules into the cell from the rest of the solution. Expanding the free energy in a Taylor's series, we find

$$\Delta A = (dA/dN)_0 \Delta N + \tfrac{1}{2}(d^2A/dN^2)_0 (\Delta N)^2 + \cdots \tag{E-16}$$

where $(dA/dN)_0$ is the change in the total free energy when one molecule is transferred into the cell from the solution outside the cell and when the system is at equilibrium ($\Delta N = 0$). If A_c is the free energy of the cell and A_s is that of the rest of the solution, then

$$(dA/dN)_0 = (dA_c/dN)_0 - (dA_s/dN)_0 = \mu_c - \mu_s$$

where μ_c and μ_s are the chemical potentials of the solution inside and outside of the cell, respectively. Since the system is at equilibrium when $\Delta N = 0$, then μ_c and μ_s must have the same value, so the first term on the right in (E-16) vanishes. The second term in (E-16) is

$$(d^2A/dN^2)_0 = d\mu_c/dN - d\mu_s/dN = (d\mu/dC)\,(dC_c/dN - dC_s/dN)$$

where dC_c/dN is the change in concentration inside the cell when one molecule is added, and dC_s/dN is the corresponding change in the solution outside the cell. From Equation (E-14)

$$dC_c/dN = m/V$$

and

$$dC_s/dN = m/(1 - V)$$

Since the volume of a cell, V, is much smaller than unity, dC_s/dN can be neglected in comparison with dC_c/dN, so that

$$\Delta A = kTa\,(\Delta N)^2$$

where

$$a = m(d\mu/dC)/(2VkT)$$

Therefore the mean square fluctuation in the dielectric constant is

$$\overline{\Delta\varepsilon^2} = \left(\frac{d\varepsilon}{dC}\right)^2 \frac{m^2}{V^2} \int_{-\infty}^{\infty} (\Delta N)^2\, e^{-a(\Delta N)^2} d(\Delta N) \Big/ \int_{-\infty}^{\infty} e^{-a(\Delta N)^2} d(\Delta N)$$

$$= \left(\frac{d\varepsilon}{dC}\right)^2 \frac{m^2}{V^2}\,\frac{1}{2a} = \frac{m}{V}\,kT \left(\frac{d\varepsilon}{dC}\right)^2 \left(\frac{d\mu}{dC}\right)^{-1}$$

Substituting this in (E-7b), we find that the light scattered by unit volume of solution into a solid angle $d\Omega$ is

$$I_s = \frac{\pi^2}{2\lambda^4}(1+\cos^2\chi)I_0\,d\Omega\,mkT\left(\frac{d\varepsilon}{dC}\right)^2\left(\frac{d\mu}{dC}\right)^{-1} \tag{E-17}$$

Since the dielectric constant is related to the index of refraction by the equation $\varepsilon = n^2$, we have $d\varepsilon/dC = 2n(dn/dC)$. Furthermore, if the solution is ideal, $\mu = \mu_0 + kT \ln C$, so that $d\mu/dC = kT/C$. Substituting these results in (E-17), we find

$$I_s = \frac{2\pi^2}{\lambda^4}(1+\cos^2\chi)I_0\,d\Omega\,n^2\left(\frac{dn}{dC}\right)^2 mC \tag{E-18}$$

It is evident that for a given weight concentration of solute, C, the intensity of the scattering is proportional to the molecular mass, m, of the solute. As Debye has emphasized, light scattering can therefore be used to determine molecular weights of solutes. One need merely determine the variation of the refractive index of the solution with solute concentration, dn/dC, and the energy scattered from a dilute solution into a known solid angle $d\Omega$ at some value of the angle χ (usually 90° to the incident beam). A more convenient procedure is to integrate (E-18) over all directions, thus obtaining the total scattered intensity. This is equated to the decrease in intensity of the incident beam after it passes through the solution (obtained through the measurement of the turbidity of the solution). Alternatively, if the molecular weight of the substance is known, Equation (E-17) makes it possible to study the deviations of the solution from ideality.

The use of light scattering to study solutions has been reviewed by Doty and Edsall (*9*).

Exercise. When 0.1 gram of a certain polymer is dissolved in 100 cc. of water, the refractive index is changed by 0.00019. If a beam of light of wave length 4360 Å. is passed through 1 cc. of the resulting solution, the total energy of the scattered light is 0.001% of the energy in the incident light. What is the molecular weight of the polymer?

e. X-ray diffraction by crystals

If the scattering material is crystalline and the incident radiation has a wave length of the same order of magnitude as the distance between the atoms, scattering in certain well defined directions will take place. This, of course, is the situation when x-rays pass through crystals. It is best understood by considering first the scattering from a single row of atoms spaced at equal intervals, Δ. Suppose that the row is normal to the incident beam. The wavelets scattered from each atom in the row will merge to form new wave fronts in certain directions; in other directions they will interfere destructively (see Fig. 15-13).

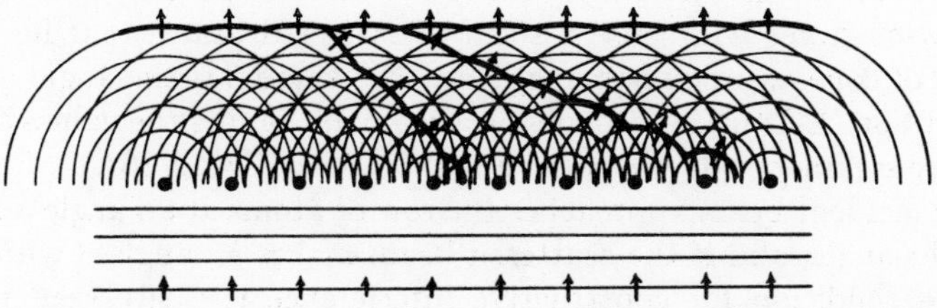

FIG. 15-13. Scattering by a row of atoms. The wave length of the light is of the same order of magnitude as the interatomic spacing.

The total scattering intensity in a given direction can be found by adding the electric fields along a normal to that direction. Suppose that the scattered beams make an angle β with a very long row of atoms $a, b, c, \cdots$ (see Fig. 15-14). Then unless the field strength is the same at $A, B, C, \cdots$ along the

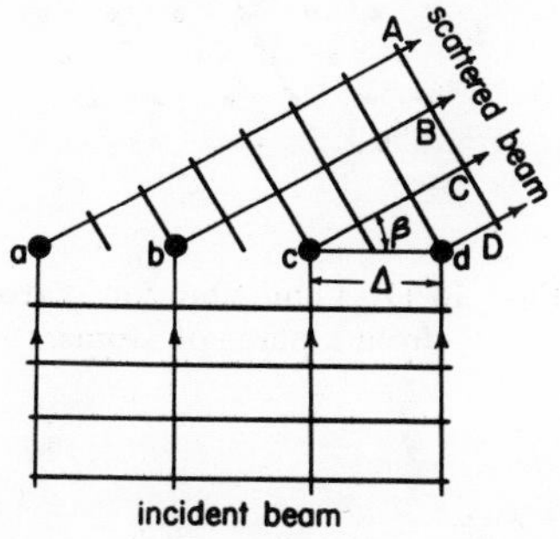

FIG. 15-14. Condition for scattering by a row of atoms; incident beam normal to the row.

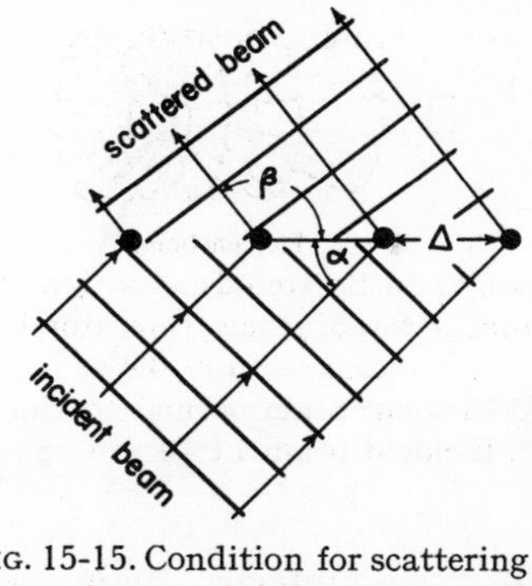

FIG. 15-15. Condition for scattering by a row of atoms; incident beam oblique to the row.

normal to the scattering direction, the rays in this direction from the different atoms will interfere destructively when they fall on a silver bromide crystal or some other detecting device. Constructive interference will occur only if the distance from atom b to point B on the presumed wave front is a whole number of wave lengths greater than or less than the distance from its neighboring atom a to point A. The condition for this is

$$\Delta \cos \beta = \pm n\lambda \tag{E-19}$$

where n is an integer. Thus a beam of radiation impinging at a right angle on a row of atoms will cause a sheaf of scattered beams that lie along the surfaces of cones whose apex angles are 2β, where β is the angle defined by the condition (E-19). These angles occur in pairs having the values β, $180°-\beta$, which are symmetrically disposed about 90° (see Fig. 15-16A).

If the incident beam approaches the row of atoms at an angle α instead of at a right angle, and if the scattered beam makes an angle β with the row, then the condition for constructive interference is easily seen to be (see Fig. 15-15)

$$\Delta(\cos \alpha - \cos \beta) = \pm n\lambda \tag{E-20}$$

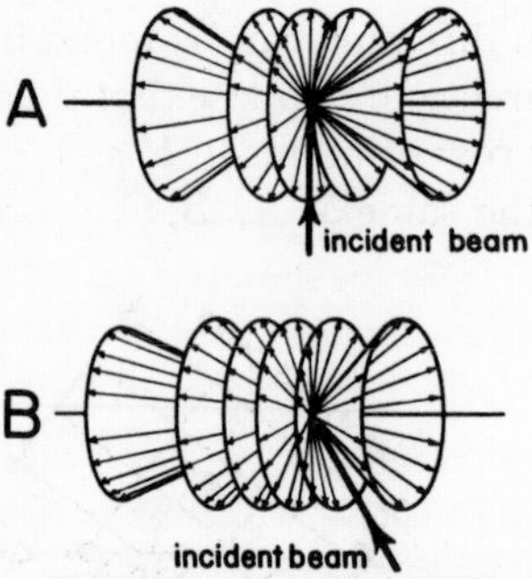

FIG. 15-16. Scattered cones of radiation from a row of atoms. Interatomic spacing $= \Delta = 0.4\,\lambda$.
A. Incident beam normal to the row.
B. Incident beam at 60° to the row.

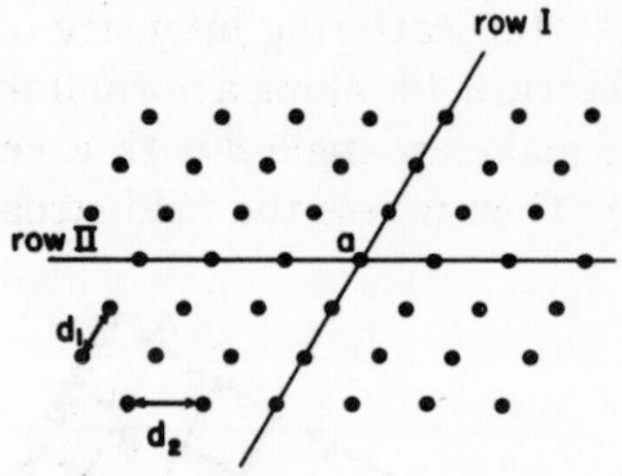

FIG. 15-17. Condition for scattering from a plane of atoms.

where n is an integer. Again the scattering is along cones with apex angles 2β, but the angles β are now subject to the condition (E-20), and they no longer occur in pairs that are symmetrically disposed about an angle of 90° (see Fig. 15-16B).

Exercise. Find the directions of scattering by a row of atoms for which $\Delta = 4\lambda$, first taking $\alpha = 90°$ and then taking $\alpha = 45°$. Draw sketches showing the conical surfaces generated by the scattered rays in the two cases.

Let us consider next the scattering by a plane of regularly spaced atoms. This can be thought of as an arrangement of parallel rows of atoms, each of which would scatter along certain conical surfaces if it were by itself. Being together, however, the scattered rays in the different conical surfaces interfere in such a way that scattering can occur in only a few directions.

Thus the atom *a* in Fig. 15-17 can be considered to be a member of both row I and row II. If the incident beam makes an angle α_1 with row I, then the scattered ray from *a* must satisfy the condition

$$d_1(\cos \alpha_1 - \cos \beta_1) = \pm n_1 \lambda \tag{E-21a}$$

in order not to be destroyed by interference with the rays scattered by the other atoms in row I. But at the same time any scattered ray from atom *a* must also satisfy the conditions for scattering from row II

$$d_2(\cos \alpha_2 - \cos \beta_2) = \pm n_2 \lambda \tag{E-21b}$$

where α_2 is the angle between row II and the incident beam. If these two conditions are fulfilled at atom *a*, then they will hold for any other atom in the plane. Thus the scattered beams from a regular planar arrangement of atoms must fall along the intersections of the two sets of cones defined by the conditions (E-21a) and (E-21b). It is significant that one of these scattered beams obeys the law of specular reflection (see Exercise 2, below). That is, a layer of atoms acts like a partially silvered mirror, reflecting a portion of the incident beam. As usual, one of the scattered beams lies in the same direction as the incident beam.

Exercises. (1) A large number of atoms are arranged in a plane in a simple square array. A beam of radiation impinges on the plane, making an angle of 45° with the plane and an angle of 90° with one of the sides of the squares. The sides of the squares are three times the wave length of the incident beam. Find the directions of all of the scattered beams, and show that one of them obeys the ordinary law of reflection from a plane mirror (angle of incidence equals the angle of reflection).

(2) Setting $n_1 = n_2 = 0$ in Equations (E-21a) and (E-21b), one finds that there must always be a scattered beam for which

$$\cos \alpha_1 = \cos \beta_1$$
$$\cos \alpha_2 = \cos \beta_2$$

Sketch the two cones so defined and show that they intersect along two directions, one of which obeys the law of reflection from a plane mirror lying in the same plane as the atoms, while the other is in the direction of the incident beam.

Finally, let us consider what will happen when a beam of radiation falls on a set of identical regular planes of atoms stacked on top of each other at constant intervals. Each atom must be considered to be a member of three independent rows of atoms not lying in the same plane. If the spacings of these rows are d_1, d_2, and d_3, and if the incident beam makes angles α_1, α_2, and α_3, respectively, with these rows, it is clear that any beam scattered by the crystal as a whole must satisfy simultaneously the conditions

$$
\begin{aligned}
(\cos\alpha_1 - \cos\beta_1) &= \pm\, n_1\lambda/d_1 \\
(\cos\alpha_2 - \cos\beta_2) &= \pm\, n_2\lambda/d_2 \\
(\cos\alpha_3 - \cos\beta_3) &= \pm\, n_3\lambda/d_3
\end{aligned}
\qquad \text{(E-22)}
$$

where n_1, n_2, and n_3 are integers. In addition, if the three rows are perpendicular, the direction cosines of the scattered beam must satisfy the trigonometric identity

$$\cos^2\beta_1 + \cos^2\beta_2 + \cos^2\beta_3 = 1 \qquad \text{(E-23)}$$

If the rows are not perpendicular, a similar but more complicated identity must be satisfied. Thus we have four equations relating the three quantities β_1, β_2, and β_3, which define the direction of the scattered beam. It is clear that these four equations cannot be satisfied simultaneously for all values of α_1, α_2 and α_3. In other words, a three dimensional crystal can only scatter radiation when it has certain definite orientations relative to the incident beam, such that conditions (E-22) and the identity (E-23) (or its counterpart) are fulfilled. Even in this orientation only a few (usually only one) directions of scattering are possible.

Exercise. Assume that a simple cubic crystal has $d_1 = d_2 = d_3 = d$. Find all of the sets of angles α_1, α_2, and α_3 for which scattering of a beam of radiation is possible when $\lambda = d$. Give the directions of the scattered beams for each set.

The scattering from a crystal is more commonly treated by the simple Bragg theory, which involves the assumption (proven above) that the planes in a crystal reflect x-rays according to the laws of specular reflection. The theory developed above is more general, however, and emphasizes the basic cause of x-ray diffraction as a scattering phenomenon. The criteria (E-22) and (E-23) are exactly equivalent to the simple and familiar Bragg condition,

$$n\lambda = 2d \sin\theta$$

where d is the spacing of a set of equivalent planes and θ is the angle between the incident beam and the reflecting plane. A simple demonstration of this equivalence will be found in Buerger (*10*, pp. 43-47).

f. Small-angle scattering by particles

The radiation scattered by a crystal or other object in the same direction as the incident beam always interferes constructively with itself, whereas that scattered in other directions is very weak, except perhaps in a few directions when the wave length is of the same order of magnitude as the interatomic spacing. What can we say about the scattering in the transition region between the forward direction and these other directions? This scattering is called "small-angle scattering" and we shall see that it is determined essentially by the dimensions of the particles through which the radiation travels.

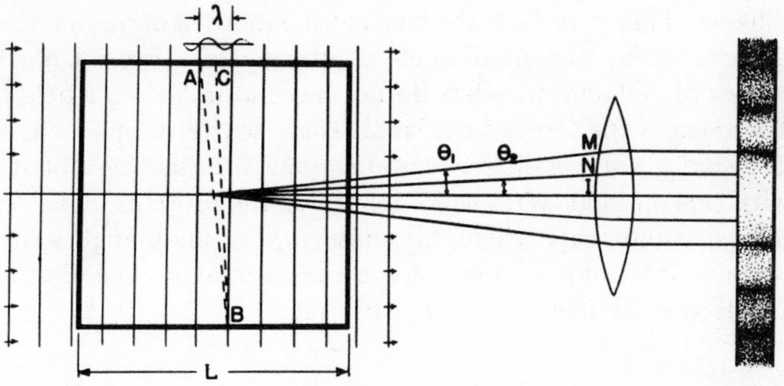

FIG. 15-18. Small angle scattering by a cube.

Consider a cube of matter having a side of length L. A beam of light of wave length λ is allowed to pass through the cube, as shown in Fig. 15-18. Draw a line AB making an angle $\theta_1 = \tan^{-1}(\lambda/L)$ with the wave fronts of the incident beam. The light scattered by the atoms on AB, in the direction M normal to AB, contains all phases from 0 to 2π, and hence will interfere destructively with itself when brought to a focus at M by the lens. This is not the case, however, for the light scattered in the direction N by atoms along the line BC, which makes an angle $\theta_2 = \tan^{-1}(\lambda/2L)$ with the wave fronts of the incident beam. These atoms all vibrate with phases in the same half cycle of the incident beam, so destructive interference does not occur.

It is readily seen that if the wave length is much smaller than the dimensions of the cube, the light scattered in the forward direction will give a diffraction pattern of the character shown on the right in Fig. 15-18, when it is brought to a focus by a lens. Most of the intensity will be concentrated in the central peak, whose edges subtend an angle

$$\varphi = 2\theta_1 \simeq 2\lambda/L$$

The exact form of the small-angle diffraction pattern will, of course, vary with the orientation of the cube relative to the incident beam. It will also be different for particles of other shapes – such as spheres, rods, etc. But it should generally be true that the "wings" added to the incident beam by the small-angle scattering of the particles will subtend an angle of the order of magnitude of the ratio of the wave length to some significant linear dimension of the particle.

One would therefore expect that by analysis of the small-angle scattering information might be gained not only about particle sizes, but also about particle shapes. This is, in fact, the basis of the method of finding the sizes of colloidal particles by the small-angle scattering of x-rays. Unfortunately most samples of colloidal powders do not have a uniform particle size, and this fact influences the form of the small-angle scattering to such an extent that it is usually not possible to obtain much information about particle shapes with this method. With colloids such as the proteins, however, where the particle size and shape are highly uniform, the small-angle scattering of x-rays can give information about shape and even about the distribution of matter inside the particle (see *11*, p. 616; *12*).

The scattering of visible light by powders occurs by an entirely different process from that described above. Partial reflection occurs every time light enters and leaves a particle. The surfaces of most particles are not normal to the incident beam, so this results in diffusion of light in all directions, which completely obscures the small-angle scattering. It is only because x-rays are feebly reflected by surfaces that the small-angle scattering can be observed with them.

g. The refraction of light as a consequence of scattering

A polarized plane light wave moving in the positive z-direction has an electric field

$$\boldsymbol{E} = \boldsymbol{E}_0 \cos \omega(t - z/c) \tag{E-24}$$

Let it fall perpendicularly onto a uniform plane layer of atoms of thickness Δz containing N atoms per cubic centimeter, Δz being small compared with the wave length, λ, of the light. If λ does not lie in any of the absorption bands of the atoms, the field $\boldsymbol{E}$ will induce a dipole moment

$$\boldsymbol{m} = a\boldsymbol{E}_0 \cos \omega(t - z/c) \qquad \text{(E-25)}$$

in each atom, where a is the in-phase polarizability. This oscillating dipole moment will give rise to a scattered spherical wavelet from each atom in the layer. The magnitude of the field in the wavelet from an atom at a distance r from the atom is (*cf.* Equation (E-1))

$$\begin{aligned} |\boldsymbol{E}_s| &= -\,[(a\omega^2/rc^2)\ \sin\theta\cos\,\omega(t - r/c)]\,|\boldsymbol{E}_0| \\ &= -\,[(4\pi^2 a/r\lambda^2)\sin\theta\cos\,\omega(t - r/c)]\,|\boldsymbol{E}_0| \qquad \text{(E-26)} \end{aligned}$$

where θ is the angle between $\boldsymbol{E}_0$ and the line from the atom to the point at which $\boldsymbol{E}_s$ is measured

Note that if the polarizability is positive (which will be the case if the incident light does not consist of x-radiation or of light that is strongly absorbed by the atom), then the field in this scattered wavelet is opposite in sign to the field in the incident beam. We may say that the phase of the scattered spherical wavelets has been retarded by 180°. The physical reason for this is that the accelerations of the charges in an oscillating dipole are opposite in direction to the displacements of the charges; the electric field in the scattered radiation tends to have the same sense as the acceleration (see footnote on page 552) whereas the displacement is in phase with the field of the incident light when the polarizability is positive.

The wavelets from all of the atoms will tend to cancel one another in all directions except the forward and backward directions. The wave scattered in the forward direction produces an important optical effect, as we shall now see. Let us consider the resultant of the wavelets from all of the atoms at some distance from the layer. Since the atoms in the layer are close together compared with the wave length of the incident light, a construction similar to that given in Fig. 15-13 will show that, far from the layer, the scattered wavelets will coalesce to give plane waves moving away from the layer in the forward and backward directions (*i.e.*, in the same direction as the incident beam and in the opposite direction to this). It is shown below that the electric field in the scattered forward-moving plane wave is

$$\boldsymbol{E}'_s = -\gamma\Delta z\, \boldsymbol{E}_0 \sin \omega(t - z/c) \qquad \text{(E-27)}$$

where

$$\gamma = 4\pi^2 N a/\lambda$$

and where $\gamma\Delta z$ is a number much smaller than unity. Thus the merging of the spherical scattered wavelets into a plane wave changes the cosine term in (E-26) into a sine term. That is, the phase of the scattered plane wave is 90° ahead of the phase of the spherical scattered wavelets. This means, in turn, that the phase of the forward-scattered plane wave is 90° behind that of the incident light wave.

Therefore, after the incident beam has passed through the layer of atoms, it finds itself accompanied by another (much weaker) plane wave that is 90° behind it in phase. The total electric field in the light wave after passing through the layer is

$$\boldsymbol{E} = \boldsymbol{E}_0 \left[\cos \omega(t - z/c) - \gamma\Delta z \sin \omega(t - z/c)\right] \tag{E-28a}$$

A time τ can be defined such that $\tan \omega\tau = \gamma\Delta z$, so that (E-28a) can be written in the form

$$\begin{aligned}\boldsymbol{E} &= \boldsymbol{E}_0 \left[\cos \omega(t - z/c) - \tan\omega\tau \sin \omega(t - z/c)\right] \\ &= \boldsymbol{E}_0 \left[\cos \omega\tau \cos \omega(t - z/c) - \sin\omega\tau \sin \omega(t - z/c)\right]/\cos \omega\tau \\ &= \boldsymbol{E}_0 (\cos \omega\tau)^{-1} \cos \omega(t + \tau - z/c)\end{aligned}$$

Since $\gamma\Delta z$ is small, $\omega\tau$ must be small, so that $\cos\omega\tau = 1$ very nearly, giving

$$\boldsymbol{E} = \boldsymbol{E}_0 \cos \omega(t + \tau - z/c) \tag{E-28b}$$

This result is interesting because it shows that the action of the layer of atoms on the incident light beam is the same as if the light beam had been delayed by a time τ in its passage through the layer. This delay is proportional to the thickness of the layer, so the transmitted beam acts as though its velocity had been reduced while it was passing through the layer. If the apparent velocity of light in the layer is given by

$$c' = c/n \tag{E-29}$$

where n is a number larger than unity, then the apparent time-delay would be

$$\tau = \frac{\Delta z}{c'} - \frac{\Delta z}{c} = (n - 1)\Delta z/c = (\gamma/\omega)\Delta z \tag{E-30}$$

giving

$$n - 1 = c\gamma/\omega = \lambda\gamma/2\pi$$

or

$$n - 1 = 2\pi N a \tag{E-31}$$

It is interesting to note that if the polarizability is negative (as is the case for an incident beam of x-rays, for instance), then τ is negative and the scattered radiation seems to effect an increase in the speed of light in its passage through the layer.

The phenomenon of refraction is commonly explained by assuming that the velocity of light in matter is different from that in a vacuum, the ratio of velocities being called the *index of refraction*. This ratio is simply the number n that we have just calculated in terms of the polarizability of the atoms. We see that the light scattered in the forward direction is responsible for the apparent change in velocity, and it therefore gives rise to the phenomenon of refraction.

The scattering in the backward direction by a slab of finite thickness can be shown to lead to the observed laws of reflection at an interface (see *13*, *14*). Furthermore, if the back-scattered light is included in the treatment of a thick layer, it is found that an extra factor of $2/(n+1)$ is obtained on the right hand side of (E-31),

$$n - 1 = 4\pi N\alpha/(n+1)$$

or

$$n^2 - 1 = 4\pi N\alpha \tag{E-31a}$$

If, in addition, the mutual electrical interactions of the induced dipole moments of the atoms in the layer are taken into account, the field of the incident light wave in the layer is increased by $(n^2+2)/3$ (the "Lorentz factor") as compared with the field of the same light wave in a vacuum (*1*, pp. 138, 305; *15*, p. 197). As a result the induced dipole moment is increased by the same factor, and (E-31a) becomes

$$n^2 - 1 = (4\pi/3)\,(n^2+2)\,N\alpha \tag{E-31b}$$

which is the well-known *Lorentz-Lorenz equation*. If this equation is written in the form

$$R = \frac{n^2-1}{n^2+2}\,\frac{M}{d} = \frac{4\pi}{3}\,\alpha\,\frac{NM}{d} = \frac{4\pi}{3}\,N_0\alpha \tag{E-31c}$$

where M is the molecular weight, d is the density, and N_0 is Avogadro's number, then the quantity R is called the *molar refraction*.

Evidently Equation (E-31) is valid only when n is close to unity. A factor $[(n^2+2)/3]^2$ of similar origin should also appear on the left hand side of Equation (D-36); since the electric field of a light wave acting on an absorbing atom is larger than the field in the same light wave *in vacuo*, the electronic motions occur with larger amplitude and the dissipation is correspondingly increased.

The dependence of the refractive index on the frequency of the incident light can be found by substituting for the polarizability from Equation (D-15). If Equation (E-31) is used for the sake of simplicity, we find

$$n = 1 + (Ne^2/2\pi m_e) \sum f_i/(\nu_i^2 - \nu^2) \qquad \text{(E-31d)}$$

Evidently the frequency dependence of the refractive index should resemble that of the polarizability. If one allows the frequency to approach the value characteristic of an absorption band, the index of refraction increases, the increase being most pronounced for strong bands (that is, bands having large oscillator strengths, f_i). This is the phenomenon of *dispersion.* It is interesting to note that at very high frequencies the denominators in the sum in (E-31 d) are negative, so that the refractive index is less than unity. Indices of refraction slightly less than one are in fact invariably observed for x-rays passing through matter. Thus glass prisms deflect x-rays in the opposite direction to that found with visible light (see Compton and Allison (*5,* Chapter 4)).

The frequency dependence of the refractive index will be discussed further in Chapter 16 (pages 688*ff*.).

Proof of Equation (E-27). Let us now calculate the net field at a point P located a distance R from a thin layer of atoms that scatter radiation according to Equation (E-26). The layer will first be divided into concentric zones in the following way (see Fig. 15-19A). Let the zones be centered at the point O where OP is normal to the layer. The diameters of the zones are so chosen that the distance from P to the outer edge of the nth zone is $r_n = R + n\lambda/2$. Then the total electric field at P due to the scattered radiation from the 1st, 3rd, 5th, $\cdots$ zones will be opposite in sign to the electric fields from the 2nd, 4th, 6th, $\cdots$ zones, because the phases will be exactly opposite. Let ϱ, φ be the polar coordinates of a point in the layer, the origin being at O, and the direction $\varphi = 0$ being parallel to the direction of polarization of the incident beam. (Recall that the propagation direction of the incident beam has been assumed to be normal to the layer of atoms, so that at a given instant all atoms in the layer are subject to the same electric field and therefore have the same induced dipole moment.)

The electric field, $\boldsymbol{E}_s$, whose magnitude is given by (E-26), has a direction normal to the line joining the point P with the atom from which it arises. It can be seen from Fig. 15-19B, however, that only the vertical component of $\boldsymbol{E}_s$ (*i.e.*, the component parallel to the plane of atoms) need be considered; the non-vertical components of $\boldsymbol{E}_s$ arising from an atom at the point (ϱ, φ)

will exactly cancel these components from an atom at $(\varrho, \varphi + \pi)$ (see Fig. 15-19B). The vertical component in question is $|\boldsymbol{E}_s| \sin \theta$ where θ is the same angle as that appearing in Equation (E-26). Thus the magnitude of the

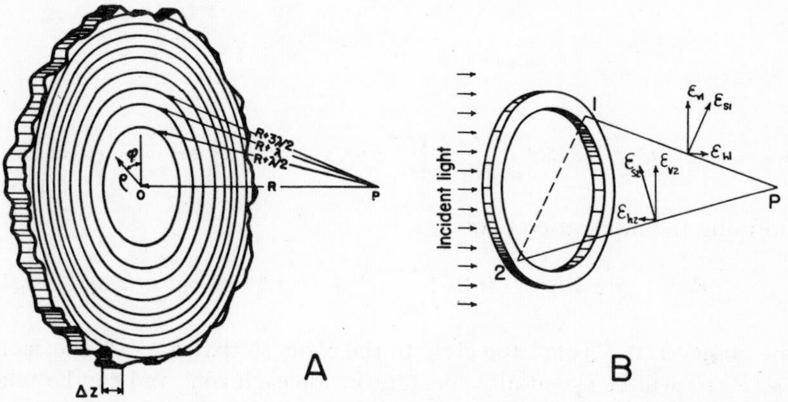

FIG. 15-19. Calculation of the net scattered radiation from a plane of atoms. I. A. Division of the plane into zones. B. Scattering from a single zone. Points 1 and 2 lie on opposite ends of a diameter. The electric fields in the scattered radiation from these two points are $\boldsymbol{E}_{s_1}$ and $\boldsymbol{E}_{s_2}$. Each of these fields can be resolved into two components; the one component, $\boldsymbol{E}_{h_1}$ and $\boldsymbol{E}_{h_2}$, respectively, lies along the direction of propagation of the incident light beam, and the other component, $\boldsymbol{E}_{v_1}$ and $\boldsymbol{E}_{v_2}$, respectively, is parallel to the direction of the electric fields in the indident light. At location P, $\boldsymbol{E}_{v_1}$ and $\boldsymbol{E}_{v_2}$ point in the same direction and interfere constructively, whereas $\boldsymbol{E}_{h_1}$ and $\boldsymbol{E}_{h_2}$ point in opposite directions and cancel one another.

resultant electric field of the radiation scattered by all of the atoms in the nth zone is

$$|\boldsymbol{E}_n| = \int_0^{2\pi} \int_{\varrho_{n-1}}^{\varrho_n} N \, \Delta z \, |\boldsymbol{E}_s| \sin \theta \, \varrho \, d\varrho \, d\varphi \tag{E-32a}$$

where N is the number of atoms per unit volume in the layer, Δz is the thickness of the layer, and ϱ_{n-1} and ϱ_n are the outer radii of the $(n-1)^{\text{st}}$ and n^{th} zones. Substituting for $\boldsymbol{E}_s$ from (E-26), we find

$$\boldsymbol{E}_n = -\boldsymbol{E}_0(a\omega^2/c^2)N \, \Delta z \int_0^{2\pi} \int_{\varrho_{n-1}}^{\varrho_n} (1/r) \sin^2 \theta \cos \omega(t - r/c) \varrho d\varrho \, d\varphi \tag{E-32b}$$

r being the distance from the point P to (ϱ, φ). It is readily shown that

$$\sin^2 \theta = (\sin^2 \varphi) + (R^2/r^2) \cos^2 \varphi$$

Furthermore

$$r^2 = R^2 + \varrho^2$$

so that

$$\varrho d\varrho = r\, dr$$

and

$$\boldsymbol{E}_n = -(a\omega^2/c^2)N\, \Delta z\, \boldsymbol{E}_0 \int_0^{2\pi} \int_{r_{n-1}}^{r_n} \left[\frac{R^2}{r^2} \cos^2 \varphi + \sin^2 \varphi\right] \cos \omega(t - r/c) dr\, d\varphi \tag{E-32c}$$

Performing the integration over φ

$$\boldsymbol{E}_n = -\pi(a\omega^2/c^2)N\, \Delta z\, \boldsymbol{E}_0 \int_{r_{n-1}}^{r_n} [1 + (R/r)^2] \cos \omega(t - r/c) dr \tag{E-32d}$$

If the observer at P is not too close to the zone, so that $R \gg \lambda$, the factor $[1 + (R/r)^2]$ will be essentially constant inside each zone and can be taken out of the integral. Writing r_n for the mean distance from the nth zone to P, we obtain for the scattering from those zones which lie completely inside the layer

$$\boldsymbol{E}_n = \pi(a\omega^2/c^2)N\, \Delta z\, \boldsymbol{E}_0 \frac{c}{\omega} \left[1 + \left(\frac{R}{r_n}\right)^2\right] \left[\sin \omega \left(t - \frac{R + n\lambda/2}{c}\right) - \sin \omega \left(t - \frac{R + (n-1)\lambda/2}{c}\right)\right] \tag{E-32e}$$

For those outermost zones that include only parts of the layer, the field will be smaller than this. Since $\lambda = 2\pi c/\omega$ and $\sin(x + n\pi) = (-1)^n \sin x$, we find

$$\boldsymbol{E}_n = (-1)^n (2\pi a\omega/c)N\, \Delta z\, \boldsymbol{E}_0 \left[1 + \left(\frac{R}{r_n}\right)^2\right] \sin \omega(t - R/c) \tag{E-32f}$$

The zone having the largest value of $\boldsymbol{E}_n$ is the innermost one, for which the term $(1 + (R/r_n)^2$ has its largest value. For this zone we have $R = r_n$ very nearly, so that

$$\boldsymbol{E}_1 = -(4\pi a\omega/c)\, N\, \Delta z\, \boldsymbol{E}_0 \sin \omega(t - R/c) \tag{E-32g}$$

The radius of the central zone is $\sqrt{(R + \lambda/2)^2 - R^2}$, which is very nearly equal to $\sqrt{R\lambda}$ if $R \gg \lambda$. The number of atoms in this zone is thus $N\Delta z\, \pi R\lambda$.

If all of the radiation scattered from the first zone in the direction of P arrived at P with the same phase, the resultant electric field at P would be

$$-(N\,\Delta z\,\pi R\lambda)\,(a\omega^2/Rc^2)\boldsymbol{E}_0 \cos\,\omega(t - R/c) = -(2\pi^2 a\omega/c)N\,\Delta z\,\boldsymbol{E}_0 \cos\,\omega(t - R/c)$$

The different times of arrival at P of the scattered rays from different parts of the innermost zone therefore result in a decrease in field strength by a factor $(2/\pi)$, and an advance in phase by one quarter cycle, as is seen from, the value given in (E-32g). This is the origin of the phase shift that we have found to be essential to the explanation of refraction in terms of scattering.

As we pass from one zone to the next, the contribution $\boldsymbol{E}_n$ gradually decreases in magnitude because of the factor $(1 + (R/r_n)^2)$. Therefore, in calculating the total effect at P

$$\boldsymbol{E}'_s = \sum_{\substack{\text{all}\\ \text{rings}}} \boldsymbol{E}_n \tag{E-33}$$

we must find the sum of a series in which successive terms have opposite signs and gradually decreasing magnitudes. Such a sum is readily evaluated by a graphical method that will now be described.

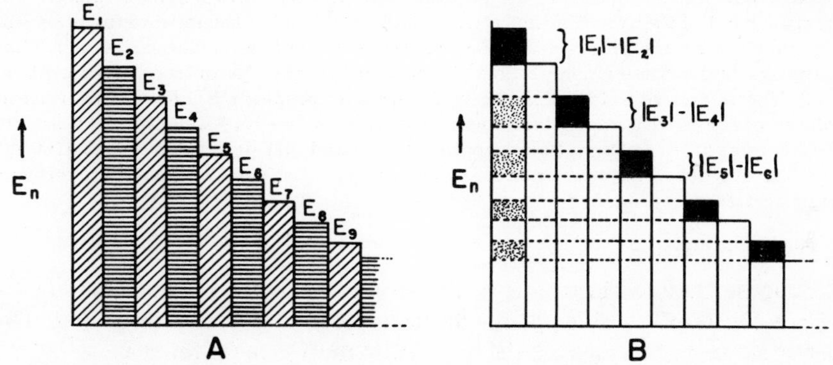

FIG. 15-20. Calculation of the net scattered radiation from a plane of atoms. II.

Figure 15-20A is a plot of $|\boldsymbol{E}_n|$ *vs.* n. The sum in (E-33) is evidently numerically equal to the areas of the shaded rectangles in this figure, the rectangles with horizontal shading being considered as having the opposite signs to those with oblique shading. Let us write (E-33) in the form

$$|\boldsymbol{E}'_s| = (|\boldsymbol{E}_1| - |\boldsymbol{E}_2|) + (|\boldsymbol{E}_3| - |\boldsymbol{E}_4|) + \cdots \tag{E-33a}$$

Since each $\boldsymbol{E}_n$ in this series is larger than the term following it, every one of the terms in parentheses has the same sign. In fact, each term in parentheses is numerically equal to the area of one of the solid black blocks in Fig. 15-20B. Now if $\boldsymbol{E}_n$ varies slowly with n, these solid black blocks clearly total in area to one half of the area of the first rectangle, $\boldsymbol{E}_1$, whose value is given in (E-32g).

Thus we have proven the remarkable result that at a distance R from the layer the total effect of the scattered radiation from a plane of atoms is simply one half the effect of that portion of the plane included in the innermost circle drawn in Fig. 15-19A; that is

$$\begin{aligned} \boldsymbol{E}'_s &= -(2\pi N\alpha\omega/c)\Delta z \sin \omega(t - R/c)\, \boldsymbol{E}_0 \\ &= -(4\pi^2 N\alpha/\lambda)\Delta z \sin \omega(t - R/c)\, \boldsymbol{E}_0 \end{aligned} \qquad \text{(E-34)}$$

which is the result used in Equation (E-27).

We may note that for most substances in the solid state, $N\alpha$ is approximately unity, and for gases at ordinary pressures and temperatures, it is very much smaller than this. We have assumed that $\Delta z/\lambda$ is very much less than unity. Thus $(4\pi^2 N\alpha/\lambda)\Delta z$ is also a number much smaller than unity. The scattered beam is therefore much less intense than the incident beam.

Exercises. (1) Show that if the incident light has its wave length within an absorption band of the atoms in the layer, the light scattered in the forward direction will not merely alter the phase of the transmitted beam, but its amplitude as well. Thus scattering under these circumstances can account for the phenomenon of absorption.

(2) The refractive index of air at 0° C. and 1 atmosphere is 1.000293. A vertical column of the earth's atmosphere whose cross-section is 1 cm.^{2} weighs 1 kg. Compute the fractions of the red light (wave length 6500 Å.) and blue light (wave length 4100 Å.) from an overhead star that are able to get through the earth's atmosphere without being scattered.

h. Double refraction

Suppose that an electron is bound to a molecule in such a way that its "force constant" is different for displacements in different directions. The potential energy of such an electron, in its most general form, is

$$V = a_{11}x^2 + a_{12}xy + a_{13}xz + a_{22}y^2 + a_{23}yz + a_{33}z^2$$

where x, y, and z are the displacements from the equilibrium position along directions parallel to the axes of a Cartesian coordinate system, and the a_{ij} are constants. By a rotation of the coordinate axes, this potential can always be reduced to the form

$$V = \tfrac{1}{2}AX^2 + \tfrac{1}{2}BY^2 + \tfrac{1}{2}CZ^2$$

where X, Y, and Z are displacements referred to new coordinate axes, called the *principal axes*, and A, B, and C are the force constants for displacements in the X, Y, and Z directions, equivalent to the force constant k that has been used elsewhere in this chapter. Such an electron is said to be *anisotropically bound* if two or more of the constants A, B and C are unequal.

From a classical point of view the electrons in all molecules ought to be anisotropically bound. This should be especially true of electrons involved in or close to chemical bonds, the force constants for displacements parallel to the bond direction being somewhat smaller than those for displacements in directions perpendicular to the bond [see, for example, the Silberstein theory of molecular polarizability, as outlined by Stuart (*16*, p. 364)].

If plane polarized light falls onto a layer of molecules containing anisotropically bound electrons, the response will depend on the orientation of the molecules with respect to the plane of polarization. Let E_X, E_Y, and E_Z be the X, Y, and Z components of the amplitude of the electric field in the light wave. Then the components of the dipole moment induced by this field will be

$$m_X = \alpha_X E_X$$
$$m_Y = \alpha_Y E_Y$$
$$m_Z = \alpha_Z E_Z$$

where the α's are the polarizabilities in the three directions. For a single electron we will have

$$\alpha_X = \frac{e^2}{A} \frac{\nu_{0X}^2}{\nu_{0X}^2 - \nu^2}; \qquad \nu_{0X}^2 = \frac{1}{4\pi^2} \frac{A}{m_e}$$

with similar expressions for α_Y and α_Z. If there is more than one electron, it is necessary to introduce oscillator strengths and to sum over the contributions of all absorption lines in the molecule, as has already been explained

$$\alpha_X = \frac{e^2}{4\pi^2 m_e} \sum \frac{f_{iX}}{\nu_{0i}^2 - \nu^2}$$

and similarly for α_Y and α_Z.

In an anisotropic molecule α_X, α_Y, and α_Z will have different values. If the incident light is polarized in a direction parallel to one of the three principal axes, so that two of the components E_X, E_Y, and E_Z are zero, the induced dipole moment will also be parallel to this principal axis. The scattered light will then have the same plane of polarization as that of the incident light, and retardation of the incident wave will take place in the normal fashion.

The intensity of the forward scattering, and hence the refractive index, will of course be different for the three principal directions of polarization.

If, on the other hand, the electric field in the incident light is parallel to neither of the three principal axes, and if the three principal polarizabilities are unequal, the induced dipole moment will not be parallel to the direction of polarization of the incident light. This means that the light scattered by the induced dipoles will have a different plane of polarization from that of the incident beam. Since the scattered light is a quarter of a cycle out of phase with the incident waves, the light that has passed through an anisotropic layer under these conditions will no longer be plane polarized, but instead will be elliptically polarized. If an anisotropic crystal is placed between crossed polaroids or crossed Nicol prisms in such a way that none of its principal axes are parallel to the plane of polarization of the incident light, the crystal will therefore appear to be luminous. This is the origin of the striking optical polarization phenomena associated with anisotropic crystals.

Behavior of this kind is particularly noticeable in crystalline solids, in which one or at most a few orientations of molecules are repeated throughout each crystallite. For instance, the carbonate ion is known to be planar, with the carbon–oxygen bonds lying in the plane of the ion, so that its polarizability is much greater in the plane than in a direction perpendicular to it. In crystals of calcite (which belongs to the trigonal system) the carbonate ions are all parallel to each other and perpendicular to the trigonal axis. Therefore the refractive index of calcite is much greater for polarized light whose plane of polarization is perpendicular to the trigonal axis than for light polarized parallel to this axis ($n_\perp = 1.658$, $n_\parallel = 1.486$ for the Na D-line). The same situation occurs in sodium nitrate, which has the same crystal structure as calcite ($n_\perp = 1.587$, $n_\parallel = 1.336$). In anthracene and other aromatic hydrocarbons of high molecular weight, the flat aromatic rings are arranged in the crystal nearly parallel to one another, and there is a similar very large birefringence. On the other hand, fibrous materials such as cellulose, nylon, and long-chain hydrocarbons consist of molecular chains running parallel to one another. Such substances usually have a higher refractive index in the direction of the fiber axis than in the direction perpendicular to this axis.

F. Other Modes of Producing Radiation

In our discussions up to this point, we have studied the relationships between radiation and the oscillating electric dipoles in atoms and molecules.

Certain types of electronic motion are conceivable, however, in which there is no change in the net electric dipole moment of the atom or molecule. We shall now show that some of these motions can produce radiation, though of much weaker intensity than dipole radiation.

a. Quadrupole radiation

Consider for instance the arrangement shown in Fig. 15-21A, in which two identical dipoles, I and II, are placed on opposite sides of the point P, both

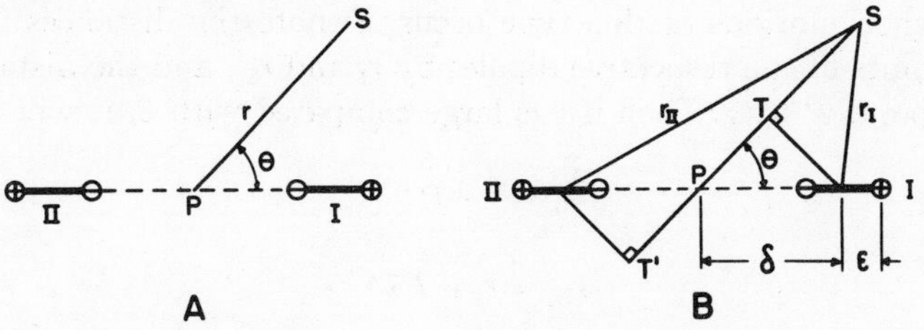

FIG. 15-21. An oscillating quadrupole.

having their negative ends pointing toward P. Such an array is known as a quadrupole. It has no net dipole moment but gives an electrostatic field that, at large distances r from the point P, has the potential (*cf.* Fig. 3-18 and pages 96 and 484)

$$V = \frac{Q}{r^3}\left(\cos^2\theta - \frac{1}{3}\right) \tag{F-1}$$

where θ is the angle between r and the axis of the array, and Q is the quadrupole moment of the array. Since the electric field strength is the rate of change of the potential with distance, the electric field strength will vary as the inverse fourth power of the distance, r.

The quadrupole moment of this array, Q, is given by

$$Q = \sum_{\substack{\text{all}\\ \text{charges}}} e_i(3x_i^2 - r_i^2) = \sum 2e_i r_i^2 \tag{F-2}$$

where r_i is the distance of the ith charge from P, and x_i is the projection of

r_i on the axis of the array. If δ is the distance from the midpoint of each dipole to P and if the charge separation in each dipole is 2ε, then

$$Q = e\,[2(\delta + \varepsilon)^2 - 2\,(\delta - \varepsilon)^2] = 8e\varepsilon\delta = 4m\delta \tag{F-2a}$$

the dipole moment of each dipole being given by

$$m = 2e\varepsilon \tag{F-2b}$$

Suppose now that the two negative charges in the array oscillate in opposite directions with equal amplitudes. The net dipole moment of the array will continue to be zero at all times, but the quadrupole moment will vary with time. Let us find the electric field in the light that is radiated to a point S when motions of this type occur. Denote the distances from S to the midpoints of the respective dipoles by r_{I} and r_{II}, and the distance from S to the point P by r. Then if r is large compared with δ, it will be nearly true that

$$r_{\mathrm{I}} = r - PT$$

and

$$r_{\mathrm{II}} = r + PT'$$

T and T' being the ends of the projections of r_{I} and r_{II} on SP, respectively (see Fig. 15-21B). The lengths of PT and PT' are equal and will be denoted by γ, which is given by

$$\gamma = \delta \cos\theta = PT = PT'$$

The magnitude of the field at S caused by dipole radiation from I is (cf. Equation (E-1).

$$|\boldsymbol{E}_{\mathrm{I}}| = -\frac{m_0}{r-\gamma}\,\frac{\omega^2}{c^2}\sin\theta\cos\omega[t - (r-\gamma)/c] \tag{F-3a}$$

and that caused by II is

$$\boldsymbol{E}_{\mathrm{II}}| = \frac{m_0}{r+\gamma}\,\frac{\omega^2}{c^2}\sin\theta\cos\omega[t - (r+\gamma)/c] \tag{F-3b}$$

where m_0 is the amplitude of the oscillating dipole moment. The net field magnitude is

$$|\boldsymbol{E}| = |\boldsymbol{E}_{\mathrm{I}}| + |\boldsymbol{E}_{\mathrm{II}}| = \frac{m_0\,\omega^2 \sin\theta}{c^2(r^2-\gamma^2)}\left[2r\sin\omega\left(t - \frac{r}{c}\right)\sin\omega\frac{\gamma}{c} - 2\gamma\cos\omega\left(t-\frac{r}{c}\right)\cos\omega\frac{\gamma}{c}\right] \tag{F-4}$$

If the separation of the two dipoles is small compared with the wave length, $\lambda(=2\pi c/\omega)$, then $\omega\gamma/c \ll 1$, and sin $\omega\gamma/c$ may be replaced by $\omega\gamma/c$ Furthermore, r may be assumed to be large compared with the wave length, so the second term in the brackets of Equation (F-4) can be neglected in comparison with the first. This gives

$$|\boldsymbol{E}| = 2\gamma m_0 \frac{\omega^3}{c^3 r} \sin\theta \sin\omega\left(t - \frac{r}{c}\right)$$

$$= 2\delta m_0 \frac{\omega^3}{c^3 r} \sin\theta \cos\theta \sin\omega\left(t - \frac{r}{c}\right) \qquad \text{(F-5a)}$$

Writing

$$Q_0 = 4\delta m_0 \qquad \text{(F-5b)}$$

where Q_0 is the amplitude of the oscillation of the quadrupole moment we have

$$|\boldsymbol{E}| = \frac{1}{2} Q_0 \frac{\omega^3}{c^3 r} \sin\theta \cos\theta \sin\omega\left(t - \frac{r}{c}\right) \qquad \text{(F-5c)}$$

Thus an oscillating quadrupole emits radiation whose field strength varies inversely with the distance. This dependence on distance is the same as that found for dipole radiation. The dependence on direction is, however, not the same for the two types of radiation (note the factor $\sin\theta\cos\theta$ in the quadrupole radiation expression, (F-5c), whereas the dipole radiation expression (A-5b) contains the factor $\sin\theta$). Because of this difference in dependence on direction, it is possible to determine experimentally the nature of the oscillations responsible for the light emitted by a system. A simple and ingenious interferometric method of detecting this difference for fluorescent and phosphorescent substances has been devised by Selenyi (see Fried and Weissman (*17*) for a description of the method, along with an application and further references). Dipole and quadrupole radiations also differ in the dependence of the field strength on the frequency: with dipole radiation, the field strength is proportional to the square of the frequency, whereas the field strength of quadrupole radiation is proportional to the cube of the frequency.

It is interesting that in spite of the fact that the *electrostatic field* from a quadrupole varies, as we have seen, in proportion to the inverse *fourth* power of the distance, the *radiation field* varies in proportion to the inverse *first* power of the distance. The radiation field from a quadrupole therefore extends over much greater distances than the electrostatic field.

Exercise. Two identical oscillating dipoles are located with their midpoints on the x-axis and separated by a distance d (Fig. 15-22). The origin is midway between them.

The dipoles are parallel to the z-axis, but one has its positive end pointing upward whereas the other has its positive end pointing downward. The net dipole moment of the pair is always zero. Show that they produce radiation at a distant point whose polar coordinates are r, θ, φ, and that the magnitude of the electric field of the radiation is

$$|\boldsymbol{E}| = Q'_0 \frac{\omega^3}{c^3 r} \sin^2 \theta \cos \varphi \sin \omega\left(t - \frac{r}{c}\right) \quad \text{(F-6a)}$$

where Q'_0 is the amplitude of the oscillation of the quadrupole moment of the array; according to page 96 and Fig. 3-18, the quadrupole moment here is

$$Q = \sum e_i x_i z_i = md \quad \text{(F-6b)}$$

so that

$$Q'_0 = m_0 d \quad \text{(F-6c)}$$

where m_0 is the amplitude of the oscillation of the individual dipole moments.

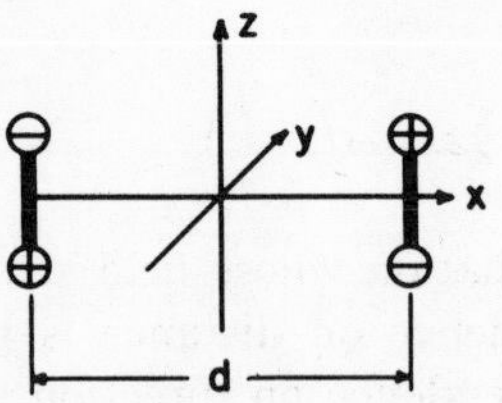

FIG. 15-22. Another type of quadrupole.

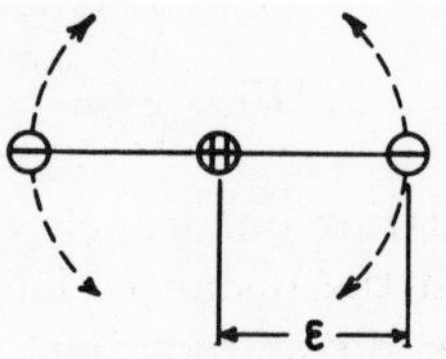

FIG. 15-23. An oscillating magnetic dipole.

b. Magnetic dipole radiation

Another possible arrangement of oscillating charges is that shown in Fig. 15-23, in which two electrons on opposite sides of a positive charge oscillate in a direction perpendicular to the line joining them. Not only is the net dipole moment zero at all times, but the magnitudes of the individual dipole moments are also constant; only the directions of the individual dipoles change with time.

This arrangement is equivalent to part of a circular coil in which there is an oscillating current. At any instant such a coil sets up a magnetic field similar to that produced by a magnetic dipole located at the center of the coil with its axis perpendicular to the coil. The "current" in this case is given by

$$i = ev/2\pi\varepsilon \quad \text{(F-7)}$$

where v is the velocity of the electron and ε is its distance from the positive charge. As we have seen in Equation (D-4) of Chapter 8, the magnitude of the resulting magnetic dipole moment is

$$|\boldsymbol{\mu}| = ev\varepsilon/2c \tag{F-8a}$$

If the displacement of each electron is

$$x = x_0 \cos \omega t \tag{F-8b}$$

the velocity will be

$$v = -\,\omega x_0 \sin \omega t \tag{F-8c}$$

and the magnetic dipole moment will be

$$\boldsymbol{\mu} = -\,\boldsymbol{\mu}_0 \sin \omega t \tag{F-8d}$$

where

$$|\boldsymbol{\mu}_0| = ex_0\varepsilon\omega/2c \tag{F-8e}$$

According to Assumption (4b) on page 554, an oscillating magnetic dipole must radiate in just the same way as does an electric dipole. Therefore this arrangement will produce radiation having an electric field whose magnitude is

$$|\boldsymbol{E}| = -\frac{ex_0\varepsilon\omega^3}{rc^3} \sin\theta \sin\omega(t - r/c) \tag{F-9}$$

where r and θ have their usual meanings. Radiation arising in this way is known as *magnetic dipole radiation.*

c. Relative orders of magnitude of the intensities of dipole, quadrupole, and magnetic dipole radiation

Comparing Equations (A-5b), (F-5c), and (F-9) we see that the amplitudes of the field strengths at a given distance for dipole, quadrupole, and magnetic dipole radiation are in the approximate ratio,

$$|\boldsymbol{E}_{\text{dipole}}| : |\boldsymbol{E}_{\text{quad}}| : |\boldsymbol{E}_{\text{mag}}| \cong \left(m_0 \frac{\omega^2}{c^2}\right) : \left(\frac{3}{2} Q_0 \frac{\omega^3}{c^3}\right) : \left(e\varepsilon x_0 \frac{\omega^3}{c^3}\right)$$

$$\cong m_0 : \left(\frac{3}{2}\,\delta m'_0 \frac{\omega}{c}\right) : \left(e\varepsilon x_0 \frac{\omega}{c}\right)$$

If the distances through which the electrons move are of the same order of magnitude for the three types of radiation, we can say that

$$m_0 \approx m'_0 \approx ex_0$$

Furthermore, $\omega/c = 2\pi/\lambda$ so that

$$|\boldsymbol{E}_{\text{dipole}}| : |\boldsymbol{E}_{\text{quad}}| : |\boldsymbol{E}_{\text{mag}}| \approx 1 : (3\pi\delta/\lambda) : (2\pi\varepsilon/\lambda)$$

Now ε and δ are distances of the order of magnitude of atomic dimensions, say 1 Å., and λ for visible light is of the order of 5000 Å. Thus

$$|\boldsymbol{E}_{\text{dipole}}| : |\boldsymbol{E}_{\text{quad}}| : |\boldsymbol{E}_{\text{mag}}| \approx 1 : \frac{1}{500} : \frac{1}{700}$$

Since the intensity is proportional to the square of the field strength, we have for the order of magnitude of the ratio of the intensities,

$$I_{\text{dipole}} : I_{\text{quad}} : I_{\text{mag}} \approx 1 : 10^{-5} : 10^{-5}$$

Thus we see that the visible light generated by quadrupole or magnetic dipole radiation from atomic or molecular systems is generally much less intense than that generated by electrical dipole radiation.

G. Optical Rotatory Power

Certain substances—for instance, crystalline quartz and solutions of sucrose—have the power of rotating the plane of polarization of a light beam that passes through them (this rotation is not to be confused with the changes from plane to elliptical polarization that occur with anisotropic substances in the manner described on page 610). The phenomenon depends on the ability of a changing magnetic field to induce an electric dipole moment in a molecule, and on the ability of a changing electric field to induce a magnetic dipole moment in a molecule. That is, if $\boldsymbol{m}$ is the induced electric dipole moment in a molecule and $\boldsymbol{\mu}$ is the induced magnetic dipole moment, then optical rotatory power will result only if we can write

$$\boldsymbol{m} = -(\beta/c)\partial\boldsymbol{H}/\partial t \tag{G-1}$$

or

$$\boldsymbol{\mu} = (\gamma/c)\partial\boldsymbol{E}/\partial t \tag{G-2}$$

where β and γ are constants determined by the structure of the molecule, c is the velocity of light and $\boldsymbol{H}$ and $\boldsymbol{E}$ are the magnetic and electric fields acting on the molecule at time t. The minus sign has been introduced into the right side of Equation (G-1) because this leads to slightly simpler equations in the results of the theory. The signs in (G-1) are consistent with the notation that is generally used.

We shall now give a qualitative description of a simple molecular model that is capable of producing relationships such as (G-1) and (G-2). We shall then show, first qualitatively and then quantitatively, how it comes about that any substance whose molecules obey these equations must rotate the plane of polarization of light. Finally, the molecular basis of (G-1) and (G-2) will be discussed in more detail.

a. Qualitative discussion of the origin of optical rotatory power

(*i*) *Origin and Effect of an Electric Dipole Moment Induced by a Changing Magnetic Field.* Let us suppose that there exists a plane layer of molecules, each of which contains electrons that are constrained to move along a right-handed helical path. Let the axes of all of the helices be parallel. Now

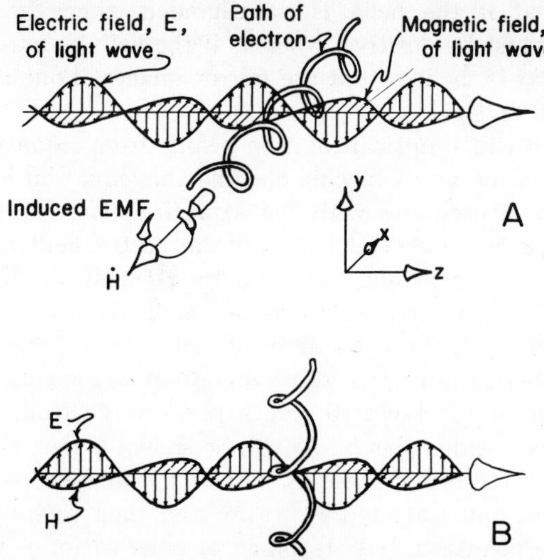

FIG. 15-24.
A. Mechanism of inducing an electric dipole moment in a helix by the changing magnetic field of a light wave.
B. Mechanism of inducing a magnetic dipole moment in a helix by the changing electric field in a light wave (see page 619 for explanation).

shine light on this layer so that the direction of propagation is normal to the plane of the layer and the magnetic field in the light wave is parallel to the axes of the helices (see Fig. 15-24A). The magnetic field, of course, continually changes with time. According to the laws of electromagnetic induction, a changing magnetic field induces an electromotive force in any conductor that encircles the changing magnetic lines of force. Therefore an e.m.f. is induced in the helix, tending to make the electrons move along the helix. This electronic motion produces a change in the molecule's electric dipole moment: the components of the changing dipole moment normal to the axis of the helix change in sign and magnitude each time the electrons negotiate one turn of the helix, while the component of the dipole moment parallel to the axis of the helix increases or decreases monotonically as the electrons move along the helix in a given direction. Consequently the motion of the electrons along the helix results in an electric dipole moment whose average direction is parallel to the axis of the helix. In this way the time variation of the magnetic field in the light wave makes a contribution to the induced electric dipole moment of the molecule, and the direction of this moment is parallel to the direction of the changing magnetic field. Therefore molecules such as these must obey Equation (G-1). It is evident, furthermore, that if the helix is right-handed, then the constant β in equation (G-1) must be positive, whereas if the helix is left-handed β must be negative. That is, helices that are mirror images of one another give β's that are opposite in sign.

The manner in which optical rotation results from Equation (G-1) is not difficult to see. When any molecule obeying this equation is acted on by a light wave, $(\beta/c)\,\dot{\boldsymbol{H}}$ oscillates with the same frequency as the light wave, but its direction is perpendicular to that of the electric field in the light wave. The oscillating electric moment induced by $\dot{\boldsymbol{H}}$ must produce a scattered wavelet in which the electric vector is perpendicular to the electric vector in the incident light. If there is a layer of molecules all obeying (G-1), then the scattered spherical wavelets from the individual molecules will merge at some distance from the layer to give a plane wave front parallel to the wave front of the incident beam. As will be shown below, the phase of this scattered plane wave is the same as that of the incident wave. Therefore, when the incident and scattered waves are combined vectorially, the resultant is still plane polarized, but its plane of polarization is rotated slightly with respect to the plane of the incident light (see Fig. 15-25). Thus we see that whenever light is passed through a layer of molecules that behave according to Equation (G-1), the plane of polarization must be rotated. That is, optical rotatory power must arise whenever an electric dipole moment can be induced in a molecule by a changing magnetic field.* Furthermore, the field in the scattered wave from the right-handed helix is equal in magnitude but opposite in sign to the field from the mirror image left-handed helix. Therefore the rotation of the plane of polarization is equal and opposite in direction in the two cases.

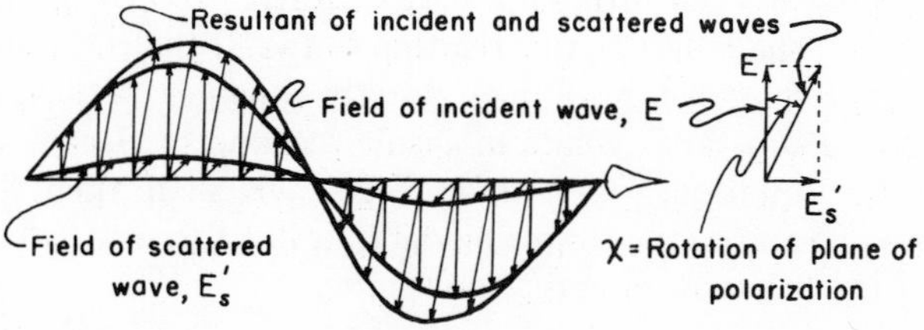

FIG 15-25. Origin of optical rotatory power from light scattered by an electric dipole that has been induced by the changing magnetic field of a light wave.

(*ii*) *Origin and Effect of a Magnetic Dipole Moment Induced by a Changing Electric Field.* Next, let us suppose that the axes of the molecular helices described above are parallel to the electric field in the light wave (that is, with the orientation shown in Fig. 15-24B). The changing electric field in the light beam causes the electrons to move along the helix because of the force that the electric field exerts on the electrons. The helical motion of the electrons produced in this way gives rise to a magnetic field similar to that produced by any flow of current through a solenoid. Since the magnetic field around a solenoid is the same as that around a magnet, we see that under these circumstances a magnetic dipole moment is induced in the molecule. This magnetic dipole is parallel to the electric field in the light wave. The more rapidly the electric field changes, the larger the magnetic moment. Thus we obtain Equation (G-2), where γ is positive if the helix is right-handed and negative if it is left-handed.

Optical rotation is produced in the following way by molecules obeying Equation (G-2). The oscillating magnetic dipole resulting from (G-2) emits radiation in the manner described on page 614. The magnetic field in the scattered wavelet lies in the same plane as the electric field in the incident

* In Section Eh it was shown that a similar nonparallel scattered component can also be produced from anisotropic molecules when they are suitably oriented. The scattered plane wave produced in this case is, however, retarded in phase by one quarter cycle behind the incident wave. Therefore, when the scattered and incident waves are combined, the resultant is no longer plane polarized.

light. Therefore the electric field in the scattered wavelet is parallel to the electric field in the wavelet produced by Equation (G-1); it also has the same phase if γ and β have the same sign. Thus we may use the same arguments as those just given for the induced electric moment, and we find once again that rotation of the plane of polarization must occur. The rotation is equal and opposite in direction for right- and left-handed helices.

We shall now proceed with a more quantitative treatment of the phenomenon. First we shall derive the relation between the rotation of the plane of polarization, the constants β and γ, and the thickness of the layer, assuming that we have a substance whose molecule respond to electromagnetic fields according to Equations (G-1) and (G-2). We shall then determine the magnitudes of β and γ for a simple model that illustrates some of the essential features of the behavior of actual molecules.

b. Relationship between β and γ and the magnitude of the optical rotation

A plane polarized light wave moving in the positive z-direction has an electric field $\boldsymbol{E} = \boldsymbol{E}_0 \cos \omega(t - z/c)$; $\boldsymbol{E}_0$ will be assumed to point in the positive y-direction. The magnetic field in the light wave is $\boldsymbol{H} = -\boldsymbol{H}_0 \cos \omega(t-z/c)$, where $\boldsymbol{H}_0$ points in the positive x-direction (see Assumption 2 on page 548). Let the wave pass through a layer of molecules that obey Equations (G-1) and (G-2), there being N of the molecules per unit volume, and the thickness of the layer being dz. The Lorentz field from the neighboring molecules increases both $\boldsymbol{E}$ and $\boldsymbol{H}$ by a factor $(n^2 + 2)/3$ inside the layer. Because of Equation (G-1), the electric moment induced in the molecules by the light has a contribution perpendicular to $\boldsymbol{E}$ amounting to

$$\boldsymbol{m}_x = -(\beta/c)\,\dot{\boldsymbol{H}} = -\frac{\beta}{c}\frac{(n^2+2)}{3}\,\omega\,\boldsymbol{H}_0 \sin \omega(t - z/c) \tag{G-3}$$

This induced moment produces a scattered wavelet in the usual way. Because of Equation (G-2), a magnetic dipole moment is also induced parallel to $\boldsymbol{E}$

$$\boldsymbol{\mu}_y = (\gamma/c)\,\dot{\boldsymbol{E}} = -\frac{\gamma}{c}\frac{(n^2+2)}{3}\,\omega\,\boldsymbol{E}_0 \sin \omega(t - z/c) \tag{G-4}$$

According to Assumption 4b (page 554), this oscillating magnetic dipole produces a scattered wavelet in which the magnetic field is coplanar with $\boldsymbol{E}_0$. Therefore the electric field in the radiation produced by $\boldsymbol{\mu}_y$ is parallel to that produced by $\boldsymbol{m}_x$; its phase is also the same as that of the radiation produced by $\boldsymbol{m}_x$.

Note that if β and γ are positive, the phases of $\boldsymbol{m}_x$ and $\boldsymbol{\mu}_y$ are 90° behind the phase of $\boldsymbol{H}$. The spherical wavelets generated by the oscillating dipoles merge at some distance from the layer of molecules to give a plane scattered wave front in the manner described on pages 601 and 604*ff*. The phase of the electric field in the plane scattered wave front is 90° behind the phase of $\boldsymbol{m}_x$ and $\boldsymbol{\mu}_y$; therefore, it is not only parallel to the magnetic field of the incident beam, but it is also 180° out of phase with it. Using the same procedure as was employed in going from Equation (E-25) to (E-27), we find that the magnitude of the electric field in the plane, forward-scattered wave front is

$$\boldsymbol{E}'_{\mathrm{s}} = \frac{4\pi^2 N}{\lambda}\, dz\, \frac{n^2+2}{3}\,(\beta+\gamma)\,\frac{\omega}{c}\,\boldsymbol{H}_0 \cos\omega(t - z/c) \tag{G-5}$$

Since the magnetic field in a light wave is numerically equal to the electric field (page 548) we may replace $|\boldsymbol{H}_0|$ by $|\boldsymbol{E}_0|$. On replacing ω/c by $2\pi/\lambda$, we then obtain

$$|\boldsymbol{E}'_{\mathrm{s}}| = \frac{8\pi^3 N}{\lambda^2}\, dz\, \frac{n^2+2}{3}\,(\beta+\gamma)\,|\boldsymbol{E}_0| \cos\omega(t-z/c)$$

$$= \frac{8\pi^3 N}{\lambda^2}\, dz\, \frac{n^2+2}{3}\,(\beta+\gamma)\,|\boldsymbol{E}| \tag{G-6}$$

When the electric fields in the scattered and incident wave fronts are combined, they give a plane wave whose plane of polarization is rotated by a small angle χ given in radians by (*cf*. Fig. 15-25)

$$\chi \cong \tan\chi = |\boldsymbol{E}'_{\mathrm{s}}|/|\boldsymbol{E}| = \frac{8\pi^3 N}{\lambda^2}\,\frac{n^2+2}{3}\,(\beta+\gamma)\, dz \tag{G-7}$$

The angle χ is defined by this expression in such a way that if the plane of polarization is rotated in a clockwise sense when one looks toward the source of the incident light, then χ is positive. This conforms with the experimental convention used in measuring optical rotations.

The experimentally observed optical rotatory power of a substance is usually expressed in terms of the specific rotation, $[\alpha]$. For a pure substance $[\alpha]$ is defined as*

$$[\alpha] = \alpha/l\varrho \tag{G-8}$$

* The wave length of the light used in measuring the optical rotation is often indicated by adding a subscript, and the temperature is sometimes indicated as a superscript. Thus $[\alpha]_{\mathrm{D}}^{20}$ stands for the rotation measured at 20° C using the D-line of sodium.

where α is the rotation in degrees resulting when plane polarized light passes through a layer of the substance l decimeters in thickness, and ϱ is the density of the substance in gm./cc. If the substance is in solution, ϱ is replaced by the concentration, c, in grams per cubic centimeter. The experimentally determined quantity $[\alpha]$ can be related to the calculated angle χ in the following way. For a pure substance

$$N = N_0\varrho/M \tag{G-9}$$

where N_0 is Avogadro's number and M is the molecular weight. In this case, setting $\alpha = 180\,\chi/\pi$ and $l = dz/10$ (dz being measured in centimeters) we have

$$[\alpha] = \frac{180}{\pi} \cdot \frac{10}{\varrho} \cdot \frac{\chi}{dz} \tag{G-10}$$

For a substance in solution

$$N = N_0c/M \tag{G-11}$$

In this case

$$[\alpha] = \frac{180}{\pi} \cdot \frac{10}{c} \cdot \frac{\chi}{dz} \tag{G-12}$$

Substituting either (G-9) and (G-7) into (G-10) or (G-11) and (G-7) into (G-12) gives

$$[\alpha] = \frac{14400\pi^2}{\lambda^2} N_0 \frac{n^2 + 2}{3} \frac{(\beta + \gamma)}{M} \tag{G-13}$$

Another commonly used unit of optical rotation is the *molecular rotation*, defined as

$$[M] = \frac{M}{100} [\alpha] \tag{G-14}$$

$$[M] = \frac{144\pi^2}{\lambda^2} N_0 \frac{n^2 + 2}{3} (\beta + \gamma) \tag{G-15}$$

Evidently the molecular property that determines the magnitude of the optical rotation is expressed through the parameters β and γ. Let us find the order of magnitude of these parameters for typical optically active substances. Assuming that the molecular rotation has the rather large value of 200° when measured with light of wave length 6000 Å., one finds from

(G-15) that $(\beta + \gamma) \cong 8 \times 10^{-34}$. Although larger values than this may be encountered, especially if one uses light whose wave length is close to that of an absorption band of the optically active substance, this may be taken as a rough upper limit to the value of $(\beta + \gamma)$ for most optically active molecules. It is interesting to use this value to compare the size of the dipole moment induced in the optically active molecule by the changing magnetic field with the ordinary induced moment, $\boldsymbol{m} = \alpha \boldsymbol{E}$, where α is the polarizability defined in Equation (D-6a) and $\boldsymbol{E}$ is the electric field. The constant α is generally of the order of magnitude 10^{-22} cc. We shall find that β and γ are usually equal in magnitude, so we shall take $\beta \simeq 4 \times 10^{-34}$. The ratio of the two induced moments is therefore

$$\frac{(\beta/c)\,|\,\dot{\boldsymbol{H}}\,|}{\alpha\,|\,\boldsymbol{E}\,|} \simeq \frac{\omega\beta\,|\,\boldsymbol{H}_0\,|}{c\alpha\,|\,\boldsymbol{E}_0\,|} = \frac{2\pi\beta}{\lambda\alpha} = \frac{2\pi \times 4 \times 10^{-34}}{6000 \times 10^{-8} \times 10^{-22}} = 4 \times 10^{-7}$$

The component of the induced moment that gives rise to the optical rotation is therefore very small compared with the component that is responsible for the ordinary refraction.

Note that if $(\beta + \gamma)$ *is positive, the substance is dextrorotatory, and if* $(\beta + \gamma)$ *is negative, the substance is levorotatory.*

c. Evaluation of β and γ for a pair of charges moving in helical paths

In order to gain more insight into the molecular basis of the phenomenon of optical rotatory power, we shall now find an expression for the parameters β and γ for a helical model similar to that discussed in Section a, above.

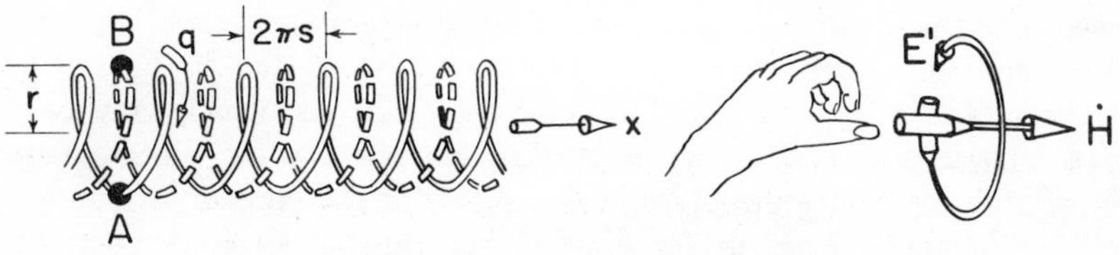

FIG. 15-26. A pair of electrons, A and B, forced to move along their helical paths because of a changing magnetic field, $\dot{\boldsymbol{H}}$.

Although this model may seem to be rather artificial, it emphasizes and illustrates the essential feature of electronic motion responsible for optical rotatory power, which is the movement of electric charge along crooked pathways under the influence of light within the molecule.

It is convenient to assume that each molecule contains two electrons, A and B, which move along two identical helices having a common axis, the motion of the electrons being such that they are always diametically opposite each other across the common axis. (As a result of this simplifying assumption, any induced moment resulting from the motion of the electrons will be strictly parallel to the axis of the helices.)

(*i*) *Evaluation of* β. Assume for the time being that the direction of propagation of the light is normal to the axes of the helices, which will be taken as the x-direction (Fig. 15-26; the helices will be given other orientations in due course). The locations of the electrons along the helices can be specified by the distance q, measured along the helix from the equilibrium positions of the electrons. A positive value of q means that, if one looks down the axis of the helix in the direction of positive x, the electrons have moved in a clockwise sense around the x-axis. If the radius of the helices is r and if $2\pi s$ is the distance between successive turns of a helix, measured along the x-axis, then the x-coordinate of the electrons is $x = qs/\sqrt{r^2 + s^2}$. The quantity $2\pi s$ is the pitch of the helix. If the helix is right-handed, s is taken to be positive and x increases when q increases. If the helix is left-handed, s is negative and x decreases as q increases.

We shall assume that both electrons are bound to their equilibrium positions by identical forces and that if an electron is displaced from its equilibrium position the force tending to return it to its resting position is

$$f_b = -kq$$

where k is a constant. Both electrons are also subject to the inertial force

$$f_i = -m_e\ddot{q}$$

where m_e is the electronic mass and $\ddot{q}$ is the acceleration.

Let there be a magnetic field parallel to the x-axis that changes at a rate $\dot{\boldsymbol{H}}$. This will exert an additional force on the electrons, which may be computed in the following way. According to the law of electromagnetic induction, a difference in potential is induced at the ends of any path that encircles magnetic lines of force when the number of lines enclosed by the path changes with time. If the potential is measured in e.s.u. and the magnetic field is measured in e.m.u., then the difference in potential is (see *3*, p. 298)

$$\Delta V = (a/c) \mid \dot{\boldsymbol{H}} \mid \tag{G-16}$$

where c is the velocity of light and a is the area enclosed by the projection of the path onto a surface normal to the direction of $\dot{\boldsymbol{H}}$. In the present example, $a = \pi r^2$, so there is a difference in potential in each turn of the helix of

$$\Delta V = (\pi r^2/c) \mid \dot{\boldsymbol{H}} \mid$$

Each turn of the helix has a length $2\pi\sqrt{r^2 + s^2}$ as measured along the helix, so the electric field acting along the helix is

$$\mid \boldsymbol{E}' \mid = \frac{\Delta V}{2\pi\sqrt{r^2 + s^2}} = \frac{r^2}{2c\sqrt{r^2 + s^2}} \mid \dot{\boldsymbol{H}} \mid \tag{G-17}$$

The direction of this field is given by Lenz's law: any current it tends to produce in the loop must cause a magnetic field opposite in direction to $\dot{\boldsymbol{H}}$. It can be seen (Fig. 15-26) that the direction of $\boldsymbol{E}'$ about the x-axis follows the left-hand rule: when the thumb of the left hand points in the direction of increasing $\dot{\boldsymbol{H}}$, the fingers of the closed left hand point in the direction of $\boldsymbol{E}'$. Thus the induced electric field exerts a force on each electron

$$f_e = e \mid \boldsymbol{E}' \mid = \mp \frac{er^2}{2c\sqrt{r^2 + s^2}} \mid \dot{\boldsymbol{H}} \mid \tag{G-18}$$

where the upper sign is used if $\dot{\boldsymbol{H}}$ increases in the positive x-direction and the lower sign is used if the increase is in the opposite direction.

We shall assume that the three forces, f_b, f_i and f_e, are the only ones acting on the electrons. This means that we shall neglect damping forces and also any forces arising directly from electric fields other than that induced by the changing magnetic field. The equation of motion of the electrons is then

$$m_e\ddot{q} + kq \pm \frac{er^2}{2c\sqrt{r^2 + s^2}} \mid \dot{\boldsymbol{H}} \mid = 0 \tag{G-19}$$

(1) This equation will be solved first for the case in which the magnetic field is changed at a constant rate. The electrons will now try to reach new equilibrium positions, q_e, determined by setting $\ddot{q}, = 0$, giving

$$q_e = \mp \frac{er^2}{2ck\sqrt{r^2 + s^2}} \mid \dot{\boldsymbol{H}} \mid$$

As a result of this change in the equilibrium position the dipole moment of the molecule is changed. Since the two electrons are assumed always to be on opposite sides of the x-axis, the induced electric moment is parallel to the x-axis because the components in other directions cancel. Thus the induced electric moment is parallel to $\dot{\boldsymbol{H}}$ and it is given by

$$\boldsymbol{m} = 2ex\boldsymbol{i} = -\frac{e^2}{ck}\frac{r^2 s}{r^2 + s^2}\dot{\boldsymbol{H}}$$

where $\boldsymbol{i}$ is the unit vector in the x-direction. Referring to (G-1) we see that under the conditions of a constant rate of change of the magnetic field, the parameter β is given by

$$\beta = \frac{e^2}{k}\frac{r^2 s}{r^2 + s^2} \tag{G-20}$$

(2) Let us next solve the differential equation (G-19) when the varying magnetic field is produced by a light wave moving in any direction normal to the axis of the helices and polarized with its electric field also normal to the axis of the helices. Then

$$\dot{\boldsymbol{H}} = \omega \boldsymbol{H}_0 \sin \omega t \tag{G-21}$$

where $\boldsymbol{H}_0$ points in the positive x-direction. This field will make the electrons oscillate back and forth along the helices. After the light has been shining for some time the electronic motions will attain a steady state consisting of an oscillation of constant amplitude having the same frequency as that of the light. That is, we may assume

$$q = q_0 \sin \omega(t + \delta) \tag{G-22}$$

where δ is a phase whose magnitude is to be determined. Substituting (G-21) and (G-22) into (G-19) we obtain

$$(k - \omega^2 m_e) q_0 \sin \omega(t + \delta) + \frac{\omega e r^2}{2c\sqrt{r^2 + s^2}} \mid \boldsymbol{H}_0 \mid \sin \omega t = 0 \tag{G-23}$$

which can be satisfied only if

$$\delta = 0 \tag{G-23a}$$

and

$$q_0 = -\frac{e r^2}{2 m_e c \sqrt{r^2 + s^2}} \frac{1}{\omega_0^2 - \omega^2} \omega \mid \boldsymbol{H}_0 \mid \tag{G-23b}$$

where ω_0 is the natural frequency of oscillation of the electrons given by

$$\omega_0^2 = k/m_e \tag{G-23c}$$

The variation in q produces an oscillating dipole moment parallel to $\dot{\boldsymbol{H}}$

$$\boldsymbol{m} = 2ex\mathbf{i} = 2eqs\mathbf{i}/\sqrt{r^2 + s^2} = -\frac{e^2}{m_e c}\frac{r^2 s}{r^2 + s^2}\frac{1}{\omega_0^2 - \omega^2}\dot{\boldsymbol{H}} \tag{G-24}$$

Comparison with (G-1) shows that

$$\beta = \frac{e^2}{m_e}\frac{r^2 s}{r^2 + s^2}\frac{1}{\omega_0^2 - \omega^2} \tag{G-25}$$

This equation reduces to (G-20) when $\omega << \omega_0$. Note, however, that because of the factor λ^2 in the denominator of (G-13) the optical rotation vanishes at low frequencies (long wave lengths) even though β remains finite.

(3) Up to this point we have assumed that the axes of the helices are normal to the incident light beam. If the helices are located in molecules of a gas or a liquid, however, their axes will be randomly oriented in space. This will produce two complications. In the first place, the value of β varies as the orientation of the molecule is changed. In the second place, the magnetic dipole moment effects that have so far been neglected will have to be considered.

Let us find the average value of β taking account of the randomness of orientation of the helices. If the axis of a helix makes an angle ε with $\dot{\boldsymbol{H}}$, then the area a in equation (G-16) becomes $\pi r^2 \cos \varepsilon$. The component of the induced dipole moment along the axis of the helix must also be multiplied by $\cos \varepsilon$ in order to obtain the component of the induced electric moment parallel to $\dot{\boldsymbol{H}}$; if the helices are randomly oriented in space, the components of the moment normal to $\dot{\boldsymbol{H}}$ will cancel and on the average, the induced moment will be parallel to $\dot{\boldsymbol{H}}$. Thus

$$\boldsymbol{m} = -(\beta_{av}/c)\,\dot{\boldsymbol{H}} \tag{G-26a}$$

where

$$\beta_{av} = (\cos^2\varepsilon)_{av}\beta \tag{G-26b}$$

and β is given by equation (G-25). For random orientations the average value of $\cos^2 \varepsilon$ is 1/3, so that we have for the final value

$$\beta_{av} = \frac{e^2}{3m_e}\frac{r^2 s}{r^2 + s^2}\frac{1}{\omega_0^2 - \omega^2} \tag{G-27}$$

(*ii*) *Evaluation of* γ. Let us investigate next the magnetic dipole moment induced in a helix by a changing electric field. We shall begin by assuming that the helices are oriented with their axes parallel to the electric field, $\boldsymbol{E}$. (Note that β vanishes for this orientation.) A force $e\boldsymbol{E}$ acts on the electrons in the helices. The component of this force along the paths of the electrons is $e \mid \boldsymbol{E} \mid s/\sqrt{r^2 + s^2}$, so the equation of motion of the electrons is

$$m_e\ddot{q} + kq \mp (es/\sqrt{r^2 + s^2}) \mid \boldsymbol{E} \mid = 0 \qquad \text{(G-28)}$$

where the upper sign is used if $\boldsymbol{E}$ points in the positive x-direction and the lower sign is used if $\boldsymbol{E}$ points the other way.

(1) If the electrons are not accelerated, so that $\ddot{q} = 0$, then

$$q = \pm \frac{es}{k\sqrt{r^2 + s^2}} \mid \boldsymbol{E} \mid$$

This displacement produces, of course, an induced electric dipole moment parallel to $\boldsymbol{E}$, but since this moment does not result in any rotation of the plane of polarization it will be disregarded. Of more interest is the magnetic moment parallel to $\boldsymbol{E}$ that results when the displacement, q, changes with the time. According to Equation (D-4) of Chapter 8, the magnetic dipole moment arising from the motion of a charged particle about an axis is equal to $evr/2c$, where r is the distance of the particle from the axis and v is the component of its velocity in a direction normal to the axis and normal to the direction of r. In the present example, $v = r\dot{q}/\sqrt{r^2 + s^2}$, so (remembering that there are two electrons in our model) the induced magnetic moment has the magnitude

$$\mid \boldsymbol{\mu} \mid = \frac{er^2\dot{q}}{c\sqrt{r^2 + s^2}}$$

If the pitch of the helices, $2\pi s$, is positive, the induced magnetic moment will have the same direction as the changing field, $\dot{\boldsymbol{E}}$. Thus we may write

$$\boldsymbol{\mu} = \frac{e^2}{ck}\frac{r^2 s}{r^2 + s^2}\dot{\boldsymbol{E}} \qquad \text{(G-29)}$$

Comparing this with Equation (G-2) we see that

$$\gamma = \frac{e^2}{k}\frac{r^2 s}{r^2 + s^2} \qquad \text{(G-30)}$$

This is identical with the value of β given in Equation (G-20) for a helix oriented with its axis parallel to a changing magnetic field.

(2) If the helix is subjected to the oscillating electric field of a light wave, with $\boldsymbol{E} = \boldsymbol{E}_0 \cos \omega t$, and if we assume a steady state solution of (G-28) of the form

$$q = q_0 \cos \omega(t + \delta)$$

we find

$$(k - m_e \omega^2) q_0 \cos \omega(t + \delta) - \frac{es}{\sqrt{r^2 + s^2}} \,|\, \boldsymbol{E}_0 \,|\, \cos \omega t = 0 \qquad \text{(G-31)}$$

which is satisfied only if $\delta = 0$ and

$$q_0 = \frac{es}{m_e \sqrt{r^2 + s^2}} \frac{1}{\omega_0^2 - \omega^2} \,|\, \boldsymbol{E}_0 \,| \qquad \text{(G-32)}$$

This gives for the induced magnetic moment

$$\boldsymbol{\mu} = \frac{er^2 \dot{q} \boldsymbol{i}}{c\sqrt{r^2 + s^2}} = \frac{e^2}{cm_e} \frac{r^2 s}{r^2 + s^2} \frac{1}{\omega_0^2 - \omega^2} \dot{\boldsymbol{E}} \qquad \text{(G-33)}$$

and comparison with (G-2) shows that

$$\gamma = \frac{e^2}{m_e} \frac{r^2 s}{r^2 + s^2} \frac{1}{\omega_0^2 - \omega^2} \qquad \text{(G-34)}$$

This expression is identical with the value of β given in (G-25) for helices whose axes are parallel to the magnetic field in a light wave.

Finally, let us find the average value of γ when the helices are allowed to assume all orientations at random. It is apparent that the same geometrical factors must be taken into account in averaging γ as were required in averaging β. Thus

$$\gamma_{\text{av}} = \frac{e^2}{3m_e} \frac{r^2 s}{r^2 + s^2} \frac{1}{\omega_0^2 - \omega^2} \qquad \text{(G-35)}$$

and

$$(\beta + \gamma)_{\text{av}} = \frac{2e^2}{3m_e} \frac{r^2 s}{r^2 + s^2} \frac{1}{\omega_0^2 - \omega^2} \qquad \text{(G-36)}$$

We see that $(\beta + \gamma)_{\text{av}}$ varies with the frequency of the incident light in much the same way as does the polarizability (*cf.* Equation (D-15)). Substituting (G-36) into Equation (G-15) we find for the molecular rotation of a gas or liquid composed of molecular helices of this type

$$[M] = 24\, N_0 \frac{n^2+2}{3} \frac{e^2}{m_e c^2} \frac{r^2 s}{r^2+s^2} \frac{\lambda_0^2}{\lambda^2-\lambda_0^2}$$
$$= 4.06 \times 10^{12} \frac{n^2+2}{3} \frac{r^2 s}{r^2+s^2} \frac{\lambda_0^2}{\lambda^2-\lambda_0^2} \qquad \text{(G-37)}$$

We note that right-handed helices ($s > 0$) should be dextrorotatory, whereas left-handed helices ($s < 0$) should be levorotatory.

Equation (G-37) was derived assuming that no dissipative forces act on the electrons. As long as the frequency of the incident light differs from the natural frequency of the electrons this assumption will not be serious. At frequencies within an absorption band, however, dissipative effects will become important. They will shift the phase of the electronic oscillation in a manner similar to that found for the linear oscillator in an absorption region (pages 570*ff*), and it is easy to see that the dipole moment induced by the changing magnetic field will consist of two parts: One part is in phase with $\dot{\boldsymbol{H}}$ and $\dot{\boldsymbol{E}}$ and the other part is a quarter cycle out of phase with $\dot{\boldsymbol{H}}$ and $\dot{\boldsymbol{E}}$. The scattered wave front that results from the dipole moments induced by $\dot{\boldsymbol{H}}$ and $\dot{\boldsymbol{E}}$ under these circumstances is no longer in phase with the incident beam, and as a result the plane polarized incident beam becomes elliptically polarized on passing through a layer of the molecules. Furthermore, when λ becomes smaller than λ_0 in (G-37), the contribution of the motion to the optical rotation changes sign. These phenomena are known as the *Cotton effect.* The equation for the ellipticity is not difficult to derive using an approach similar to that outlined above, but the details will be left to the student.

d. The interpretation of the observed optical rotation of molecules in terms of the helical model

The wave length λ_0 in Equation (G-37) would be expected to correspond to an absorption band of the molecule. If the optically active molecule contains several electrons moving in different helical paths, the optical rotation should therefore be expressible as a series of terms,

$$[M] = \sum_{\substack{\text{all absorp-}\\ \text{tion bands}}} B_i/(\lambda^2 - \lambda_i^2) \qquad \text{(G-38)}$$

where the constants B_i are determined by the shapes of the paths and each term is associated with an absorption band (wave length λ_i) belonging to one type of path. This relationship between the rotation and the wave length was proposed by Drude and is known as the Drude formula.

The observed optical rotations of optically active molecules are, in fact, accurately reproduced by expressions of this form, provided, of course, that the wave length λ used in measuring $[M]$ does not lie in any of the absorption regions of the molecule. For instance, Lowry and Cutter (*18*) found that the optical rotation of camphor for light of wave lengths between 7600 Å. and 3600 Å. is given very accurately by the expression

$$[M] = \frac{44.701}{\lambda^2 - \lambda_1^2} - \frac{30.636}{\lambda^2 - \lambda_2^2} \qquad \text{(G-39)}$$

where the wave length is measured in Ångstrom units and $\lambda_1 = 2960$ Å. and $\lambda_2 = 2330$ Å. The wave length λ_1 corresponds to the known weak absorption band of the carbonyl group. (Its oscillator strength is only 5×10^{-4}.) The wave length λ_2 probably represents the weighted average of many weak and strong bands farther in the ultraviolet. Here we have an example of a very important characteristic of the optical rotatory power: weak absorption bands are often found to make just as important contributions to the optical rotation as do strong absorption bands. This is in marked contrast to the behavior of the polarizability and the refractive index, where the contribution of a band is directly proportional to the intensity of absorption (page 582).

It is interesting to determine the dimensions of a helix that will produce an optical rotation of the same order of magnitude as the values that are actually observed. Let us suppose that an absorption band with $\lambda_0 = 3000$ Å. makes a contribution of 100° to the molecular rotation, $[M]$, using light of wave length $\lambda = 6000$ Å. to make the measurement. Assuming that the Lorentz factor is of the order of magnitude of unity, we find from (G-37) that

$$r^2s/)r^2 + s^2) = 8 \times 10^{-11} \text{ cm.}$$

This value might conceivably arise in either of two ways: (a) It could be that $r \gg s$, in which case $s = 8 \times 10^{-11}$ cm. This would mean that we are concerned with a helix of very small pitch, so that the electrons move in nearly circular paths (see Fig. 15-27A). For small oscillations of the electrons we would have a situation similar to that described in connection with magnetic dipole radiation (Fig. 15-23) except that the two charges do not move in quite the same plane. Since this kind of motion results in a very small change of the electric dipole moment of the molecule, its oscillator strength is small, and the motion must give rise to weak absorption. This

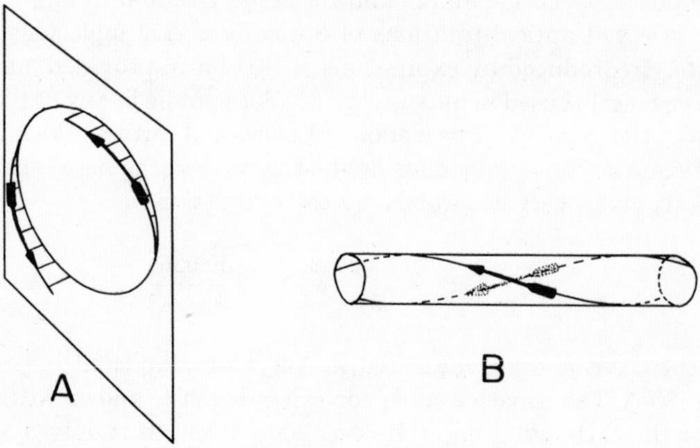

FIG. 15-27.

A. Classical paths of charges that give rise to a weak absorption band, yet make a large contribution to the optical rotation.

B. Classical paths of charges that give rise to a strong absorption band and make a large contribution to the optical rotation.

corresponds to a large contribution to the optical rotation arising from a weak absorption band. (b) On the other hand, it might be that $s \gg r$. In this case we have $r^2/s = 8 \times 10^{-11}$ cm. If s is of the order of an atomic dimension (10^{-8} cm.), then $r = 9 \times 10^{-10}$ cm., and we have a helix with a diameter much smaller than its pitch–that is, essentially a slightly twisted linear oscillator. Alternatively, if r is of the order of an atomic dimension, then $s = 1.3 \times 10^{-6}$ cm., and once again we have essentially a slightly twisted linear oscillator. Thus the case $s \gg r$ corresponds to the electrons moving in nearly straight lines (Fig. 15-27 B). The oscillator strength of such a motion is large and it will give rise to strong absorption. This is therefore an illustration of the manner in which optical rotations can be associated with strong absorption bands.

Evidently the electronic motions in the 2960 Å. absorption band of the carbonyl group can be said to resemble those of the first type ($r \gg s$). In the absence of other groups in the environment of the carbonyl group, the carbonyl electrons move in more or less circular paths, so the electric dipole moments tend to cancel. If the carbonyl group is in an asymmetric molecule such as camphor, however, the environment provided by the asymmetric molecule distorts the paths very slightly so that they acquire a small amount of helical character, and an important contribution is made to the optical rotation. Similarly, the electronic motions responsible for strong absorption bands occur more or less along straight lines, and in an asymmetric molecule the environment twists these linear paths slightly, giving them a slight helical character and causing optical rotation. The interactions between different groups in an asymmetric molecule that distort electronic paths in this way and cause optical rotation are known as *vicinal actions*.

Kuhn (*19, 20*) has discussed optical rotatory power from the point of view of the so-called *coupled oscillator model*, which appears at first sight to be rather different from the helical model described here, but is actually closely similar in its most important respects. Kuhn assumed that any optically active molecule contains at least two linear oscillators located at different positions in the molecule (see Fig. 15-28). These oscillators must not be

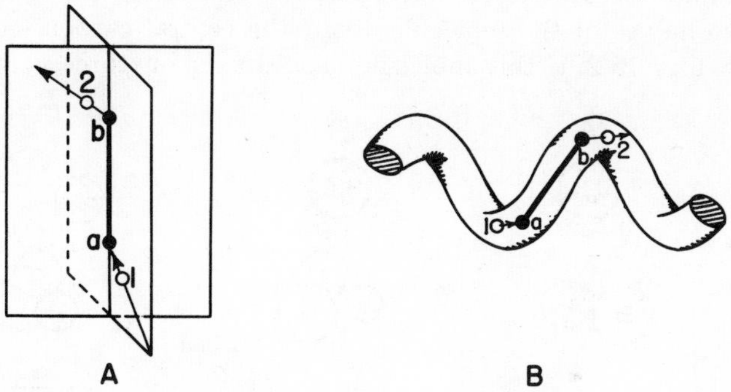

FIG. 15-28.

A. The coupled oscillator model of an optically active molecule.

B. Helical character of the motion of the charges in the coupled oscillator model.

coplanar. Kuhn also assumed that the oscillators are coupled in such a way that when the electron in one of the oscillators is displaced, a force is exerted on the electron in the other oscillator. Kuhn was able to show that such a system would rotate the plane of polarized light and would give dispersion equations similar to (G-38). In this model, the vicinal actions are those forces responsible for the coupling that causes the optical rotation.

It is not hard to see that Kuhn's model results in a helical motion of charges that is similar to the motions postulated in the helical model.

Suppose, for instance, that in Fig. 15-28 the motion of electron no. 1 toward point *a* tends to make electron no. 2 move away from point *b* because of the electrostatic repulsion of the electrons. Such a pair of oscillators is equivalent to a part of a turn of a helix along which charge is constrained to move, and it produces an optical effect similar to that described using the helical model. Because the coupling force is usually rather weak, however, the effective charge, *e*, appearing in Equation (G-37) is much smaller than an electronic charge.

From the discussion that has been given up to this point, it appears that the general features of optical rotatory power can be understood without any great difficulty in terms of relatively simple molecular models. When one attempts to construct a detailed theory relating rotatory power and structure, however, the problem becomes much more difficult. In order to see some of the reasons for this, let us consider the problem of predicting the sign of the rotation of the typical optically active molecule, secondary butyl bromide. Because of the energy barrier restricting the relative motion of the two halves of the molecule about the central carbon–carbon bond (C_b–C* in Fig. 15-29), this molecule should exist in three conformations.

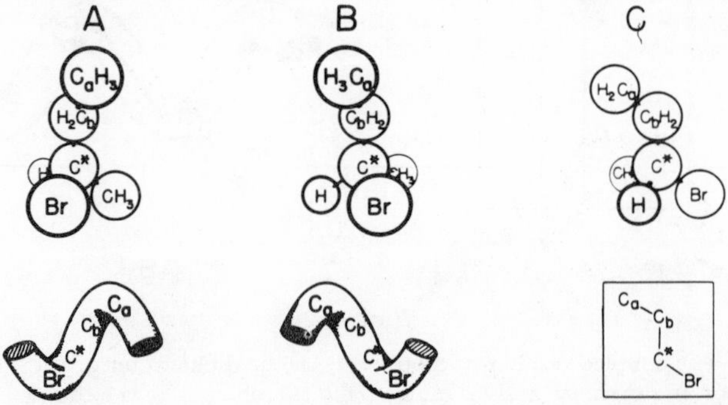

FIG. 15-29. The three conformations of the secondary butyl bromide molecule, CH_3–CH_2–CHBr–CH_3. If the electrons tend to move most easily along the chain of atoms, C_a–C_b–C*–Br, then conformation *A* is equivalent to a portion of a right handed helix, and conformation *B* is equivalent to a portion of a left handed helix. Conformation *C* has no helical character because these four atoms lie in the same plane.

The C–Br bond and the two C–CH_3 bonds are probably somewhat more polarizable along the bond than in the direction normal to the bond. It also seems likely that when the electrons in the C–Br bond move toward the carbon atom, the electrons in the C_a–C_b bond would tend to move toward the methyl group because of the electrostatic repulsions between them. The movements of charge along the chain of atoms C_a–C_b–C*–Br in conformation *A* of Fig. 15-29 therefore resembles those in a right-handed helix, whereas in configuration *B* they resemble those in a left-handed helix. In configuration *C* these four atoms are coplanar, so they have no helical character. On the other hand, the chain of atoms C_a–C_b–C*–C_d resembles a left-handed helix in conformation *A*, and a right-handed helix in conformation *C*; in conformation *B* these atoms are coplanar. Thus if one wanted to predict the sign of the rotation of the secondary butyl bromide molecule whose absolute configuration is shown in Fig. 15-29, one would have to know: (1) The relative amounts of the three conformations *A*, *B*, and C present. (If all three were present in equal amounts the rotation should vanish because of the cancellation of the effects of the right- and left-handed helices that occur in the different conformations.) (2) The relative effectiveness of the helices derived from the C_a–C_b–C*–Br chain and the C_a–C_b–C*–C_d chain. (Presumably the electrons in the former chain would be somewhat more mobile, because of the presence of the more polarizable bromine atom, so this chain should have the larger effect.) (3) One would also want to be sure that the electronic motions had been analyzed correctly. (It is by no means certain that the most effective motions of the electrons are along the chains of atoms in the molecule.) The first of these uncertainties can be avoided by working only with molecules having perfectly rigid skeletons in which there can be no uncertainty as to the relative positions of the atoms. Since relatively little is known about the detailed motions of electrons under the influence of light, however, the third problem is much more difficult. Because of the many failures of the classical theory of optics in describing the detailed electronic motions in atoms and molecules, it seems advisable to turn to the quantum mechanical theory of the phenomenon before further investigation of this difficulty (see page 703).

REFERENCES CITED

1. H. A. Lorentz, "The Theory of Electrons," 2nd ed., Teubner, Leipzig, 1916. (Reprinted by Dover, New York, 1952).
2. M. Abraham and R. Becker, "The Classical Theory of Electricity and Magnetism," Blackie and Son, London, 1937.
3. G. Joos, "Theoretical Physics," Stechert, New York, 1934.
4. R. W. Wood, "Physical Optics," 3rd ed., Macmillan, New York, 1934.
5. A. H. Compton and S. Allison, "X-rays in Theory and Experiment," 2nd ed., Van Nostrand, New York, 1935.

6. H. Margenau and W. W. Watson, *Revs. Mod. Phys.* **8,** 22, 398 (1936).
7. R. S. Mulliken and C. A. Rieke, *Repts. Progr. Phys.* **8,** 231 (1941).
8. C. I. Carr and B. H. Zimm, *J. Chem. Phys.* **18,** 1616 (1950).
9. P. Doty and J. T. Edsall, *Advances in Protein Chem.* **6,** 35 (1951).
10. M. J. Buerger, "X-ray Crystallography," Wiley, New York, 1952.
11. J. T. Edsall, *in* "The Proteins" (H. Neurath and K. Bailey, eds., Vol. 1, Part B) Academic Press, New York, 1953.
12. A. J. C. Wilson, "X-ray Optics," Methuen, London, 1949.
13. P. P. Ewald, *Fortschr. Chem. Physik. u. physik. Chemie,* **18,** 494 (1925).
14. K. L. Wolf and K. F. Herzberg, *in* "Handbuch der Physik" (H. Geiger and K. Scheel, eds.), Vol. 20, Springer, Berlin, 1928.
15. C. J. F. Böttcher, "Theory of Electric Polarization," Elsevier, New York, 1952.
16. H. A. Stuart, "Die Physik der Hochpolymeren," Vol. 1, Springer, Berlin, 1952.
17. S. Fried and S. Weissman, *Phys. Rev.* **60,** 440 (1941).
18. T. N. Lowry and J. O. Cutter, *J. Chem. Soc.* **127,** 604 (1925).
19. W. Kuhn, *in* "Stereochemistry" (K. Freudenberg, ed.), Book I, Chapter 8. Deuticke, Leipzig, 1932.
20. W. Kuhn and K. Freudenberg, *in* "Hand- und Jahrbuch der Chemischen Physik" (A. Eucken and K. L. Wolf, eds.), Vol. 8, No. 3, Akademische Verslagsges., Leipzig, 1932.

GENERAL REFERENCES

P. Drude, "The Theory of Optics" (translated by C. R. Mann and R. A. Millikan). Longmans, Green, New York, 1902.

H. A. Lorentz, reference 1.

L. Rosenfeld, "Theory of Electrons," Interscience, New York, 1951. Chapter 6 contains a discussion of scattering as a cause of refraction that is much more rigorous than the one given in this chapter.

M. Born, "Optik," Springer, Berlin, 1933.

R. W. Wood, reference 4.

Chapter 16

The Interactions of Matter with Light II Quantum Mechanical Aspects

Although the classical electron theory discussed in Chapter 15 is very successful in describing many of the phenomena of optics, serious difficulties exist in the assumptions on which this theory is founded. The difficulties are of two kinds. In the first place it is known that the classical description of radiation (as given in Assumptions 2, 3, and 4 on pages 548–554) is incomplete because it does not account for the existence of light quanta, for which there is much incontrovertible experimental proof (photoelectric effect, Planck radiation law, etc.). In the second place the basic model for atomic structure (as given in Assumption 5 on page 554) could hardly be farther from the truth. Electrons are not embedded in a diffuse cloud of positive electricity in which they undergo normal-mode oscillations in response to some kind of elastic forces. Instead they move in orbits about a very small positively charged nucleus, undergoing accelerations that do not produce any radiation despite the predictions of the classical theory (Assumption 4 on page 551).

In other chapters of this book, we have seen how quantum theory deals in a revolutionary way with the second of these difficulties—*viz.*, the problem of the structure of atomic and molecular systems. Strictly speaking we should now describe the quantum mechanical approach to the first difficulty—*viz.*, the radiation problem. This, however, would entail a more extensive excursion into the physical theory than the average chemist would be prepared to undertake. Fortunately it is possible to avoid such an excursion. The classical theory of radiation that we have presented can be amalgamated with the quantum mechanical theory of atoms and molecules to give a useful and—within limits—sound theoretical structure.

It turns out that the classical equations have the correct form, but that the constants appearing in them must be given a different interpretation from the classical one. The present chapter is concerned with this reinterpretation of the classical concepts.

A. General Theory

a. Basic equations

Suppose that we are given a molecule* whose steady state wave functions in the absence of external perturbing fields are $\phi_0, \phi_1, \phi_2, \cdots$, the corresponding energies being $E_0, E_1, E_2, \cdots$. Let us put the molecule in some definite state ϕ_0 and then shine plane-polarized light on it. Usually ϕ_0 will be the ground state of the molecule, but this need not necessarily be the case. For the moment we shall assume that the wave length of the light is so great compared with the dimensions of the molecule that the electric field of the light can be taken to be the same at all points in the molecule at a given instant. The electric field acting on the molecule will therefore be written

$$\boldsymbol{E} = \boldsymbol{E}_0 \cos \omega t \tag{A-1}$$

The field will be assumed to be parallel to the x-axis. The light wave perturbs the molecule because of the force that its electric field exerts on the charged particles of which the molecule is composed. If the mth charged particle in the molecule (i.e., a nucleus or an electron) possesses a charge e_m and if its x-coordinate is x_m, then the perturbation-energy operator arising from the action of the electric field on the molecule will be

$$\begin{aligned} \mathrm{H}' &= -\sum e_m x_m \, |\boldsymbol{E}| \\ &= -m_x \, |\boldsymbol{E}_0| \cos \omega t \end{aligned} \tag{A-2}$$

where m_x is the operator for the x-component of the dipole moment of the molecule

$$m_x = \sum_{\substack{\text{all}\\ \text{partides}}} e_m x_m \tag{A-3}$$

Substituting (A-2) into Equation (B-10) of Chapter 14 and writing $\cos \omega t$ in the form $(e^{i\omega t} + e^{-i\omega t})/2$, we find that under the influence of the electric field in the light wave, the wave function of the molecule becomes

$$\Psi = \sum a_j(t) \, \phi_j \tag{A-4}$$

where

$$a_0(t) = e^{-iE_0 t/\hbar} \tag{A-4a}$$

* In the discussion that follows, we shall refer to molecules, but it should be obvious that all of the conclusions apply equally well to atoms.

$$a_j(t) = \frac{|\mathbf{E}_0|\, m_{xj0} e^{-iE_j t/\hbar}}{2\hbar} \left[\frac{e^{i(\omega_{j0}+\omega)t} - 1}{\omega_{j0} + \omega} + \frac{e^{i(\omega_{j0}-\omega)t} - 1}{\omega_{j0} - \omega} \right] \quad (j \neq 0) \tag{A-4b}$$

$$\omega_{j0} = (E_j - E_0)/\hbar \tag{A-4c}$$

$$m_{xj0} = \int \phi_j^* (\textstyle\sum e_m x_m) \phi_0 d\tau \tag{A-4d}$$

The square of the absolute value of the coefficient $a_j(t)$ in Equation (A-4) gives the probability that at time t the molecule has been excited to state ϕ_j by the perturbation introduced by the light.

According to the basic postulates of quantum mechanics as outlined in Chapter 5, all of the physically measurable properties of the molecule in the light wave can be found from the wave function Ψ, given by Equation (A-4). Evidently the response of the molecule to the light depends on the difference between ω, the frequency of the light, and the constants ω_{j0}, determined by the energy levels of the unperturbed molecule. Two different situations can be recognized:

Case I. If the frequency of the incident light, ω, happens to be exactly the same as one of the ω_{j0}, say ω_{k0}, and if $E_k > E_0$, then ω_{k0} is positive and the denominator of the second term in the brackets in Equation (A-4b) for the corresponding coefficient a_k will vanish. Since

$$\lim_{\omega \to \omega_{k0}} \frac{e^{i(\omega_{k0}-\omega)t} - 1}{\omega_{k0} - \omega} = it \tag{A-5}$$

we see in this case a_k will increase continuously with time throughout the exposure of the molecule to the light, and the probability of finding the molecule in the state ϕ_k can become very large. All the remaining a_j (except for a_0) will always remain negligibly small, as we shall show below. Since the state ϕ_k was assumed to have a higher energy than the initial state ϕ_0, energy must be taken up by the system as it undergoes transitions from ϕ_0 to ϕ_k. This energy must come from the light (there being no other possible source). We must therefore conclude that this is the well-known phenomenon of *light absorption*.

On the other hand, suppose that ϕ_0 is an excited state and that ϕ_k is a state with lower energy. Then ω_{k0} will be negative and the denominator of the first term in brackets in (A-4b) vanishes. The system again undergoes transitions from ϕ_0 to ϕ_k, but in this case energy is given up under the influence of the incident radiation. The energy thereby released is found to appear as radiation. This is the much less well-known phenomenon of *induced emission*.

Evidently both absorption and induced emission result in an exchange of energy between the molecule and the radiation, in increments whose magnitude is

$$E_k - E_0 = \hbar\omega_{k0} = h\nu_{k0} \tag{A-6}$$

where ω_{k0} is the circular frequency of the radiation and ν_{k0} is the ordinary frequency ($\omega_{k0} = 2\pi\nu_{k0}$). It thus appears that the radiant energy exchanged between light and matter comes in bundles of finite size—or as they are more often called, quanta. Here we have an indication of the origin of the quantization of radiant energy. Equation (A-6) is of course one of the assumptions made by Bohr in setting up the pre-quantum mechanical form of quantum theory. In the present theory, this condition for light absorption does not have to be assumed, since it is a logical consequence of the postulates of the theory.

Case II. If ω is not close to any of the ω_{j0}, then all of the a_j except a_0 remain small at all times: Even the most intense light sources have a field strength of less than one volt per centimeter (i.e., 0.003 e.s. volts/cm.). The integrals m_{xj0} are of the order of magnitude of an atomic or molecular dimension (say 10^{-7} cm.) multiplied by an electronic charge (about 5×10^{-10} e.s.u.). If $\omega_{j0} \pm \omega$ has the order of magnitude 10^{15} sec.$^{-1}$ (corresponding to visible light) and if we note that the maximum values attainable by the numerators of the fractions in the brackets of Equation (A-4b) is 2, we find $|a_j| < 10^{-7}$. Under these conditions the probability of finding the molecule in an excited state, even when extremely intense light shines on it, is negligible ($|a_j|^2 < 10^{-14}$). This is true regardless of the duration of the exposure to light. Furthermore, no particular one of the a_j is likely to get much larger than all of the others, so in evaluating Ψ, many of the terms in the sum $\sum a_j\phi_j$ must be considered.

Let us now discuss these two cases in more detail.

b. Case I. The absorption and induced emission of light

If ω is equal to ω_{k0} for some particular state, then as we have seen, a transition to that state becomes increasingly probable as the duration of the exposure to the light increases. Retaining the last term in Equation (A-4b) and making use of (A-5) we find

$$a_k = -\frac{i|\mathbf{E}_0|m_{xk0}}{2\hbar}\, te^{-iE_k t/\hbar} \tag{A-7a}$$

and the probability of a transition is

$$|a_k|^2 = \frac{|\mathbf{E}_0|^2\,|m_{xk0}|^2}{4\hbar^2}\, t^2 \tag{A-7b}$$

According to this result, the probability of absorption of light by a system in its ground state should increase in proportion to the square of the time, so that the *rate* of absorption of radiant energy ought to increase linearly with the duration of the exposure. This does not agree with our everyday experience, however, for we usually find that the absorption cœfficients of dyes and other colored materials are either constant in time or decrease due to bleaching.

The reason for this discrepancy is that Equations (A-7) are based on exposure to strictly monochromatic light, whereas laboratory sources are never even approximately monochromatic to the degree required here. Neglecting the first term in (A-4b) we find that for frequencies in the vicinity of ω_{k0}

$$|a_k|^2 = \frac{|\boldsymbol{E}_0|^2 \, |m_{xk0}|^2}{4\hbar^2} \left[\frac{2 - e^{i(\omega_{k0}-\omega)t} - e^{-i(\omega_{k0}-\omega)t}}{(\omega_{k0}-\omega)^2} \right]$$

$$= \frac{|\boldsymbol{E}_0|^2 \, |m_{xk0}|^2}{\hbar^2} \, \frac{\sin^2 \frac{1}{2}(\omega_{k0}-\omega)t}{(\omega_{k0}-\omega)^2} \tag{A-8}$$

When plotted against the incident frequency ω, this expression has the form of a rapidly damped oscillation, whose "frequency" increases linearly with the time (see Fig. 16-1). The function has a maximum at ω_{k0} whose height increases as the square of the time, as expected from Equation (A-7b). If

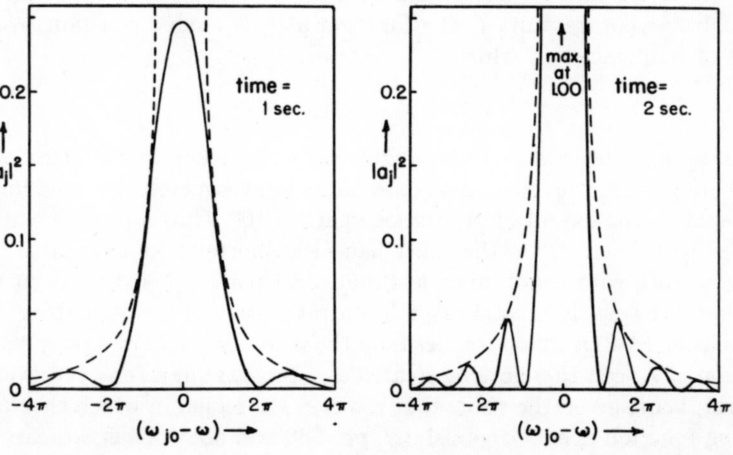

FIG. 16-1. Plot of Equation (A-8) for two values of the time t, one of which is twice as large as the other. The dashed curves give the variation of the factor $1/(\omega_{j0}-\omega)^2$. The transition probability, $|a_j|^2$, is expressed in units of $|\boldsymbol{E}_0|^2 \, |m_{xj0}|^2/\hbar^2$.

Equation (A-7b) is to be applicable, however, the incident radiation must be sufficiently monochromatic for its entire range of frequencies to fall well within the central peak. That is, all of the incident frequencies ω must be such that

$$| \omega - \omega_{k0} | << \frac{\pi}{2t} \approx \frac{1}{t} \tag{A-9}$$

where t is the exposure time. In a flash of light lasting t seconds, however, it is impossible in practice to specify the frequency more accurately than to within about $1/t$ sec.$^{-1}$ (*cf.* page 241). Therefore it is impossible to fulfill Equation (A-9), and there can be no light source sufficiently monochromatic to permit the use of Equation (A-7b).

We are therefore forced to turn to Equation (A-8) and make allowance for the actual distribution of frequencies in our necessarily polychromatic light source. This distribution is conveniently described in terms of the radiation energy density function, $\varrho(\nu)$, introduced on page 583. According to Equation (D-37) of Chapter 15, the value of $|\boldsymbol{E}_0|^2$ for the radiation in any small frequency interval $d\nu$ can be replaced by $8\pi\varrho(\nu)d\nu$.

The transition probability can therefore be written

$$| a_k |^2 = \frac{2\, | m_{xk0} |^2}{\pi \hbar^2} \int_0^{\infty} \varrho(\nu) \, \frac{\sin^2 \pi(\nu_{k0} - \nu)t}{(\nu_{k0} - \nu)^2} \, d\nu \tag{A-10}$$

It is easy to see that for any actual light source and for any experimentally attainable exposure time t, the function $\varrho(\nu)$, is almost constant over the range of frequencies in which

$$[\sin^2 \pi\, (\nu_{k0} - \nu)t]/(\nu_{k0} - \nu)^2$$

has an appreciable value. Visible light has a frequency of the order of 10^{15} sec.$^{-1}$ and even the best monochromatic light sources are uncertain in frequency to the extent of at least one part in 10^8. Thus $\varrho(\nu)$ covers at least a range of 10^7 sec.$^{-1}$. On the other hand the shortest exposure time one is likely to work with is not much less than a microsecond, so that even under the most extreme circumstances, the central portion of the oscillatory factor in Equation (A-8) is much smaller than the width of $\varrho(\nu)$. The slowly varying function, $\varrho(\nu)$, can therefore be treated as a constant and removed from the integral, being given the value that it has in the region in which the rapidly varying function does not vanish (*cf.* pp. 580 and 583). Thus we can write

$$| a_k |^2 = \frac{2\, | m_{xk0} |^2 \, \varrho(\nu_{k0})}{\pi \hbar^2} \int_{-\infty}^{\infty} \frac{\sin^2 \pi(\nu_{k0} - \nu)t}{(\nu_{k0} - \nu)^2} \, d(\nu - \nu_{k0}) \tag{A-10a}$$

(The lower limit of the integral has been written $-\infty$ instead of $-\nu_{k0}$ for obvious reasons.) Since

$$\int_{-\infty}^{\infty} \frac{\sin^2 ax}{x^2}\, dx = a\pi$$

we have

$$|a_k|^2 = \frac{2\pi\, |m_{xk0}|^2\, \varrho(\nu_{k0})}{\hbar^2}\, t \qquad \text{(A-11)}$$

Evidently when polychromatic light falls on an atom, the transition probability increases in direct proportion to the time of exposure, as we should expect.

Up to this point, we have assumed that the incident light has definite directions of polarization and propagation with respect to some chosen direction in the molecule. If, however, the molecule is in the vapor or liquid state and is free to orient in all directions, or what is equivalent, if the radiation passes through the molecule from all directions and with all directions of polarization, then the y- and z-components of the dipole moment operator, $\boldsymbol{m}$, must be considered. In this case we must write Equation (A-11) in the form

$$|a_k|^2 = \frac{2\pi\, [|m_{xk0}|^2 + |m_{yk0}|^2 + |m_{zk0}|^2]}{3\hbar^2}\, \varrho(\nu_{k0})t \qquad \text{(A-11a)}$$

Since the total dipole moment of a molecule has the vector operator

$$\boldsymbol{m} = \boldsymbol{i}m_x + \boldsymbol{j}m_y + \boldsymbol{k}m_z \qquad \text{(A-12)}$$

and since

$$|\boldsymbol{m}_{k0}|^2 = \left| \int \phi_k{}^* \boldsymbol{m} \phi_0 d\tau \right|^2 = |m_{xk0}|^2 + |m_{yk0}|^2 + |m_{zk0}|^2 \qquad \text{(A-12a)}$$

we may write

$$|a_k|^2 = \frac{2\pi\, |\boldsymbol{m}_{k0}|^2}{3\hbar^2}\, \varrho(\nu_{k0})t \qquad \text{(A-12b)}$$

Each time a molecule in its ground state undergoes a transition, a quantity of energy $(E_k - E_0)$ is absorbed. Thus if there are n molecules per unit volume, the rate of energy absorption is

$$n(E_k - E_0)\frac{|a_k|^2}{t} = \frac{8\pi^3 n}{3}\frac{|\boldsymbol{m}_{k0}|^2\, \nu_{k0}}{h}\, \varrho(\nu_{k0}) \qquad \text{(A-13)}$$

This result is to be compared with the analogous classical result for the rate of energy dissipation, Equation (D-39) of Chapter 15; we see that the two can be related if we replace

$$\frac{8\pi^3 \nu_{k0} \mid \boldsymbol{m}_{k0} \mid^2}{3h} \text{ by } \pi \frac{fe^2}{m_e}$$

where f is the oscillator strength of an absorption band. That is, we can make use of the classical expression for absorption if we define the oscillator strength by means of the relation

$$f = \frac{8\pi^2 \nu_{k0} m_e \mid \boldsymbol{m}_{k0} \mid^2}{3he^2} \tag{A-14}$$

We therefore see that according to quantum mechanics the atomic or molecular property that determines the intensity of the light absorption of a substance is the quantity $\boldsymbol{m}_{k0}$.

It will be convenient to define a transition probability per unit time and per unit of radiation energy density for the transition ϕ_0 to ϕ_k

$$B_{0k} = \mid a_k \mid^2 / t\varrho(\nu_{k0}) = 2\pi \mid \boldsymbol{m}_{k0} \mid^2 / 3\hbar^2 \tag{A-15a}$$

Similarly, we may have a transition probability, B_{k0}, for the reverse transition, $\phi_k \rightarrow \phi_0$, in which the molecule is initially placed in the state ϕ_k(a_k initially equal to unity) and then exposed to radiation of frequency ν_{k0} and intensity $\varrho(\nu_{k0})$. Evidently

$$B_{k0} = \mid a_0 \mid^2 / t\varrho(\nu_{k0}) = 2\pi \mid \boldsymbol{m}_{0k} \mid^2 / 3\hbar^2 \tag{A-15b}$$

Since $\boldsymbol{m}$ is a Hermitian operator, $\mid \boldsymbol{m}_{k0} \mid^2 = \mid \boldsymbol{m}_{0k} \mid^2$ so that

$$B_{k0} = B_{k0} \tag{A-16}$$

Thus for a given value of the radiation density, the induced transition $\phi_0 \rightarrow \phi_k$ has the same probability as the induced transition $\phi_k \rightarrow \phi_0$. That is, the probabilities of induced emission and of absorption are equal for any pair of states.

The quantities B_{k0} and B_{0k} are known as *Einstein transition probabilities for induced emission and absorption.*

c. Case II. Calculation of the induced dipole moment and the polarizability

According to the classical electron theory, if a molecule is exposed to light that it does not absorb, the primary response of the molecule is to acquire an induced dipole moment. All of the optical effects observed in transparent objects are a result of the scattered wavelets generated by the oscillations of this induced moment, as we have shown in Chapter 15. We shall now show that the irradiated molecule described by the wave function Ψ given by Equation (A-4) also has an induced dipole moment that oscillates in phase with the incident light. This induced moment must be responsible for the optical properties of transparent bodies.

According to the postulates of quantum mechanics (Law III, page 159) the component of the dipole moment in the direction of the field of the light wave (here the x-direction) is

$$\overline{m_x} = \int \Psi^* m_x \Psi d\tau \Big/ \int \Psi^* \Psi d\tau \tag{A-17}$$

where m_x is the dipole moment operator defined in Equation (A-3). Substituting (A-4) into (A-17), neglecting terms in the numerator that involve cross products of the coefficients a_j with all other coefficients except a_0, and replacing the denominator by unity, we find

$$\overline{m_x} = \int \phi_0^* m_x \phi_0 d\tau + a_0 \sum a_j^* \int \phi_j^* m_x \phi_0 d\tau + a_0^* \sum a_j \int \phi_0^* m_x \phi_j d\tau$$

$$= m_{x00} + \frac{|\boldsymbol{E}_0|}{2\hbar} \sum e^{i\omega_{j0}t} m_{x0j} m_{xj0} \left[\frac{e^{-i(\omega_{j0}+\omega)t} - 1}{\omega_{j0} + \omega} + \frac{e^{-i(\omega_{j0}-\omega)t} - 1}{\omega_{j0} - \omega} \right]$$

$$+ \frac{|\boldsymbol{E}_0|}{2\hbar} \sum e^{-i\omega_{j0}t} m_{x0j} m_{xj0} \left[\frac{e^{i(\omega_{j0}+\omega)t} - 1}{\omega_{j0} + \omega} + \frac{e^{i(\omega_{j0}-\omega)t} - 1}{\omega_{j0} - \omega} \right]$$

$$= m_{x00} + \frac{|\boldsymbol{E}_0|}{2\hbar} \sum |m_{xj0}|^2 \left[\frac{e^{i\omega t} + e^{-i\omega t} - e^{i\omega_{j0}t} - e^{-i\omega_{j0}t}}{\omega_{j0} + \omega} \right.$$

$$\left. + \frac{e^{i\omega t} + e^{-i\omega t} - e^{i\omega_{j0}t} - e^{-i\omega_{j0}t}}{\omega_{j0} - \omega} \right]$$

$$= m_{x00} + \frac{2|\boldsymbol{E}_0|}{\hbar} \cos \omega t \sum \frac{|m_{xj0}|^2 \omega_{j0}}{\omega_{j0}^2 - \omega^2} - \frac{2|\boldsymbol{E}_0|}{\hbar} \sum \frac{|m_{xj0}|^2 \omega_{j0} \cos \omega_{j0} t}{\omega_{j0}^2 - \omega^2} \tag{A-18}$$

The first term in this expression is independent of the time and represents the x-component of the permanent dipole moment of the molecule. The third term represents oscillating dipole moments whose frequencies are entirely different from that of the incident light wave and are of no interest here. The second term oscillates in phase with the light wave and therefore represents the induced dipole moment that we seek.

The polarizability, α, is the ratio of the induced moment to the applied field (here, $\boldsymbol{E}_0 \cos \omega t$). Thus the quantum mechanical expression for the polarizability of a molecule as a function of the frequency of the applied field is

$$\alpha = \frac{2}{\hbar} \sum \frac{\omega_{j0} \mid m_{xj0} \mid^2}{\omega_{j0}{}^2 - \omega^2}$$

$$= \frac{2}{h} \sum \frac{\nu_{j0} \mid m_{xj0} \mid^2}{\nu_{j0}{}^2 - \nu^2} \tag{A-19a}$$

If the molecules are randomly oriented in space, then $\mid m_{xj0} \mid^2$ must be replaced by the average of the squares of the three components

$$\tfrac{1}{3} \left[\mid m_{xj0} \mid^2 + \mid m_{yj0} \mid^2 + \mid m_{zj0} \mid^2 \right] = \tfrac{1}{3} \mid \boldsymbol{m}_{j0} \mid^2$$

so that

$$\alpha = \frac{2}{3h} \sum \frac{\nu_{j0} \mid \boldsymbol{m}_{j0} \mid^2}{\nu_{j0}{}^2 - \nu^2} \tag{A-19b}$$

Comparing this with the classical expression for the polarizability (Equation (D-15) of Chapter 15) we see that $2\nu_{j0} \mid \boldsymbol{m}_{j0} \mid^2/3h$ takes the place of $e^2 f_j/4\pi^2 m_e$. Thus we can identify the oscillator strength, f_j, with the quantity

$$\frac{8\pi^2 m_e \nu_{j0} \mid \boldsymbol{m}_{j0} \mid^2}{3he^2}$$

This is identical with the expression for the oscillator strength found for Case I (Equation (A-14)).

Thus the classical electron theory of absorption and polarizability can be taken over into quantum mechanics provided we interpret the oscillator strengths according to Equation (A-14) and provided we interpret the frequencies of absorption, ν_{j0}, in terms of the Bohr assumption, $\nu_{j0} = \mid E_j - E_0 \mid / h$, rather than as the normal-mode frequencies of oscillating charges.

It is interesting to note that if the state ϕ_0 is an excited state, some of the ν_{j0} in Equation (A-19b) will be negative. The corresponding terms in the polarizability expression therefore give negative oscillator strengths and make negative contributions to the polarizability. Experimental evidence for this has been obtained by Ladenburg (*1*).

d. Spontaneous emission of light from excited molecules

Having deduced the behavior of a molecule in a light beam from the quantum mechanical laws that were set up in Chapter 5, let us next see what happens when a molecule or atom is placed in a definite excited state in the complete absence of radiation. According to the procedures of Chapter 5, the Hamiltonian of such an atom or molecule depends only on the coordinates of the particles making up the system, and on derivatives with respect to these coordinates. Since the radiation is explicity assumed to be absent, the Hamiltonian should not contain the time. Under these conditions we have shown (corollary II, page 162) that the square of the wave function of an energy eigenstate is independent of time. Therefore, according to this formulation, if an atom gets into an excited state, it should remain there as long as it is undisturbed.

This result is, of course, not correct, since it is well known that systems in excited energy states will emit light spontaneously, undergoing transitions to states of lower energy. Apparently our procedure in this instance has led to a wrong answer.

The reason for this error lies in the assumption that the Hamiltonian depends only on the coordinates of the particles in the system. The charged particles in an atom or molecule do not interact only with each other. They also produce electric and magnetic fields in the space about them—that is, they interact with the "ether". If the Hamiltonian is to be complete, the energy associated with these interactions should be included. These interactions are very weak, but Dirac has shown that when they are included, spontaneous transitions take place from states of high energy to states of lower energy, the difference in energy appearing in the space around the system as a photon. In this way the spontaneous emission of light is quantitatively accounted for.

The detailed arguments of Dirac's theory are beyond the scope of this book (for relatively brief accounts of Dirac's theory see Slater (*2*, pp. 460*ff.*) and Mott and Sneddon (*3*, pp. 259*ff.*). A more complete discussion will be found in Heitler's book (*4*).) Fortunately it is possible to find other ways of obtaining Dirac's expression for the rate of spontaneous emission of light by excited atoms and molecules. We shall consider two different methods of doing this. The first is a simple, if somewhat indirect argument proposed by Einstein. This procedure depends on the requirement that the results of quantum mechanics must be consistent with the Planck black body radiation formula

$$\varrho(\nu) = \frac{8\pi h\nu^3}{c^3} \Big/ (e^{h\nu/kT} - 1) \tag{A-20}$$

where $\varrho(\nu)$ is the density of radiant energy (*cf.* page 583) in a cavity in thermal equilibrium with its walls (which have the temperature T), c is the velocity of light, and k is the Boltzmann constant.

We have seen that a system in a state m can be induced to undergo a transition to another state, n, by radiation whose frequency is $\nu_{mn} = | E_m - E_n |/h$, where E_m and E_n are the energies of the two states. It does not matter which of the two energies is the greater; if the initial state has a lower energy than the final state, we speak of "absorption," and if the initial state has the higher energy we speak of "induced emission." In addition we must now recognize the possibility of "spontaneous emission."

Let the state m be assumed to have the higher energy, so that B_{mn} is the Einstein probability of induced emission and B_{nm} is the Einstein probability of absorption (*cf.* page 644). Furthermore, let A_{mn} be the probability that an atom in state m will in unit time undergo a spontaneous transition to the lower energy state n. Let there be a large number of atoms or molecules in thermal equilibrium, with N_m atoms or molecules in state m and N_n in state n. The thermal energy density will be $\varrho(\nu)$, given by Equation (A-20). From the Boltzmann expression

$$N_m/N_n = \exp\,[-(E_m - E_n)/kT] = \exp\,(-h\nu_{mn}/kT) \tag{A-21}$$

The number of atoms undergoing transitions in unit time from state m to state n by means of processes involving the emission of a photon is

$$N_m(A_{mn} + \varrho(\nu_{mn})B_{mn})$$

and the number undergoing the reverse transition through the absorption of a photon is

$$N_n B_{nm} \varrho(\nu_{mn})$$

Since the system is in thermal equilibrium, these two rates must be equal.* Thus

$$N_m [A_{mn} + B_{mn}\varrho(\nu_{mn})] = N_n B_{nm} \varrho(\nu_{mn}) \tag{A-22}$$

or since $B_{mn} = B_{nm}$

$$\varrho(\nu_{mn}) = \frac{A_{mn}}{B_{mn}} \left[\frac{N_n}{N_m} - 1 \right]^{-1}$$

$$= \frac{A_{mn}}{B_{mn}} \Big/ (e^{h\nu_{mn}/kT} - 1) \tag{A-23}$$

Comparing (A-23) with the Planck radiation formula, Equation (A-20), we see that

$$A_{mn}/B_{mn} = 8\pi h \nu_{mn}^3/c^3 \tag{A-24a}$$

Substituting for B_{mn} from Equation (A-15) we find

$$A_{mn} = 64\pi^4 \nu_{mn}^3 \mid \boldsymbol{m}_{mn} \mid^2/3hc^3 \tag{A-24b}$$

This is the expression found by Dirac. We may note that once again the integral of the dipole moment operator between a pair of states makes its appearance.

A second simple method is available for deducing this expression for the spontaneous transition probability, A_{mn}. It depends on the comparison of the decay of radiation from an excited oscillator with the decay expected from a group of excited atoms. Suppose that N_m atoms are in state m and that they undergo spontaneous transitions exclusively to state n, the energy released appearing in the form of photons. Then from the definition of A_{mn}

$$dN_m/dt = -A_{mn}N_m \tag{A-25a}$$

or

$$N_m = N_{m0}e^{-A_{mn}t} \tag{A-25b}$$

* We do not exclude the possibility that transitions between m and n may occur without the emission or absorption of a single photon — for instance by collision, or by nonadiabatic processes (crossing of potential energy surfaces) or by simultaneous emission of more than one photon. According to the statistical mechanical principle of detailed balance (also known as the principle of microscopic reversibility), however, the rate of a transition at equilibrium must be exactly equal to the rate of its inverse for each separate conceivable mechanism of transition. Therefore, in writing Equation (A-22) we are justified in disregarding other mechanisms of transition even if they have greater rates.

The rate of emission of energy by these atoms is

$$\mathscr{W} = -(E_m - E_n)dN_m/dt$$
$$= A_{mn}(E_m - E_n)N_{m0}e^{-A_{mn}t} \tag{A-26}$$

On the other hand the classical theory gave the result that the rate of emission of a group of oscillators excited to a given amplitude, x_0, is proportional to the square of the amplitude (Equation (A-7a) of Chapter 15)

$$\mathscr{W} = Kx_0^2$$

According to Equation (B-3) of Chapter 15, the amplitude decays with time according to the relation

$$x_0 = x_{00}e^{-\omega' t}$$

giving

$$\mathscr{W} = x_{00}^2 K e^{-2\omega' t} \tag{A-27}$$

If all of the energy of the oscillator appears as radiation (*i.e.*, if there is no viscous damping) we have (*cf.* Equation (D-19) of Chapter 15)

$$\omega' = f\frac{e^2\omega^2}{m_e c^3} \tag{A-28}$$

Comparing Equations (A-27) with Equation (A-26) we see that we may identify A_{mn} with $2\omega'$, so that

$$A_{mn} = 8\pi^2 e^2 \nu^2 f/m_e c^3 \tag{A-29}$$

Substituting the quantum mechanical expression for the oscillator strength, (A-14), we obtain

$$A_{mn} = 64\pi^4 \nu_{mn}^3 \mid \boldsymbol{m}_{mn} \mid^2/3hc^3 \tag{A-30}$$

which is identical with (A-24b).

The reciprocal of the transition probability A_{mn} is the average time spent by the system in the state m before undergoing a transition to state n (usually called the *mean lifetime* for the transition). If the oscillator strength of the transition is unity and the emitted light is in the visible range, say $\nu = 5 \times 10^{14}$ sec.$^{-1}$ (corresponding to $\lambda = 6000$ Å.), then Equation (A-29) leads to a mean lifetime of about 0.5×10^{-8} sec. Since the oscillator strengths of transitions emitting visible light are rarely greater than unity, this represents a lower limit to the lifetimes of such excited states.

e. Quantum mechanical proof of the Kuhn-Thomas sum rule

We have mentioned (page 576) that according to the classical theory of radiation the oscillator strengths should obey the rule

$$\sum_{\substack{\text{all} \\ \text{absorptions}}} f_j = N \tag{A-31}$$

where N is the number of electrons in the atom or molecule. We shall now show that this rule is valid in the quantum mechanical theory. By definition

$$\sum_{\substack{\text{all} \\ \text{transitions, } j}} f_j = \frac{8\pi^2 m_e}{3he^2} \sum_{\substack{\text{all} \\ \text{states, } j}} \nu_{j0} \left[\mid m_{xj0} \mid^2 + \mid m_{yj0} \mid^2 + \mid m_{zj0} \mid^2 \right] \tag{A-32}$$

Consider the quantity $\nu_{j0} m_{xj0}$. This can be written as

$$\nu_{j0} m_{xj0} = \frac{e}{h} \sum_{\substack{\text{all} \\ \text{electrons, } n}} (E_j - E_0) \int \psi_j^* x_n \psi_0 d\tau$$

$$= \frac{e}{h} \sum_n \left[\int (H\psi_j^*)(x_n\psi_0) d\tau - \int (\psi_j^* x_n)(H\psi_0) d\tau \right] \tag{A-33}$$

Since the Hamiltonian operator H is Hermitean and real the first integral can be written

$$\sum_n \int (H\psi_j^*)(x_n\psi_0) d\tau = \sum_n \int \psi_j^* H x_n \psi_0 d\tau$$

$$= -\frac{h^2}{8\pi^2 m_e} \sum_{n'} \sum_n \int \psi_j^* \nabla_{n'}^2 x_n \psi_0 d\tau + \sum_n \int \psi_j^* V x_n \psi_0 d\tau \tag{A-34}$$

Now if $n' \neq n$

$$\nabla_{n'}^2 x_n \psi_0 = x_n \nabla_{n'}^2 \psi_0 \tag{A-35a}$$

and if $n' = n$

$$\nabla_n^2 x_n \psi_0 = \left(\frac{\partial^2}{\partial x_n^2} + \frac{\partial^2}{\partial y_n^2} + \frac{\partial^2}{\partial z_n^2} \right) x_n \psi_0$$

$$= 2 \frac{\partial \psi_0}{\partial x_n} + x_n \nabla_n^2 \psi_0 \tag{A-35b}$$

so that (recalling that V commutes with all of the coordinates because it is an ordinary function of the coordinates)

$$\nu_{j0} m_{xj0} = -\frac{eh}{4\pi^2 m_e} \sum_n \int \psi_j^* \frac{\partial \psi_0}{\partial x_n} d\tau$$

$$= -\frac{ei}{2\pi m_e} \sum_n \int \psi_j^* p_{xn} \psi_0 d\tau \tag{A-36}$$

where p_{xn} is the operator for the component of momentum of the nth particle in the x-direction. Multiplying by m_{x0j}, adding over all states and writing

$$\sum_n p_{xn} = P_x \text{ and } \sum_n x_n = X$$

we obtain

$$\sum_j \nu_{j0} \mid m_{xj0} \mid^2 = -\frac{e^2 i}{2\pi m_e} \sum_j (\psi_0^* \mid X \mid \psi_j)(\psi_j^* \mid P_x \mid \psi_0) \tag{A-37}$$

where the expression $(\psi_0^* \mid X \mid \psi_j)$ stands for the integral

$$\int \psi_0^* X \psi_j d\tau$$

and similarly for the expression $(\psi_j^* \mid P_x \mid \psi_0)$. But for any pair of operators, α and β, it is true that*

$$\sum_j (\psi_0^* \mid \alpha \mid \psi_j)(\psi_j^* \mid \beta \mid \psi_0) = (\psi_0^* \mid \alpha\beta \mid \psi_0) \tag{A-38}$$

Therefore,

$$\sum_j \nu_{j0} \mid m_{xj0} \mid^2 = -\frac{e^2 i}{2\pi m_e} (\psi_0^* \mid X P_x \mid \psi_0) \tag{A-39}$$

* Expand $\beta\psi_0$ in terms of the complete set of functions, ψ_j

$$\beta\psi_0 = \sum_j (\psi_j^* \mid \beta \mid \psi_0)\psi_j$$

Operate on this expression with α, multiply by ψ_0^* and integrate. This gives

$$(\psi_0^* \mid \alpha\beta \mid \psi_0) = \sum_j (\psi_0^* \mid \alpha \mid \psi_j)(\psi_j^* \mid \beta \mid \psi_0)$$

Alternatively, we may take the complex conjugate of Equation (A-36) and obtain

$$\nu_{j0} m_{x0j} = \frac{ei}{2\pi m_e} \sum_n (\psi_j \,|p_{xn}^*|\, \psi_0^*)$$

$$= \frac{ei}{2\pi m_e} (\psi_0^* \mid P_x \mid \psi_j) \qquad \text{(A-40)}$$

which, on multiplying by m_{xj0} and summing over all states gives

$$\sum_j \nu_{j0} \mid m_{xj0} \mid^2 = \frac{e^2 i}{2\pi m_e} \sum_j (\psi_0^* \mid P_x \mid \psi_j)\, (\psi_j^* \mid X \mid \psi_0)$$

$$= \frac{e^2 i}{2\pi m_e} (\psi_0^* \mid P_x X \mid \psi_0) \qquad \text{(A-41)}$$

Adding Equations (A-39) and (A-41) we find

$$2 \sum_j \nu_{j0} \mid m_{xj0} \mid^2 = \frac{e^2 i}{2\pi m_e} (\psi_0^* \mid P_x X - X P_x \mid \psi_0) \qquad \text{(A-42)}$$

But

$$P_x X - X P_x = \sum_n (p_{xn} x_n - x_n p_{xn})$$

$$= \sum_{\substack{\text{all} \\ \text{electrons}}} h/2\pi i = Nh/2\pi i \qquad \text{(A-43)}$$

because of the commutation rules for position and momentum (see page 172). Thus

$$\sum \nu_{j0} \mid m_{xj0} \mid^2 = \frac{Ne^2 i}{4\pi m_e} \frac{h}{2\pi i} (\psi_0^* \mid 1 \mid \psi_0) = N \frac{he^2}{8\pi^2 m_e} \qquad \text{(A-44)}$$

It is easy to see that the same procedure leads to the result

$$\sum_j \nu_{j0} \mid m_{yj0} \mid^2 = \sum_j \nu_{j0} \mid m_{zj0} \mid^2 = N \frac{he^2}{8\pi^2 m_e} \qquad \text{(A-45)}$$

When these are substituted in Equation (A-32), we obtain the Kuhn-Thomas sum rule, Equation (A-31).

f. Other mechanisms of transition

In the discussion given above, it has been assumed that the wave length of the light that interacts with the molecule is long compared with the dimensions of the molecule. Furthermore, the interaction with the magnetic field in the light wave has been neglected. This means, in effect, that we have disregarded the magnetic dipole and electric quadrupole interactions discussed from the classical point of view on pages 610*ff*. When these effects are taken into account it is found (see *5*, pp. 93, 96) that small additional terms occur in the expression for the transition probability. The Einstein coefficient for spontaneous emission then has the form

$$A_{mn} = \frac{64\pi^4\nu_{mn}{}^3}{3c^3h}\left\{ |(\psi_m{}^* \mid e\mathbf{r} \mid \psi_n)|^2 + \left|\left(\psi_m{}^* \mid \frac{e}{2m_ec}\mathbf{r}\times\mathbf{p} \mid \psi_n\right)\right|^2 + \frac{3\pi^2}{10}\frac{\nu_{mn}{}^2}{c^2}|(\psi_m{}^* \mid e\mathbf{rr} \mid \psi_n)|^2\right\} \quad \text{(A-46)}$$

where

$$|(\psi^*_m \mid e\mathbf{rr} \mid \psi_n)|^2 = \frac{e^2}{2}\left[\frac{1}{3}\left|\int \psi_m{}^* \sum (3z_i{}^2 - r_i{}^2)\,\psi_n d\tau\right|^2 + 4\left|\int \psi_m{}^* \sum x_iy_i\psi_n d\tau\right|^2 + 4\left|\int \psi_m{}^* \sum x_iz_i\psi_n d\tau\right|^2 + 4\left|\int \psi_m{}^* \sum y_iz_i\psi_n d\tau\right|^2 + \left|\int \psi_m{}^* \sum (x_i{}^2 - y_i{}^2)\,\psi_n d\tau\right|^2\right]$$

and where the first term arises from electric dipole interactions, the second term from magnetic dipole interactions and the third term from electric quadrupole interactions. The ratios of these terms for typical transitions are of the same order of magnitude as was found for the similar terms in the classical theory (see page 615). It sometimes happens, however, that the dipole moment integral, $(\psi_m{}^* \mid e\mathbf{r} \mid \psi_n)$, vanishes and the weak radiation produced by the quadrupole and magnetic dipole terms can be observed. In general the nature of the interaction may be determined experimentally by means of the interferometric method mentioned on page 613 and also from the behavior of the spectrum in electric and magnetic fields (see *6*, p. 156; *7*).

B. Selection Rules for Dipole Transitions

The intensity of absorption and emission resulting from an electric dipole transition between a given pair of states a and b depends, as we have seen, on the magnitude of the integral

$$\int \psi_b{}^* \boldsymbol{m}\psi_a d\tau = \boldsymbol{m}_{ba}.$$

This integral vanishes for many pairs of states of actual molecules, so that the corresponding transition can occur only by the much less intense multipole or magnetic dipole mechanisms, if at all. The conditions under which the integral fails to vanish for the various possible pairs of states of a given system are known as *selection rules*. A transition between a pair for which $\boldsymbol{m}_{ba}$ vanishes is said to be *forbidden*. In some instances, a first approximation to the correct wave functions a and b will give $\boldsymbol{m}_{ba} = 0$, whereas a higher approximation will lead to a nonvanishing (though usually small) value of the moment integral. Such transitions are said to be *approximately forbidden*. If the precise wave functions give $\boldsymbol{m}_{ba} = 0$, the transition is said to be *strictly forbidden*, even though it may be possible through a magnetic dipole or electric quadrupole mechanism.

The quantity $|\boldsymbol{m}_{ba}|^2$ is called the *line strength* of the transition between states a and b.

We shall now investigate the selection rules for some typical systems.

a. The harmonic oscillator

Suppose that a charged particle of mass M is constrained to move in one direction at the end of a spring whose force constant is K. We have seen (Chapter 6, especially Equation (E-26)) that the normalized wave functions are

$$\psi_n = \left(\frac{\sqrt{\beta/\pi}}{2^n n!}\right)^{1/2} e^{-\beta x^2/2} H_n(\sqrt{\beta}x) \tag{B-1}$$

and the energy levels are

$$E_n = (n + \tfrac{1}{2})h\nu_0 \tag{B-2}$$

where x is the displacement of the particle, $\beta = 2\pi\sqrt{MK}/h$, ν_0 is the frequency of oscillation, and $H_n(\xi)$ is the nth Hermite polynomial. According to the Bohr postulate, the absorption frequencies should be

$$\nu_{mn} = \frac{E_m - E_n}{h} = (m - n)\nu_0 \tag{B-3}$$

so we might expect absorption and emission at frequencies that are whole multiples of the natural frequency of the oscillating charge. On the other hand, the intensity of the dipole radiation is determined by the value of the integral

$$|\boldsymbol{m}_{mn}| = |(\psi_m | \boldsymbol{m} | \psi_n)|$$

$$= e\sqrt{\frac{\beta}{\pi}}\,(2^{n+m}m!n!)^{-1/2} \int_{-\infty}^{\infty} xe^{-\beta x^2} H_m(\sqrt{\beta}x) H_n(\sqrt{\beta}x)dx \tag{B-4}$$

According to Equation (E-25) of Chapter 6

$$\xi H_n(\xi) = nH_{n-1}(\xi) + \tfrac{1}{2} H_{n+1}(\xi) \tag{B-5}$$

Thus,

$$|\, \boldsymbol{m}_{mn} \,| = e(\pi 2^{m+n} m!\, n!)^{-1/2} \int_{-\infty}^{\infty} e^{-\beta x^2} [nH_{n-1}(\sqrt{\beta} x) + \tfrac{1}{2} H_{n+1}(\sqrt{\beta} x)] H_m(\sqrt{\beta} x) dx$$

$$= \frac{e}{\sqrt{2\beta}} \left[\sqrt{n} \int_{-\infty}^{\infty} \psi_{n-1} \psi_m dx + \sqrt{n+1} \int_{-\infty}^{\infty} \psi_{n+1} \psi_m dx \right] \tag{B-6}$$

Since the ψ_n are orthonormal, it is obvious that $\boldsymbol{m}_{mn} = 0$ unless $m = n \pm 1$. If $m = n + 1$

$$|\, \boldsymbol{m}_{mn} \,| = e \sqrt{\frac{n+1}{2\beta}} = \frac{e}{2} \left[\frac{(n+1)\, h}{\pi} \right]^{1/2} (MK)^{-1/4} \tag{B-7a}$$

If $m = n - 1$

$$|\, \boldsymbol{m}_{mn} \,| = e \sqrt{\frac{n}{2\beta}} = \frac{e}{2} \left[\frac{nh}{\pi} \right]^{1/2} (MK)^{-1/4} \tag{B-7b}$$

Therefore the selection rule for the harmonic oscillator limits transitions to those between adjacent states. Since

$$E_{n+1} - E_n = E_n - E_{n-1} = h\nu_0 \tag{B-8}$$

all transitions produce or absorb radiation of the same frequency, and this frequency is that of the oscillator itself. For the oscillator strength we obtain in absorption

$$f_{n \to n+1} = \frac{2\pi}{3} \sqrt{\frac{M}{K}}\, (n+1)\nu_0 = \frac{n+1}{3} \tag{B-9a}$$

and in emission

$$f_{n \to n-1} = -\frac{2\pi}{3} \sqrt{\frac{M}{K}}\, n\nu_0 = -\frac{n}{3} \tag{B-9b}$$

The total oscillator strength of a system in state n is, from (A-14)

$$f_{n \to n+1} + f_{n \to n-1} = 1/3 \tag{B-9c}$$

which is the same as the classical value for a linear oscillator.

Exercise. Effect of perturbations on selection rules.

Suppose that the potential energy of a charged particle is given by

$$V = \tfrac{1}{2} Kx^2 + kx^3$$

where x is the displacement and k is a small constant. Using first order perturbation theory show that the wave function of the mth state is of the form

$$\psi_m = \psi^0{}_m + k(A\psi^0{}_{m-3} + B\psi^0{}_{m-1} + C\psi^0{}_{m+1} + D\psi^0{}_{m+3})$$

where the $\psi^0{}_l$ are harmonic oscillator wave functions and A, B, C, and D are constants. Show that in addition to the transition with $\Delta m = \pm 1$, transitions are also possible whose probabilities are proportional to k with $\Delta m = \pm 2$ and ± 4. The frequencies of the light emitted and absorbed in the latter transitions are close to $2\nu_0$ and $4\nu_0$, where ν_0 is the frequency of the oscillator in the absence of the anharmonic term kx^3 in the potential.

Thus the anharmonicity makes possible transitions that are not possible for the harmonic oscillator itself. Since the frequencies associated with these new transitions are approximately whole multiples of the frequency of the unperturbed oscillator, they are called *overtones*. It is interesting to note that overtones of the infra-red vibration frequencies of H_2O occur in the red end of the visible spectrum and are probably responsible for the blue color of pure water.

This is an example of a change in the selection rules of a transition by a perturbation—a rather common occurrence, as we shall see.

b. The particle in a box

The calculation of the oscillator strengths and the selection rules for the transitions of a particle in a box is a perfectly straightforward matter of substitution of the wave functions given in Equation (B-4a) of Chapter 6 into the appropriate expressions. The results are quoted below, and the proofs are left as exercises for the student.

Exercise. Show that for a charged particle moving in a one dimensional box, the oscillator strength of a transition from a state of quantum number m to one of quantum number n is

$$f_{nm} = \frac{64}{3\pi^2} \frac{m^2 n^2}{(n^2 - m^2)^3} \qquad \text{(B-10a)}$$

if $(m + n)$ is an odd number and

$$f_{nm} = 0 \qquad \text{(B-10b)}$$

if $(m + n)$ is an even number. Prove by numerical substitution or otherwise that when $m = 1$, and also when $m = 2$

$$\sum_n f_{nm} = {}^1/_3 \qquad \text{(B-10c)}$$

Note that the oscillator strength is independent of the charge and mass of the particle. Also, if $n = m + 1$

$$f_{m+1,\, m} = \frac{64}{3\pi^2} \frac{m^2 (m+1)^2}{(2m+1)^3} \tag{B-10d}$$

so that as m increases

$$f_{m+1,\, m} \rightarrow \frac{8m}{3\pi^2} \tag{B-10e}$$

Values of f_{nm} larger than unity are thus possible if the charged particle is initially in an excited state. This does not contradict the sum rule because the f_{nm} are negative when $m > n$.

c. The linear rotator

Consider a particle of mass M and charge e that can rotate about a fixed point at a distance R. According to Equation (D-8) of Chapter 6, the wave function is

$$\psi = K_{lm} P_l^{|m|} (\cos\theta) e^{im\varphi} \tag{B-12}$$

where K_{lm} is the normalizing constant. The energy is

$$E = \frac{\hbar^2}{2MR^2} l(l+1) \tag{B-13}$$

The dipole moment operators in polar coordinates are

$$m_x = eR \sin\theta \cos\varphi$$

$$m_y = eR \sin\theta \sin\varphi$$

$$m_z = eR \cos\theta$$

Thus the integrals determining the transition probabilities are

$$m_{xlm,\, l'm'} = eRK_{lm}K_{l'm'} \int \sin\theta\, P_l^{|m|} P_{l'}^{|m'|} \sin\theta d\theta \int \cos\varphi e^{i(m'-m)\varphi} d\varphi$$

$$m_{ylm,\, l'm'} = eRK_{lm}K_{l'm'} \int \sin\theta\, P_l^{|m|} P_{l'}^{|m'|} \sin\theta d\theta \int \sin\varphi e^{i(m'-m)\varphi} d\varphi$$

$$m_{zlm,\, l'm'} = eRK_{lm}K_{l'm'} \int \cos\theta\, P_l^{|m|} P_{l'}^{|m'|} \sin\theta d\theta \int e^{i(m'-m)\varphi} d\varphi$$

Making use of the relations

$$\cos\varphi = \tfrac{1}{2}(e^{i\varphi} + e^{-i\varphi}),\ \sin\varphi = \frac{1}{2i}(e^{i\varphi} - e^{-i\varphi})$$

and Equations (B-72), (B-73), and (B-75) of Chapter 3, which give

$$\sin\theta P_l^{|m|} = [P_{l+1}^{|m|+1} - P_{l+1}^{|m|+1}]/(2l+1)$$

$$= [(l+|m|+1)(l+|m|)P_{l-1}^{|m|-1}$$

$$-(l-|m|+2)(l-|m|+1)P_{l+1}^{|m|-1}]/(2l+1)$$

$$\cos\theta P_l^{|m|} = [(l-|m|+1)P_{l+1}^{|m|} + (l+|m|)P_{l-1}^{|m|}]/(2l+1)$$

it is easy to obtain the selection rules

$$m_{xlm,\,l'm'} \text{ and } m_{ylm,\,l'm'} = 0 \text{ unless } m = m' \pm 1 \text{ and } l = l' \pm 1 \qquad \text{(B-14a)}$$

$$m_{zlm,\,l'm'} = 0 \text{ unless } m = m' \text{ and } l = l' \pm 1 \qquad \text{(B-14b)}$$

Since in either case $\Delta l = \pm 1$, the frequency for absorption when the particle is in a state with quantum number l is

$$\nu = (E_{l+1} - E_l)/h = \frac{\hbar}{2\pi MR^2}(l+1) \qquad \text{(B-15a)}$$

and the frequency for emission from the same state is

$$\nu = (E_l - E_{l-1})/h = \frac{\hbar}{2\pi MR^2}l \qquad \text{(B-15b)}$$

d. The hydrogen atom

Since the wave function of the hydrogen atom contains the same angular dependence as that of the linear rotator, the selection rules must be identical as far as changes in l and m are concerned. Thus a hydrogen atom can undergo only transitions of the type $s \longleftrightarrow p$, $p \longleftrightarrow d$, $d \longleftrightarrow f$. Apparently a hydrogen atom in a 2s-state cannot return directly to the ground state by dipole radiation. The 2s-state is therefore metastable.

It is found that there is no limitation on the change that the principal quantum number, n, can undergo, though in general small changes in n are the most probable. Detailed formulae are given by Bethe (*8*, p. 440).

The oscillator strengths of some of the transitions of the hydrogen atom are given in Table 16-1.

TABLE 16-1

OSCILLATOR STRENGTHS OF SOME TRANSITIONS OF THE HYDROGEN ATOM (TAKEN FROM BETHE (*8*, PAGE 443))

Initial state Final state	1s np	2s np	2p ns	2p nd
$n = 1$	—	—	−0.139	—
2	0.4162	—	—	—
3	0.0791	0.425	0.014	0.694
4	0.0290	0.102	0.003	0.122

Exercise. Violation of the Selection Rules of an Atom, Caused by an Electric Field. If a hydrogen atom is placed in a small, uniform electric field $|\mathbf{E}_0|$ acting in the z-direction, there is a perturbation on the electron

$$V = -e|\mathbf{E}_0|z = -e|\mathbf{E}_0|R\cos\theta$$

Show that according to the first order perturbation theory, the wave function of the 1s-state acquires the form

$$\psi = 1s + A2p_z + B3p_z + \cdots$$

and the wave function of the 2s-state acquires the form

$$\psi' = 2s + A'2p_z + B'3p_z + \cdots$$

where $A, A', B, B', \cdots$ are constants proportional to $|\mathbf{E}_0|$. Show that the selection rule $\Delta l = \pm 1$ breaks down in this case, so that transitions between the 1s- and 2s-states are possible, with a probability proportional to $|\mathbf{E}_0|$.

In gases at high pressures and in solutions in polar solvents, the electric fields of the permanent molecular dipoles and of ionic charges in the vicinity of an atom may cause the ordinary selection rules, as derived for the isolated atom, to break down in a manner similar to that found above.

e. Atoms other than hydrogen

The "one-electron spectra" of the alkali atoms would be expected to be similar to those of the hydrogen atom. Inspection of Fig. 9-4 will verify that this is indeed the case. The observed strong lines in the spectrum of lithium are invariably of the type $^2S \longleftrightarrow {}^2P$, $^2P \longleftrightarrow {}^2D$, $^2D \longleftrightarrow {}^2F$, etc. Thus it is not possible to go from an S-state to a D-state of sodium or potassium by the emission or absorption of dipole radiation. In atoms containing more than one outer electron, however, additional factors must be taken into account.

(*i*) *The Laporte Rule.* All atomic wave functions can be written in such a way that they are either symmetric ("even") or antisymmetric ("odd) with respect to the operation of inversion of the electrons through the nucleus (that is, a change of the coordinates of each electron from x, y, z to $-x, -y, -z$). Now the operation of inversion on the dipole moment operator, $\boldsymbol{m} = e(\boldsymbol{i}x + \boldsymbol{j}y + \boldsymbol{k}z)$ results in a change in the sign of $\boldsymbol{m}$. In order that the integral

$$\int \psi_a{}^* \boldsymbol{m} \psi_b d\tau$$

be different from zero, the operation of inversion on the integrand, $\psi_a{}^* \boldsymbol{m} \psi_b$, must leave its sign unchanged (see Theorem II page 141). Therefore the product $\psi_a{}^* \psi_b$ must be odd with respect to inversion, which is possible only if one of the two states is even and the other is odd. That is, there is a strict selection rule for dipole radiation from atoms and other systems having a center of symmetry, requiring that transitions be between odd states and even states. This is the *Laporte rule.*

It is obvious from inspection that the hydrogenic wave functions for s, d, g, etc. states ($l = 0, 2, 4, \cdots$) are even, whereas those for p, f, h, etc. states ($l = 1, 3, 5, \cdots$) are odd. In an atom whose wave function can be written as the product of one-electron orbitals, the state is even if the sum of the l-values of the occupied orbitals ($\sum l_i$) is an even number, and the state is odd if this sum is an odd number. We can therefore readily see, for example, that for iron any wave function of the configuration $1s^2 2s^2 2p^6 3s^2 3p^6 4s^2 3d^6$ is even, whereas the configuration $1s^2 2s^2 2p^6 3s^2 3p^6 4s^2 3d^5 4p$ can have only odd wave functions. Thus according to the Laporte rule, transitions between these two configurations should be possible, but transitions between the multiplets within either configuration are forbidden.

According to the Laporte rule dipole transitions are impossible between states having the same electronic configuration. Such transitions can occur, however, through quadrupole and magnetic dipole radiation; as can be seen from (A-46) these transitions involve the operators $\boldsymbol{r} \times \boldsymbol{p}$ and $\boldsymbol{rr}$, neither of which change sign when inverted through the origin. Therefore quadrupole and magnetic dipole radiation can occur only between states which are either both odd or both even.

An interesting example of the failure of the Laporte rule, caused by quadrupole and magnetic dipole radiation, is to be found in the transitions between the ^2P, ^2D, and ^4S-states of the $2p^3$-configuration of the O^+ ion and also in the transitions between the ^1S, ^1D, and ^3P-states of the $2p^2$-configuration of the O^{2+} and N^{3+} ions. Such transitions are forbidden by the Laporte rule. Nevertheless strong lines corresponding to these transitions are observed in the spectra of nebulae.

Since these transitions are never observed in terrestrial sources, they were once supposed to be caused by a new element, which was given the name nebulium. Bowen, however, pointed out that because of the Laporte rule, the excited states of these configurations are metastable and hence have very long lifetimes. Under terrestrial conditions (frequent collisions with other atoms and ions and with the walls of the container, etc.), the metastable ions are able to find ways of getting to the ground state without giving off radiation. In the extremely rarefied conditions of interstellar space, however, it is a very long time (of the order of a day) between collisions. Given such a long time, emission by the quadrupole or magnetic dipole mechanisms is by far the most rapid method of getting to the ground state from one of the metastable states.

(ii) Restrictions on Changes in the Orbital Angular Momentum (Russell-Saunders Coupling). To the extent that Russell-Saunders coupling is valid (that is, when the spin–orbit coupling is weak), it can be shown that the selection rule for dipole radiation is $\Delta L = 0$ or ± 1, except that states with $L = 0$ cannot undergo a transition to another state with $L = 0$ (L is the total orbital angular momentum quantum number). Furthermore, to the extent that (a) the atomic wave function can be described as a product of one-electron wave functions and (b) the transition involves a single electron, the orbital quantum number l of the electron that undergoes the transition must change by unity ($\Delta l = \pm 1$). Both of these rules can, however, be violated, the former when Russell-Saunders coupling breaks down, and the latter when different electronic configurations become mixed (configuration interaction, see page 325).

(iii) Restrictions on Changes in the Spin. We have seen (Chapters 10 and 11) that if the spin–orbit interaction can be neglected, atomic and molecular wave functions can be written as eigenfunctions of the total spin operators, S^2 and S_z. Furthermore, the dipole moment operator $\boldsymbol{m}$ is completely independent of the spin and therefore commutes with the spin operators (this is also true of the operators for the quadrupole and magnetic dipole moments). Making use of Theorem V of Chapter 4 (page 144) we see that if ϕ_a and ϕ_b are eigenfunctions of S^2 and S_z with eigenvalues $S_a(S_a + 1)$, $S_b(S_b + 1)$, $\mathscr{S}_a$, and $\mathscr{S}_b$, respectively, then

$$\int \phi_a{}^* \boldsymbol{m}\, \phi_b d\tau = 0$$

unless $S_a = S_b$ and $\mathscr{S}_a = \mathscr{S}_b$. That is, in the absence of spin–orbit interaction, we have the selection rules $\Delta S = 0$ and $\Delta \mathscr{S} = 0$. The same selection rules hold for quadrupole and magnetic dipole transitions.

The conditions under which this rule is most likely to be obeyed are found in the helium atom, where the spin–orbit coupling is very small. This explains the experimental observation that no transitions are observed spectroscopically between any of the triplet states of helium and any of the singlet states. The spectrum of helium therefore appears to be produced by a mixture of two substances. This was, in fact, noticed before the role of the spin in atomic structure had been elucidated, and the names ortho-helium and para-helium were given to what we now know are the triplet and singlet sets of states, respectively, the implication being that there were two different kinds of helium. The lowest state of ortho-helium is $1s\,2s\,{}^3S$, which has a considerably higher energy than the lowest state of para-helium, $1s^2\,{}^1S$, but since the transition ${}^3S \rightarrow {}^1S$ is forbidden, the ground state of ortho-helium cannot be reached through the emission of radiation. Thus the ground state of ortho-helium is metastable. There are, of course, other ways in which a ${}^3S$ helium atom can return to the ground state (e.g., by interaction with a paramagnetic atom or with another atom of ortho-helium).

As we go further into the periodic table, spin–orbit interaction becomes increasingly important, and the selection rule $\Delta S = 0$ breaks down. For instance, the line corresponding to the transition $ns\,np\,{}^3P \rightarrow ns^2\,{}^1S$ in the group II metals (Mg, Ca, Sr, Ba, Zn, Cd, Hg) becomes more and more intense as the atomic number increases. In mercury this transition is responsible for the very intense 2537 Å. line that is so frequently used in photochemical investigations. Lines arising from transitions involving a change in the multiplicity are known as *intercombination lines.*

(*iv*) *Restrictions on the Changes in the Total Angular Momentum.* It was shown in Chapter 10 that when the spin–orbit interactions are included in the Hamiltonian operator H of an atom, the operator J_e for the total angular momentum still commutes with H, even though the operators for spin and orbital angular momentum fail to commute with H. It is found (see *5*, Chapter 3) that the selection rules for the total angular momentum quantum number J arising from these operators are as follows:

For electric and magnetic dipole radiation: $\Delta J = 0$ or ± 1, except that a state with $J = 0$ cannot go to another state with $J = 0$.

For quadrupole radiation: $\Delta J = 0, \pm 1$, or ± 2, except that (a) $J = 0$ cannot go to either $J = 0$ or $J = 1$, (b) $J = 1/2$ cannot go to $J = 1/2$, and (c) $J = 1$ cannot go to $J = 0$.

f. Selection rules for molecules

Because of the Born-Oppenheimer principle, the wave function of a molecule can, to a good approximation, be written in the form

$$\psi = \phi_e \phi_v \phi_r$$

where ϕ_e is a function of the positions of the electrons and the nuclei, ϕ_v pertains to the much slower vibrational motions of the nuclei and depends on the internuclear distances alone, and ϕ_r describes the rotation of the molecule as a whole and depends on the orientation of the molecule in space. The function ϕ_r is independent of internuclear distances.

(*i*) *Selection Rules for Changes in the Angular Momentum.* In diatomic molecules and in linear polyatomic molecules ϕ_e can be written as an eigenfunction of the electronic orbital angular momentum about the axis of the molecule. Selection rules can be derived that specify the changes in this component of the angular momentum as well as the changes in the angular momentum of the nuclear framework associated with ϕ_r. Depending on the nature of the coupling of these two types of rotation with each other and with the electron spin, one can obtain various alternative sets of selection rules. In principle the basis of these rules is similar to the basis of the selection rules for the different types of coupling found in atoms, but the possibility of over-all molecular rotation makes for greater complexity because of the larger number of coupling possibilities. The student is referred to Herzberg (*9*, pp. 240*ff*.) for details.

(*ii*) *Selection Rules for Changes in the Vibrational State. The Franck-Condon Principle.* The effect of the vibrational motions of the nuclei on the selection rules is of much more general interest and will be discussed in more detail. The internuclear distance occurs in the molecular wave function in two places: in ϕ_e and in ϕ_v. Especially in states of large vibrational energy, ϕ_v has the oscillatory character imparted by the Hermite polynomials, whereas ϕ_e varies much more slowly with the internuclear distances. Let us consider the dipole moment integral,

$$\int \mathrm{I}_i{}^* \boldsymbol{m}\, \mathrm{II}_j d\tau$$

where the symbols I and II refer to the electronic states of a molecule, and the subscripts i and j refer to the vibrational levels of the molecule in the respective electronic states. (The volume element $d\tau$ includes both nuclear and electronic coordinates.) For simplicity we shall assume that the molecule is diatomic and that the potential energy curves of the two electronic states are as shown in Fig. 16-2. Let us compare the magnitudes of the integrals for transitions between the lowest vibrational level, $i = 0$, of the lower electronic state, I, to various vibrational levels, $j = 0, 1, 2, \cdots$, of the upper state, II. The wave functions of these two states will have the form $\phi^{\mathrm{I}}{}_e\phi^{\mathrm{I}}{}_v$ and $\phi^{\mathrm{II}}{}_e\phi^{\mathrm{II}}{}_v$. The approximate shapes of $\phi^{\mathrm{I}}{}_v$ and $\phi^{\mathrm{II}}{}_v$ have been superimposed on each vibrational level in Fig. 16-2.

It is evident that the magnitude of

$$\int \mathrm{I}_i{}^* \boldsymbol{m}\, \mathrm{II}_j d\tau$$

depends largely on the behavior of the product $\phi^{\mathrm{I}}{}_v\phi^{\mathrm{II}}{}_v$, since this product varies most rapidly with the internuclear distance, R. For a transition from the vibrational level $i = 0$ of state I to the vibrational level $j = 0$ of state II, this product is very small because $\phi^{\mathrm{I}}{}_v$ is large in regions where $\phi^{\mathrm{II}}{}_v$ is small and *vice versa* (that is, the two vibrational functions overlap only slightly). Therefore such a transition is very unlikely when the potential energy curves are arranged as in Fig. 16-2.

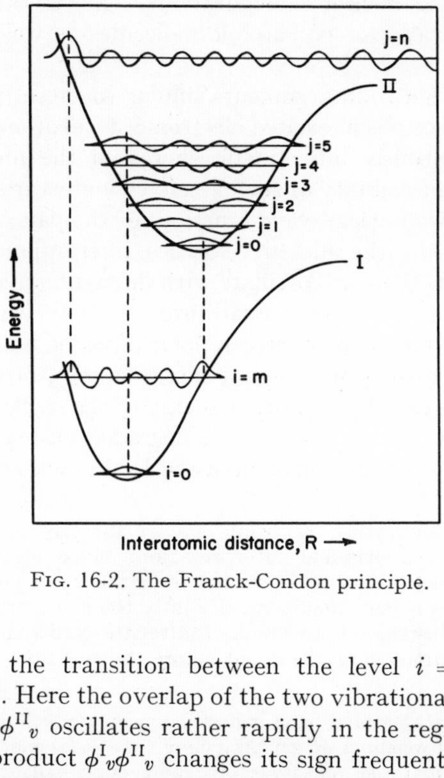

FIG. 16-2. The Franck-Condon principle.

Consider next the transition between the level $i = 0$ and the level $j = n$ in Fig. 16-2. Here the overlap of the two vibrational functions is good, but the function $\phi^{\mathrm{II}}{}_v$ oscillates rather rapidly in the region of the overlap. As a result, the product $\phi^{\mathrm{I}}{}_v\phi^{\mathrm{II}}{}_v$ changes its sign frequently as R varies, the positive values tending to cancel the negative ones, so that the integral as a whole tends to be small. Thus the transition probability between these two levels is small.

Finally consider the transitions between $i = 0$ and the levels $j = 3$, 4, and 5. In this case ϕ^{II}_v does not oscillate at all in the region where ϕ^{I}_v is large, and ϕ^{I}_v is small in the range of values of R for which ϕ^{II}_v does oscillate. Therefore the integral of the product $\phi^{I}_v\phi^{II}_v$ for these two levels will tend to be large, so that a transition between these levels is possible. For similar reasons the transitions from $i = m$ to $j = 0$ and n will be possible, whereas transitions from $i = m$ to other levels of electronic state II will be less likely to occur.

The general conclusions of the reasoning outlined above can be summarized in the following statement: Transitions tend to occur between vibrational levels of two different electronic states for which either the maximum or the minimum values of the internuclear distance in the potential energy diagram occur with the same nuclear configurations. This is the *Franck-Condon principle* and it is valid for polyatomic molecules as well as for diatomic molecules.

It is also easily seen from arguments similar to those just outlined that the energy of the electrons in excited electronic states of molecules cannot be converted instantaneously into kinetic energy of the nuclei by external perturbations. The probability of such a conversion invariably involves the integration of the two nuclear wave functions of the pair of states involved in the transition, and if the nuclear velocities differ appreciably in the two states, these two functions will oscillate with different wave lengths, so that their product will tend to give a small integral. Only when electrons and nuclei are able to interact for relatively long times or under special conditions, such as the crossing of potential energy curves, predissociation, and internal conversion, can the excitation energy of the electrons be converted directly into nuclear motion and thereby degraded ultimately into thermal energy. This important principle is discussed by Franck and Levi (*10*).

The difficulty of converting electronic energy directly into nuclear kinetic energy can be given a classical interpretation. Since electrons are several thousand times lighter than atomic nuclei, the transfer of energy between the two types of particles is very inefficient. Exactly the same principle is observed when a ping pong ball strikes a bowling ball: after the collision the ping pong ball has practically the same energy as it did before the collision.

Exercise. Suppose that a ping pong ball weighing m gm. collides head-on with a stationary billiard ball weighing M gm. Assuming that linear momentum and energy are conserved, show that the velocity of the ping pong ball after the collision is $(M - m)/(M + m)$ times the velocity before the collision. Thus if $M = 10{,}000\ m$, the velocity changes by only 0.02 %, and the energy lost by the ping pong ball is only 0.04 % of the initial energy.

The Franck-Condon principle also has a classical interpretation. When a molecule vibrates, the probability of finding a given atom at a certain point is inversely proportional to its velocity when it is at that point. Therefore the atoms in a vibrating molecule spend most of their time in configurations in which the kinetic energy is low — that is, the configurations in which the potential energy is nearly identical with the total energy, or at the intersection of the vibrational energy level with the potential energy surface of the molecule. Thus the photon is most likely to be absorbed when the nuclei are stationary or are moving slowly. Furthermore, the excitation resulting from the absorption of the photon cannot be transferred immediately to the nuclei. The nuclei will therefore tend to continue moving slowly immediately after the absorption process. Thus in the excited state the nuclear configuration also tends to be close to the intersection of the vibrational energy level with the excited potential energy surface. Therefore transitions tend to take place between vibrational levels in which the nuclear configurations are the same in both states, and they tend to occur when the nuclear kinetic energies are small. This, however, is merely another statement of the Franck-Condon principle.

For a more extended discussion of the Frank-Cordon principle, see Herzberg (*9*, pp. 200 *ff*.).

(*iii*) *The Effect of Molecular Symmetry on Selection Rules.* If a molecule possesses symmetry – as for example is the case in benzene and ethylene – this symmetry is reflected in the electronic and vibrational wave functions, which can be written as eigenfunctions of the symmetry operations belonging to the molecule. Since the coordinate system can always be chosen so that the components of the dipole moment operator are also eigenfunctions of the symmetry operations and since an integral such as

$$\int \psi_a^* x \psi_b d\tau$$

must vanish unless the integrand is unchanged by all symmetry operations, the symmetry properties of a state introduce a new factor in the selection rules for transitions. An illustration of this has already been discussed in Chapter 12 in connection with the interpretation of the spectrum of benzene. The general treatment is best given in terms of group theory (see *11*, Chapter 11).

(*iv*) *The Effect of Electron Spin.* If a molecule is made of light atoms for which spin – orbit interactions are small, its wave functions will be eigenfunctions of the spin operators to a good approximation, and the selection rule $\Delta S = 0$ will operate for the same reason that it applies in atoms under the same conditions. Intercombinations, and in particular, transitions between singlet and triplet states of molecules, will be unlikely. The introduction of heavy atoms, as for instance the substitution of a hydrogen atom by bromine in benzene, will increase the probability of such transitions very considerably, however.

Recent investigations by Lewis and Kasha (*12*) indicate that the phosphorescence of dyes and other organic materials probably involves intercombinations between triplet and singlet states of the molecules. This work will be discussed in more detail below (pages 700*ff*).

C. Absorption Spectra and Color

a. General discussion of the factors influencing color

A substance is colored if its absorption bands happen to fall partially or entirely in the visible range of the spectrum (roughly from 4000 Å. to 7500 Å.) Three factors can be mentioned that influence the color of a substance:

(*i*) *The frequencies of the electronic transitions that can occur in the substance.* These frequencies are, of course, determined by the locations of the electronic energy levels of the excited states through the Bohr condition $\nu_{ba} = (E_b - E_a)/h$.

(*ii*) *The intensity of the absorption.* For dipole transitions (which are the most important cause of light absorption), the intensity is determined by the dipole moment integrals,

$$\int \psi_a{}^* \boldsymbol{m} \psi_b d\tau$$

(*iii*) *The shape of the absorption band and the extent to which it overlaps the visible range of the spectrum.* This factor is actually not entirely independent of the other two factors just mentioned. It depends in part on the contributions of the molecular vibrations to the dipole moment integral. In some instances (e.g., the 2300–2600 Å. absorption band of benzene) the entire absorption depends on vibrational contributions. The shape may also depend on the more or less random shifts of the electronic levels brought about by the perturbing fields of the molecules which surround the colored molecule.

Vibrational structure in an absorption band arises in the following way. The equilibrium interatomic distances in a molecule are generally increased when the molecule goes into an excited electronic state, because the bonding between the atoms in the molecule is weakened. Thus the potential energy curves tend to be related to one another as indicated in Fig. 16-3A, the minimum in the excited state occurring at a larger separation than in the ground state. If the substance is not at too high a temperature, most molecules will be in the first few vibrational levels of the ground state. Therefore the Franck-Condon principle tends to cause transitions to a group of higher vibrational levels of the excited electronic state that have their

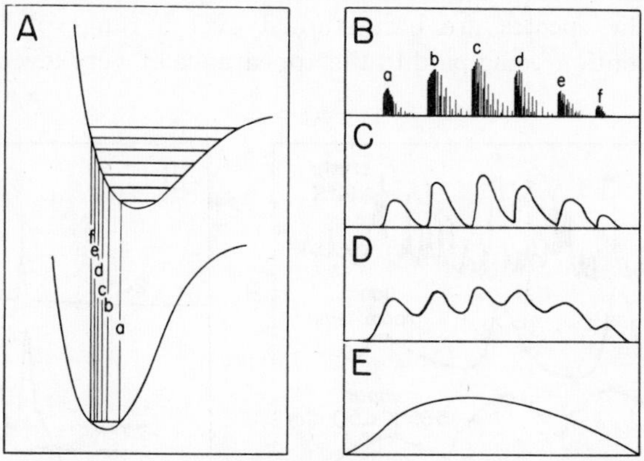

FIG. 16-3. Effect of intermolecular forces on an idealized absorption band.

A. Potential energy curves involved in the transition.
B. Absorption spectrum of the dilute vapor.
C. Absorption spectrum of the vapor at moderate pressures (rotational structure smeared out).
D. Absorption spectrum in liquid with weak intermolecular forces (vibrational structure present, but less pronounced than in vapor and somewhat shifted in frequency).
E. Absorption spectrum in liquid with strong intermolecular forces (vibrational structure smeared out).

end points almost directly above the minimum of the ground state curve. In the dilute vapor, each of the vibrational levels consists of many closely spaced rotational levels and the absorption spectrum under high resolution has the general appearance shown in Fig. 16-3B. As the pressure of the vapor is increased, collisions between molecules broaden the rotational levels until a continuum is produced that still shows, however, a group of distinct peaks reflecting the vibrational energy level spacings. In the liquid state the interactions with neighboring molecules cause further broadening (Fig. 16-3C), and if this is sufficiently strong the vibrational structure may even disappear (Fig. 16-3D).

In the crystalline state, especially at low temperatures, the surroundings of a given molecule tend to assume less random arrangements and the banded appearance of the spectrum may become more pronounced (see

Fig. 16-4). Furthermore, fewer vibrational levels of the ground state are populated at low temperatures, so that fewer transitions between vibrational levels take place and the spectrum is simplified. Therefore in the analysis of the vibrational structure of the absorption spectra of molecules in liquids and solids, the spectra are best studied over a range of temperatures, particular attention being paid to the appearance at very low temperatures.

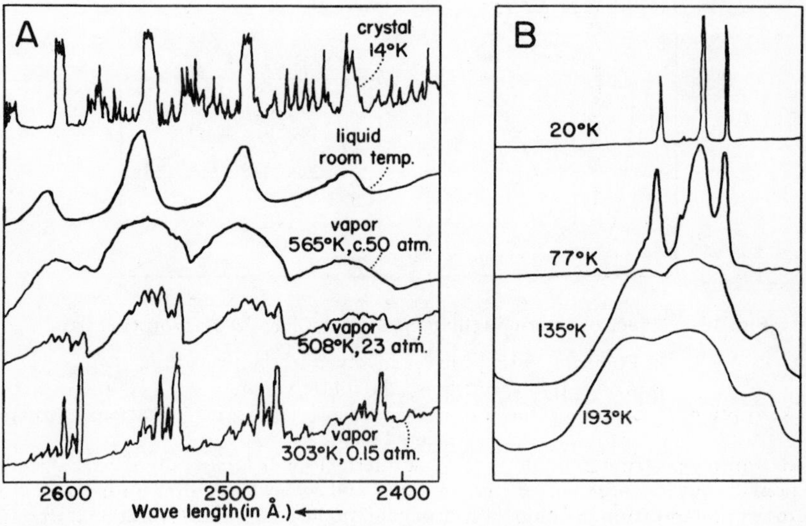

FIG. 16-4.

A. Absorption spectra of benzene under various conditions. (The spectrum of crystalline benzene is taken from P. Kronenberger, *Z. Physik* **63**, 494 (1930), and of liquid and vapor from S. Sambursky, A. Halperin, and H. Henig, *J. Chem. Phys.* **21**, 2041 (1953). The middle curve corresponds approximately to the critical point of benzene, 561° K and 47.7 atm.)

B. Effect of temperature on the fluorescence spectrum of crystalline $CsUO_2(NO_3)_3$ in the region 5400–5500 Å. (After G. H. Diecke and A. B. F. Duncan, "Spectroscopic Properties of Uranium Compounds," p. 110, McGraw-Hill, New York, 1949).

Strong interactions with solvents and with vibrational modes are responsible for the broad absorption bands observed in all organic compounds and many inorganic compounds in solution. This broadening has a very important effect on the color. For instance, a substance whose absorption spectrum consists of a few narrow bands in the visible range would be weakly colored even if the oscillator strengths associated with the bands were large. In spite of the fact that the wave lengths falling inside the bands would be strongly absorbed, most of the white light falling on such a substance would not be absorbed.

It is also interesting to note that the colors of many substances, especially those of yellowish or brownish tint, are caused by absorption bands whose peaks are in the near ultra-violet, but whose edges extend into the visible. Similarly, the colors of solutions containing Cu^{2+} and Ni^{2+} ions are determined by the wings of absorption bands whose peaks lie in the near infra-red and near ultra-violet.

Absorption spectra in the visible and near ultra-violet can be classified conveniently according to the distance through which the electrons move when they become excited. In many substances, particularly the rare earth ions and the simple hydrated transition metal cations, the wave functions concerned in the absorption are fairly completely localized on a single atom or ion. In other substances, wave functions in the vicinity of two atoms are changed. In still other substances, the wave function of almost the entire molecule may be affected. These types of spectra will now be discussed in more detail.

b. Spectra caused by transitions localized on single atoms

(*i*) *The Rare Earth Ions.* The trivalent rare earth ions (except for La^{3+}, Ce^{3+}, Yb^{3+}, and Lu^{3+}) have spectra in aqueous solution that consist of numerous relatively sharp, weak lines. The oscillator strengths of these lines are of the order of magnitude 10^{-6} and smaller (*13*). Their wave lengths are only slightly influenced by the surroundings of the ions; for instance, complexing of Nd^{3+} with ethylenediamine tetracetic acid hardly affects its absorption spectrum (*14*).

These lines undoubtedly originate from transitions between different multiplets of the f^n-configurations that characterize all of the rare earth ions. This would account for the absence of lines in the spectra of La^{3+}, Ce^{3+}, Yb^{3+}, and Lu^{3+}, which have the configurations f^0, f^1, f^{13}, and f^{14}, respectively, none of which show any multiplet structure. It would also account for the weakness of the absorption, since transitions between multiplet levels are forbidden by the Laporte rule. That the transitions occur at all must be a consequence either of quadrupole or magnetic dipole mechanisms or of deviations from centro-symmetry resulting from perturbations by neighboring molecules in the solutions and in the crystalline state. The sharpness of the lines and their insensitivity to their surroundings are a result of the feeble interactions of the circular and deeply buried $4f$-orbitals with other atoms – the same factor that is responsible for the small chemical effects of these orbitals.

In spite of the fact that the environment has only a slight effect on the spectra of the rare earths, this small effect provides a method of studying the interactions of these ions with their surroundings (see (*15*, Chapter 3) for a review). In solutions and in crystals the fields of the surrounding molecules split the electronic states of the rare earth ions into a number of distinct levels, thus causing small but readily measurable displacements and splitting of the absorption lines. The observations in aqueous solution and crystalline hydrates are more or less consistent with an octahedral arrangement of water molecules around the ions, but there are some puzzling inconsistencies in the case of samarium. With colored substances other than the rare earths in condensed phases, the interaction of the chromophoric electron with neighboring atoms is so strong that a smeared out band is usually obtained in the absorption spectrum. Such a band is much less amenable to quantitative study than are the sharp lines of the rare earth ions.

Among the actinide metal ions, the uranyl ion, UO_2^{2+}, has a great many weak but rather sharp absorption lines throughout the visible spectrum. The strongest of these are around 4600–4700 Å. and have oscillator strengths of only 10^{-8} to 10^{-7}.

(*ii*) *The Transition Metal Cations.* The visible absorption bands of the transition metal cations are much less sharp than those of the rare earths, and their oscillator strengths are somewhat larger (usually of the order of 10^{-4} for the hydrated ions). They are also affected much more by the formation of complexes. This accounts for some of the striking color changes that make the chemistry of these elements so fascinating. For instance, the red band of Cu^{2+} extends from 6000 Å. to beyond 12,000 Å. with a peak at 8100 Å. and an oscillator strength of 2×10^{-4}. Formation of the $Cu(NH_3)_4^{2+}$ complex increases the absorption coefficient about five-fold and shifts the peak to about 6000 Å. Co^{2+} has a band at 4000–6000 Å. with an oscillator strength of about 4×10^{-5} when it is hydrated, but in the presence of a high concentration of chloride ion, the absorption region shifts to 5500–7500 Å. and the oscillator strength is 100 times greater. (This shift is presumably responsible for the well-known change in the color of $CoCl_2$ from blue to pink when it becomes moist.)

The colors of these cations are apparently caused by transitions between the multiplet levels of the d^n configurations. Although these transitions are forbidden by the Laporte rule, they are made possible by the perturbations of the solvent or crystal lattice. Furthermore, Bethe (*16*) has shown that perturbation by the solvent or crystal shifts the relative positions of the multiplets and splits them into several components (the number of components depending on the symmetry of the environment and the angular momentum of the multiplet level). The hydrated cupric ion is interesting in this connection, because Cu^{2+} has the ground state configuration d^9, which has the single multiplet term ^2D. This term is split into two levels by the surrounding water molecules, and the familiar blue color of cupric salts is caused by a transition between these two components of the $d^9\ {}^2D$ term.

c. Spectra caused by transitions localized in small groups of atoms

The permanganate ion, MnO_4^-, has a rather strong absorption band extending from 4200 to 7000 Å. with an oscillator strength of 0.03, and the chromate ion, $CrO_4^=$, absorbs in the near ultra-violet between 3100 and 4500 Å. with an oscillator strength of 0.09. (The tail end of this band is responsible for the color of the chromates.) The transitions causing these bands probably involve bonding orbitals between the metal and the oxygen atoms in the anion (see (*17*) for a quantum mechanical calculation of the spectra of these ions). The nitrate ion, NO_3^-, has a weak band (oscillator strength about 10^{-4}) between 2600 and 3200 Å. that also probably involves the bonding orbitals.

Molecules containing unconjugated double bonds generally have fairly distinctive absorption regions in the ultraviolet. For instance, the carbonyl group, C=O, in aldehydes and ketones invariably gives rise to a weak band near 2900 Å. with an oscillator strength of about 5×10^{-4}. This is believed to be caused by a transition from a nonbonding oxygen electron to an antibonding orbital between the oxygen atom and the carbon atom (see (*18*)). Similarly, substances containing NO_2, NO, C=C, C=S, CO–OH, COO- etc. each absorb in typical regions of the ultra-violet regardless of the nature of the rest of the molecule, as long as the double bonds in these groups are not conjugated with double bonds in the rest of the molecule. The absorption bands of these molecules at wave lengths above about 2000 Å. are believed to be caused by transitions involving bonding and antibonding orbitals in the double bonds of the groups.

d. Spectra caused by transitions involving entire molecules

In many of the most important colored substances (especially dyestuffs and pigments) the electronic transitions responsible for light absorption involve changes in orbitals that extend over most of the molecule. This is true of the aromatic compounds such as benzene, naphthalene, and anthracene and their derivatives, but the most interesting examples are found in molecules containing conjugated double bonds arranged in a long chain. We shall discuss these molecules in some detail since they offer interesting demonstrations of the usefulness, and also of the limitations, of various simple quantum mechanical approaches.

We shall be especially interested in the polyene skeletons

$$-\underset{|}{N}-(CH{=}CH)_{N-1}-\underset{|}{C}{=}\underset{|}{\overset{+}{N}}- \qquad \text{(A)}$$

and

$$-\underset{|}{C}{=}\underset{|}{C}-(CH{=}CH)_{N-2}-\underset{|}{C}{=}\underset{|}{C}- \qquad \text{(B)}$$

containing N double bonds in the conjugated chain. Molecules containing these skeletons have intense absorption bands in the visible and infrared, with molecular extinction coefficients of the order of 10^5 and higher, equivalent to oscillator strengths not very far from unity. The spectra of these compounds will now be discussed from three different points of view: the free electron molecular orbital model (also often called the "box model"), the LCAO molecular orbital model, and the valence bond model.

(*i*) *Polyene Spectra and the Free Electron Molecular Orbital Model.* Bayliss (*19*), Kuhn (*20*), and Simpson (*21*) have independently made use of a particularly simple and attractive form of molecular orbital that has been quite successful in the interpretation of the spectra of some of the polyenes. They suggest that the π-electrons in the extended conjugated chain move along the chain in an approximately constant potential field and that their molecular orbitals therefore resemble those of a one-dimensional electron gas in their dependence on position along the chain. (This simplifying assumption was also made earlier by Pauling (*22*), Lonsdale (*23*), and Schmidt (*24*) in connection with the wave functions of the electrons in aromatic hydrocarbons.) Thus the skeleton (A) contains $(2N + 2)$ π-electrons (two for each double bond plus two on the neutral nitrogen atom), which should occupy the first $(N + 1)$ molecular orbitals

$$\psi_k = A \sin (k\pi\, x/L) \qquad \text{(C-1)}$$

having energies

$$E_k = h^2k^2/8m_eL^2 \qquad \text{(C-2)}$$

where k is a positive integer, x is the position of the electron measured along the chain, L is the total distance from one end of the chain to the other through which the π-electrons are free to move, and m_e is the mass of an electron. Kuhn measures the distance L along the zig-zag path formed by the individual bonds. Because of the resonance of (A) with the form

$$-\overset{+}{\underset{|}{N}}=CH-(CH=CH)_{N-1}-\underset{|}{N}- \qquad (A')$$

which is identical in energy with (A), the carbon–carbon bonds in this skeleton all have 50% double bond character, so Kuhn assumes a bond length that is the same as that observed in benzene (1.39 Å.). He also assumes that the ends of the "box" are one bond length beyond each terminal nitrogen atom. This means that

$$L = 2(N + 1)\cdot 1.39 \text{ Ångstrom units} \qquad \text{(C-3)}$$

The first excited state of (A) should result from the transition of an electron from the highest occupied molecular orbital (i.e., that with $k = N + 1$) to the lowest unoccupied orbital (i.e., that with $k = N + 2$). The energy is thereby increased by

$$\Delta E = \frac{h^2}{8m_eL^2}\,[(N + 2)^2 - (N + 1)^2] = \frac{h^2\,(2N + 3)}{8m_e \times 1.39^2 \times (2N + 2)^2} \times 10^{16} \qquad \text{(C-4)}$$

The corresponding absorption should occur at a wavelength of

$$\lambda = hc/\Delta E = 2550(N + 1)^2/(2N + 3) \text{ Ångstrom units} \qquad \text{(C-5)}$$

The observed wave lengths of maximum absorption for the compounds

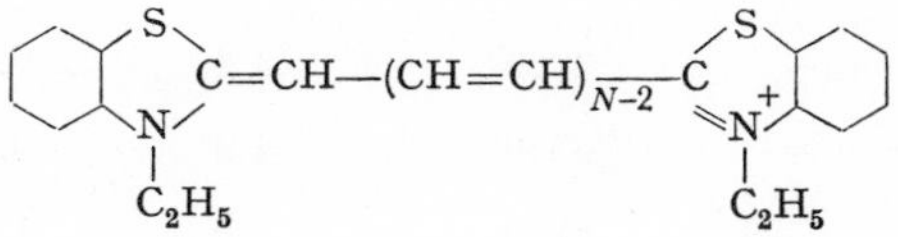

(C)

and

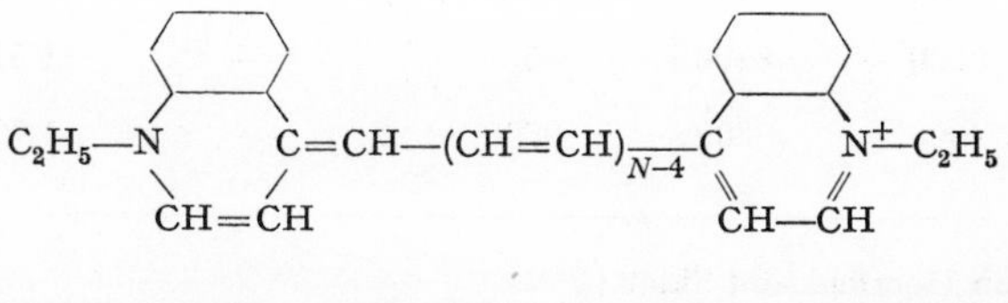

(D)

are compared with the wave lengths calculated from Equation (C-5) in

Table 16-2. The agreement is surprisingly good, considering the great simplicity of the model.

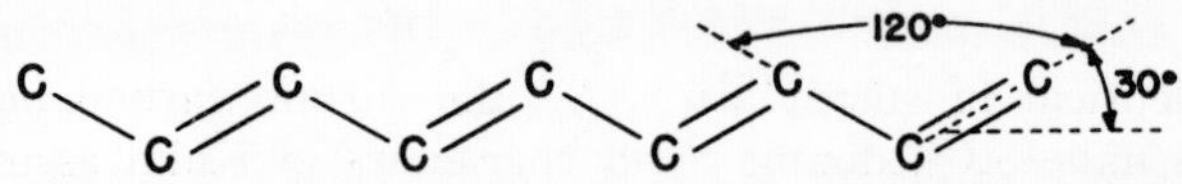

FIG. 16-5. The all-trans configuration of the polyene chain.

The oscillator strengths may also be calculated using Equation (B-10a). Kuhn assumed that the chain is stretched out, all the double bonds having the *trans* configuration as shown in Fig. 16-5. A factor must be included to take account of the fact that the polyene chain is kinked, the component of the dipole moment parallel to the molecular axis being less than the total length L of the chain by the factors cos 30°. (The components normal to the molecular axis are, of course, zero.)

TABLE 16-2

ABSORPTION CONSTANTS OF POLYMETHENE DYES

	Wave length of maximum absorption			Oscillator strengths		
	Observed		Calculated	Observed	Calculated	
N	Compound C* (Å.)	Compound D† (Å.)	Equation (C-5) (Å.)	Compound C†	Equation (C-6)	Equation (C-19)
2	4250	—	3280	1.2	0.69	7.6
3	5600	—	4540	1.2	0.89	12.4
4	6500	5900	5800	1.6	1.10	17.2
5	7600	7100	7060	1.9	1.31	22.1
6	8700	8200	8330	—	1.51	27.0
7	9900	9300	9600	—	1.72	31.9

* Data from Herzfeld and Sklar (*25*).
† Data by Brooker, as reported by Sklar (*26*).

The resulting expression for the oscillator strength is *

$$f = \frac{64}{3\pi^2} \frac{(N+1)^2 (N+2)^2}{(2N+3)^3} \cos^2 30^\circ \tag{C-6}$$

Values of the oscillator strengths calculated from this equation are compared with the observed values in Table 16-2. The agreement is again reasonably good considering the simplicity of the model. It should be noted that if some of the double bonds in the chain have the *cis* configuration the calculated f-values would be even smaller than those given (see *27, 28* for discussions of the effects of *cis-trans* isomerism on the absorption spectra of the polyenes).

This simple approach accounts remarkably well for the positions and intensities of the absorption bands of the polymethene dyes having skeletons of type (A). It is significant, however, that when the same approach is used with the type (B) skeletons much poorer agreement is obtained. Consider for instance the carotenoid series

$$\rangle C{=}C{-}(CH{=}CH)_{N-2}{-}C{=}C\langle \tag{C}$$

for which the experimentally observed maxima and oscillator strengths are shown in Table 16-3.

TABLE 16-3

ABSORPTION CONSTANTS OF CAROTENOIDS [DATA FROM BAYLISS (*29*)]

	Wave Lengths		Oscillator Strengths	
N	Observed absorption maximum (Å.)	Calculated from Equation (C-5a) (Å.)	Observed	Calculated from Equation (C-6a)
1	1600	870	—	0.24
2	2170	2080	0.53	0.47
3	2600	3360	0.62	0.68
4	3020	4650	—	0.90
5	3460	5940	1.05	1.10
6	3690	7230	—	1.30
11	4510	13700	2.49	2.33
12	4750	15000	—	2.53
15	5040	18900	—	3.14

* Kuhn also introduces a factor of two into this expression, arguing that there are two electrons in the state $k = (N + 1)$, each capable of undergoing the transition. When the spin factors are included in the wave function, however, it is seen that only one singlet excited state exists to which the singlet ground state can make the postulated transition $(N + 1) \rightarrow (N + 2)$. The other excited state with the same configuration is a triplet, to which transitions from the ground state are forbidden. Therefore the transition is made between two nondegenerate states and no additional factor is called for.

In this case, the length of the chain is slightly different from the value given in (C-3). Assuming that the double bond distance is 1.35 Å. and the single bond distance is 1.46 Å. (as observed in butadiene), and, for simplicity, that the "box" extends half a single bond length beyond each terminal carbon atom, then

$$L = 2.81\,N \text{ Ångstrom units} \tag{C-3a}$$

and the wave length of the absorption becomes

$$\lambda = 2610\,N^2/(2N + 1) \text{ Ångstrom units} \tag{C-5a}$$

The expression for the oscillator strength is given by

$$f = \frac{64}{3\pi^2}\,\frac{N^2\,(N + 1)^2}{(2N + 1)^3}\cos^2 30^\circ \tag{C-6a}$$

The calculated absorption wave lengths and oscillator strengths are given in Table 16-3. Although there is excellent agreement with the observed oscillator strengths, the calculated wave lengths deviate markedly from the observed values. It is especially noteworthy that the formula predicts that the wave length should vary in proportion to the length of the chain when the chain is long, whereas the observed data seem to indicate a much more gradual variation.

Kuhn has suggested an interesting explanation for this discrepancy. He points out that although the two resonating structures of skeleton (A) are exactly equivalent in energy, so that all carbon–carbon bonds in the chain are equivalent, the resonating forms of skeleton (B)

—C=C—(CH=CH)$_{N-2}$—C=C— and —C—C=(CH—CH)$_{N-2}$=C—C—

do not have the same energy. Therefore alternating carbon–carbon bonds in (B) should have different bond lengths, some being more like single bonds, while others would be more like double bonds. Since the single bonds are longer than the double bonds, the electrons would be expected to have a higher potential energy when they were in a single bond than when they were in a double bond. This introduces a perturbation on their motion that can be described approximately by writing the potential energy inside the "box" in the form

$$V = \text{constant} + V_0 \sin \frac{2\pi\,(x - 1.35)}{l} \tag{C-7}$$

where l is the length of the C–C=C repeating unit (1.35 + 1.46 Å. in butadiene), x is the distance measured from one end of the chain, and V_0 is the amplitude of the fluctuation of the potential due to the alternating single and double bond character along the chain. It is found that the oscillating potential shifts the energy levels of the free electron into groups (*cf.* solutions of Mathieu's equation, page 209). If there are N double bonds, then the N lowest levels tend to be drawn together toward one common value, the next N levels are drawn together toward another common value, and so on (see Fig. 16-6). As a result, the energy

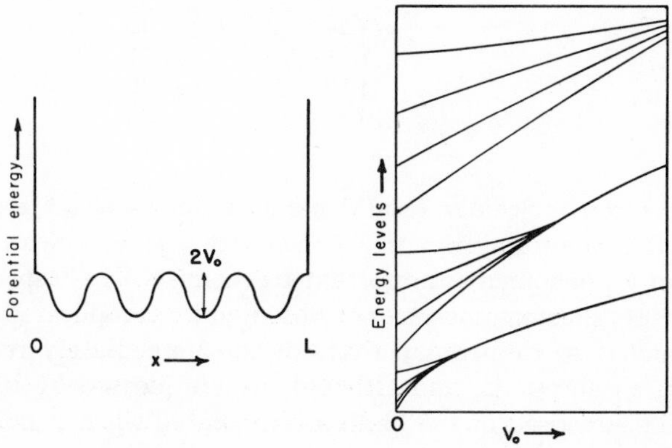

FIG. 16-6. Schematic diagram of the energy levels of a particle in a box with infinite potential walls at $x = 0$ and $x = L$, subject to an additional sinusoidal potential of amplitude V_0. For purposes of illustration the sinusoidal potential has been given four minima inside the box. When $V_0 = 0$, the energy levels are those of a particle in a box of length L ($E = N^2h^2/8mL^2$, where $N = 1, 2, 3, \cdots$). When V_0 is very large, the levels gather together in groups of four, with energies proportional to $(M + \frac{1}{2})\sqrt{V_0}$, where $M = 0, 1, 2, \cdots$ (*cf.* page 210).

difference between the Nth and $(N + 1)^{\text{st}}$ levels is increased. Since the first absorption band arises from a transition between just these two orbitals, this makes the absorption take place at shorter wave lengths than are expected from the free electron model. Kuhn finds that a value $V_0 = 2$ electron volts results in calculated energy level spacings agreeing very well with those actually observed. Kuhn also found that the value of V_0 required to give the observed absorption wave length decreases slightly as N increases ($V_0 = 2.55$ e.v. when $N = 2$; 2.06 when $N = 8$; 1.89 when $N = 17$). This is qualitatively in accord with the conclusions of Lennard-Jones (*30*), who, using the LCAO molecular orbital model for a conjugated chain, found that the alternating bonds should become more nearly of the same length as the number of double bonds in the chain increases. Lennard-Jones' calculations indicated, however, that the difference in bond lengths between alternating bonds is already quite small when $N = 8$ (see Table 16-4). It is therefore somewhat surprising that Kuhn would find such a marked alternation persisting when $N = 17$.

TABLE 16-4

LENGTHS OF "SINGLE" AND "DOUBLE" BONDS IN

$$—C{=}C—(CH{=}CH)_{N-2}—C{=}C\langle$$

(LENNARD-JONES (*30*))

N	Bond Lengths	
	"Single" bond (Å.)	"Double" bond (Å.)
1	(1.54)	1.33
2	1.41	1.34
8	1.39	1.36
∞	1.38	1.38

The free electron molecular orbital model is an excellent example of a semiquantitative quantum mechanical theory that gives a clear and simple explanation of a phenomenon important to chemists. It is especially interesting because its shortcomings are not obscured by complicated mathematics. The model is so elementary that we can immediately recognize the severity of its assumptions, and although we are pleased by its successes, we need not be surprised or too badly disappointed when it fails. Furthermore, the calculations that it requires, at least in its first approximation, are very simple, so that we need not be very much disturbed at having done some paper work for nothing when the results of the calculations fail to agree with experimental observations. Unfortunately, this cannot be said to be true of all quantum chemical theories.

Exercises. (1) One would expect that even in skeleton (A), in which all of the carbon–carbon distances are the same, there would be a periodic potential due to the atoms along the chain. The wave length of this perturbation is, however, just half of that used by Kuhn to explain the difference between skeletons (A) and (B) (that is, the distance l in Equation (C-7) is 1.39 Å. instead of 2.81 Å.). Explain why this potential should have less effect on the observed spectrum than the constant potential used by Kuhn. [Hint: Consider the case of V_0 very large, with $2N$ atoms in conjugation. How many levels would tend to coalesce into the lowest cluster? Is the gap between the highest occupied level of the ground state and the lowest unoccupied level increased by this clustering?]

(2) Dyes of the type

$$R—(CH{=}CH)_j—CH{=}(CH—CH)_j{=}CH—R'$$

absorb at longer wave lengths than analogous dyes of the type

$$R—(CH{=}CH)_j—N{=}(CH—CH)_j{=}CH—R'$$

when j is an odd number. On the other hand, when j is even, the nitrogen-containing analogue absorbs at longer wave lengths. Explain this. In neither type of compound do the groups R and R′ resonate with the conjugated system of double bonds. [Hint: Nitrogen has a higher electron affinity than carbon. When N is substituted for CH, those orbitals that have large values at the nitrogen atom are therefore stabilized to a greater extent than are orbitals with a node at the nitrogen atom. See Kuhn (*31*).]

(*ii*) *Polyene Spectra and the LCAO Molecular Orbital Model.* In the conventional LCAOMO approach (*cf.* page 378) to the problem of the spectra of the polymethenes the molecular orbitals for the π-electrons in skeleton (B) are written in the form

$$\psi = \sum_{i=1}^{2N} a_i \phi_i \tag{C-8}$$

where ϕ_i is the $2p\pi$ atomic orbital on the ith carbon atom in the chain. Using the variation method to solve for the constants a_i we obtain in the usual way the $2N^{\text{th}}$ order determinantal equation

$$\begin{vmatrix} (\alpha - E) & \beta & 0 & 0 & 0 \cdots 0 & 0 \\ \beta & (\alpha - E) & \beta & 0 & 0 \cdots 0 & 0 \\ 0 & \beta & (\alpha - E) & \beta & 0 \cdots 0 & 0 \\ \vdots & \vdots & \vdots & \vdots & \vdots \quad \vdots & \vdots \\ 0 & 0 & 0 & 0 & 0 \cdots \beta & (\alpha - E) \end{vmatrix} = 0 \tag{C-9}$$

where α is the Coulombic integral,

$$\int \phi_i^* H \phi_i \, d\tau$$

for an electron in a given $2p\pi$-orbital, and β is the exchange integral,

$$\int \phi_i^* H \phi_{i+1} \, d\tau$$

for an electron in two neighboring $2p\pi$-orbitals. In obtaining (C-9), all exchange integrals between nonadjacent $2p\pi$-orbitals have been neglected, as have all of the overlap integrals,

$$\int \phi_i^* \phi_j \, d\tau$$

for which $i \neq j$. It is also assumed that all bond lengths along the chain are equal.

Equation (C-9) has the roots (see Exercise 4 on page 49)

$$E = \alpha + 2\beta \cos\left(\pi \frac{l}{2N+1}\right) \tag{C-10}$$

where $l = 1, 2, \cdots, 2N$. For each value of l there is a set of coefficients a_j of the form

$$a_{lj} = \sin\left(\pi j \frac{l}{2N+1}\right) \tag{C-11}$$

which give the form of the lth molecular orbital. There are thus just as many molecular orbitals as there are electrons, but since each orbital can be occupied by two electrons, only those orbitals for which $l \leqslant N$ are filled in the ground state. In the first excited state, an electron is transferred from the Nth orbital to the $(N+1)^{\text{st}}$ orbital, giving an increase in energy

$$\Delta E = 2\beta\left[\cos\left(\pi \frac{N+1}{2N+1}\right) - \cos\left(\pi \frac{N}{2N+1}\right)\right]$$

$$= -4\beta \sin \frac{\pi}{2(2N+1)} \tag{C-12}$$

which leads to a calculated wave length for the first absorption band

$$\lambda = -\frac{hc}{4\beta \sin \dfrac{\pi}{2(2N+1)}} \tag{C-13}$$

(Note that β is a negative quantity, so λ is positive.) For large values of N this gives

$$\lambda \simeq -hcN/\pi\beta \tag{C-13a}$$

Thus we see that, just as in the free electron model, we should expect a linear dependence of the wave length on the length of the chain—which is not observed. Furthermore, the values of the exchange integral, β, that are required to account for the observed positions of the absorption bands range from about 100 kcal./mole to 300 kcal./mole as N varies from 1 to 15. It will be recalled that the same exchange integral appears in the calculation of the resonance energy of benzene, and that its value there was deduced to be of the order of 15–20 kcal./mole, which is much less than the value required by the spectra. Thus the account of the positions of the absorption bands of the polymethenes of type (B) given by the simple LCAOMO theory can hardly be said to be satisfactory.

The oscillator strengths predicted from the wave functions derived from the coefficients in (C-11) are in good agreement with the observed values. This is easily seen by writing these coefficients in the form

$$a_{lj} = \sin(\pi l x_j/L) \tag{C-14}$$

where $x_j = jx_0$, x_0 being the mean distance between successive atoms in the chain, and L is the total length of the chain. These coefficients have the same form as the free electron molecular orbitals in (C-1), except that they are defined only at the positions of the carbon atoms. The molecular orbitals constructed from them have the appearance shown in Fig. 16-7, and their similarity to the free electron orbitals is evident. Because of this similarity, it is obvious that the integral

$$\int \psi_{N+1}^* \, x\psi_N d\tau$$

should not be far from the value found for the free electron gas model.

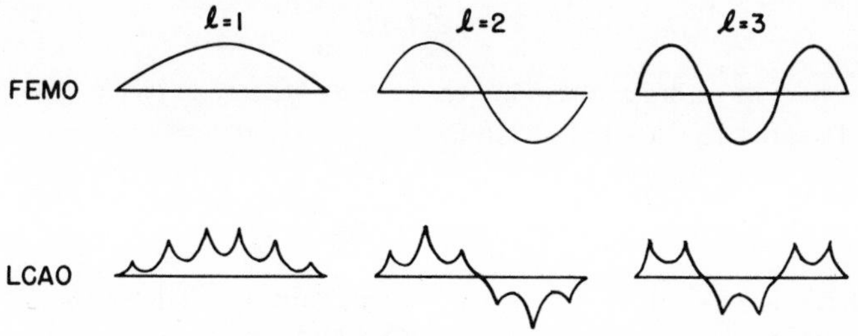

FIG. 16-7. Comparison of the free electron molecular orbital (FEMO) and LCAO wave functions along a polyene chain.

The LCAO method has been applied to the skeleton of type (A) by Herzfeld and Sklar (*25, 32*). A determinantal equation similar to (C-9) is obtained, except that the exchange and Coulombic integrals involving the nitrogen atoms are different from the rest. This complicates the arithmetic, but it is found that for long chains the wave length of the first transition varies linearly with the chain length, as is observed. Again, the value of the exchange integral is rather large, being of the order of 60 to 70 kcal./mole. This value is, however, much more reasonable than the value found for systems having skeleton (B). In view of this improvement, it is likely that if alternations in the bond lengths of the (B) skeleton were taken into account in the LCAO method, a somewhat better result would be obtained, just as was found with the free electron model.

(*iii*) *Polyene Spectra and the Valence Bond Model.* Let us apply the valence bond method to the skeleton

$$\begin{array}{c}\overset{+}{\text{—N}}\text{=CH—CH=CH—N—} \\ \;|\qquad\qquad\qquad\quad\; | \end{array} \qquad \text{(I)}$$

which has the resonance form of identical energy

$$\begin{array}{c}\text{—N—CH=CH—CH=}\overset{+}{\text{N}}\text{—} \\ \;|\qquad\qquad\qquad\quad\; | \end{array} \qquad \text{(II)}$$

If the wave functions of these two structures are ϕ_{I} and ϕ_{II}, then the wave function of the ground state of the molecule should be

$$\psi = \phi_{\text{I}} + \phi_{\text{II}} \qquad \text{(C-15a)}$$

and there should be an excited state with the approximate wave function

$$\psi' = \phi_{\text{I}} - \phi_{\text{II}} \qquad \text{(C-15b)}$$

The difference in energy between these two states is twice the exchange energy between these two structures

$$\Delta E = 2 \int \phi_{\text{I}} H \phi_{\text{II}} d\tau \qquad \text{(C-16)}$$

It is easy to show that this integral is very small. Writing the $2p$-orbitals along the chain in order as a, b, c, d, e (a and e being located on the nitrogen atoms) we see that ϕ_{I} consists of determinants having two electrons in orbital e, of the form (see page 420)

$$|(a\alpha)_1(b\beta)_2(c\alpha)_3(d\beta)_4(e\alpha)_5(e\beta)_6|$$

and ϕ_{II} consists of determinants in which two electrons are in orbital a, of the form

$$|(a\alpha)_1(a\beta)_2(b\alpha)_3(c\beta)_4(d\alpha)_5(e\beta)_6|$$

The integral

$$\int \phi_{\text{I}} H \phi_{\text{II}} d\tau$$

is therefore composed of a sum of terms such as

$$\int a_1 b_2 c_3 d_4 e_5 e_6 \, H a_1 a_2 b_3 c_4 d_5 e_6 d\tau$$

which contain multiple exchanges of electrons and which must therefore all be very small. The two wave functions ψ and ψ' are therefore not capable of explaining the positions of the absorption bands of this compound.

Let us consider next the oscillator strength for the transition $\psi \to \psi'$. The dipole moment integral for the component parallel to the length of the stretched out chain is

$$m_x = e \int \psi(\textstyle\sum x_i)\psi' d\tau$$

$$= e \int (\phi_{\mathrm{I}} + \phi_{\mathrm{II}}) (\textstyle\sum x_i) (\phi_{\mathrm{I}} - \phi_{\mathrm{II}}) d\tau$$

$$= e \left[\int \phi_{\mathrm{I}}(\textstyle\sum x_i) \phi_{\mathrm{I}} d\tau - \int \phi_{\mathrm{II}}(\textstyle\sum x_i)\phi_{\mathrm{II}} d\tau \right]$$

$$= e\, [\overline{X}_{\mathrm{I}} - \overline{X}_{\mathrm{II}}] \qquad \text{(C-17)}$$

where $\sum x_i$ is the sum of the x-coordinates of all electrons, x being measured along the axis parallel to the chain, $\overline{X}_{\mathrm{I}}$ is the mean value of the x-coordinate for all electrons in structure (I) and $\overline{X}_{\mathrm{II}}$ is the same quantity for structure (II). These two mean values may be estimated by assuming that in each structure the electrons in a double bond are, on the average, located at the midpoint of the double bond. Writing x_0 for the bond length, $x_0 \cos 30°$ for the projection of x_0 on the x-axis, and taking $x = 0$ at the nucleus of the nitrogen atom on the left end of the molecule, we see that in structure (I) two electrons are at $x = \frac{1}{2}x_0 \cos 30°$, two more are at $x = \frac{5}{2}x_0 \cos 30°$ and two more are at $x = 4x_0 \cos 30°$. Thus

$$\overline{X}_{\mathrm{I}} = 2(\tfrac{1}{2}\, x_0 + \tfrac{5}{2}\, x_0 + 4x_0) \cos 30° = 14\, x_0 \cos 30°$$

Similarly for structure (II)

$$\overline{X}_{\mathrm{II}} = 2(0 + \tfrac{3}{2}\, x_0 + \tfrac{7}{2}\, x_0) \cos 30° = 10\, x_0 \cos 30°$$

Thus

$$m_x = 4ex_0 \cos 30° \qquad \text{(C-18a)}$$

For compounds of the type

$$-\overset{+}{\underset{|}{N}}{=}CH{-}(CH{=}CH)_{N-2}{-}\underset{|}{N}-$$

one finds in a similar fashion

$$m_x = 2Nex_0 \cos 30° = eD \qquad \text{(C-18b)}$$

where D is the distance between the terminal nitrogen atoms. Using this result to calculate the oscillator strength, one finds

$$f = \frac{8\pi^2 m_e c}{3h} \frac{D^2}{\lambda} = 1080\ D^2/\lambda = 6250\ N^2/\lambda \qquad \text{(C-19)}$$

where D and λ are in Angstrom units, and where we have taken $D = 2 \cdot 1.39\ N \cos 30°$. The oscillator strengths calculated from this equation are given in Table 16-2. The values are much too large.

Thus the simple valence bond theory based on structures I and II is quite unable to account for the observed spectra of these compounds. It is sometimes said that resonance is "responsible" for color in organic dyestuffs. Although this may be true in a very broad sense, we see that the color does not arise from transitions involving only the most stable resonating forms.

Herzberg and Sklar (*32*) have extended the valence bond method to include structures of the type

—N—C⁺H—CH=CH—CH=CH—· · ·

and

—N—ĊH—C⁺H—ĊH—CH=CH—· · ·

These structures give rise to much larger exchange integrals. The determinantal equation to which they give rise on applying the variaton method is of the same form as that obtained in the LCAOMO method for the same compounds, but only the two lowest roots are utilized. It is found that the dependence of the wave length of the absorption on the number N of double bonds, is not correctly given. The inclusion of the new structures should, however, decrease the size of the calculated oscillator strength very considerably, because these structures move the charge away from the ends of the molecule.

On the whole it appears that the valence-bond method does not give as satisfactory an account of the spectroscopic properties of these compounds as do the free electron and LCAOMO methods. Herzfeld and Sklar (*32*) have shown, however, that the effect of introducing different groups on the ends of the conjugated chain (which has been studied experimentally by Brooker (*33*)) is more adequately explained by the valence bond method than by the LCAOMO method (see also Kuhn (*20*) for a discussion of the effects of substituting different groups on the ends of the chain from the point of view of the free electron model).

e. Electron transfer spectra

In Chapter 14, it was shown that the ground states of the alkali halide molecules in the vapor are essentially ionic in structure. There is also an excited state that is essentially covalent at the interatomic distance characteristic of the normal molecule in its ground state. The transition between these two states results in strong absorption bands of alkali halide vapors in the ultra-violet around 2500–3000 Å. This transition might be described by the equation

$$M^{+}\ddot{X}^{-} \rightarrow M:X$$

but the covalent state, M: X, is so weak that it tends to fall apart into atoms with very little energy absorption. The transition might therefore better be symbolized as

$$M^{+}\ddot{X}^{-} \rightarrow M\cdot + \cdot X$$

That is, it can be regarded as the result of a movement of an electron from the anion to the cation. The absorption resulting from such a movement of electrons from one atom to another is known as an *electron transfer spectrum.** These spectra probably play an important part in the photochemical behavior of many substances (detailed reviews given by Rabinowitch (*34*) and Orgel (*35*)).

Because of the weakness of the bond in the covalent excited state of the alkali halides, the atomic energy levels of the halogen atoms are not much affected by the proximity of the metal atoms. The halogen atom formed by the charge transfer may be in either the $^2P_{3/2}$ or $^2P_{1/2}$-states. The difference in energy between these two states in the free atom is 880 cm.$^{-1}$ in Cl, 3700 cm.$^{-1}$ in Br, and 7600 cm.$^{-1}$ in I. It is therefore strong evidence for the concept of electron transfer spectra, that the absorption spectra of NaBr and NaI vapors each show two peaks whose separations (3300 cm.$^{-1}$ for NaBr and 7200 cm.$^{-1}$ for NaI) are nearly the same as the separations in the free atoms (the separation in NaCl would be too small to be observed). Somewhat similar sets of peaks in crystalline alkali halides may be explained in the same way (*34*).

* Early papers referred to these spectra as *electron affinity spectra.* Mulliken (*36*) uses the symbol $N \rightarrow V$ to denote a transition from the normal, covalent state (N) of a molecule to a state (V) that is characterized in the Heitler-London approximation by an ionic structure: in the MO approximation $N \rightarrow V$ refers to transitions between bonding and anti-bonding orbitals. These $N \rightarrow V$ spectra are therefore essentially the same as electron transfer spectra.

Using Equation (C-19), it is easy to make an estimate of the distance moved by an electron in an absorption that takes place at a wave length λ

$$D = \sqrt{f\lambda/1080} \tag{C-20}$$

where D is a measure of the distance (in Ångstrom units) moved by the electron in a transition whose oscillator strength is f, the wave length being measured in Ångstrom units. If $f = 1$ (implying a molar extinction coefficient of about 5×10^4), and if the absorption band is at 2000 Å., then $D \cong 1.4$ Å. A movement of this magnitude would not, of course, be remarkable in a conjugated polyene, but if it occurred in a simple ion, such as a halide ion, it would seem to require a transfer of charge to another atom.

The ions Cl^-, Br^-, and I^- have absorption bands in aqueous solution whose maximum extinction coefficients reach values of 10^4 near 2000 Å. This implies the movement of an electron through a distance so large that it is extremely difficult to see how the electron can avoid becoming intimately associated with the molecules of water that surround the ions. It has therefore been suggested that the absorption bands of the halide ions in water are charge-transfer spectra involving the movement of an electron to the water molecule, with the formation of a free halogen atom plus either an H_2O^- ion or a decomposition product such as $H + OH^-$. The unstable products formed in this way would account for the photochemical changes that occur in these solutions on irradiation with ultraviolet light (see *37*). Similarly the absorption spectra of cations such as Tl^+, Ag^+, Cu^{2+}, Hg^{2+}, Pb^{2+}, Fe^{3+}, and Ce^{4+} show absorption peaks with maximum extinction coefficients close to 10^4 in the far ultraviolet, near 2000 Å. When these ions are complexed with halide ions or with OH^-, CNS^-, etc., the intense absorption bands move toward the visible. It is probable that electron transfer to the solvent and to the complexing ion, respectively, takes place in connection with these absorption spectra, and this process is undoubtedly important in the photochemical behavior of these solutions. For a detailed discussion of the photochemical aspects of charge transfer spectra in solutions of inorganic ions, see the review by Uri (*38*).

D. The Relationship of Absorption and Dispersion

According to Equations (E-31b) and (E-31c) of Chapter 15, the refractive index of a substance may be related to the polarizability through the molar refraction

$$R = \frac{M}{d}\,\frac{n^2 - 1}{n^2 + 2} = \frac{Ne^2}{3\pi m_e} \sum \frac{f_i}{\nu_i^2 - \nu^2} \tag{D-1}$$

Thus the refractive index is determined by contributions from all of the absorption bands of the substance. Two factors determine the magnitude of the contribution by a given band: (*1*) the oscillator strength, f_i, of the band, and (*2*) the proximity of the frequency of absorption of the band, ν_i, to the frequency of the light used, ν.

Because of their small f-values, weak absorption bands make small contributions to the refractive index at all frequencies not close to that of the weak band. Except for organic molecules containing conjugated double bonds, most substances begin to show strong absorption (that is, bands with oscillator strengths close to unity and molar extinction coefficients of the order of 10^4 to 10^5 at wave lengths below about 2000 Å.). Therefore practically all of the molar refraction for visible light arises from bands in the far ultraviolet. That this is actually the case in a typical example can be seen from Fig. 16-8, where the dependence of the refractive index of butanone and cyclohexanone on the wave length is shown. Both of these substances have

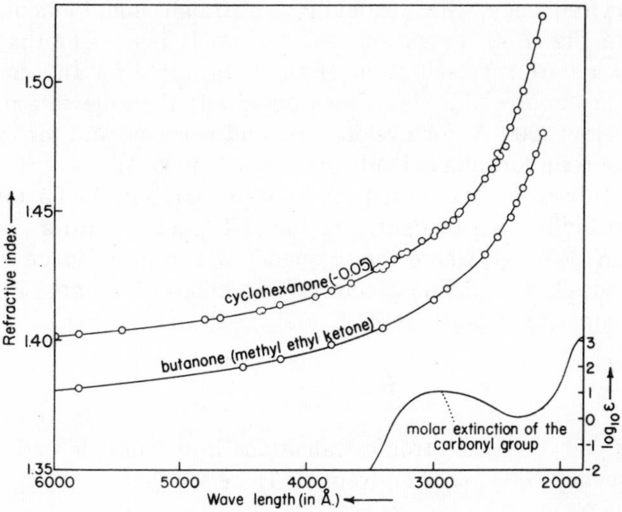

FIG. 16-8. Negligible effect of a weak absorption band on the refractive index. In passing through the weak carbonyl absorption band at 3000 Å. the refractive index (and hence the polarizability) fails to show any appreciable anomalous dispersion of the type shown in Fig. 15-10. On the other hand, the refractive index (and hence the polarizability) increases rapidly as the wave length approaches the strong bands below 2000 Å. (data from H. Voellmy, *Z. physik. Chem.* **A 127,** 305 (1927); T. M. Lowry and C. B. Allsopp, *Proc. Roy. Soc.* **A133,** 26 (1931); **A 143,** 622 (1933); **A146,** 300 (1934)). The refractive indices of cyclohexanone plotted here have been reduced by 0.05 from their actual values.

a weak band ($\varepsilon_{max} = 10$, $f = 3 \times 10^{-4}$) at about 2900 Å. In neither case is there evidence of any "kink" in the curve of n *vs.* wave length, such as would be expected if the carbonyl band made an appreciable contribution to the refraction.

The dispersion of many substances in the visible and near ultraviolet can be represented accurately by the empirical one-term dispersion formula

$$R = \frac{n^2 - 1}{n^2 + 2} \frac{M}{d} = \frac{K}{\nu_0^2 - \nu^2} \tag{D-2}$$

where K and ν_0 are constants. If this equation is turned upside down, we obtain

$$\frac{1}{R} = \frac{\nu_0^2}{K} - \frac{1}{K} \nu^2 \tag{D-2a}$$

so that a test of this relationship is provided by plotting $(1/R)$ against the square of the frequency, which should give a straight line. Plots of this kind are shown in Fig. 16-9 for various substances. It is evident that (D-2) is usually an accurate representation of the data. Only for the unsaturated compounds are appreciable deviations observed; these deviations become noticeable below 3000 Å. for cyclohexene and benzene, and for cyclohexadiene the one term formula is inadequate below 4000 Å.

For many substances the contributions of the bands in the far ultra-violet to the polarizability can evidently be lumped together into a single term of the form $K/(\nu_0^2 - \nu^2)$, where the frequency ν_0 is a mean value for all the effective bands. Let us define a composite oscillator strength, F, of these bands by comparing Equations (D-1) and (D-2) and writing

$$K = \frac{Ne^2}{3\pi m_e} F \tag{D-3}$$

Suppose that all of the electronic transitions in a molecule had the same natural frequency, $\nu_i = \nu_0$. Then from (D-1) we see that

$$R = \frac{Ne^2}{3\pi m_e} \frac{\Sigma f_i}{\nu_0^2 - \nu^2}$$

so that $F = \Sigma f_i$. But the Kuhn-Thomas sum rule (page 576) tells us that Σf_i is equal to the total number of electrons present in the molecule. In the more realistic case that all natural frequencies are not the same, we may therefore interpret the empirical constant F (determined from the slope of a plot of $1/R$ vs. ν^2 using (D-2a) and (D-3)) as an effective number of electrons responsible for the polarizability and its dispersion at long wave lengths. The empirical constant ν_0 (or the corresponding wave length, $\lambda_0 = c/\nu_0$) is an indication of the spectral region in which are to be found the absorption bands chiefly responsible for the polarizability at long wave lengths. It is interesting to inspect typical empirical values of F and λ_0 from this point of view.

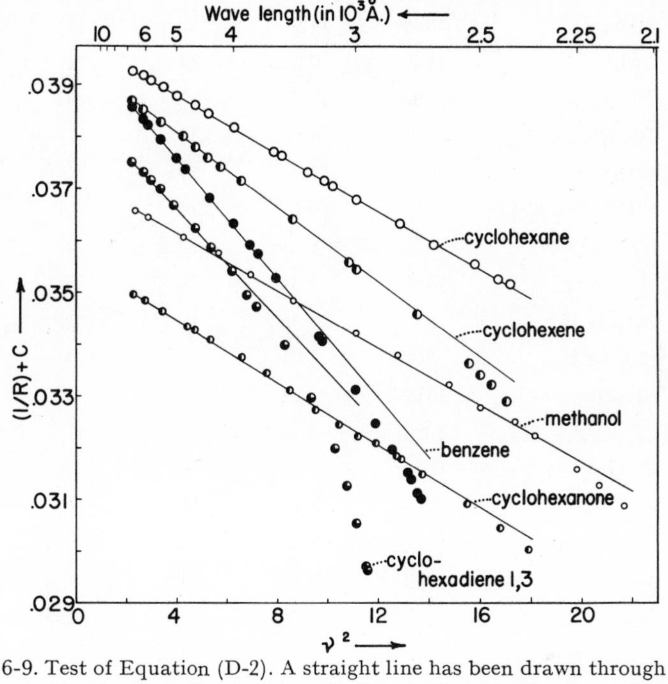

FIG. 16-9. Test of Equation (D-2). A straight line has been drawn through the points at long wave lengths for each substance. In the abscissa the frequency is in units of 10^4 cm.$^{-1}$ To avoid crowding, the points for some of the substances have been shifted vertically by an amount C ($C = 0.3$ for cyclohexane, 0.15 for cyclohexene, and -0.1 for cyclohexanone). The molar refractions of methanol have been multiplied by one third. Data for methanol from Voellmy; remaining data from Lowry and Allsopp (see Fig. 16-8 for references.).

Table 16-5 will show the reader that in general the mean absorption wave length of the bands responsible for the refraction lies very far in the ultraviolet, in the vicinity of 1000 Å. Furthermore, F is considerably smaller than the total number of electrons in the molecule. It is not surprising that the electrons in the inner shells of the atoms in these molecules do not make much of a contribution to the polarizability in the visible; this would be expected from the tightness with which they are bound (that is, the large

TABLE 16-5

Substance	λ_0 (Å.)	F	Number of electrons in outer shells	Reference (below)
$SiCl_4$	997	15.8	32	(1)
$SiBr_4$	1335	11.9	32	(1)
BCl_3	1125	8.82	24	(1)
BBr_3	1188	11.3	24	(1)
$COCl_2$	1040	8.53	24	(1)
$(CH_3)_2CO$(vapor)	800	13.9	24	(1)
$(CH_3)_2CO$(liquid)	917	10.2	24	(2)
$(C_2H_5)_2O$	785	19.8	32	(1)
$HC{\equiv}CH$	1155	3.52	10	(1)
$H_2C{=}CH_2$	1077	4.94	12	(1)
$H_3C{-}CH_3$	803	7.81	14	(1)
CH_3Cl	1072	5.39	14	(1)
CH_2Cl_2	1024	8.46	20	(1)
CH_2Br_2	1010	11.25	20	(1)
CH_3OH	866	5.89	14	(2)
cyclohexane	866	19.8	36	(2, 3)
cyclohexene	899	17.85	34	(2, 4)
benzene	1160	10.2	30	(2, 3)
cyclohexanone	901	18.45	40	(5)
diamond	715	2.27	4	(6)

1. H. Lowery, *Proc. Roy. Soc.* **A 133,** 188 (1931).
2. H. Voellmy, *Z. phys. Chem.* **A 127,** 305 (1927).
3. T. M. Lowry and C. B. Allsopp, *Proc. Roy. Soc.* **A 133,** 26 (1931).
4. C. B. Allsopp, *Proc. Roy. Soc.* **A 143,** 622 (1933).
5. C. B. Allsopp, *Proc. Roy. Soc.* **A 146,** 305 (1934).
6. F. Peter, *Z. Physik* **15,** 358 (1923).

values of ν_i). It is interesting, however, that F tends to be only one-half to one-third of the number of electrons in the outer, valence shells. Therefore many of the strongly absorbed frequencies associated with these outer electrons must be extremely high, so that they make small contributions to the polarizability as measured with visible light.

It is well known that one can calculate the molar refraction (and hence the polarizability) of many substances by adding up the contributions of individual atoms or groups of atoms in the molecule (*39*). Since this procedure does not always work, it is interesting to look for the conditions under which additivity should be expected, especially in the light of Equation (D-2). This question is best answered by giving a simple example. Let us calculate the molar refraction of the CH_2 group in a hydrocarbon, first by adding the atomic refraction of the carbon atom found from the index of refraction of diamond ($n_D = 2.4173$, $d = 3.51$, $M = 12.01$, giving $R = 2.115$) to the value for the refraction of two hydrogen atoms found from the difference between *n*-hexane (C_6H_{14}, $n_D = 1.37536$, $d = 0.6603$, giving $R = 29.88$) and cyclohexane (C_6H_{12}, $n_D = 1.42900$, $d = 0.7791$, giving $R = 27.72$). This difference is found to be 2.16. The refraction of a CH_2 group should therefore be $2.115 + 2.16 = 4.28$. On the other hand, the refraction of a CH_2 group as found by dividing the value of R for cyclohexane by 6 is 4.93, and the values found by subtracting the R-values of successive members of various homologous series is 4.5 to 4.6. The agreement between these values, while not perfect, is reasonably good. It means that the ν_0-values of these compounds must be fairly similar and that the F-values are additive.

The calculation of molar refractions by adding the contributions of individual atoms should break down if, in bringing two atoms together, one shifts ν_0 appreciably by moving a strong absorption band toward the visible. This is just what happens when a double bond is introduced into a molecule and to an even greater extent when a conjugated system of double bonds is introduced. For this reason, additional numbers ("exaltations") must be added to the calculated molecular refraction when such structural elements are present in a molecule.

E. The Return of Excited Molecules to their Ground States

When a colored substance absorbs light, its molecules are raised to an excited state. Since most colored substances retain their color even when illuminated for a long time, it is obvious that the excited molecules must be able to return to the ground state in some way. The processes by which they do this are interesting, complex, and only partially understood. A single example will show that this process is also of tremendous practical importance, for the mechanism of photosynthesis — certainly one of the most important terrestrial reactions from the point of view of all living things — is intimately concerned with the return of the photoexcited chlorophyll molecule to its ground state.

a. Return to the ground state by the direct re-emission of a light quantum

When an excited molecule reverts to the ground state, the excitation energy may appear in the form of light ("fluorescence," "phosphorescence," "luminescence"), chemical energy ("photochemical change"), or heat. Of these alternatives, the simplest would seem to be the direct re-emission of the same light quantum that was initially absorbed. Therefore it is somewhat surprising that this mechanism is not very often used; most of the colored substances that we encounter convert the light energy that they absorb into heat. And even when light is re-emitted, its color is usually different from that of the light that was originally absorbed.

The reason for the change in color is not hard to understand. Consider

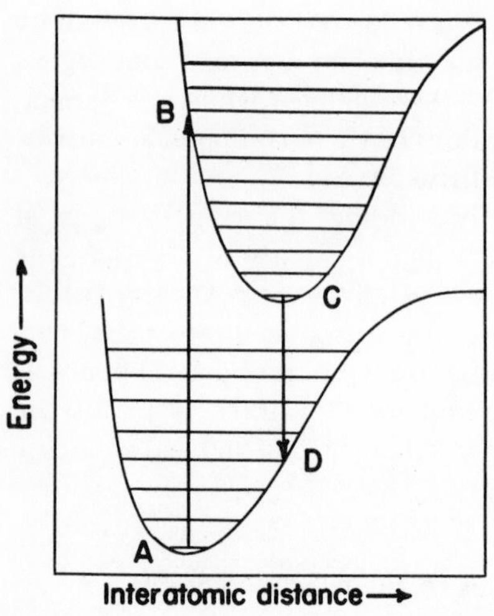

Fig. 16-10

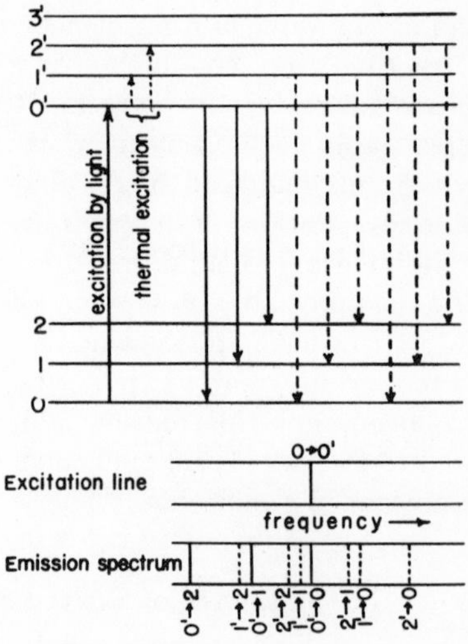

Fig. 16-11

FIG. 16-10. Typical mechanism of fluorescence.

FIG. 16-11. Origin of anti-Stokes fluorescence. The numbers with primes refer to vibrational levels of an excited electronic state, and numbers without primes refer to vibrational levels of the lowest electronic state. It is assumed that the system is irradiated with light whose wave length corresponds to the transition from level 0 of the lowest state to level 0′ of the excited state. Thermal energy then raises a few of the excited molecules to higher vibrational levels. By returning directly to the level 0, these thermally activated molecules are able to emit a photon having more energy than the photon absorbed in the initial excitation process.

the result of the absorption of light by a diatomic or polyatomic molecule originally in the lowest vibrational level, *A*, of the ground state, I, bringing it into the level *B* of the excited state II (see Fig. 16-10). In order that the molecule be able to re-emit light of the same frequency as that absorbed in going from *A* to *B*, it must remain in the vibrational level *B* for at least 10^{-8} sec., since this is the minimum mean lifetime of a level under the most favorable circumstances (oscillator strength $\sim$ 1, see p. 650). According to the kinetic theory, however, if a molecule is in the vapor, it undergoes $\bar{u}/l$ collisions per second, where $\bar{u}$ is the mean velocity and l is the mean free path. For typical molecules at room temperature the time between collisions comes to about $(10^{-10}/p)$ sec., where p is the pressure in atmospheres. At ordinary pressures, therefore, the excited molecule undergoes at least one hundred collisions in the vapor before it is able to reradiate. If the molecule exists in a liquid or in a solid, the time between collisions will be several powers of ten smaller than this. Even if the efficiency of transfer of vibrational energy is rather low, the excited molecule should have little difficulty in transferring its vibrational energy to other molecules in less than the time required to reradiate, especially if it is dissolved in a liquid. It will therefore usually find its way to the lowest vibrational level, *C*, of the excited electronic state II in Fig. 16-10 long before it is able to get rid of the absorbed energy by reradiation. If the situation is as shown in Fig. 16-10, the molecule will then remain in level *C* until it can lose the rest of its energy by radiation. By the Franck-Condon principle, it will fall to level *D* of the ground state. The transition from *D* back to the original vibrational level, *A*, then occurs through collisions with other molecules in the manner just described. Thus in this case the absorption of light of energy *AB* is accompanied by re-emission of light of energy *CD*. The difference between these two energies appears, of course, as heat.

Because of this tendency of excited molecules to cascade into their lowest vibrational levels before emitting, and because the interatomic distances in excited states generally differ from those in the ground state, the frequency of the emitted light is almost always less than that of the light that was initially absorbed. This frequency rule was discovered empirically by Stokes in 1852. Only when thermal energy puts appreciable numbers of molecules into higher vibrational levels of the ground state or of the excited state, is it possible for the frequency of the emitted light to be greater than that of the light that was absorbed (see the dashed lines in Fig. 16-11). Such lines must, however, always disappear on cooling to 0° K. The temperature-dependent lines that disobey the Stokes rule are known as *anti-Stokes fluorescence.*

b. The mechanism of the conversion of excitation energy into heat; "internal conversion"

The complete conversion of absorbed radiant energy into heat is a much more widespread phenomenon than the partial conversion that has just been described. The complete conversion is believed to occur by the following mechanism.

Suppose that the potential energy surfaces for the ground state and an excited state of a molecule cross one another, so that the nuclear skeletons have some configurations in common. Then it will be possible for the excited state to undergo a nonadiabatic transition to the ground state. Since polyatomic molecules have so many degrees of freedom of nuclear motion, and since the complete potential energy surfaces may be quite complex, it is easy to believe that these cross-overs are not unusual in molecules. An illustration of a hypothetical situation is shown in Fig. 16-12, using two nuclear degrees of freedom in order to indicate the increased complexity that can be expected as compared with a diatomic molecule.

If such a polyatomic molecule absorbed a photon, being in this way raised to a high vibrational level of the excited electronic state II, it would execute a complicated vibration that could easily pass again and again through the intersection of the surfaces of the states I and II. There would thus be frequent opportunities for radiationless transitions to occur to high vibrational levels of state I, and once such a transition had occurred, collisions would readily bring the molecule back down to the lowest vibrational level of state I. This process is known as *internal conversion*. Unfortunately, next to nothing is known about the details of this process in any actual example, in spite of its being a very widespread phenomenon.*

Internal conversion is also apparently quite common among the higher excited states of most substances, and is responsible for cascading from highly excited states down to lower states even when the final return to the ground state can only take place through fluorescence. Thus the fluorescence spectra observed from many substances are not affected by the wave length of the light used to excite the fluorescence. This means that, regardless of the electronic state to which the molecule was lifted in the initial absorption process, it soon finds itself in the lowest vibrational state of some particular low-lying excited state from which the characteristic fluorescence takes

* Teller (*40*) has pointed out that in polyatomic molecules the "slow" potential energy surfaces of two electronic states can actually cross even if the two states interact strongly. This makes it much easier in general for internal conversion to occur in polyatomic molecules than in diatomic molecules.

place. A typical arrangement of excited potential energy curves that would account for such behavior is shown in Fig. 16-13.

It is, of course, possible for the excited molecule produced by the initial absorption step to dissociate or rearrange into chemically active fragments, or for radiationless transitions to occur into states that can in turn dissociate or rearrange. This is one way in which light energy may be converted into chemical energy and can lead to photochemical change. Another possibly even more important method of promoting chemical change through light absorption will be mentioned below in connection with the discussion of the phosphorescence of organic molecules in glassy solvents.

The process of internal conversion and its relationship to predissociation are discussed in detail by Franck and Sponer (*41*).

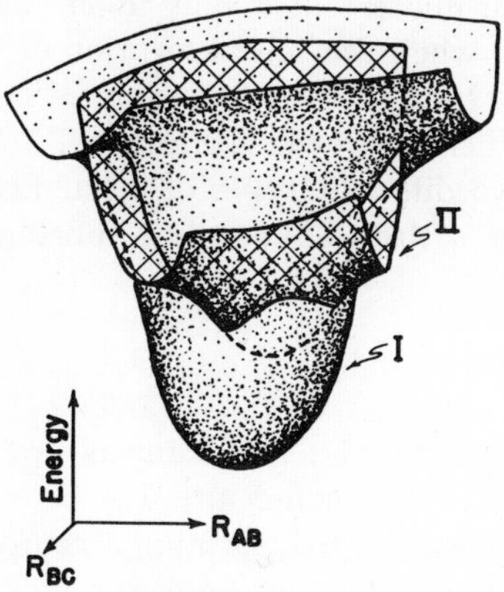

Fig. 16-12

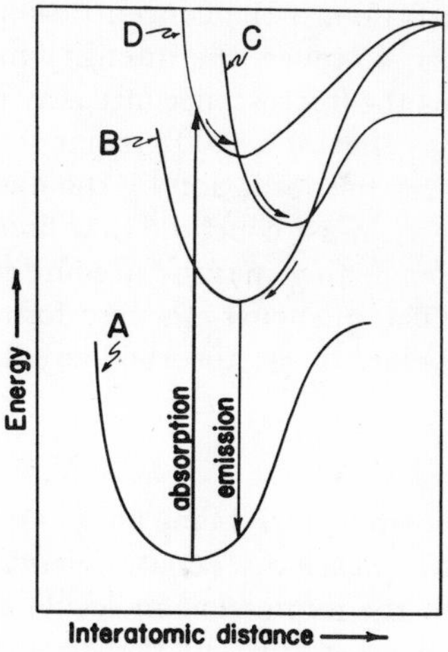

Fig. 16-13

FIG. 16-12. Typical potential energy surfaces that might give rise to internal conversion in a polyatomic molecule. It is assumed that there are three atoms, *A*, *B*, and *C*, lying along a line. The surfaces represent the energy as a function of the distances between *A* and *B* and between *B* and *C*. Surface I belongs to the electronic ground state. Surface II (which is cross-hatched) represents an excited electronic state. Radiationless transitions can occur at the intersection of the two surfaces, so that a molecule in state II could return to the ground state without emitting light.

FIG. 16-13. Role of internal conversion in fluorescence. A molecule in the ground state *A* is brought to the excited state *D* by absorption of light. It passes to state *C* and then to state *B* by radiationless transitions at the intersections of the potential energy curves. From state *B* it returns to the ground state with emission of a photon. The same photon is emitted if a different exciting wave length is used, so that the initial absorption process brings the molecule directly to either states *B* or *C*.

c. Quenching

It is found that when foreign substances such as amines, oxygen, and certain salts are added to a solution of a fluorescent substance such as quinine or fluorescein, the intensity of the fluorescence is reduced. The added substances are said to *quench* the fluorescence. The quenchers are themselves chemically unaffected by their activity: they merely catalyze the conversion of the excitation energy into heat. Furthermore, the quenching does not occur by the direct absorption of the radiation involved in the fluorescence by the quenchers, since as a rule they are transparent to the wave lengths concerned. It is found that fluorescent substances themselves are among the most potent quenching agents. For instance, the intensity of the fluorescence of aqueous solutions of fluorescein is proportional to the concentration, if the concentration is below 0.002 molar. At higher concentrations, however, the intensity increases less rapidly than the concentration. Thus the fluorescence intensity of 0.01 molar fluorescein is only about three times that of a 0.001 molar solution, even when the self-absorption of the fluorescent radiation by the fluorescein has been taken into account. Therefore we must conclude that fluorescein quenches its own fluorescence.

Quenching may be accounted for by two different types of mechanism: (1) The quencher, Q, may form a complex with the fluorescent substance, F, before it has absorbed any light

$$Q + F \rightleftharpoons QF$$

the complex QF being nonfluorescent when excited. This mechanism is called *static quenching*. (2) In *dynamic quenching* the quencher interacts directly with the fluorescent molecule after it has been excited, forming a complex that can return to the ground state without emitting light through an internal conversion process

$$h\nu + F \rightarrow F^* \rightarrow F + h\nu'$$

$$F^* + Q \rightarrow FQ^* \rightarrow F + Q + \text{heat}$$

In some instances it may happen that FQ^* dissociates to give unexcited F plus excited quencher Q^*, which may then be able to cause chemical reactions. This would be an example of *photocatalysis*, or *photosensitization* by F.

There is a simple and interesting way of distinguishing between these two general mechanisms of quenching. Since F^* has a very short mean lifetime, the collisions between F^* and Q must occur very soon after F^* is formed if dynamic quenching is to take place. If, however, the viscosity of the solvent is sufficiently high, molecules of Q will not be able to move about with sufficient rapidity to make the necessary inactivating collision with the F^* molecule before it reradiates. Therefore dynamic quenching should become negligible in solvents of high viscosity. On the other hand, the equilibrium constant for the complexing of F and Q to form FQ, and therefore the static quenching, should not depend on the viscosity of the solvent, all other things being equal. Furthermore, low temperatures should favor association and should therefore tend to increase static quenching, whereas dynamic quenching should be decreased at low temperatures because of the increased viscosity.

It is found that quenching by many foreign substances becomes negligible in solvents of very high viscosity, showing that the dynamic mechanism is operative here. On the other hand, there is evidence that self-quenching in many instances is caused by the static mechanism, the fluorescent molecules forming dimers and polymers in solution that are unable to fluoresce, presumably because of an increased probability of internal conversion in the polymer. The reason for this increased probability is not known.

Very little is known about the detailed mechanism of quenching and in spite of the availability of many data, the relationship between the chemical nature of a quencher and its quenching ability is not understood. Excellent reviews of the present state of our knowledge will be found in the books by Foerster (*42*) and Pringsheim (*43*).

There are various experimental methods for following the decay of fluorescence following exposure to a pulse of incident radiation. In most of these methods, the fluorescent substance is set between a pair of shutters that can be opened at different times. One of the shutters allows light from an exciting source to fall on the sample, while the other permits the observer to view the sample a short time after the first shutter has been closed. Mechanical shutters can be employed for time intervals down to 10^{-5} sec.; for times down to 10^{-9} sec. Kerr cells and diffraction by ultrasonic waves have been used (see (*43a*) for a very rapid device). By these means, it is possible to measure directly the lifetime of the excited molecule, as well as the effect of quenchers on this lifetime. It should be noted that this provides another means of distinguishing between static and dynamic quenching. In static quenching, variation in the concentration of quencher merely changes the number of potentially fluorescent molecules, without altering their lifetime, whereas in dynamic quenching an increase in the concentration of quencher increases the rate of deactivating collisions, and should therefore decrease the lifetime.

d. Phosphorescence of organic substances in solvents of high viscosity

We have seen that quenching tends to be reduced in solvents having a high viscosity because of the slow diffusion and hence the infrequent collisions between the excited molecules and the quenchers. Apparently high viscosity has a similar effect on the rate of internal conversion, because a great many normally nonluminescent organic molecules become fluorescent when they are dissolved in very viscous media such as glucose and boric acid glasses at room temperature and supercooled mixtures of alcohol, isopentane, and ether at liquid air temperatures. Even more striking, however, is the fact that these solutions also emit light from states having much longer lifetimes (up to several seconds) than the values of around 10^{-8} sec. that are expected for molecules emitting by normal dipole mechanisms. This long-lived emission will be tentatively referred to as *phosphorescence* (the distinction between the terms fluorescence and phosphorescence will be discussed in more detail in Section e, below). Evidently internal conversion in these molecules requires movements of the molecular framework that are large enough to be prevented by encasing the molecule in a rigid solvent.

The temperature dependence of the phosphorescence of these solutions is most interesting, and has led Lewis and Kasha (*12, 44*) to important generalizations about the excited states of organic molecules. It is found that when almost any unsaturated or aromatic compound is dissolved in a rigid solvent and excited with suitable radiation, the spectral characteristics of the phosphorescence (that is, of the long-lived emission) change in a characteristic way with the temperature. At high temperatures the spectrum of the phosphorescent emission is identical with that of the fluorescence, but the duration of the emission is much longer than the lifetimes typical of fluorescence (about 10^{-8} sec.), and it varies strongly with the temperature. As the temperature is lowered, a range of temperatures is reached in which the emitted light occurs at longer wave lengths. At still lower temperatures, this new spectral distribution completely replaces the high-temperature, fluorescent-like emission, and the duration becomes independent of the temperature.

Jablonski (*45*) showed that this behavior can be explained if there are two excited states, F and P, whose potential energy surfaces are related to one another in the manner indicated schematically in Fig. 16-14. The transition from the ground state N to the excited state F, and its reverse, are assumed to proceed with a probability corresponding to reasonably strong absorption, so that the mean lifetime of a molecule in state F is of the order of 10^{-8} sec. Transitions between N and P are, however, forbidden. The lowest vibrational level of P is assumed to have a lower energy than the lowest vibrational level of F. Under these conditions, the exposure of a molecule in the ground state N to radiation will excite it to state F. If the cross-over point F and P is suitably located, and if the nonadiabatic transition from F to P at this cross-over point is not too strongly forbidden,

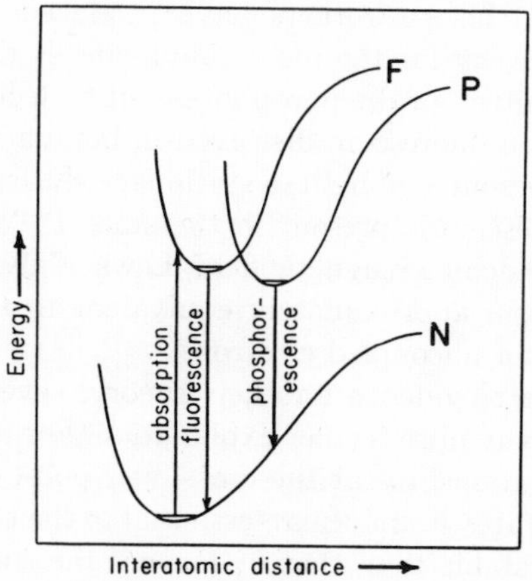

FIG. 16-14. Jablonski's scheme for explaining the temperature variation of the fluorescence and phosphorescence of organic molecules.

some of the excited molecules will turn up in the ground state of P before they have a chance to return from F to N by ordinary fluorescent emission. The molecules thus finding themselves in P have a choice of two pathways for returning to the ground state. They can wait in P for the long time required for the forbidden transition $P \rightarrow N$ to take place, or by acquiring some thermal energy they can return to F and drop to N by emitting light having the same spectral character as the normal fluorescent light. At high temperatures, the second mechanism will be favored, whereas at low temperatures the return to the ground state can occur only by means of the forbidden transition.

The fact that a great many organic compounds behave in a manner consistent with this scheme, indicates that a pair of excited states having these characteristic relationships and properties must be present in nearly all unsaturated and aromatic compounds. The nature of the state P and the reason for the low probability of the transition $P \rightarrow N$ are therefore matters of considerable general interest. Franck and Livingston (*46*) suggested that this state is a tautomer of the ground state. Lewis and Kasha proposed that P

is a triplet or diradical state in which a pair of electron spins has been uncoupled. Since the ground states of the molecules concerned are all singlets, and transitions in which $\Delta S \neq 0$ are strongly forbidden in compounds containing atoms of low atomic number, this would explain the low probability of the $P \rightarrow N$ transition.

Although Lewis and his co-workers have produced much evidence in support of their theory, by far the most convincing is the direct measurement of the paramagnetism of the phosphorescent molecules in the state P. It was shown that when fluorescein dissolved in boric acid glass is exposed to a sufficiently intense source of light, a stationary state is reached in which nearly all of the fluorescein is present in the state P. When the exposure to intense light was made in a magnetic field, Lewis, Calvin, and Kasha (*47*) observed paramagnetism approximately equivalent to that expected of a substance with a pair of uncoupled electrons.

In spite of this direct evidence for their theory, several features of the Lewis-Jablonski theory require further explanation. For instance, the proper spatial and energy relationships at the cross-over point and for the lowest vibrational levels of states F and P are essential to the theory. If the lowest level of F were more stable than that of P, or if the surfaces for P and F failed to cross, or if they crossed only at very high energies, the observed phosphorescent behavior would not be obtained. It is somewhat surprising that the required relationships are found in such a wide variety of substances. Furthermore, it is not at all clear why the same prohibition that restricts the $P \rightarrow N$ transition should not be equally effective in inhibiting the transitions between F, a singlet state, and P, a triplet state.

e. Definitions of the terms fluorescence and phosphorescence

Several definitions of the terms fluorescence and phosphorescence have been proposed by workers in this field. These definitions are unfortunately not at all consistent with one another, and it is important to realize this in order to avoid unnecessary confusion in reading papers in the field.

Pringsheim (*43*, pp. 5, 290) defines phosphorescence as emission that occurs indirectly because it involves passage of the excited molecule through a metastable state. Thus in the example discussed in the previous section, Pringsheim would give the name fluorescence to the light emitted immediately after the primary absorption process, with the mechanism indicated by the following reactions

$$N + h\nu \rightarrow F$$
$$F \rightarrow N + h\nu' \qquad \text{(I)}$$

He gives the name "slow fluorescence" to the process

$$P \rightarrow N + h\nu'' \tag{II}$$

The delayed emission produced indirectly by the sequence of reactions

$$P \rightarrow F$$
$$F \rightarrow N + h\nu' \tag{III}$$

he would call "phosphorescence."

Lewis and Kasha, on the other hand, would restrict the term phosphorescence to emission involving forbidden transitions between states of different multiplicity. Any emission resulting from a transition between a pair of singlet states (or between a pair of triplet states) would be called fluorescence. Thus Lewis and Kasha would call process (II), above, phosphorescence and processes (I) and (III), fluorescence.

Perrin (*42*, p. 12) proposed that emission whose lifetime is more or less independent of temperature be called fluorescence, while emission whose duration increases with decreasing temperature be called phosphorescence. In the example given above, this terminology agrees with that of Pringsheim.

Leverenz (*48*) defines fluorescence as the radiation that appears within about 10^{-8} sec. after the absorption of the light quantum by the molecule; phosphorescence is defined as the radiation emitted over longer periods of time. According to this definition, the light produced in process (I) would be fluorescence, whereas that obtained in (II) and (III) would be called phosphorescence. This definition has the advantage of being much easier to use in experimental work.

F. The Quantum Mechanical Basis of Optical Rotatory Power

In Chapter 15 we found that a substance will rotate the plane of polarization of light if an electric dipole moment $\boldsymbol{m}$ can be induced in its molecules by a changing magnetic field $\dot{\boldsymbol{H}}$, and if a magnetic dipole moment $\boldsymbol{\mu}$ can be induced by a changing electric field $\dot{\boldsymbol{E}}$

$$\boldsymbol{m} = -(\beta/c)\,\dot{\boldsymbol{H}} \tag{F-1a}$$

$$\boldsymbol{\mu} = (\gamma/c)\,\dot{\boldsymbol{E}} \tag{F-1b}$$

where β and γ are constants. It was shown (Equations (G-13) and (G-15) of Chapter 15) that the optical rotatory power is proportional to the sum of the quantities β and γ, which are determined in turn by the structure of the molecule. Chapter 15 also contained a description of a classical mechanical model for which values of the parameters β and γ could be calculated. The optical rotation of this classical model showed certain important characteristics actually observed in optically active molecules. For instance, the model could explain the variation of the optical activity with the wave length of the light used in the measurement. It also explained the fact that the electronic motions associated with weak absorption bands may make important contributions to the optical rotation. On the other hand, since classical mechanics is known to fail badly on the molecular level, one would hardly take this model seriously as a basis for a detailed theory of the relationship between molecular structure and optical rotatory power. Such a theory must be based on quantum mechanical expressions for the parameters β and γ.

Expressions (F-1) may be derived using a procedure analogous to that followed earlier in this chapter, where an expression for the polarizability, α, was obtained. First the perturbation of a wave function by a magnetic field is determined. The perturbed wave function is then used to find the electric moment of the molecule. The resulting expression contains a term proportional to the rate of change of the magnetic field, and the coefficient of this term may be equated to $(-\beta/c)$ in Equation (F-1a). A similar calculation of the magnetic moment resulting from a wave function perturbed by an electric field gives a term dependent on the rate of change of the electric field; the coefficient of this term may be equated to (γ/c) in Equation (F-1b).

a. The electric moment induced by a magnetic field

When a molecule is exposed to a magnetic field, $\boldsymbol{H}$, its energy is perturbed by

$$H' = -\boldsymbol{\mu} \cdot \boldsymbol{H} \tag{F-2}$$

where $\boldsymbol{\mu}$ is the magnetic moment of the molecule (see Equation (D-8) of Chapter 8). The wave function, Ψ, of the perturbed molecule can be expanded in terms of the complete set of the wave functions, ψ_k, of the unperturbed molecule (*cf.* Equations (B-1) and (B-7) of Chapter 14)

$$\Psi = [\sum b_k e^{-i\omega_{k0}t}\psi_k]e^{-iE_0t/\hbar} \tag{F-3}$$

where $\omega_{k0} = (E_k - E_0)/\hbar$, E_k being the energy of the kth state and ψ_0 being the ground state. If the molecules are in their ground state, Equation (B-10) of Chapter 14 gives for the coefficients b_k of the excited states ($k \neq 0$)

$$b_k = (1/i\hbar) \int_0^t H'_{k0} e^{i\omega_{k0}t} dt \tag{F-3a}$$

$$H'_{k0} = \int \psi_k^* H' \psi_0 d\tau = - \int \psi_k^* (\boldsymbol{\mu} \cdot \boldsymbol{H}) \psi_0 d\tau \tag{F-3b}$$

Let the system be exposed to a light wave in which the magnetic vector points in the z-direction. If the unit vector in the z-direction is **k**, we may write

$$\boldsymbol{H} = \boldsymbol{H}_0 \cos \omega t = \tfrac{1}{2} |\boldsymbol{H}_0| \, \mathbf{k} \, (e^{i\omega t} + e^{-i\omega t}) \tag{F-4}$$

If the z-component of $\boldsymbol{\mu}$ is μ_z, then

$$H' = - \tfrac{1}{2} \mu_z |\boldsymbol{H}_0| (e^{i\omega t} + e^{-i\omega t}) \tag{F-5}$$

On substitution into (F-3) and integration with respect to the time, this gives

$$b_k = - \frac{1}{2i\hbar} |\boldsymbol{H}_0| \int \psi_k^* \mu_z \psi_0 d\tau \left[\frac{e^{i(\omega + \omega_{k0})t}}{i(\omega + \omega_{k0})} - \frac{e^{-i(\omega - \omega_{k0})t}}{i(\omega - \omega_{k0})} + \frac{2i\omega_{k0}}{\omega_{k0}^2 - \omega^2} \right] \tag{F-6}$$

The electric dipole moment of the molecule in the perturbed ground state is (neglecting cross products of the small quantities b_k)

$$\overline{\boldsymbol{m}} = \int \Psi^* \boldsymbol{m} \Psi d\tau = \int \left(\sum_k b_k^* \psi_k^* e^{i\omega_{k0}t} \right) \boldsymbol{m} \psi_0 d\tau$$

$$+ \int \psi_0^* \boldsymbol{m} \left(\sum_k b_k \psi_k e^{-i\omega_{k0}t} \right) d\tau \tag{F-7}$$

Writing

$$\boldsymbol{m}_{k0} = \int \psi_k^* \boldsymbol{m} \psi_0 \, d\tau \text{ and } (\mu_z)_{0k} = \int \psi_0^* \mu_z \psi_k \, d\tau \tag{F-8}$$

and substituting for b_k from (F-6), we obtain*

* The time-independent term inside the brackets in (F-6) has been neglected because it does not produce oscillatory contributions to $\overline{\boldsymbol{m}}$ that have the same frequency as the magnetic field and its time derivative.

$$\overline{\boldsymbol{m}} = \frac{|\boldsymbol{H}_0|}{2\hbar} \sum_k \left\{ \boldsymbol{m}_{k0}(\mu_z)_{0k} \left[\frac{e^{-i\omega t}}{\omega + \omega_{k0}} - \frac{e^{i\omega t}}{\omega - \omega_{k0}} \right] \right.$$

$$\left. + \boldsymbol{m}_{0k}(\mu_z)_{k0} \left[\frac{e^{i\omega t}}{\omega + \omega_{k0}} - \frac{e^{-i\omega t}}{\omega - \omega_{k0}} \right] \right\}$$

$$= -\frac{|\boldsymbol{H}_0|}{2\hbar} \sum_k \left\{ \boldsymbol{m}_{k0}(\mu_z)_{0k} \left[\frac{(\omega + \omega_{k0})e^{i\omega t} - (\omega - \omega_{k0})e^{-i\omega t}}{(\omega + \omega_{k0})(\omega - \omega_{k0})} \right] \right.$$

$$\left. + \boldsymbol{m}_{0k}(\mu_z)_{k0} \left[\frac{(\omega + \omega_{k0})e^{-i\omega t} - (\omega - \omega_{k0})e^{i\omega t}}{(\omega + \omega_{k0})(\omega - \omega_{k0})} \right] \right\}$$

$$= -\frac{|\boldsymbol{H}_0|}{\hbar} \sum_k \left\{ \boldsymbol{m}_{k0}(\mu_z)_{0k} \left[\frac{i\omega \sin \omega t + \omega_{k0} \cos \omega t}{\omega^2 - \omega_{k0}^2} \right] \right.$$

$$\left. + \boldsymbol{m}_{0k}(\mu_z)_{k0} \left[\frac{-i\omega \sin \omega t + \omega_{k0} \cos \omega t}{\omega^2 - \omega_{k0}^2} \right] \right\} \qquad \text{(F-9)}$$

The quantity $\boldsymbol{m}_{k0}(\mu_z)_{0k}$ will in general be complex and may be written in the form

$$\boldsymbol{m}_{k0}(\mu_z)_{0k} = \boldsymbol{A}_k + i\boldsymbol{B}_k \qquad \text{(F-10a)}$$

where $\boldsymbol{A}_k$ and $\boldsymbol{B}_k$ are real quantities.* The quantity $\boldsymbol{A}_k$ may also be written $\mathrm{Re}[\boldsymbol{m}_{k0}(\mu_z)_{0k}]$ ("the real part of $\boldsymbol{m}_{k0}(\mu_z)_{0k}$") and $\boldsymbol{B}_k$ may be written as $\mathrm{Im}[\boldsymbol{m}_{k0}(\mu_z)_{0k}]$ ("the imaginary part of $\boldsymbol{m}_{k0}(\mu_z)_{0k}$"). Since the operators $\boldsymbol{m}$ and μ_z must be Hermitian

$$\boldsymbol{m}_{k0} = \boldsymbol{m}_{0k}^*$$

and

$$(\mu_z)_{k0} = (\mu_z)_{0k}^*$$

so that

$$\boldsymbol{m}_{0k}(\mu_z)_{k0} = [\boldsymbol{m}_{k0}(\mu_z)_{0k}]^* = \boldsymbol{A}_k - i\boldsymbol{B}_k \qquad \text{(F-10b)}$$

Substituting (F-10a) and (F-10b) into (F-9), we obtain

* Note that if ψ_k and ψ_0 are both real, then $\boldsymbol{A}_k$ must vanish, because the operator for the electric moment

$$\boldsymbol{m} = \sum e_j \boldsymbol{r}_j$$

is real, whereas the operator for the magnetic moment (*cf.* page 262),

$$\boldsymbol{\mu} = (e/2m_e c) \sum (\boldsymbol{r}_j \times \boldsymbol{p}_j)$$

is imaginary, because of the occurrence of $\sqrt{-1}$ in the operators for linear momentum $\boldsymbol{p} = (\hbar/i)\,[\boldsymbol{i}\partial/\partial x + \boldsymbol{j}\partial/\partial y + \boldsymbol{k}\partial/\partial z]$.

$$\overline{m} = \frac{|\boldsymbol{H}_0|}{\hbar}\sum_k \frac{(\mathbf{A}_k + i\mathbf{B}_k)(i\omega \sin \omega t + \omega_{k0}\cos \omega t) + (\mathbf{A}_k - i\mathbf{B}_k)(-i\omega \sin \omega t + \omega_{k0}\cos \omega t)}{\omega_{k0}^2 - \omega^2}$$

$$= \frac{2|\boldsymbol{H}_0|}{\hbar}\sum_k \frac{\mathbf{A}_k \omega_{k0}\cos \omega t - \mathbf{B}_k \omega \sin \omega t}{\omega_{k0}^2 - \omega^2} \qquad \text{(F-11)}$$

It is interesting to note that at this point all imaginary quantities have disappeared, so that the expression for $\overline{m}$ is real, as it must be for any observable property (Corollary V, page 169). Since $|\boldsymbol{H}_0|\cos \omega t = |\boldsymbol{H}|$ and $-\omega \boldsymbol{H}_0 \sin \omega t = |\dot{\boldsymbol{H}}|$ and since $\mu_z|\boldsymbol{H}_0|\cos \omega t$ can be written as $\boldsymbol{\mu}\cdot|\boldsymbol{H},|$ we have

$$\overline{m} = \frac{2}{\hbar}\sum_k \frac{\mathbf{A}_k \omega_{k0}}{\omega_{k0}^2 - \omega^2}|\boldsymbol{H}| + \frac{2}{\hbar}\sum_k \frac{\mathbf{B}_k}{\omega_{k0}^2 - \omega^2}|\dot{\boldsymbol{H}}|$$

$$= \frac{2}{\hbar}\sum_k \frac{\omega_{k0}\mathrm{Re}[\boldsymbol{m}_{k0}(\boldsymbol{\mu}_{0k}\cdot\boldsymbol{H})]}{\omega_{k0}^2 - \omega^2} + \frac{2}{\hbar}\sum_k \frac{\mathrm{Im}[\boldsymbol{m}_{k0}(\boldsymbol{\mu}_{0k}\cdot\dot{\boldsymbol{H}})]}{\omega_{k0}^2 - \omega^2} \qquad \text{(F-12)}$$

According to this expression, the direction of $\overline{m}$ need not be the same as the direction of $\boldsymbol{H}$, because it is determined in a complex way by the directions of the integrals $\boldsymbol{m}_{k0}$. Thus the quantity β in (F-1a) is not necessarily an ordinary number, since it may be able to alter the direction of $\dot{\boldsymbol{H}}$ as well as its magnitude, in transforming $\dot{\boldsymbol{H}}$ into $\overline{m}$. The optical rotation is, however, almost always measured on substances in the liquid state, so that the molecules are free to orient at random relative to the direction of the perturbing magnetic field, $\boldsymbol{H}$. Under these conditions the components of $\overline{m}$ normal to $\dot{\boldsymbol{H}}$ cancel on the average, so that $\overline{m}$ has the same direction as $\dot{\boldsymbol{H}}$, and β is an ordinary number. Let us now evaluate β for this condition of random orientation.

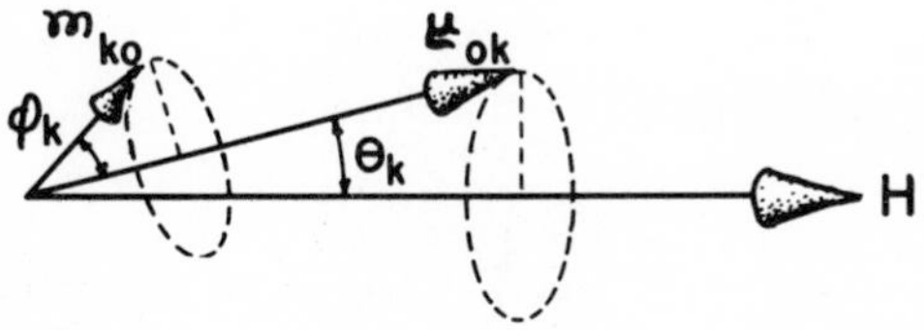

FIG. 16-15. Averaging of $\boldsymbol{m}_{k0}(\boldsymbol{\mu}_{0k}\cdot\dot{\boldsymbol{H}})$ for randomly oriented molecules.

Each of the vectors $\boldsymbol{m}_{k0}$ and $\boldsymbol{\mu}_{0k}$ is, of course, fixed inside the molecule, and it rotates with the molecule as the orientation of the molecule in space is changed. Let the angle between $\boldsymbol{m}_{k0}$ and $\boldsymbol{\mu}_{0k}$ be φ_k (see Fig. 16-15); this angle is not changed when the molecule is rotated. First let us turn the molecule about an axis parallel to $\boldsymbol{\mu}_{0k}$. The components of $\boldsymbol{m}_{k0}$ normal to

$\boldsymbol{\mu}_{0k}$ cancel on the average; the mean value of $\boldsymbol{m}_{k0}$ has the same direction as $\boldsymbol{\mu}_{0k}$ and has the magnitude $|\boldsymbol{m}_{k0}| \cos \varphi_k$. This mean value of $\boldsymbol{m}_{k0}$ parallel to $\boldsymbol{\mu}_{0k}$ will be called $\boldsymbol{m}'_{k0}$. Let the angle between $\boldsymbol{\mu}_{0k}$ and $\boldsymbol{H}$ be called θ_k. Then $(\boldsymbol{\mu}_{0k} \cdot \boldsymbol{H}) = |\boldsymbol{\mu}_{0k}|\,|\boldsymbol{H}| \cos \theta_k$. If θ_k is held constant and the molecule is rotated about an axis parallel to $\boldsymbol{H}$, the components of $\boldsymbol{m}'_{k0}$ normal to $\boldsymbol{H}$ cancel on the average; the mean value of $\boldsymbol{m}'_{k0}$ is then parallel to $\boldsymbol{H}$ and has the magnitude $|\boldsymbol{m}'_{k0}| \cos \theta_k = |\boldsymbol{m}_{k0}| \cos \varphi_k \cos \theta_k$. Thus for each value of θ_k, the average value of $[\boldsymbol{m}_{k0}(\boldsymbol{\mu}_{0k} \cdot \boldsymbol{H})]$ lies in the direction of $\boldsymbol{H}$ and has the magnitude

$$|\boldsymbol{m}_{k0}|\,|\boldsymbol{\mu}_{0k}|\,|\boldsymbol{H}| \cos \varphi_k \cos^2 \theta_k$$

Finally, this quantity must be averaged over all values of the angle θ_k. Since the average value of $\cos^2 \theta_k$ is 1/3, and since $|\boldsymbol{m}_{k0}|\,|\boldsymbol{\mu}_{0k}| \cos \varphi_k = \boldsymbol{m}_{k0} \cdot \boldsymbol{\mu}_{0k}$, we have for the average electric moment induced by the changing magnetic field

$$\langle \overline{\boldsymbol{m}} \rangle = \frac{2}{3\hbar} \sum_k \frac{\omega_{k0} \mathrm{Re}(\boldsymbol{m}_{k0} \cdot \boldsymbol{\mu}_{0k})}{\omega_{k0}^2 - \omega^2} \boldsymbol{H} + \frac{2}{3\hbar} \sum_k \frac{\mathrm{Im}(\boldsymbol{m}_{k0} \cdot \boldsymbol{\mu}_{0k})}{\omega_{k0}^2 - \omega^2} \dot{\boldsymbol{H}} \qquad \text{(F-13)}$$

The first term in (F-13) gives the direct effect of the magnetic field in inducing an electric moment in the molecule; it has the same phase as the electric field of the light wave and therefore makes a contribution to the polarizability, but not to the rotation of the plane of polarization. This term may therefore be disregarded for our present purposes. (Its contribution to the polarizability turns out to be very small in general.*) Comparison with Equation (F-1a) shows that for molecules in the liquid and vapor states

$$\beta = -\frac{2c}{3\hbar} \sum_k \frac{\mathrm{Im}(\boldsymbol{m}_{k0} \cdot \boldsymbol{\mu}_{0k})}{\omega_{k0}^2 - \omega^2} \qquad \text{(F-14a)}$$

Since $\mathrm{Im}(\boldsymbol{m}_{k0} \cdot \boldsymbol{\mu}_{0k}) = -\mathrm{Im}(\boldsymbol{m}_{0k} \cdot \boldsymbol{\mu}_{k0})$, this may also be written

$$\beta = \frac{2c}{3\hbar} \sum_k \frac{\mathrm{Im}(\boldsymbol{m}_{0k} \cdot \boldsymbol{\mu}_{k0})}{\omega_{k0}^2 - \omega^2} \qquad \text{(F-14b)}$$

Exercise. Suppose that a molecule is subjected to a magnetic field that changes at a constant rate, $\dot{\boldsymbol{H}}_0$, so that the field at time t is

$$\boldsymbol{H} = \dot{\boldsymbol{H}}_0 t$$

* If ψ_k and ψ_0 are both real, $\boldsymbol{m}_{k0} \cdot \boldsymbol{\mu}_{0k}$ will be a pure imaginary, so $\mathrm{Re}(\boldsymbol{m}_{k0} \cdot \boldsymbol{\mu}_{0k}) = 0$ and the contribution of this term to the polarizability vanishes.

Using this in the operator for the perturbation energy, Equation (F-2), find the perturbed wave function by means of Equations (F-3), (F-3a), and (F-3b) and show that the electric moment induced in the molecule is

$$\overline{\boldsymbol{m}} = \frac{2}{\hbar}\sum_k \left[\frac{\mathrm{Re}[\boldsymbol{m}_{k0}(\boldsymbol{\mu}_{0k}\cdot\boldsymbol{H})]}{\omega_{k0}} + \frac{\mathrm{Im}[\boldsymbol{m}_{k0}(\boldsymbol{\mu}_{0k}\cdot\dot{\boldsymbol{H}}_0])}{\omega_{k0}^2} \right]$$

Show that if the molecules are oriented at random relative to $\dot{\boldsymbol{H}}_0$, the parameter relating the induced moment to the rate of change of the magnetic field is

$$\beta = \frac{2c}{3\hbar}\sum_k \frac{\mathrm{Im}(\boldsymbol{m}_{0k}\cdot\boldsymbol{\mu}_{k0})}{\omega_{k0}^2}$$

Note that this result is the same as (F-14b) if we set $\omega = 0$.

b. The magnetic moment induced by an electric field

If a molecule is in an electric field, $\boldsymbol{E} = \boldsymbol{E}_0 \cos \omega t$, the perturbation energy is

$$H' = -\,\boldsymbol{m}\cdot \boldsymbol{E}$$

This perturbation results in a slightly altered ground state wave function whose magnetic moment can be calculated in a manner similar to that used in the previous section. It is found that the contribution to the magnetic moment due to the electric field is

$$\overline{\boldsymbol{\mu}} = \frac{2}{\hbar}\left[\sum_k \frac{\omega_{k0}\mathrm{Re}[\boldsymbol{\mu}_{k0}(\boldsymbol{m}_{0k}\cdot\boldsymbol{E})]}{\omega_{k0}^2-\omega^2} + \sum_k \frac{\mathrm{Im}[\boldsymbol{\mu}_{k0}(\boldsymbol{m}_{0k}\cdot\dot{\boldsymbol{E}})]}{\omega_{k0}^2-\omega^2}\right] \qquad \text{(F-15)}$$

The first term in this expression can be disregarded because it does not contribute to the optical rotation. If the molecule is able to orient at random, an averaging process similar to that used in finding $\langle \overline{\boldsymbol{m}} \rangle$ leads to

$$\langle \overline{\boldsymbol{\mu}} \rangle = \frac{2}{3\hbar}\sum_k \frac{\mathrm{Im}(\boldsymbol{\mu}_{k0}\cdot\boldsymbol{m}_{0k})}{\omega_{k0}^2-\omega^2}\,\dot{\boldsymbol{E}} \qquad \text{(F-16)}$$

Referring to (F-1b), we see that the parameter γ is given by

$$\gamma = \frac{2c}{3\hbar}\sum_k \frac{\mathrm{Im}(\boldsymbol{\mu}_{k0}\cdot\boldsymbol{m}_{0k})}{\omega_{k0}^2-\omega^2} \qquad \text{(F-17)}$$

This is identical with the expression for the parameter β given in (F-14b) so we find that for molecules in the liquid and vapor states,

$$\beta = \gamma \tag{F-18}$$

It will be recalled that the same result was obtained for the randomly oriented classical model described in Chapter 15.

Exercise. Prove (F-15) and (F-16).

c. General discussion of the quantum mechanical expression for the optical rotatory power

Making use of Equation (G-15) of Chapter 15, we may write the following expression for the molecular rotation of a substance in solution or in the vapor state, measured with light of wave length λ

$$[M]_\lambda = \frac{48N_0}{\hbar c}\frac{n^2+2}{3}\sum_{\substack{\text{all}\\ \text{transitions}}}\frac{\lambda_{k0}{}^2\,\mathrm{Im}(\boldsymbol{m}_{0k}\cdot\boldsymbol{\mu}_{k0})}{\lambda^2-\lambda_{k0}{}^2} \tag{F-19}$$

where N_0 is Avogadro's number, n is the refractive index, and λ_{k0} is the wave length associated with the frequency ω_{k0}. This is the basic quantum mechanical equation for the optical rotation.

Equation (F-19) has essentially the same form as Drude's equation, (G-38) of Chapter 15. Therefore the optical rotation can be regarded as the sum of contributions from all of the absorption bands of the molecule. In this respect the quantum mechanical theory agrees with the discussion from the classical point of view given in Chapter 15.

The important quantity in (F-19) that determines the differences in the magnitudes of the contributions of an absorption band to the optical rotatory power in different substances is $\mathrm{Im}(\boldsymbol{m}_{0k}\cdot\boldsymbol{\mu}_{k0})$. Condon (*49*) gave this quantity the symbol R_{k0} and called it the *rotatory strength* of the transition $0 \rightarrow k$. It plays a central role in the phenomenon of optical rotation, analogous to the role of the *line strength*, $S_{k0} = (\boldsymbol{m}_{0k}\cdot\boldsymbol{m}_{k0}) = |\,\boldsymbol{m}_{k0}\,|^2$, in the phenomena of polarization and refraction. Thus the quantum mechanical interpretation of optical rotatory power resolves itself into the discussion of the rotatory strengths of the absorption bands of substances.

The factors that influence the contribution of an absorption band to the optical rotation are evidently not the same as those that determine its contribution to the polarization and refraction. True, the electric moment integral, $\boldsymbol{m}_{0k}$, appears in the equations for both phenomena, but for optical rotation the magnetic moment integral, $\boldsymbol{\mu}_{k0}$, is also important. Furthermore, the directions of these two vectors relative to each other are important, because if $\boldsymbol{\mu}_{k0}$ and $\boldsymbol{m}_{0k}$ are perpendicular, $\boldsymbol{m}_{0k} \cdot \boldsymbol{\mu}_{k0} = 0$ and the optical rotation vanishes.

Consider the sum of the rotatory strengths of all of the absorption bands of a molecule

$$\sum_k R_{k0} = \mathrm{Im}[\sum_k (\boldsymbol{m}_{0k} \cdot \boldsymbol{\mu}_{k0})]$$

According to Equation (A-38),

$$\sum_k \boldsymbol{m}_{0k} \cdot \boldsymbol{\mu}_{k0} = \int \psi_0^* (\boldsymbol{m} \cdot \boldsymbol{\mu}) \psi_0 \, d\tau$$

The quantity on the right in this equation is the mean value of the scalar product of the electric and magnetic moments of the ground state. This is an observable property of the state and must be a real number (Corollary V, page 169). Therefore its imaginary part must vanish and we have

$$\sum_k R_{k0} = 0 \tag{F-20}$$

This means that all the absorption bands of a molecule cannot rotate the plane of polarization in the same direction, so that to some extent, at least, the contributions of different bands must tend to cancel. Furthermore, at wave lengths much shorter than those of all of the absorption bands important in determining the optical rotation, we may replace $(\lambda^2 - \lambda_{k0}^2)$ by $-\lambda_{k0}^2$ in all of the denominators in (F-19), so that

$$[M] = -\frac{48N_0}{\hbar c}\frac{n^2+2}{3}\sum \mathrm{Im}(\boldsymbol{m}_{0k} \cdot \boldsymbol{\mu}_{k0}) = 0 \tag{F-20a}$$

That is, the optical rotation vanishes at very short wave lengths. According to (F-19), the optical rotation also vanishes at very long wave lengths. Thus we reach the interesting conclusion that *optical rotatory power tends to go to zero at both ends of the spectrum.*

d. The optical rotation of a molecule and its mirror image

Let ψ_{kd} be the wave function of the kth state of a molecule and let ψ_{kl} be the wave function of the corresponding state of the same molecule reflected in a mirror. Since the Hamiltonian operator of any molecule commutes with the operation of reflection, ψ_{kd} and ψ_{kl} must have exactly the same energy, and the absorption bands of the two forms must occur at identical wave lengths. Therefore the wave lengths λ_{k0} in (F-19) are the same for both forms, and any difference in their optical rotations must come entirely from the rotatory strengths.

The numerical values of the integrals

$$\int \psi_{0d}^* \boldsymbol{m} \psi_{kd}\, d\tau \quad \text{and} \quad \int \psi_{kd}^* \boldsymbol{\mu} \psi_{0d}\, d\tau$$

cannot depend on the coordinate system used in evaluating the integral (Theorem I, page 138). Consider, then, the effect of the operator σ_{xy}, which denotes reflection in the x - y plane. This transformation has no effect on the x- and y-coordinates of the electrons and nuclei in the molecule, but it changes all of the z-coordinates into-z. The electric moment operator is

$$\boldsymbol{m} = \sum_n e_n(x_n\mathbf{i} + y_n\mathbf{j} + z_n\mathbf{k}) = m_x\mathbf{i} + m_y\mathbf{j} + m_z\mathbf{k} \tag{F-21}$$

where x_n, y_n, and z_n are the Cartesian coordinates of the nth particle, and e_n is its charge. Reflection in the x-y plane gives

$$\sigma_{xy}\boldsymbol{m} = \sum_n e_n(x_n\mathbf{i} + y_n\mathbf{j} - z_n\mathbf{k})$$

$$= m_x\mathbf{i} + m_y\mathbf{j} - m_z\mathbf{k} \tag{F-21a}$$

Similarly, for the magnetic moment

$$\boldsymbol{\mu} = \frac{1}{2c}\sum_n \frac{e_n}{m_n}\mathbf{r}_n \times \mathbf{p}_n = \frac{\hbar}{2ci}\sum_n \frac{e_n}{m_n}[(y_n\partial/\partial z_n - z_n\partial/\partial y_n)\mathbf{i}$$

$$+ (z_n\partial/\partial x_n - x_n\partial/\partial z_n)\mathbf{j} + (x_n\partial/\partial y_n - y_n\partial/\partial x_n)\mathbf{k}] \tag{F-22}$$

$$= \mu_x\mathbf{i} + \mu_y\mathbf{j} + \mu_z\mathbf{k}$$

so that

$$\sigma_{xy}\boldsymbol{\mu} = -\mu_x\mathbf{i} - \mu_y\mathbf{j} + \mu_z\mathbf{k} \tag{F-22a}$$

Furthermore

$$\sigma_{xy}\psi_{kd} = \psi_{kl}$$

Thus

$$(\boldsymbol{m}_{0k}\cdot\boldsymbol{\mu}_{k0})_d = \left(\int \psi_{0d}^* \boldsymbol{m}\psi_{kd}\, d\tau\right)\cdot\left(\int \psi_{kd}^* \mu\psi_{0d}\, d\tau\right)$$

$$= \left[\int \sigma_{xy}(\psi_{0d}^* \boldsymbol{m}\psi_{kd})\, d\tau\right]\cdot\left[\int \sigma_{xy}(\psi_{kd}^* \boldsymbol{\mu}\psi_{0d})\, d\tau\right]$$

$$= \left[\left(\int \psi_{0l}^* m_x\psi_{kl}\, d\tau\right)\mathbf{i} + \left(\int \psi_{0l}^* m_y\psi_{kl}\, d\tau\right)\mathbf{j} - \left(\int \psi_{0l}^* m_z\psi_{kl}\, d\tau\right)\mathbf{k}\right]\cdot$$

$$\left[-\left(\int \psi_{kl}^* \mu_x\psi_{0l}\, d\tau\right)\mathbf{i} - \left(\int \psi_{kl}^* \mu_y\psi_{0l}\, d\tau\right)\mathbf{j} + \left(\int \psi_{kl}^* \mu_z\psi_{0l}\, d\tau\right)\mathbf{k}\right]$$

$$= -\left(\int \psi_{0l}^* \boldsymbol{m}\psi_{kl}\, d\tau\right)\cdot\left(\int \psi_{kl}^* \boldsymbol{\mu}\psi_{0l}\, d\tau\right) = -(\boldsymbol{m}_{0k}\cdot\boldsymbol{\mu}_{k0})_l \tag{F-23}$$

That is, *the rotatory strengths, R_{k0}, of two mirror image molecules are equal in magnitude but opposite in sign.* This is the basis of the fundamental principle of stereochemistry discovered by Pasteur: molecules that are mirror images of one another rotate the plane of polarization in equal and opposite directions.

It is obvious that if a molecule is identical with its mirror image, its optical rotatory power must vanish, because in that case, $R_{k0} = -R_{k0}$, and this is only possible if $R_{k0} = 0$. This may happen either because $\boldsymbol{m}_{0k} = 0$, or because $\boldsymbol{\mu}_{k0} = 0$, or because the two moments are perpendicular.

e. Application of the quantum mechanical formula to the helical model

In Chapter 15 a model was considered in which two electrons were constrained to move along helical paths about a common axis. The particles were connected in such a way that they were at all times located at opposite ends of a common diameter of the helices (Fig. 15-26). Equation (G-27) of Chapter 15 gives the value of β for an assembly of randomly oriented helices of this kind. It is not difficult to derive the same expression from the quantum mechanical relation, (F-19).

Let q be the displacement of each of the electrons along the helices measured from their equilibrium positions. Let k be the force constant for the displacement. The potential energy of each electron is then $\frac{1}{2}kq^2$, and the potential energy of the entire system is $V = kq^2$. Since the electrons are constrained to move together, the effective mass of the system is double the mass, m_e, of each electron. Thus the Hamiltonian operator of the system is

$$H = -(\hbar^2/4m_e)d^2/dq^2 + kq^2 \tag{F-24}$$

The eigenfunctions of this operator are those described in Chapter 6 in connection with the harmonic oscillator

$$\psi_n = \left[\frac{\sqrt{\beta/\pi}}{2^n n!}\right]^{\frac{1}{2}} H_n(\xi)e^{-\xi^2/2} \tag{F-24a}$$

where

$$\beta = 2\sqrt{mk}/\hbar \text{ and } \xi = \sqrt{\beta}\, q$$

and $H_n(\xi)$ is the Hermite polynomial of degree n.

It is apparent from the symmetry of the system that both the electric moment integral, $\boldsymbol{m}_{ab}$, and the magnetic moment integral, $\boldsymbol{\mu}_{ba}$, are parallel to the axes of the helices. Therefore the magnitude of the electric moment integral between the states a and b is

$$|\boldsymbol{m}_{ab}| = \int \psi_a(2ex)\psi_b\, dq \tag{F-25a}$$

where x is the component of the displacement of the electron parallel to the axis and e is the charge on each electron. Since $x = (s/\sqrt{r^2 + s^2})q$, where $2\pi s$ is the pitch of the helices and r is their radius (see Fig. 15-26), we have

$$| m_{ab} | = (2es/\sqrt{r^2 + s^2}) \int_{-\infty}^{\infty} \psi_a q \psi_b \, dq \tag{F-25b}$$

It was shown on page 656 that the integral

$$\int_{-\infty}^{\infty} \psi_a q \psi_b \, dq$$

vanishes unless $a = b \pm 1$. Assuming that the electrons are in the ground state, with $a = 0$, so that $b = 1$, we have from equation (B-7b) of this chapter

$$\int_{-\infty}^{\infty} \psi_0 q \psi_1 dq = \sqrt{1/2\beta} \tag{F-25c}$$

$$| m_{01} | = \frac{2es}{\sqrt{r^2 + s^2}} \sqrt{\frac{1}{2\beta}} \tag{F-25d}$$

The component of the magnetic moment of the two electrons parallel to the axis of the helix is (*cf.* Equation (D-4) of Chapter 8)

$$\mu_x = \frac{er}{c} v$$

where v is the component of velocity in a plane normal to the axis of the helix and normal to the line joining the two electrons. Let p_q be the momentum associated with the coordinate q. Remembering that the effective mass of the system is double the mass of each electron, so that the velocity along the helix is $p_q/2m_e$, we see that

$$v = (p_q/2m_e)\,(r/\sqrt{r^2 + s^2})$$

Thus the operator for μ_x is

$$\mu_x = \frac{er^2}{2m_e c\sqrt{r^2 + s^2}} \frac{\hbar}{i} \frac{d}{dq} \tag{F-26}$$

and the magnetic moment integral of interest is

$$(\mu_x)_{10} = \frac{e\hbar r^2}{2im_e c\sqrt{r^2 + s^2}} \int_{-\infty}^{\infty} \psi_1 (d\psi_0/dq)\, dq \tag{F-27a}$$

The integral is evaluated in the following manner

$$\int_{-\infty}^{\infty} \psi_1(d\psi_0/dq)dq = \int_{-\infty}^{\infty} \psi_1(d\psi_0/d\xi)\, d\xi$$

$$= \sqrt{\beta/2\pi} \int_{-\infty}^{\infty} H_1(\xi)e^{-\xi^2/2}(de^{-\xi^2/2}/d\xi)\, d\xi$$

$$= -\sqrt{\beta/2\pi} \int_{-\infty}^{\infty} \xi H_1 e^{-\xi^2}\, d\xi \qquad \text{(F-27b)}$$

But $H_1(\xi) = 2\xi$, so

$$\int_{-\infty}^{\infty} \psi_1(d\psi_0/dq)dq = -\sqrt{2\beta/\pi} \int_{-\infty}^{\infty} \xi^2 e^{-\xi^2} d\xi = -\sqrt{\beta/2}$$

Thus

$$(\mu_x)_{10} = -\frac{e\hbar r^2}{2im_e c\sqrt{r^2 + s^2}} \sqrt{\frac{\beta}{2}} \qquad \text{(F-27c)}$$

and

$$\boldsymbol{m}_{01} \cdot \boldsymbol{\mu}_{10} = |\boldsymbol{m}_{01}|\, (\mu_x)_{10} = i\, \frac{e^2\hbar r^2 s}{2m_e c(r^2 + s^2)} \qquad \text{(F-27d)}$$

so that

$$\text{Im}(\boldsymbol{m}_{01} \cdot \boldsymbol{\mu}_{10}) = \frac{e^2\hbar r^2 s}{2m_e c(r^2 + s^2)} \qquad \text{(F-28)}$$

Substitution into (F-14a) gives, for the contribution to β of the $0 \to 1$ transition of the normal mode in which both electrons move along the helix in the same direction

$$\beta = \frac{e^2 r^2 s}{3m_e(r^2 + s^2)} \frac{1}{\omega_0^2 - \omega^2} \qquad \text{(F-29)}$$

where ω_0 is 2π times the frequency corresponding to the transition to the first vibrational state. This expression is identical with the classical expression, (G-27) of Chapter 15.

f. Contributions by weak and strong absorption bands

It is interesting to estimate the order of magnitude of the contribution to the optical rotation in the most favorable hypothetical case of a transition in which $\boldsymbol{m}_{0k}$ and $\boldsymbol{\mu}_{k0}$ are parallel and both large. If there is strong absorption,

$\boldsymbol{m}_{0k}$ will be of the order of an electronic charge multiplied by a bond length – say 1.5 Å. Similarly, $\boldsymbol{\mu}_{k0}$ might be of the order of a Bohr magneton, $e\hbar/2m_ec$. Then

$$R_{k0} \approx (1.5 \times 10^{-8}\, e)\, (e\hbar/2m_ec) = 0.6 \times 10^{-37} \text{ c.g.s. units}$$

Suppose that the absorption band corresponding to excitation to state k from the ground state is at 3000 Å., and suppose that the optical rotation is measured with light whose wave length is 6000 Å. Then substitution in (F-19) leads to a value of 10,000° for the contribution of the absorption band to $[M]$. This is ten or twenty times larger than the largest contributions that are ordinarily observed under these conditions. Therefore we must conclude that in most optically active substances either (a) $\boldsymbol{m}_{0k}$ is large but $\boldsymbol{\mu}_{k0}$ is much less than a Bohr magneton, (b) $\boldsymbol{\mu}_{k0}$ is large but $\boldsymbol{m}_{0k}$ is much smaller than the value corresponding to strong absorption, or (c) $\boldsymbol{m}_{0k}$ and $\boldsymbol{\mu}_{k0}$, though both large, are very nearly perpendicular. Situations (a) and (c) correspond to optical rotations arising from strong absorption bands. Situation (b) will lead to a relatively large contribution by a weak absorption band. This is the quantum mechanical explanation of the experimental observation that weak absorption bands often make just as important contributions to the optical rotation as do strong absorption bands (*cf.* p. 632). Of course, if neither $\boldsymbol{m}_{0k}$ nor $\boldsymbol{\mu}_{k0}$ is large for a band, the contribution will invariably be small. Such bands are, in fact, known to occur.

g. The one-electron theory of optical rotation

The absorption spectra of organic compounds are determined by isolated groups of atoms in the molecule, known as chromophoric groups. Thus, all ketones and aldehydes have a weak absorption band at about 2950 Å. if the carbonyl group is not conjugated with any other double bonds that may be present in the molecule. Similarly, nitro, phenyl, and other groups each give rise to characteristic absorption bands in substances in which they occur. As one changes the rest of the molecule, the spectral characteristics of a given chromophore (that is, the shape, wave length, and oscillator strength of its absorption band) do not change very much. We can therefore say that, to a good approximation, the transition responsible for the absorption is localized on the atoms of the chromophoric group. Furthermore, the transitions responsible for absorption bands occurring at wave lengths down to at least 2000 Å. invariably involve the transfer of a single electron from one of the occupied orbitals of the ground state to one of the less stable unoccupied orbitals.

These well established principles form the basis of the one-electron theory of optical rotation, originally proposed by Condon, Altar, and Eyring (*50*) and later refined by Eyring and co-workers (review in (*51*)). Chromophoric groups by themselves are invariably identical with their mirror images. Therefore the rotatory strengths of the transitions on an isolated chromophoric group must vanish. If, however, the group is placed in an asymmetric molecule, the environment provided by the molecule will perturb the wave functions of the ground and excited states of the chromophore in such a way that the rotatory strength no longer vanishes. These perturbations of the chromophore by the remaining groups in the molecule are the counterparts of the vicinal actions that Kuhn introduced in his discussion of optical rotation from a classical point of view (page 633). The perturbations involve electrostatic interactions of the same type as those responsible for van der Waals forces and chemical bonding. They may also involve the fields due to dipoles and ionic charges in the vicinity of the chromophore. They are discussed in detail in the review by Kauzmann, Walter, and Eyring (*51*).

As an illustration of the approach used by the one-electron theory, let us consider the contribution of the 2950 Å. absorption band of the carbonyl group to the optical rotation of a substance such as camphor. According to McMurry (*18*), this absorption band is caused by a transition from a $2p$ nonbonding orbital on the oxygen (one of the "lone pair" oxygen electrons) to the σ antibonding orbital of the C–O bond.* Considered as a one-electron transition, approximate wave functions can be written for the ground state

$$\psi_0 = 2p_y$$

and for the excited state

$$\psi_1 = (2p_z - t_c)/(2 - 2S)$$

where t_c is a trigonally hybridized atomic orbital on the carbon atom of the carbonyl (t_c points toward the oxygen atom), S is the overlap integral,

$$S = \int t_c 2p_z d\tau$$

and $2p_y$ and $2p_z$ are hydrogen-like atomic orbitals on the oxygen atom. The z-axis is taken along the C–O bond, and the y-axis lies in the plane defined by

* Unfortunately there is evidence, from the study of the circular dichroism of the carbonyl band in camphor by Kuhn and Gore (*52*), that this band is actually the superposition of two bands, involving two different electronic transitions. Therefore McMurry's analysis and the discussion given here may be incomplete.

the carbonyl group and the two atoms to which it is attached. Using these wave functions, McMurry found that the electric moment integral,

$$\boldsymbol{m}_{01} = \int \psi_0 \boldsymbol{m} \psi_1 \, d\tau$$

has a small component in the z-direction, which accounts for the weakness of the absorption. On the other hand, the magnetic moment integral,

$$\boldsymbol{\mu}_{10} = \int \psi_1 \boldsymbol{\mu} \psi_0 \, d\tau$$

has a large component in the x-direction (see Exercise 1, below). Since $\boldsymbol{m}_{01}$ and $\boldsymbol{\mu}_{10}$ are perpendicular, the rotatory strength vanishes in this approximation (as it must, since the carbonyl group is superposable on its mirror image).

In the one-electron theory, the remaining groups in an asymmetric molecule containing a carbonyl group are assumed to perturb the carbonyl in such a way as to introduce a small component of electric moment parallel to the large magnetic moment. This is accomplished (see Exercise 2, below) by writing for the excited state

$$\psi'_1 = \psi_1 + C3d_{xy}$$

where $3d_{xy}$ is a hydrogenic orbital on the oxygen atom and C is a constant determined by means of perturbation theory. This wave function results in an electric moment component along the x-axis and leads to an optical rotation proportional to the constant C. An approximate calculation of C is described in (*51*). (There is an error in sign in the expression for R_{ba} on page 369 of this reference, which results in an assignment of the wrong sign to the rotation of 3 methyl cyclohexanone.) It is found that optical rotations of the observed order of magnitude can be accounted for in this way, showing that the effects considered in the one-electron theory are real and must be taken into account in any theory. (We shall see, however, that other important effects are not included in this theory.)

According to the one-electron theory, the large contribution of the weak carbonyl absorption band to the optical rotation is explained by the large magnetic moment integral associated with this band. Evidently the classical analogue of the electronic path in the carbonyl transition is a helix with a very small pitch ($r >> s$ in Equation (F-29)). If a weak absorption band does not have a large magnetic moment, it will fail to contribute appreciably to the optical rotation. This appears to be the case with the weak absorption band of the phenyl group at 2500–2700 Å. Experimental observation of the rotatory dispersion of asymmetric molecules containing phenyl groups invariably shows that this band does not contribute a term to the Drude formula. This finding is consistent with the description of the corresponding benzene transition given on pages 472*ff*, since both the magnetic moment integral and the electric moment integral vanish to a first approximation for a $^1A_{1g} \rightarrow {}^1B_{2u}$ transition (see Exercise 3, below).

Exercises. (1) Using the expression (F-22) for $\boldsymbol{\mu}$ and the hydrogen wave functions for $2p_y$ and $2p_z$, show that the one-electron transition $2p_y \rightarrow 2p_z$ has a magnetic moment equal to one Bohr magneton and parallel to the x-axis.

(2) Show that out of all the s, p, d, and f hydrogen-like atomic orbitals, only d_{xy} is capable of giving an electric moment with a component along the x-axis for a transition whose ground state is $2p_y$. [Hint: Investigate the symmetry properties of the various integrands with respect to reflection in the xy, xz and yz planes.]

(3) Using the symmetry principles discussed in Section Cc of Chapter 12, show that the magnetic moment for the transition ${}^1A_{1g} \rightarrow {}^1B_{2u}$ in benzene vanishes. [Hint: Consider the behavior of the two wave functions and of the magnetic moment operator under the inversion operation, i.]

h. Kirkwood's polarizability theory

Kirkwood has pointed out (*53*) that since most of the electrons in optically active molecules can be assigned to definite groups of atoms in the molecule (such as the methyl group, the ethyl group and the hydroxyl group in the case of secondary butyl alcohol, CH_3–CH_2–CHOH–CH_3), it is possible to express the electric and magnetic dipole moment operators in the form

$$\boldsymbol{m} = \sum_g \boldsymbol{m}^g \tag{F-31}$$

and

$$\boldsymbol{\mu} = \frac{e}{2m_e c} \sum_g [\mathbf{R}^g \times (\sum_i \mathbf{p}^g{}_i) + \sum_i \mathbf{r}^g{}_i \times \mathbf{p}^g{}_i] \tag{F-32a}$$

where $\boldsymbol{m}^g$ is the electric moment operator for all of the electrons in the gth group, $\mathbf{R}^g$ is the vector drawn to the center of gravity of the gth group from some fixed origin, $\mathbf{r}^g{}_i$ is the vector drawn from the center of gravity of the gth group to the ith electron in that group, and $\mathbf{p}^g{}_i$ is the linear momentum of the ith electron in the gth group. Let us write

$$\boldsymbol{\mu}^g = \frac{e}{2m_e c} \sum_i \mathbf{r}^g{}_i \times \mathbf{p}^g{}_i$$

and

$$\mathbf{P}^g = \sum_i \mathbf{p}^g{}_i$$

where $\boldsymbol{\mu}^g$ is the magnetic moment operator for the gth group relative to its center of gravity and $\mathbf{P}^g$ is the total momentum of the electrons in the gth group. Then

$$\boldsymbol{\mu} = \frac{e}{2m_e c} [\sum_g \mathbf{R}^g \times \mathbf{P}^g] + \sum_g \boldsymbol{\mu}^g \tag{F-32b}$$

Using (F-31) and (F-32b) to calculate the rotatory strength of the transition to state n from the ground state, we find

$$R_{n0} = \mathrm{Im}(\boldsymbol{m}_{0n} \cdot \boldsymbol{\mu}_{n0}) = R^{(0)}{}_{n0} + R^{(1)}{}_{n0} + R^{(2)}{}_{n0} \tag{F-33}$$

where

$$R^{(0)}{}_{n0} = \frac{e}{2m_e c} \sum_{g \neq k} \mathrm{Im}[\boldsymbol{m}^g{}_{0n} \cdot (\mathbf{R}^k \times \mathbf{P}^k{}_{n0})]$$

$$R^{(1)}{}_{n0} = \sum_{g \neq k} \mathrm{Im}(\boldsymbol{m}^g{}_{0n} \cdot \boldsymbol{\mu}^k{}_{n0})$$

$$R^{(2)}{}_{n0} = \sum_{g} \mathrm{Im}(\boldsymbol{m}^g{}_{0n} \cdot \boldsymbol{\mu}^g{}_{n0})$$

One can therefore consider that the optical rotation arises from terms of three general types. Terms of the type $R^{(2)}{}_{n0}$ come from the asymmetry inherent in the groups themselves. In a certain sense, the one-electron theory can be regarded as an attempt to evaluate this term. Kirkwood assumed that this type of term was small and could be neglected, but the justification of his assumption is questionable, since the one-electron calculations show that it can be quite large, especially in compact molecules where the environment close to the chromophoric group is asymmetric.

The term $R^{(1)}{}_{n0}$ represents an interaction between the magnetic moment on one group and the electric moment on another group. As an illustration of a situation in which it might be important, consider an asymmetric molecule containing a carbonyl group. The transition responsible for the 2950 Å. absorption band of the carbonyl group undoubtedly results in movements of the electrons on other groups elsewhere in the molecule—for instance, through a change in the inductive effect that the carbonyl group exerts on atoms close to it. This movement gives rise to electric moment integrals, $\boldsymbol{m}^g{}_{0n}$, on other groups in the molecule, and the scalar products of these $\boldsymbol{m}^g{}_{0n}$ with the large magnetic moment $\boldsymbol{\mu}_{n0}$ located on the carbonyl will appear in $R^{(1)}{}_{n0}$. Kirkwood assumed that this effect would be small, but it is by no means obvious that this is the case, especially in compact molecules. Unfortunately the detailed evaluation of $R^{(1)}{}_{n0}$ is not easy. (This term is also neglected in the one electron theory.)

Kirkwood considered that all of the optical rotation of an asymmetric molecule arises from the terms of type $R^{(0)}{}_{n0}$. He showed that $R^{(0)}{}_{n0}$ could be transformed to

$$R^{(0)}{}_{n0} = \sum_{g \neq k} \omega_{n0} \,\mathrm{Re}\, [\mathbf{R}^g \cdot (\boldsymbol{m}^g{}_{0n} \times \boldsymbol{m}^k{}_{n0})] \tag{F-34}$$

This expression can be interpreted in the following way. Suppose that a transition involves predominantly an electron on group G, and that the electric moment integral for the transition on this group is $\boldsymbol{m}^G{}_{0n}$. Because of the movement of charge in G, the charges on other groups in the molecule are also slightly displaced in the excited state—though generally to a much smaller extent than the displacements that occur on G as a result of the transition. Consequently the transition on G gives rise to small electric moment integrals on the other groups. It is the interaction of these in accordance with (F-34) that gives rise to $R^{(0)}{}_{n0}$. In calculating the values of $\boldsymbol{m}^k{}_{n0}$ induced in groups other than G by the transition on G, Kirkwood assumed that the interaction between groups can be approximated by a dipole-dipole potential similar to that used by London in discussing the origin of van der Waals forces (Chapter 13). (Since this potential is accurate only if the groups are relatively far apart, a serious error may be introduced in this way if the method is applied to compact asymmetric molecules.) In this way Kirkwood was able to deduce a relatively simple closed formula for the optical rotatory power in terms of the polarizabilities, anisotropies, and spatial positions of the groups in the molecule. (In Kirkwood's original paper there is an error in sign, which is corrected in (*54*)).

When applied to a pair of linear (that is, completely anisotropic) oscillators whose polarizabilities are α_1 and α_2, Kirkwood's formula for the rotatory parameter β reduces to

$$\beta = \frac{1}{6}\,\alpha_1\alpha_2\left[\mathbf{b}_1\cdot\mathbf{b}_2 - 3\,\frac{(\mathbf{b}_1\cdot\mathbf{R}_{12})\,(\mathbf{b}_2\cdot\mathbf{R}_{12})}{|\,\mathbf{R}_{12}\,|^2}\right]\frac{\mathbf{R}_{12}\cdot\mathbf{b}_1\times\mathbf{b}_2}{|\,\mathbf{R}_{12}\,|^3} \qquad \text{(F-34)}$$

where $\mathbf{b}_1$ and $\mathbf{b}_2$ are unit vectors pointing in the same directions as the two oscillators, and $\mathbf{R}_{12}$ is the vector joining oscillator 1 to oscillator 2. If there are more than two oscillators in the molecule, then β is made up of a sum of terms each of the form (F-34), there being one term for each pair of oscillators. If the oscillators are not completely anisotropic, their contribution to β is proportionately reduced. If either oscillator of a pair is completely isotropic, the pair can make no contribution at all to β in this approximation.

According to Kirkwood's formula, the magnitude of the contribution of a given pair of interacting groups to the optical rotation depends, among other things, on the product of the polarizabilities of the two groups. Since weak absorption bands make small contributions to the polarizability (page 582), this means that weak absorption bands should always make small contributions to the optical rotation. We have seen, however, that this is often not the case for absorption bands in the near ultraviolet. Therefore it appears that Kirkwood's formula (F-34) does not include all of the factors responsible for the optical rotation. The important contributions neglected by Kirkwood are presumably those that have large magnetic moments and small electric moments. These transitions would not contribute appreciably to $R^{(0)}{}_{n0}$, but would contribute to the terms $R^{(1)}{}_{n0}$ and $R^{(2)}{}_{n0}$, so Kirkwood was clearly not justified in neglecting at least one of these two types of terms.

Exercises. (1) Show from (F-34) that the pair of oscillators in Fig. 15–28A should be levorotatory. Assume that $\mathbf{b}_1 = (\mathbf{i} + \mathbf{k})/\sqrt{2}$, $\mathbf{b}_2 = (-\mathbf{j} + \mathbf{k})/\sqrt{2}$, and $\mathbf{R}_{12} = d\mathbf{k}$, where $\mathbf{i}$, $\mathbf{j}$, and $\mathbf{k}$ are a right-handed system of unit vectors with $\mathbf{k}$ parallel to the line ab, and d is the distance between a and b. (Note that if we regard this pair of oscillators as a part of a left-handed helix, as in Fig. 16–28B, the classical considerations given on page 630 also lead us to expect that this system is levorotatory.)

(2) Find the absolute configuration of dextrorotatory

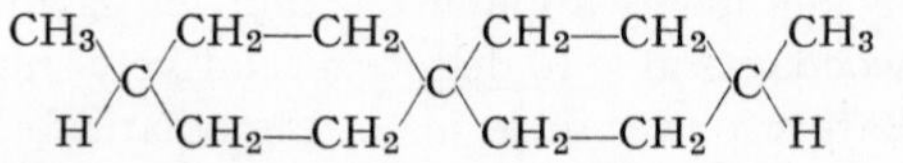

according to Kirkwood's theory. (Assume that both cyclohexane rings are planar and that the two C–CH_3 groups may be replaced by linear oscillators.)

i. Current status of theories of optical rotatory power

Much effort has gone into the calculation of the optical rotations of asymmetric molecules, the chief purpose being the determination of the spatial arrangements of the groups in these molecules. A reliable method of calculating these spatial relationships from the sign and magnitude of the optical rotation would provide a very useful tool for studying molecular structure and for interpreting the stereochemistry of chemical reactions. Unfortunately none of the calculations performed to date can be said to be very reliable.*

The reason for lack of confidence in the absolute configurations derived from classical theories of optical rotatory power is the incorrect detailed picture of the electronic motions on which the classical theories are based.

* Recently a method of determining absolute configurations by means of x-ray diffraction has been successfully applied by Bijvoet (*55*), using the phase shift occurring when the x-rays are absorbed by one of the atoms in the crystal. It is probably fortuitous that the absolute configurations found by Bijvoet agree with those predicted to date by Kuhn, by Kirkwood, and by the one-electron theory. In any case, it is safe to say that in the future the results of the x-ray method will prove most useful as a guide for the development of more reliable theories of optical rotatory power.

As to the quantum mechanical calculations, a number of difficulties arise that have not yet been adequately dealt with. We have seen that Kirkwood's theory is incomplete because it neglects the contributions of weak absorption bands, which are known to be important. Although this shortcoming is not implicit in the one-electron theory, this theory has other difficulties to contend with. In order to apply the one-electron theory to an optically active compound, one must have a relatively detailed description of the orbitals involved in some transition whose contribution to the optical rotation is known experimentally. Unfortunately, in none of the one-electron calculations that have yet been made is the nature of the electronic transitions well enough known to justify much confidence in the assignment of the absolute configuration. Furthermore, the one-electron theory does not deal effectively with the terms $R^{(0)}{}_{n0}$ and $R^{(1)}{}_{n0}$ in (F-33).*

In spite of these shortcomings, all of these theories provide a basis for a useful phenomenological approach to the study of optical rotatory power, and it is probably in this direction that the most important applications of the theories can be expected. Thus it is possible to explain many empirical observations concerning optical activity in terms of general concepts common to all of the theories discussed here. For instance, it has long been known that if an asymmetric carbon atom occurs in a ring, the optical rotation tends to be much larger than when the asymmetric carbon is in an open chain. Furthermore, the optical rotations of simple open chain compounds almost always decrease when the temperature is raised. Both of these observations can be accounted for if the optical rotation of a molecule is the sum of the interactions of no more than two groups taken together at a time (*56*). This assumption of the pairwise interaction of groups is common to the Kuhn, Kirkwood, and one-electron theories. For a detailed discussion, see (*51*) and (*56*).

REFERENCES CITED

1. R. Ladenburg, *Revs. Mod. Phys.*, **5**, 243 (1933).
2. J. C. Slater, "Quantum Theory of Matter," Mc Graw-Hill, New York, 1951.
3. N. F. Mott and I. N. Sneddon, "Wave Mechanics and Its Applications," Oxford U.P., New York, 1948.

* Quite aside from any inherent unreliability of the theories, many of the calculations have been made with molecules such as secondary butyl alcohol, which contains a flexible chain and a hydroxyl group whose spatial conformations are not known with much certainty (*cf.* page 634).

4. W. Heitler, "The Quantum Theory of Radiation," Oxford U.P., New York, 1954.
5. E. U. Condon and G. Shortley, "The Theory of Atomic Spectra," rev. ed., Cambridge U.P., New York, 1950.
6. G. Herzberg, "Atomic Spectra and Atomic Structure," Prentice-Hall, New York, 1937. (Reprinted by Dover, New York).
7. A. Rubinowicz, *Repts. Progr. Phys.*, **12,** 233 (1948–9).
8. H. Bethe, *in* "Handbuch der Physik," (H. Geiger and K. Scheel, eds.), 2nd ed., Vol. 24, Part 1, Springer, Berlin, 1933. (Reprinted by Edwards Brothers, Ann Arbor, Mich., 1943).
9. G. Herzberg, "Spectra of Diatomic Molecules," 2nd ed., Van Nostrand, New York, 1950.
10. J. Franck and H. Levi, *Z. physik. Chem.*, **B27,** 409 (1935).
11. H. Eyring, J. Walter, and G. E. Kimball, "Quantum Chemistry," Wiley, New York, 1944.
12. M. Kasha, *Chem. Revs.*, **41,** 401 (1947).
13. Articles by F. Franzen, C. J. Gorter, J. Hoogschagen, S. Kruyer, T. G. Scholte, A. P. Snoek, and J. P. M. Woudenberg, *Physica*, **9,** 217, 936 (1942); **10,** 365, 693 (1943); **11,** 504, 513, 518 (1946).
14. T. Moeller and J. C. Brantley, *J. Am. Chem. Soc.*, **72,** 5447 (1950).
15. D. M. Yost, H. Russell, and C. S. Garner, "The Rare Earth Elements and Their Compounds", Wiley, New York, 1947.
16. H. Bethe, *Ann. Phys.* [5], **3,** 133 (1929).
17. M. Wolfsberg and L. Helmholz, *J. Chem. Phys.*, **20,** 837 (1952).
18. H. L. McMurry, *J. Chem. Phys.*, **9,** 231 (1940).
19. N. S. Bayliss, *Quart. Revs. (London)*, **6,** 319 (1952).
20. H. Kuhn, *J. Chem. Phys.*, **17,** 1198 (1949).
21. W. T. Simpson, *J. Chem. Phys.*, **16,** 1124 (1948).
22. L. Pauling, *J. Chem. Phys.*, **4,** 673 (1936).
23. K. Lonsdale, *Proc. Roy. Soc.*, **A159,** 149 (1937).
24. O. Schmidt, *Z. physik. Chem.*, **B39,** 59 (1938); **B42,** 83 (1939); **B44,** 185, 194 (1939); **B47,** 1 (1940).
25. K. F. Herzfeld and A. L. Sklar, *Revs. Mod. Phys.*, **14,** 294 (1942).
26. A. L. Sklar, *J. Chem. Phys.*, **10,** 521 (1942).
27. R. S. Mulliken, *J. Chem. Phys.*, **7,** 364 (1939).
28. L. Zechmeister, *Chem. Revs.*, **34,** 295 (1944).
29. N. S. Bayliss, *J. Chem. Phys.*, **16,** 287 (1948).
30. J. Lennard-Jones, *Proc. Roy. Soc.*, **A158,** 280 (1937).
31. H. Kuhn, *Helv. Chim. Acta*, **34,** 2371 (1951).
32. K. F. Herzfeld and A. L. Sklar, *J. Chem. Phys.*, **10,** 508, 524 (1942).
33. L. G. S. Brooker, *Revs. Mod. Phys.*, **14,** 275 (1942).
34. E. Rabinowitch, *Revs. Mod. Phys.*, **14,** 112 (1942).
35. E. Orgel, *Quart. Revs. (London)*, **8,** 422 (1954).
36. R. S. Mulliken, *J. Chem. Phys.*, **7,** 14 (1939).
37. R. Platzman and J. Franck, *in* L. Farkas Memorial Volume, *Research Council Israel (Jerusalem), Spec. Publ.* **No. 1,** (1952).
38. N. Uri, *Chem. Revs.*, **50,** 375 (1952).
39. J. R. Partington, "An Advanced Treatise on Physical Chemistry", Vol. 4, pp. 42–78, Longmans, Green, London, 1953.
40. E. Teller, *J. Phys. Chem.*, **41** 109 (1937).

41. J. Franck and H. Sponer in "Contribution à l'étude de la structure moléculaire" (V. Henri commemorative volume), page 169, Maison Desoer, Liége, 1947–8.
42. T. Foerster, "Fluoreszenz organischer Verbindungen," Van der Hoeck and Ruprecht, Göttingen, 1951.
43. P. Pringsheim, "Fluorescence and Phosphorescence," Interscience, New York, 1949.
43a. E. A. Bailey and G. K. Rollefson, *J. Chem. Phys.*, **21**, 1315 (1953).
44. G. N. Lewis and M. Kasha, *J. Am. Chem. Soc.*, **66**, 2100 (1944).
45. A. Jablonski, *Z. Physik*, **94**, 38 (1935).
46. J. Franck and R. Livingston, *J. Chem. Phys.*, **9**, 184 (1941).
47. G. N. Lewis, M. Calvin, and M. Kasha, *J. Chem. Phys.*, **17**, 804 (1949).
48. H. W. Leverenz, "An Introduction to Luminescence of Solids," Wiley, New York, 1950.
49. E. U. Condon, *Revs. Mod. Phys.*, **9**, 432 (1937).
50. E. U. Condon, W. Altar, and H. Eyring, *J. Chem. Phys.*, **5**, 753 (1937).
51. W. Kauzmann, J. E. Walter, and H. Eyring, *Chem. Revs.*, **26**, 339 (1940).
52. W. Kuhn and H. K. Gore, *Z. physik. Chem.*, **B12**, 389 (1931).
53. J. G. Kirkwood, *J. Chem. Phys.*, **5**, 479 (1937).
54. W. W. Wood, W. Fickett, and J. G. Kirkwood, *J. Chem. Phys.*, **20**, 561 (1952).
55. J. M. Bijvoet, *Endeavour*, **14**, 71 (1955).
56. W. Kauzmann and H. Eyring, *J. Chem. Phys.*, **8**, 41 (1940).

GENERAL REFERENCES

On the fundamentals of the quantum theory as applied to light and other radiation:
W. Heitler, reference 4.

On the theory of the relationship of absorption, polarizability, dispersion, and oscillator strengths:
R. S. Mulliken and C. A. Rieke, *Repts. Progr. Phys.*, **8**, 231 (1941).
A. Maccoll, *Quart. Revs.* (*London*), **1**, 16 (1947).

For a comprehensive discussion of the theory and applications of the free electron molecular orbital model for conjugated systems, along with further references:
K. Ruedenberg and C. W. Scherr, *J. Chem. Phys.*, **21** 1565 (1953).
C. W. Scherr, *J. Chem. Phys.*, **21** 1582 (1953).

On the quantum mechanical theory of optical rotation:
E. U. Condon, reference 49.
W. Kauzmann, J. E. Walter, and H. Eyring, reference 51.
E. Hückel, *Z. Elektrochem.*, **50**, 13 (1944).
J. P. Mathieu, "Les théories moléculaires du pouvoir rotatoire naturel," Gauthier-Villars, Paris, 1946.

Summaries of experimental observations on optical rotatory power and its relation to absorption:
T. M. Lowry, "Optical Rotatory Power", Longmans, Green, London, 1935.
P. A. Levene and A. Rothen, *in* "Organic Chemistry" (H. Gilman, ed.), 1st ed., Vol. 2, p. 1779, Wiley, New York, 1938.

APPENDIXES

I. Atomic Units

Quantity	Size of atomic unit in c.g.s. units	Values of important properties in atomic units
Mass	$m_e = 9.108 \times 10^{-28}$ gm.	Mass of electron = 1 Mass of proton = 1836.1
Length	$a_0 = \hbar^2/m_e e^2$ $= 0.52917 \times 10^{-8}$ cm.	Radius of innermost Bohr orbit = 1
Time	$\tau_0 = a_0\hbar/e^2$ $= 2.4189 \times 10^{-17}$ sec.	Time for one circuit of innermost Bohr orbit = 2π
Frequency	$\nu_0 = 1/\tau_0$ $= 4.1341 \times 10^{16}$ sec.$^{-1}$	Rydberg frequency = $1/4\pi$
Velocity	$e^2/\hbar = 2.1877 \times 10^8$ cm./sec.	Velocity of electron in innermost Bohr orbit = 1 Velocity of light = 137.037
Force	$e^2/a_0^2 = 8.2377 \times 10^{-3}$ dynes	Electrostatic force between electron and proton in first Bohr orbit = 1
Action	$h = 6.625 \times 10^{-27}$ erg-sec.	Planck's constant = 1
Angular momentum	$\hbar = h/2\pi$ $= 1.0544 \times 10^{-27}$ erg-sec.	Component of orbital angular momentum about any axis = 0, 1, 2, $\cdots$
Energy	$e^2/a_0 = 4.3592 \times 10^{-11}$ erg	Ionization energy of hydrogen = 1/2
Electric charge	$e = 4.8029 \times 10^{-10}$ e.s.u. $= 1.6021 \times 10^{-20}$ e.m.u.	Charge on electron = − 1
Electrical potential	$e/a_0 = 9.0762 \times 10^{-2}$ statvolts = 27.219 Internat. volts	Potential in innermost Bohr orbit = 1
Electric field	$e/a_0^2 = 1.7152 \times 10^7$ statvolt/cm. $= 5.1436 \times 10^9$ Int.volt/cm.	Field strength in innermost Bohr orbit = 1
Magnetic moment	$e\hbar/2m_e c = 0.9273 \times 10^{-20}$ erg/gauss = 1 Bohr magneton	Magnetic moment due to orbital motion in innermost Bohr orbit = 1
Magnetic field	$eh/m_e c a_0^3 = 7.8640 \times 10^5$ gauss	Magnetic field at the nucleus due to orbital motion of electron in innermost Bohr orbit = 1

[Based on values of fundamental constants given by E. R. Cohen, J. W. M. DuMond, T. W. Layton and J. S. Rollett, *Revs. Mod. Phys.* **27** 363 (1955)].

II. Conversion Factors for Energy Units

	Atomic units	ergs	electron volts	cm.$^{-1}$
1 atomic unit =	1	4.3592×10^{-11}	2.7210×10^{1}	2.1947×10^{5}
1 erg =	2.2940×10^{10}	1	6.2420×10^{11}	5.0348×10^{15}
1 electron volt =	3.6752×10^{-2}	1.6021×10^{-12}	1	8.0660×10^{3}
1 cm.$^{-1}$ =	4.5563×10^{-6}	1.9862×10^{-16}	1.2398×10^{-4}	1
1° K =	3.1668×10^{-6}	1.3804×10^{-16}	8.6167×10^{-5}	6.9502×10^{-1}
1 kcal. =	9.5982×10^{20}	4.1840×10^{10}	2.6116×10^{22}	2.1066×10^{26}
1 kcal./mole =	1.5931×10^{-3}	6.9446×10^{-14}	4.3348×10^{-2}	3.4964×10^{2}
1 gram =	2.0618×10^{31}	8.9876×10^{20}	5.6100×10^{32}	4.5251×10^{36}

	°K	kcals.	kcal./mole	grams
1 atomic unit =	3.1578×10^{5}	1.0419×10^{-21}	6.2771×10^{2}	4.8502×10^{-32}
1 erg =	7.2441×10^{15}	2.3901×10^{-11}	1.4400×10^{13}	1.1126×10^{-21}
1 electron volt =	1.1605×10^{4}	3.8290×10^{-23}	2.3069×10^{1}	1.7825×10^{-33}
1 cm.$^{-1}$ =	1.4388×10^{0}	4.7471×10^{-27}	2.8601×10^{-3}	2.2099×10^{-37}
1° K =	1	3.2993×10^{-27}	1.9878×10^{-3}	1.5359×10^{-37}
1 kcal. =	3.3009×10^{26}	1	6.0249×10^{23}	4.6553×10^{-11}
1 kcal./mole =	5.0307×10^{2}	1.6598×10^{-24}	1	7.7268×10^{-35}
1 gram =	6.5107×10^{36}	2.1481×10^{10}	1.2942×10^{34}	1

[Based on values of fundamental constants given by Cohen *et al.* (see Appendix I) and F. D. Rossini, F. T. Gucker, H. L. Johnston, L. Pauling and G. W. Vinal, *J. Am. Chem. Soc.* **74**, 2699 (1952)].

III. Hydrogen Atom Wave Functions

$n = 1$ Energy $= -Z^2/2$

$$1s = \frac{Z^{3\,2}}{\sqrt{\pi}} e^{-Zr}$$

$n = 2$ Energy $= -Z^2/8$

$$2s = \frac{Z^{3/2}}{\sqrt{32\pi}} (2 - Zr)\, e^{-Zr/2}$$

$$2p_0 = \frac{Z^{3/2}}{\sqrt{32\pi}} Zr\, e^{-Zr/2} \cos\theta$$

$$2p_{\pm 1} = \frac{Z^{3/2}}{\sqrt{64\pi}} Zr\, e^{-Zr/2} \sin\theta\, e^{\pm i\varphi}$$

$n = 3$ Energy $= -Z^2/18$

$$3s = \frac{Z^{3/2}}{81\sqrt{3\pi}} (27 - 18Zr + 2Z^2r^2)\, e^{-Zr/3}$$

$$3p_0 = \frac{Z^{3/2}\sqrt{2}}{81\sqrt{\pi}} (6Zr - Z^2r^2)\, e^{-Zr/3} \cos\theta$$

$$3p_{\pm 1} = \frac{Z^{3/2}}{81\sqrt{\pi}} (6Zr - Z^2r^2)\, e^{-Zr/3} \sin\theta\, e^{\pm i\varphi}$$

$$3d_0 = \frac{Z^{3/2}}{81\sqrt{6\pi}} Z^2r^2\, e^{-Zr/3} (3\cos^2\theta - 1)$$

$$3d_{\pm 1} = \frac{Z^{3/2}}{81\sqrt{\pi}} Z^2r^2\, e^{-Zr/3} \sin\theta \cos\theta\, e^{\pm i\varphi}$$

$$3d_{\pm 2} = \frac{Z^{3/2}}{162\sqrt{\pi}} Z^2r^2\, e^{-Zr/3} \sin^2\theta\, e^{\pm 2i\varphi}$$

$n = 4$ Energy $= -Z^2/32$

$$4s = \frac{Z^{3/2}}{1536\sqrt{\pi}} (192 - 144Zr + 24Z^2r^2 - Z^3r^3)\, e^{-Zr/4}$$

$$4p_0 = \frac{Z^{3/2}\sqrt{5}}{2560\sqrt{\pi}} (80Zr - 20Z^2r^2 + Z^3r^3)\, e^{-Zr/4} \cos\theta$$

$$4p_{\pm 1} = \frac{Z^{3/2}\sqrt{5}}{2560\sqrt{2\pi}} (80Zr - 20Z^2r^2 + Z^3r^3)\, e^{-Zr/4} \sin\theta\, e^{\pm i\varphi}$$

$$4d_0 = \frac{Z^{3/2}}{3062\sqrt{\pi}} (12Z^2r^2 - Z^3r^3)\, e^{-Zr/4} (3\cos^2\theta - 1)$$

$$4d_{\pm 1} = \frac{Z^{3/2}\sqrt{3}}{1536\sqrt{2\pi}} (12Z^2r^2 - Z^3r^3)\, e^{-Zr/4} \sin\theta \cos\theta\, e^{\pm i\varphi}$$

$$4d_{\pm 2} = \frac{Z^{3/2}\sqrt{3}}{3072\sqrt{2\pi}} (12Z^2r^2 - Z^3r^3)\, e^{-Zr/4} \sin^2\theta\, e^{\pm 2i\varphi}$$

$$4f_0 = \frac{Z^{3/2}}{3072\sqrt{5\pi}} Z^3r^3\, e^{-Zr/4} (5\cos^3\theta - 3\cos\theta)$$

$$4f_{\pm 1} = \frac{Z^{3/2}\sqrt{3}}{6144\sqrt{5\pi}} Z^3r^3\, e^{-Zr/4} \sin\theta (5\cos^2\theta - 1)\, e^{\pm i\varphi}$$

$$4f_{\pm 2} = \frac{Z^{3/2}\sqrt{3}}{3072\sqrt{2\pi}} Z^3r^3\, e^{-Zr/4} \sin^2\theta \cos\theta\, e^{\pm 2i\varphi}$$

$$4f_{\pm 3} = \frac{Z^{3/2}}{6144\sqrt{\pi}} Z^3r^3\, e^{-Zr/4} \sin^3\theta\, e^{\pm 3i\varphi}$$

Index to Symbols

List of the more important symbols used more than once, with pages where defined or discussed

Subject Index

M

N

S

W

X

Z